Agricultural insect pests of temperate regions and their control

For Melinda

Agricultural insect pests of temperate regions and their control

DENNIS S. HILL
M.Sc., Ph.D., F.L.S., C.Biol., M.I.Biol.

Consultant in Entomology, Ecology, and Plant Protection, and Honorary Research Associate, Department of Zoology, University of Hull
Associate Professor of Crop Protection, Department of Plant Science, University of Agriculture, Alemaya, Ethiopia

The right of the University of Cambridge to print and sell all manner of books was granted by Henry VIII in 1534. The University has printed and published continuously since 1584.

CAMBRIDGE UNIVERSITY PRESS
Cambridge
London New York New Rochelle
Melbourne Sydney

Published by the Press Syndicate of the University of Cambridge
The Pitt Building, Trumpington Street, Cambridge CB2 1RP
32 East 57th Street, New York, NY 10022, USA
10 Stamford Road, Oakleigh, Melbourne 3166, Australia

First published 1987

Printed in Great Britain by
Hazell Watson & Viney Limited,
Member of the BPCC Group,
Aylesbury, Bucks

British Library cataloguing in publication data

Hill, Dennis S.
Agricultural insect pests of temperate regions and their control.

1. Field crops – Diseases and pests. 2. Insect control
I. Title
632′.7′0912 SB601

Library of Congress cataloging in publication data

Hill, Dennis S., 1934–
Agricultural insect pests of temperate regions and their control.

Bibliography: p.
Includes index.
1. Insect pests – Control. 2. Agricultural pests – Control.
I. Title.
SB931.H45 1985 632′.7 85-5890

ISBN 0 521 24013 1

SE

Contents

Preface

The original project was to produce a textbook for teaching agricultural entomology in the tropics (initially in Uganda) as at the time no suitable text was available. The accumulation of information for that compilation was generally regarded as successful, and it was suggested that a complementary text be produced along the same lines for use in the temperate regions of the world.

In the UK, Europe and North America there are various textbooks available in English, but none gives an overall (international) view of the subject and none has quite the same approach as this, where large amounts of information have been incorporated into a summarized form for easy assimilation.

For reasons of space and cost, only insect and mite pests are included in this book; many of the bird, mammal, mollusc, and other pests are best viewed regionally in any case, because of their variation from region to region.

The sources of information are many, and are listed in the bibliography; some are referred to in the text. In a number of cases the original publication was not seen; the information was taken from a review article or from an abstract.

Specimens for drawing were either personally collected or loaned from various institutions or collections, especially from the British Musuem (Natural History) through the Keepers of Entomology (Dr P. Freeman, and Dr L. Mound), and the Trustees are thanked. Drawings were made by Hilary Broad, Karen Phillipps, and Alan Forster; a few were from other sources. Photographs were mostly taken by the author, but a few were from other sources and have appropriate acknowledgement under the plate.

Identifications of insect specimens were made by staff of the Commonwealth Institute of Entomology and the Department of Entomology, British Museum (Natural History), who were also sources of general information and advice.

Initial support for the project was made by the Rockefeller Foundation through a grant to the Faculty of Agriculture, Makerere University, Uganda.

General facilities were made available by the Department of Zoology, University of Hong Kong, and the Department of Zoology, University of Hull, for which I am grateful.

The successful completion of this project would not have been possible without the help of many colleagues, especially those from ADAS, and the Harpenden Laboratory of MAFF; also from Rothamsted Experimental Station, from FAO (Rome) and from many chemical companies.

I would like to take this opportunity to thank specifically the following for their help in many different ways: Dr D.V. Alford, Mr R. Bardner, Dr V.F. Eastop, Susan D. Feakin, Mr C. Furk, Mr R. Gair, Marion Gratwick, Dr D.J. Greathead, Gwyneth Johnston, Mr A. Lane, Dr Lee, Hay Yue, Mr R.J.A.W. Lever, Dr Li, Li-ying, Dr W. Linke, Professor B. Lofts, Professor J.L. Nickel, Professor J.G. Phillips, Mr G. Rose, Dr K.A. Spencer, Dr D.L. Struble, Dr J.D. Sudd and Mr R. Wong.

Dennis S. Hill
April, 1984

'Haydn House'
20 Saxby Avenue
Skegness
Lincs. PE25 3LG
England

1 *Introduction*

This book is intended for use as a student text for courses in 'Applied entomology', 'Crop pests' and 'Crop protection', at both undergraduate and post-graduate level. It presupposes a basic knowledge of entomology to the level of that in Imms, A.D. (1960) *A general textbook of entomology*, or alternatively Borror, D.J. & D.M. Delong (1971) *An introduction to the study of insects*. In other words, the reader should be acquainted with the major groups of insects and their characteristics, which may mean order, suborder or superfamily in some cases, but in the more economically important orders this would mean familiarity with superfamilies or families, for example in the Hemiptera, Lepidoptera, Coleoptera, and Diptera.

There is more information in the 10th Edition of 'Imms' (Richards & Davies, 1977), but this is now so expensive that most students are probably still using the previous edition. This book is intended to be complementary to *Agricultural insect pests of the tropics and their control*, so that the total coverage is, in fact, worldwide. There is a certain deliberate overlap between the two books, such as the principles of control and chemicals which are much the same in both regions, and there are cosmopolitan crops and pests that have to be in both books, but wherever possible the emphasis is different in each book.

It is hoped that some sections of the book will also serve for reference purposes, as these sections represent the distillation of much information acquired by extensive experience and detailed literature searching.

Although the title is *Agricultural insect pests of temperate regions*, there are crops and pests included that are found in many warmer parts of the world, namely in S. Africa, Australia, N. New Zealand, China, Korea, Japan, S. USA, other parts of Asia and the Mediterranean region. Certain tropical crops, such as rice and citrus fruit, can be cultivated in countries outside the tropics. A recent trend in many countries is to make the effort to diversify local agricultural crops: in the tropics many temperate crops are now being grown in cooler locations, and in temperate countries some tropical crops are being successfully grown both in greenhouses and in the open. The breeding of new varieties of crops has made the more widespread cultivation more feasible.

Because of the escalating costs of publication the overall size of the book is limited and so the number of pests and crops studied in detail is less than desired. In an attempt to compensate for this the pest section is aimed at generic level rather than individual species, where possible; some pests are Holarctic (or Palaearctic) at the genus level, occurring as several distinct allopatric species. Where an important group of pests is only sparsely represented then a brief review of the group is presented in a couple of pages of text.

To make it as clear as possible which insect species are being referred to in the text both the scientific name (genus and species) and a suitable common name are used in conjunction. Unfortunately there is a lack of international agreement over the use of names, despite the efforts of the *International Code of Zoological Nomenclature*. Similarly with the different taxa used: what is often regarded as a distinct family in one country may be relegated to subfamily status in another. In this respect the present text shows a bias towards the classification used in the UK, and particularly that employed in 'Imms' *General textbook of entomology*, 10th edition, by Richards & Davies (1977), with occasional modification from Kloet & Hincks where it seems warranted.

In a text such as this, where the aim is an international coverage, inevitably some names given here will differ from those used in other parts of the world, but hopefully the identity of the pest will not be in doubt. The scientific names given in the book are those used by the

Commonwealth Institute of Entomology (CIE) on their distribution maps, and those listed in Seymour (1979) and Kloet & Hincks (2nd edition – revised, 1964–78) for the UK, Werner (1982) for the USA, and the checklists for China, Japan, Australia, etc., listed in the bibliography. Clearly, some of these major publications are already out of date, and some major name changes have taken place very recently, within the last year or two.

The question as to whether very recent name changes for insect pests should be followed in a student text is very vexing. The taxonomic purists will, of course, insist that all name changes be strictly adhered to. But in many parts of the world the news of such name changes is slow to arrive, and so far as students are concerned the great majority of their reference sources, if not all, will be using the previous name for the pest, if not even earlier ones. As a practical entomologist I am loathe to see well-established names being changed, unless really necessary, because of the confusion that will ensue. But of course, if there is good reason for the name change then it must be accepted. In chapter 9, where individual pest species are described, if there has been a recent name change the former name is included in parentheses as a synonym, or whatever. Older previous names are not included, for many of the widespread crop pests have lists of synonyms and misidentifications of interminable length.

The common names used are from the same sources generally as the scientific names (with the exception of Kloet & Hincks, as there are no common names included in this checklist). There are considerable divergences in usage of common names; for example in Europe it is traditional to refer to the adult insect (such as Onion Fly) whereas in the New World the damaging stage is referred to (that is Onion Maggot); but in these cases the identity is fairly obvious. In some cases though, for the purpose of this text, an arbitrary choice of common names has had to be made as to the more appropriate when used internationally.

It should be stressed that some records are taken from local or regional publications, and sometimes there are complications in that a particular pest may either have been misidentified or else have been identified correctly but referred to by an invalid name. Sometimes the use of an invalid scientific name is obvious and the record can be rectified, but with some less well-known species it may not be evident, and so some incorrect names will inevitably be included.

Finally it should be remembered that the writing of many book manuscripts takes several years, and then actual publication generally takes 1–2 years to complete, so it is inevitable that the published book will be out of date scientifically, both with regard to names used and also pesticides and their recommendations, even on the day of release. But hopefully by the combined use of scientific and common names for each pest it will be clear as to the identity of the organism concerned.

The distributions given are summarized from the maps produced by the CIE, and in the cases where a map has not been produced for a particular species the appropriate distribution data have been made available from the CIE card index system (now computerized). Reference to the CIE map, where one is available, is made at the end of each summary of distribution.

In the section on control, emphasis has been placed on methods of cultural control whenever these are available, and so far as pesticides are concerned no details as to rates, etc., are included. Pesticide recommendations vary extensively from country to country, and also from season to season, so only the barest details of pesticide recommendations are included. For full details of these for local crops in each country, the appropriate Ministry or Department of Agriculture or Regional Entomologist should be consulted. It would be quite impossible to provide adequate pesticide detail suitable for practical use in all the different parts of the temperate world.

When considering some aspects of the basic principles underlying the study of crop pests and their control, some of the examples given are from tropical situations. They are used because they are particularly suitable as examples, and are usually very well-known pests.

The section on pesticides was compiled from data published by Martin (1970), from Worthing (1979) and from various original data sheets provided by the firms concerned, and that part dealing with application equipment largely from G. Rose (1963) and Matthews (1979). It is not feasible to generalize extensively about persistence, efficiency, pre-harvest intervals, toxicity, and tolerance levels, for not only do these characteristics vary considerably according to local climatic conditions, but each country has its own requirements with regard to residues and toxicity. Some countries are more concerned with operator safety, whereas others regard consumer hazards the more important. Thus the same chemical may have a pre-harvest interval of seven days in one country and as many as 28 in another; or alternatively an approved pesticide in one country may be banned in another.

In chapter 9 on pest descriptions, biology and control measures, the original scheme was to illustrate all the important stages of the major insect pests and to show the damage done to the host. But it was not possible to provide all stages and damage for more than 205 major pests, and so in some cases only the adult insect is drawn. Unfortunately, some of the earlier drawings were designed more to give an impression of the pest and the crop plant rather than accurate detail of the insect. In the more recent drawings by Hilary Broad, Alan Forster, and Karen Phillipps we have endeavoured to reproduce morphological details which are taxonomically specific.

The species here designated as major pests have, in a few instances, been chosen for academic reasons or to demonstrate a point of particular biological interest rather than always being primarily economic pests. I have attempted to include a well-balanced range of pests, most of which are important on major crops, and widely distributed throughout the temperate parts of the world. The denotation of the term 'major pest' to a species is necessarily somewhat arbitrary when dealing with 75 crops grown throughout the cooler parts of the world. However, this term has usually only been applied to species which are economically important over a wide part of the range in which the crop is cultivated.

According to figures provided by Dr R.G. Fennah for Wilson (1971) it can be said that there are some 30 000 insect pest species, but Fletcher (1974) referred to there being only about 1000. Later in the book he mentioned that the total number of insect and mite pests species recorded from several major crops ranges from 1400 on cocoa and cotton to 838 on coffee. It seems reasonable to assume that on a worldwide basis there is something in the region of 1000 species of 'serious' crop pest species, including pests of forests and ornamentals, and maybe up to 30000 minor pest species. In chapter 10, under the headings of the 75 crops considered, are listed the more important major pests, many of which were included in chapter 9, and in addition a selection of the minor pests recorded from each crop. In some of the more restricted crops the number of recorded pest species is very small, whereas the widespread crops may have more than 1000 recorded minor pests. In these cases the list of minor pests has been restricted to the more important, more interesting, or more widespread of the minor pest species.

2 *Pest ecology*

The information provided in this chapter has been separated off in an attempt to emphasize the need for a greater understanding of the complex ecological relationships between the insects and plants in the agricultural context. Here are also included various aspects of basic biology that have broad ecological relevance. There is clearly overlap between this and the next chapter, as various factors in the consideration of basic principles relating to pest control are aspects of the pest/crop ecology.

Ecology and pest control

The earliest recorded attempts at pest control were often basically concerned with the biology of the pests and their ecology, and attempts were made to make the environment less favourable for the pests by various physical and cultural means. With the recent disillusionment with pesticides and with the increased awareness of the importance of ecological aspects of the pest/crop situation, as now defined by most integrated pest management (IPM) programmes, there has been a reversal of approach to basic ecological aspects.

Ecology

The complex and interacting system comprising all the living organisms of an area, and their physical environment (soil, water, climate, shelter etc.) is termed the *ecosystem*, and the study of ecosystems is called *ecology*. Definitions of ecology vary according to the speciality of the definer; botanists often have a different viewpoint from zoologists, and agriculturalists may have a third view. In its simplest form ecology can be defined as 'the total relationships of the plants and animals of an area (habitat) to each other, and to their environment'.

Environment has been defined by Andrewartha & Birch (1961) as being composed of four main factors: weather, food, other animals and plants, and shelter (a place in which to live).

For convenience it is customary to lump together environmental factors into two broad categories, biotic (i.e. organic) and physical (i.e. abiotic or inorganic). Weather and shelter (usually) are clearly physical factors, although shelter for a parasite could be regarded as biotic. Other animals and plants clearly constitute a biotic factor. Food is a biotic factor for animals which are holozoic (heterotrophic) in their feeding habits, but could possibly be more suitably described as physical for plants, which are holophytic (autotrophic) in their nutrition.

The *environmental factors* can be further defined as follows.

Weather

(a) Temperature – ranges defined as tropical, temperate, arctic or boreal
(b) Humidity – ranges from moist, moderate, to dry conditions
(c) Water – includes groundwater, rainfall, etc.
(d) Light – intensity important for many organisms
(e) Wind – important for dispersal, and drying effects

Food

(a) For animals
    (i) Organic remains – detritivores
    (ii) Plant material – herbivores (phytophagous)
    (iii) Other animals – carnivores and parasites
(b) For plants
    (i) Organic remains – saprophytes (mostly fungi and bacteria)

(ii) Other plants – parasites and pathogens
(iii) Animals – insectivores (carnivores)
(iv) Sunlight, water, carbon dioxide, minerals, chlorophyll (autotrophs)

Other animals and plants (i.e. the community)
(a) Competition – intraspecific (within the species)
– interspecific (between different species)
(b) Predation
(c) Parasitism
(d) Pathogens causing diseases

Shelter (a place in which to live; habitat)
(a) For animals (insects) and pathogens – frequently a plant, and often a specific location on the plant, e.g. in the cases of a leaf-miner, stem borer, bollworm and leaf-roller. Some insects are soil-dwellers (e.g. termites, crickets, beetle larvae), and adult, winged insects may not be very habitat-specific (i.e. eurytopic).
(b) For plants – usually a physical location (habitat), including the soil (e.g. hilltop, valley, field) together with the other plants that constitute the community.

Two basic ecological terms should perhaps be included here for reference.

Habitat. The place where the plants and animals live; usually with a distinctive boundary, e.g. a field, pond, stream, sand-dune or rocky crevice. Often initially broadly subdivided into terrestrial, marine, and freshwater habitats.

Community. The collection of different species and types of plants and animals, in their respective niches, within the common habitat, e.g. a lake community, mangrove community or ravine community. The basic plan for all communities is the same, i.e. they are composed of saprophytes, autotrophic plants, detritivores, herbivores, carnivores, parasites, etc.

With the general disillusionment that followed the widespread continual use of synthetic chemical pesticides, especially the early organochlorine compounds, the situation has changed so that now attention is focussed on biological and ecological understanding, linked with careful application of selected pesticides. This approach was initially called *integrated control* but was more recently redefined as *pest management* (PM), and is now finally referred to as *integrated pest management* (IPM).

As indicated above, the number of different factors operating in an insect pest/host plant relationship is large and hence the different possibilities available for ecological manipulation are considerable. But, of course, a vital prerequisite is a detailed knowledge of the insect's life-history and biology, and especially its relationship with the host plant.

As mentioned later (p.7), an insect species is only a pest (that is an economic pest) at or above a certain population density, and in any pest ecosystem any one (or more) aspect of the environment may be of over-riding importance. In the study of pest populations the key to control will inevitably lie in the understanding of the complex of environmental factors and their relative importance. However, our knowledge at present of most pest situations falls short of this ideal, and much basic ecological study is still required. Too frequently pest control still consists of hastily and ill-considered applications of chemical pesticides, which sometimes wreak ecological havoc, especially in the tropics, often without controlling the pest at which they were aimed. Progress is gradually being made though, as evidenced by the ever-growing number of IPM programmes for different crops in different parts of the world.

Agroecosystems

An agroecosystem is basically the ecosystem of an area as modified by the practice of agriculture, horticulture or animal rearing. Agriculture consists of methods of soil management and plant cultivation so as to maintain a continuous maximum yield of crop produce, in the shortest time possible. This is achieved by manipulation of the environment so as to make growing conditions for the crop plants as near ideal as possible, and also to minimize damage to the crop by pest and disease attacks. Obvious manipulations are listed below, under the appropriate environmental headings.

Weather

(a) Temperature control by shading (lowering) or use of greenhouse (raising)
(b) Humidity control by spraying or altering plant density
(c) Irrigation (below or above-ground) and drainage
(d) Light increased by use of ultra-violet lamps, or reduced by shade trees, shelter, etc.
(e) Wind protection by growing shelter-belts, windbreaks, tall trees and hedgerows

Food

(a) Animal feedstuff, grazing leys, dietary supplements
(b) Plants are 'fed' by addition of fertilizers, minerals, and trace elements; sometimes increased radiation by use of extra illumination.

Competition – intraspecific, reduced by careful crop spacing
– interspecific, reduced by weeding and use of herbicides

Predation, parasitism and disease reduced by crop protection procedures

Shelter

(a) Animal houses
(b) Windbreaks, shelter-belts, greenhouses, polythene shelters, protected seedbeds, etc. Also soil improvement by drainage, irrigation, liming, fertilizers, deep ploughing, hardpan breaking, manuring etc.

Thus it is clear that every aspect of the environment can be (and usually is) manipulated in the course of modern sophisticated agriculture. Generally though, only practices that show a definite economic profit are indulged.

The major ecological modifications that are made during the process of agriculture (*sensu lato*) that affect pest populations are as follows.

Monoculture – the extensive growth of a single plant species, with a simplification of the flora, partly by weed destruction.

Increased edibility of crop plants – the crop plants are more succulent, larger and generally more attractive to pests than the wild progenitors.

Multiplication of suitable habitats – the habitat and the microclimate becomes uniform over a large area.

Loss of competing species – may lead to the formation of new pests.

Change of host/parasite relationships – will lead to the development of secondary pests.

Spread of pests by man – as crops are grown in more parts of the world, the pests are also eventually spread around by accident.

These and other topics will be looked at in more detail in later sections of this chapter.

It should be stressed at this point that the vast majority of crop pests are in fact human-created through the ancient practice of agriculture. Completely 'natural' serious crop pests are very few and are limited to locusts, possibly a few tropical armyworms, and some defoliating caterpillars and sawfly larvae that occur in the extensive natural semi-monocultures of the northern taiga in N. Europe, N. Asia, and N. America and the northern deciduous forests.

Pest populations

The important point to remember about any pest is that it is only an economic pest at or above a certain population density, and that usually the control measures employed against it are designed only to lower the population below the density at which the insect is considered to be an economic pest; only very rarely is complete eradication of the pest aimed at. The schematic representation of the growth of a population in fig. 1 (adapted from Allee *et al*, 1955) has had four separate population levels indicated; these are represented by the numbers 1 to 4. These population levels indicate purely hypothetical densities at which any particular insect species may be designated an economic pest. Population level 1 might well represent an economic pest level for such an insect as Rosy Apple Aphid (*Dysaphis plantaginea*), for in this case control measures are recommended when the population density reaches one aphid per tree (at bud-burst), and similarly for a pest such as Colorado Beetle.

Fig.1. The growth of populations (after Allee *et al.*, 1955).
Stage I Period of positive, sigmoid growth; population increasing
A Establishment of population
B Period of rapid growth (exponential growth)
II Equilibrium position (asymptote); numerical stability
III Oscillations and fluctuations
A Oscillations – symmetrical departures from equilibrium
B Fluctuations – asymmetrical departures
IV Period of population decline (negative growth)
V Extinction

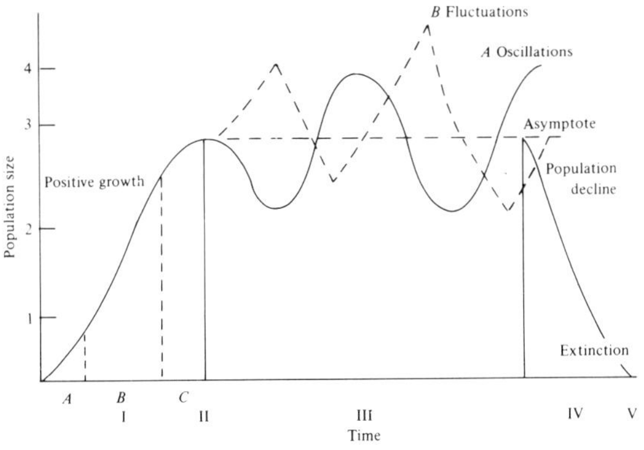

At the other extreme population level 4 could well apply to insects such as various cutworms which are only economic pests in Europe at irregular intervals at times of population irruption. Most of the more common pests would come into the categories which reach pest density at population levels 2 and 3. The growth of a population can be expressed very simply in the equation:

$$P_2 \rightleftharpoons P_1 + N - M \pm D,$$

where P_2 = final population, P_1 = initial population, N = natality, M = mortality, D = dispersal.

To simplify this equation, natality can be regarded as synonymous with birthrate, mortality with deathrate, and dispersal is either regarded as movement out of the population (emigration), or movement into the population from outside (immigration). The object of pest control is to lower P_2, which quite clearly can be done by either lowering the birthrate of the pest, increasing the deathrate, or inducing the pest to emigrate away from the area concerned.

Four hypothetical pest populations are illustrated graphically in Stern, Smith, van den Bosch & Hagen, (1959), in relation to their equilibrium position, economic threshold, and economic injury levels; these graphs are illustrated in fig. 2.

Life-table

The examination of a pest population and its separation into the different age-group components, i.e. eggs, larvae, pupae and adults, enables a life-table for that pest population to be compiled. The construction of a life-table for a pest species is an important component in the

Fig.2. Schematic graphs of the fluctuations of four theoretical arthropod populations in relation to their general equilibrium position, economic threshold and economic injury levels (from Stern *et al.*, 1959).

1. Non-economic species whose general equilibrium position and highest fluctuations are below the economic threshold, e.g. *Aphis medicaginis* on alfalfa in California, USA.

2. Occasional pest whose general equilibrium position is below the economic threshold, but whose highest population fluctuations exceed the economic threshold, e.g. *Cydia molesta* on peaches in California, USA.

3. Perennial pest whose general equilibrium position is below the economic threshold, but whose population fluctuations frequently exceed the economic threshold, e.g. *Lygus* spp. on seed alfalfa in western USA.

4. Serious pest whose general equilibrium position is above the economic threshold, usually requiring insecticide application to prevent economic damage, e.g. *Musca domestica* in milking sheds of dairy farms.

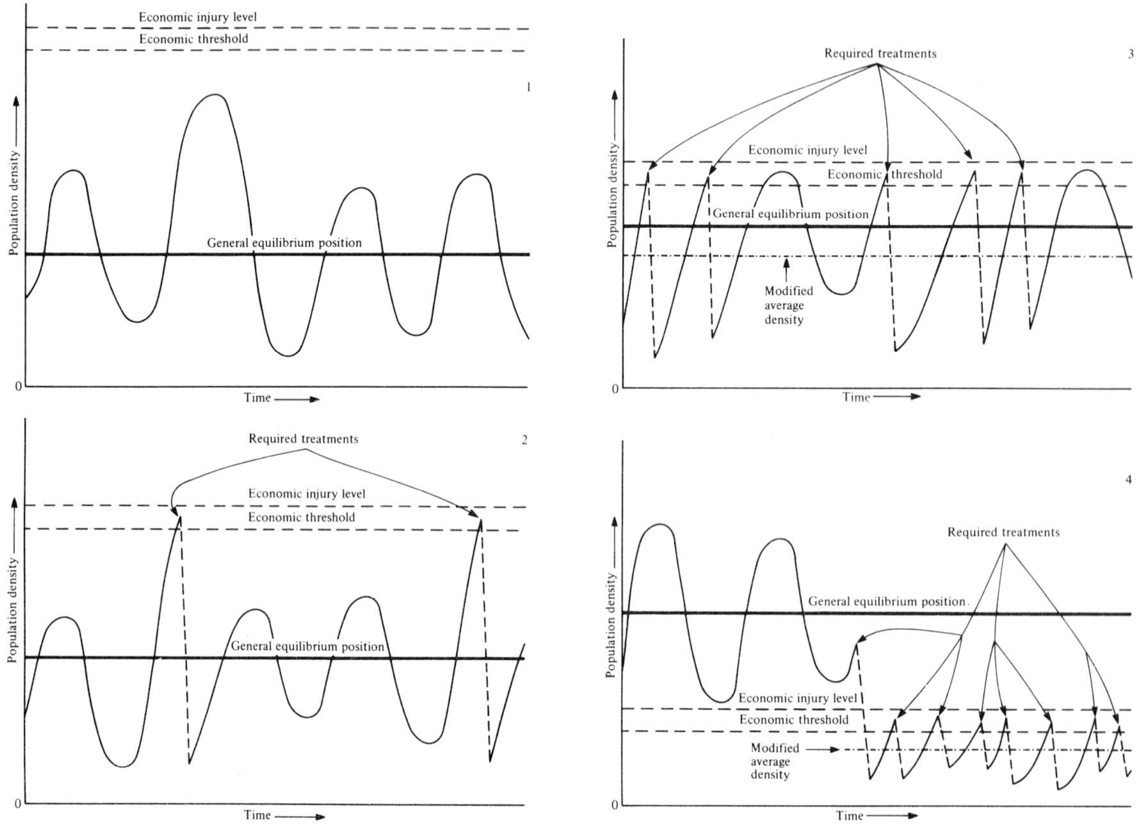

understanding of its population dynamics, particularly in relation to natural predation and mortality, and is, in point of fact, a vital part of any IPM programme. The growth of an insect population, especially the recruitment and the survival of the different stages, varies considerably according to the type of insect concerned. One result of this variation is that there are half-a-dozen different methods for the construction of a *budget* (for further details see chapters 10 and 11 in Southwood (1978)). As pointed out by Harcourt (1969), it is necessary to be careful in the choice of the appropriate method for compiling a life-table budget when planning the sampling methods to be used.

Resurgence

The term resurgence is used to express a sudden increase in population numbers. One type occurs when the target species, which was initially suppressed by the insecticidal treatment, undergoes rapid recovery after the decline of the treatment effect.

It may also occur as a result of the development of a new biotype of the pest, or if the insecticide treatment kills a disproportionate number of the natural enemies of the pest species.

Population dynamics theory

(After Southwood, 1977.) Applied biologists have for a long time been concerned with two basic aspects of animal numbers: firstly that population numbers may change greatly, as pointed out by Andrewartha & Birch (1954), and secondly that most animal populations are relatively stable in comparison with their prodigious powers of increase. It now seems that certain species of animals belong to the one category and others to the second. Southwood pointed out that the change of the population fluctuation to a state of stability is associated with an increasing duration of stability in the habitat, and may be conveniently equated with the *r–K continuum*. *r-strategists* are opportunists, living in temporary (ephemeral) habitats and adapted to obtain maximum food intake in a short time; they are generally small, mobile and migratory, and have a short generation time. *K-strategists* live in stable habitats, often in crowded conditions, with their population size near the carrying capacity of their habitat; they are usually larger in size, less migratory and have a long generation time.

Fig. 3 is a synoptic population model. Three regions can be recognized. First the r-strategists, whose habitats are ephemeral and whose numbers are characteristically 'boom and bust': this strategy is dominated by large-scale migration, massive population losses, and new populations continually developing from a handful of colonizers. Secondly, the K-strategists represent the other extreme, maintaining a steady population at or near the carrying capacity of the habitat, basically in equilibrium with their resources; recruitment, mortality, and migration are low, so there is less opportunity to adapt to changed environments. These animals are specialized to their particular environment, and if their numbers are reduced to a low

Fig.3. The synoptic population model (after Southwood & Comins, 1976). K = carrying capacity (as in K-selected)
Note: equilibrium points only occur where an 'east-facing slope' cuts a zero population growth contour, as only here does negative feedback occur.

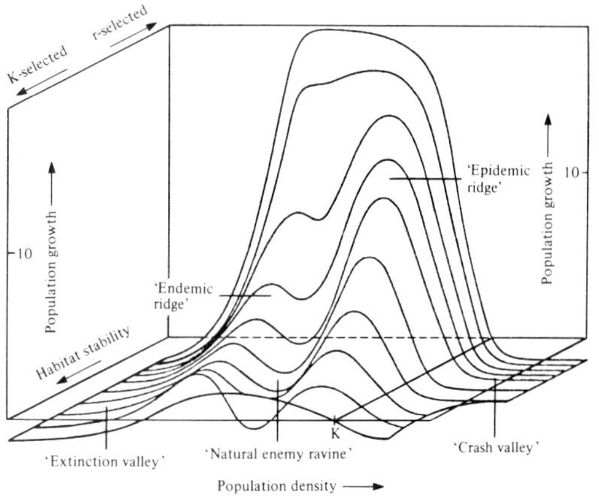

level they are liable to become extinct.

Finally, the middle region recognized is the 'natural enemy ravine'. Both kinds of strategists have a stable equilibrium point, the upper one at the population density of the carrying capacity of the habitat. Where the 'natural enemy ravine' dips below the zero population growth contour there is a second equilibrium point, and where it rises through the contour on the other side of the ravine is the release, or escape, point from natural enemies. Above this point, in the absence of density-independent catastrophes, the population rises to the upper equilibrium point where intraspecific competition mechanisms (disease, etc.) operate. These two levels have been referred to as the endemic (lower) level and the epidemic (upper) level.

r-pests include species such as locusts, armyworms, leafhoppers, aphids, planthoppers, many flies, and, in plants, the ruderals (weeds) belong to this category. *K-pests* include elephants in Africa, tapeworms, Codling Moth, ants, tsetse flies, and many beetles. Obviously the r- and K-pests represent the extremes of a continuum, and there is correspondingly a large group of *intermediate pests*. It is with this large group that natural enemies have most population impact.

Applied biologists generally appreciate that habitat characters are important indicators for IPM strategies, and Conway (in May, 1976) has shown that as the r–K continuum is related to habitat characteristics it is relevant to decisions on the choice of a particular control strategy.

Insect pest diversity (competitive exclusion)

Many zoology students have been taught the idea that no two animal species can occupy the same ecological niche without one species (the 'stronger') replacing the other (the 'weaker') over a period of time. Supporting evidence was usually an experiment carried out by Gause in 1934: he kept *Paramecium aurelia* and *P. caudatum* together in nutritive fluid in a small container; after about three weeks the latter species was exterminated. This idea is generally referred to as 'Gause's hypothesis', or 'the principle of competitive exclusion'. By definition this refers to the exclusion of one species by another when they compete for a common resource (often food) that is in limited supply; the principle being that two species with identical ecological requirements cannot coexist indefinitely.

It would seem that the basic principle of competitive exclusion is clearly valid, and it may be a factor of importance in evolution. But the *Paramecium* experiment was clearly a very simple case where both species ate the same simple food within an enclosed habitat, and as such represented a very artificial situation. For most phytophagous insects either the food sources are not limited, or else the ecological requirements of the different insect species are not identical, even though they may be quite similar.

When a student finds a particular crop pest species *in situ* on a host plant, frequently the assumption is made that 'that niche is clearly occupied so there will not be another species in that microhabitat'! In practice the converse situation prevails, as is generally recognized by most experienced field biologists: if one particular insect species is found on a plant at a particular location on the plant body, then the student should expect to find other species at the same location. It should never be assumed that an infestation is a single-species population; it is actually preferable to assume that each infestation may be a mixed population of several closely related (or otherwise) species, until proven otherwise. Many ecological studies have failed because of the inability of the observer/recorder to recognize a mixed-species population.

Thus it should be expected that many natural animal populations are likely to be composed of several (often closely-related) species, sometimes very similar in appearance and occasionally indistinguishable morphologically. Common examples in agricultural entomology include the following: the stalkborer complex on maize, rice, sorghum and sugarcane (Lep., Pyralidae, Noctuidae); scale insects on *Citrus*; mealybugs on sugarcane; aphids on lettuce; aphids on potato; leafminers on apple;

chafer grubs eating the roots of sugarcane; weevil larvae eating roots of strawberry.

On many agricultural crops, food for an insect pest can be regarded as virtually unlimited; also most closely related insects have slight differences in their basic ecology or diet, so in most insect/crop plant situations the concept of competitive exclusion does not apply.

Insect pheromones in relation to pest control

Pheromones, originally referred to as ectohormones, are complex chemical compounds, basically long-chain hydrocarbons such as alcohols, esters, ketones, aldehydes and sometimes ethers. Many have now been successfully identified, and some synthesized; a number are now available commercially from some chemical companies, for pest monitoring or control purposes.

They are secretions from several different types of glands, on different parts of the insect body, which open directly to the exterior, the secretory product is usually airborne for its distribution. Their basic function is for communication of a specific type between individuals of the same species, and the chemical elicits a specific reaction in the receiving individual. Within the large, complex, social colonies of ants, bees, wasps and termites, apparently quite sophisticated systems of communication have evolved, mostly based upon the use of pheromones.

The term 'pheromone' is usually regarded in a behavioural context in that it is a chemical or chemical complex that elicits a specific behavioural response in the receiving insect. Apparently some glands secrete a single chemical, whereas others secrete several which appear to act in concert. Generally there has been no overall agreement for a scheme of classification, but most workers favour the basis to be the type of behaviour released in the receiving insect. One approach is to consider them to be of two basic types, those which give a releaser effect (this entailing a more or less immediate and reversible effect on the behaviour of the recipient), the others having a primer effect (this starting a chain of physiological events in the receiving insect). The latter group are usually gustatory in operation and typically control the social behaviour in Hymenoptera and Isoptera. Behaviour-releasing pheromones are typically odorous and their action is direct upon the central nervous system of the recipient, usually through the chemoreceptors in the antennae.

Recent work has demonstrated that most insect pheromones are in fact a complex of chemicals, and the individual chemicals are referred to as *components*. For example, the Smaller Tea Tortrix Moth sex pheromone has four components, two are major components and two are minor. Current opinion is that further research will reveal that almost all pheromones are actually chemical complexes of several compounds, and that the original idea of there being only one chemical present is completely incorrect, and originated because at the time the methods of chemical analysis were insufficiently sensitive to detect the minor components.

It is now apparent that much of the early work on insect pheromones is largely worthless (or, at best, of limited value) in that the researchers did not appreciate that almost invariably each pheromone was in fact a complex of major and minor components acting in concert on the receiving insect. Experimentation using only some of the chemical components of a particular pheromone inevitably led to anomalous results.

The types of behaviour (pheromones) used by Shorey (1976) were aggregation (including aerial and ground trail-following), dispersion, sexual, oviposition, alarm and specialized colonial behaviour. Some pheromones appear to have more than one function so these categories are of somewhat limited application.

Aggregation

The reasons for aggregation are numerous and varied, and include collection around or to a food source, a shelter site, a site for oviposition or colonization, recruitment of a sexual partner, and aggregation for swarming or dispersal purposes.

One of the most obvious cases can be seen by watching ants on their foraging trails, where the scouts are clearly laying scent trails. Trail-following by ants has been well studied. The scouts, after having located a food source, deposit droplets of pheromone on the ground and this stimulates trail-following behaviour amongst other workers. While the food source persists the ant workers continually reinforce the scent trail, but after depletion trail reinforcement ceases and the trail eventually disappears. Some trails are ephemeral but others persist for weeks or months. The Imported Fire Ant (*Solenopsis saevissima*) in the USA uses short-lived trails near the nest, but some species of Leaf-cutting Ant (*Atta* spp.) lay persistent trails to leaf sources up to 100 m distant which may last for months. Similar trails are laid by termites when foraging.

Bees and wasps (*Vespa vulgaris*) leave a scent trail from their feet which is important in delimiting the entrance to their nests.

Bark beetles (Scolytidae) use aggregation pheromones to designate host trees suitable for colonization, as these beetles only flourish when present in quite dense populations. The pheromones are released from the hind-gut of the beetle mixed with the various terpenoid compounds of the host tree which initially attracted the first invaders to the tree. These aggregation pheromones can be released by either sex and serve to attract individuals of both sexes. Colonization is usually succeeded by mating which involves the use of sexual pheromones. This type of aggregation can also be seen in the Japanese Beetle (*Popillia japonica*) and the Cotton Boll Weevil (*Anthonomus grandis*).

Certain mosquitoes release pheromones into the water at oviposition which attract other females. Sheep Blowfly (*Lucilia cuprina*) females apparently use an aggregation pheromone to form dense populations at sheep carcases for oviposition.

Aggregation at a suitable resting site has been demonstrated for the Bed Bug (*Cimex lectularis*) and some other cryptozoic species.

The mechanism for aggregation at a chemical source is usually chemotaxis where the insect can detect the gradient of odour molecules, and it often involves orientation by anemotaxis, that is positive orientation to air currents, particularly in the case of flying insects.

Dispersal

Dispersal is clearly the opposite of aggregation, but it is not encountered very often. However, some bark beetle males produce pheromones after mating which repel other males, and some female beetles release a repelling pheromone when they are unwilling to mate. *Tribolium confusum* females release a pheromone in the foodstuff they infest which repels other females and ensures a uniform population distribution throughout the available space. It is thought that the female Apple Fruit Fly leaves a pheromone trace on the apple surface after oviposition, for she can be seen to drag her ovipositor over the surface and generally other females do not lay their eggs in the same fruit.

It seems that some dispersal secretions are the same as the defence secretions; for instance, nymphs of *Dysdercus* produce stinking coxal gland secretions when disturbed (thought to be a defence against predators) which causes the gregarious bug nymphs to scatter. Certain species of ants have alarm pheromones which in some circumstances (in the nest) induce aggregation but under other conditions (away from the nest) result in dispersal.

Sexual behaviour

Sex pheromones may be produced in either sex and stimulate a series of behavioural sequences that usually results in mating. Typically, there appears to be a hierarchy of behavioural responses with increasing stimulation by sex pheromones. Once the two sexes are in proximity there is usually a close-range series of behavioural reactions, referred to as courtship behaviour.

The most usual situation is that a receptive virgin female insect will announce her availability through release of aerial sex pheromones, known as 'calling', and these cause a flight response and approach by receiving males. The night-flying moths (especially Saturniidae, Geometridae, and Noctuidae) are best known for their nocturnal emission of sex pheromones, which reputedly can attract males from as far as 5 km downwind. Males may produce a pheromone (sometimes called an 'aphrodisiac') when in the immediate vicinity of the female, which operates by inhibiting the female's tendency to fly away.

Sex pheromones are commonly referred to as 'sex attractants' or 'sex lures', which is misleading in that it implies that the odorous chemicals simply cause attraction, which is a great oversimplification. As will be discussed later, the male response to a female pheromone is complicated and sequential, involving half-a-dozen or more separate stages.

Most of the sex pheromones that have been isolated, identified and synthesized are from the Lepidoptera, and include Red Bollworm, Spiny Bollworm, Pink Bollworm, *Heliothis* spp., *Spodoptera littoralis*, *Chilo* spp., *Prays citri*, *Prays oleae*, Gypsy Moth, *Bombyx mori*, Cabbage Looper, Codling Moth, and Honey Bee queen. Pheromones of some of the most important fruit flies (Tephritidae) such as *Dacus* and *Ceratitis* spp. have also been synthesized, as have some for Scarab Beetles and Scolytidae. Many of the sex pheromones are either difficult to synthesize or else expensive to produce, and this has led to the development of chemical pheromone mimics for large-scale management programmes. These chemicals are discussed in the section on 'attractants' which follows later.

Oviposition

As already mentioned, Sheep Blowfly females release an aggregation pheromone when they oviposit on a suitable sheep carcase in Australia, and the result is the formation of a dense population. Bark beetles (Scolytidae) aggregate on suitable trees as a result of use of aggregation pheromones, but the ultimate purpose of the aggregation is for oviposition and breeding. Thus, functionally many of the aggregation pheromones are also used to stimulate oviposition upon the correct host plant. This point is mentioned later under the heading of 'plant odours', as there appears to be probable interaction between plant volatiles and pheromones connected with the oviposition of many phytophagous insects.

Alarm behaviour

This is characteristic of social Hymenoptera and Isoptera, and may be seen most dramatically when field workers disturb a nest of Paper Wasps (*Polistes* spp.) or arboreal ants in plantation trees. The wasps (and bees) produce alarm pheromones both when they sting and when gripping with their mandibles. The pheromone is released from glands in the stinging apparatus and from the mandibular glands, but apparently the worker can open the sting chamber and emit the pheromone without the necessity of stinging.

It has recently been demonstrated that aphids secrete alarm pheromones from their siphunculi when distressed. Recent work at Rothamsted has shown that a wild potato (*Solanum berthaultii*) produces a chemical mimic of this pheromone from special secretory hairs on the foliage which appears to repel aphids from its leaves.

Attractants

Initial experimental studies on female sex pheromones were conducted using live virgin females that had been laboratory-reared, or else a chemical extract was made from the abdomen tips of many young females and this was used instead of the live insects. It was soon apparent that these sex pheromones have considerable potential application in pest management programmes and so many organic chemists in government and industrial establishments started work in this field.

One approach was to carefully analyze the tiny quantities of natural pheromone produced by virgin females, and, when the chemical components were isolated and identified, to attempt to synthesize the same chemical compounds in the laboratory.

The second approach was to try and synthesize closely related chemical compounds which might possess the behavioural qualities of the natural pheromone, but were easier and cheaper to manufacture. In this way pheromone homologues and analogues have been produced commercially. A *pheromone homologue* is a very closely related compound, which differs only from the natural pheromone by chain lengthening or shortening following, for example, addition or removal of a methylene group. A *pheromone analogue* is a less-closely related compound that has major basic differences in structure, such as the change of a functional group or its position, for example an alcohol, ester or ketone.

The third approach was to use a vast range of organic chemicals, which it was thought might possibly function in a manner similar to sex pheromones, and to use them in laboratory and field trials in a purely empirical manner to see if they did possess such qualities. The organic chemicals that have been found to be successful in attracting certain male insects are collectively referred to as *sex attractants*. Obviously for monitoring purposes it does not matter how the chemical attracts the insects so long as it does attract them sufficiently well, and a great deal of research effort is being expended in this field at present. Sometimes these chemicals are termed *pheromone mimics*, for the obvious reason that they produce a similar reaction in the receiving male insect. 'Hexalure' is a chemical attractant produced commercially in the USA for use with Pink Bollworm on cotton, in a disruptive technique to prevent mating.

Recent work has demonstrated that many pheromone complexes are subjected to *synergism* in one way or another. In some cases it appears that some of the minor components have a synergistic effect on the major components or else on the pheromone complex as a whole. Sometimes the synergist may be a chemical released by the host plant; this may be of more importance in aggregation or oviposition behaviour, for example in scolytid infestations of forest trees various terpenoids are released by the injured tree which interact with the aggregation pheromones released by the beetles. Empirical chemical testing has discovered a number of synergists for use with sex pheromones that appreciably enhance their performance.

The sex pheromones of insects must, by their very nature, be quite specific to each species but some of the attractants have a very useful and much broader response, for example 'Cu-lure', developed initially for Melon Fly (*Dacus cucurbitae*), and methyl eugenol both attract all species of *Dacus* and some other fruit flies in addition, which makes these chemicals very useful for survey studies.

Sex attraction in Lepidoptera

Most of the work on sex pheromones and attraction has been done on the Lepidoptera, and the greatest potential for pheromones in pest management is in this group.

The 'calling' female moth emits her sex pheromones from the genital opening on the abdomen tip and the chemical complex is carried downwind as a plume. The odour plume is basically cone-shaped, but is flattened ventrally if the moth is close to the ground; the plume widens and the chemicals disperse as they are carried from the source. The shape of the plume is clearly controlled by wind speed and direction, the contours, and the presence of tall vegetation such as trees. Most virgin female moths 'call' at dusk or at night when there may be strong temperature gradients, or even inversions, over the ground, and these will obviously have effects on the spread of the pheromone.

The male moth responds to the pheromone by anemotaxis, in that it flies upwind in a zig-zag pattern. At first (assuming the male to be some distance from the female) the flight pattern of the male moth diverges from the plume of pheromone quite often, but as it approaches the female the pheromone concentration increases and the flight of the male moth becomes more direct.

The patterned response by a resting male moth to female sex pheromone can generally be described in half-a-dozen sequential stages, as follows:

(a) reception – antennal elevation or twitching,
(b) activation – wing fanning or fluttering,
(c) active flight,
(d) orientation to the source of pheromone – i.e. anemotaxis,
(e) alighting – landing in the immediate vicinity of the female moth,
(f) courtship – including gland extrusion and release of male pheromone,
(g) mating.

Thus it is clear that the behavioural response by a male moth to a 'calling' female is a complex sequential series of events and is not just a simple attraction.

The initial responses are made to a low concentration of pheromone, but the later events require an ever-increasing concentration of pheromone. It is now thought that the different components of the sex pheromone are responsible for different parts of the behaviour sequence. Thus, if in an experiment one minor component is missing, this will result in the behaviour sequence being broken, and the experimental results confusing.

With *Adoxophyes orana* it has been demonstrated that the female sex pheromone has two components and that there are three different types of chemoreceptors on the male antennae (Den Otter, 1980). One type of sensillum reacts to the first component in electroantennagram studies, the second type reacts to the second component in the female pheromone, and it is suspected that the third type of sensillum may react to the male pheromone at close range, but this has not yet been demonstrated.

Insect feeding on plants

Insect feeding

The feeding process in animals involves different aspects, all of which have some importance in relation to control of pests. The main aspects include:

(a) Recognition of food
    (i) distant
    (ii) proximal
    (iii) contact
(b) Manipulation of food
(c) Ingestion

(a) Recognition of food

(i) *Distant*. This is usually (for most animals) a sight reaction to shape and colour; at what distance host/food recognition is achieved is probably very variable. With some birds of prey it can be up to a kilometre, but at the other extreme for mammals such as moles it is probably only a few centimetres. For many haematophagous Diptera, host/prey movement is very important for distant recognition, followed by heat radiation when closer; but these factors are scarcely applicable to plant hosts. With herbivorous insects long-range recognition is probably a combination of sight and smell. A recent review on this topic is by Prokopy & Owens (1983). Presumably with a large crop, odour recognition from a distance may be achieved; this must give some crop pests a distinct advantage over insects feeding on wild hosts in natural (mixed) vegetation.

Yellow coloration is certainly an attraction to many insects, and it is thought that this is basically because the young and very old leaves have more available foodstuffs in the tissues than the mature leaves: there would certainly be soluble sugars in such leaves as opposed to insoluble starch deposits, and apparently a higher than usual nitrogen content usually confers (in young leaves) a yellowish coloration.

(ii) *Proximal*. At closer quarters sight may still be important, as some plants do have what appear to be quite definite recognition signals for searching insects (although most of these are connected with the need to attract pollinators). Scent recognition is presumably of prime importance, utilizing both natural volatile elements and also metabolic by-products.

(iii) *Contact* (host-plant testing). If the host plant looks and smells correct then the insect alights and contact is effected. The aura of the plant will then be reinforced or superseded by the taste; these stimuli being received through chemoreceptors on the fore-tarsi and the palpi of the mouthparts.

(b) Manipulation of food. This involves the cutting up of pieces of plant material by the mandibles, tasting them, and their manipulation by the various mouthpart structures into a position for ingestion. For sap-sucking insects this involves the insertion of the proboscis into the correct site for food ingestion, sometimes into the xylem vessels (Cercopidae, etc.), usually into phloem tubes (Aphididae, etc.), and sometimes just into mesophyll tissues or a ripening fruit.

(c) Ingestion. For insects with biting and chewing mouthparts the addition of saliva to the chewed fragments is important for lubrication to avoid damage to the oesophagus. The Hemiptera apparently all inject saliva and/or regurgitate stomach enzymes when feeding. Precisely why this is done with plants is not clear, but obviously with blood-sucking and predacious bugs their saliva contains an anti-coagulant which permits them to feed without the blood clotting in the proboscis. The predacious forms also practise external digestion of part of their prey in order to be able to render their food liquid enough to be imbibed through their proboscis. The injection of saliva and/or enzymes into the host plant when feeding is of importance agriculturally as the plant reacts to the presence of these alien substances by growth distortion, or necrosis of tissues.

Plant odours

Most plants release volatile odorous chemicals into the atmosphere (although the majority are undetectable by human sense) and phytophagous insects react to these chemical stimuli when locating host plants. Monophagous and oligophagous phytophagous insects usually react to specific volatile odorous chemicals in, and emitted from, the host plant. It is thought that polyphagous insects either have no olfactory chemical response, or else react to general plant chemicals. The olfactory chemoreceptors are mostly situated in the antennae, but in some Diptera they are located on the tarsi (feet).

It has very recently been demonstrated that very strong plant odours can inhibit sex pheromone reception; the interaction of plant odours and sex pheromones is thought to be complementary under natural conditions, so that mating is more likely to be successful on appropriate host plants, whereas the chances of mating taking place on inappropriate host plants are reduced. As this is a very recent discovery, as yet little work has been carried out, but future studies might well give rise to a greater understanding of the general phenomenon of host-specificity in phytophagous insects.

The chemical attractants in plants, when identified, are usually a mixture of many different compounds, for example, cruciferin in the brassicas is a complex mixture of glucosides, amines, and other chemicals. Biochemical research has shown that most of the more important volatile chemicals in plants are secondary metabolites.

Work at Wellesbourne (NVRS) has demonstrated that Cabbage Root Fly are clearly attracted to some of the volatile chemical components released by plants of the Cruciferae, and some stimulate increased egg-laying. It was shown that gravid females could move at least 24 m upwind to a brassica crop in response to the odour stream. Volatile hydrolysis products are constantly released, at low concentrations, from Cruciferae during normal growth and development, resulting from damage or death of cells, and by the endogenous enzyme system. More than 23 different compounds were obtained from cultivated crucifers at NVRS (Cole, 1980); the actual component constitution varied with the species, the age and stage of development of the plants. Only a few of the 23 compounds actually elicited a response from the flies when used in isolation. The overall situation with regard to plant production of volatile chemicals is clearly complicated, but a great deal of research effort is being expended on this subject worldwide, and gradual elucidation is to be expected. Other relevant publications include Wallbank & Wheatley (1979), Ellis, Cole, Crisp & Hardman (1980), and Crowson (1981). An especially useful book is *Insect Herbivory* by Hodkinson & Hughes (1982).

Ferns as a group are little attacked by insects, and this is thought to be because they contain considerable quantities of repellent/toxic chemicals such as ecdysones, glycosides, phenols, sesquiterpenes, tannins, thiaminase, etc.; the group is ancient and has presumably been grazed extensively, particularly in the days before the evolution of the flowering plants. One striking feature in ferns is that the chemical production is at a peak very early in the growing season so that the youngest fronds are usually the most toxic as well as containing the most available protein.

Studies at Rothamsted have shown that a type of wild potato (*Solanum berthaultii*) from Peru, has two types of 'hairs' (trichomes) on its foliage: one type is short with a sticky head, and the secretion can trap insects on the leaves and stems by adhesion; the other hairs are longer and secrete fluid containing (E)-B farnesene, which is the main ingredient of the alarm pheromone in most species of Aphididae. Experiments showed that *Myzus persicae* aphids were reluctant to invade the foliage of this plant.

In addition to the attractive odours, plants also produce volatile chemicals that function as repellents (from a distance) or feeding inhibitors (at close range), and these form part of the plant defence mechanism against insect attack.

Plant resistance to insect feeding

A very recent and exciting development in insect/plant relationships is the concept of rapidly induced anti-insect defences in plants; this is mentioned at the end of this section. It has even been postulated that there can be communication between adjacent trees through airborne chemicals, so that neighbouring trees increase their defences before being attacked by the insects. However, it should be stressed that this line of research is very much in its infancy and the available data may have alternative interpretation (see Fowler & Lawton, 1984).

The basic resistance exhibited by plants to insect attack is partly physical and partly chemical. The physical properties include:

(a) Thickened cuticle
(b) 'Hairy' epidermis (trichomes may be hooked, secretory, or just physically close together)
(c) Hardening of tissues by general sclerenchymatization
(d) Increasing the extent of natural silica deposits in the tissues
(e) Spiny leaf margins (e.g. holly) may deter some leaf-margin eaters (the thorns and spines developed presumably to deter vertebrate grazers and browsers are not effective against insects; in fact some insects mine spines!)

The chemical defences include:

(a) Absence of specific attractants or feeding stimulants that would otherwise normally be present
(b) Presence of repellent odours to deter insects from alighting on the plant
(c) Presence of distasteful or poisonous chemicals in the tissues to deter feeding
(d) Absence of certain chemicals (often amino acids) required for normal development of the immature insects
(e) Presence of chemicals that mimic insect alarm pheromones, for example the wild potato already mentioned and the aphid alarm pheromone mimic

The major chemical repellents in plants seem to be terpenes, tannins and various alkaloids. Tannins are mostly found in horsetails, ferns, gymnosperms, and some angiosperms, and they are quite antibiotic to many pathogens. Alkaloids are mostly found in angiosperms and are thought to be of more recent origin. It is thought that tannins were developed initially in the process of evolution as a deterrent to grazing reptiles, and the alkaloids similarly evolved as a protective mechanism in angiosperms to repel grazing mammals. It seems unlikely that the Insecta were at all involved in the evolution of feeding repellents in plants, although these may now be of considerable importance with respect to phytophagous insect feeding behaviour. The insect biotypes that come to feed on 'repellent' varieties of crop plants (and other plants) usually develop biochemical detoxification mechanisms, so that the poisonous compounds are broken down into non-toxic degradation products.

It has been suggested that 'dominant' (termed 'apparent') species of plants have chemical defences that tend to cause digestive difficulties and retard development of insects feeding on their foliage, rather than actually causing death (see Crowson, 1981; chapter 18); and also that the pests of these plants tend to be polyphagous (and possibly the more ancient) species. Good examples of such dominant plants include the oaks (*Quercus* spp.) and beech (*Fagus* spp.); their main defences against insect herbivores appear to be a combination of sclerification of the leaf tissues and accumulation of tannins. After about a week or two from leaf-unfolding, the young leaves become quite inimical to the insects; larval development slows, mortality increases, fewer eggs are laid, etc. Many insects have adjusted their life cycles so that they are able to feed on the young leaves during the short period of time while they are palatable. The less-dominant plants (termed 'unapparent'), which are usually the more recently evolved, tend to be the ones to develop actual poisons (alkaloids, etc.) as their chemical defence.

It is now clearly established through recent research that some plants respond to insect feeding damage by active production of deterrent chemicals, so that their final concentration in the leaf tissues is considerably greater than before. In some cases the whole tree produces more chemicals, not just the damaged leaves. The entire subject of insect/plant relationships has recently been evoking a great deal of interest and much has been published of late.

As mentioned at the start of this section, it has been said that insect-injured plants emit volatile chemicals that stimulate neighbouring trees to produce defensive chemicals irrespective of whether they are attacked or not. But the data at present are not really conclusive, and are usually open to other interpretation. It might be expected that the levels of chemical secreted into the air would be too low a concentration to elicit such a response from a receiving tree.

Food sources for adult insects

When considering the subject of insect pest feeding, it is invariably assumed that it is the pest stage of the insect feeding on the cultivated plant. With groups such as Orthoptera, Hemiptera, and some Coleoptera (e.g. Chrysomelinae) both adult and immature insects are found side-by-side on the crop and both cause damage (often identical, sometimes different). However, within the Diptera and Lepidoptera it is the larval stages that are agricultural pests; the adults are free-living and with a very few exceptions (such as fruit-piercing moths) not crop pests, although some females may cause damage to the plants when ovipositing. Most of these adult insects (females anyway) require food prior to ovulation or egg development, and in temperate regions the spring/summer emergence of adults is often closely synchronized with the flowering of various local wild herbs and shrubs.

A striking example of this dependence upon local vegetation is seen with many flies (Muscidae, Anthomyiidae, Psilidae, etc.) where the newly emerged adults congregate and feed upon the flowers of wild Umbelliferae common on headlands and in hedgerows. The most abundant and widespread species of Umbelliferae concerned as nectar sources as follows:

Common name	*Scientific name*	*Flowering period*
Cow Parsley	*Anthriscus sylvestris*	April–June
Hemlock	*Conium maculatum*	June–July
Hogweed	*Heracleum sphondylium*	June–September
Upright Hedge Parsley	*Torilus japonica*	July–August
Wild Angelica	*Angelica sylvestris*	July–September
Fennel	*Foeniculum vulgare*	August–Sept.

One reason for the importance of this group, apart from their widespread distribution, is the sequential flowering periods which result in a continuous flower availability in hedgerows from early April until the end of September, or even into October. Thus there is nectar available for each successive generation of adult flies as they emerge.

For many groups of insects there is no obvious link with a particular group of nectar-producing plants upon which they are dependent. Many moths are quite opportunistic in their feeding and will take nectar from many different sources. The early spring in Europe is characterized by a paucity of flowers: only a small number of plants have flowers at this time so natural sources of nectar for early-emerging flies (and moths) in April are quite limited. Later, by June and July, the countryside is a profusion of flowers, and nectar sources are numerous.

Adult insects that emerge and oviposit over the winter period, such as Winter Moth and other hibernal Geometridae, use their fat body for nutrients and they lay their eggs without feeding at all; but, of course, their pupal period was relatively brief.

Many of the muscoid flies, whose maggots attack vegetables in the soil, or cereal seedlings, feed predominantly on the flowers of the wild Umbelliferae listed above, and in most cases it appears that the nectar feed is necessary for egg development. These flies typically emerge from overwintering pupae in April, in the UK; they include Cabbage Root Fly, Onion Fly, Bean Seed Fly, and Carrot Fly amongst others.

Knowledge as to the feeding requirements/preferences of the adult insects can be ecological information of great use in survey studies and in population monitoring, and might on occasion be used for an adult insect control programme. This would be a line of research that might be profitably pursued for a number of important crop pests.

Abundance and richness of insect (arthropod) faunas on host plants

Another aspect of insect–plant relationships receiving attention recently is the analysis of the factors seemingly responsible for the abundance (numbers of individuals of each species) and richness (numbers of different species) of the insect fauna (including phytophagous mites) on trees. Most of the published studies refer to indigenous trees, but the basic ecological concepts involved should have application to the study of long-term orchard and plantation crops and their pests. Clearly these studies are, at present, confined to more or less permanent hosts such as trees, for the study of annual plants would present additional complications reflecting their ephemeral nature.

In a recent paper by Kennedy & Southwood (1984) (which includes a comprehensive bibliography) they investigated the following factors with reference to insects on British trees: host-tree abundance, time, taxonomic isolation, tree height, leaf size, and two more nebulous characters termed ‘coniferousness’ and ‘deciduousness’. The most important factors were the first two. *Host-tree abundance* refers to the overall area of habitat available for colonization; a larger area is also likely to provide more different microhabitats and thus a wider range of niches for a larger group of associated species. *Time* refers to the evolutionary age of the tree species, and would seem to be positively correlatable to the species richness of the insect community, although there has been some controversy on this point. Of the other five factors considered by Kennedy & Southwood all, with the exception of ‘coniferousness’, apparently made significant contributions to the present recorded diversity.

Banerjee (1981) made an analysis of tea pest species and reported that time (measured as the age of the plantation) was the major factor in relation to pest recruitment; pest diversity reaching maximum at a plantation age of about 35 years. With other tropical plantation crops it seems that other factors may be more important in relation to pest diversity.

A number of useful papers on insect–plant relationships are presented in the Royal Entomological Society of London, Symposium Number Six, edited by H.F. van Emden (1972).

Insect (pest) distributions

The distribution of animals and plants throughout the world is controlled by many different factors. With cultivated plants clearly the most dominant factor has been the deliberate transport by man. For insects and other animals the evolutionary history of the area is of importance, but probably the overriding factor is the climate. Of the climatic factors temperature is the most important (and most easily measured). As insects are poikilothermic they have only a little influence over their body temperature over which the various bodily functions operate most efficiently, and they have a heat death point and a cold death point at which they die. From the point of view of distribution globally the cold death point is the most important, and the insects can be divided into three main groups on this basis, as follows:

(a) Tropical insects – cold death point circa 10–15 °C
(b) Temperate insects – cold death point at 0 °C (death because of ice crystal formation in the cells/tissues)
(c) Boreal (Arctic) insects – death point well below 0 °C (−20–30 °C often); body fluids supercool and freeze to glass

A few well-known species are clearly *eurythermal* and are able to have a worldwide distribution as they can function over a wide range of ambient temperatures. Some others are *stenothermal* and only thrive in a narrow range of temperatures, either low, high or intermediate.

Pest organisms are species renowned for their biologically aggressive and opportunist nature in relation to hosts, and on the whole their distribution tends to be ever-increasing to the limits of suitable environmental conditions; these limits are often climatic.

Knowing the *optimum* conditions of temperature and relative humidity for the development/activity of an insect species, and its preferred range of conditions, it is possible to plot a *climatograph* with different areas of suitability/abundance for the species. This is sometimes used in making prediction assessment of the climatic suitability of an area for an outbreak/invasion of a particular pest; knowing the zones of suitability in regard to climate for the pest, if the monthly means of temperature and humidity are plotted on to the graph a polygonal diagram results, and the placement of the diagram indicates the likelihood of climatic suitability. In fig. 4 is shown the climatograph for the Medfly (*Ceratitis capitata*) in relation to (A) Orlando, Florida, (B) Naples, Italy and (C) Ankara, Turkey (from Edwards & Heath, 1964).

Fig.4. Climatograph for the Mediterranean Fruit Fly. A = Orlando, Florida; B = Naples: C = Ankara (from Edwards & Heath, 1964).

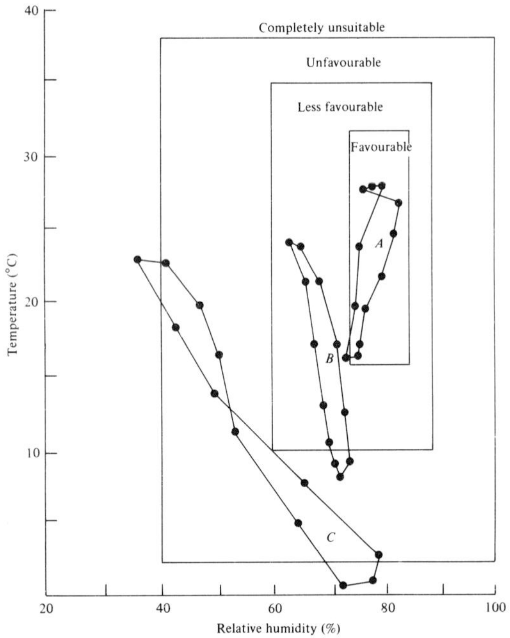

If temperature and rainfall are used as criteria the diagram is called a *hythergraph*.

After extensive laboratory studies and field observations it is possible to designate three fairly distinct zones of abundance for each insect (pest) species, as follows.

(A) (Endemic) zone of natural abundance. Here the pest species is always present, often in large numbers, and regularly breeding. Environmental conditions are generally optimal for this species, and in this zone the species is regularly a pest of some importance.

(B) Zone of occasional abundance. Here the environmental conditions are either less suitable (i.e. drier, cooler, etc.) or else with pronounced variation (often seasonal), having periods of suitable conditions alternating with unsuitable. The population is kept low by the overall climatic conditions; some breeding does occur, but only occasionally does the population rise to pest proportions. Sometimes climatic conditions are sufficiently severe to destroy the entire population, which then has to be re-established by dispersal from the endemic zone.

(C) Zone of possible abundance. This is essentially a zone into which adult insects spread (disperse) from zones (A) and (B). The immigrant population may survive for a time, and may actually be a pest for a while, until changing climate destroys the organisms. Breeding in this location is rare, but permitted occasionally by a period of mild weather. Occupation of this zone is strictly ephemeral (short-lived).

In fig. 4, the three boxes on the graph (A, B, C) could be regarded as corresponding to the three natural zones of abundance.

The basic nature of an insect population is to increase, and unless it is controlled by changing climate, heavy predation or parasitism, or artificial control measures (i.e. insecticide spraying) there is usually dispersal of part of the population to alleviate the competition pressure for food or other limited resources. Thus many pest species increase in numbers in zone (A), and when the population density is high some disperse into zones (B) and (C) from time to time. These three zones are not necessarily constant in their demarcation, depending in part upon the nature of the limiting factors controlling the distribution of the pest organism. Often the main limiting factor is available food, and if the host crop becomes more widely cultivated then many pests may follow the crop into the new regions.

The dispersal success of a pest organism depends upon several factors, including the effectiveness of the precise method of dispersal (e.g. insect flight, wind-carried fungal spores, transport on agricultural produce, etc.) and the adaptability of the pest. Many of the most successful pests have a eurythermal physiology and a polyphagous diet.

As an example of the interaction between temperature and relative humidity the diagram made by Uvarov (1931) for the Cotton Boll Weevil can be used (fig. 5). It may be generally regarded that conditions optimal for rate of development are also optimal for the whole organism and its general well-being. Strictly speaking this may not

Fig.5. Time of development of the Cotton Boll Weevil (in days) in relation to two climatic factors (after Uvarov, 1931).

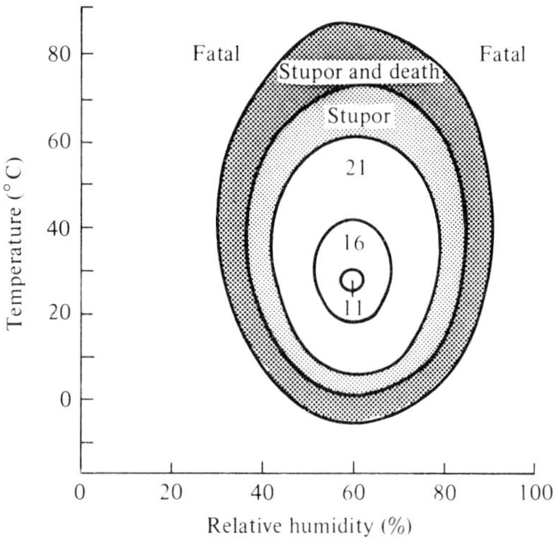

be true, for in some species the different physiological processes have slightly different optima.

Changing distributions

In the past many of the major changes in the distribution of a pest species were made through human agency, for example the Gypsy Moth (page 34) or Colorado Beetle (page 44), either intentionally or accidentally. But occasionally an animal drastically increases its distributional range under its own powers of dispersal; the reason for the spread is generally not understood. These sudden changes are usually termed *invasions* so far as the new countries are concerned. There have been two quite recent and interesting invasions in the UK. Firstly, the Collared Dove (*Streptopelia decaocto*), an Asiatic species, normally resident (i.e. non-migratory), which invaded Europe through Turkey early this century and was first recorded in the UK in 1951 in East Anglia. This is now widespread and locally abundant, and has even bred in Iceland. It is somewhat urban in habits and regarded as a pest species by chicken rearers as it takes grain fed to the chickens in open runs. A recent insect pest to invade the UK is the American Lupin Aphid (*Macrosiphum albifrons*), first recorded in West London in 1981. It withstood the statutory eradication measures that were immediately implemented, and is now abundant and widespread in England and Wales as far north as Yorkshire. This pest is confined to lupins, so far as is known, and infested plants often die, unless control measures are applied. In both the UK and in many parts of Europe there is recent interest in lupins both as a break crop in cereals, and some annual species are grown for their seeds which have a protein content higher than soybean. Some perennial species are also being used very successfully as pioneer colonizers on open-cast mining reclamation sites. Thus the Lupin Aphid, which first seemed to be only serious to gardeners in destroying their flowers, now appears to have much more serious economic significance. At present it has not been recorded from Europe.

Pests that are *native*, or *endemic*, to a region are referred to as *autochthonous*, the implication being that they have evolved locally. Species found locally, but which are thought to have originated elsewhere, are termed *allochthonous*; they are usually immigrants of one type or another.

Pest distribution

When considering the different insect and mite pests in an area or on a crop it is sometimes necessary to regard them in a broader context of their overall distribution. To elucidate the terminology used in biogeography, fig. 6 shows a map of the World with the generally accepted major biogeographical subdivisions.

Zoogeographically a somewhat different terminology is used, as shown in fig. 7.

Dispersal

This is the natural spread of part of a population away from its source (origin) at a time of high population density. With birds and mammals, the dispersal is often partly to seek new territories, and sometimes it is a population survival mechanism to ensure that, on dispersal of the first brood, the parents find sufficient food in their territory to raise a second brood of offspring. With insects, dispersal sometimes appears to be obviously in response to dwindling food supplies, or a reduction in the availability of suitable food (such as progressive drying of leaves, etc.), and sometimes it appears more as a behavioural quirk which coincides with certain weather conditions. The overall effect is clearly beneficial from the point of view of survival of the species. Dispersal appears to be very important for the overall survival of the species as it enables diminished populations to be replenished, and the spread of genetic material through the entire population is advantageous. Also newly available habitats (such as new agricultural crops) can be colonized. In the same way that all animal and plant populations have an innate tendency to increase, they display a similar innate tendency to dispersal.

Fig.6. General geographical/biological subdivisions of the World.

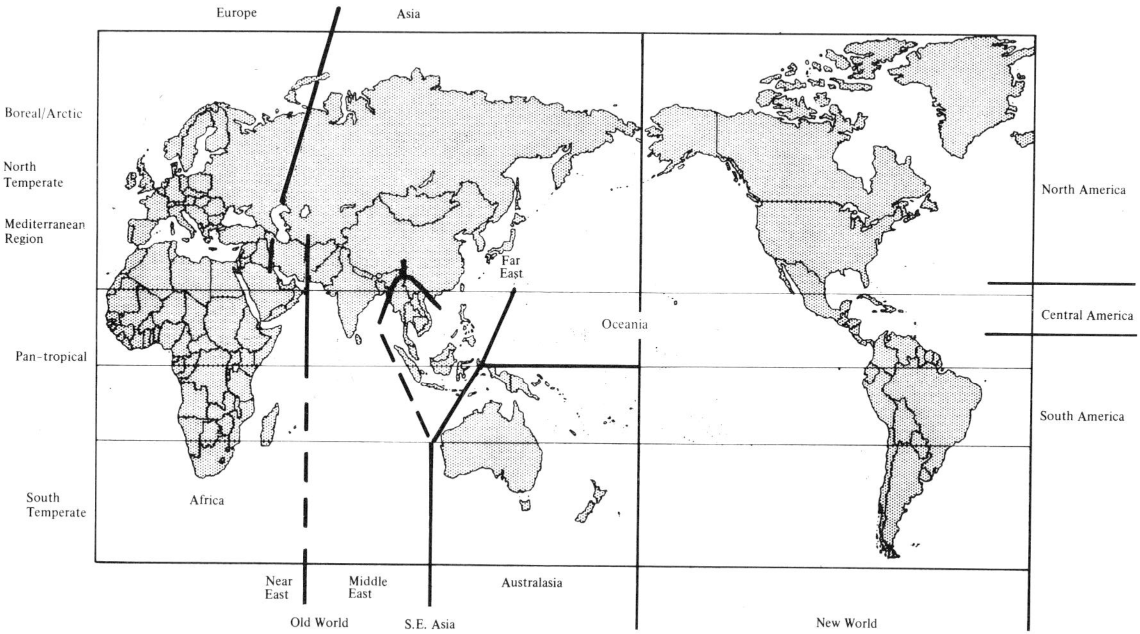

Fig.7. Zoogeographical regions of the World.

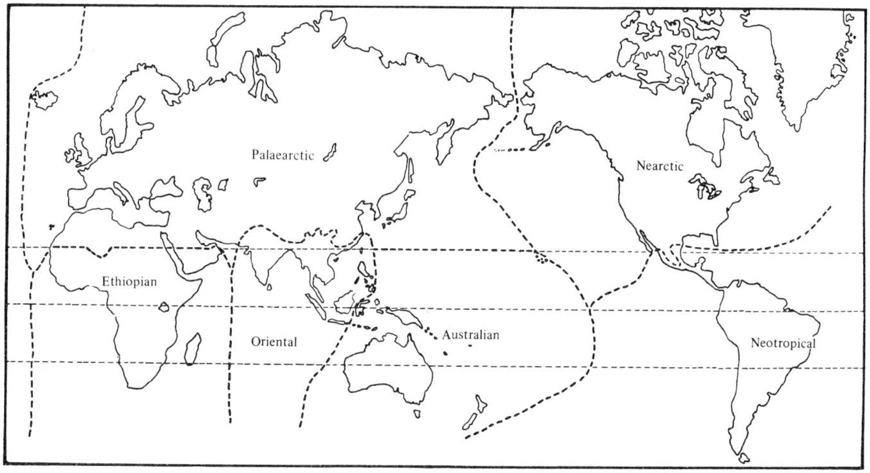

Migration

The movement of animal populations, or individuals, from one area to another on a larger scale than merely a local dispersal, can be defined in three different ways.

(a) Immigration – This is the movement of animals into a region.
(b) Emigration – This is the movement of animals out of a region.
(c) Migration – In the strict sense this applies to a definite double journey, firstly out of one region to another, and then the return to the original region. In the literature the term migration is sometimes incorrectly used to denote just a lengthy dispersal movement. It should be stressed that sometimes it is not clear to what extent a return movement occurs when pest migrations are discussed.

Flying birds, and migratory bats, are clearly in control of their direction of dispersal, although some species appear to prefer to fly into prevailing wind systems, whereas others apparently fly downwind for long distances. Most insects are small in size, and recent research does indicate that even with locusts, which are of moderate size, most movements are completely controlled by prevailing winds. It appears that the insects use their wings primarily to remain in the air and they are then carried quite passively by the wind or air currents to wherever the wind blows; when they stop flying they lose their buoyancy and descend. This is the situation for many long-distance dispersals; local dispersal and food/mate seeking is clearly an active procedure under individual control by the insect(s) concerned.

An interesting short review, titled *Dispersal and movement of insect pests* was published recently (Stinner *et al.*, 1983).

Immigration is the movement of a pest population into an area from elsewhere, and in certain parts of the world is a very important source of major pests. In most of the tropics there is little insect migration (apart from locusts) because of the general stability of climate. Migration is typically an animal phenomenon of the colder parts of the world, where animals move away from northern areas after the short warm summer, and before the onset of the long cold winter. There are some tropical migrations, including the spectacular ungulate migrations of eastern Africa, where animals move to new food sources away from arid areas suffering from their annual dry season, they are basically following seasonal rains and the new grass growth that is promoted across part of the continent.

Animal migrations and dispersal movements are natural phenomena characteristic of all phyla in the animal kingdom and many species now regarded as pests will have generally dispersed during the millenia from their endemic areas (areas of origin). This innate tendency will still be present in all animal populations, but in some cases it operates slowly and may not be at all obvious. Flying insects can effect their own dispersal, influenced by winds and air currents of course. Some of the more novel means of dispersal include riding on floating flotsam (for rats, insects etc.), concealment in the feathers of migrating birds, and aerial transport by typhoon and hurricane. Migration and natural succession was clearly demonstrated on the island of Krakatoa after its total devastation by a volcanic eruption.

Locusts and armyworms (*Spodoptera* spp.) are migratory tropical species of considerable economic importance; they breed in areas of hazardous climate which compels them to disperse on their sometimes lengthy journeys.

The Brown Planthopper of Rice (BPH) (*Nilaparvata lugens*) has recently become established as a major pest of rice in Japan through its migratory behaviour. The climate of S. Japan (Kyushu) is sub-tropical or tropical in the summer, but the winter is cold with snow and ice and the BPH cannot survive the winter. The White-backed Planthopper (*Sogatella furcifera*) is likewise an annual migrant into Japan. These planthoppers (Delphacidae) live along the coastal regions of E. Asia (and elsewhere in India and S.E. Asia) and each spring they migrate northwards, usually entering Kyushu in Japan in the period mid-June to mid-July when they are caught in wind traps along the coast. They then breed on the rice crops and usually have four generations each summer in Japan before dying out in the cooler weather of the autumn after harvest.

Some of the countries most affected by regular pest migrations include Canada, Japan, North China, Fennoscandia, and to a lesser extent the UK. In Canada, particularly, there is quite a large number of insect pests that cannot survive the very cold winter; they arrive in early summer from the USA and breed during the hot Canadian summer/early autumn and then die later in the fall when temperatures plummet.

Pollination

The value of insect pollination of crops to man is really inestimable, although the annual yield of insect-pollinated crops in the USA has been estimated to be in the region of US $5000 million (at present values).

The crops pollinated by insects (entomophilous) include top fruit (e.g. citrus, apple, pear, peach, plum, cherry, almond, mango); bush and cane fruits (e.g. currants, raspberry, blackberry, gooseberry); ground fruit (e.g. strawberry); some Leguminosae (e.g. pulses, clovers); all Cruciferae (brassicas and other vegetables); other vegetables such as Cucurbitaceae and onions; cotton, cocoa, tea, some coffees; most flowers and some trees (e.g. lime).

The crops that are anemophilous (wind-pollinated) are mostly the Gramineae (cereals and sugar-cane), and many trees, particularly the Gymnospermae (pines, spruces and other conifers). A few crop plants are partly entomophilous and partly anemophilous, such as beet, spinach, carrot, parsnip, white mustard, charlock and chrysanthemum. With a surprisingly large number of crops there is apparently uncertainty as to the precise manner of natural fertilization. Some crops are grown in cultivated clones with fruit set by parthenocarpy (e.g. banana, and Smyrna fig), and others are propagated vegetatively like sugarcane and potato. Some crops are self-pollinating, such as *Coffea arabica*, pea, groundnut, some beans and many Solanaceae, but apparently sometimes cross-pollination by insects is effected.

One result of widespread use of chemical insecticides has been the great reduction of insect pollinator populations (mostly bees) in many parts of the world, and at the present time many crops are grown under conditions of inadequate pollination. However, for some crops which are at least partially self-pollinating the precise value of increased insect pollination is not known (Free & Williams, 1977). As mentioned later (page 75), some crops typically over-produce flowers and fruit so increased pollination of these crops may be of little value; but it does seem quite likely that many crops are now suffering from under-pollination. An example of the difference in yield that can be achieved by increased pollination in some crops was shown on red clover in Ohio (USA) where the average yield reported was about 0.1 m^3/ha of seed; after an increase in the local bee population the field yield rose to about 0.4 m^3/ha, and when a plot was enclosed and subjected to maximum pollination by bees the yield was raised to 1.1 m^3/ha.

Bee destruction in Japan has been serious since World War II due to the very intensive nature of agriculture there, and in some orchards, by 1980, growers had to resort to hand-pollination – both time-consuming and labour-expensive. It had been estimated that about 25% of labour-time in fruit orchards in Japan was spent on pollination of the crop. An interesting development in Japan has been to use mason bees (in particular *Osmia cornifrons*) in fruit orchards for pollination. Artificial nest sites (usually consisting of small bundles of open canes or narrow tubes, about 5–6 mm diameter) are manufactured and situated in suitable sheltered locations in the orchards, often under the eaves of the house and buildings.

These sites are readily colonized by the mason bees. Research has shown the *Osmia* bees to be good pollinators; they forage in Japan from 08.00 h to 18.00 h daily, visiting some 15 flowers per minute. A local population of 500–600 bees per hectare will give adequate pollination without recourse to outside pollinators, and there will be a 50% fruit-set within 65 m of the nest site (Maeta & Kitamura, 1980).

The main groups of insects responsible for flower pollination are:

(a) Hymenoptera

  Apidae – Honey Bee (*Apis mellifera*), cosmopolitan

  – Bumble Bees (*Bombus* spp.), native to the Holarctic only (now in New Zealand)

  Megachilidae – Leaf-cutting Bees (*Megachile* spp.)

  – Mason Bees (*Osmia* spp.)

(b) Diptera

  Syrphidae – Hover Flies

  Muscidae – House Fly, Bluebottles, etc.

Some Lepidoptera (butterflies and moths) may be of importance, but mostly for ornamentals with flowers having a long tubular corolla. A few beetles and some thrips pollinate some crops, and a few somewhat bizarre tropical plants (e.g. orchids) are pollinated by hummingbirds or bats.

The most important wild pollinators are probably bumble bees and flies, but *Bombus* is mainly Holarctic in distribution. In the tropics, honey bees, Megachilidae and flies are probably the most important pollinators, but flies (Muscidae) can only pollinate the open-type flowers such as in the Cruciferae. In the cooler temperate regions *Apis mellifera* is domesticated and kept in hives which are easily handled and may be transported to orchards specifically for crop pollination. In the tropics however *Apis mellifera*, although domesticated and kept in 'hives', also occurs widely as wild colonies nesting in hollow trees, etc.

After the early catastrophes, when insecticide spraying in orchards and crops in flower resulted in large-scale bee destruction, most chemical companies are now including toxicity testing against bees as part of their regular pesticide screening programmes, and in the UK there are elaborate arrangements made to ensure that apiarists (bee-keepers) are warned before any major local insecticide applications are made. Also care is taken to avoid spraying particular crops in flower, so that bumble bee populations are safeguarded. In the tropics, however, there is seldom any warning given prior to spraying, but in these regions the majority of bees are wild anyway.

In the UK work on plant pollination is being conducted at Rothamsted Experimental Station, Herts., and from a bibliographical point of view by the International Bee Research Association (for example see Crane & Walker, 1983). An interesting recent review paper is by Kevan & Baker (1983).

3 *Principles of pest control*

Definition of the term 'pest'

Before contemplating taking any control measures against an insect species in a crop, the species must be correctly identified; then, presuming its biology is known, it should be clearly established that the species in this particular context is a pest, and that it could be profitable to attempt population control.

In this section the various terms used to describe pests are defined. It should be noted that some terms are more or less synonymous, but they are all well established in the literature. For example a *major* pest is very often a *serious* pest, and all *economic* pests are serious.

Pest. The definition of a pest can be very subjective, varying according to many criteria; but in the widest sense any animal (or plant) causing harm or damage to man, his animals, his crops or possessions, even if just causing annoyance, qualifies for the term pest. From an agricultural point of view, an animal or plant out of context is regarded as a pest (individually) even though it may not belong to a pest species. Thus a deer on a farm is a pest, but next-door in a game park it is not, and is in fact a valuable national asset there. Similarly, volunteer cabbage plants growing in a field with onions have to be regarded as 'weed' pests.

Many insects belong to generally accepted *pest species*, as listed in chapter 10, but individual populations are not necessarily always pests; that is, of course, not necessarily *economic pests*.

As pointed out by Norton & Conway (in Cherret & Sagar, 1977), we are often somewhat over-preoccupied at the present time with the state of the 'pest' population, whereas probably the most important aspect of a pest species is the damage (or illness) caused by the pest and the value placed upon these consequences by human society.

Economic pest. On an agricultural basis, we are concerned when the crop damage caused by insects leads to a loss in yield or quality, resulting in a loss of profits by the farmer. When the yield loss reaches certain proportions the pest can be defined as an economic pest. Clearly the value of the crop is of paramount importance in this case, and it is difficult to generalize, but as a general guide for most crops it is agreed that most species reach *pest status* when there is a 5–10% loss in yield. Obviously a loss of 10% of the plant stand in a cereal or rape field (note that this is not the same as a 10% loss in yield!) is not particularly serious, whereas the loss of a single mature tree of *Citrus*, apple or peach is important.

Economic damage. This is the amount of damage done to a crop that will financially justify the cost of taking artificial control measures, and will clearly vary from crop to crop according to its basic value, the actual market value at the time and other factors. In practice, many peasant farmers engaged in subsistence farming feel that they cannot justify use of pesticides at all.

Economic injury level (EIL). This is the lowest population density that will cause economic damage, and will vary between crops, seasons and areas. But it is of basic agricultural importance that it is known for all the major crops in an area.

Economic threshold. It is desirable that control measures be taken to prevent a pest population from actually causing economic injury. So the economic threshold (Stern *et al.*, 1959) is the population density of an increasing pest population, at which control measures should be started to prevent the population from reaching the economic injury level (see fig.2).

Pest complex. The normal situation in a field or plantation crop is that it will be attacked by a number of insects, mites, birds and mammals, nematodes and pathogens which together form a complicated interacting *pest complex*. The control of a pest complex is complicated and requires careful assessment, especially as to which are the *key pests*, and careful integration of the several different methods of control which may be required. This, of course, makes the process of evaluation difficult, and generally, in the past, much money was wasted on uneconomic pest control, either through carelessness or lack of knowledge.

Pest spectrum. This is the total range of different types and species of pests recorded attacking any particular crop, and especially of concern in one particular area. The total number of insects (and mites) recorded from the major crop species are considerable; these records, incidentally, are of insects feeding or egg-laying on the plant, and do not include casual observations when the insect might just be resting, or when, for example, caterpillars have crawled up on to the plant just to pupate (e.g. Large White Butterfly). Simmonds & Greathead (in Cherrett & Sagar, 1977) listed the numbers of pest species, on a world basis, recorded from sugarcane as 1300, cotton 1360, coffee 838, and cocoa 1400. Fortunately, for the practising entomologist, and the farmers, these numbers reflect the situation globally and many of these pests are restricted geographically to one part of the world. For example, wherever apple is grown it will be attacked by a tortricid complex (Lep., Tortricoidea) but the actual species differ from region to region. Only a few major pests are completely cosmopolitan (e.g. *Myzus persicae*, *Agrotis ipsilon*) or pantropical (e.g. *Maruca testulalis*) in distribution (in point of fact, their widespread distribution is one reason for their being regarded as major pests).

Pest load. This is the actual (total) number of different species (and numbers of individuals) of pests found on either a crop or an individual plant at any one time, and, as already mentioned, this would usually be a pest complex, but could also be a monospecific population, although this would be rare.

Key pests. In any one local pest complex it is usually possible to single out one or two major pests that are the most important; these are defined as key pests, and are usually perennial and dominate control practices. A single crop may have one or more key pests, which may or may not vary between areas and between seasons. It is of course necessary to establish economic thresholds for these key pests in order to be certain when to apply control measures, for it has been often observed that the mere presence of a few individuals of a key pest species in a crop may cause undue alarm and lead to unnecessary pesticide treatment. Key pests owe their status to several factors, including their usually high reproductive potential, and the type of damage they inflict on the host plant (e.g. Codling Moth on apple).

Serious pest. This is a species that is both a major pest and an economic pest of particular importance, being very damaging and causing considerable harm to the crop plants and a large loss in yield. It almost invariably occurs in large numbers.

Major pest. In this book these are the species of insects and mites that are either serious pests of a crop (or crops) in a restricted locality, or are economic pests over a large part of the distributional range of the crop plant(s). Thus the species here regarded as major pests usually require controlling over a large part of their distributional (geographical) range, most of the time. As mentioned in chapter 1 however, some species of insects have been included as 'major pests' in this book because of their widespread and frequent occurrence, biological interest, wide range of host plants, or other aspects of academic interest. In any one crop, in one location, at one time, there is usually only a rather small number (say 4–8) of major pests in the complex that actually require controlling. For example, although the pest spectrum for cotton worldwide is 1360 species, on any one cotton crop there will probably only be about five species requiring population control. Usually for most crops in most localities the major pest species remain fairly constant from year to year, but several entomologists have commented recently that in some areas they have observed that the major pest species complex has been gradually changing over a long period of time. Soehardjan (1980) reported that in Indonesia the 8–10 major pests of rice have largely changed over the period of time 1929–79, although there are some differences within different parts of the island of Java. As mentioned already, the Brown Planthopper (BPH) of rice has risen in 10 years from obscurity to becoming the most serious pest of rice in most parts of Asia. And it is reported from IRRI that there are now two new major rice pests in tropical Asia (Pathak, 1980): the Sugarcane Leafhopper and Rusty Plum Aphid. So over a period of some 10–50 years it is expected that the complement of major pests for a crop will change. It must be remembered that evolution continues all the time, though it is not often obvious, and that in an artificial environment, such as agriculture, it can be expected that evolution will be accelerated.

Minor pests. These are the species that are recorded feeding or ovipositing on the crop plant(s) but usually do not inflict damage of economic importance; often their effect on the plant is indiscernible. They may be confined to particular crop plants or may prefer other plants as hosts. Many (but not all) pests listed as minor pests are potentially major pests (viz. BPH of rice). Many species that are major pests of one crop will occur in a minor capacity on other crops. And sometimes a major pest of a particular crop in one part of the world (e.g. Europe) will be a minor pest on the same crop in a different part (e.g. Australia or the New World).

Potential pest. This term is used occasionally in the literature and refers to a minor pest species that could become a major pest following some change in the agroecosystem. Only a relatively small proportion of the species listed as minor pests are really potential pests in this sense, because of their basic biology.

Secondary or sporadic pest. Defined by Coaker (in Cherret & Sagar, 1977) as a species whose numbers are usually controlled by biotic and abiotic factors which occasionally break down, allowing the pest to exceed its economic injury threshold.

Pest populations. A most important point to remember is that an insect is only an actual pest (in practice) at or above a certain population density, and most control measures are aimed only at reducing this population to a lower level. So insect pest population studies are vitally important, and although mentioned here briefly the topic is included in chapter 2 (page 7), as an aspect of pest ecology.

Pest species accumulation. Long-term stable habitats generally exhibit an extensive species diversity, both in host plants and in phytophagous arthropods. This is shown typically in old forests, and also some plantation crops such as cocoa, rubber, sugarcane and tea (Banerjee, 1981). The pest species accumulation in the monocultures is in part a reflection of the area under cultivation; other important factors are the type of plant, the geographical location of the habitat (area) and its natural species richness, and the age of the actual plants and of the community (crop). It appears that in some cases the age of the plant community is not particularly important, but Banerjee observed that in tea plantations (from the Old World tropics, excluding China) age appeared to be a major factor and pest species saturation was apparently reached (in plantations in N.E. India) at the plantation age of about 35 years. After this age there was no further increase in pest species numbers.

Pest species recruitment. As mentioned elsewhere, each major crop has a more or less clearly defined area of origin in one particular part of the world, and during historical times most of the crops have gradually been transported by travellers and commerce to other parts of the world where suitable climatic conditions prevail.

With some crops, such as *Citrus*, it appears that most of the species in the pest spectrum in most parts of the world are allochthonous in that they have originated in the S. China/Indo-China region where *Citrus* is endemic, and the pests have been gradually spread to the new areas of cultivation. Thus at the present time most of the important *Citrus* pests are pantropical in distribution.

Other crops such as rubber, sugarcane and tea (Banerjee, 1981) have apparently been subjected to autochthonous pest recruitment in their new areas of cultivation. Tea, according to Banerjee, has only about 3% of its total number of pest species common to the different areas of cultivation, the other pests being recruited locally and thus being different in each area.

As would be expected, it appears that with many crops pest recruitment has been in part allochthonous and partly autochthonous. The origin of the crop pest spectrum may have an important effect so far as pest control strategy is concerned, particularly with regard to natural and biological control.

Development of pest status

This takes place through a number of different agencies that fall into two main categories, as follows.

(a) Ecological changes – as already mentioned, these are largely results from the artificial innovations of widespread agriculture.

(b) Economic changes – which are related to the human social context.

An important publication relating to this problem is the 18th Symposium of the British Ecological Society (April 1976) on *'Origins of pest, parasite, disease and weed problems'*, edited by J.M. Cherrett & G.R. Sagar (1977).

Ecological changes

There are really only three aspects of importance in considering crop pests, and these are:

(a) state of the pest population (i.e. numbers of pests present);

(b) nature of the damage done to the crop;

(c) value of the damage as assessed by human society.

Thus each aspect of the way in which pests arise through ecological changes will relate to population numbers, for by definition a species is only a pest at or above a particular population density.

Increase in numbers. This is the most common way in which an insect species attains pest status and it usually follows because of one of the factors listed below. But often insect populations can be seen to change, sometimes drastically from a few individuals per plant to literally thousands on the same plant the following season, for no obvious reason at all; probably the most spectacular population fluctuations of this sort are seen with some Diaspididae, Aleyrodidae and Aphidoidea. Presumably there is a subtle population control being effected, but often we are not able to understand it.

Most insects species will have a natural phenological cycle whereby the relatively small number of overwintering (or equivalent) or immigrant adults lay large numbers of eggs and the population develops in a regular, stepped cycle, increasing through each generation of the growing season; finally, there will be a population 'crash' or large-scale emigration at the time of harvest, or the onset of winter (or dry season). The population growth cycle is 'stepped' because of natural predation on the different instars, the earliest instars usually being the most heavily preyed upon. See also the section on pest populations in chapter 2 (page 7).

Population resurgence. The effect of most chemical pesticides is short-lived nowadays, and once the suppressive effect declines the pest population will naturally resurge, possibly up to the economic injury level again.

There may also be resurgence of a secondary pest due to insecticidal destruction of natural enemies, or from some other ecological upset.

Migration. In northern temperate regions a number of pest species regularly migrate into the country in the spring and early summer from farther south where conditions are milder and warmer. In the national pest spectra of the UK, Fennoscandia, Japan, and Canada there are several (or more) quite important migratory species that do not survive the winter locally, but regularly immigrate into the country each year from farther south. This topic is dealt with in more detail in chapter 2 (page 20).

Character of food supply. Plants grown for agricultural purposes have usually been selected for their nutritive value and are typically large and succulent with especially large fruits, leaves and lush foliage. Thus cabbage and lettuce are far more attractive as food for caterpillars than are wild crucifers (such as *Capsella* and *Cardamine*), and *Solanum nigrum* hardly compares with egg-plant, tobacco or tomato as a food source; similarly oats and wheat are far more attractive than wild grasses. A large field containing such a crop represents, literally, an inexhaustible food supply (for at least the duration of the crop) for a potential pest species, so it is little wonder that pests evolved concurrently with the practice of cultivation.

This topic is scarcely distinguishable from the following one in practical terms, especially concerning the use of new varieties and clonal propagation. The growing of new varieties of crop plant can lead to the development of a new pest species. In Asia new high-yielding varieties of rice were introduced some years ago from IRRI, and in order to achieve the high yields possible the farmers had to use extra nitrogenous fertilizers; the use of these fertilizers produced a lush vegetative growth which seems to be more favourable for BPH reproduction. Thus, as already mentioned, *Nilaparvata lugens* rose in status from a minor pest in the 1960s to its first record as a major pest in 1970, to a generally serious pest in 1975 and a widespread serious pest by 1980. In many parts of Asia it is now regarded as the 'number one' rice pest. This new-found fecundity has also led to a very rapid biotype formation and resistance to many pesticides and resistance-breaking of many new rice varieties.

Monoculture. The growing of a single crop species over a large area provides an unlimited source of food for pest species, especially when the crop plant is particularly succulent. The present worldwide tendency is towards mechanized agriculture which requires larger fields and fewer hedgerows which traditionally delimited each field, and so the crop monoculture becomes even more extensive. This practice will encourage some insect species to become more abundant, and hence important as pests, as was observed about a century ago with Colorado Beetle on potatoes in the USA (see fig. 8).

As well as monoculture becoming more extensive, some new crop varieties have become more specialized in their growing conditions, and there is a tendency to reduce the extent of crop rotation.

As mentioned before, the taiga of northern Asia and America with its coniferous forest dominated by only a few tree species, and the northern temperate oak/beech forests, are the natural equivalents of an agricultural monoculture. Thus it is not surprising to learn that *Quercus* is attacked by 1000 different species of insects!

The extensive increase in rape as a commercial crop

Fig.8. Schematic graph of the change in general equilibrium position of the Colorado Beetle (*Leptinotarsa decemlineata*) following the development of widespread potato culture in the USA (from Stern *et al.*, 1959).

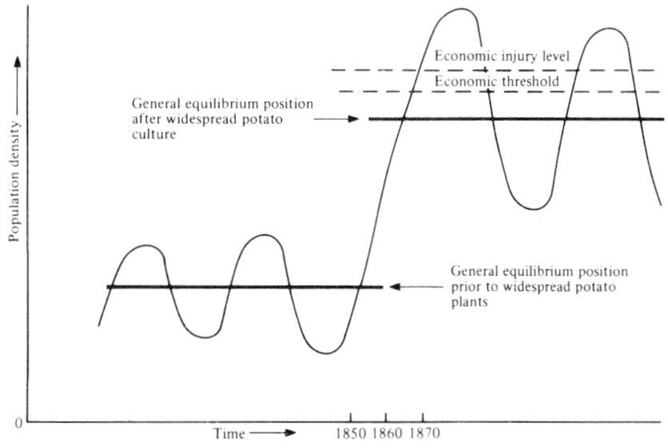

in many parts of Europe has resulted in a number of pests (see page 568) that attack the flowers and pods becoming far more abundant and widespread, and thus putting *Brassica* seed crops more 'at risk' than formerly.

Farmers also often specialize in particular crops for reasons of convenience, both agriculturally and economically, so that, for example, in parts of eastern England and eastern Canada where the fenland peats (and mucklands) occur, it is common practice to grow carrots, rotated with celery and parsnip, because they all require similar soil conditions, but from a pest point of view this is scarcely a rotation as all belong to the Umbelliferae and have a similar pest spectrum. The more extensive a monoculture becomes then the greater is the pest problem, generally.

Certain regions of Canada, the USA and the USSR are referred to as the 'wheat belts', and here the annual wheat crops are measured in square kilometres or square miles, rather than hectares. This practice has evolved in part because of the national need for cereals, but also because the climatic and edaphic factors prevailing limit crop production to plants that require little water.

A recent trend in agriculture is towards extreme standardization of product, for a number of different reasons connected with both crop production and product utilization. A good example is rubber in Malaysia. After World War II it was thought that synthetic rubbers (produced from oil) had destroyed the natural rubber (NR) market. However, it was found that natural rubber possesses qualities that have not been matched by synthetic rubber. Then in 1965 new cultivation methods and processing produced the new Standard Malaysia Rubber (SMR), constant in quality to tight specifications. There is now a tremendous world market for natural rubber, and this industry is booming throughout S.E. Asia, mostly through new improved methods of production.

These new production methods include *cloning*, the vegetative mass-production of ideal genetic stock through mist-propagation and other techniques. Clonal propagation of oil palm in Malaysia and S.E. Asia is proceeding rapidly at present as this industry expands throughout the region. To have large areas of genetically identical trees is very useful from both crop production and product-processing points of view, but these trees are vulnerable to disease and pest epidemics by virtue of their uniformity, and the most serious threat would be from a new major pest or disease which might arise. Clonal cultivation clearly represents the most extreme form of monoculture.

Oddly enough, in the very large-scale monocultures, such as the wheat fields on the Canadian prairies, there is the rather anomalous situation whereby the wheat monoculture is regarded as a highly stable habitat with a diminished pest problem. On reflection this situation is not so strange because the extent of the wheat monoculture is very great. It seems that each extreme of habitat diversity (from tropical rain forest to prairie wheat) is stable with a reduced pest problem, whereas it is the intermediate stages in the process of simplification that suffer the greatest pest problems.

Mixed stands (*mixed cropping*) are being recognized as a cultural method of alleviating pest situations (Way, in Cherret & Sagar, 1977), and it is being developed successfully for use in barrios in the Philippines by IRRI. For large-scale agriculture, with the present reliance on mechanization, mixed cropping is generally not feasible, but for small peasant farmers in the tropics and for smallholders it can be a valuable means of combating pest infestations. In the tropics by far the greatest area of crops is actually grown on smallholdings and not the huge agricultural estates characteristic of the northern temperate regions.

Continuous cropping. The plantation and orchard crops are all very long-term and because of this they suffer from particular pest problems, but in compensation their pests are to some extent controlled by natural enemies. Field crops are typically characterized by their short duration (an ephemeral habitat) as at most they are annuals, and sometimes by careful timing the crop may be grown before the pest population catches up. However, there is pressure of late to produce a continuous supply of vegetables, and new varieties tolerant of different climates have been bred. In the vicinity of large urban conurbations many vegetables are now being grown almost all year round. For example in S. China and S.E. Asia, by varying the varieties grown, it is possible to produce onions, tomatoes, and several species of *Brassica* continuously for the local markets. This encourages the build-up of the pest populations; for instance throughout Asia, Diamond-back Moth and Turnip Mosaic Virus have now both become serious pests of *Brassica* crops.

Minimum cultivation techniques. A recent agricultural technique in ground preparation is known as minimum cultivation, and it consists essentially of a chemical destruction of old crop remains and weeds, followed by a subsequent planting of the new crop into the undisturbed soil. Ploughing and harrowing normally reduce the population of soil pests by exposing them to sunlight and desiccation, and to predators and parasites. Many Coleoptera, Lepidoptera and Diptera feed on the aerial parts of plants as larvae but pupate in the soil. These and other soil-inhabiting pests would normally be depleted in numbers by normal cultivation methods, and so in areas where minimum cultivation techniques are employed there is often a build-up in numbers of soil pests.

Multiplication of suitable habitats. Farming leads to a simplification of the flora by a selection of plants suitable for husbandry. Thus insects associated with these plants have a more attractive and concentrated food supply, as well as a greater total host number. The most outstanding example of this is shown in the storage of grain and foodstuffs; many storage pests exist in small populations in the field but increase enormously in numbers in the favourable micro-climate and abundant food of the grain store. Examples are to be seen in *Sitophilus* on maize cobs, *Sitotroga* on sorghum, and the Bean and Cowpea Bruchids.

Loss of competing species. Under conditions of monoculture an area harbours fewer insect species than under 'natural conditions', and many insects now become pests which were not pests under the natural conditions. Sometimes specific pest control measures may remove one pest, but another insect released from competitive pressure may increase in numbers and become a new pest. Once the new pest is established mutations make the relationship between it and the crop even closer, for as numbers increase more mutants appear and can be selected to consolidate their niche as pests.

Change of host/parasite relationships. Many insects are kept in check by their predators and parasites, although when a pest species increases in number there is typically a time-lag between its increase and the parasite and predator numbers. Parasites are generally quite specific but predators less so, and the time-lag of the parasite population build-up is generally less than that for a predator. The greater the time-lag between the pest population increase and that of the parasite or predator, then the more likely is the species to be a serious pest. Generally agricultural operations involving large-scale insecticide applications may affect parasites and predators more than the pests. One of the classical cases is that of the Fruit Tree Red Spider Mite (*Panonychus ulmi*) which became a more serious pest on fruit trees after widespread use of DDT in orchards throughout Europe and North America.

Spread of insects and crops by man. Almost every major crop was once endemic only to one particular part of the world, but during the centuries of world exploration and trade most crops were distributed throughout the regions which had a suitable climate for their cultivation. This state of affairs is generally highly desirable, especially since some crops flourish in their new habitats (e.g. rubber in Malaysia, coffee in S. America and potatoes in Europe) and most countries now aim at agricultural diversification. Unfortunately some of the endemic crop pests were accidentally (or otherwise) distributed along with the crops (or afterwards) and became serious pests in the new locations, for they were usually without the restraint of their endemic natural enemies. Several aspects of this problem are reviewed in the book by Elton (1958). The main means of pest dispersal by human agency are as follows.

Dispersal by unknown means. Some pests (and crops) appear to have been dispersed since antiquity, though it is not clear how this dispersal was effected. It must be remembered that in some parts of the world there have been human population movements ever since the early days of agriculture. Also, of course, all animal species have an innate tendency to disperse, or migrate, of their own accord. However, there are numerous records of crop pests having been spread by man, either accidentally or intentionally.

Transport on the host (plant). The symbionts that normally live with man and his animals have been transported to all parts of the world, and likewise the transport of live plants provides an easy mechanism for the dispersal of many pests, especially scale insects, mealybugs, and aphids on the plants, and beetles, nematodes, moth pupae, etc. in the soil around the roots. It is only recently that phytosanitation measures have been adopted. Classical cases include the spread of *Icerya purchasi* around the Mediterranean on nursery stock, and also several scale insects on *Citrus* and other fruit stocks from S. China to California, Hawaii and Australia. Hessian Fly (*Mayetiola destructor*) is thought to have been transported to N. America from Europe in wheat straw. Colorado Beetle was brought from the USA to Europe (France) on stored potatoes.

Dispersal through trade. Rats, many weeds, and some insects have been dispersed over the international trade routes, usually accidentally, in either produce (as above), containers or ballast. A recent case of note is the spread of Rice Water Weevil from the USA into Japan, where unfortunately it is now established. It is thought that weevil pupae were present in hay exported to Japan for use by dairy farmers, and they were then carried into rice fields in the manure from dairy farms. About half the immigration records for Colorado Beetle in the UK are from produce (seldom potatoes) in ships arriving at coastal ports in southern and eastern England.

Stowaways in air transport. With the recent development of international airways and rapid transport from one part of the world to another, there is a constant danger of insect pests being transported live from one region to another. This threat applies also to medical pests (especially vectors) which is why aeroplane cabins are sprayed with aerosol insecticides prior to passenger departures. Several of the records of Colorado Beetle introductions into the UK are clearly cases where the beetles were stowaways in the undercarriage of the planes from Spain and France, and the beetles were dropped along the airport approach line as the undercarriage was lowered.

Deliberate introduction by man. The most notorious examples are without doubt the introduction of prickly pear cactus (*Opuntia* spp.) and rabbit into Australia. Many other examples are listed by Simmonds & Greathead (in Cherrett & Sagar, 1977) and include the Gypsy Moth (*Lymantria dispar*) imported into the USA in about 1870 by an amateur entomologist to test its value for silk production. Some moths escaped and became established as one of the most serious forest pests in the north-eastern USA.

Economic changes

The definition of a pest relates clearly to the value of the damage done by the insect (or animal) as assessed by human society, and so any changes in the value of the crop will affect the importance of the pest. Damage that is not important when prices are low can be very serious when prices are high. Sometimes the converse situation is true, if an important food crop is in short supply then some pest damage may be tolerated.

Change in demand. If some crops are replaced by others, the pests of the former crops become less important. Greater demand for a crop increases its value and the incentive to grow it. The demand may be for increased quantity and quality, both of these factors affecting the importance of the pests. If the crop is in short supply the consumers are less selective than if it is abundant. Wireworms do not greatly affect the yield of potatoes, but their tunnels spoil the appearance and keeping qualities. If the supply is short, consumers overlook a little damage, but with the recent demand for packaged vegetables such potatoes are generally unsaleable, thus making wireworms a much more serious pest.

Change in production costs. A pest may become economically important when agricultural practices change. If a new high-yielding variety is developed, minor pests which attack it may become of economic importance.

The case of the Brown Planthopper becoming a serious pest on new rice varieties is complicated in that, in order to achieve the high yield possible from these new varieties, it is necessary to use more fertilizer on the crop. The extra fertilizer produces a lush foliage which induces the BPH to a higher level of fecundity, making it a serious pest on a now more expensive crop.

Crops for export are more valuable than those for local consumption. In Europe it is now possible to buy fresh peppers, aubergines, avocados, and many other tropical fruits and vegetables that are being air-freighted from Africa, the Middle East and other tropical regions. These crops are of course now much more expensive to produce (and sell) because of air freight but are sold at correspondingly higher prices.

Pest damage

This is considered in more detail later (chapter 5) but is mentioned here briefly because of its implications in connection with choosing a pest control strategy. The single most important aspect is the relation between the damage done by the pest and the part of the plant harvested. *Direct damage* is when the part of the plant to be harvested is the part attacked, such as the leaves of tobacco, fruits of tomato and apple, tubers of potato and sweet potato; in these cases clearly the damage is more important. *Indirect damage* is when the part of the plant damaged is not the part to be harvested. Examples include the roots of tobacco and wheat, leaves of sugarbeet, potato and apple. In these cases it is usually possible to ignore quite surprisingly high levels of pest infestation, as these infestations may only have a marginal effect on the crop yield.

Direct effects of insect feeding

Biting insects may damage plants as follows.

(a) Reduce the amount of leaf assimilative tissue and hinder plant growth; examples are leaf-eaters, such as adults and nymphs of locusts and *Epilachna* and larvae of *Plutella*, *Pieris*, *Plusia* (Lepidoptera) and sawfly larvae.

(b) Tunnel in the stem and interrupt sap flow, often destroying the apical part of the plant; these are stem borers and shoot flies, such as *Zeuzera* in apple branches, *Cephus* in wheat, *Ostrinia* in maize, *Atherigona* in maize and sorghum.

(c) Ring-bark stems, for example some Cerambycidae.

(d) Destroy buds or growing points and cause subsequent distortion or proliferation, as with Fruit Bud Weevils (*Anthonomus* spp.) on shoots of apple, pear, etc.

(e) Cause premature fruit-fall, as with Cherry Fruit Fly, Codling Moth, Apple Sawfly.

(f) Attack flowers and reduce seed production, as with the blossom beetles (*Meligethes* spp.) and Japanese Beetle.

(g) Injure or destroy seeds completely, or reduce germination due to loss of food reserves; examples are Hazelnut Weevil, Maize Weevil, Pea and Bean Bruchids, Pea Pod Borers, and Bean Pod Borers.

(h) Attack roots and cause loss of water and nutrient absorbing tissue, as with wireworms and various chafer larvae (Scarabaeidae) and other beetle larvae in the soil.

(i) Remove stored food from tubers and corms, and affect next season's growth; examples are cutworms and wireworms in potato, and Potato Tuber Moth larvae.

Insects with piercing and sucking mouthparts may damage plants as follows.

(a) Cause loss of plant vigour due to removal of excessive quantities of sap, in extreme cases wilting and foliage distortion results, as in the stunting of cotton by *Bemisia* (Whitefly), and aphids on many plants.

(b) Damage floral organs and reduce seed production, for example capsid bugs (Miridae) and other Heteroptera (Wheat Shield Bugs, Chinch Bugs, etc.).

(c) Cause premature leaf-fall, as do many diaspidid scales.

(d) Inject toxins into the plant body, causing distortion, proliferation (galls) or necrosis; examples are seen in capsid damage on bean leaves and shoots, and the stem necrosis on plants by *Helopeltis* and other Heteroptera.

(e) Provide entry points for pathogenic fungi and bacteria, as does *Dysdercus* on cotton bolls (for fungus *Haemospora*) and other bugs.

Indirect effects of insects on crops

(a) Insects may make the crop more difficult to cultivate or harvest; they may distort the plant and cause the plant to develop a spreading habit which makes weeding and spraying more difficult. They may delay crop maturity, as do the bollworms on cotton, and grain in cereals may become distorted or dwarfed.

(b) Insect infestation results in contamination and loss of quality in the crop; the quality loss may be due to reduction in nutritional value or marketability (lowering of grade). Loss of yield in a crop is obvious but a nutritional quality loss is easily overlooked; this is the type of damage done to stored grain by *Ephestia cautella* and *Tribolium*. A more common loss of quality is the effect of insects on the appearance of the crop, for example skeletonized or discoloured cabbages have a lower market value than intact ones. Attacked fruit is particularly susceptible to this loss in quality, as seen by skin blemishes and hard scales on citrus fruit and capsid damage on apples. Contamination by insect faeces, exuviae, and corpses all reduce the marketability of a crop, as do black and sooty moulds growing on the honeydew excreted by various homopterous bugs. A major problem in the tropics is 'stickiness' of cotton lint caused by honeydew from Cotton Whitefly; the sticky cotton is difficult to gin, and its value is diminished.

(c) *Transmission of disease organisms*

(1) Mechanical transmission, also termed passive transmission, takes place through feeding lesions in the cuticle. Sometimes the pathogen (usually fungi or bacteria) is carried on the proboscis of the bug or sometimes it is on the body of the tunnelling insects. Examples are seen in the case of the *Scolytus* beetles which transmit Dutch Elm and other fungal diseases.

(2) Biological transmission. Most viruses depend upon the activity of an insect vector for transmission. The vector is usually also an intermediate host, as is the case with most aphid and whitefly hosts. Diseases transmitted in this manner include Cucumber Mosaic, Tobacco Mosaic and Turnip Mosaic.

Economics of pest attack and control

Economics in this sense is the relation of crop losses to production costs. Sometimes due to pest outbreaks a particular crop may be scarce and sell for a higher price, so that the individual producer may not suffer financially. From the economic point of view, however, one must consider the overall (national) picture, not just individual producers.

Decision to take action

This decision is based upon an accurate estimation of the cost of control measures in relation to possible profits likely to accrue as a result of control. The *economic threshold* in pest control is the point at which a particular pest can be controlled at a cost of less than the expected market value of the expected yield increase. Often the cost of control may be known very accurately but the possible profit is a subjective assessment based upon past experience and guesswork. Because many pest control costs are relatively low, against a serious pest attack it is usual for no calculations to be made; none are needed! However, sometimes costs are critical, for some pesticides are quite expensive and their use at the recommended rates may be prohibitive in cost to the small farmers and generally can only be afforded by the large estates. Impressions of the degree of infestation of a crop are often wrong or else very misleading; temperate examples include the infestation of Mangold Fly on sugarbeet, where the damage is often striking and unsightly, but experiments have shown that 70% defoliation reduces the yield by only 5%, and a 50% defoliation has no measurable yield loss. Similarly Black Bean Aphid infestation on spring-sown field bean stems is misleading in that an early infestation which looks only light can reduce the yield by as much as 30%; in this case damage is more serious than would appear at first sight.

In some areas certain crops are always *at high risk* from particular pests, as shown by years of empirical observations, and in these cases preventative pesticide applications are usually made, often in the form of granule application at the time of sowing or else seed dressings.

There is a definite danger though that in some areas the mere sight of a particular key pest in a crop will precipitate panic action by a farmer who will immediately spray with insecticides when in fact their application might not be necessary.

Cost/potential benefit ratio in decision

In practice, cost/potential benefit ratios are only known for valuable crops like citrus and apples, and usually then apply only to the benefit likely to result from control of the heaviest infestations that may occur. Generally little account is taken of differences in infestation and prices; thus spray programmes planned using this concept are only applicable to comparatively steady rates of infestation and cash return. When the ratio is high, uncertainty about the likelihood of attack can be ignored. Seed dressings are a cheap form of protection and can give a cost/potential benefit ratio as high as 1 : 10, when used against severe attacks of Wheat Bulb Fly or wireworms. Doubt as to whether they are needed on each occasion wheat is planted after fallow (danger from Wheat Bulb Fly) or ploughed-up pasture (wireworm) is usually discounted because the ratio of 1 : 10 need only be reached once in ten times for the operation to break even with regard to costs. Surveys for use of insecticides against wireworms on potato lands in the British Isles by the National Agricultural Advisory Service (now ADAS) have shown that in only 25% of fields is there likelihood of wireworm damage. Therefore the cost/potential benefit ratio must be at least 1 : 4 for treatment to be profitable on average, but attacks are seldom severe enough to give a ratio of 1 : 10, and so an average of 1 : 4 or more for treated fields is unlikely. Thus treatment of all potato fields is not only unnecessary three times out of four, but the total effect is to increase the cost of production of potatoes and probably to decrease growers' profits. Surveys have shown that after the decline of wireworms as serious pests some years ago, many farmers continued to apply soil insecticides in the UK (despite the low levels of damage recorded) which was quite unnecessary. Experience has shown that often growers continue a treatment long after the need for it has finished.

Relation between yield increase and pesticide dosage

If a graph is drawn showing the relationship between rate of application of pesticide and percentage increase in crop yield, it is apparent that the curve is sigmoid but slightly asymmetrical. At the lower application rates the yield increases are low; at higher application rates the increase becomes high and then gradually decreases; finally, the yield increase decreases rapidly. It is possible to calculate a dosage response curve for a particular pest at different population densities, which shows what control measures are necessary, and also shows the rate of application at which profits no longer increase. However, if the dosage response curve is to be used to decide upon the correct pesticide use, the relationship between yield and infestation must be known, and the populations likely to cause economic damage must be predicted. Unfortunately, this relationship and prediction cannot yet be made very accurately for most insect pests.

Crop yields. When referring to crop yield in relation to pest infestation and damage levels, it should be borne in mind that in practice there are various levels of crop yield, as shown in fig. 9 (CAB/FAO, 1971). *Primitive* yield is the lowest, and refers to the situation where none of the modern agricultural inputs (improved seeds, mechanized tillage, irrigation, fertilizers, and crop protection) are used. *Actual* yield is the crop yield actually attained through partial use of these inputs. *Attainable* yield is what could be achieved under optimal conditions with maximum input. *Economic* yield is that which gives the highest return financially on investment, and is usually lower than the attainable yield. *Theoretical* yield is the highest, and it represents the upper limit of crop production calculated by crop physiologists.

Fig.9. Interrelation between crop yield, crop loss and crop protection profits.

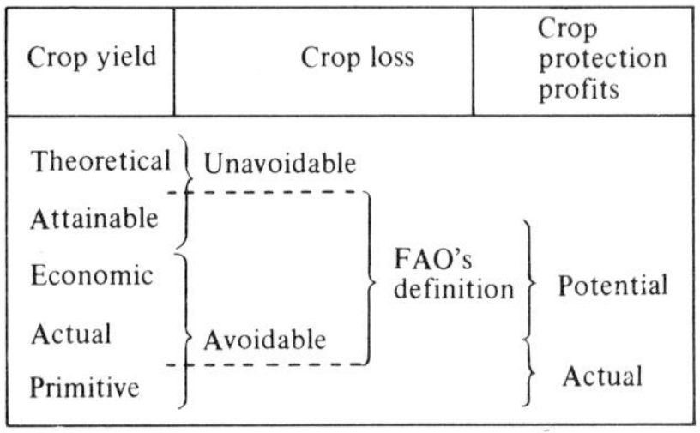

Economics of national pest control programmes

Most of the cost of pest control programmes is borne by the consumer in the price which he pays for food. However, the cost of quarantine, eradication of imported pests, biological control, and much of the cost of research into pest biology is met through government expenditure. The cost of the UK quarantine laws against Colorado Beetle is probably in the order of £50000 per year, or 4p per hectare to be added to the cost of the potato crop; this is a very good investment when compared with the losses which a heavy attack would cause. Attempts to eradicate a newly established pest by chemical control can be very costly. After the Mediterranean Fruit Fly was accidentally introduced into Florida a total of 2 million hectares of citrus land and adjoining land was sprayed at a total cost of some £3–4 million, yet the entire operation cost no more than about 5% of the annual value of the citrus crop and saved the industry.

The cost of biological control programmes is sometimes difficult to estimate, but DeBach (1964) said that, for the Department of Biological Control of the University of California for the period 1923–59, a total expenditure of £1½ million has resulted in a saving on five projects of £40 million, with a recurrent benefit of £10½ million. Thus it can be seen that the cost of biological control does compare very favourably with chemical control.

Pest assessment

In sampling and assessment programmes it is found that the relationship between pest numbers and damage is often logarithmic rather than linear, and it is usually more effective to take many samples and separate them quickly into different, easily distinguished, categories of infestation, than to carefully count numbers of pests on a few samples, if the results are to be statistically analysable.

It is highly advisable to seek advice from a biological statistician, when planning a sampling and assessment programme, in order to ensure that the results obtained will be suitable for statistical treatment. The publication by FAO/CAB (1971) includes details of damage assessment for about 80 major pests of 27 crops (see chapter 5 for more detail). In the UK pest assessment details for some crops are available from local ADAS offices.

Efficiency of pest control measures

Methods of pest control tend to operate with a characteristic *efficiency* in that they reduce pest populations by a fixed proportion, more or less regardless of the number of pests involved. The efficiency of a treatment can be expressed as the percentage reduction of the pest population, or of the damage. It therefore follows that the numbers of a pest surviving a treatment depend upon the numbers attacking the crop, as well as the efficiency of the control measure. The efficiency of an insecticide treatment invariably declines with time since application, and is closely related to residue persistence, the growth of the crop, type of soil, sunshine, and moisture level.

The *effectiveness* of a control measure, as distinct from its efficiency, is reflected by the number of pests surviving after treatment, or, from a practical point of view, the amount of damage occurring after treatment. Thus a control measure will be most effective when dealing with relatively light infestations. Consequently, any actions which reduce a pest population will automatically improve the effectiveness of a control treatment, even though the efficiency of that treatment remains unchanged. Conversely, any practice encouraging a high population density in an area will immediately reduce the effectiveness of the control treatment, often giving the false impression that the treatment no longer works. The kinds of action which encourage a pest population build-up are destruction of predators and parasites, insufficient crop rotation, presence of weeds and alternative hosts in the vicinity of the crop, a low standard of crop hygiene, etc.

Forecasting pest attack

The ultimate goal of pest control programmes is the accurate forecasting of pest attacks before they actually take place, so that control measures can be planned with maximum efficiency. Successful forecasting techniques should be as simple as possible, and will be based upon detailed knowledge of the biology and ecology of the pests concerned. The types of detailed studies required to give the basic information are as follows.

Quantitative seasonal studies which must be made over several years to determine seasonal range, variability in numbers, and geographical distribution. Such studies must use sampling methods appropriate to the pest and its abundance, and the seasonal counts should be related to climate and topographical data.

Life history studies, to find the length of the life cycle, number of eggs laid, amount of food eaten, maturation period for the females, are aspects that can be studied both in the field and in the laboratory. Behaviour of different larval instars and possible number of generations under different conditions can most suitably be studied in the laboratory. The expected range in relative humidity, temperature, etc., can be considered in relation to limits of survival of the insect under study.

Field studies to find the effects of weather on the pest. Either directly or indirectly climatic factors control pest numbers, affecting not only the pests themselves but also their predators and parasites. The ways in which pests spread from crop to crop are largely influenced by weather. Until population dynamics of insects are more fully understood, accurate forecasting is very difficult, and generally the many forecasting schemes in different countries have met with widely differing successes. The National Agricultural Advisory Service (now ADAS) in Great Britain is usually quite good at making pest outbreak forecasts. The essential point of forecasting is to predict the timing of critical pest populations, or populations reaching the economic threshold. In this section the term forecasting is used in the wide sense to include simple spray warnings based upon first occurrence records, as well as more complicated forecasting by prediction.

Emergence or occurrence warnings (pest monitoring)

In temperate regions these are basically emergence warnings as the first of the overwintering eggs hatch, or the first adults emerge from the overwintering pupae. Because of the climatic regulation most emergences take place over a relatively short period of time and are not too difficult to monitor. In the tropical parts of the World, where weather conditions permit continuous breeding of the pests most of the time, the warning is basically for the first occurrence of the pest in the crop, or sometimes the recording of immigrants from an adjoining area. In southern Japan on the island of Kyushu there are several serious rice pests which are unable to overwinter and which arrive from E. China each summer as aerial immigrants. In the areas at risk from these pests large wind-traps are mounted along the coast to sample the anemoplankton (small insects carried on the wind or air currents) to detect the arrival of these visitors. In these different cases, this type of warning is applicable only where the pest is known to be a serious one, and when economic damage is to be expected, based upon years of previous empirical experience. Such a crop is referred to as being at *high risk*.

In its simplest form, this type of warning consists of a visual record of adult insects in the vicinity of the crop, otherwise the use of different types of traps is required. To make this method more reliable it is preferable to use a trap that actively attracts the insects and so will be effective at lower pest densities; if electricity is available in the field then an ultra-violet (u.v.) mercury-vapour light-trap (sometimes called 'black-light') can be used for some species such as moths and leafhoppers. Black-light traps are used to monitor the arrival of migratory rice leafhoppers in Japan, and are used to such an extent throughout China for pest monitoring that electricity cables run throughout all the agricultural areas of eastern China. In some cases the u.v. traps are of a sufficient density to be also used for the '*trapping-out*' of several pest species. Otherwise pheromone traps are very good, and quite a few pheromones are now commercially available in most countries. Codling Moth warnings are based upon both pheromone and u.v. light trap catches; fruit flies of the genus *Dacus* are easily trapped using 'Cu-lure' or methyl eugenol baited traps. Warnings of danger from locusts and armyworms usually rely on sight records of hopper or adult swarms. Suction traps are suitable for small flying insects such as aphids and midges, but the more active fliers can elude the trap easily. Sticky traps are cheap to make and easy to handle, and by adjusting the level of stickiness it is possible to catch only the correct-sized insects. But the efficiency of such traps is rather dubious and may require some trial runs to establish whether it is a suitable method for any particular pest. For some small insects water traps can be quite effective. Some traps such as sticky and water traps may be more effective if coloured yellow, as it seems that a number of pests find the colour yellow attractive.

Emergence traps fixed on the soil between the crop plants are very suitable for species that pupate in the soil or litter, such as some Diptera, Lepidoptera and Thysanoptera, but care has to be taken for such traps may heat the soil and cause precocious emergence.

The use of traps as an ecological tool for the assessment of population size is highly problematical, and requires great care and forethought, but for forecasting use most of the time all that is required is a record of the first emergence or occurrence in the crop. In some temperate situations empirical studies over many years can link certain levels of trap catches (e.g. Codling Moth) with particular levels of infestation and crop damage. For details concerning methods of trapping see Southwood (1978) and in McNutt (1976) various types of traps are illustrated.

Spray warnings based on such trapping methods have to make due allowance for the female maturation period, rate of oviposition and time required for egg development.

Forecasting by sampling

Strictly speaking there is no real difference between this category of forecasting and the previous one, for in both cases the pest population is being sampled (and monitored), but for practical purposes there is some merit in treating them as distinct. The study of the development of a pest population is now often referred to as *pest monitoring*.

By sampling immature stages of insect pests it is possible to arrive at approximate estimations of numbers expected in later stages. The further back in the life history one samples, the less reliable will be the method, because of the difficulty in estimating natural mortality rates. Thus, it is rather unreliable to sample fly pupae in the soil in the autumn (for example in the UK); it is more useful to sample eggs in the soil (around the plants) in the following spring. In the UK the taking of soil cores for insect eggs of Carrot Fly and Cabbage Root Fly is quite successful for estimating later population of root maggots, and hence determining whether or not to apply insecticides.

With pests that have alternative hosts, they may be sampled while on the other host, so that an estimate of their probable pest density on the crop can be made; this method has most application in temperate regions where the pest overwinters as eggs on another plant, e.g. Peach–potato Aphid; Black Bean Aphid is often sampled as overwintering eggs on spindle trees.

With many Lepidoptera it may be feasible to determine the best spraying date by the finding of eggs on the crop. This is done with various pulse moths (Pea Moth, etc.), some stem borers and many of the bollworms of cotton. In many parts of Africa the major cotton bollworms are *Diparopsis* and *Heliothis* and crops at risk are examined either in the field or else samples taken back to the laboratory to search for eggs and/or young larvae.

Forecasting by prediction

Temperature. Temperature is the single most important factor controlling insect development and hence population numbers. A simple method using mean temperatures for two months has been developed to predict the date of emergence of the adult of the Rice Stem Borer (*Chilo simplex*) in Japan. It has also been used in the USA for the prediction of outbreaks of the European Corn Borer (*Ostrinia nubilalis*). A base temperature of 10 °C is used, and the amount of development the pest attains daily is indicated by the number of degrees above 10 °C each daily mean temperature reaches. The accumulation of these departures from the base temperatures in one season is expressed as *degree-days*. Observations over several years have given the number of degree-days required and have equated these with each stage of development.

Rice Leaf-roller has recently become a serious pest of rice in Japan, and here it is recorded that there are five larval instars on young plants and six instars on the ripening crop, usually with three generations each season in Japan. The second generation requires 210 degree-days for development, whereas the later, third generation requires 300 degree-days.

The developmental requirements, in terms of degree-days, are now known for a large number of pests in several different countries, mostly in Canada, USA, Japan and Europe (including the UK). Information on this matter may be obtained from the appropriate national research station in most countries.

Rainfall. Rainfall has also been used to forecast the likelihood of pest attack. In Tanzania outbreaks of the Red Locust (*Nomadacris septemfasciata*) have been forecast from an index of the previous year's rainfall, and in the Sudan the amount of pre-sowing rains frequently enables jassid damage to the cotton crops to be predicted.

Generally rainfall-based predictions are tenuous at best, and this line of research does not appear at present to be very profitable.

Climate. A climatograph (see fig. 4) will show the relationship between the basic pest requirements and the prevailing conditions in the area in question. This will give an indication as to the likelihood of the pest becoming established in that area. Such studies have formed the basis for many of the international phytosanitary regulations, when it has been shown that there are areas where a particular pest could become established because the climatic conditions are suitable and if introductions were permitted.

The final type of prediction is based upon observation of climatic areas, for on this basis some areas where critical infestations are likely to occur can be predicted for some pests. The principal factors controlling a pest population build-up may be climatic, biotic and topographical, although a combination of temperature and relative humidity (or rainfall) is probably the most important.

Knowing the temperature and relative humidity requirements for the different instars of an insect species and knowing the climatic conditions of an area, enables the likelihood of an outbreak of that pest in the area to be estimated.

The effects of climate upon insect pests is discussed in more detail in chapter 2 (page 20).

4 *Methods of pest control*

There is a large number of different methods of pest (including disease) control available to the crop protectionist, but careful deliberation is required in making a choice of methods. In general the orientation of the control project is towards the crop plant population rather than individual plants, so that low levels of pest infestation are acceptable provided that damage levels are low. Obviously with some expensive horticultural crops the welfare of each individual plant is of concern.

The choice of method(s) to be used depends on several factors.

(a) Degree of risk – some crops in some fields are at *high risk*, because in that area serious pests are invariably present in large populations. In such situations *preventative* measures (sometimes called insurance measures) may be justified. Similarly a high risk area may be predicted by sampling and forecasting techniques.
(b) Nature of pest and disease complex – usually several (or many) different pests and pathogens will be interacting on the crop in the form of a pest complex. Key pests will dominate the control strategy. With many crops between four and eight major pests will require control at any one time. Ideally the method(s) used will control several pests (and sometimes pathogens also) simultaneously.
(c) Nature of the crop and agricultural system – e.g. height of crop and spacing.
(d) Economic factors – e.g. cost of chemicals and specialized equipment.
(e) Ecological factors – e.g. extent and type of natural control, and availability of water.

Obviously it is vitally important that the pests be correctly identified and that their general biology be known.

The system of classifying control measures which follows is based upon the mode of action, and is widely used by plant pathologists.

(a) Exclusion – including quarantine, use of disease-free seed and planting material; designed to keep (new) pests and diseases out of an area or crop.
(b) Avoidance – uses cultural control methods and sites free of infection, and resistant crops.
(c) Protection – use of chemicals mostly, as protectants, therapeutants and disinfectants; physical protection may be included; anticipated protective measures are termed preventative.
(d) Eradication – for an outbreak of a pest or disease in a new area; uses soil sterilization, fumigation, heat treatment, insecticidal saturation, etc.

Generally it is more useful, from a pest viewpoint, to regard control measures according to their basic nature, as follows:

(a) legislative methods
(b) physical methods
(c) cultural control
(d) crop plant resistance to pest attack
(e) biological control
(f) chemical control
(g) integrated control
(h) integrated pest management (pest management)
(i) eradication.

Legislative methods

These are obviously methods of control where government legislation (laws) has been passed so that certain control measures are mandatory, with failure to comply being a legal offence. These are extreme measures and only apply to certain very serious pest situations of national importance.

Phytosanitation (quarantine)

When the major crops were distributed around the world from their indigenous areas, initially they may have been free of their native pests and diseases (particularly crops grown from seed). But over the years many reintroductions have been made and gradually the native pests and pathogens have spread also, until now when some pests and diseases are completely sympatric with their host crops. However, there is still a large number of pests and diseases that have not yet spread to all parts of the crops' areas of cultivation. For example, the early success of coffee in Brazil (S. America) was due to the absence of Coffee Rust, Antestia Bugs, and Coffee Berry Borer, which remained behind in E. Africa; but this advantage has now been lost as in recent years both Rust and Berry Borer have become established in S. America. Similarly the success of rubber as a crop in S.E. Asia is due in part to the absence of its native South American Leaf Blight.

FAO (Food and Agriculture Organization of the United Nations) have organized a system of international plant protection with respect to the import and export of plant material. The world is divided into a number of different geographical zones for the basis of phytosanitation; each zone has its own regional organization for co-ordination. There is now an International Phytosanitary Certificate, an essential document required for importation of plant material into almost every country of the world.

Disease-free and pest-free plants can usually be imported, provided they are accompanied by appropriate documentation from the country of export, but certain plants (and fruits) are completely prohibited because of the extreme likelihood of their carrying specific noxious pests or pathogens. Other categories of plants are allowed to be imported (and exported) after routine treatment to eradicate possible pests; such treatment usually consists of fumigation (e.g. fumigation of fruit in the USA using ethylene dibromide). Sometimes the plants have to be kept in quarantine isolation for a period of time to check that no symptoms develop (as with domestic pets and rabies quarantine). Specific import regulations vary from country to country according to the nature of the main agricultural crops; thus California and Florida are very concerned about *Citrus* fruit import, and Hawaii has rigorous regulations concerning orchids. Importation of pome fruit into many countries from Asia and America is very rigorously controlled because of the danger of importation of San José Scale (*Quadraspidiotus perniciosus*); this is potentially the most destructive orchard pest and caused tremendous damage to commercial orchards in the southern USA at the turn of the century, attacking all types of pome fruit, plums and cherries, in addition to shade trees and ornamentals. The terminal stages of infestation usually resulted in the death of the trees. California Red Scale (*Aonidiella aurantii*) is a serious pest of *Citrus* that is not yet quite pantropical, so that citrus fruit importation is subjected to careful scrutiny in several tropical countries.

For further information on international phytosanitation see Caresche *et al.* (1969) and Hill & Waller (1982).

Prevention of spread

A very serious temperate pest is Colorado Beetle (*Leptinotarsa decemlineata*), first observed in the semi-desert areas of western USA in 1824, feeding on the foliage of a wild *Solanum* species. When the cultivated potato was introduced into this area the beetle found a new and more desirable source of food, and in a short period of time both potato cultivation became widespread in the USA and the Colorado Beetle spread rapidly. By 1874 it had reached the Atlantic seaboard of the USA and both Ontario and Quebec in Canada. The first record of Colorado Beetle in Europe was in Germany in 1876, and in 1877 a single beetle was found in the Liverpool docks. In the UK it was realized that this was potentially a very serious pest and that measures were required to prevent its establishment in Britain; in 1877 the Destructive Insects Act was established, the first British legislation concerned with plant health. It referred only to the Colorado Beetle, and it controlled the importation of potato plants, or tubers, and other vegetables from anywhere outside the UK. It authorized destruction of a crop on which the beetle was found, and stipulated various control measures to be taken in the event of finding an infestation. It was changed to the Colorado Beetle Order in 1933.

The first established colony of Colorado Beetle in the UK was found in Kent in 1901, but it was successfully eradicated; since then there have been more than 120 breeding colonies found in parts of Kent and S.E. England, but prompt action by MAFF staff has ensured that none became established.

Similar introductions occurred in Europe, and in 1921 the pest became established near Bordeaux in France. It gradually spread throughout much of Europe, aided in part by the confusion caused by World War II. Fig. 10 shows the recorded spread of infestation of Colorado Beetle in Europe from 1921–61 (MAFF, 1961). Since then it has spread further in Continental Europe, and recently successfully invaded southern Sweden (see map on page 293), though after rigorous preventative measures it was contained and finally eradicated so that Sweden is once again free from this pest.

Colorado Beetle is restricted to plants of the Solanaceae as hosts, and it is particularly damaging to potato crops; infested crops seen in Europe often suffer complete defoliation if insecticides are not used.

An important pest of stored grains is *Trogoderma*

Fig. 10. Approximate lines of advance of the Colorado Beetle in Europe since 1921.

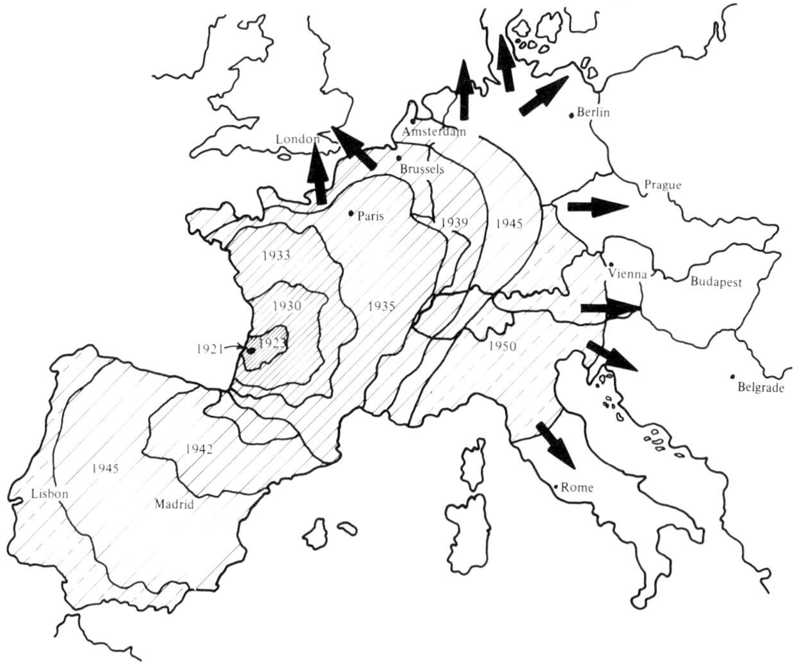

granarium (Khapra Beetle, see page 273), now almost worldwide in distribution but still not established in parts of S. and E. Africa. Regular accidental introductions occur, usually from cargo ships carrying infested grain, but very strict examinations and control measures are taken to ensure that there is no spread of infestation beyond the dockside areas, and also that the infestation is soon destroyed. This pest was established in the USA in 1952, when it rapidly spread throughout the country, but it is now largely contained at a very low level by regular produce inspection and fumigation.

Fruit flies (*Dacus* and *Ceratitis* spp.; Tephritidae) are very important pests of most tropical fruits and a serious threat to *Citrus* (and other fruit) cultivation in both Florida and California (USA). There are regular accidental introductions of Medfly (*Ceratitis capitata*) into both areas, and on each occasion legislation has ensured an immediate eradication programme involving compulsory intensive widespread insecticide spraying both from the ground and from the air. To date, these campaigns have proved successful and the Medfly (and some other Tephritidae) have been denied establishment. The cost of eradication of a pest such as the Medfly (caught before it has spread very far afield) is quite infinitesimal in comparison with the value of the $14000 million fruit industry in California.

An important publication on this topic is *Plant Health* (The scientific basis for administrative control of plant diseases and pests), edited for the Federation of British Plant Pathologists by Ebbels & King (1979).

Physical methods

These refer to methods of mechanical removal or destruction of the pest, and are usually unimportant in most countries owing to the high cost of labour. They are still of use in some countries and for particularly valuable crops.

Mechanical

Hand-picking of pests was probably one of the earliest methods of pest control and is still a profitable method for the removal of some caterpillars from young fruit trees. The killing of longhorn beetle larvae boring in branches of some bushes is recommended by the pushing of a springy wire (e.g. bicycle spoke) up the bored hole and spiking the insect. The use of mechanical drags, which crush insects on the ground, has been made against armyworms (*Spodoptera* larvae), but this practice is now generally outmoded. Banding on fruit trees is particularly effective against caterpillars and ants, which gain access to the tree by crawling up the trunk. Spray-banding of the apple trunks is practised against the wingless females of Winter Moth and other Geometridae which typically climb the lower parts of the trunk from the soil. An earlier method of locust control was to herd the hoppers into a large pit which was afterwards filled in with soil. Armyworms when on the move can often be trapped in trenches dug across their line of marching, and when trapped they are easily destroyed by burning, filling the trench or spraying with chemicals.

In parts of Asia, especially in gardens and on smallholdings, it is common practice to place bags around large fruits (e.g. grapefruit, pomelo, pomegranate and jackfruit) to deter fruit flies (Tephritidae) from oviposition. In rural areas and in the forests the bag is usually woven from grass or raffia, but in more suburban situations paper bags and polythene may be employed. Occasionally a single fruit may be left unprotected so that the local fruit-flies will concentrate their egg-laying upon this one fruit which can later be destroyed.

In temperate smallholdings and gardens it is a common practice to build a large cage of wire or string (now often plastic) netting over a fruit bed, but this protection is usually against fruit-eating birds rather than insects; such a cage may be portable and temporary or may be a permanent construction. This idea is now being developed on quite a large scale throughout the world, particularly for the protection of seed-beds and also for particularly valuable crops that may be at risk locally (e.g. orchids, etc.). Fig. 11 shows a protected area of seed beds, about 5 × 10 m, in E. England, for the rearing of virus-free *Brassica* seedlings and other crops. The netting is a fine plastic mesh and will keep insects as small as aphids at bay; if the mesh is too small however, plant growth may be retarded somewhat due to reduced sunlight levels. Small structures such as this are ideal for rearing seed crops of vegetables, and then insects such as blowflies (Calliphoridae) can be introduced inside the netting and maximum pollination may be easily achieved. These structures also provide some physical protection from strong winds, frost, heavy rain and extreme insolation, but do allow free air circulation which prevents over-heating, which makes these structures very useful for the tropics. They are particularly effective for rearing virus-free seedlings, for with care the aphid vectors (and others) can be completely excluded. Fungal spores will, of course, pass freely through the mesh.

Greenhouses in temperate situations provide complete physical protection for a crop, but the main purpose is to modify the temperate climate to provide a suitable hot/moist microclimate for the cultivation of exotic crops or local crops out of season. Nowadays the use of polythene tunnels has become very widespread (fig. 12) as these are effective and much cheaper to erect than the traditional glasshouses. The cultivation of protected crops has now become very sophisticated and in many respects is a separate branch of horticulture. In the UK the main source of expertise and information lies with GCRI (Glasshouse Crops Research Institute) at Littlehampton, Sussex. (There are similar institutions in other countries.) It is in these places where biological control of pests is widely practised, within the controlled microhabitat, with great success.

As long ago as a hundred years it was reported that discs made of tarred-felt could protect cabbage transplants from Cabbage Root Fly attack, and recent work at NVRS has shown that foam rubber (carpet underlay) discs of 12 cm diameter, placed around the stem of transplants, significantly reduced the level of Root Fly damage. This reduction in damage is due partly to the reduced numbers of eggs laid (only half the number) on the discs, and in part due to aggregation of predatory

Fig.11.

Fig.12.

ground beetles under the discs, which consequently eat proportionally more fly eggs.

Use of physical factors

The use of lethal temperatures, both high and low, for insect pest destruction is of importance in some countries. (Often high temperatures are lethal for temperate pests, and low temperatures for tropical pests.) The use of cool storage in insulated stores for grain is practised in Asia and Africa. The purpose of this method is not the actual destruction of the pests, which may occur when lethal temperatures are employed, but the drastic retardation of development following the reduction of the metabolic rate.

Kiln treatment of timber for control of timber pests is very widely practised in many countries. Plant bulbs are often infested with mites, fly larvae (Syrphidae) or nematodes, and hot-water treatment (dipping) can be a very successful method of control if carefully carried out. The drying of grain, which is widely practised for a reduction in moisture content, usually results in lower infestation rates by most pests. The heating of cotton seed to kill the larvae of Pink Bollworm (*Pectinophora gossypiella*) is an effective control. In different parts of the World hermetic storage of grain is being developed as a standard long-term storage method. The stores are now of several different basic types and generalization is not feasible; the principle involved is that only a small quantity of air is enclosed within the sealed bin, the oxygen in which is quickly used up by the respiration of the pests and the subsequent carbon dioxide accumulation quickly results in the death of all contained pests, both arthropod and microbial.

On-farm storage of grain is being carried out in some areas using butyl silos; the addition of small quantities of diatomite fillers increases the effectiveness of this control, as the abrasive effect removes the outer waxy covering of the epicuticle of the insects, resulting in greater water loss (and possible dehydration) and greater ease of insecticide penetration through the cuticle.

Use of electromagnetic energy

The radio-frequency (long wavelength radiations) part of the spectrum has been extensively studied in the development of radio communications, radar, etc., and it has been known for a long time that absorption of radio-frequency energy by biological material results in heating of the tissues. Control of insect pests by such heating is only practicable in enclosed spaces of small or moderate size (food stores, warehouses, timber stores). The nature of absorption of radio-frequency energy by materials in a high-frequency electrical field is such that for certain combinations of hosts and insect pests their dielectric properties are favourable for differential absorption of energy, hence the insects can be killed without damaging the host material. Timber beetles in wood blocks have been killed in this manner, but whether this treatment offers any real advantage over normal kiln treatment is doubtful.

Use of infrared radiation for heating purposes is very much in its infancy.

Many insects show distinct preferences for visible radiation of certain wavelengths (i.e. certain colours), as well as the long recognized attraction of ultra-violet radiation for various nocturnal insects, especially Lepidoptera. Ultra-violet light traps have, on occasions, significantly lowered pest populations in various crops, but have also failed when used against other pests. Mercury vapour lamps are mostly being used for pest monitoring, especially for various night-flying Lepidoptera (Codling Moth, Pea Moth, rice stem borers, etc.), and for this purpose their use is now very widespread. In some countries, such as China, electricity lines are laid extensively throughout agricultural areas and ultra-violet light traps are used in large numbers (fig. 13).

Aphids and some other plant bugs are attracted to yellow colours; this is possibly because most aphids feed either on young or senescent leaves, presumably because these are the plant parts where active transport of food material occurs. The young leaves are photosynthesizing rapidly and the sugars formed are transported away in the phloem system to be stored as starch grains in older leaves, tubers, etc. As leaves become senescent, the stored starch is reconverted into soluble sugars for transportation prior to leaf dehiscence. Senescent leaves are usually yellowish in colour and young foliage is often a pale yellowish-green. So the attraction of aphids to yellow colours seems fairly obvious, but why so many flies (especially Anthomyiidae) and some moths are similarly attracted to yellow is not obvious.

This attraction for the colour yellow by so many insects is exploited in the use of many different types of trap; yellow water traps consistently catch more aphids and more Cabbage Root Fly adults; sticky traps coloured yellow, and even pheromone traps for Tortricoidea and Tephritidae, are often more effective than white ones. In a monitoring programme, it is of course necessary to make allowance for any enhanced trapping effect due to the colour of the trap.

Conversely, many flying insects are repelled by blue colours and by reflective material. This has been exploited by using strips of aluminium foil, or metallicized plastic, between the rows and around the periphery of the crop, the result being that fewer flying aphids and other insects settle in the crop than would otherwise. Thus, reflective strips also result in crops having far less aphid-borne virus diseases. Another effective method is to surround the seed-beds by flooded furrows, as practised in S. China with vegetable crops.

Fig.13.

The ionizing radiations (X-rays, γ-rays) are sterilizing at lower dosages but lethal at higher. The use of these radiations in controlling stored product pests, particularly in grain, is being quite extensively studied in various countries.

Cultural control

These are regular farm operations, that do not require the use of specialized equipment or extra skills, designed to destroy pests or to prevent them causing economic damage. Often these are by far the best methods of control since they combine effectiveness with minimal extra labour and cost.

Most of the cultural methods do not give high levels of pest control, and in the recent past, when reliance was placed almost entirely on chemical control (using organochlorines), these methods received little attention. However, with the recent interest in integrated pest management there is a revival of interest in the use of cultural methods for incorporation into management programmes. Often the present-day scheme is to use several different methods in conjunction, each method achieving a certain level of control, so that in concert the desired level is achieved with minimal ecological disruption etc.

Optimal growing conditions

A healthy plant growing vigorously has considerable natural tolerance to pests and diseases (as with a healthy animal), both physically and physiologically. Good plant vigour is a result of sound genetic stock and optimal growing conditions. Obviously the farmer attempts to provide such growing conditions so that the crop yield will be maximal. Many diseases are more severe, and the damage by pests more serious, if the plant is suffering from water-stress (drought), unfavourable temperature, imbalance of nutrients or nutrient deficiency, etc. This *predisposition* to pest attack and disease can be very serious when crops are grown on marginal land. This is one main reason, together with reduced yield, why the cultivation of marginal land is generally not very successful.

This point is of particular interest at the present times, for various countries are endeavouring to increase their national agricultural yield, partly to produce more food crops to feed the ever-increasing population, and partly to increase the cash crops as a source of national revenue. But the great majority of countries are already utilizing all their high-quality agricultural land and the only land available for agricultural development is marginal tracts with a very limited potential. In some countries this situation is exacerbated in that land has to be taken from agriculture in order to provide sites for new towns, airports, and the general ever-increasing sprawl of urbanization; such land is almost invariably choice agricultural land, and may be the finest such land in the country (that being the main reason for the historical siting of the town in that location).

Photosynthetic efficiency

When aiming at a maximum level of sustained crop yield it is advantageous to understand the physiology of the crop plant in order that cultivation be practised in the most effective manner. This is also important in some countries where cultivation of marginal land is being undertaken, for only certain crops give adequate yields on poor land. Plants struggling in poor growing conditions are invariably more susceptible to damage by pests.

Some plants are now known to have a more efficient photosynthetic activity than others. They are referred to as C_4 plants – so named because of the chemical intermediary, the four-carbon oxaloacetic acid. The majority of plants, which could be called the 'normal' plants, are referred to as C_3 plants. C_4 plants generally have a high productivity in situations of high temperature and low humidity (and the concomitant low carbon dioxide concentration). They have a high level of stomatal resistance, which conserves water at high temperatures, but of course restricts carbon dioxide entry, but they are able to carry out photosynthesis very effectively under low carbon dioxide concentrations. As a group these plants are generally the most productive agriculturally. Examples of C_4 plants include maize, sugarcane and other tropical Gramineae as well as many weed species in the Chenopodiaceae, Euphorbiaceae, etc.

The more effectively functioning plants are generally those with the highest agricultural yields, and are also often more tolerant of insect attack and less susceptible to damage.

The importance of shade has been somewhat misunderstood in the past; several crops have in the past been thought to require shade for best growth and production; in some cases because they are naturally occurring forest understorey plants and to be regarded ecologically as skiophytes (i.e. shade-tolerant). A striking example is tea – for some obscure reason tea was thought to need shade, and many plantations bear mute testimony to this previous practice in the large number of dead trees to be seen standing throughout the plantation. Recent experiments in Indonesia have shown that cocoa grown as a heliophyte (i.e. fully exposed to sunlight) gives a greater yield than when grown as usual under the shade of a tree cover. In the wild this shrub is a skiophyte adapted for life as a rain forest understorey shrub; but this adaptation to a low light intensity does not necessarily imply that it prefers shaded conditions. And the practice of growing cocoa as a forest-edge crop, under the shade of the trees, frequently exacerbates the pest situation in that many of the more serious pests live on (or in) the forest trees and continually invade the crop plants from this vast natural reservoir. In parts of Africa the most serious pests to ripening cocoa are the local monkeys, and in the forest these animals are almost impossible to control economically.

In Europe it has long been realized that strawberry plants grow and yield best when fully exposed to sunlight, even though they are found wild as woodland ground flora.

To minimize pest damage to agricultural crops it might be advantageous to review cultivation methods to determine that each crop is really being grown under optimum conditions, and not just to continue cultivation practices on the grounds of historical precedent.

Avoidance

Empirical observations will reveal that certain areas (and fields) are constantly 'at risk' from particular pests and conversely others are pest-free. Clearly, if a crop can be grown in areas of the latter category it can be expected to remain free from that particular pest. This practice is particularly effective against certain soil-borne diseases and nematodes, but less so against most insects because of their greater mobility. This is one of the advantages of shifting cultivation, so widely practised in parts of the tropics. Soil insects, such as root maggots (Anthomyiidae), wireworms, chafer grubs, rootworms (USA) and swift moth caterpillars, can to some extent be avoided by the planting of non-susceptible crops and the growing of the vulnerable crops at some distance away. Such practice is to be highly recommended, but is sometimes difficult to achieve as there may be strong agricultural, or other, reasons for growing the crops in those areas. Thus in the UK the East Anglian fenlands are ideal for growing carrots, celery and parsnip, but many of these areas have endemic Carrot Fly populations on the native hemlock and other Umbelliferae, and have long been areas highly 'at risk' from Carrot Fly. In this situation it is possible to minimize the risk from this pest by using other cultural methods such as crop rotation, etc., and of course chemical protection has to be used.

Time of sowing

By sowing early (or sometimes late) it may be possible to avoid the egg-laying period of a pest, or else the vulnerable stage in plant growth may have passed by the time the insect numbers have reached pest proportions. Early sowing is regularly practised against Cotton Lygus (*Taylorilygus vosseleri*) and Sorghum Midge (*Contarinia sorghicola*) in Africa. In N. Thailand it has been shown that early transplanting of paddy rice reduced the level of Rice Gall Midge attack appreciably. Another important aspect of the time of sowing is that of simultaneous sowings of the same crop over a wider area, to avoid successive plantings which often permit the build-up of very large pest populations.

In Europe the recent trend towards autumn-sowing cereals (winter wheat, winter barley) has reduced the risk of aphid damage to seedlings to a negligible level. At the same time this practice does increase the risk from Wheat Bulb Fly. Autumn-sown field beans are generally not at risk at all from Black Bean Aphid, whereas the spring-sown plants may be severely damaged.

With Wheat Bulb Fly only the very early-sown spring wheat is at risk, so if spring wheat is sown later there is little risk from this pest; but if sowing is delayed too much then crop growth is impaired!

Deep sowing (planting)

Some seeds are less liable to damage and pest attack if rooted deep, but of course if planted too deep germination will be impaired. Many root crops are also less liable to attack by pests if they are deeper in the soil. This is true for sweet potato in that the deeper tubers always have fewer weevils (*Cylas* spp.) boring inside, and the deeper potatoes have fewer infestations of Tuber Moth larvae. Even if no special attempts at deeper planting of root crops are made, then care should be taken to ensure that no tubers are allowed to grow too close to the soil surface; earthing-up should be done when required. Exposed root crops may be damaged by pheasants, other birds, and by rodents and rabbits grazing.

Time of harvesting

Prompt harvesting of maize and beans may prevent these crops from becoming infested by Maize Weevil

(*Sitophilus zeamais*) and Bean Bruchid (*Acanthoscelides obtectus*) respectively. Both of these pests infest the field crops from neighbouring stores, but are generally not able to fly more than about half a mile; so an added precaution is to always grow these crops at least half a mile away from the nearest grain store.

New varieties of crops which mature early may enable a crop to be harvested early, before pest damage is serious. This is one of the qualities that many plant breeders are constantly seeking in a very wide range of different crops. This approach requires detailed knowledge of the ecology and life history of the local serious pests so that crop development might be desynchronized in relation to pest population development.

Close season

In E. Africa legislation has been passed to ensure that there is a close season for cotton growing in order to prevent population build-up of Pink Bollworm (*Pectinophora gossypiella*) which is oligophagous on Malvaceae. This legislation stresses that all cotton plants should be uprooted and destroyed (or burned) by a certain date and quite clearly no seed would be planted until the following rains arrive. However, it is clear that many farmers do not bother to destroy the old plants by the appointed date and so in some areas there is considerable survival of diapausing Pink Bollworm larvae.

This approach to pest control tends to be more applicable to the tropics where insect development and crop production may be more or less continuous. In temperate regions there is already established very firmly a close season for virtually all crops, namely winter!

Deep ploughing

Many Lepidoptera (particularly Noctuidae, Hepialidae, Sphingidae and Geometridae), Coleoptera and Diptera pupate in the soil, and a large number of their larvae live there. The bulk of the soil insect population lie in the top 20 cm of the soil (most are in the top 10 cm). Deep ploughing will bring these insects to the surface, to be exposed to hot sunlight (insolation), desiccation and predators. In many tropical areas a farmer ploughing a field will be followed by a flock of cattle egrets, little egrets, crows or starlings, in a seaside locality there might be a flock of gulls; all these birds will feed on the exposed worms, slugs and insects. In temperate regions a plough is usually accompanied by a large flock of birds (gulls, crows, etc.), and although the birds are also eating earthworms their consumption of insect larvae and pupae (many or most of which are pests) must be prodigious, as often a hundred or more birds may follow a single tractor (fig. 14).

Fallow

Allowing a field to lie fallow almost invariably reduces pest and pathogen populations, but care must be taken to ensure that there are no volunteer crop plants or important secondary host weed species. Fallowing may be done as *bare fallowing* when the soil surface is left bare, or *flood fallowing* when the field is flooded with water for a while. Sometimes, as an alternative, a cover crop of legumes is grown as *green manure* which is then ploughed under.

Fig. 14.

Crop rotation

In olden times a period of fallow was an essential part of all crop rotations, but nowadays economic pressures mean that fields can seldom be left fallow. Instead, basic crop rotation is usually practised for the obvious reasons that continuous cultivation of one crop depletes the minerals and trace elements in the soil quite rapidly, and also induces disease and pest build-up. However, some agricultural crops require rather specialized growing conditions, and these, combined with the practice of large-scale cultivation, result in some areas growing crops such as sugarcane, wheat, pineapple, maize or potato, almost continuously. Obviously the orchard crops (apple, plum, pear, citrus, peach, olive, etc.) are also very long-term monocultures, as are vineyards.

The alternation of completely different crops in a field has very obvious advantages from the pest and disease control aspects. But in a rotation it is necessary to remove a particular crop quite a distance away; having the same crop in an adjacent field is not really 'rotation' so far as active insects are concerned, although it might be adequate for soil nematodes and some soil-borne diseases. So, for effective pest control crop rotation has to separate crops both spatially and in time (temporally).

Against monophagous and oligophagous pests crop rotation can be effective, especially with beetle larvae that may take a year or more to develop, but it is not effective against migratory pests or those with effective powers of dispersal. The alternation of cereals with non-cereals may be an important method of curtailing *Nematocerus* weevils in Africa. A common type of rotation is the alternation of a legume crop with cereals; this is effective against some pests, but others (e.g. *Colaspis*, *Diabrotica*) can utilize both host types. In Europe a combination of potato, wheat and rape is currently popular and successful.

Secondary hosts

Most pests are not monophagous and so will live on other plants in addition to the crop. Sometimes, in point of fact, the crop itself is not the preferred host! For example, Turnip Aphid is often more abundant on *Cardamine* and charlock, etc., than on the *Brassica* plants. In many cases the pests build up their numbers on wild hosts and then invade the crop when the plants are at the appropriate stage of development. Many Cicadellidae and Delphacidae that are crop pests (on rice and other cereals) feed and breed on wild grasses in the vicinity of the paddy fields so that when the young rice is planted out there is a large bug population waiting to infest the young tender shoots of the rice. The destruction of alternative hosts (or the control of insects on them) may be an important part of an IPM programme. The alternative hosts are usually native plants (but may be introduced), and may be trees, shrubs, or herbaceous plants; they may be cultivated species, wild plants or weeds. Solanaceous weeds, for example, are important alternative hosts for pests of tomato, tobacco, eggplant and potato. In many cases the permanent pest population in an area is maintained not on crop plants but on wild plants belonging to the same family; there are many wild Cruciferae that support *Brassica* pests and the number of wild Leguminosae, Chenopodiaceae, Rosaceae, etc., is very large.

In temperate situations a number of pests and diseases have an alternation of generations on quite different hosts, e.g. *Myzus persicae* on peach and potato. The removal of the alternative host can effectively reduce such pests to insignificance, but in the case cited, of course, the alternative host is itself a crop plant. Black Bean Aphid overwinters on spindle trees, and in some areas attempts have been made to remove these trees locally, but usually without much success. *Pemphigus bursarius* overwinters on poplar trees where it makes petiole galls in the spring; later generations migrate to lettuce where they encrust the roots.

With monophagous pests it is important to remove any host plants between crops (if feasible for the species concerned). A monophagous insect would normally be confined to the species of a single genus for host, though it would be unusual for it to be restricted to a single species of one genus. Thus, for sugarbeet pests it is not feasible to attempt to remove all wild species of Chenopodiaceae as they are too abundant. But, for the Brown Planthopper of Rice (*Nilaparvata lugens*) which is restricted to rice (*Oryza* spp.) as a host plant, the alternative host plants are either wild rice or volunteer rice; by destruction of wild rice weeds, volunteer plants and crop residues, this pest can be partially controlled.

The situation with regard to secondary/alternative hosts tends to be more important in the tropics where usually there are more wild plants in the immediate neighbourhood of the crop, especially where land is cultivated along the edges of tracts of native forest ('jungle').

Weeds

Most cultivation practices require destruction of weeds because of their competition with the crop plants, and their interference with different aspects of cultivation. But weeds can also be important from the viewpoint of pests and diseases; sometimes the weeds may be alternative hosts, as already mentioned. Weeds in a particular crop often belong to the same family as the crop plant because of the selective nature of many post-emergence herbicides. Some pests seem to prefer weeds as oviposition sites; for example, some cutworms (*Agrotis* spp.) and some beetles (Scarabaeidae) lay most of their eggs on, or in the immediate vicinity of, weeds in the crop. Weed removal at the appropriate time may result in many potential pests being destroyed. Some weeds are important as natural reservoirs (alternative hosts) of both pathogens and invertebrate pests.

Trap crops

The use of trap plants to reduce pest infestation of various crops is based upon the knowledge that many pests actually prefer feeding upon plants other than those on which they are the most serious pests. This preference may be exploited in two different ways: either the pests are just lured from the crop on to the trap plants where they stay and feed, or else, because of the greater concentration of pests on the trap plants, only the trap plants need be sprayed with pesticides. In the latter case, since the trap plants are either grown as a peripheral band or else interplanted at about every fifth to tenth row, the saving of insecticide represented is considerable.

This is a technique used mostly in warmer parts of the world where insect pest breeding is more or less continuous; in the cooler temperate regions, with the cold dormant winter period, it is of less applicability. The practice of intercropping is based upon the same premise, and this has long been a regular method of reducing levels of pest damage in Europe and N. America.

Intercropping

This practice obviously has various drawbacks for large-scale agriculture, but can be of particular application for the small farmers, who often use little insecticide. Intercropping can certainly reduce a pest population on a crop, and without doubt reduces the visual and olfactory stimuli that attract insects to a particular crop species. As a method of control it is most effective against exogenous pests, such as locusts, which enter the crop for only part of their life-cycle. In N.E. Thailand in 1980 groundnuts were inter-cropped with maize, and nymphs of Bombay Locust (*Patanga succincta*) were induced to leave the maize (particularly on hot days) for the lower foliage of the groundnuts where they were eaten by ducks. Work is in progress on different schemes of intercropping at IRRI in the Philippines, for use particularly in that country with its large proportion of peasant farmers.

In Europe, on smallholdings, it has long been a regular practice to intercrop, often using plants with very strong odoriferous qualities in the hope that the odours released would confuse the host-seeking female insects, which it does appear to do. Onions and crucifers, for example, would be used to shield carrot crops. Often wild-type plums are grown around the periphery of apple orchards where they act partly as a wind shield and also as a diversion for a number of different insect pests.

Crop sanitation

This is a rather general term, and usually is used to include the following different aspects of crop cultivation.

(a) Destruction of diseased or badly damaged plants – the roguing of such plants is an important agricultural practice, but of course requires hand labour. For agricultural crops this is sometimes not feasible, but some farmers do make the necessary effort; with horticultural crops such as fruit orchards, flowers and some vegetables, destruction (preferably by burning) of infected and infested branches etc., is an important aspect of any control programme, as it is easier to remove the foci of infection than to kill the organism with pesticides.

(b) Removal and destruction of rubbish – old crop remnants, fallen leaves, branches, dead trunks, also weeds, etc. Some pests use rubbish heaps for breeding purposes, for example scarab larvae are to be found in rotting vegetation and soil, especially in rubbish heaps. The tropical *Oryctes* beetles are good examples, as the larvae are found in rotting palm trunks and rubbish heaps.

(c) Removal and destruction of fallen fruits – for the control of many fruit flies and boring caterpillars this is important, as the insects will continue to develop in the fallen fruit and will pupate either there or in the soil. This is, in fact, still a successful method for reducing the numbers of Codling Moth and other fruit-boring caterpillars, as well as for many different fruit flies (Tephritidae). It is still one of the best methods of control for Coffee Berry Borer, where labour permits.

(d) Destruction of crop residues – this is often vitally important in order to kill the resting stages (pupae, etc.) of many pests after harvest. Many stalk borers (Pyralidae, Noctuidae) pupate in the lower parts of the cereal stems and as such will be left in the stubble even if the main parts of the stalks are removed. With some crops of maize, sorghum and millets most of the actual stem is left. The stems should be burnt immediately after harvest. The rotting residues of crops such as turnip, parsnip and *Brassica* generally, after being ploughed in attract ovipositing female Bean Seed Fly and other Muscoidea. For diseases, crop residue destruction may be even more important. Ploughing in of crop stubble may kill a small proportion of pupae but most will survive; burning is most effective. European Corn Borer traditionally overwinters in maize stubble, but most farmers are now aware of this and so the stubble is destroyed.

The recommended method of destruction of all weeds, crop residues, rubbish, rogued plants, fallen fruits, etc., is by collection and burning; other methods may not kill the pests.

Crop sanitation tends to be a term mostly used by plant pathologists rather than entomologists, as it aims mostly at the removal of sources (foci) of disease infection.

Crop plant resistance to pest attack

In most growing crops it may be observed that some individual plants either harbour far fewer pests than the others or else show relatively little sign of pest damage. These individuals usually represent a different genetic variety from the remainder of the crop, and this variety is said to show *resistance* to the insect pest. Also when different varieties of the same crop are grown side by side, differences in infestation level may be very marked. Resistance to pest attack is characterized by the resistant plants having a lower pest population density, or fewer damage symptoms, than the other plants which are termed *susceptible*. Conversely, there will be some plants that appear to be preferred by the pests and these especially susceptible plants will bear very large pest populations. Frequently these plants will actually be destroyed by the pests and so will not breed and pass on their disadvantageous genetic material.

The main use of resistant varieties of crop plants in agriculture has been against plant diseases (Russell, 1978), and in general a high level of success has been achieved. Plant-parasitic nematodes (eelworms) in some ways behave rather like soil pathogens, and the development of resistant varieties of potato and wheat have been very successful in combating Potato Cyst Eelworm and Cereal Root Eelworm. Against insect pests plant breeding for resistance has not had the same success, but in some instances good control has been achieved with enough success to encourage further work in this area. On the whole it can be said that many resistant varieties of crop plants have given quite good control of insect pests, albeit only partial, against a very wide range of insect species. Many varieties of crop plants showing good resistance to important pest species have not been fully exploited because their yield is less, or of inferior quality, than the usual susceptible varieties.

In January, 1984, the Royal Entomological Society of London had a paper presented on the topic of 'Hypotheses concerning the evolution of plant resistance to insects' by Dr P.P. Feeny, illustrated mostly by reference to swallowtail butterflies and their food plants.

Varietal resistance to insect pests was broadly classified by Painter (1951) into three categories: non-preference, antibiosis, and tolerance; but Russell (1978) suggests the use of a fourth category: pest avoidance. Some workers restrict the use of the term varietal resistance to antibiosis, but this view is rather narrow and not practical. In fact it is often very difficult to distinguish between some cases of non-preference and antibiosis.

Types of resistance

(a) Pest avoidance
(b) Non-preference (= non-acceptance)
(c) Antibiosis
(d) Tolerance

The basis of these types of resistance is slight variations in genetic material; as defined by Russell (1978), 'Resistance is any inherited characteristic of a host plant which lessens the effect of parasitism.' The term *parasitism* is used in a broad sense to include the attack of insect pests, mites, vertebrates, nematodes, and pathogens (fungi, bacteria, viruses) on the host plant. The feeding of phytophagous insects on plants is not generally regarded as parasitism by most zoologists, but rather as ecological grazing.

Genetically there are three main types of a resistance. *Monogenic* resistance is controlled by a single gene, usually a *major gene* which has a relatively large effect. This type of resistance (often biochemical, involving phytoalexins) is fairly easily incorporated into a breeding programme, and it usually gives a high level of resistance; unfortunately this resistance is just as easily 'broken' by new pest 'biotypes' (new races or strains).

Oligogenic resistance is the term used when the character is controlled by several genes acting in concert.

Polygenic resistance is the result of many genes, and is clearly more difficult to incorporate into a plant breeding programme. It may be either morphological or biochemical, and it is generally less susceptible to biotype resistance ('breaking'). Many of the genes will be *minor genes* which individually only have a small effect genetically.

In epidemiological terms, resistance is classified as either *horizontal resistance* (alternatively, durable resistance), with a long-lasting effect and effective against all genetic variants of a particular pest, or *vertical resistance* (alternatively, transient resistance), effective for a short period and against certain variants only.

There are a few other terms which are in use in plant breeding for pest resistance. *Field resistance* is the term used commonly to describe resistance which gives effective control of a pest under natural conditions in the field, but is difficult to characterize in laboratory tests; usually it is a complex kind of resistance giving only partial control. *Passive resistance* is when the resistance mechanism is already present before the pest attack, for example an especially thick cuticle, or hairy (pubescent) foliage.

Active resistance is a resistance reaction of the host plant in response to attack by a parasite more usually applicable to attack by pathogens rather than pests (insects etc.); for example, the formation of phytoalexins or other antibiotics (antifungal compounds) by some host plants in response to attack by some pathogenic fungi. This reaction is not unlike the human production of antibodies in response to foreign matter in the blood or tissues. *Qualitative resistance* applies when the frequency distribution of resistance and susceptible plants in the crop population is discontinuous, and the plants are individually easily categorized as either resistant or susceptible. *Quantitative resistance* is the term used when a crop shows a continuous gradation between resistant plants and susceptible plants within the population, with no clear-cut distinction between the two types.

Varietal resistance has been shown in a broad range of crop plants ranging from cereals and herbaceous plants to trees, against an equally broad range of insect, mite and vertebrate pests. The insect groups involved include Orthoptera (locusts), Hemiptera (bugs), Lepidoptera (caterpillars), Diptera (fly maggots), Thysanoptera (thrips), and Coleoptera (beetles), in both temperate and tropical parts of the world.

In many respects crop plant breeding for pest resistance is the most ecologically desirable method to be employed in pest management programmes. The time factor is quite considerable in that a breeding programme is slow to conduct, and the cost may be exorbitant, but the end result could be very long-term control without the use of pesticides and their associated problems. At present there are a number of plant breeding programmes in operation in different parts of the world for different crops and different pests, some at major international research institutes. Some of the better known stations include IRRI, Philippines (rice), CIAT, Colombia (cassava, etc.), IITA, Nigeria (many tropical crops), Cotton Growing Corporation, Africa, India, etc. (cotton), PBI, England (potatoes, wheat), NVRS, England (vegetables), and many other stations specializing in fruit (UK, USA), sorghum (East Africa), maize (India), pulses (E. Africa and UK), and so on. The scope of some of these plant breeding programmes is often very large, for example at IRRI all 10000 varieties of rice, representing the world germ plasm collection, have been screened for resistance to rice stalk borers. More than 20 varieties show high levels of resistance to *Chilo* caterpillars by non-preference and/or antibiosis mechanisms, and some of these varieties are also resistant to leafhoppers and planthoppers. Some of these varieties of rice show tolerance towards these pests, as well as definite antibiosis and non-preference.

It is, however, now clear that, in response to the development of new resistant plants, it must be expected that new insect biotypes will arise to which the existing plants are not resistant. Fortunately, it appears that most insect resistance in plants is complex and polygenic in nature, which may help to discourage development of insect biotypes.

Pest avoidance

This is when the plant escapes infestation by the plant by not being at a susceptible stage when the pest population is at its peak. Some varieties of apple escape infestation by several different pest species in the spring by having buds which do not open until after the main emergence period of the pests, thus reducing the final amount of damage inflicted.

Non-preference

The term 'non-acceptance' has been proposed as a more suitable alternative, but has never gained general acceptance. Insects are noticeably reluctant to colonize some individual plants, or some particular strain of host-plant, and these plants seem to be less attractive to the pest by virtue of their texture, colour, odour or taste. Non-preference usually is revealed when the bug or caterpillar either refuses to feed on the plant or takes only very small amounts of food, or when an ovipositing female insect refrains from laying eggs on the plant. In the Philippines *Chilo supressalis* females laid about 10–15 fewer egg masses on resistant rice varieties than on susceptible ones. Also at IRRI, the Brown Planthopper of Rice (*Nilaparvata lugens*) punctures the tissues of a certain rice variety but apparently feeds only little, probably because of the reduced amount of a particular amino acid (asparagine) in the sap of that variety. In Uganda, experiments at Namulonge (Stride, 1969) showed that Cotton Lygus Bugs (*Tayloriligus vosseleri*) found some red-coloured varieties of cotton less acceptable for feeding purposes than the usual green plants, and this difference in preference was attributed to the presence of certain aromatic compounds in the sap of the red-coloured plants.

Non-preference based upon the absence or presence of certain chemicals in the plant tissues or sap may equally well be categorized as a form of antibiosis.

Two temperate examples of this type of resistance include Raspberry Aphid (*Amphorophora idaei*) which when placed on to leaves of resistant plants exhibit a reaction so strong that they will quickly walk off the plants completely. On sugarbeet aphids do not actually walk off the resistant plants, but they do feed for noticeably shorter periods of time and are quite restless whilst on resistant plants. The precise details of the resistance in these cases is not yet known, but the insect reactions are easily observed.

Antibiosis

In this case the plant resists insect attack, and has an adverse effect on the bionomics of the pest by causing the death of the insects or decreasing their rate of development or reproduction. The resistant plants are generally characterized by anatomical features such as thick cuticle, hairy stems and leaves, a thickened stem (cereals), a narrower diameter of the hollow pith in cereal stems, compactness of the panicle in sorghum, tightness of the husk in maize, and tightness of leaf sheaths in rice. Biochemical aspects usually involve the presence of various toxic or distasteful chemicals in the sap of the tissues of the plant which effectively repel feeding insects, sometimes to the extent that the odour is sufficient to completely deter them from feeding. Alternatively, there may be a chemical which normally functions as a feeding stimulant missing from the body of the resistant plant, or else at a sufficiently low concentration that it fails to stimulate the insect into feeding behaviour.

Cotton Jassids (*Empoasca* spp.) have ceased to be important pests of cotton in Africa and India since the post-war development of pubescent strains which the bugs find quite unacceptable as host plants. In a similar manner, hairy-leaved varieties of wheat in N. America are attacked significantly less often by the Cereal Leaf Beetle (*Oulema melanopa*); the females lay fewer eggs on the leaves and, of the larvae that hatch, fewer survive. It is also recorded that pubescent foliage apparently deters oviposition by many species of Lepidoptera, but this situation is complicated in that some bollworms will apparently lay more eggs on the foliage of some pubescent varieties of cotton.

The tightness of the husk in some maize varieties will deter feeding on the cobs by larvae of *Heliothis zea* (Corn Earworm) in the USA, and should also apply to field infestations of the drying grain by Maize Weevil (*Sitophilus zeamais*). At IRRI it has been shown that the tightness of the leaf cleavage in rice varieties is closely correlated with resistance to stem borers. If the leaf sheath is tight and closed and covers the entire stem internode, the young caterpillars usually fail to establish themselves between the leaf sheath and the stem where they would normally spend some six days feeding before boring into the stem. Varieties of sorghum in Africa, with an open panicle, suffer far less damage by False Codling Moth (*Cryptophlebia leucotreta*) and other caterpillars. Wheat varieties with solid stems (i.e. very reduced pith) are noticeably resistant to Wheat Stem Sawfly (*Cephus cinctus*) in N. America in that growth and development of the larvae are retarded. Some species of pyralid and noctuid stemborers are closely restricted to cereal hosts with stems of a particular thickness, and varieties with thicker or thinner stalks may harbour fewer caterpillars.

An anatomical factor of considerable importance is the development of silica deposits in leaves and stems of various graminaceous crops. The Gramineae as a group are basically semixerophytic and so most species have silica deposits in the leaves and stems; not only the truly xerophytic species such as marram grass (*Ammophila* spp.), but also species which have become secondarily adapted as hydrophytes such as rice, still possess some deposits. A number of cereals are indigenous to semi-arid areas, and one example is sorghum. In these plants silica deposits are apparently formed in some resistant varieties at about the fourth leaf stage. Up to this stage most varieties of sorghum are attacked by Sorghum Shoot Fly (*Atherigona soccata*), but in the resistant varieties infestations are not usually recorded once these silica deposits are formed, although infestations will still occur in susceptible varieties up to the sixth leaf stage. This is a simple type of resistance which is easily incorporated into all sorghum breeding programmes. A recent account of the role of silica in protection of Italian ryegrass from Frit and other shoot flies is given in Moore (1984).

The Hessian Fly (*Mayetiola destructor*) was introduced into the USA in the mid-18th century and rapidly became a major pest of wheat; the larvae attack the wheat stem, and either destroy the shoot or weaken the stem so that lodging subsequently occurs. Heavy attacks cause a serious drop in wheat yield. Since 1914 a search for resistance was carried out in Kansas. Several resistance varieties were developed, and since 1965 the widespread growing of resistant varieties has resulted in the virtual extinction of Hessian Fly in Kansas. Some 25 different resistant varieties of wheat are being grown in the USA now, and the annual value of the wheat crop has increased by many millions of dollars. The mechanism of resistance has not been established, but it is known that these resistant wheats have unusually large deposits of silica in the leaf sheaths.

At IRRI rice varieties have been developed with a high silica content in the leaves and stems, and these are resistant to larvae of *Chilo suppressalis* in that the caterpillars' mandibles become worn down by the abrasive nature of the silica deposits.

The biochemical factors involved in plant resistance arise from differences in the chemical constituents of the plants. The differences may be restricted to different parts of the plant body and/or particular stages in the growth of the plants. It is thought that in some resistant plants the pest concerned suffers nutritional deficiencies resulting from the absence of certain essential amino acids. Some maize varieties show direct physiological inhibition of larvae of the European Corn Borer (*Ostrinia nubilalis*) for they possess biochemical growth inhibitors at various stages. Some pests are influenced in their host selection by aromatic compounds present in the plant tissues, as well as various sugars, amino acids, and vitamins in the sap. Resistance to the Brown Planthopper of Rice by a rice variety at IRRI was attributed to a low concentration of asparagine, thought to be a feeding stimulant for this pest. Cabbage Aphid (*Brevicoryne brassicae*) is stimulated by sinigrin (a mustard oil glucoside) in the leaves, and it has been shown that a high level of resistance to Cabbage Aphid is associated with low foliar concentrations of sinigrin. Some cotton varieties have a poisonous polyphenolic pigment, gossypol, in subepidermal glands, and they show a strong resistance to several different insect pests (Pink Bollworm, Spiny Bollworm, etc.), although a direct causal relationship has not been established.

Members of the family Solanaceae often have peculiar glandular hairs on the leaves and stems, and recently new potato varieties have been obtained by crosses with various wild stocks from S. America which have an abundance of these glandular hairs. The sticky exudates easily trap small insects like aphids, which soon die, and even such large insects as larvae of Colorado Beetle have been trapped. This easily inherited character is hoped to be of value in combatting potato aphid attack, for these sap-sucking bugs are vectors of several major virus diseases. It was reported in 1982 that breeding programmes had been established at both PBI and GCRI in the UK to try to incorporate the 'hairy' attributes of the Peruvian varieties into the usual domestic varieties.

Many plants have evolved biochemical as well as physical methods of protection against grazing and browsing herbivores; this topic is reviewed more extensively in chapter 2 on pest ecology (page 16) but should be mentioned here. One common method of deterring grazers is shown clearly by both oak and beech in that the leaves are soft and tender for only a short time after budburst; after about a week sclerification is evident within the leaf tissues and at the same time various chemical substances (tannins) are accumulating in the tissues (page 18). The result is that various leaf-mining weevils (in beech) and caterpillars, such as Winter Moth (on oak), find the older leaves less suitable as food and larval mortality increases quite significantly (Bale, 1984). Apparently individual trees vary quite extensively in this respect, and may often be easily categorized as 'susceptible' and 'resistant' trees.

Tolerance

Tolerance is the term used when host plants suffer little actual damage in spite of supporting a sizeable insect pest population. This is characteristic of healthy vigorous plants, growing under optimal conditions, that heal quickly and show compensatory growth. In fact most plants bear more foliage than they actually need, and can usually suffer a fair amount of defoliation with no discernible loss in crop yield.

Tolerance is frequently a result of the greater vigour of a plant, and this may result from the more suitable growing conditions rather than from the particular genetic constitution of the plant. For example, sorghum growing vigorously will withstand considerable stalk borer damage with no loss of yield. Some varieties of crop plant (e.g. rice) may show both tolerance to a pest as well as antibiosis; this is true for several stalk borers.

Sometimes pest attack on a tolerant variety can actually increase the crop yield; this occurs quite frequently with the tillering of cereals following shoot fly, stem borer, or cutworm destruction of the initial shoot in the young seedling.

From a pest management point of view the use of a tolerant variety could in theory be a disadvantage in that it could support a larger population of the pest and so encourage a local population build-up rather than a decline.

Many cases of clear-cut resistance to insect pests have been recorded, but they have not been investigated sufficiently for the mechanism of resistance to be evident. This is particularly the case in respect to the aphids *Myzus persicae* and *Aphis fabae* on sugar beet, and to Carrot Fly (*Psila rosea*) on carrots (Hill, 1974*b*; Ellis *et al.*, 1980).

Breaking of host plant resistance

In some agricultural situations there develop physiological races of the insect pest, known as *biotypes*, some of which are not susceptible to the host plant resistance. These are *resistance-breaking biotypes*. In nematology this type of variant is known as a *pathotype*, in virology it is a *strain* and, applied to fungi, it is a *race*.

The development of resistance-breaking biotypes has been known for a long time in the Hessian Fly, and several biotypes can attack wheat varieties that are quite resistant to other biotypes. The Brown Planthopper of Rice (BPH) in S.E. Asia has recently become notorious in that for several reasons it has changed status from a minor to a serious major pest on rice. However, rice varieties resistant to this bug were developed at IRRI, and have been widely grown throughout the area; they have given such good control that sometimes insecticides have not been required. In some localities, however, resistance-breaking biotypes of BPH have developed to such an extent as to threaten local rice production. The present situation is that as fast as the research workers at IRRI produce resistant varieties of rice to BPH, the insects correspondingly produce new resistance-breaking biotypes. Detailed biosystematic and ecological studies of the biotypes of *Nilaparvata lugens* have very recently been initiated at IRRI and through the auspices of the ODA in the UK.

The breaking of host plant resistance is generally less common amongst insect pests than pathogens. This is thought to be because insects produce far fewer propagules than the fungi, bacteria and viruses, and thus far less genetic variation can be expressed.

Biological control

In the broad sense (*sensu lato*) this can include all types of control involving the use of living organisms, so that, in addition to the use of predators, parasites, and disease-causing pathogens (biological control – *sensu stricta*), one can include sterilization, genetic manipulation, use of pheromones, and the use of resistant varieties of crop plant.

As already indicated in this book, the use of resistant crop varieties is being dealt with separately as it is an aspect of control of such importance, and plant breeding is a very specialized subject in its own rights. In chapter 6 biological control (*sensu stricta*) is considered in more detail, and here it is only intended to be introduced in its broadest aspects.

The main attraction of biological control is that it obviates the necessity (or at least reduces it) of using chemical poisons, and in its most successful cases gives long-term (permanent) control from one introduction. This method of control is most effective against pests of exotic crops which often do not have their full complement of natural enemies in the introduced locality. Then the most effective natural enemies usually come from their native locality, for the local predators/parasites/pathogens are usually in a state of delicate ecological balance in their own environment and cannot be expected to exercise much population control over the introduced pests. On rare occasions a local predator or parasite will successfully control an introduced pest, but this is rare!

Natural control

This is the existing population control already being exerted by the naturally occurring predators and parasites (and diseases) in the local agroecosystem, and it is vitally important in agriculture not to upset this relationship. Because it is not readily apparent, the extent of natural control in most cases is not appreciated. It is only after careless use of very toxic, broad-spectrum, persistent insecticides which typically kill more predators and parasites than the less sensitive crop pests, and which is then followed by a new, more severe pest outbreak, that the extent of the previously existing natural control may be appreciated. In summary, the importance of natural control of pests in most agroecosystems cannot be overemphasized.

Predators

The animals that prey and feed on insects are very varied, as are their effects on pest populations. The main groups of entomophagous predators are as follows:

Mammalia – (man), Insectivora, Rodentia

Aves – Passeriformes (many families), many other groups, especially ducks, game birds, egrets and herons, hawks

Reptilia – small snakes, lizards, geckos, chamaeleons

Amphibia – most Anura (frogs and toads)

Pisces – *Gambusia* etc. (control mosquito larvae)

Arachnida – spiders, harvestmen, chelifers, scorpions, etc.

Acarina – mostly family Phytoseiidae

Insecta – Odonata (adults, and nymphs in water), Mantidae, Neuroptera, Heteroptera (Miridae, Anthocoridae, Reduviidae, Pentatomidae), Diptera (Cecidomyiidae, Syrphidae, Asilidae, Therevidae, Conopidae, etc.), Hymenoptera (Vespidae, Scoliidae, Formicidae), Coleoptera (Cincindelidae, Carabidae, Staphylinidae, Histeridae, Lampyridae, Hydrophilidae, Cleridae, Meloidae and Coccinellidae)

A few predators are quite prey-specific; for example, the larvae of Meloidae feeding on the egg-pods of

Acrididae in soil, and Scoliidae feeding on scarab larvae in soil and in rubbish dumps. But most predators are not particularly confined to any specific prey. Some of the predators live in rather specialized habitats; for example, all the fish are aquatic, as are some insect larvae (e.g. Odonata), and so only prey on aquatic insects (such as mosquito larvae); some live in soil or leaf litter so their prey is restricted to certain types of insects (more details are given in chapter 6).

Parasites

These are almost entirely other insects and belong to two large groups (Diptera and Hymenoptera) and one small group (Strepsiptera), together with a number of important species of entomophilic nematodes. The Diptera include the large family Tachinidae which parasitize Lepidoptera (larvae), Coleoptera, Hemiptera and Orthoptera. Other parasitic families include Phoridae, Pipunculidae, Bombylinidae, and there are some parasites in the Sarcophagidae and Muscidae.

The Hymenoptera include the very important Chalcidoidea and Ichneumonoidea, and some Bethylidae, Scelionidae and Proctotrupidae. Some groups of parasitic wasps are miniscule (being the tiniest insects known, about 0.2mm in body length) and all are egg-parasites. Almost all groups of insects (as well as spiders and ticks) are parasitized by the Hymenoptera Parasitica, at all stages of development from egg to adult. Many species of parasitic wasps lend themselves to exploitation in biological control projects.

Pathogens

Control by pathogens is sometimes referred to as *microbial control*. There are three main groups concerned: bacteria, fungi and viruses, and some other groups of entomophagous micro-organisms which are rather obscure and little studied. There are several types of *Bacillus*, which are specific to caterpillars or beetle larvae, responsible for natural epizootics, and several species are now commercially formulated and very important in pest control projects.

Fungi are responsible for producing antibiotics and apparently about 300 antibiotics do show some promise as pesticides; these act directly as killing agents or inhibitors of growth or reproduction.

Viruses are quite commonly found attacking insects in wild populations of caterpillars and beetle larvae, as well as some temperate sawfly larvae. They have long been used as biological insecticides, by finding dead larvae in the field and making an aqueous suspension of their macerated bodies. But now a few commercial preparations are available.

Some of these new biological insecticides using insect pathogens are, however, only easily available in the USA as yet, although others are commercially available in Europe and parts of Asia.

Sterilization

This usually refers to the sterilization of males by X-rays or γ-rays and is called the *sterile-male technique*; control of a pest by this technique is termed *autocide*. Sterilization can be effected by exposure to various chemicals and this practice is called *chemo-sterilization*. The rationale behind this method is that male sterilization is effective in species where females only mate once and are unable to distinguish or discriminate against sterilized males. The classical case was in about 1940 on the island of Curaçao against Screw-worm (*Callitroga*) on goats: the male flies were sterilized by exposure to γ-rays, and dropped from planes at a rate of 400/square mile/week. The whole pest population was eradicated in 12 months. The life-cycle took only about four weeks to complete, and the females only mated once in their lifetime. Generally, autocide is most effective when applied to restricted populations (islands, etc.), but can be effective on parts of the continents. The Screw-worm eradication campaign was extended to the southern part of the USA where the pest is very harmful to cattle. In Texas 99.9% control was achieved in only three years. Male sterilization trials were effective against Mediterranean Fruit Fly (*Ceratitis capitata*) on part of the island of Hawaii in 1959 and 1960, but immigration from untreated parts of the island prevented control from being long-lived.

Chemosterilization has now advanced from a theoretical technique to a practical one, a variety of chemicals have been demonstrated to interrupt the reproductive cycles of a large number of insect species, see Curtis (1985).

Genetic manipulation

In reality this method is an extension of the previous one, in that the electromagnetic radiations (X-rays, γ-rays) induce *dominant lethal mutations* in the germ cells of the insects. These mutations in insect sperm have been used successfully in several eradication programmes. Lethal mutations are not lethal to the treated cell, they are lethal to its descendant in that the zygote fails to develop to maturity. These mutations arise as a result of chromosome breakages in the treated cells.

Potential uses for pheromones in pest control

Pheromones are reviewed in more detail in chapter 2 (page 11), but here they are considered in their actual roles in pest control.

The two obvious ways in which pheromones may be used in a pest control programme are firstly, in pest population surveys or for population monitoring (for emergence warnings, and spray warnings), and secondly for direct behavioural modification control. It is clear now from the work that has been done in recent years that pheromone traps are extremely useful in monitoring projects and this use is likely to be increased in the future. But to date there has not yet been a good example of pheromone use actually achieving a significant level of population control in a pest management programme; though, as already mentioned (page 13), there have been behaviour disruption trials on several different crops with very encouraging results.

Insect population monitoring. The presence or absence of a particular insect species in an area can be established through the use of attractant pheromones, so that control measures may then be exercised, if necessary, with precise timing. Previously field population monitoring relied largely on either light-trapping, which requires a source of electricity, or the finding of eggs on the crop plants. The finding of the first eggs on a particular crop is a very tedious and time-consuming process requiring a great deal of labour and is not particularly efficient. The examination of a few pheromone traps for the presence of male insects is relatively very easy, and much more efficient. Alternatively, the pheromone traps can be used to monitor the effectiveness of a pest control programme, even though not directly employed in the programme themselves.

Emergence of male Codling Moth in apple orchards in the spring in Europe and N. America, and Pea Moth, is now regularly monitored by the use of small paper (waterproof) pheromone traps with a sticky interior. Pink Bollworm, and other bollworms, on cotton crops in many parts of the tropics are likewise monitored with the use of these sticky pheromone traps, with considerable success. Various species of fruit flies (*Dacus* spp.) are monitored, sometimes using sticky pheromone traps and sometimes in traps with insecticides inside, in citrus and peach orchards to determine whether or not insecticide spraying is required, and if so just when. As more and more sex pheromones are being synthesized, and more chemical attractants are being discovered, it seems likely that the use of these chemicals in monitoring programmes will increase and will play a constant role in many pest management programmes.

Insect behavioural control. Bark beetles have been induced to fly to inappropriate host trees by aggregation pheromones; the host tree was either resistant and killed the boring beetles or they were unable to breed successfully. Carefully designed traps, incorporating the required visual stimuli, have been very successfully used in destroying bark beetle populations in forests in N. America.

Orientation to a trap baited with sex pheromones (*trapping-out*) has proved a feasible method of reducing populations of various moths and some fruit flies (Tephritidae). Destruction of males responding to a female sex pheromone-baited trap will, however, only effectively control a pest population if enough males respond and are destroyed to result in most females not being inseminated. Using theoretical population models it is suggested that such a trapping technique could be effective at low population densities, but would be unlikely to be effective at high densities. Such traps can kill the male insects either by adhesion (sticky traps) or by insecticides.

Male nocturnal moths have been observed to be additionally attracted to light in the presence of female pheromones. An experiment in America in 1966 using Cabbage Looper moths showed that when a cage of virgin females was placed on an ultra-violet (black-light) light trap the catch of males overnight was increased twenty-fold! With the recent synthesizing of Cabbage Looper sex pheromone this method of population reduction could be feasible agriculturally.

The *communication disruption technique* uses pheromones or sex attractants to prevent orientation of males to virgin females. This method could give direct population control of a pest species in a crop, but as yet no really convincing results have been obtained although some trials were fairly successful. The basic idea is to saturate the air with pheromone or sex attractant, or at least to make the concentration high enough so that the pheromone released by the wild females is imperceptible to the males; thus the males would not find the females in the crop and the females would not get inseminated. A recent experiment in the New World gave encouraging results. On cotton crops grown in southern USA, C. and S. America, Pink Bollworm is a major pest which builds up through six generations during the warmer part of the year, the last two or three generations generally being above the economic threshold for this crop and causing economic damage. A sex attractant, 'Gossyplure', was used in the communication disruption technique to saturate the air over the cotton fields with pheromone mimic so that the males would be disoriented by the odour and would fail to find and mate with many of the virgin females. This technique was timed to be used against the fourth and fifth Pink Bollworm generations. Following the use of 'Gossyplure', the level of cotton boll attack was significantly reduced, demonstrating that this technique can be agriculturally successful (Brooks, 1980). Similar experiments on other crops have given encouraging results.

It has been shown by experiments that most male moths show adaptation in their response to sex pheromones as it is weakened following previous recent exposure. In addition, it was shown that some non-pheromone chemicals may react on the antennal sense cells or central nervous system to cause adaptation to the natural pheromone, but conclusive studies have not yet been carried out, although there would seem to be promise for eventual agricultural use.

Recent work by Cherrett and others have involved the addition of trail pheromones to baits for Leaf-cutting Ants in C. and S. America; this renders the bait pellets more attractive to the foraging ants which pick them up and take them back to the nest, making the poison bait more effective.

Pheromone release. The precise method of release of the pheromone, or mimic, is of importance, since for trapping purposes a steady, controlled release over a known period of time is most desirable. Some preparations are sold as impregnated rubber or polythene caps, or as a 'wick', each usually containing 1 mg of the chemical, which are generally effective for several (often four) weeks. However, experimentation has shown that there does appear to be a considerable difference in success of pheromone release, sometimes when a different matrix is used. This is a factor that requires consideration when experimentation or monitoring programmes are planned, rather like the importance of the base (matrix) of a bait in a baiting project.

A recent development described by Brooks (1980) is a hollow fibre formulation for pheromone release. Each tiny fibre is $1\frac{1}{2}$ cm in length, sealed at one end, and with the lumen filled with pheromone which evaporates from the open end. In the experiment against Pink Bollworm two pheromone components were used in a 1 : 1 ratio and the release was steady over a period of 2–3 weeks.

Another method of delaying evaporation of the chemicals is by microencapsulation, but after some encouraging results ICI found that pheromones in gelatine capsules used in experiments in the Mediterranean region rapidly degraded in daylight because of the ultraviolet radiation. However, several successful uses of microencapsulated pheromones have now been reported (Matthews, 1979).

One of the companies now supplying a range of insect pheromones commercially for use in monitoring and control programmes is International Pheromones Ltd., with its main branches in the UK and Norway. Some of their products are for use in temperate situations and others are designed for use against tropical pests.

Chemical methods

Pesticides

Insecticides and their methods of application will be dealt with more fully in chapter 7. Only rarely does chemical application kill all the pests, and the few which survive usually soon give serious problems by the development of resistance. Chemical control is essentially repetitive in nature and has to be applied anew with each pest outbreak. However, this method is very quick in action and, for the majority of pest outbreaks, chemical control remains the method by which the surest and most predictable results are obtained. The different modes of action of insecticides are briefly listed below.

(a) Repellants – designed to keep the insects away; usually employed against mosquitoes and other medical pests.

(b) Antifeedants – certain chemicals block part of the feeding response in some phytophagous insects, and they can be used for plant protection.

(c) Fumigants – volatile substances that vaporize and the toxic gases kill pests within enclosed containers (food stores), greenhouses, or in soil.

(d) Smokes – finely divided insecticidal powders mixed with a combustible material; the insecticide is dispersed as 'smoke'; only of use in greenhouses and other enclosed spaces.

(e) Stomach poisons – have to be ingested to be toxic; either sprayed on to foliage (for foliage eaters) or mixed with a bait to encourage ingestion.

(f) Contact poisons – usually absorbed directly through the cuticle:
  (i) ephemeral – short-lived; usually a foliar application.
  (ii) residual – persistent (long-lived); soil or foliage application.

(g) Systemic poisons – watered into the soil, sprayed on to the plant, or applied to the trunk; absorbed and translocated by the plant and effective against sapsuckers especially. May be applied as sprays or granules, to either soil or foliage.

Pesticide application still remains the major weapon in the pest war, for obvious reasons, but we now find that often, sometimes usually, there are three definite post-application effects.

(a) Normal resurgence of treated pest – this always occurs as the target species initially suppressed by the insecticidal treatment shows a rapid population recovery after the decline of the treatment effect.

(b) Resurgence of the target species due to either development of a resistant biotype (as with the Brown Planthopper of Rice) and/or destruction of natural enemies.

(c) Outbreak of a secondary pest or pests, due to the alteration of the agroecosystem, usually by the destruction of natural enemies.

Insect resistance to pesticides

This topic is now of sufficient importance that a small section of text be devoted to it, for it is a major factor to be considered in many tropical and temperate pest management programmes.

In the 1940s the many new synthetic organic pesticides, mostly the organochlorine group, became widely available throughout the world for crop protection. Pesticides such as DDT, BHC(HCH) and dieldrin were in many respects thought to be ideal for crop protection: they were highly toxic to insects (most groups), broad-spectrum and persistent. Control levels (kills) of major insect pests were consistently at the level of 98–99% or even higher. Many serious pests were controlled effectively for the first time, and a heavy dependence upon those pesticides resulted. Invariably, though, despite a high kill there was always a small proportion of individuals that possessed a natural resistance to such poisons, and of course during successive generations this genetically based, inherited, natural resistance spread throughout the insect (and mite) populations. After a number of years (generations) this resistance became manifestly obvious and a cause for concern. Eventually the entire local pest population became resistant to the chemical involved. These groups of individuals, with slight genetic differences from the main stock of the insect species, are known as *biotypes*. In disease-causing pathogens these races are known as *pathotypes*, and in a few cases a widespread pathogen is known to have as many as 20–30 pathotypes throughout its geographical range. Insects generally have far fewer biotypes, but the Brown Planthopper of Rice probably has a dozen biotypes in the area from India, through S.E. Asia, up to Japan. *Myzus persicae* occurs worldwide as a large series of biotypes, many of which show resistance to a large number of different pesticides.

Sometimes resistance to one chemical compound leads to resistance to other closely related compounds in the same group. For instance, after initial development of resistance to DDT, it rapidly developed to most of the other organochlorines. There followed resistance to the organophosphorous compounds, and now there is also resistance to some carbamates and some pyrethroids.

To date, more than 300 major pest species (insects and mites) have developed resistance to one or more major pesticides. An extreme case is that of Diamond-back Moth in Asia which shows very marked resistance to most of the pesticides used generally against caterpillars, with the exception of BTB; and it has now developed resistance to many of the new pyrethroids. One of the most spectacular cases of biotype development is that shown by the Brown Planthopper of Rice (*Nilaparvata lugens*) in India and S.E. Asia. Prior to the widespread use of diazinon on paddy rice it was only a very minor pest. Diazinon was the most widely used insecticide for general rice pest control in S.E. Asia for many years, and in 1970 this planthopper was first recorded as a major rice pest in a few localities; by 1975 it was recorded as a serious pest, and by 1980 was both serious and widespread. Recent work at IRRI has shown that the situation now is such that if diazinon is used on rice it invariably causes a resurgence of the planthopper.

It is generally thought that some 10–15 generations are required for the development of manifest resistance, and, as already mentioned, a widespread crop pest may show the development of several distinct biotypes over different parts of its distributional range. In temperate regions where many pests are uni- or bi-voltine, resistance develops slowly (6–10 years), but in the tropics, with the higher temperatures and no cold winter, insect breeding may be more or less continuous, and resistance may be strongly apparent after only 2–4 years in extreme cases. The Brown Planthopper of Rice can apparently develop a new biotype in 18 months on rice, that is about 18 generations.

To date, field resistance when encountered has been dealt with by just increasing the amount of pesticide applied, or using available chemical substitutes. The former remedy is generally useless (except for a very short time) and only adds to the general ecological disturbance as well as causing residue problems; it may also accelerate resistance development. Resistance problems are expanding and intensifying to such an extent that they are outstripping the development of new chemical pesticides.

Most cases of resistance have been shown by the

Arthropoda, but in recent years resistance has been also shown by rats, some fungi, and some weeds.

Antifeedants

Certain chemicals possess the properties of inhibiting the feeding of insect pests; these can be classed as antifeedants. The first chemicals noticed with these properties were initially referred to as repellents, but they are not repellents in that they do not merely drive the insect away to another plant but actually inhibit feeding on that plant. In laboratory tests insects have remained on treated plants indefinitely and eventually starved to death without eating the leaves. In field tests the insects were free to wander elsewhere seeking food; they either found weed plants to feed on or died of predation and starvation.

The most successful source of natural antifeedant is the Neem tree (Indian lilac, *Azadirachta indica*) and the closely related Persian lilac (*Melia azedarach*). Extracts from the Neem have long been known to have germicidal properties, but for the last ten years research in India and Germany has concentrated on its pesticide potential. Extracts from leaves and fruits (Neem oil) act as an antifeedant for many different insects (locusts, caterpillars and Hemiptera) and they also upset development and inhibit gravid females from oviposition. The active ingredient is a chemical called azadirachtin, and it is thought to be related to the ecdysonoids. The first Neem Conference was held in Germany in 1980 and was reported in *International Pest Control* (1981) No. 3, pp. 68–70 (25 references are listed).

The earliest antifeedant used in agriculture was ZIP (a complex compound zinc salt) used to keep rodents and deer from feeding on the bark and twigs of trees in the winter. The first recognized antifeedant for use in insect pest control was introduced by Cyanamid in 1959. Since then a number of compounds have been shown to possess antifeedant properties, but a commercially successful antifeedant has not yet been produced, although work is still progressing along this line, and it appears that Neem oil may be the first such product.

This line of research is being pursued in many different countries and it now appears that quite a number of plants possess chemicals in their tissues that repel phytophagous insects and act as antifeedants. Another such plant is the sub-tropical shrub *Ajuga remota* (Labiateae) found in India. The active ingredient is called ajugarin (there are several) and is known to be particularly effective against African Armyworm. The ajugarins belong to the clerodanes which form part of the large group known as the terpenes (page 17). Ajugarin I (the main active component) has now been synthesized at Imperial College, London (1983).

Integrated control

The original concept of integrated control was developed to stress the need for understanding the complicated and antagonistic relationship between biological control and the use of chemical pesticides. Early pest control measures, up until about the turn of the century, were mainly concerned with the biology and the ecology of the pests, and in particular those aspects relating to their population numbers. Attempts were made to make the crop environment less favourable to the pests by a combination of cultural and biological methods. The chemical poisons available at that time were rather simple, such as kerosene, sulphur and some inorganic salts (lead, arsenic and mercury), and were not very effective as insecticides. The earliest attempts at biological control were in the early 1900s when various insect predators and parasites were imported into California and Hawaii to try to control some of the pests of the newly established *Citrus* industry. It was soon appreciated that the use of chemical poisons was basically inimical to biological control in that the poisons were not selective, and in fact the natural enemies were almost invariably more susceptible to poisons than the biologically very robust pests. In 1940 DDT was discovered to have insecticidal properties and was soon available commercially. It was capable of killing a broad range of insects and mites in small doses, and had a long-lasting residual activity. Literally almost overnight insect pest control was revolutionized, and a series of new synthetic organic insecticides was rapidly discovered and made available commercially. The organochlorine compounds (DDT, HCH, dieldrin, etc.) gave consistently high kills (98–99%) of a wide range of pests, as seed dressings, sprays and powders; persistence was lengthy and application was not difficult.

But there was no universal panacea after all; it soon transpired that pest problems often continued, and indeed sometimes worsened, and many undesirable side-effects were noted. These were the accidental destruction of natural enemies of the pests, and the development of resistance to the insecticides by the pests, and soon various other ecological disruptions became evident. Now, after considerable ecological damage and widespread resistance has developed, the culmination has been the extensive banning of the organochlorines in most parts of the World, particularly the total ban imposed by the Environmental Protection Agency (EPA) in the USA. The cycle of events is now completed as we are turning back to the original approach whereby biological and ecological understanding assumes predominance. The judicious use of the synthetic organic pesticides is a most important weapon in pest control, but must be used in conjunction with other appropriate methods.

The term *integrated control* was originally coined to describe the combining of biological control with compatible chemical application. (In this sense biological control included natural control and biological control *sensu stricta*.) The basic idea was to use chemical pesticides judiciously so as to avoid disrupting the existing natural control by killing the predators and parasites in the crop community. This can be done in several ways; by using specific, carefully screened pesticides only, by careful timing of the treatment, using minimal dosages, by reducing spray drift, and so on. This attitude developed because these two basic approaches to pest control are our primary resources, and as used in many instances in the past they have been in direct conflict with each other (Smith & van den Bosch, in Kilgore & Doutt, 1967).

Eventually accumulated experience and logic made it clear that it is necessary to integrate not only chemical and biological control, but all available techniques and procedures, into a single pattern aimed at profitable crop production together with minimal environmental disturbance. This realization led to the concept of *pest management* (PM) with its broad ecological approach to pest situations.

During the last decade the literature on pest control has been very confusing for in some instances integrated control was regarded in its original context and in many others it was virtually synonymous with pest management. After all, there is no basic qualitative difference between the two approaches, the difference is essentially quantitive. This confusion in terminology was widespread, but finally there appears to have been some international agreement to remove this confusion and the new term PM is now accepted as *integrated pest management* (IPM) as formulated by Glass (1975) on behalf of the Entomological Society of America. The term integrated control is regarded as an historical term, now superseded.

Some of the nicest examples of integrated control have come from tropical parts of the world, and include the control of Coffee Leaf Miners (*Leucoptera* spp.) in Africa and elsewhere, and the oil palm pest complex (bagworms and other defoliators) and cocoa pests in Sabah and Malaysia (Conway, 1972). It became apparent in these situations that the local natural control was of paramount importance and that the pesticides used formerly were killing the natural enemies of the pests and so chemical application was followed by extensive pest resurgence.

In temperate regions probably the best example of integrated control was that practised in glasshouses, where the red spider mites were controlled by predacious mites, the Glasshouse Whitefly controlled by parasitic wasps (Aphelinidae), and the aphids controlled with the aid of nicotine smokes, or other judicious insecticide application. Nowadays the glasshouse pest control situation is even more complex in that aphids can be controlled using a fungus, and mealybugs using a predacious ladybird beetle.

The importance of biological control as a supplement to chemical use has long been appreciated in China, and they were in fact practising integrated control centuries before the term was coined (page 108).

A series of publications providing basic information on integrated control was produced by FAO in 1966 following a symposium on integrated control held in Rome in 1965: *Proceedings of the FAO Symposium on Integrated Pest Control,* Volume 1 (pp. 91), Volume 2 (pp. 186), Volume 3 (pp. 129).

Integrated pest management (IPM) (formerly pest management (PM))

In 1967 the FAO panel of experts on integrated pest control defined *integrated control* as 'a pest management system that, in the context of the associated environment and the population dynamics of the pest species, utilizes all suitable techniques and methods in as compatible a manner as possible and maintains the pest population at levels below those causing economic injury'.

This definition incorporates the concept of pest management as defined by the Entomological Society of America, now expressed as IPM (Glass, 1975).

The concept of (I)PM is now well established. One of the earliest definitions was by Rabb & Guthrie (1970); they commented that originally integrated control generally referred to the modification of insecticidal control in order to protect and enhance the activities of beneficial insects (predators and parasites). Subsequently, however, integrated control interpretations have become more comprehensive until, now, some definitions of integrated control embody most of the essentials of pest management. Rabb preferred the term pest management because it connotes a broader ecological basis and a wider variety of opinions in devising solutions to pest problems.

Pest management can be defined as the reduction of pest problems by actions selected after the life-systems of the pests are understood and the ecological as well as economic consequences of these actions have been predicted, as accurately as possible, to be in the best interests of mankind. In developing a pest management programme, priority is given to understanding the role of intrinsic and extrinsic factors in causing seasonal and annual changes in pest populations. Such an understanding implies a conceptual model of the pests' life-system functioning as a part of the ecosystem involved. Ideally such a model would be mathematical, but a word or pictorial model may be useful in predicting effects of environmental manipulations.

Five of the most characteristic features of the population management approach to pest problems are as follows, after Rabb & Guthrie (1970).

(a) The *orientation* is to the entire pest population, or a relatively large portion of it, rather than to localized infestations. The population to be managed is not contiguous to an individual farm, county, state or country, but is more often international; hence a high degree of co-operation, both nationally and internationally, is a prerequisite for success.

(b) The *immediate objective* is to lower the population density of the pest so that the frequency of fluctuations, both spatially and temporally, above the economic threshold is reduced or eliminated.

(c) The *method*, or combination of methods, is chosen to supplement the effects of natural control agents where possible and is designed to give the maximum long-term reliability of protection, the minimum expenditure of effort and money, and the least objectionable effects on the ecosystem.

(d) The *significance* is that alleviation of the problem is general and long-term rather than localized and temporary, and that harmful side-effects are minimized or eliminated.

(e) The *philosophy* is to manage the pest population rather than attempt to eradicate it. The real significance of the concept is seen in relation to serious pest problems which defy solution through the more traditional approaches.

As previously stated, this broad ecological approach to pest control problems is still rather new in concept and, as yet, is more developed theoretically than in practice, particularly with regard to the use of computer models.

There is now quite an extensive list of publications concerned with IPM on a worldwide basis, many edited by the pioneers in this field, such as Apple & Smith (1976), Rabb & Guthrie (1970), Metcalf & Luckmann (1975), Watson, Moore & Ware (1975). In some of these books are published specific IPM programmes for a range of particular crops. Other IPM programmes for specific crops have been published by Rabb, Todd & Ellis (1976) for tobacco in N. Carolina, Hensley (1980) for sugarcane in Louisiana and Chu (1980) for sugarcane in Taiwan. Many such papers are being published regularly in *Ann. Rev. Entomol.* as can been seen from the Bibliography. A series of papers dealing with IPM for major crops has been produced by the FAO as Plant Production and Protection Papers: *Guidelines for integrated control of rice insect pests* No. 14, pp. 115 (1979); *maize pests* No. 18, pp. 91 (1979); *sorghum pests* No. 19, pp. 159 (1979).

A recent publication of interest called *Integrated Pest Management* was written for the US Council on Environmental Quality by Bottrell (1979), and it views the subject in its broadest aspects.

Clearly some of these programmes are in the category of first attempts, and as yet rather crude, whereas others are quite elaborate, very sophisticated and obviously very effective. Presumably, the basic problems with a particular crop remain much the same wherever the crop is grown, but the pest and disease spectra will be different in each locality where the crop is important (i.e. site-specific). The ultimate aim for crop protectionists is clearly an IPM programme for all the more important crops in all the major agricultural regions of the world.

Results of recently implemented IPM programmes in Canada, USA, and Europe have shown a pesticide usage reduction of 35–80%; a general overall reduction of 50% appears possible for many crops.

A recent trend in the literature is to refer to IPM meaning 'insect pest management', but the precise meaning is the same as the original use of 'integrated pest management.'

Eradication

In most cases, pest control is undertaken to reduce the population density of an insect to a point at which the damage done is not of economic significance; very rarely is complete eradication the goal, and even more seldom is it achieved. The more usual cases of complete eradication are directed against pests of medical importance, and very successful campaigns have been carried out in many areas against diseases such as malaria, yellow fever and dengue. Against agricultural pests about the only time eradication is aimed at is when a new pest, which is potentially very serious, has been introduced into a country and has not spread too far. The campaign is often very costly and difficult but can be won if the pest is still restricted to a relatively small area.

Some eradication programmes have been successful against Screw-worm (*Callitroga*) and fruit flies (*Drosophila* spp.), when sterile-male techniques were employed. This technique is particularly suitable for eradication programmes, especially in restricted locations.

California and Florida in the USA have multimillion dollar *Citrus* industries, as well as for peach, fig, guava and many other tropical and sub-tropical fruits. Some very damaging fruit flies are still not established here, as are some other pests such as San José Scale and California Red Scale (Florida). Some years ago Medfly (*Ceratitis capitata*) was accidentally introduced into Florida; a total of 0.7 million hectares of *Citrus* orchards and adjoining land was sprayed with insecticides, at a total cost of US $8 million; but the project was successful and the Medfly population exterminated. The total value of the Florida fruit industry was at that time about $180 million. Very recently (1981) Medfly was accidentally established in California and presented the Governor with a difficult decision. With a $14 000 million fruit industry in California at stake it was imperative that an aerial spraying eradication programme be started quickly, but in recent years the environmentalists in California have become a very powerful political lobby and they objected strongly to the idea of aerial spraying of suburban homes and gardens. The Medfly, was, however, too serious a threat so the eradication project went ahead.

Colorado Beetle is accidentally introduced into the UK in most years, but this very damaging pest of potatoes is usually destroyed quite rapidly. Since the 1933 Colorado Beetle Order all farmers and growers are aware of this pest and the danger it presents, and the odd infestation is usually reported promptly. On receipt of a report, the Ministry of Agriculture regional staff take immediate action to collect and kill adults and larvae on the potato foliage, and also to fumigate the soil in the infested area in order to kill any pupae that might be present.

Several forestry pests in North America have been the subjects of eradication programmes because of the tremendous damage potential they represented in both the deciduous and coniferous forests; these include Gypsy Moth, Winter Moth, Larch and Pine Sawflies; the list is quite long! Some of these pests were eradicated in some regions following extensive aerial insecticide spraying; others succumbed to introduced parasites (page 106), but others, although some measure of control was achieved, still remain a problem.

The decision to eradicate a pest is a grave biological responsibility and should not be made unless careful study, involving diverse perspectives, has produced convincing evidence that the benefits to be accrued more than balance the ecological impoverishment represented by removing the pest species.

The present state of insect pest management

In Europe and N. America there has been extensive publicity given to recent record harvests, and general agricultural over-production has led to the so-called 'grain mountains', and other produce surpluses, that are embarrassing the EEC administrators. This recent high level of production is due partly to the advanced level of agriculture being practised, particularly large-scale mechanization and the abundance of agrochemicals. This publicity leads to a general assumption that in temperate countries crop pest problems are greatly diminished now, and the main pest problems are only to be found in the tropics.

But in a very recent article by Pimental (1985) he points out that these record harvests are misleading in that in the USA (probably the heaviest user of insecticides in the world) during the last 35 years there has been a ten-fold increase in insecticide use, and over this period crop losses have actually doubled (7% to 13%).

Integrated pest management (IPM) for insects has now been studied and discussed for more than 20 years, but up to the present time there has still been more discussion than action in regard to its implementation. One major problem has been the lack of a truly interdisciplinary approach to the basic research. Another has been the reluctance of farmers to appreciate and accept the principles of IPM and thus to adopt this long-term approach to control: they usually still prefer frequent and repeated applications of insecticides.

Enough IPM programmes have been carried out in N. America and Europe to demonstrate the effectiveness of this approach, and if an overall reduction in insecticide usage of 50% can be achieved the future for IPM is surely bright.

5 *Pest damage to crop plants*

The agricultural entomologist has a distinct advantage over his colleagues in nematology and plant pathology in that the damaging organisms with which he is concerned are relatively large and usually to be found in the vicinity of the damage on the crop plant. This helps to make the identification of insect pests a relatively simple matter, at least to the level of family and genus.

For many control purposes pest identification to family, or preferably genus, is often adequate, for most members of most insect (and mite) families produce the same type of damage on the crop plant and are likewise controlled by the same practices. After all, this is to be expected, since the object of insect systematics is to place closely related species together in the same taxa. However, there are occasions when the specific identity of the pest is important, especially in the case of aphids and other Homoptera that are virus vectors. For example, *Toxoptera citricidus* is the vector of Citrus Die-back disease (Tristeza) in Africa and S. America, whereas the partially sympatric and morphologically very similar *T. aurantii* is apparently not a vector. The genus *Ceutorhynchus* is an example where several closely related species have quite different biologies. At such times a high level of taxonomic expertise is demanded, but these occasions are relatively few. Generally however, it is necessary to identify the pest correctly (to the appropriate systematic level) in order that the economic significance of the infestation be accurately assessed, and the most appropriate control measures be applied.

On occasions only the damaged plant may be found, or else there may be several similar pests on the crop and it may not be evident which insects are responsible for which damage. Some pests are nocturnal or crepuscular in habits and during daylight hours remain hidden in the plant foliage or the soil, others drop off the plant and hide in the soil litter when anyone approaches. The damage inflicted on the various parts of the plant body is, however, sometimes characteristic of a specific pest, or a group of closely related pests. Then the experienced entomologist will usually be able to make a fairly accurate determination (identification) of the identity of the damaging animal.

In practice the situation is often more complicated. Several quite unrelated insect groups produce almost identical attack symptoms on the plants; for example, leaf miners may be the larvae of Diptera (Anthomyiidae, Ephydridae, Agromyzidae) or Lepidoptera (Gracillariidae, Phyllocnistidae, Nepticulidae, Tischeriidae, Lyonetidae) or Coleoptera (Chrysomelidae, Hispinae, Halticinae). 'Dead-hearts' in graminaceous seedlings may be produced by larvae of Lepidoptera (Pyralidae, Noctuidae) or Diptera (Muscidae, Anthomyiidae) or Coleoptera (Scarabaeidae). The more generalized defoliation resulting from the browsing and grazing of herbivores can be very difficult to identify without extensive practice, and even then it is often not possible. Leaf-eating is the normal method of feeding of some mammals (ungulates, pigs, rabbits, rodents), birds (sparrows, ducks, etc.), molluscs (slugs, snails), most grasshoppers, locusts, caterpillars, and some beetles (both larvae and adults). If one is presented with such damaged plant material in the laboratory or regional 'Plant Clinic', then identification of the causal organism may be quite impossible. But if the plants are examined *in situ* in the field various clues such as footprints, scats, hair, feathers, slime trails, etc. may permit the entomological detective to identify the culprit. Detailed ecological knowledge is of great assistance in such cases, especially knowledge of local animal migrations and behaviour.

The situation is actually even more complicated in reality because damage similar to that produced by some pests may also result from adverse weather conditions, such as drought, flooding or waterlogging of the soil, lightning-strike, hail, frost, sun-scorch and strong wind. Excessive fertilizer application, herbicide damage and use of insecticide on susceptible crops also produce distortions and damage to the plant body. Disease organisms (bacteria, fungi, viruses) and nematodes (eelworms) produce some symptoms reminiscent of insect and mite attack, and mineral deficiencies (manganese, magnesium, etc.) result in symptoms similar to various virus infections.

The growing crop will be subjected to the vagaries of the weather, prevailing soil conditions, and the ravages of maybe a vast pest and disease complex, which altogether will influence the basic genetic control of the development of the plant body. All of these constituent ecological factors will be interacting, and the various agricultural advisers (agronomist, horticulturalist, pathologist, entomologist) will be confronted by the end-result of this complicated interaction. Obviously, in some situations the adviser cannot hope to rectify the situation, but only aim at minimizing damage in an attempt to secure a crop yield of economic proportions.

It should finally be noted that various pests and disease organisms interact by *predisposition*, so that an infected plant may be more susceptible to pest attack, and a slight infestation may produce unusually severe symptoms. This also applies to climatic and edaphic factors; a water-stressed plant is invariably more susceptible to pest attack.

Pest damage assessment and crop yields

The ultimate aim of agriculture is to produce a sustained economic yield of crop produce, so it becomes of prime importance to understand the effect of the insect pest population on the subsequent yield or harvest. Obviously, if the pests are causing no crop loss their presence on the plants and the damage they cause may be ignored, and in the context of ecological stability they should be left alone! However, most pest populations produce some damage of significance, but the damage assessment in relation to possible or expected yield loss is difficult. The total number of interacting factors responsible for determining crop yield is quite overwhelming, and any decision as to the probable effect of any single factor, such as the population of one insect pest species, is problematical. However, the gradual accumulation of empirical data over many years has resulted in our being able to make various generalizations about some pest populations and their probable effect on crop yield. These results are used to define economic injury levels (and economic thresholds) for some pests on some crops in different parts of the World. But in general, many more data are required for many more pests on the more important crops, especially those in the tropics.

Part of plant body damaged

Some types of damage are obviously more important than others, depending upon the part of the plant body damaged and the part harvested; if the two are the same then clearly the damage is more serious. A single Codling Moth larva can effectively destroy a single apple or peach, and a relatively small number can ruin an entire crop. On the other hand an apple tree can accommodate a large number of foliage-eating caterpillars, sap-sucking bugs, and root-eating beetle larvae, with no discernible loss of yield. In general, root crops can stand considerable leaf damage without appreciable yield loss; pulses and most cereals can likewise tolerate leaf-eating, root-eating and sap-sucking (with a few exceptions) at a moderate level. Vegetables such as cabbage, lettuce, celery, etc., may have their outside leaves removed at harvest, prior to sale, so damage to the outer leaves is relatively unimportant.

It has long been known that many crop plants can tolerate partial defoliation without discernible loss of yield; two well-documented cases are cucumbers attacked by Red Spider Mite, where 30% of leaf cover has to be damaged before there is an effect on yield, and sugarbeet, where young crops suffer little loss of yield unless defoliation exceeds about 50%.

Fig.15. Examples of five closely related crops (*Brassica* spp.) which have different parts of the plant body harvested, and also a generalized seedling.

1. *Brassica* seedling
2. Broccoli (*B. oleracea* var. *botrytis*), grown for flower heads
3. Brussels sprouts (*B. oleracea* var. *gemmifera*), grown for lateral buds
4. Cabbage (*B. oleracea* var. *capitata*), grown for 'heart', i.e. telescoped main shoot
5. Turnip (*B. rapa*), grown for swollen root
6. Rape (*B. napus*), grown for seeds as a source of oil

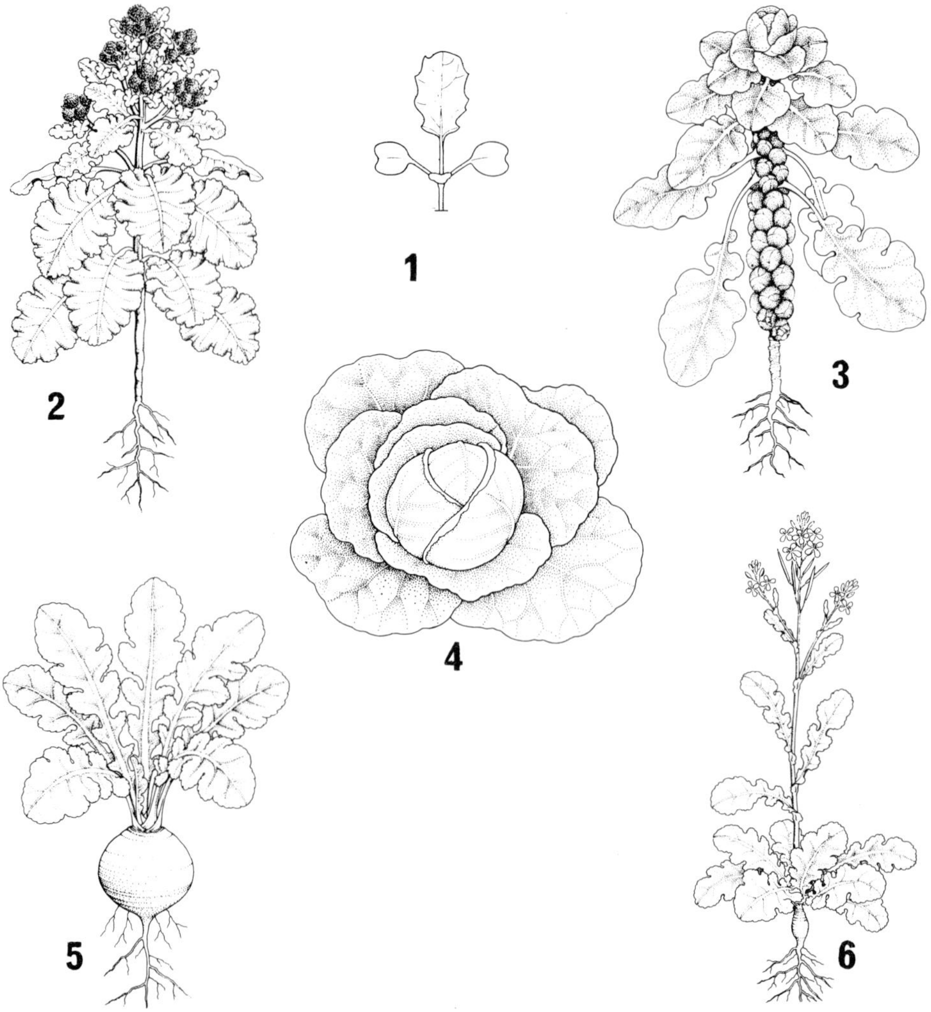

In summary, damage which can be ignored on one crop may be of considerable economic importance on another (even closely related) crop, so damage assessment is different for each crop grown. For example fig. 15 shows five different *Brassica* crops together with a seedling; should the seedling be killed by Cabbage Root Fly, cutworm, or white grub, then the damage is usually serious for the entire future plant is lost and there is a large gap left in the field which encourages weed development (except rape). With Broccoli the flower heads are eaten, so damage to lower leaves, and some root damage can be ignored. Brussels sprouts are lateral buds harvested over the winter period, so late caterpillar defoliation in the autumn may be of no consequence, but a single Cabbage Root Fly inside a button for freezing is serious damage. The many types of Cabbage are grown for the 'heart', so all the outer leaves may be damaged without affecting the saleability of the heart. Turnip is one of the cruciferous root crops and will tolerate considerable leaf damage, but even slight Cabbage Root Fly damage may spoil the appearance of the root. Rape is becoming more and more important in many regions as a source of seed for oil extraction, and in this crop it is the flower and pod pests that are important (as they are also for the other crops grown to seed); small numbers of plants destroyed have no effect on final yield because of the density of the crop.

Details of sale procedure may be relevant; sprouts grown for freezing have to be totally free of Cabbage Root Fly maggots; turnips sold with soil adhering to the root can have surface tunnelling by Root Fly maggots, but, if washed and prepacked in plastic, maggot tunnelling may affect sale (now some retailers are trimming off the roots prior to sale, and also removing surface tunnels). There will always be some unusual cases causing problems. Some cabbages are unsaleable not because of the caterpillar damage but because of the extensive frass accumulation between the leaves. Broccoli heads have also been unsaleable when infested with Diamond-back Moth pupae inside their flimsy silken cocoons, although there was no actual damage.

Compensatory growth

Most plants produce far more leaf cover than actually required, and of course grazing by herbivores is to be expected; the plant responds to grazing by compensatory growth, either in the form of extra leaves, or in Gramineae by tillering if the apical shoot is killed. The *leaf area index* (l.a.i.) of many crop plants may be as high as 3, 4, or 5, so that much of the lower foliage is shaded by the upper leaves, and with closely spaced crops the lower parts of the plant will be shaded by adjacent plants.

Some plants produce far more flowers and fruits than the plant can actually sustain, and there is a regular natural flower-fall and fruit-fall. In temperate crops one of the most noticeable cases is with apple, where there is a flower-fall followed by an early fruit-fall and sometimes a later fruit-fall. With a number of crops there is a tendency towards overproduction and the loss of later flowers or buds may be beneficial (e.g. by Pea Midge), leading to a smaller yield of better fruits and more simultaneous ripening which makes harvesting easier. It must be remembered that in all terrestrial ecosystems the Spermatophyta, and in fact all members of the Plant Kingdom, are the main primary producers and so heavy grazing of all crop plants is to be expected. It will be equally expected that the plants have evolved various methods of counteraction.

With some crops the inter-row spacing and the intra-row spacing habitually employed is often a compromise between yield per plant, overall yield, canopy density (to reduce weed competition) and other factors. If a row of crop plants is destroyed by pests then the adjacent plants usually would grow larger and bear more/larger fruit so that the overall crop yield does not necessarily diminish.

Beneficial effects of pests

One anomalous aspect of pest infestations is that sometimes the overall effect is agriculturally beneficial; occasionally the yield is actually increased, sometimes it is 'quality', and sometimes the effect is more subtle. The most usual effect is one of 'pruning' whereby suppression of growth of one part of the plant body results in the increase in size of another part, usually the 'fruits' or harvested storage organ. Some cereal crops after early shoot fly attack may show an increased yield at harvest because of increased tillering in response to pest damage. Defoliation of some crops of potato and turnip has resulted in small yield increases, though the precise way in which this came about is not clear. Some legumes after pest attack produce fewer seeds but larger ones, and this could be regarded as an increase in quality. Small infestations of *Aphis fabae* may increase yields of field beans because of their suppressing effect on apical growth. Similarly Pea Midge can increase the marketable yield of peas for freezing by effectively destroying the apical shoot (causing a 'nettle-head') and encouraging simultaneous ripening of the peas in the pods already formed. (For the UK freezer-market a pea plant with four simultaneously ripe trusses is aimed at.) Potato tubers of very large size are difficult to market, and it has been noticed that infestations of *Aphis nasturtii* have decreased the size but not the number of tubers, so although total yield weight might be reduced the actual value of the crop at market could be increased. And as already mentioned, sometimes a pest attack destroys young plants which are closely spaced, and adjacent plants take advantage of this reduced competition pressure for space by increasing in size and yield, and occasionally an overall yield increase has been recorded.

Repelling of insect pests

Some plant species and some varieties of crop plants possess properties, either physical or biochemical, that deter insects from either feeding on their tissues or from ovipositing. These topics have been included in the section on plant resistance to pests, and in chapter 4 on page 56.

Pest infestations

There are two basic ways in which pest infestation (or damage) can be assessed. The *incidence* of the pest (or damage symptoms) is generally the proportion of plants in a sample which are host to the pest (or which show damage symptoms), and is usually expressed as a percentage. The *severity* of the infestation is a measure of the size of the pest population on the plants, or the extent of the damage done, and is often measured as so many insects per plant, per bush, per 10 leaves, egg masses per plant, etc. And of course the total damage caused to a crop is a combination of severity of infestation together with *duration* (time).

In ecological studies of populations of both plants and animals, several methods of assessment are employed. These are based upon the proportion of area covered within the habitat (for plants), or the number of animals seen or sampled, in relation to area, or proportion of plants examined (sampled). Botanists are able to use more precise systems for plant population assessment because of the immobility of the organisms, and the three most widely used methods employ between 4 (Raunkiaer, 1934), 6 (Braun-Blanquet, 1927) and 11 (Domin) abundance categories. But for small, highly mobile insects the latter level of precision is not feasible, especially when different recorders are being used. In the UK the *Biological sites recording scheme* advocates the use of four abundance/frequency categories for population size assessment, without the use of lengthy or detailed sampling procedures, and this approach would seem to be appropriate for assessing field populations of insect pests on crops. These categories of abundance are as follows:.

abundant	(a)	=	very common	(VC)
frequent	(f)	=	common	(C)
occasional	(o)	=	uncommon	(U)
rare	(r)	=	rare	(R)

The alternative categories in the right-hand column would appear preferable as their designation is somewhat more obvious. Further details on pest population assessment can be found in the publications by Southwood (1978), Bardner & Fletcher (1974), and FAO/CAB (1981).

Damage assessment

It was suggested by Bardner & Fletcher (1974) that the term *injury* be used for slight (i.e. non-damaging) effects of insects feeding or other activities on the growth or appearance of crop plants; and that *damage* is injury resulting in a measurable loss of yield or reduction in quality. This distinction would be most useful, but to date in the literature these two terms are used more or less synonymously, and of course the term 'economic injury level' is well established.

The extent of crop damage is usually proportional to the numbers of insects present, and would accordingly be rated as follows:

very severe	(VS)	= 1 or	1, 2
severe	(S)	= 2	3
mild	(M)	= 3	4
very mild	(VM)	= 4	5, 6

In some systems of recording a numerical categorization is used, and a six-point scale seems to be quite popular; presumably in this case number 6 (and 'very mild') come under the 'injury' category of Bardner & Fletcher (1974) and would be detectable but not of any economic importance. With simple damage, such as leaf lamina being eaten or apples infested with Codling Moth larvae, damage can be expressed easily as proportion of lamina destroyed or percentage of fruits infested per tree. In some systems numbers of pests present are correlated empirically against expected loss of yield (percentage), on a scale of no loss (0%) to total loss (100%).

Because each crop has its own growth characteristics (see below), and the vast diversity of types of pests and pest damage and all the other factors involved with crop production, it is not possible to generalize extensively. Damage assessment will remain different for each crop and sometimes also for each major locality as the pest complex usually varies regionally. It is however generally agreed that for most purposes a damage assessment scale of not more than six levels is preferable, for easy recognition in the field by non-experts, and it is recommended that a large number of small (easily categorizable) samples be taken rather than a small number of large samples; this also caters better for the uneven pest distribution within the crop, which is usual.

Plant age/stage of development

This could alternatively be termed *crop vulnerability*. Casual inspection of pest infestations can be very misleading in relation to actual damage done. Pests such as aphids may at times be very damaging, but at other times a heavy infestation may be very conspicuous and unsightly (especially if associated with sooty moulds) whereas the actual damage might be slight.

Many crops are only vulnerable to certain pests at a particular time in their growth (development). Generally the two most stressful times for plants are the time of establishment and the time of flowering/fruit development, and at these vulnerable times pest damage is often most serious. Cereals are only susceptible to Shoot Fly attack when very young; by the third or fourth leaf stage they are no longer attacked. Bean Fly only kills seedlings of *Phaseolus*, on larger (older) plants the infestation occurs in the leaf petioles where they have little effect on crop yield.

There have long been established well-defined growth stages for most of the major crops as a necessity for convenience in cultivation practices. Fig. 16 shows the growth stages for wheat (and other temperate cereals), and for comparison fig. 17 shows the growth stages for rape and fig. 18 for sugar beet; all are taken from FAO/CAB (1971).

With wheat, and other cereals, aphid infestation is most damaging at the young seedling stage (at establishment) and later on the flag leaf and head at the start of flowering; the flag leaf is mainly responsible for the photosynthetic activity providing foodstuff for grain formation. With autumn sowing of many cereals now

Fig.16. Growth stages in wheat, oats, barley and rye (modified from drawings by E.C. Large, 1954).

Stage 0 Pre-emergence

1 One sprout (number of leaves can be added) = 'grairding'
2 Beginning of tillering
3 Tillers formed, leaves often twisted spirally; in some varieties of winter wheats, plants may be 'creeping' or prostrate
4 Beginning of the erection of the pseudostem, leaf sheaths beginning to lengthen
5 Pseudostem (formed by sheaths of leaves) strongly erected

} Tillering

6 First node of stem visible at base of shoot
7 Second node of stem formed, next-to-last leaf visible
8 Last leaf visible, but still rolled up, ear beginning to swell
9 Ligule of last leaf just visible
10 Sheath of last leaf completely grown out, ear swollen but not yet visible

} Stem extension

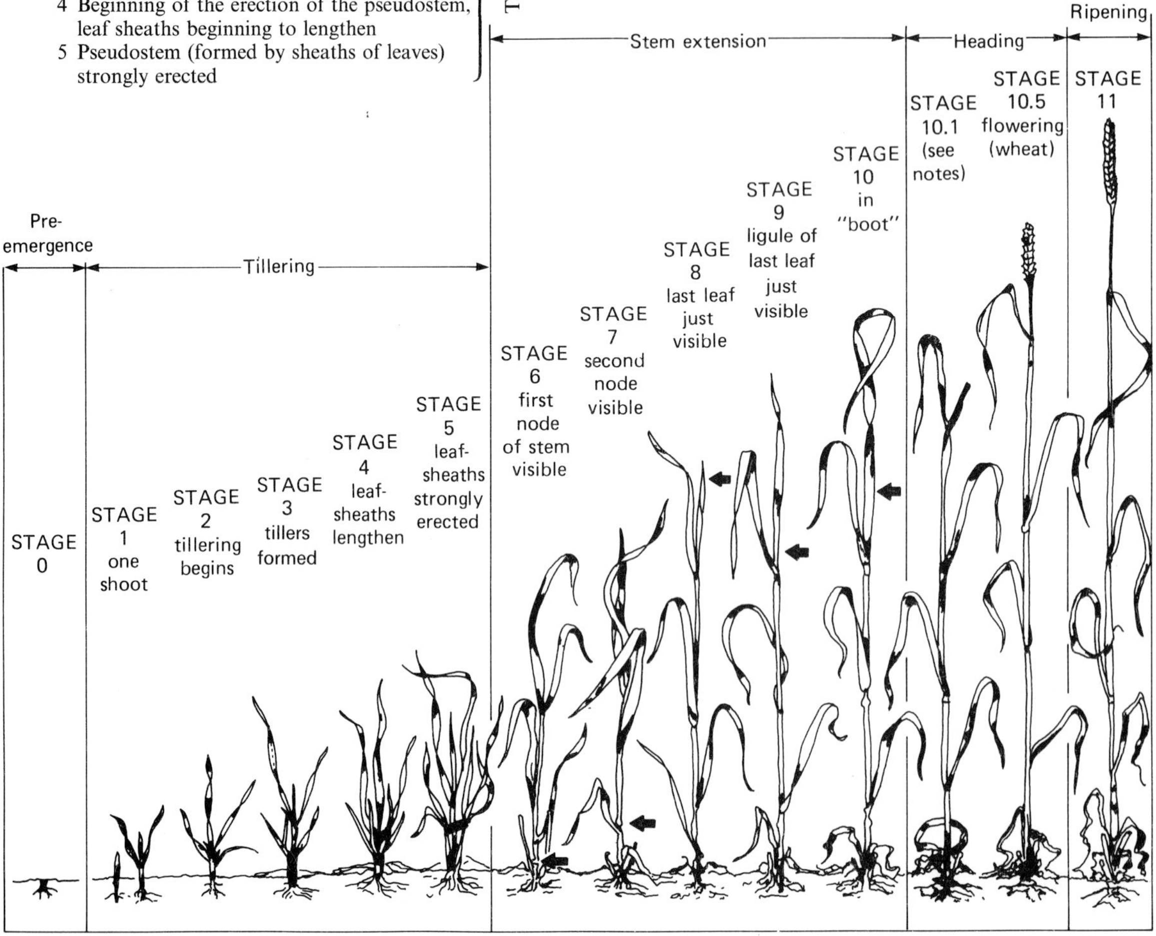

caption 10 (*cont.*)

10.1. First ears just visible (awns just showing in barley, ear escaping through split of sheath in wheat or oats)
10.2. One-quarter of heading process completed
10.3. One-half of heading process completed
10.4. Three-quarters of heading process completed } Heading
10.5. All ears out of sheaths
10.5.1 Beginning of flowering (wheat)
10.5.2. Flowering complete to top of ear
10.5.3. Flowering over at base of ear
10.5.4. Flowering over, kernel watery ripe } Flowering (wheat)

11.1 Milky ripe
11.2 Mealy ripe, contents of kernel soft but dry
11.3 Kernel hard (difficult to divide by thumbnail)
11.4 Ripe for cutting; straw dead } Ripening

widely practised in the UK, the plants are long past the susceptible stage (to aphids) by the time the aphids become abundant in the spring and early summer. Similarly, field beans sown in the autumn are generally not at risk from *Aphis fabae* in the spring, as the vulnerable stage has passed.

Most annual crops are grown directly from sown seed in the field, sometimes precision-drilled (e.g. sugar-beet), sometimes furrow-drilled and requiring thinning (now mostly superseded by precision-drilling), and sometimes furrow-drilled quite densely (cereals and rape), and sometimes grown in rather dense beds (some carrots, some onions). With all these crops the tiny seedlings are exposed in the field to all the local pests. Some crops, such as many brassicas, tobacco, etc., are first grown in seed-beds and then transplanted into the field. In the seed-beds the plants may be intensively protected, but at the time of transplanting and subsequent establishment the seedlings are very vulnerable.

Pest thresholds

The need to establish international co-operation for studies on crop damage assessment, and for publication of data, to enable pest control measures to be applied on a more rational basis led to the FAO (Rome) convening a *Symposium on crop losses* in October 1967, which was attended by representatives from 36 countries. One recommendation was to prepare a manual of methodology; this was published for FAO by CAB in 1971, with supplement No. 1 in 1973, No. 2 in 1977, and No. 3 in 1981. The *Manual* includes loss-assessment methods for a total of about 80 pests on 27 crops, about half of which are tropical and half temperate. The assessment techniques here presented are extremely varied, both direct and indirect; in some cases infestation levels/yield losses are clearly evident and simply expressed, in others regression analysis of sampling data is required.

In addition to the 40-odd temperate examples included in the *Manual* there are threshold recommendations locally available in most countries; a few of the more useful examples available in the UK, through local ADAS (MAFF) offices, are listed below.

Some insect pest thresholds for crops in the UK (ADAS, MAFF).

(1) Wheat
– Cereal Grain Aphid (*Sitobion avenae*): five aphids per ear at the start of flowering, with weather conditions fine and settled.
– Rose–grain Aphid (*Metopolophium dirhodum*): 30+ aphids per flag leaf.

(2) Rape
– Blossom beetles (*Meligethes* spp.): 15–20 adult beetles per plant, at flower-bud stage.
– Cabbage Seed Weevil (*Ceutorhynchus assimilis*): one or more weevils per plant at flowering (spray at end of flowering).

Fig.17. Growth stages in rape (modified from Berkenkamp, 1973, and Harper, 1973).
Descriptions are based on the main stem

Stage 0 Pre-emergence
1 Seedling
2 Rosette
2.1. First true leaf expanded
2.2. Second true leaf expanded (add 0.1 for each additional leaf)
3 Bud
3.1. Inflorescence visible at centre of rosette
3.2. Inflorescence raised above level of rosette
3.3. Lower buds yellowing
4 Flower
4.1. First flower open
4.2. Many flowers opened, lower pods elongating
4.3. Lower pods starting to fill
4.4. Flowering complete, seeds enlarging in lower pods
5 Ripening
5.1. Seeds in lower pods full size, translucent
5.2. Seeds in lower pods green
5.3. Seeds in lower pods green-brown mottled
5.4. Seeds in lower pods brown
5.5. Seeds in all pods brown, plant senescent

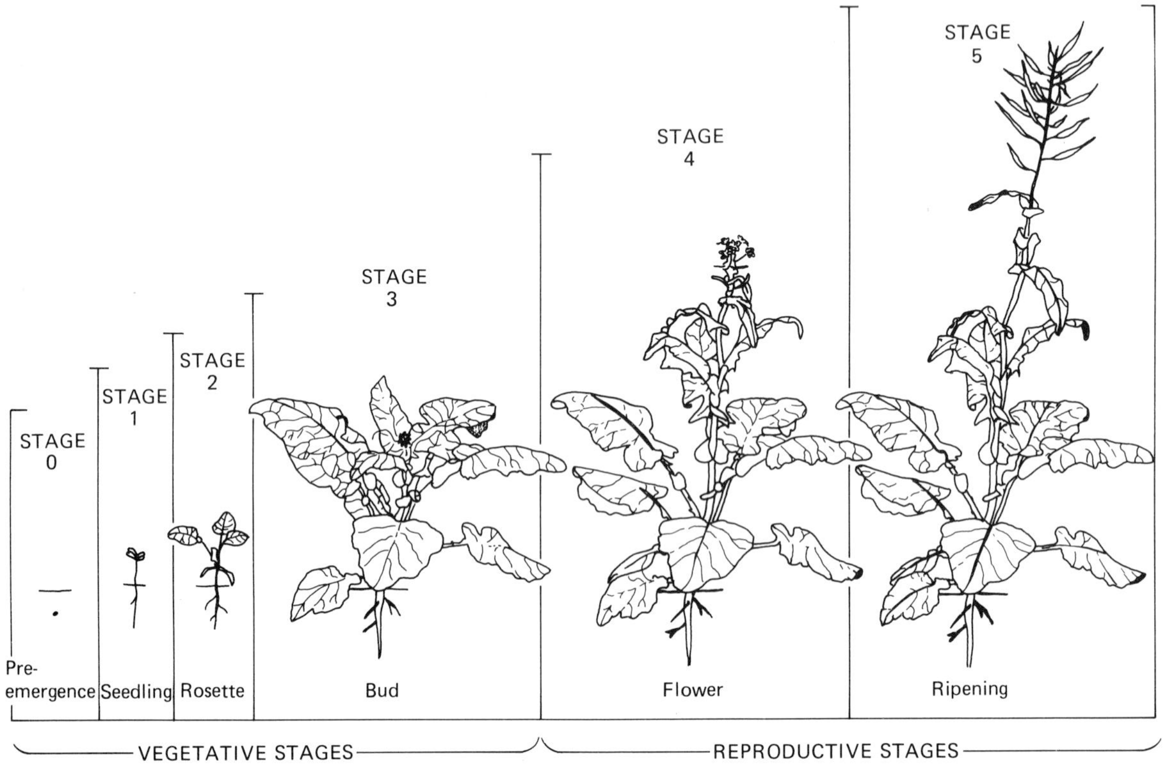

Fig.18. Growth stages of the sugar beet root crop (after G.D. Heathcote on behalf of IIRB).

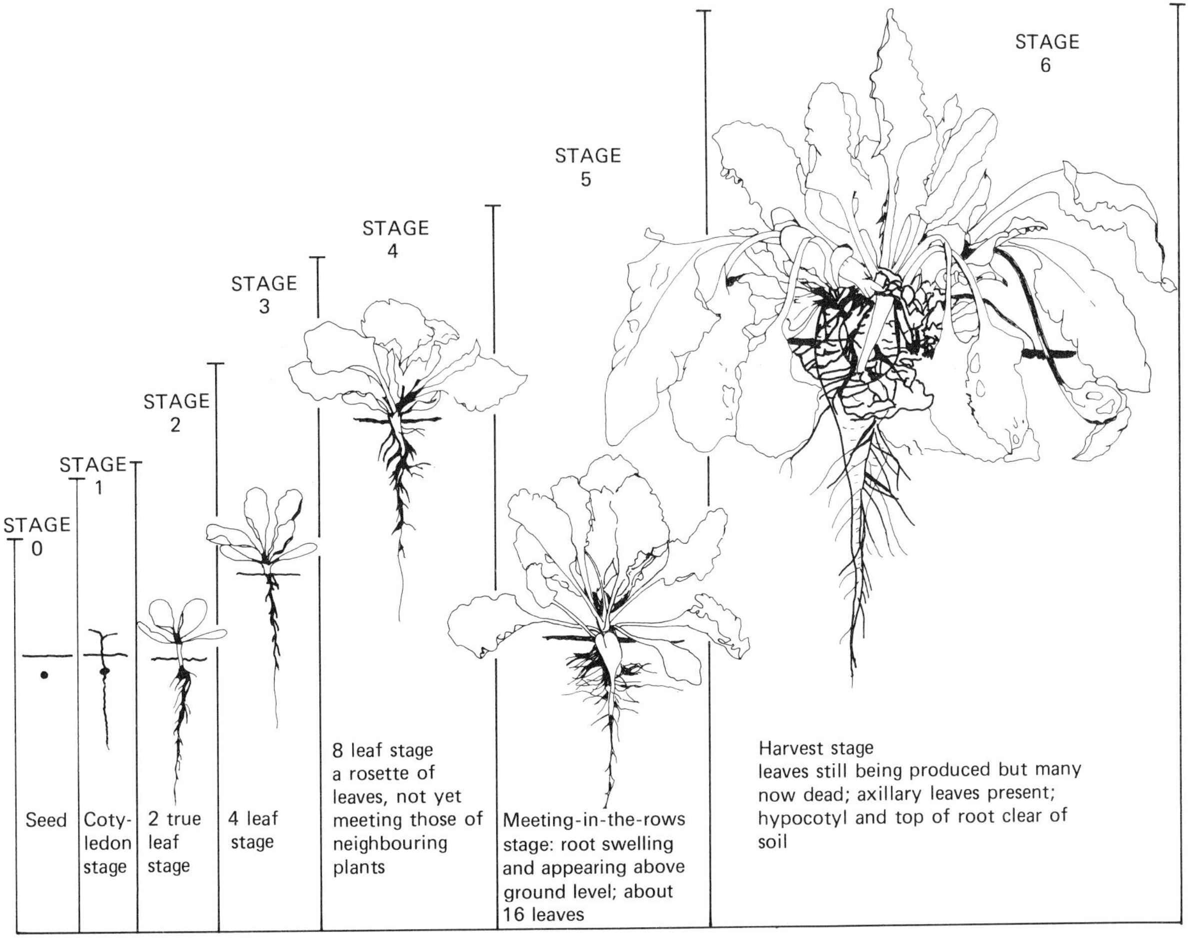

(3) Potato (ware crops) – Aphids: average 3–5 aphids per true leaf, in a sample of 30 each of top, middle, and lower leaves taken across the field. (On retarded crops two aphids per leaf may be appropriate).

(4) Pea – Pea Moth: 10+ moths per pheromone trap on two consecutive two-day periods (wait 3–4 days then apply sprays).
– Pea Aphid (*Acyrthosiphon pisum*): 5–10% of the growing tips infested.

(5) Field bean – Black Bean Aphid (*Aphis fabae*): if 5% or more of the plants on S.W. headland are infested (spring-sown crops only; winter-sown crops are not at risk).

(6) Fruit trees – Fruit Tree Red Spider Mite (*Panonychus ulmi*): when more than seven leaves in a 50-leaf sample carry more than four mites per leaf.

(7) Apple – Winter Moth (*Operophtera brumata*): larvae on 10% of trusses at bud-burst.
– Apple–grass Aphid (*Rhopalosiphum insertum*): aphids on 50% of trusses at bud-burst.
– Rosy Apple Aphid (*Dysaphis plantaginea*): any aphids present on trusses at bud-burst.

It has been shown for some crops that the effect of a particular pest infestation level will vary according to the usual yield of the crop, a high-yielding variety of crop suffering relatively less reduction in yield than a low-yielding variety. Crops such as pulses, sorghum and some other cereals, where varietal differences reflect in large differences in expected yield, should for assessment purposes be distinguished as high, medium, or low-yielding varieties. This tends to be more of a problem in some tropical countries rather than in most temperate ones, as crop yields there are more often at the 'primitive' level (page 38).

Crop loss profile

One of the first objects in an IPM programme is to establish a *crop loss profile* for the crop in question; this defines the *pest complex* and in particular the *key pests* which are in the main responsible for yield reduction in that crop. When the main sources of crop loss are identified it is necessary to describe and quantify the pests. The assembling of these data has been called the *preliminary portfolio* (Large, 1966). Such assemblage has been done for some temperate crops, such as cereals, but to date mostly for the disease complex rather than the animal pests. After the compilation of the preliminary portfolio a *damage function* can be designed; this is an equation relating the level of pest infestation to yield loss, usually expressed as a percentage. Damage functions are now in use for many crops worldwide, but mostly in relation to diseases and nematodes rather than insect and mite pests. A major problem in designing damage functions is that loss is calculated (estimated) in relation to a *reference expected yield*, but the derivation of this reference expected yield for most crops is problematical at best. The information on crop losses is usually acquired through national and international surveys (e.g. FAO yearbook) and other sources of indirect data, and the reliability of such data are really highly suspect at times.

Types of pest damage to crop plants

In order to break down this section for convenient handling it is necessary to subdivide according to the part of the plant body being attacked by the pest. In practice this is not too satisfactory because of overlap; for example a shoot consists basically of a short terminal portion of stem surrounded by young leaves; similarly a rhizome or tuber, although technically a stem, is attacked more by root pests than those usually associated with stems.

For convenience of handling, the data on damage are grouped under the following headings:

Damaged leaves (1) – mostly by biting insects
Damaged leaves (2) – by sap-sucking insects and mites
Damaged shoots and buds (including flowers)
Damaged stems
Damaged fruits and seeds
Damaged roots and tubers (and bulbs)
Damaged seedlings and sown seeds

Damaged leaves (1) – mostly by biting insects (fig. 19)

(a) Margin with regular notches (1, 2): many adult broad-nosed weevils (Coleoptera: Curculionidae) feed on the leaf margins of many different plants throughout the World, producing characteristic notches.

(b) Margin with large, clean, semicircular (or sub-circular) pieces of lamina missing: adults of leaf-cutting ants (Formicidae: Attinae) in C. and S. America, and leaf-cutting bees of the genus *Megachile* (7).

(c) Margin irregularly eaten (3, 4, 5, 6): the commonest form of leaf damage by defoliating pests, caused by grasshoppers, locusts, many caterpillars, leaf-beetles, sawfly larvae, some slugs and snails, some birds; in a severe attack the entire leaf lamina may be eaten away, sometimes leaving only the main veins intact (4).

(d) Lamina skeletonized (11, 12): caterpillars of several families (Epiplemidae, Bombycidae, Noctuidae etc.) especially when young, eat part of the way through the leaf lamina but leave the veins and one epidermis intact; some beetles (adults and larvae of *Epilachna* (11)), adult Citrus Flea Beetle and some slug sawflies (Tenthredinidae).

(e) Lamina holed (8, 9, 10): many tiny holes are made by adult flea beetles (Chrysomelidae, Halticinae); fewer larger holes are made by adults of tortoise beetles and some other Chrysomelidae, and some caterpillars (8).

(f) Lamina windowed (8): a window is a small hole in the lamina with one epidermis left intact, but after a while the thin epidermis dries and ruptures leaving a small hole; such damage is characteristic of larvae of the Diamond-back Moth.

(g) Leaves and shoot webbed: some caterpillars (Tortricidae, Pyralidae, Lasiocampidae, etc.) web leaves and shoots with silk and make a 'tent' or web in which they live and eat the leaves, often the web traps a large number of faecal pellets.

(h) Lamina mined (19, 20, 21): the mines may be either tunnel mines or blotches, sometimes starting as a tunnel and ending as a blotch mine; tunnel mines with a central line of faecal pellets usually belong to caterpillars (Lepidoptera: Gracillariidae, Lyonetidae, etc.); tunnels without evident faecal pellets are made by maggots (Diptera: Agromyzidae, Anthomyiidae, Ephydridae) and beetle larvae (Hispinae, Halticinae).

(i) Lamina cut and rolled or folded (22): caterpillars of Hesperiidae (skippers), some Tortricidae and Pyralidae cut the lamina and make a leaf-roll, binding the edges with strands of silk; commonest examples are Cotton Leaf-roller (*Sylepta derogata*) on cotton and *Hibiscus*, banana skippers (*Erionota* spp.) and the several rice skippers; the leaf lamina is eaten within the protection of the roll.

(j) Cereal leaf with a linear series of small regular holes (13): made by young larvae of stem borers (Lepidoptera: Pyralidae, Noctuidae) before they penetrate the stem.

(k) Cereal or grass leaf with longitudinal scarification (15): feeding scars of some beetles (adults and larvae of *Oulema*, etc.); larval mines in such leaves (and Palmae) made by larvae of Hispinae (Coleoptera, Chrysomelidae), or Agromyzidae.

Fig.19. Damaged leaves (1) – mostly by biting insects

1. Leaves of Field Bean notched around the margin by adult *Sitona* weevils.
2. Generalized leaf with margin notched by adult broad-nosed weevils, e.g. *Systates*, *Hypera*, etc.
3. Brassica leaf with large irregular holes and margin eaten away by cabbage caterpillars (*Mamestra*, *Pieris*, etc.).
4. Hazel leaf eaten by gregarious sawfly larvae (*Croesus septentrionalis*); they are leaf-margin feeders and usually eat the entire leaf, leaving only the main veins; this damage also typical of grasshoppers (*Zonocerus*, etc.).
5. Leaf with irregular holes in both lamina and around the margin, typical of many larger caterpillars (Sphingidae, Noctuidae, some grasshoppers, some adult chafer beetles).
6. Leaf lamina eaten between the veins, some very irregular damage to leaf margin; many small veins not eaten completely making a ragged edge; typical of *Phyllobius* and some other adult leaf weevils.
7. Leaf margin with neat circular (or subcircular) pieces of lamina removed; by adults of *Megachile* (leaf-cutting bees), or in tropical America by worker leaf-cutting ants (*Atta*, *Acromyrmex* spp.). If subcircular, they could be made by some broad-nosed weevils (fig. 19–2).
8. Cabbage leaf with smallish holes/windows in the lamina, made by larvae of Diamond-back Moth and a few other small caterpillars.
9. Leaf with numerous very small holes in lamina (shot-hole effect); often leaf of *Brassica*, cotton, sweet potato; feeding holes made by adult flea beetles (Chrysomelidae, Halticinae).
10. Leaf lamina with larger, often quite regular-sized holes; made by adults and some larvae of leaf beetles, especially tortoise beetles (Chrysomelidae, Cassidinae).
11. Leaf lamina with ladder-like windowing (leaving small veins intact); done by adults and larvae of *Epilachna* (Coleoptera, Coccinellidae) (mostly sub-tropical).
12. Leaf lamina extensively skeletonized, by larvae of various Lepidoptera (Epiplemidae, Bombycidae, etc.); many leafworms are gregarious when newly hatched and at first skeletonize the leaf surface (*Mamestra*, *Spodoptera*, *Pieris brassicae*; Lepidoptera, Noctuidae and Pieridae).
13. Graminaceous plants with a series of small holes along the leaf; made by small stalk borers (Lepidoptera, Noctuidae) biting through the folded young leaf, prior to penetration of the actual stem.
14. Leaf rolled longitudinally and lamina eaten by enclosed larva; Rose Leaf-rolling Sawfly (*Blennocampa pusilla*); also many Tortricidae and some Pyralidae are called 'leaf-rollers'.
15. Cereal (graminaceous) leaf with longitudinal external scarification, by adult and larvae of *Oulema* and some other Chrysomelidae; elongate leaf mines made by larvae of hispid beetles (Chrysomelidae, Hispinae) (also in palm leaves) and larvae of leaf-miners (Agromyzidae).
16. Onion leaves with elongate holes, made by various cutworms and leafworms (Lepidoptera, Noctuidae); the caterpillars live inside the hollow leaves and eat the lamina; also Leek Moth larvae on onion.
17. Leaves (apple, top fruit, etc.) partly folded and held with silken strands, or adjacent leaves cut and webbed together, often with faecal pellets evident; characteristic of various fruit tree (and other) tortricids (Lepidoptera, Tortricidae).
18. Leaf with interveinal area eaten out, but dorsal epidermis intact; often pupal remains inside; typical of some Tortricoidea and some Microlepidoptera, also some sawflies and Agromyzidae (Diptera).
19. Leaf with long narrow linear mine, often with puparium terminally; made by leaf miners (Diptera, Agromyzidae); on the left edge are shown the feeding scars left by the female fly, she punctures the epidermis with her ovipositor and sucks up the exuding sap; often many mines per leaf.
20. Blotch leaf mine in leaf of sugar beet, made by larvae of *Pegomya hyoscyami* (Diptera, Anthomyiidae); a large blotch only in large leaves, often with several large maggots inside (pupation in the soil).
21. Leaf mine starting as a linear mine but ending up in a blotch; quite small; made by some Gracillariidae (Lepidoptera), and some obscure groups of Diptera; also larvae of Nepticulidae.
22. Graminaceous leaf longitudinally folded or rolled, sometimes held by silken strands; larvae or pupae of skippers (Lepidoptera, Hesperiidae); sometimes an egg-site for some Pyralidae, or spiders.
23. Leaves with part of the lamina eaten and with small or tiny cases dangling by silken strands; these include a few tiny species of bagworms (Lepidoptera, Psychidae), some Microlepidoptera (Coleophoridae, casebearer moths); the illustration is actually of Rice Caseworm (*Nymphula* spp., Pyralidae).
24. Leaves with lamina eaten, and nearby dangling cases, attached with silken strands, the cases made of silk covered with pieces of plant material, some as

long as 5 cm; made by bagworms (Lepidoptera, Psychidae), most abundant in warmer countries; most on trees and bushes, but may be on herbs and Gramineae also.

25. Leaf(let) folded, swollen and reddish; galled by larvae of Cecidomyiidae (in this case *Dasineura rosarum*, Rose Leaf Midge; many species fold and distort leaflets of clovers and other Leguminosae); some leaves folded by aphids; some folded by thrips (*Gynaikothrips ficorum*, Phlaeothripidae).

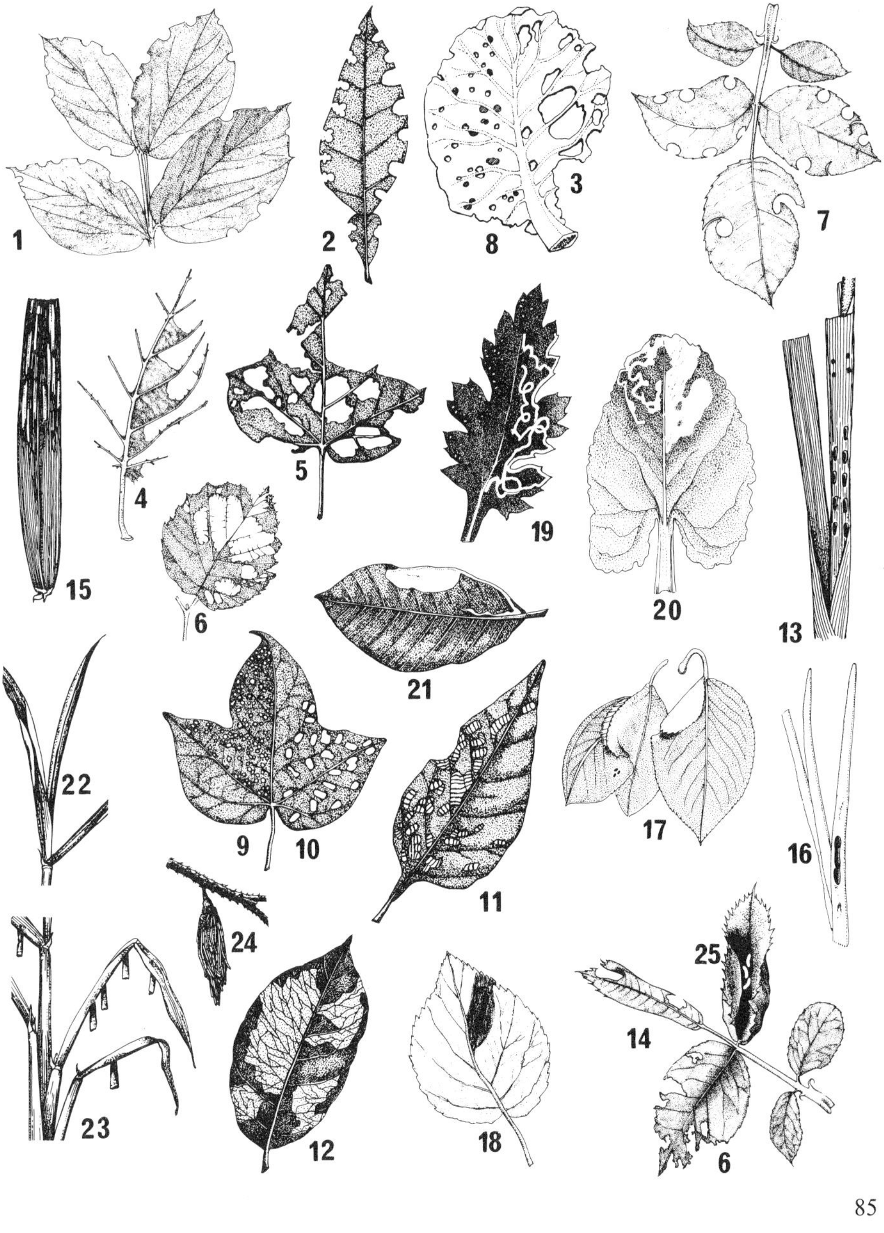

(l) Onion leaf with holes and internal feeding scars (16): made by various leaf-eating caterpillars (Noctuidae, etc.).

(m) Fruit tree leaves cut and folded, or webbed together, often with faecal pellets evident (17): made by larvae of various fruit tree tortricids.

(n) Tree leaf with interveinal area mined, tunnelled or skeletonized (18): caused by a single caterpillar of some Tortricidae, or by single sawfly larva (Tenthredinidae).

(o) Tiny or small bagworm cases (made of silk and plant fragments) adjacent to leaf lamina damage (holing or skeletonization) (23): made by caterpillars of some Microlepidoptera (Coleophoridae, etc.), some Pyralidae (*Nymphula* spp.) and some small species of Psychidae.

(p) Large bagworm cases, dangling from twigs, adjacent to leaves with lamina extensively eaten (24): large bagworms (Lepidoptera, Psychidae), most abundant in warmer climates.

(q) Leaflet rolled longitudinally and eaten (14): leaf-rolling sawfly larvae (*Blennocampa* spp.).

(r) Leaflet folded longitudinally, sometimes swollen and discoloured (galled) (25): larvae of some Cecidomyiidae (several on clovers), and some leaf-rolling thrips (Phlaeothripidae).

Damaged leaves (2) – by sap-sucking insects and mites (fig. 20)

(a) Leaf curled under (1) or generally distorted, sometimes completely distorted into a bunchy lump of tissue (11): the former damage is done by Aphididae, Psyllidae, Aleyrodidae, Cicadellidae and some other Homoptera, the latter damage is by thrips of the family Phlaeothripidae, and some aphids.

(b) Leaf folded (2) by aphids; folded and galled (5) by *Trioza* psyllids; folded or rolled longitudinally (3) by nymphs of some Aleyrodidae, some thrips, or spiders making a nest.

Fig.20. Damaged leaves (2) – by sap-sucking insects and mites

1. Leaves with curled-under edges, somewhat distorted, sometimes discoloured; caused by sap loss from feeding aphids, psyllids, cicadellids and some thrips.
2. Leaf of pear with characteristic curled shape, caused by feeding of Pear–Bedstraw Aphid (*Dysaphis pyri*); many fruit trees show specific distortion of leaves by various aphids. (See also 21–5.)
3. Leaf edge rolled dorsally, by nymphs of some Aleyrodidae, thrips (some Phlaeothripidae), some aphids; spiders make nests sometimes in rolled leaves.
4. Leaf with many small distinct blisters, bright red in colour, caused by colonies of Redcurrant Blister Aphid on the underside; slightly similar small/tiny bright red galls are caused by some gall mites (Eriophyidae), but mostly on shade and forest trees.
5. Leaf with edge rolled, distorted and coloured (reddish), forming a definite gall; nymphs of *Trioza alacris* on Bay leaves, and similar leaf galls are made by some other species of *Trioza* (Psyllidae). A few similar leaf galls are made by Cecidomyiidae.
6. A spittle mass under or on the leaf surface, made by certain spittle bugs (Cercopidae), in this case *Philaenus spumarius* on strawberry. (Large frothy clumps on plant foliage in the tropics are usually the egg masses of tree frogs!)
7. Leaves with many small necrotic black spots, usually causing death of the leaf; the drawing shows leaves of Tea after feeding by *Helopeltis* bugs (Heteroptera, Miridae); many Heteroptera with their toxic saliva cause necrotic spots where they feed, especially some Pentatomidae, Coreidae, Lygaeidae, Miridae, etc.; some adult Cercopidae also cause leaf spotting.
8. Leaf with lamina tattered; caused by feeding of Heteroptera on young leaf, toxic saliva causes necrotic spots that become holes and as the young leaf expands the holes tear and enlarge, causing 'tattering'; this drawing shows damage to leaf of Runner Bean by Common Green Capsid.
9. Leaf with covering of black sooty mould growing on the honeydew excreted by various Homoptera; usually an indication of infestation higher on the plant by aphids, mealybugs, whiteflies, or soft scales.
10. Leaf surface scarified, with a shiny silvery appearance and often wilting somewhat; damaged by feeding thrips (Thripidae). (See also 20–15).
11. Leaf sometimes rolled longitudinally and/or else very severely distorted;

colonies of some thrips in the family Phlaeothripidae; also a few aphids can cause such distortion (e.g. Plum Leaf-curling Aphid) (see fig. 21–5).

12. Onion (and other monocotyledons) leaf speckled and scarified and eventually severely wilting, due to feeding of Onion Thrips (*Thrips tabaci*).
13. Leaf petiole galled; this is a poplar leaf galled by Lettuce Root Aphid (*Pemphigus bursarius*) spring generation. Leaf petioles are bored or mined by various caterpillars (Lepidoptera, Gelechiidae, etc.) and some beetle larvae (Coleoptera, Curculionidae, etc), but most are to be found on wild plants.
14. Petiole of bean leaf galled by second generation Bean Fly larvae (Agromyzidae).
15. Leaf surface scarified, often with a silvery or a bronzed appearance, usually wilting somewhat, and often with a film of silk webbing over the surface; typical Red Spider Mite damage (*Tetranychus* spp. etc., Tetranychidae); often moulted 'skins' (exuviae) on the leaf surface.
16. Young leaf severely distorted and crinkled in appearance; dwarfed; typical damage by Strawberry Mite (*Stenotarsonemus pallidus*, Tarsonemidae) and other tarsonemids.
17. Warty outgrowths, usually on underside of leaves but projecting

Fig. 20 (*cont.*)
dorsally, termed erinia, made by gall mites (Acarina, Eriophyidae); usually many mites per erinium, but on Sycamore the tiny red galls contain only single mites. Many different sizes/shapes/colours are to be found on leaves galled by Eriophyidae.

18. Leaf with definitely shaped galls on either upper or lower surface, sometimes distinctively coloured (often red); the economic species are mostly gall midges (Diptera, Cecidomyiidae), but spectacular woodland (temperate) species include the oak leaf galls made by Cynipidae, and the red willow leaf-galls made by *Pontania* (Hymenoptera, Tenthredinidae).

19. Leaves, both monocotyledons and dicotyledons, with striking discoloration between the veins indicating that the plant is suffering a virus infection, usually transmitted by a homopterous bug (aphid, whitefly, leafhopper, etc.) in its feeding activity; the coloration pattern is usually linear in monocotyledons and reticulate on broad leafed dicotyledons. Deficiency of some minerals may produce a similar effect on some plants.

(c) Leaf lamina with red blisters projecting dorsally (4): Red Currant Blister Aphid, and some gall mites (Eriophyidae).

(d) Lamina pitted: some Psyllidae in the group Triozinae cause ventral leaf-pits at the sites where the nymphs sit and feed, young leaves sometimes may be considerably deformed (mostly tropical).

(e) Lamina scarified (10, 15): adults and nymphs of some thrips (Thysanoptera) and some spider mites (Tetranychidae) make tiny epidermal feeding lesions which give the lamina a silvery, bronzed or scarified appearance; soft leaves will be caused to wilt, as shown by Onion Thrips on onions (12).

(f) Lamina with erinia (17): on the lower surface (usually) of some leaves are found wart-like outgrowths (erinia) inhabited by microscopic gall mites (Eriophyidae).

(g) Leaf galls (18): round or elongate galls, small or large, few or numerous, found on either upper or lower leaf surface, sometimes arranged randomly, sometimes distributed peripherally or alongside veins; made by feeding larvae of gall midges (Diptera: Cecidomyiidae), gall mites (Acarina: Eriophyidae), gall wasps (Hymenoptera: Chalcidoidea, Cynipoidea and Symphyta) and some Psyllidae.

(h) Lamina tattered with irregular holes and tears (8): margin usually intact; many Miridae and other Heteroptera feed on young leaves, and their toxic saliva results in small necrotic spots. As the leaf grows and expands the dead areas enlarge and tear, resulting in the characteristic tattering of the expanded leaf.

(i) Leaf with numerous small necrotic dark spots (7): caused by toxic saliva of *Helopeltis* and other Miridae.

(j) With bubble-froth (6) either on leaves or in leaf axils: spittle mass built by nymphs of Cercopidae (spittle bugs) for protection.

(k) With black sooty mould (9): infestations by aphids, mealybugs, and soft scales are often associated with sooty moulds that develop on the sugar excreted as 'honey-dew'.

(l) Leaf petiole galled: several species of temperate woolly aphids (Pemphigidae) make large galls in the leaf petioles of various woody shrubs and trees (13); also Bean Fly larvae mine and swell petioles of beans (14).

(m) Leaf dwarfed, crinkled and distorted (16): Strawberry Mite, and other Tarsonemidae.

(n) Cereal leaves with longitudinal streaking (adjacent pale and dark areas) (19), and dicotyledonous leaves with a conspicuous regular blotching: these are symptoms of virus infection; all plant viruses are transmitted by feeding Hemiptera, or sometimes by nematodes, or occasionally by beetles.

Damaged shoots and buds (including flowers) (fig. 21)

(a) Shoot killed distally (brown and wilted) (1): by boring larva of *Earias* spp. in cotton (spiny bollworms), larvae of longhorn beetles (Cerambycidae), caterpillars of goat and leopard moths (Cossidae); sometimes killed by ovipositing cicadas or Tettigoniidae; also killed by toxic saliva of some sap-sucking bugs (Coreidae, Pentatomidae, etc.) (3).

(b) Cereal or grass seedling with 'dead-heart' (2): caused by boring larva of shoot flies (Agromyzidae, Anthomyiidae, Muscidae, etc.); also by caterpillar of some Pyralidae and Noctuidae (stem borers).

(c) Shoot and apical leaves completely distorted, crinkled and small (5): caused by Plum Leaf-curling Aphid and Green Apple Aphid; similar damage caused by some leaf-rolling thrips (Phlaeothripidae).

(d) Petals gnawed (12, 13): adult blister beetles (Meloidae) chew petals of many plants, often common on Malvaceae; adult flower beetles (Scarabaeidae) make small holes in petals, *Popillia* being especially injurious.

(e) Petals scarified: flowers of Leguminosae, Compositae, etc. inhabited by adults and nymphs of thrips (Thripidae) which scarify the bases of the petals.

(f) Flowers inhabited by tiny black beetles, making feeding scars at the base of the petals: legume flowers inhabited by *Apion* weevils; worldwide.

(g) Anthers eaten: pollen beetles (*Coryna* spp. etc.) feed on the anthers of many flowers, especially Malvaceae, destroying the pollen sacs.

(h) Maize tassels eaten: by grasshoppers, or Maize Tassel Beetle in East Africa, or Japanese Beetle in the USA.

(i) Flowers inhabited by tiny maggots: gall midge larvae (Diptera: Cecidomyiidae), either white, yellow, orange or red in colour, usually causing shoot and flower deformation (nettlehead).

(j) Buds bored (8): caterpillars of some Tortricidae bore into large buds of shrubs and trees (called 'budworms' in the USA).

(k) Flower buds (rape) bored by blossom beetles (*Meligethes* spp.) (9): larvae develop inside unopened buds.

(l) Grass or cereal shoot galled and swollen (7): caused by some grass gall flies (*Chlorops* spp. etc.).

(m) Flower bud (apple) 'capped' (11): caused by adult Apple Blossom Weevil, larvae develop inside unopened bud, feeding on the flower.

(n) Buds gnawed, with large holes (6): eaten by large caterpillars (e.g. Cotton Semi-looper, *Anomis flava*, Noctuidae); sometimes by long-horned grass hoppers (Orthoptera: Tettigoniidae).

(o) Buds webbed and gnawed: some Tortricidae have caterpillars that feed on opening buds which they cover with a fine silk webbing.

(p) Buds pierced and dying: feeding adult and nymphal jassids (Miridae) and other Heteroptera, with their toxic saliva.

(q) Buds enlarged and swollen (10): gall mites of the family Eriophyidae infest buds of some shrubs causing a disorder called 'big-bud'.

(r) Buds and shoots stunted/wilting (4): heavy infestations of aphids, scales, mealybugs, can stunt young shoots during active growth; Hibiscus Mealybug is unusual in causing severe shoot telescoping and often shoot death; 'nettleheads' caused by larvae of Cecidomyiidae in many cultivated plants.

(s) Woody shoot cut off (14): twig cut by adult twig-cutter weevils (*Rhynchites* spp.) after ovipositing in distal part of shoot.

Damaged stems (fig. 22)

These fall into four main categories for practical purposes: shoots of graminaceous seedlings; stems of cereals and grasses; shoots and twigs of shrubs; trunks and branches of trees (and large woody shrubs).

(a) Cereal shoots with 'dead-hearts' (fig. 21–2): caused by boring larvae of shoot flies (Agromyzidae, Muscidae, Anthomyiidae), and caterpillars of Pyralidae.

(b) Cereal stems galled and distorted: larvae of Diptera (Opomyzidae, Chloropidae); cereal stem gall midges (Cecidomyiidae) are mostly temperate in distribution.

(c) Cereal and grass stems bored (2): caterpillars of the family Pyralidae generally bore rice and grass stems, while the larger caterpillars of the Noctuidae bore stalks of maize, sorghum and other larger species of Gramineae; tunnels in sugarcane are usually very short because the stem is solid and pithless. Stem sawflies (*Cephus*) bore apically in temperate cereals, especially wheat.

(d) Herbaceous stems and woody shoots bored, swollen often with a single emergence hole: larvae of gall weevils (Curculionidae) (5), also Sisal Weevil, and the temperate Cabbage Stem Weevil and Turnip Weevil (6).

(e) Banana pseudostem bored: Banana Stem Weevil (*Odoiporus longicollis*) larvae make extensive tunnel galleries in which they pupate and adults may also be found.

(f) Herbaceous stems, including Sweet Potato stem (vine) bored (3): caterpillars of some clearwing moths (Sesiidae) and plume moths (Pteryphoridae), also some weevils and some sawfly larvae.

(g) Twigs galled (7): made by feeding larvae of gall midges (Diptera: Cecidomyiidae), some gall wasps (Cynipidae, Eurytomidae, Torymidae), old galls have multiple exit holes; a few twig galls are made by weevil larvae and some woolly aphids (Pemphigidae).

(h) Stems (often Gramineae) with frothy spittle mass in the axils (1); made by nymphs of Cercopidae (spittlebugs).

(i) Seedling stem gnawed at about ground level, commonly sugarbeet gnawed by adults of Pygmy Mangold Beetle (*Atomaria linearis*) (4); similar damage done by wireworms (Coleoptera, Elateridae), symphylids (Myriapoda), and some other soil pests.

Fig.21. Damaged shoots and buds (including flowers)

1. Distal part of woody shoot wilting and dying: some longhorn beetle larvae (Coleoptera, Cerambycidae); spiny bollworms (*Earias* spp.) on cotton, boring inside the stem; also larvae of goat moths and leopard moths (Lepidoptera, Cossidae); shoot also killed by ovipositing cicadas.
2. Cereal or grass shoot dead and brown ('dead-heart'): shoot growing point eaten by larvae of shoot flies (Anthomyiidae, Muscidae, etc.); or stem borer caterpillars (Lepidoptera, Pyralidae, Noctuidae).
3. Woody or herbaceous stem with apical part wilting and killed: various Heteroptera with toxic saliva (mostly Coreidae, also Pentatomidae, Lygaeidae, etc.).
4. Apical shoot distorted, dwarfed, telescoped, often referred to as a 'nettlehead' condition: the illustration shows the effect on a *Hibiscus* shoot of the mealybug *Maconellicoccus hirsutus* (unusual in having toxic saliva), but only a light infestation; such shoot deformation may be produced by some aphids, but more commonly by Cecidomyiidae (Pea Midge, Hawthorn Shoot Gall Midge, etc.), as a result of the saliva of the feeding larvae; the 'nettlehead' is usually a tight rosette of small leaves.
5. Apical shoot and young leaves completely distorted, crinkly, small and deformed: Plum Leaf-curling Aphid is one of the aphids that has toxic saliva; shoot may die.
6. Flower bud eaten into, by various grasshoppers, some caterpillars (Noctuidae, etc.), and some beetles (*Popillia* etc., Scarabaeidae).
7. Grass or cereal shoot galled, swollen and telescoped: caused by larvae of Gout Fly (*Chlorops* spp.), and a few other grass flies; Psyllidae of the genus *Livia* cause similar galls on *Juncus* spp. but not on cereals.
8. Flower bud bored by small caterpillars belonging to the Tortricidae, collectively called 'budworms', often frass outside the hole.
9. Rape flower buds bored by adult blossom beetles (*Meligethes* spp.) larvae feed on developing flower internally. Other pollen beetles may feed on exposed anthers after flower has opened.
10. Bud swollen and distended but does not open; this is the 'big-bud' condition produced on various shrubs by mites of the genera *Phyllocoptella* and *Cecidophyopsis* (on Hazel and Blackcurrant).
11. 'Capped' flower bud, turning brown, and does not open, produced by larva of Apple Blossom Weevil (*Anthonomus pomorum*).

12. Flower with small holes in the petals, sometimes numerous; made by adult flower beetles, especially *Popillia* (Scarabaeidae, Rutelinae); also some adult and larval tortoise beetles (Chrysomelidae, Cassidinae).
13. Flower with large pieces of perianth removed: eating by adult blister beetles, especially *Mylabris* and *Epicauta* (Meloidae), sometimes the entire flower is destroyed.
14. Woody twig cut through and apical shoot fallen or dangling: this is done by adult weevils after laying an egg in the terminal part of the stem, the larva develops inside the fallen piece of stem; seen on apple by Twig Cutter (*Rhynchites caeruleus*). Some shoot weevil larvae live inside the intact stem, which is killed distally by their feeding activity (e.g. Bamboo Shoot Weevil); but most stem weevils do not actually kill the plant distally.

(j) Tree trunk and branches bored, sometimes bark eaten externally: timber borers belong to two orders of insects, the Lepidoptera and Coleoptera; Metarbelidae feed externally at night on the bark under a silken tube coated with frass but retire during the day to a deep tunnel in the heartwood (tropical); larvae of Cossidae and some Sesiidae bore in the heartwood of trunks and usually down the centre of smaller branches (8); the circular adult emergence hole typically contains the pupal exuvium projecting. Tree-boring beetles belong mainly to the families Cerambycidae, Buprestidae, Bostrychidae, and Scolytidae, with a few weevils (Curculionidae) and a few small families such as the Lymexylidae. Cerambycidae are longhorn beetles and may be very large (*Batocera* spp.) with larvae that bore in trees for up to 3–4 years before pupating; they mostly eat sapwood just under the bark and may leave frass holes to the exterior; the emergence hole is circular (8, 9); the adults sometimes chew patches of bark. Buprestidae are called jewel beetles, and the larvae, known as flat-headed borers, and the emerging adults leave an oval exit hole in the tree bark (11). Bostrychidae are black, cylindrical beetles, completely circular in cross-section and the tunnels are bored by the adult beetles (10) called black borers. Scolytidae (shot-hole and twig borers) belong to the group of ambrosia or fungus beetles; the adults bore into trees (11) and make extensive breeding galleries under the bark (12); they carry fungus spores on their bodies with which to inoculate the new fungus galleries where the larvae feed; some species bore down the centre of twigs on bushes such as tea.

Damaged fruits and seeds (fig. 23)

(a) Cereal panicle with few grains developed, or grains small and distorted: larvae of Sorghum Gall Midge (*Contarinia sorghicola*; Diptera: Cecidomyiidae); small distorted grains of wheat, sorghum, etc.,

Fig.22. Damaged stems
The apical parts of stems, shoots, have been included with buds (and flowers) on fig. 21; there is similar overlap with roots (fig. 24), some borers may attack any part of the plant, from apical tip down to the upper part of the root system.

1. Leaf axil with spittle mass, inhabited by nymphs of Cercopidae, but no overt signs of damage usually.
2. Graminaceous stem bored internally, sometimes with frass/emergence hole; plant wilts, may be stunted, and sometimes breaks; leaves sometimes with characteristic rows of holes: stem borers are larvae of Noctuidae and Pyralidae (Lepidoptera), the latter usually in the narrower stems of rice, wheat and grasses. Worldwide many genera are stem borers.
3. Herbaceous stem bored, may be vines (prostrate) or short upright stems such as in brassicas: Sweet Potato vines bored by larvae of Sesiidae, Pterophoridae; also several species of stem weevils (Curculionidae); some boring sawflies; usually there is no external sign of the stem infestation, although distally the plant may be wilting.
4. Seedling with stem gnawed at about ground level: this Sugar Beet seedling shows damage by adult Pygmy Mangold Beetle (*Atomaria linearis*); similar damage is done by other soil pests, including wireworms (Coleoptera, Elateridae), symphylids (Myriapoda) and springtails (Collembola).
5. Stem swollen and galled, often with an emergence hole quite distinct: caused by stem weevil larvae, usually in woody stems or twigs (Coleoptera, Curculionidae). Pupation takes place in the soil usually.
6. Cabbage stem and gall weevils (*Ceutorhynchus* spp.); the larvae may be found in either stems or upper parts of the roots; many different weevils bore in the stems of herbaceous plants, making evident galls.
7. Woody twig with globular galls, either regular or irregular in shape, old galls with emergence holes, sometimes with remains of pupal exuviae extruding: caused by larvae of many different gall midges (Cecidomyiidae); occasionally by wasps such as *Eudecatoma* (Hymenoptera, Chalcidoidea). In forest trees such as oaks there are many different gall wasps (Hymenoptera, Cynipidae) making leaf, twig, shoot and flower galls of great diversity.
8. Woody stem or tree trunk/branches bored centrally with many frass holes to the exterior: made by the larvae of some longhorn beetles

(Coleoptera, Cerambycidae); adult emergence holes are quite circular in shape.
9. Tree trunk or branch with some large holes, with external frass, and parts of the bark eaten away: tunnelling by larvae of Cerambycidae and bark-eating by the adult beetles.
10. Woody branch or tree trunk with cylindrical hole medially (up the branch); holes completely cylindrical are made by adult black borers (Coleoptera, Bostrychidae); tunnels less cylindrical are made by larvae of Cossidae (Lepidoptera) or Sesiidae, and these tunnels usually have frass packed inside, with only a little being extruded. On some trees these tunnels may extend down into the top of the root system. With wood-boring Lepidoptera the old pupal exuvium is left projecting out of the emergence hole in a very characteristic manner.
11. Tree trunk with oval emergence hole made by young adult jewel beetle (Coleoptera, Buprestidae), such as *Agrilus*, *Chrysobothris*, etc.; the larval tunnel system is oval in cross-section and tightly packed with frass. Also on this piece of trunk are some entry/emergence holes made by adult bark borers (Coleoptera, Scolytidae).
12. Breeding gallery in the sapwood under the bark of a tree, made by adult bark ambrosia beetles (Coleoptera, Scolytidae), and later by the larvae in some cases. Some of these beetles are collectively known as twig borers and they bore in woody twigs rather than making this type of breeding gallery; *Scolytus* spp. are the most common cause of such damage shown (12).

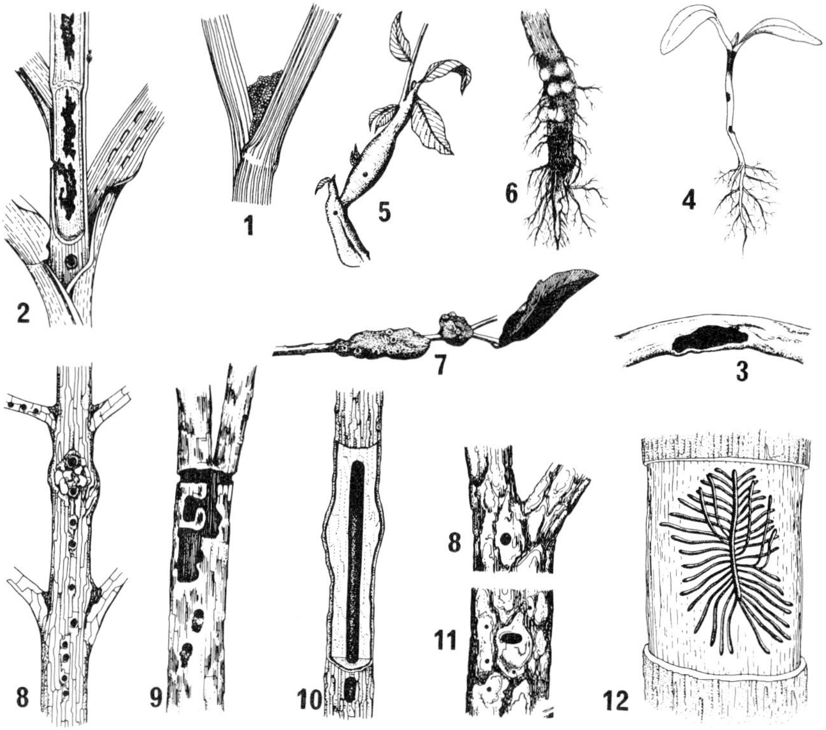

result from feeding of sap-sucking Heteroptera with toxic saliva (Miridae, Jassidae, Coreidae, etc.), and sometimes grain aphids.

(b) Cereal panicle with missing or bitten grains: young cereal heads attacked by caterpillars which eat the softy milky grains (e.g. sorghum attacked by False Codling Moth larvae in Africa); similar damage done by grasshoppers (Orthoptera: Acrididae and Tettigoniidae); riper grains eaten by birds; maize cobs grazed by *Heliothis* and other caterpillars (Noctuidae and Pyralidae) which may also bore up inside cobs (5).

Fig.23. Damaged fruits and seeds

1. Apple with corky patch, caused by feeding of a capsid bug (Miridae) with its toxic saliva causing some necrosis and some distortion of tissues; similar damage done by other Heteroptera, and also on pear fruitlets by feeding of adult froghoppers (*Cercopis* etc.).
2. Cotton boll with necrotic spots caused by feeding Heteroptera bugs; this necrotic spotting very common on cocoa pods and many other fruits, both tropical and temperate.
3. Apple with discoloured patch (reddish), hole in epidermis, and usually small pile of frass externally; internally with excavation extending into core region, extensive frass usually accompanied by moulds and rots; typical Codling Moth damage to apple, but several other fruit moths do the same damage to a wide range of temperate fruits (Lepidoptera, Tortricidae).
4. Cotton boll showing typical bollworm damage, by feeding caterpillars of many Noctuidae, and a few Tortricidae (*Cryptophlebia* etc.).
5. Maize cob bored internally, and usually with surface grazing also, by caterpillars (mostly Noctuidae) that are also stem borers; the illustration shows damage by European Corn Borer (*Ostrinia*, Pyralidae).
6. Apple fruitlet showing feeding scars by Apple Sawfly (*Hoplocampa testudinea*), initial damage is surface feeding followed by boring into the fruitlet; several closely related sawflies do similar damage on other temperate fruits.
7. Apple with surface eaten by Summer Fruit Tortrix (*Adoxophyes orana*); several Tortricidae eat the surface of fruits, usually under a leaf webbed on to the fruit surface: these species eat both leaves and the surface of fruits.
8. Apple fruitlet eaten by adult cockchafers (*Melolontha* spp.); several adult scarab beetles also regularly eat leaves.
9. Fruit (orange) with necrotic spot externally and extensive internal cavity, made by fruit fly maggots (Diptera; Tephritidae: *Ceratitis*, *Rhagoletis* etc.); usually many maggots per fruit; often accompanied by fungal and bacterial rots.
10. Apple fruitlet with feeding scars made by adult Apple Fruit Rhynchites (*Caenorhinus aequatus*); many phytophagous weevils when feeding make small surface holes in the plant tissue, some of which may also be used as oviposition sites.
11. Pea pod with surface feeding scars caused by Pea Thrips (*Kakothrips robustus*); most thrips (Thripidae) feed on leaves and scarify the epidermis, but a few are found on developing fruits (botanical sense); pupation takes place in the soil, this is typically larval damage.
12. Rape pod with emergence hole made by full-grown larva of Cabbage Seed Weevil (*Ceutorhynchus assimilis*), emerging from the fruit to pupate in the soil; many weevils have larvae that feed on the seed (kernel) inside fruits, and some on the fleshy part of fruits, both throughout the tropics and in temperate regions; pupation is invariably in the soil, so the hole is made by the full-grown larva.
13. Hazel nut with hole bored by emerging larva of Hazelnut Weevil (*Curculio nucum*); many temperate nuts are attacked by weevils (*Curculio* spp.) whose larvae eat the kernel, and then bore a hole in the shell in order to pupate in the soil (cf. 21).
14. Bean pod holed by caterpillars of American bollworms (*Heliothis* spp.) in order to feed on developing seeds internally; large caterpillars (Noctuidae) seldom actually enter the pod but just push head and thorax inside to feed; smaller caterpillars (Pyralidae) usually live inside the pod whilst feeding, although one caterpillar may visit several pods. All legume pods are attacked by a wide range of caterpillars belonging to both Pyralidae and Noctuidae.
15. Pea pod bored by small caterpillar and the seeds eaten inside the pod; several important pea-pod borers are members of the genus *Cydia* (Tortricidae) both in temperate regions and in the tropics, in pods of *Pisum* and other genera: for pods picked as a vegetable there is usually no external sign of damage as the larvae and frass are all contained inside the pod, but pupation does take place in the soil so in drying pods an emergence hole is usually evident. In pigeon pea (and some other tropical legumes) Bean Pod Fly (*Melanagromyza obtusa*) causes similar damage by the larvae eating the seeds inside the pods.
16. Ripening maize cob (in the field) showing infestation by Maize Weevil (*Sitophilus zeamais*); the holes are made by emerging young adults.
17. Pulses (seeds of Leguminosae) showing emergence holes made by young adult bruchid beetles (*Bruchus* spp. etc.); infestation may be of ripening crops in the field or else later of the dried seeds in a produce store.
18. Stored grains with larvae or pupae inside, and with emergence holes made by the young adult *Sitophilus* weevils; usually in wheat it is *S. granarius*, in maize *S. zeamais*, and in rice usually *S. oryzae*; these are the preferred foods.
19. Mushroom ‘fruit’ with stem and cap tunnelled

extensively; holes made by larvae (maggots) of several different flies; usually Sciaridae, Phoridae and Cecidomyiidae.

20. Fruit (orange) distorted during early development by infestation of Citrus Bud Mite (*Aceria sheldoni*); some other species of *Aceria* (Eriophyidae) cause fruit deformation on other fruit crops.
21. Macadamia fruit showing emergence hole made by full-grown caterpillar of *Cryptophlebia* spp. (Tortricidae) through this extremely hard nut shell; not all holes in hard nut shells are made by weevils, some are made by tortricid caterpillars.

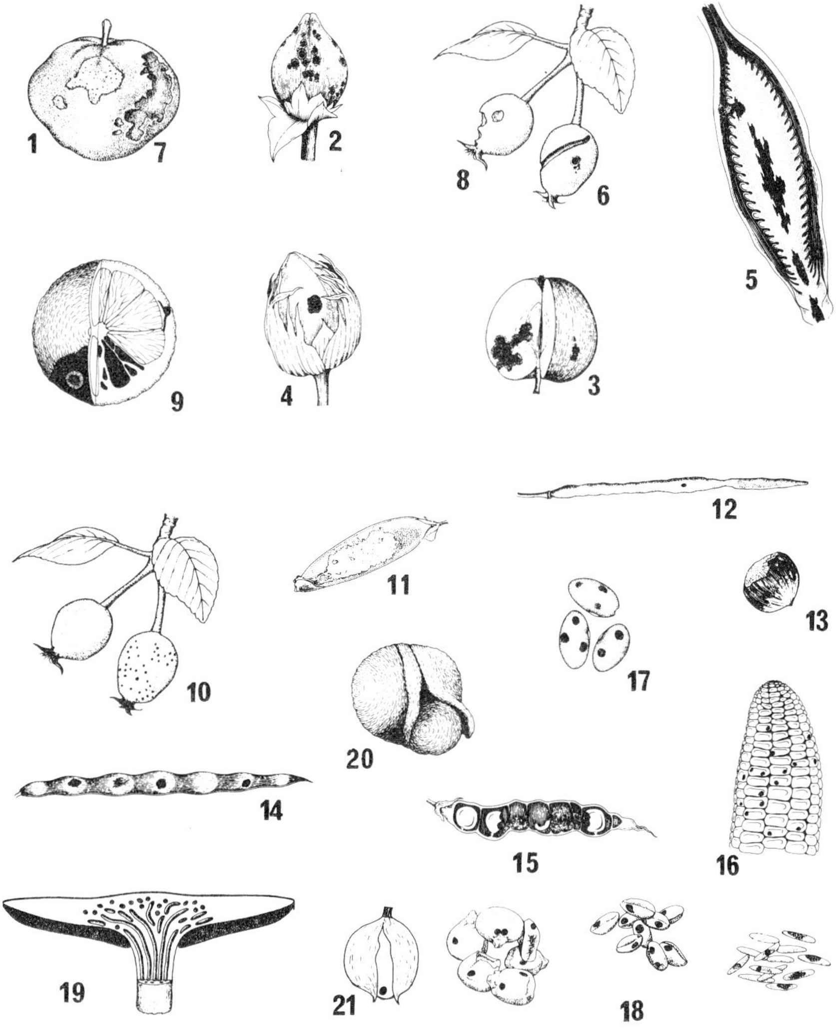

(c) Fruits with numerous small necrotic patches: heteropteran bugs of the families Miridae, Coreidae and Pentatomidae have toxic saliva, and when feeding on fruits they cause necrosis at the feeding sites, which usually become infected with fungi and bacteria, resulting in rotting, death, and premature fruit-fall, e.g. cocoa capsids, cotton stink bugs (2) Apple Capsid, and *Cercopis* adults (froghoppers), (1).

(d) Fruits, nuts and seeds bored: bored internally by weevil larvae, and hole made by larva or adult, examples are mango weevils, Hazel-nut Weevil (13), Cotton Boll Weevil, Maize Weevil (16, 18); in small ovaries one weevil larva will eat all the endosperm and thus each ovary contains only one weevil pupa (e.g. clover seed weevils: *Apion* spp.); several Maize Weevils can develop in one maize kernel.

(e) Fruits with large internal tunnels, sometimes large holes to the exterior (3, 4, 5, 6), sometimes infected with fungal and bacterial rots, sometimes frass expelled: made by caterpillars of the families Tortricidae, Pyralidae and Noctuidae; eggs are laid externally, and the tiny caterpillars (first instar) bore into the fruit; sometimes infested fruits fall prematurely; cotton bolls bored by many bollworms; top fruit bored by many different tortricid larvae (3) but Codling Moth is the most serious pest.

(f) Pulse pods bored by many different caterpillars: large Noctuidae (14) too stout to enter the pods make holes and reach into the young seeds which are eaten; pea moths (*Cydia* spp.) live inside the pods (15) and eat the seeds, as does Pea Pod Borer (*Etiella*, Pyralidae); many of the Lycaenidae have larvae feeding on or in the pods of Leguminosae.

(g) Fruits webbed and gnawed (7): some caterpillars (mostly Pyralidae) will spin silk over young fruits and feed on the fruits (Coffee Berry Moth: *Prophantis smaragdina*, Pyralidae); Sesame Webworm spins silk over the pod and shoot before boring the pod. Apples and other fruits with surface damage due to feeding caterpillars (Tortricidae), usually under a leaf webbed to the fruit.

(h) Fruitlets (apple, etc.) gnawed (18): by adult chafer beetles (Scarabaeidae) in the spring.

(i) Fruitlets bored and with surface feeding scars (6): made by feeding larvae of Apple Sawfly (*Hoplocampa* spp.).

(j) Fruits tunnelled and bored: adults of some Scolytidae bore host plants to make breeding tunnels and galleries in which the larvae develop; Coffee Berry Borer (*Hypothenemus hampei*) bore coffee berries in Africa and South America.

(k) Fruits with maggots inside, a necrotic area, and sometimes small holes (9): fruit flies of the family Tephritidae attack almost all larger types of fruit; eggs are laid under the skin of the young fruit and the fly larvae develop internally as the fruit ripens; the fruit often falls prematurely, pupation usually takes place in the soil; secondary infection of the tunnels and oviposition site by bacteria and fungi is common; typically there is neither frass hole, entrance nor exit hole evident in the skin of the fruit while still on the tree, but sometimes there may be sap exudation. Also Pear Midge, causing fruitlet fall.

(l) Cruciferous seed pods bored and with an emergence hole (12): made by emerging larva of Cabbage Seed Weevil (*Ceutorhynchus assimilis*) on its way to pupate in the soil.

(m) Pulse and cruciferous pods with internal damage made by tiny maggots, which are larvae of several pod midges (Diptera, Cecidomyiidae) and some Agromyzidae (pod flies).

(n) Fruits (apple, etc.) with small feeding scars (10): made by adults of some weevils (*Caenorhinus* spp.).

(o) Fruits (pea pods, etc.) with external scarification (11): caused by the feeding of some thrips (Thripidae).

(p) Fruit deformed (20): various phytophagous mites (Eriophyidae) feeding on flowers and young fruits cause distortion and deformation of the fruits.

(q) Mushroom fruiting body with extensive tunnelling in the cap and up the stalk (19): made by feeding larvae of the numerous 'mushroom flies' (Cecidomyiidae, Sciaridae, and Phoridae).

(r) Pulse seeds, both inside ripening and ripe pods, and in storage with tunnels (17): made by larvae of several different species of Bruchidae; pupae to be found under the 'windows'; *Bruchus* spp. are more field pests, and *Callosobruchus* more adapted to life in stores.

(s) Maize with ripening seeds eaten in the field by boring larvae (16) of *Sitophilus zeamais*, then carried into grain store where further breeding continues (18).

(t) Various ripening seeds attacked in the field, hollowed out by feeding larvae of *Araecerus fasciculatus* (Coffee Bean Weevil).

(u) Nuts (in this case Macadamia) with kernel eaten and hole bored through the shell to the exterior (21): bored by caterpillar of various Tortricidae (*Cryptophlebia* spp.).

(v) Grains in store with 'windows' and hole bored (18): *Sitophilus* species with *S. granarius* preferring wheat, *S. zeamais* in maize and *S. oryzae* preferring rice, and the latter two species more tropical than the temperate grain weevil.

Damaged roots and tubers (and bulbs) (fig. 24)

(a) Fine roots eaten (1): by subterranean caterpillars, fly larvae, beetle larvae and termites.

(b) Tubers tunnelled (2): usually infected with secondary rots; Potato Tuber Moth larvae bore down the stem into the tubers.

(c) Tubers with narrow tunnels (3): bored by wireworms (larvae of Elateridae) and some small species of slugs.

(d) Tubers with wide, sometimes shallow tunnels (4): bored by cutworms (Noctuidae), chafer grubs (Scarabaeidae), and some larger species of subterranean slugs.

(e) Yam tubers with wide tunnels: bored by adults and larvae of yam beetles (Scarabaeidae).

(f) Sweet potato tubers bored: infected with secondary rots; larvae and adults of Sweet Potato Weevils (*Cylas* spp., Apionidae).

(g) Root stock encrusted: root mealybugs form a hard layer of encrustation over plant roots; ants are often found in attendance; mostly tropical.

(h) Roots with waxy agglomeration: root aphids (usually Pemphigidae) found as a waxy or mealy agglomeration on roots of some herbaceous plants; mostly temperate.

(i) Roots galled (9, 10): larvae of several weevils (*Ceutorhynchus* spp.) and other beetles live inside root galls; *Sitona* inside the 'nodules' on roots of Leguminosae.

(j) Taproot eaten or hollowed (5, 7): cutworms (Noctuidae) and chafer grubs (Scarabaeidae), but in particular Rosy Rustic Moth larvae.

(k) Taproot tunnelled and eaten (6), symptoms are wilting and plant collapse (7): larvae of Anthomyiidae (root maggots), and other Diptera such as Carrot Fly maggots (*Psila rosae*); with root crops, plant wilting does not normally occur.

(l) Bulb bored and tunnelled (8): Onion Fly (*Delia antiqua*) larvae (temperate and sub-tropical), and the temperate narcissus flies (Syrphidae).

(m) Rhizome bored: Banana Weevil (*Cosmopolites sordidus*) larvae bore the rhizome extensively, and both pupae and adults may also be found in the tunnels; various fly maggots belonging to the Chloropidae and other families bore the rhizomes of ginger.

(n) Roots with small globular cysts (12): although nematodes are not included in this book, mention of them cannot be avoided here; small globular cysts are produced by female *Heterodera* spp. (cyst eelworms) on a wide range of host plants; probably more temperate than tropical.

(o) Roots with large swellings (13): root-knot nematodes (eelworms), *Meloidogyne* spp. cause extensive root swellings on many crop plants, especially in the tropics.

(p) Roots stunted but bushily prolific (11): various free-living nematodes.

Damage to seedlings and sown seeds (fig. 25)

(a) Cotyledons of large seeds bored and eaten (7): larvae of Bean Seed Fly (Corn Seed Maggot in the USA) (*Delia platura*) bore into cotyledons, epicotyl and hypocotyl and prevent germination; similar damage done by small species of slugs.

(b) Seeds dug up and eaten: various species of birds (sparrows etc.) and mice.

(c) Cereal seedling stem with 'dead-heart' (1, 2): maggots of various Diptera (Agromyzidae, Anthomyiidae, Muscidae, Opomyzidae, Chloropidae, and Oscinellidae) bore into the young stem, usually killing the growing point and making the apical leaf turn brown and die; some caterpillars (Crambidae, Pyralidae and Noctuidae) also bore seedling stems of graminaceous plants, but typically they attack older plants.

(d) Stem bored, usually swollen (3): several species of Agromyzidae (Diptera) have larvae that bore in the stem of various seedlings, best known is probably the Bean Fly (*Ophiomyia phaseoli*) widespread and abundant on legumes, it will also bore into leaf petioles on beans.

(e) Stem severed with shoot lying alongside (6): typical cutworm (Noctuidae) and cricket (Orthoptera: Gryllidae) damage; often several consecutive seedlings are cut through; with crickets the cut plant is left for a day to wither and the next night is pulled down into the nest; such damage may be done by surface slugs.

(f) Stem severed and plant removed: several species of

Fig.24. Damaged roots and tubers (and bulbs)

1. Root hairs, rootlets, lateral roots eaten: this general damage is done by a wide range of soil-inhabiting insects and phytophagous Myriapoda, including cutworms (Noctuidae), beetle larvae (wireworms, chafer grubs, etc.), weevil larvae, mole crickets, sometimes fly larvae, and symphylids; usually specific identification is not possible from damage symptoms alone.
2. Potato tuber eaten by larvae of Potato Tuber Moth, after having tunnelled down the stem (direct damage in storage); several other Lepidoptera, especially Gelechiidae, have larvae that tunnel in the stems of plants and eventually pass down into the main roots.
3. Potato tuber with tiny holes bored by wireworms (Coleoptera; Elateridae); also sometimes holes made by smaller species of slugs, but then usually slime trails are evident.
4. Potato tuber with large, usually shallow, hole excavated by feeding cutworm (Lepidoptera, Noctuidae); sometimes by chafer grub (Coleoptera, Scarabaeidae); sometimes by larger species of slug, but then slime trails are usually to be seen.
5. Plant (sugarbeet) with large taproot hollowed out, and often with lower parts of aerial stem or thick leaf petioles also tunnelled; typical damage by larvae of Rosy Rustic Moth and a few allied species (Lepidoptera, Noctuidae); these are the main temperate crop stem borers.
6. Carrot root with surface (and some deeper) tunnels made by larvae of Carrot Fly (*Psila rosae*); also found on (in) parsnip roots and eating out the root-base/stem of celery.
7. Plant with thick stem/root eaten into and hollowed (lettuce plant); caused by feeding cutworm (Lepidoptera, Noctuidae), or chafer grub (Coleoptera, Scarabaeidae); occasionally by a slug; a few other soil beetles may also cause such damage (wireworms, etc.).
8. Bulb (onion, tulip, *Narcissus*, etc.) with inner scales eaten away, often with internal rots; typical damage done by Onion Fly larvae, or narcissus flies (Diptera, Syrphidae); also similar damage by stem and bulb eelworms (Nematoda).
9. Root crop with carbuncle-like spherical galls on the surface: made by larvae of Turnip Gall (*Ceutorhynchus pleurostigmata*) Weevil.
10. Herbaceous root with nodular galls: cabbage plant with Turnip Gall Weevil attack; similar weevils attack a range of plants causing root galls.
11. Roots partly stunted and partly bushy: typical damage caused by several different free-living plant parasitic nematodes (*Tylenchulus* spp. etc.).
12. Roots with many small

globular cysts, made by root cyst nematodes such as female *Heterodera* etc., on many different plants.

13. Roots with large irregular swellings; caused by root-knot nematodes, *Meloidogyne* spp., mostly tropical but also sub-tropical and in some temperate regions.

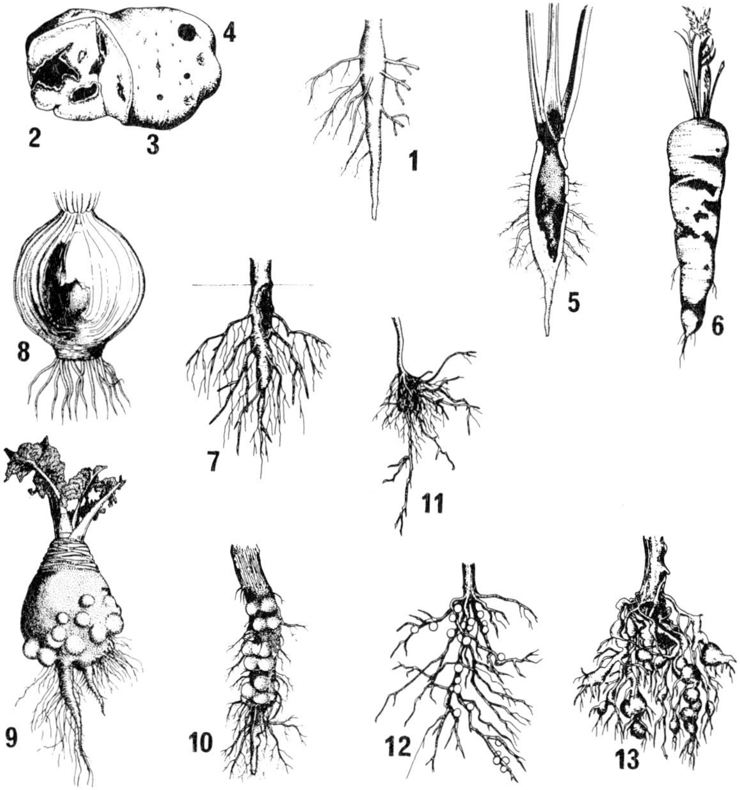

termites; can also be done by some leaf-cutting ants and harvester ants.

(g) Stem gnawed at about ground level (2) or below ground: a few beetles belonging to small families, and a few adult scarabs.

(h) Cotyledons or first leaves pitted and eaten: adult flea beetles (Halticinae) make a shot-hole effect on seedlings of Cruciferae, cotton, and other crops, frequently stunting and killing the seedling.

(i) Seedling or young plant wilting and dying (5) as a result of the underground stem being eaten: typical damage by temperate Cabbage Root Fly (*Delia radicum*), cutworms, and African Black Maize Beetles (*Heteronychus* spp.) on maize and sugar-cane seedlings (2).

(j) Woody seedling or young plant wilting, with earthtube up part of the stem; under the tube the bark is eaten away by termites (tropical), or ants.

(k) Seedling with hypocotyl gnawed at about, or below, ground level (4): by adult Pygmy Mangold Beetle, symphylids, wireworms (*Agriotes* etc.), and some other soil pests.

Damage to stored products

Damage to stored grains, seeds, and foodstuffs is of particular importance in that it occurs post-harvest, and in theory should be easy to prevent. On-farm storage is usually practised for a while after harvest, particularly on smaller farms, although this is more typical of the tropical parts of the world. In temperate regions it is becoming more usual for communal or commercial harvesting with immediate removal of produce to bulk stores or to market. After a possible period of on-farm storage, the produce is usually taken to a local or regional depot or silo, depending upon the type of crop. Once the produce is taken into warehouses and stores in the cities and towns it is no longer the purview of the agricultural entomologists, but usually the public health inspectorate.

In some parts of the world it is general practice to store root crops (potatoes especially, but also beets, carrots, parsnips, mangels, etc.) in earthen clamps at the edges of fields where they are covered with straw and earthed over to protect them from frost. However, it is now becoming more usual to keep roots and fruits overwinter in large cold stores sited strategically around the country. Stored potatoes and apples clearly have their own special pest problems, such as Potato Tuber Moth, aphids, Codling Moth, fruit flies, etc. In the general studies on stored products pests the produce is usually grain, dried pulses, nuts and the like.

Types of pests

The more important pests of stored products are a few moths in the family Pyralidae, a few mites (Acarina) and the rest are beetles belonging to several different families. In practical terms the most serious pests must include the synanthropic rats and mice, but they are excluded from the present study. On the basis of their feeding behaviour the pests can be categorized as follows.

(a) **Primary pests**: These insects are able to penetrate the intact test of grains and seeds, and include *Trogoderma*, *Sitophilus*, *Rhizopertha*, *Cryptolestes* and *Ephestia* spp.
(b) **Secondary pests**: These are only able to feed on grains already damaged by either primary pests or physically damaged during harvest; such as *Oryzaephilus* spp.
(c) **Fungus feeders**: A number of insects that are regularly found in infested stored products are actually feeding on the fungi growing on the moist produce; but a few species may be both fungus feeders and secondary pests, such as some Psocoptera.
(d) **Scavengers**: These are polyphagous, often omnivorous, casual or visiting pests (as distinct from the resident or permanent pests), such as cockroaches, crickets, some beetles, ants, etc.

Types of damage

(a) **Direct damage**: This is the most obvious and typical form of damage recorded, often measured as a direct weight loss or reduction in volume. But this is not accurate as there is an accumulation of frass, faecal matter, dead bodies, etc. All the insects and the rodents are responsible for such damage. In some flour mite infestations observed the final bulk of exuviae, faeces and dried dead bodies amounted to nearly 50% of the original food volume.
(b) **Selective eating**: Some insects show preference for the germ region of seeds and grains; thus a fairly low level of damage will severely impair germination of stored seeds, and in food grains there will be a serious loss of quality. This preference is shown by *Cryptolestes* and *Ephestia* larvae.
(c) **Heating of bulk grain**: Stagnant air becomes heated by insect metabolism and 'hot-spots' develop. The moisture from the insects' bodies condenses on the cool grains at the edge of the hot-spot, and the water causes caking, leads to fungal development, and may also cause some grains to germinate.

Fig.25. Damaged seedlings and sown seeds

1. Cereal seedling with 'dead-heart': caused by Wheat Bulb Fly larvae and other Anthomyiidae and Muscidae; also larvae of some grass and cereal flies (Opomyzidae, Chloropidae, etc.); sometimes larvae of lepidopterous stem borers (Pyralidae, Noctuidae, etc.).
2. Graminaceous seedling with stem eaten just below ground level: sometimes damage done by cutworms underground; usually by various scarab beetles, either adults or larvae. The illustration is actually showing damage to sugarcane sett by adult *Heteronychus* spp. Wireworms (Coleoptera, Elateridae) sometimes cause similar damage to a wide range of crop plants.
3. Bean (*Phaseolus* spp.) seedling with stem bored/galled by larvae of Bean Fly (*Ophiomyia phaseoli*, Agromyzidae).
4. Sugarbeet seedling with stem gnawed by adult Pygmy Mangold Beetle (*Atomaria linearis*); similar damage done by various beetles, either adult or larvae, including wireworms (*Agriotes* spp. etc.), etc.; also symphylids.
5. Seedling or young transplant wilting or dying; this is the main symptom of serious root or stem damage: typical of Cabbage Root Fly, and other Anthomyiidae; also cutworm and chafer grub damage; the uprooted seedling will show more detailed damage.
6. Seedling with stem cut through: damage typical of cutworms nocturnal surface feeding (Lepidoptera, Noctuidae); also done by crickets (*Acheta*, *Gryllus* spp. etc.; Orthoptera, Gryllidae), and by some weevils (surface weevils: *Tanymecus* spp.).
7. Sown seed and germinating seeds are dug up and eaten by many birds and by field mice (*Apodemus* spp. etc.), but seeds with cotyledons bored are usually attacked by larvae of bean seed flies (*Delia* spp., Anthomyiidae). The birds and rodents usually remove the entire seed and eat it, and signs of their attack may include footprints, scats, etc.

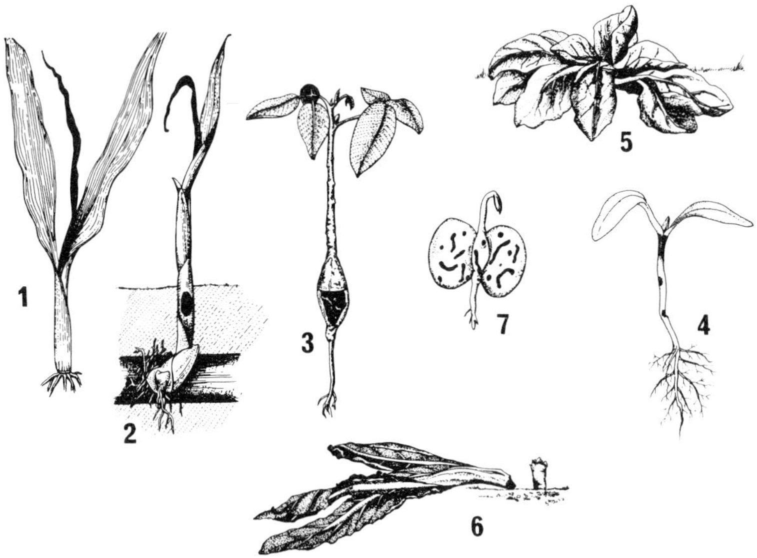

(d) **Webbing by moth larvae**: The pyralid larvae in stored products all produce silk-webbing, which if present in large quantities may clog machinery and otherwise be a nuisance.

(e) **Contamination**: The presence of insects in the produce, and dead bodies, exuviae, frass, faeces, etc., causes a general loss of quality and value; aesthetic rejection of produce becomes even more pronounced when prepacked in transparent wrappings for supermarket sale.

Cross-infestation

Possibly the single largest problem in produce storage is that of cross-infestation. It generally happens in two ways: firstly, the clean produce is brought into a store that is 'dirty' (i.e. already infested with insects and mites present in cracks and crevices and in rubbish and spilled grain). Secondly, the clean store containing uninfested produce receives a consignment of infested produce and the insects then spread into the previously uninfested material.

Some of the major regional produce stores in some countries, as well as many on-farm stores, are actually never properly 'clean', and in some stores the total pest spectrum in a miscellany of produce may be quite enormous.

Regular inspection

Stored grain should be inspected, and spear samples taken, weekly if insect attack is not to go unnoticed. In some stores ultra-violet light traps and pheromone traps are used for monitoring moth and some beetle populations. With practice trap catches can be correlated with insect populations in the stored produce. For bulk storage regular temperature monitoring is also required.

Reduction of damage to stored products

For the different ways in which damage to stored products may be minimized, and for pesticide application recommendations, see p.274.

6 *Biological control of crop pests*

The initial enthusiasm with which the first modern insecticides, the chlorinated hydrocarbons, were greeted in the late 1940s and early 1950s has long since begun to wane. This is, in part, the result of the development of resistance by so many pests to many of the pesticides, and also because of the growing awareness of the dangers involved in a gradual poisoning of the whole environment. One result is that many governments are now introducing legislation designed to curb indiscriminate use of the more toxic and persistent insecticides (mostly chlorinated hydrocarbons) and indeed to prohibit the use of the most dangerous pesticides. Another result of this awareness is the acknowledgement of the desirability of trying to control pests by use of cultural or biological means. The single most promising aspect of cultural control is the development and use of resistant varieties of crops. In some texts this is regarded as a biological method of control, but this view is not generally held, since whilst it involves the reactions of living organisms, it is normally regarded as a distinct branch of agricultural science.

The main advantages of biological control lie in:

(a) the absence of toxic effects;
(b) no development of resistance by the pests;
(c) no residues of poison in the soils and rivers, etc;
(d) no build-up of toxins in food chains;
(e) no killing of pollinators, or development of secondary pests through the destruction of their natural enemies;
(f) the permanence of successful biological control programmes where repeated application of chemicals would be required;
(g) the fact that biological control is self-adjusting and does not require the careful timing and organization which should be given to pesticide applications, and which often make it impracticable on small peasant holdings in underdeveloped areas.

Quite often natural control is effective for many pest populations most of the time, but when a population outbreak does occur and methods of control are required, care should be taken so that the *disturbance of natural control factors should be kept to a minimum*. Some of the ways in which this can be attained are:

(a) prediction of pest outbreaks so that insecticidal control can be well planned in advance and used only when necessary;
(b) use of selective insecticides – the more specific the insecticide the less damage it is likely to do to natural control;
(c) by avoiding 'blanket' treatments with insecticides as an insurance measure;
(d) the use of cultural or biological control measures wherever possible;
(e) the encouragement of the increase of natural enemies of the pest;
(f) the growing of pest-resistant crops where possible.

In this chapter biological control (BC) is viewed in the strict sense as the control (lowering) of pest populations by predators, parasites and disease-causing pathogens.

Natural control

First and foremost it must be stressed that it is not possible to overemphasize the importance of the natural control of insect populations. Under natural conditions most insect populations are regulated by a complex of predators, parasites and pathogens causing diseases, all sharing the same habitat and belonging to the same ecological community. This population regulation is sometimes referred to as the *balance of nature* or *natural control.* Under agricultural systems of extensive monoculture, conditions are created that may induce a pest population explosion, but often it is found that the natural enemies increase correspondingly, and so natural mortality rates for the pest remain high and the final pest population does not become very large after all. This is particularly true for perennial orchard and plantation crops, which are present for a long enough time to allow some stabilization of insect and mite populations. The time factor is critical in that parasite populations always lag behind the host populations in their rate of growth (increase), and a fairly lengthy time is required to reach community stability (that is some measure of pest population control). With annual crops, such as vegetables, cereals and cotton, there is generally insufficient time available for specific predator or parasite populations to build up to a level where they can exert any controlling effect, although non-specific (polyphagous) predators and parasites may be important in such situations. The spectacular control of greenhouse pests (red spider mites and whiteflies) by introduced predators and parasites on cucumbers and tomatoes in the UK, northern Europe and the USA is attributable to the greenhouses being very small (in effect tropical) and completely enclosed agroecosystems, and an unnaturally high predator or parasite population is introduced at the start of the control programme. However, although the natural enemies of pests on annual and short-lived crops may seldom completely control the pest populations, their controlling effect is always important. Many crop pests have spectacular fecundity; for example, some female noctuid moths (especially cutworms) lay as many as 2000 eggs per season, and the only reason that the world is not inundated with caterpillars is the high rate of natural mortality due largely to predators, parasites and disease. In temperate regions the fecundity of viviparous and parthenogenetic aphids is almost unbelievable!

Since natural control is present in all pest ecosystems at all times, it is difficult to appreciate the actual extent of its population-controlling effect as it is usually not really noticed until it breaks down. Indiscriminate use of highly toxic, non-specific, persistent pesticides has often resulted in the death of more natural enemies than crop pests. In point of fact, this is now the normal situation with most orchard infestations of scale insects (Coccoidea), in that the mature scales are notoriously difficult to kill with insecticides but the complex of tiny parasitic chalcid and braconid wasps are far more susceptible to the poisons. In these situations pesticide application (with chemicals such as DDT and dieldrin) is almost invariably followed by a population outbreak of the pest. Examples of this phenomenon are now legion!

A nice example of recognized natural control was recorded in northern Thailand where *Patanga succincta* is a serious maize pest; careful field studies revealed that the egg-pods in the soil were heavily parasitized by a scelionid wasp and preyed upon by *Mylabris* beetle larvae; insecticide spraying was carefully restricted to avoid disruption of this natural control. At the same time ducks were introduced into the fields and ate vast numbers of the nymphs, and local villagers were employed to walk through the crops and hand-collect the adults (most of which were later cooked and eaten!). This was a case of partial natural control supplemented by judicious biological control (*sensu stricta*), and with a minimal pesticide application a very high level of overall control was achieved.

Biological control

This is the deliberate introduction of predators, parasites, and/or pathogens into the pest/crop agroecosystem, and is designed to reduce the pest population to a level at which damage is not serious.

Two schematic representations of a pest population being controlled (economically) by natural enemies are shown (from Smith & van den Bosch in Kilgore & Doutt, 1967) in figs. 26 and 27.

The advantages of biological control are several, and have already been listed (see page 103). The predators and parasites may be local species whose natural popula-

Fig.26. Complete biological control (based on the economic criterion) of a pest by an introduced natural enemy. Note that it is not the economic threshold of the pest that is affected by the parasite, but rather its equilibrium position (e.g. long-term mean density) (after Smith & van den Bosch, 1967).

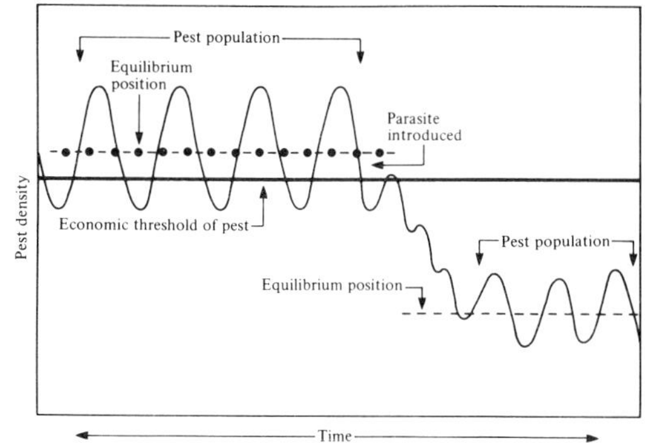

Fig.27. Three-dimensional representation of biological control of a pest species by an introduced natural enemy. The two populations depict the distributional status of the pest species and its density at particular points in time before and after parasite introduction (after Smith & van den Bosch, in Kilgore & Doutt, 1967).

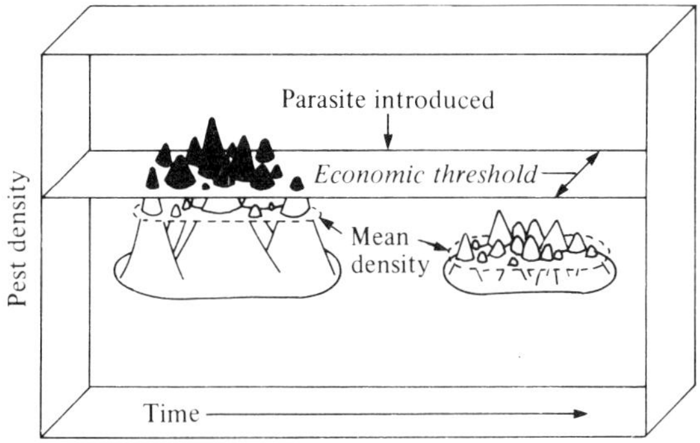

tion is being augmented by the introduction, or they may be exotic species from another country. The most successful examples of biological control are elegant in all respects, but unfortunately most pest situations are not amenable to this method of control. Suitable situations are more likely to be found in the tropics (and sub-tropics) where the climatic conditions favour continuous breeding of insect populations. To date, there are in the region of 200–300 cases of clear success of biological control (*sensu stricta*) that are well documented. The original work in this field was carried out in the USA in the early part of this century, mostly in California and Hawaii, to find means of controlling pests of introduced exotic fruits (*Citrus*, etc.). At the present time most work is being done by the various stations of the Commonwealth Institute of Biological Control (CIBC), which are situated in Africa, India, Pakistan, Switzerland, the W. Indies, and S. America. This is one of the institutes belonging to the Commonwealth Agricultural Bureaux, Slough, UK, and is a sister institute to the Commonwealth Institute of Entomology, London.

As already stated, most of the best examples of biological control are from the tropics, but there are many very nice examples from the sub-tropical regions of Florida and California in the USA, and from South China and the Mediterranean Region, all of which lie just to the north of the Tropic of Cancer, and some from Australia and New Zealand. In the cooler temperate regions of Canada, northern USA and northern Europe (including the UK) there is extensive biological control practised, but it is mostly confined to greenhouses and protected crops, although some forest defoliators have been controlled.

Most crops are indigenous to one particular region of the world, and have been widely transported by man to other regions with a suitable climate in relatively recent times. These exotic crops in their new locations may be attacked by indigenous insect pests quick to seize upon a new food source. Alternatively, their own pests may have either accompanied the crops or else followed soon after, and in the new situation their usual complement of natural enemies is absent, thus permitting a rapid population build-up. The local predators, parasites and pathogens will inevitably be in a state of delicate balance in their own environment and usually cannot be expected to exercise much control over the introduced pests.

The most successful cases of biological control are usually where the predator or parasite has been brought from the area of origin of the crop. For example, the most successful citrus scale parasites have all come from S. China where *Citrus* is indigenous. Usually specific parasites are more successful in stable habitats such as long-established orchards. If the habitat is unstable (as with many agricultural crops) then more general parasites are better suited for control as they can better survive the periods when the specific pest host is absent (between crops). An interesting case is recorded from Japan; here the rice stem borers are controlled partially by the use of *Trichogramma* spp. to parasitize the eggs. After harvest, when there are no more lepidopterous egg masses available, the wasps parasitize the eggs of flies of the genus *Sepedon* (Diptera: Sciomyzidae) in the weedy, fallow lands around the paddy fields, and so the parasite population is maintained (in a state of diapause) over winter. The fly larvae prey on aquatic snails of the genus *Lymnaea* and as such are normally part of the fauna of oriental rice paddy fields.

There is now an extensive literature dealing with biological control, in all its aspects including methodology, history, and giving many examples of successful and unsuccessful cases. The major publications include DeBach (1974), Delucchi (1976), Greathead (1971), Wilson (1960), and the various *Technical Bulletins* and *Reports* published by CIBC.

The information here presented is intended as a summary, including the more important points, such as basic principles, and with a few selected examples. In table 1 is listed some of the more successful biological control projects that have been carried out in temperate situations.

There is an interesting history of biological control practices against crop pests in China, indeed the use of some predacious insects in fruit orchards go back to the earliest recorded times. Much of this information has not previously been published internationally, and so in table 2 is listed a selection of the more successful biological control projects carried out in the cooler parts of China; these data were kindly made available by Dr Li, Li-ying in 1980.

Cost of a biological control programme

The cost of a successful introduction, giving long-term control, may be relatively low in comparison with most chemical applications, and very low in relation to the annual saving in terms of losses avoided. For example, DeBach (1964) estimated that the Department of Biological Control, University of California, over a period 1923–59 spent a total of US $3.6 million with a resulting annual saving in California of about US$100 million. A recent case by CIBC, Pakistan, was the establishment of *Apanteles ruficrus* (Hym., Braconidae) against the maize pest *Mythimna separata* (also attacking vegetables and other crops) in New Zealand in 1973–5. The cost of supplying the parasites from Pakistan was US$1700 and the costs involved in New Zealand amounted to about US$20000; the annual saving is estimated at about US$5 million, with the control still being successful. One oddity noticed was that only the Pakistan strain of parasite was particularly effective.

A similar case was the control of Winter Moth (*Operophtera brumata*) in Canada. It was accidentally introduced into Nova Scotia in the 1930s, but first noticed in 1949; it eventually increased in numbers and spread inland and the damage to deciduous hardwoods (mostly oak, but this is a polyphagous pest) soon reached economic proportions. Parasites from Europe were imported and effectively colonized over the period 1955–60 (see table 1) bringing the Winter Moth under complete population control. The overall total cost of the project was estimated at about $160000 and the annual saving to the hardwood industry was estimated at $2 million.

Table 1: *Some successful biological control projects (worldwide).*

Pest	Crop	Predator or parasite	Country	Date	Estimated annual value (US $)
Mythimna separata (Noctuidae)	Maize	*Apanteles ruficrus* (Hym., Braconidae)	New Zealand	1973–5	US $5 mill.
Phthorimaea operculella	Potato	*Apanteles subandrinus*	Zambia	1968	US $70000
Chrysomphalus ficus	Citrus	*Aphytis holoxanthus* (Hym., Aphelinidae)	Israel	1965–7	US $1 mill.
Pseudococcus citriculus	Citrus	*Clausenia purpurea* (Hym., Encyrtidae)	Israel	1940	(Many)
Eriosoma lanigerum	Apple	*Aphelinus mali* (Hym., Aphelinidae)	Australia	1923–36	(Many)
Saissetia oleae	Fruit trees	Various Chalcidoidea	Australia	1898–1905	(Many)
Operophtera brumata	Deciduous trees	*Cyzenis albicans* (Dipt., Tachinidae) *Agrypon flaveolatum* (Hym., Ichneumonidae)	Canada	1955–60	US $2 mill.
Chromaphis juglandicola	Walnut	*Trioxys pallidus* (Hym., Braconidae)	USA (California)	1959 & 1968	(Many)
		Regular biological control practices in temperate greenhouses			
Trialurodes vaporariorum	Glasshouse crops	*Encarsia formosa* (Hym., Aphelinidae)	UK, Europe, N. America, Asia, etc.	1930–	
Tetranychus urticae	Glasshouse crops	*Phytoseiulus persimilis* (Acarina, Phytoseiidae)	Europe, etc.	1960–	
Myzus persicae	Glasshouse crops	*Aphidius matricariae* (Hym., Braconidae)	UK, etc.	1970–	

See also table 3, for some of the B–C agents commercially available now, all of which are obviously economically successful.

These levels of cost in relation to savings are typical of the more spectacular biological control successes, but the less-successful projects are still quite inexpensive and the whole problem of pesticide residues and side-effects is avoided. The control of glasshouse pests using predators and parasites compares favourably with the use of chemicals, and there are no toxicity hazards.

Table 2: *Biological control of some insect pests in China* (Li, Li-ying, 1980; *in litt.*)

Pest	Crop	Predator, parasite or pathogen	Date	Locality	Success
1. *Pseudococcus* spp.	Tung	*Cryptolaemus montrouzieri* (Col., Coccinellidae)	1955–1979	Guangdong	Yes
2. *Aphis* spp.	Cotton, Wheat, etc.	*Coccinella septempunctata* (Col., Coccinellidae)	1971–	Henan	Yes
3. *Heliothis armigera*	Cotton	*Trichogramma* spp. (3)	1970–	Hubei, Shangsi	Yes
4. Corn borers	Maize	*Trichogramma* spp. (3)	1970–	N. China	Yes
5. *Grapholitha glycinivorella*	Soybean	*Trichogramma* spp. (3)	1970–	N. China	Yes
6. *Eriosoma lanigerum*	Apple	*Aphelinus mali* (Hym., Aphelinidae)	1955	Shangdon	Yes
7. *Pieris brassicae*	Cabbages	*Pteromalus puparum* (Hym., Pteromalidae)	1974–	Anhui, Guangdong	Yes
8. Cotton pest complex	Cotton	*Vespa, Polistes* spp. (Hym., Vespidae)	1975–	Henan, Jiansu	In progress
9. Various borers	Sugarcane	*Tetramorium guineense* (Hym., Myrmecinae)	1961–	Guangdong, Fujian	Yes
10. Various pests	Cotton	*Chrysopa* spp. (Neuroptera, Chrysopidae)	1975	Henan	Limited
11. *Panonychus citri*	Citrus	{ *Amblyseius newsami* *Typhlodromus pyri* (Acarina, Phytoseiidae) }	1975– 1975–	Guangdong Sichuan	Yes Yes
12. Various pests	Rice, Cotton	{ Various Argiopidae, Lycosidae, Tetragnathidae, Micryphantidae, Clubionidae, etc. (Arachnida, Araneida) }	1975–	Hunan, Zhejiang	Yes
13. Corn borers *Grapholitha glycinivorella*	Maize Soybean }	*Beauveria bassiana* (fungi)	1965–	most provinces of China	Yes Yes
14. *Trialeurodes vaporariorum*	Greenhouse vegetables	*Encarsia formosa* (Hym., Aphelinidae)	1979–	Beijing	In progress
15. Lepidopterous pests	Vegetables, rice, cotton	*Bacillus thuringiensis*	1970–	Most provinces of China	Yes

Commercial availability of biological control agents

In the last few years many commercial establishments have been incorporated and they breed and supply predators, parasites and pathogens to commercial growers and farmers. The pioneer work in this field was usually carried out on Government research stations, the various CIBC stations throughout the world, and other autonomous establishments, such as GCRI in the UK. But now bulk production is in the hands of commercial companies.

Most of the predators tend to be non-specific and so may be used against a number of different crop pests, or against a particular pest complex. Lacewings, ladybird beetles, wasps and spiders belong in this category.

The parasites are usually quite specific, but they sometimes accept hosts belonging to the same family, so some *Trichogramma* species will parasitize a wide range of lepidopterous eggs, and *Nasonia* (Hym., Pteromalidae) will accept the pupae of many different Muscoidea. It is more usual though for the parasites (Hymenoptera and Tachinidae) to be quite host specific.

Some of the biological control agents currently available from commercial suppliers in temperate regions are listed in table 3. However, it must be stressed that the situation differs in each country and so the local Ministry of Agriculture staff should be consulted for local information.

Fortuitous biological control

It should perhaps be mentioned, as pointed out by DeBach (1971), that there has been considerable, unrecognized, fortuitous biological control resulting from ecesis (accidental dispersal and establishment) of natural enemies throughout the world. This is particularly the case with some scales (especially Diaspididae) and their parasites, because the scales are tiny, sessile, inconspicuous (especially in low numbers) and easily transported on either fruits or planting material (shoots and rootstocks). DeBach uses the specific parasite, *Aphytis lepidosaphes*, of the Purple Scale (*Lepidosaphes beckii*) as a good example. Purple Scale is indigenous to the area of S. China and Indo-China, and it is highly specific to *Citrus*. During the past 100 years or so Purple Scale has gradually (accidentally) spread and invaded most of the major *Citrus*-growing areas of the world, from the Mediterranean to S. Africa, Australia, N., C. and S. America, and the W. Indies. DeBach records that *Aphytis lepidosaphes* (first discovered in China in 1949) is slowly spreading around the world, although the only deliberate introduction was into California in 1949; apparently, at that time it was (as discovered later) already established in Hawaii, presumably as a result of an earlier accidental introduction. The most recent colonizations by this specific parasite include Spain (1969) and Argentina (1970). In almost every country checked by DeBach, *Aphytis lepidosaphes* is responsible for substantial to complete biological control of Purple Scale on *Citrus*.

Predators

The importance of insect pest predators is now being reassessed in many countries because of their obvious value in IPM programmes.

The value of wild birds as insect predators is clearly demonstrated in urban situations, where tits (Paridae) and sparrows (Ploceidae) can be seen searching fruit trees, roses, etc., for aphids, caterpillars, and other insects, and thrushes (Turdidae) eat large numbers of slugs and snails. Domesticated birds are being used to control insects in parts of the tropics very successfully. For many years peasant farmers in Africa have been using chickens on their shambas in cotton plots to eat the cotton stainers and other bugs that drop to the ground (as an escape mechanism) when disturbed. About 40 chickens per acre is the accepted number of birds required for adequate control. Recently in S. China (and other parts of India and S.E. Asia) effective use of ducks has been made against rice pests. In Guangdong on the early rice crop, after transplanting and establishment of the plants, 220 000 ducklings (about half-grown) were herded slowly through the paddy fields. Each duckling was estimated to eat about

Table 3: *Some of the biological control agents commercially available in the UK, Europe, and N. America*

Organism	Description	Against	Supplied as:
Predators			
Chrysopa spp.	Lacewings (Neuroptera)	General predators	Larvae or pupae
Cryptolaemus spp.	Ladybird beetles (Col., Coccinellidae)	Mealybugs, etc.	Active adults
Amblyseius spp.	Predacious mites (Acarina, Phytoseiidae)	Thrips, etc.	Active mites
Phytoseiulus persimilis		Red Spider Mites	Active mites
Parasites			
Trichogramma spp.	Egg parasites (Hym., Trichogrammatidae)	Lepidoptera	Pupae in parasitized eggs
Encarsia formosa	Nymphal parasite (Hym., Aphelinidae)	Glasshouse Whitefly	Pupae in parasitized 'scales'
Diglyphus isaea	(Hym., Eulophidae)	Chrysanthemum–Tomato	Active adults
Dacnusa sibirica/Opius pallipes	(Hym., Braconidae)	Leaf Miners	Active adults, or parasitized pupae
Aphidius matricariae	(Hym., Braconidae)	*Myzus persicae*	Parasitized aphid mummies
Micro-organisms			
Bacillus thuringiensis	Bacterium	Caterpillars	Suspension
Bacillus spp.	Bacteria	Beetle larvae	Suspension
Verticillium lecanii	Fungus	Aphids Glasshouse Whitefly	Spore suspension as a w.p.
Beauveria bassiana	Fungus	Caterpillars	
Granulosis viruses/nuclear polyhedrosis viruses	Viruses	Caterpillars & some beetle larvae, sawflies	Viral bodies in a w.p.

A few of the best-known suppliers of biological control agents are listed below:

UK

Applied Horticulture Ltd., Billinghurst, Sussex
Bunting & Sons, Colchester, Essex
Natural Pest Control, Bognor Regis, W. Sussex

USA

Fairfax Biological Labs.
Nutrilite Products Inc.
Biological Control Supplies
Rincon–Vitova Insectaries Inc., California
International Minerals Chemical Corp.

(It must be stressed that these are just a few of the companies concerned.)

200 insects per hour, and most of these were inevitably pests. The overall effect of this predation was that the amount of chemical insecticides required on the early rice crop was reduced from 77000 kg in 1973 to 6700 kg in 1975. Following this early success, ducks are now being used as rice pest predators in many parts of the Far East. In Thailand, in 1980, adult ducks were used successfully to eat adults and large nymphs of the Bombay Locust (*Patanga succincta*) that had been enticed out of the upland maize crop into intercropped soybean foliage and also groundnuts. In this case ducklings would have been useless as the locusts were too large for them to kill. At the same time local villagers were paid to hand-collect adult locusts, and in a few days 80 tons were collected; the locusts were later either roasted and eaten by the villagers or made into a paste for culinary purposes. (There are times when man can be a direct insect predator of some consequence!)

Within the Vertebrata one amphibian has been used a few times for biological control of insects; this is the giant toad (*Bufo marinus*), which was introduced into Hawaii in 1932–3 in an attempt to control various beetles (Scarabaeidae) attacking sugarcane, and was released in Australia (Queensland) in 1935–6. It is now flourishing in both locations where it generally seems to cause more ecological disturbance than benefit; this is in part because it eats small vertebrates as well as insects, and the toads are poisonous if eaten by other vertebrates.

It should perhaps be mentioned that the main reason for the tremendous upsurge of rats (*Rattus* spp.) as urban and agricultural crop pests in India and throughout S.E. Asia is largely due to the destruction of their main predators which are snakes, especially the venomous cobra (*Naja naja*; Elapidae), and the non-venomous, but fanged and aggressive, rat snakes (*Ptyas* spp., *Elaphe* spp., etc., Colubridae). For inexplicable reasons most peasants and farmers in Africa and tropical Asia have a morbid dread of all snakes, most lizards, and salamanders, being convinced that they are all highly venomous and lethal. The consequence is, regrettably, the wholesale and widespread destruction of local snake populations, with a subsequent population explosion of suburban rats. The situation is somewhat aggravated in S.E. Asia as it is the endemic source of the frugivorous arboreal *Rattus rattus* subspecies group. It is estimated by FAO that in India there are five rats per person, and that in recent years 10% of the grain harvests have been lost due to rat attack. In S.E. Asia the main pests of oil palm (and in some areas coconut) are rats; in the Philippines the crops damaged also include sugarcane and maize.

It has long been obvious that spiders were an important part of the natural control community throughout the world, but only recently have they been studied in this role. Work at IRRI in the Philippines and in S. China has shown that many rice pests (especially Cicadellidae and Delphacidae) are heavily predated by spiders. In some experiments mortality rates of the Brown Planthopper of Rice reached 70% due almost entirely to spider predation. In parts of China (Hunan, Zhejiang), since 1975, spiders belonging to more than five different families have been reared and released on rice and cotton crops, against a broad range of insect pests, with considerable success (Li, Li-ying, *in litt.*, see table 2). Both the web-spinning and the webless wolf spiders have been used. A recent publication on spiders as biological control agents is by Riechert & Lockley (1984). At IRRI it was discovered that γ-BHC was particularly toxic to spiders, both directly by contact and indirectly by their eating poisoned prey; obviously such insecticides should be avoided in any IPM programme where spiders are important natural predators.

Also in the class Arachnida are the mites (order Acarina). In the same way that insects are heavily preyed upon by other insects, the major predators of phytophagous mites are carnivorous mites, most of which belong to the family Phytoseiidae. The three main genera used in biological control programmes are *Phytoseiulus*, *Amblyseius*, and *Typhlodromus*. The original widespread use of these predatory mites against spider mites (Tetranychidae) was in greenhouses in the UK and the USA (on cucurbits, tomatoes, etc.) with spectacular success. Later it was realized that on field and orchard crops in warmer regions they could also be successful. Various companies now provide commercial sale of phytoseiid mites for biological control purposes.

Predacious insects are important population controlling factors for many insect pests, but their precise roles in natural control have seldom been evaluated and documented. It was, however, reported from the Solomon Islands that, in 1975 and 1976, the Brown Planthopper of Rice was not a pest on rice (economic pest, that is) because of good control by a predacious mirid bug (*Cyrtorhinus*). Many members of the Miridae, Anthocoridae, Reduviidae, Pentatomidae, and other Heteroptera, are important natural predators. *Orius minutus* has been used in China since 1976 for the control of various cotton pests in Hubei and Jiansu with some success. *Platymeris laevicollis* (Reduviidae) is an important predator of Rhinoceros Beetle adults (*Oryctes* spp.) and has been introduced from Zanzibar into India, Fiji and New Caledonia (Lever, 1969); each adult bug can destroy one adult beetle per day and may live for four months.

In the Diptera there are some Cecidomyiidae that prey on Coccoidea, and there are a number of rather obscure families, in addition to the obvious Asilidae (robber flies), where the adults prey on other insects. But without doubt the most important group is the Syrphidae (hover flies) in which many species have predacious larvae. They feed on aphids, coccids, other small plant bugs and small caterpillars. Being legless, headless and blind it is rather surprising that syrphid larvae are effective predators, but the eggs are laid amidst dense populations of suitable prey so the emerging larvae manage to feed effectively with the aid of their mouth-hooks and rather simple sensilla (chemosensory and tactile). Although very important in natural control, Syrphidae are not much used in biological control programmes as their limited powers of searching make them not really suitable. One oddity is that in the UK they have been recorded as a nuisance in commercial pea (*Pisum sativa*) crops; the pea-viner also threshes and when the fresh peas (seeds) are removed from the pods and collected in the hopper they are accompanied by the small globular syrphid pupae which were dislodged from the foliage. Customers buying a packet of frozen peas are invariably annoyed to find syrphid pupae enclosed.

The Hymenoptera have a few predacious groups of importance. The Vespidae, the social wasps, are all carnivorous and feed upon other insects. In most parts of the world *Vespa* and *Polistes* are regarded as pests because they damage ripe fruits, and also they nest in crops and attack the field-workers if disturbed. But in China (and some other countries) they are being used as general predators in cotton crops. Similarly ants (Formicidae) include carnivorous species that will prey on crop pests, and *Oecophylla smaragdina* (Red Tree Ant) is successfully controlling *Rhynchocoris humeralis* on *Citrus* in Guangdong, China; this method has in fact been used in China since ancient times. The African Red Tree Ant (*O. longinoda*), when nesting in a coconut palm, keeps the palm free of Coconut Bug. But these ants sometimes guard bug colonies, and as they are quite aggressive they may attack field-workers. So, like the wasps, (*Vespa* and *Polistes*) it is a moot point whether they are more important as pests or as beneficial predators. Scoliidae prey on beetle larvae, mostly chafer grubs, in the soil, but attempts to use them for the control of *Oryctes* spp. in S.E. Asia have not been very successful.

The Neuroptera as an order are almost entirely predacious, both as adults and larvae, and clearly they are an important group in the natural control of many insect species. Green lacewings (*Chrysopa* spp.) are quite easy to rear, and are now available commercially for biological control use. As general (i.e. non-specific) predators they are useful in IPM programmes, but will only give limited control of any particular pest.

Coleoptera include many important groups of predators. In some families, both adults and larvae are fiercely predacious (e.g. Carabidae, Staphylinidae, Cicindelidae, Histeridae and Coccinellidae), but in others only the larvae are (Hydrophilidae, Meloidae, Lampyridae). The Carabidae include the Indian *Pheropsophus hilaris* which has been successfully used in Mauritius to kill larvae of the Rhinoceros Beetle. Most of the predacious beetles are litter and soil-dwellers and so are mostly used against soil-inhabiting pests, for obvious reasons. The major exception to this is, however, the Coccinellidae (ladybird beetles) which are arboreal foliage dwellers. Apart from *Epilachna*, the entire family are predators and very effective control agents, both as adults and larvae. They are used mainly against aphids and coccoids, and also kill small caterpillars. Ladybirds are easy to rear and can be bought commercially in large numbers from many establishments. Successful examples of their use in biological control are too numerous to list, but a few are included in the tables. Fig. 28 shows the effect of coccinellid predators on *Icerya purchasi* in California in 1868, one of the earliest documented cases of biological control.

Fig.28. Schematic graph of the fluctuations in population density of the Cottony Cushion Scale (*Icerya purchasi*) on citrus from the time of its introduction into California in 1868. Following the successful introduction of two of its natural enemies in 1888, this scale was reduced to noneconomic status except for a local resurgence produced by DDT treatments (from Stern *et al.*, 1959).

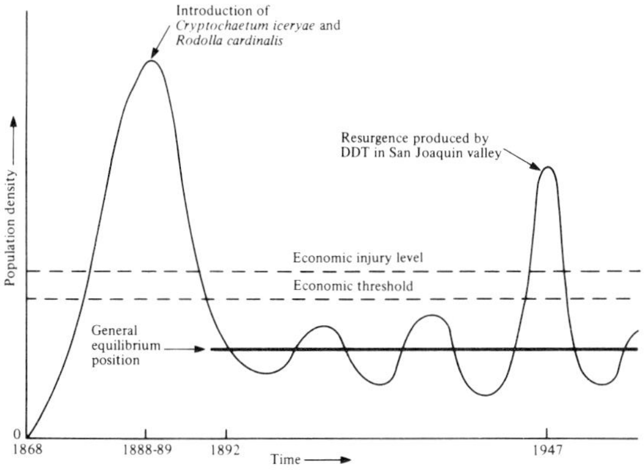

The Meloidae are important in the natural control of many locusts and grasshoppers; the triungulin larvae seek out and prey upon the egg-pods of Acrididae in the soil, and in parts of the tropics their value is considerable. As already mentioned *Mylabris* larvae exert considerable control pressure on the Bombay Locust in N. Thailand.

Soil insects, especially root maggots (Anthomyiidae) and cutworms, suffer very heavy natural predation by ground beetles (Carabidae) and rove beetles (Staphylinidae). Cabbage Root Fly regularly suffer 90–95% loss of eggs and larvae due to beetle predation (Coaker & Finch, 1970), about two-thirds of the loss being in the egg stage. The most important Carabidae are *Bembidion*, *Ferronia*, *Harpalus* and *Trechus*, all of which are large and abundant genera.

Parasites

The three outstanding groups of insect parasites are the Tachinidae (Diptera), and the Chalcidoidea and Ichneumonoidea (both Hymenoptera). They are vitally important, both in the natural control and biological control of insect crop pests. Within the Diptera are families such as Pipunculidae, Phoridae, Bombylinidae and Sarcophagidae which are of minor importance as natural parasites of insect pests. There are also some anomalous groups like the Sciomyzidae which live in oriental paddy fields and whose eggs provide alternative hosts for *Trichogramma* wasps during the period when stalk borers (*Chilo*, *Tryporyza*, etc.) are absent.

The Tachinidae parasitize a very large range of hosts worldwide, including the larvae of lepidopterans, the larvae and adults of coleopterans, and nymphs and adults of orthopterans and hemipterans.

Ichneumonidae and Braconidae parasitize a wide range of insects, ranging from the wood-boring larvae of Siricidae, to cereal stem borers, aphids and scales. Some species have long ovipositors and are able to parasitize larvae *in situ* in wood or cereal stems. Probably the most important genus is *Apanteles* (Braconidae) with its many species, which parasitize the larvae of Noctuidae and

Pyralidae. These species are generally not difficult to rear, and some are bred regularly at various CIBC stations.

The Chalcidoidea includes some 19 families, most of which (excluding the Agaonidae – fig wasps) are parasites of caterpillars, beetle larvae, fly larvae and pupae, and large numbers feed on aphids, scales, mealybugs, psyllids and other Hemiptera. A few species parasitize Orthoptera, spiders, ticks, and other insects. Several groups, especially the Trichogrammatidae, are solely egg parasites and use the eggs of Hemiptera, Lepidoptera, and some Diptera, Orthoptera and others. Species of *Trichogramma* are being widely used in many IPM programmes in different parts of the tropics, and several species are available commercially in the USA. Some species of *Trichogramma* are especially useful in their being polyphagous and parasitizing a range of similar-sized eggs. It should be mentioned that Scelionidae and Proctotrupidae are also important as egg parasites, largely of Lepidoptera; they are Parasitica but not Chalcidoidea. The total complex of parasitic Hymenoptera (including hyperparasites) that can be reared from scale insect and aphid colonies is at times quite bewildering. The most important chalcid families for biological control purposes and their insect hosts are as follows: Chalcididae (pupae of Lepidoptera and Diptera mostly); Pteromalidae (many different insects, some species are polyphagous); Encyrtidae (Coccoidea, also aphids, psyllids, some Diptera, Lepidoptera, Coleoptera; some species are hyperparasites); Eulophidae (eggs, larvae and pupae of many insects); Aphelinidae (Coccoidea, Aleyrodidae, Aphidoidea, some Psyllidae, Cicadoidea and eggs of Orthoptera and Lepidoptera); Trichogrammatidae and Mymaridae (eggs of many different groups).

The Strepsiptera are a small order of the Insecta in which the larvae and adult females are obligatory internal parasites of some bees and Hemiptera, and a few species are regarded as important in the natural control of some heteropteran bugs.

A few species of Nematoda are entomophagous. Species of nematodes, including *Romanomermis culicivorax*, are being used very successfully in controlling populations of mosquito larvae in ponds and swamps. The nematode–bacterium complex of *Neoaplectana carpocapsae* and *Achromobacter nematophilus* (known as strain DD-136 and available commercially as 'Biotrol NCS' from Nutrilite Products Inc., USA) appears to be effective against some rice stem borers (larvae and sometimes pupae), for example *Chilo*, *Tryporyza* and *Sesamia*, and seems to be a very promising biotic insecticide for use against many caterpillars and some beetle larvae (such as Colorado Beetle).

In Australia commercial use of two genera of rhabditoid nematodes has recently started (Bedding, 1984) with considerable success. *Neoaplectana bibionis* almost totally eradicated the stem borer *Synanthedon salmachus* from 0.5 million cuttings of blackcurrant used to establish new plantations. *Heterorhabditis heliothidis* is reported as successfully controlling Black Vine Weevil larvae (*Otiorhynchus sulcatus*) in potted plants, and in strawberry plantations. And *Neoaplectana feltiae* (= *carpocapsae*) is said to control the carpenterworm, *Prionoxystus robinae*, in trees of *Ficus carica*. In this paper a method of rearing insect-parasitic nematodes on a large scale is described, as also are methods of storage and transport; studies such as this will enable commercial production techniques for nematodes to be refined and costs lowered so that this method of insect pest control may be economically comparable to other present techniques.

A very interesting account, titled *Nematodes with potential for biological control of insects and weeds* was written by Nickle (1981) and published in *Biological control in crop production* (Papavizas, editor). A summary of his comments follows:

Host insects	Nematode parasites
Grasshoppers (Orthoptera)	*Mermis* spp.
	Agamermis spp.
Beetles	*Psammomermis* spp.
(Coleoptera: Scarabaeidae)	*Howardula* spp.
(Curculionidae)	*Hexamermis* spp.
Lepidoptera	*Neoaplectana* spp.
	Amphimermis spp.
	Hexamermis spp.
Diptera	*Heterotylenchus* spp.
	Howardula spp.
(Mosquito larvae, *Simulium*, *Culicoides*)	*Romanomermis* spp.; *Mermis* spp.
Homoptera (Cicadellidae)	*Agamermis* spp.

In some cases the nematodes did not kill the insect, but sterilized the females so that they lived and spread the parasites but did not breed.

Pathogens

Fungi. Many fungal antibiotics have been in commercial production for use against plant diseases for a long time, and they have generally been quite effective. It is well known that insects (and some mites) regularly suffer epizootics from fungal attack, but it is only in the last decade that the development of entomopathogenic fungi for insect pest control has been seriously contemplated, and now it is evident that the use of fungi for pest control is both feasible and should be profitable in all respects. A paper by Soper & Ward in Papavizas (1981) titled, *Production, formulation, and application of fungi for insect control* presents a summary of work on this topic. It is estimated that there are 750 species of entomopathogenic fungi, but at present only a few are being used to any extent. However it is thought that many more will be developed commercially in the not too distant future, now that the principles of epizootiology are better understood.

It was formerly thought that the fungi were unreliable pathogens in that they were too dependent upon environmental conditions, especially moisture, to be satisfactorily manipulated for pest control. For the past few years a fungus has been used to control greenhouse aphids and whitefly in the UK, very successfully, but this has been in the confines of greenhouses. However, it does appear that fungal control on field crops is likely to be quite feasible.

A few of the more important parasitic fungi are listed below:

Fungus	Host insect/mite	Country
Aschersonia spp.	Aleyrodidae; scales	USSR
Beauveria bassiana (=*globulifera*)	Colorado Beetle	USSR, USA
	Corn borers	China
	Chinch Bug	USA
Conidiobolus obscurus	Aphids	UK, USA, France
Culicinomyces clavosporus	Mosquitoes	Australia
Entomophaga grylli	Grasshoppers/locusts	USA
Entomophthora musci	Muscoid flies	Europe
Hirsutella thompsoni	Citrus Rust Mite	USA
Lagenidium giganteum	Mosquitoes	USA
Metarhizium anisopliae	Cercopidae	Brazil
	Mosquitoes	USA
	Field crickets	Australia
	Rhinoceros Beetle	S. Pacific
Nomuraea rileyi	Caterpillars	USA
Verticilium lecanii	Greenhouse aphids/whitefly	UK
Zoophthora radicans	Spruce Budworm	USA

Other entomophagous fungi recorded in UK greenhouses include *Cephalosporium aphidicola* and *Entomophthora coronata*, and they show some potential for use in control programmes.

Bacteria. There are four species of *Bacillus* that are regarded as being entomophagous, but only the one being produced commercially *in vitro*. The four species are as follows:

Bacterium	*Insect host*
Bacillus thuringiensis	Caterpillars (Lepidoptera)
B. popilliae	Beetle larvae (Scarabaeidae)
B. moritai	Some caterpillars
B. sphaericus	

B. thuringiensis is now produced commercially *in vitro* and is available under a number of different trade names in different countries. In this book it is included in chapter 8 as a biological insecticide. This bacterium produces several different toxins which are entomocidal, and it appears to occur as several distinct strains, making it a very complex species. This basic variability may account for some of the variation in results when used to control caterpillars on crops. It is often found to work well in laboratory testing, but to produce very variable results in field trials. The commercial preparations are generally short-lived, and so repeated applications may be necessary. However, with the rapidly developing resistance shown by some caterpillars (especially Diamond-back Moth), BTH is very useful insecticide against cabbage caterpillars, and others, as part of a control programme.

B. popilliae causes Milky Disease and is very effective in controlling Japanese Beetle. It has been shown to be very persistent and to give long-term control, but to date has only been produced *in vivo* in *Popillia japonica* larvae and cannot be produced *in vitro*, which means that commercial production is very limited, although it is marketed in the USA under the trade name 'Doom' by Fairfax Biological Labs.

Three important publications of relevance here are Dulmage in Papavizas (1981) *Production of bacteria for biological control of insects*, Falcon in Huffaker (1971), chapter 15, *Microbial control as a tool in integrated control programmes*, and Ferron (1978) on entomogenous fungi.

Viruses. The two main types of entomophagous viruses are those described as granulosis viruses (including Codling Moth GV, Summer Fruit Tortrix GV and Cabbage White Butterfly GV) and the polyhedrosis viruses. *Heliothis* PHV has already been mentioned as a biological insecticide (page 62) as it is now widely available commercially. Other PHVs include ones specific to *Neodiprion* larvae (sawfly), *Trichoplusia* and *Spodoptera/Prodenia* larvae. These latter viruses are still undergoing trials, and at present it is not known if they will become generally available.

A natural PHV occurs on Cotton Semi-looper (*Anomis flava*) in Queensland, Australia, and is an important population-controlling factor; at times the proportion of the looper population which was infected was as high as 50%. It was suggested by Blood & Bishop (1975), as might be expected with an epizootic, that a certain minimum population of larvae (population abundance threshold) is required for a disease outbreak proper, and they calculated from two seasons' data, using regression analysis, that for *A. flava* the population abundance threshold might be about 130000 (±10000) larvae per hectare before a significant disease outbreak would occur.

A review of recorded cases of insects with virus diseases was published by Hughes (1957).

Protozoa. In the Beltsville symposium published in 1981 (Papavizas, editor) Canning discussed *Insect control with Protozoa*, and stressed that the only groups likely to be of use in pest control were the Coccidia, Neogregarines and Microsporidia. She refers to a model developed to investigate the eradication of pest populations by microbial agents (Anderson & May, 1981); the parasite parameters in the model are pathogenicity, spore survival time, and transmission rate. Microsporidia are seldom quickly lethal, but have higher mortality rates with older larvae and pupae. Spores are not very resistant and seldom survive long in the environment, but some are transmitted transovarially which is highly desirable in a biological control agent.

The effectiveness of a protozoan parasite on insects can be seen by the well-publicized damage done to honey bee colonies by *Nosema apis*. A few of the more notable protozoan insect parasites are listed below:

Protozoan	*Host insect*
Nosema operophterae	Winter Moth (*Operophtera brumata*)
Nosema sp.	
Pleistophora operophterae	
Nosema locustae	Grasshoppers/locusts

Biological control failures

In the past many of the biological control failures appear to have resulted from careless handling of the predators and parasites in shipment, together with accidental delays in transit and distribution, and were not due to intrinsic factors in the host/parasite biology. However, simultaneously, many introductions were made without any real scientific basis for the choice of predator or parasite, except perhaps convenience! Thus, there was often little success achieved. For example, the history of biological control attempts in Australia (F. Wilson, 1960) against insect pests reveals a large number of unsuccessful cases, and a similar situation prevailed in Africa (Greathead, 1971; see Hill, 1975 & 1983).

Firstly, the identity of the pest must be clearly established, and the taxonomists need to be made aware of any biological differences, as well as morphological or anatomical differences between various pest species (e.g. *Planococcus kenyae*). Secondly, most serious pests are not indigenous to the area in which they are pests, and so it is necessary to go back to their country of origin for suitable predators and/or parasites for possible introduction (e.g. *Planococcus kenyae*). Thirdly, it is generally a waste of time to introduce parasites or predators which usually attack another genus of pest, even though the two pests may have very similar biologies (e.g. *Scyphophorus*). It must be borne in mind, however, that generally predators are less specific in their choice of prey than are parasites. Parasites of closely related species of pests from different parts of the world may be worthwhile introducing sometimes, but more often parasites are species-specific and will not attack the new host (e.g. *Leucoptera* spp. and *Mirax*). Fourthly, when the pest is indigenous to the area in which it is a nuisance, it is generally not worthwhile to attempt to breed-up large numbers of one of the local parasites for local release since the usual pest/parasite complex is more often a stable and long-established one which will not permit changes of this nature. Also, if the parasite complex is extensive, there is little chance of being able to successfully insert another species into the complex, even if it comes from another area altogether (e.g. *Leucoptera* spp.), although sometimes this is successful. Generally it is best to find out if there are any predator/parasite niches which are not filled and then attempt to introduce a species that will effectively fill that niche.

Fifthly, when making the original survey of the pest parasite complex, it is vital to do the job thoroughly to make certain that all the local parasites are found. Should an introduction be made of a parasite species which is already present in the area, but not detected owing to careless surveying, then this is likely to have little effect and prove to be a waste of effort (e.g. *Hypothenemus hampei* in Kenya).

One of the problems associated with early attempts at the introduction of parasites from other countries was the time taken for consignments of insects to be shipped from country to country, and many consignments were found to be dead on arrival. Nowadays, with rapid and regular airline services, such consignments of insects can be sent either by air-mail or air freight and arrive at their destination within only a few days.

The facilities required for mass propagation of parasites or predators are not unduly complicated but are generally beyond the scope of most agricultural and entomological research stations. However, in recent years many research stations have become equipped with such facilities. In addition there is the chain of stations and substations of CIBC scattered throughout the world. The result of these developments is that now it is generally possible to obtain suitable predator or parasite species in very large numbers, bred either in the country of origin or in the country of introduction, so that tens or hundreds of thousands of individuals may be released instead of only a few hundred, which was all that was often possible in the past. Clearly the larger the number of individuals that can be released, the greater the chances of the introduction being successful.

As pointed out by F. Wilson (1960), it has been customary to characterize attempts at biological control as either 'successful' or 'unsuccessful'. In reality this oversimplification is undesirable for it is fairly obvious that many different levels of biological control are achieved. For the sake of definition it can be said that a biological control project is 'completely successful' if other forms of control can then be dispensed with. In the oversimplified view the project is regarded as 'unsuccessful' if other methods of control still have to be utilized. This attitude is totally unrealistic; chemical applications normally give many different levels of control, as is only to be expected. Wilson summarizes attempts at biological control of insect pests in Australia by making five categories based on the levels of success achieved:

(a) pests substantially reduced in status;
(b) pests reduced in status;
(c) pests of doubtfully diminished status;
(d) pests of unchanged status;
(e) pests against which the introduced enemies failed to become established.

7 *Pesticide application*

Methods of pesticide application

In the application of pesticides to crop plants, the object is to place the chemical in or on the correct part of the plant in order that it might come into suitable contact with the insect pest. For a leaf-eating insect a pesticide should either be on the leaf surface or in the leaf tissues; for a sap-sucker the poison should be in the phloem system. For a leaf-miner the poison must penetrate into the leaf tissues to be effective. A soil-inhabiting, root-eating pest can be attacked either through the tissues of the roots or else by a contact insecticide, introduced into the soil around the roots. Many of the caterpillars (and fly larvae) which in their later instars bore into fruit, or plant stems, hatch from eggs laid on the surface of the plant or on the soil and the first instar larvae spend some time on the plant surface before burrowing into the stem or fruit. Thus, these larvae, which in later instars are almost invulnerable once they are inside plant tissues, can be attached by carefully timed and placed pesticide application. Seedling pests can be attacked by the use of a seed dressing on the sown seed. With a contact insecticide it is imperative that the chemical comes into contact with the pest during the period of its potency; this point is most important because of the increasing use of the more short-lived organophosphorous and other pesticides.

Two recent publications of considerable importance are *Pesticide application methods* (Matthews, 1979), and the entire number of *Outlook on Agriculture*, vol. 10 (7) (1981) which is devoted to recent trends in spray application technology. In recent years spray application technology has become established as a scientific speciality in its own right, and much has been written, both in the two publications mentioned above, and also now in the latest editions of *Insecticide and fungicide handbook* (5th edition) by Martin & Worthing (1976), chapter 3, and the *Pest and disease control handbook*, by Scopes & Ledieu (1979), chapter 2.

Because of all the published information now available, in this book only the more important aspects are presented in summary and the other sources are cited for further details.

Most pesticides are either crystalline solids or oily liquids in the pure state or the technical product, and they are usually effective in quite small quantities (i.e. a litre or two per hectare). In order to apply a pesticide to a crop it is usually necessary for the chemical to be contained in a carrier fluid or powder. The earliest carrier (or diluent) used was water and the chemical was either dissolved in it or, if it was water-insoluble, as a suspension or emulsion. In the case of emulsions and suspensions it is important that the mixture be stable enough to ensure the application, under practical conditions, of a solution of uniform and known concentration. The physical stability of a spray solution may be conferred by the addition of supplementary materials. Thus the sedimentation of a suspension can be delayed by the addition of protective colloids or dispersing agents. The coalescence of the scattered droplets in an emulsion may be retarded by the addition of an emulsifier.

Formulations

As previously mentioned most pesticides are crystalline solids or oily liquids and as such are usually not suitable for spraying direct on to a crop. In a few cases the technical product is soluble in water; then the pesticides can be prepared as a very *concentrated solution* (c.s.) which only requires dilution by the farmer to the appropriate strength for spraying. Usually these concentrated solutions have to have wetting agents or detergents added (see later in the chapter). This type of spray solution is typically very homogeneous and spreads a very even level of pesticide over the foliage.

Many solid substances that will not dissolve in water can be ground and formulated as *wettable powders* (w.p.). Wettable powders are powders which can easily be wetted and do not resist the penetration of water, and are miscible in water in which they pass into suspension. Others are more accurately termed water-dispersible powders in that when mixed with water they remain as individual particles in suspension for a considerable period of time. Various additives (dispersants) can be included in the formulation of wettable powders to delay the process of sedimentation.

Oils and other water-immiscible liquids, if agitated with water, break up into tiny droplets which on standing rapidly coalesce to form a separate layer. This coalescence may be retarded or prevented by the addition of auxiliary materials known as surfactants or emulsifiers. With mixtures of oil and water, two types of *emulsion* are possible. The oil may be dispersed as fine droplets suspended in water, which is then the continuous phase, giving an oil-in-water (o/w) emulsion, or the water may be the disperse phase giving a water-in-oil (w/o) emulsion. The type of emulsion generally required for crop spraying practice is the o/w emulsion which, as water is the continuous phase, is readily dilutable with water. Thus pesticides insoluble in water may be dissolved in various organic solvents forming an *emulsifiable concentrate* (e.c.) which can be diluted in water to an appropriate spray strength. The 'breaking' of an emulsion is the usual way in which the toxic dispersed phase comes into play, breaking occurring after the evaporation of most of the water. Various chemical substances can cause the 'inversion' of an emulsion which then becomes useless for spray purposes. Sometimes emulsions are caused to 'cream' (named from the analogous creaming of milk), resulting from differences in specific gravity between the dispersed and continuous phases.

Some pesticides are more suitably formulated as *miscible liquids* (m.l.). In this case the technical product is usually a liquid and is mixed (dissolved) with an organic solvent which is then, on dilution, dissolved in the water carrier.

Other pesticides may, for use against specific pests, be formulated as *seed dressings* (s.d.), both wet and dry, or *granules*, but these types will be dealt with later in the chapter.

At times it will be necessary to know precisely the quantity or proportion of the pure chemical in any formulation; sometimes this may be shown on the pesticide container, but more often is not. However, it will be available on the company data sheets, and in such sources as *The pesticide manual*. The usual method of expression of the proportion is as grams of *active ingredient* per kilogram of formulation (g a.i./kg) for powders, and grams per litre (g/l) for liquids.

Spraying

When water is the carrier, the usual method of application of the spray is by passage under pressure through special nozzles which distribute the chemical in a fine spray over the crop. In general, the type of spray application can, for convenience, be expressed as high-volume, low-volume, or ultra-low-volume according to the amount of carrier liquid.

During the last 20 years, especially when viewed worldwide, there have been many different definitions published for high- and low-volume spraying, with very little overall agreement as to rates. There has tended to be variation, not only from country to country, but sometimes according to the crop concerned, and also it appears according to the availability of local water supplies in some cases.

In the UK recently there has been a concerted effort to promote the following definitions.

		Rate (l/ha) for trees and bushes	Rate (l/ha) for ground crops
High-volume	(h.v.)	> 1000	> 700 (600)
Medium-volume	(m.v.)	500–1000	200–700

Low-volume	(l.v.)	200–500	50–200
Very-low-volume	(v.l.v.)	50–200	5–50
Ultra-low-volume	(u.l.v.)	< 50	< 5

High-volume spraying. As already mentioned, there have been many different definitions published over recent years, with little international agreement as to the rates. Even now in the UK there is not total agreement in that Matthews (1979) prefers the rate of 600 l/ha (more than) for field crops, as opposed to the more widely used 700 l/ha. An early definition of some authority was that by Maas (1971) who used the rate of 400 l/ha. However, at that time there was generally no 'medium-volume' rate. A distinction is now drawn between rates for tree crops and rates for field crops because of the need for good foliar spray cover, and the fact that the leaf area index for tree crops is usually far higher than ground crops.

With high-volume spraying the carrier is invariably water, and the usual quantity involved is in the region of 600–1200 l/ha. In the case of wanting a run-off of spray from the upper parts of the plants on to the lower parts, or on to the soil, then the water volume may be doubled up to the extent of 2400 l/ha. High-volume spraying has several major disadvantages, the first being the problem of transporting the large quantities of water required, especially in areas where piped water is not available. In many drier regions obtaining sufficient water for this purpose can be difficult. The cost of the high-volume spraying equipment is considerable, and the bulk of the equipment is such that its operation often requires a large tractor.

Medium-volume spraying. This is a somewhat arbitrary category, designated for convenience so that the other categories are more clearly defined, as with v.l.v.

Low-volume spraying. As now defined, this term applies to rates in the region of 50–500 l/ha, according to the type of crop being sprayed and the extent of its foliage cover. This technique was developed initially because of some of the problems associated with high-volume spraying.

Because of these problems alternative methods of application for the treatment of large areas have been sought. The 'scent spray' principle involves the jet of liquid being dispersed into fine droplets by the force of a copious air-flow. This method has been successful for the dispersal of DDT/petroleum mixtures from aeroplanes where the air speed alone is sufficient to break up the jet of solution into droplets which are then dispersed by the slipstream. A helicopter can be even more effective for this method using the down-draught from the rotor blades. For terrestrial use, the air-stream can be provided by a fan mounted horizontally, or by a turbine fan. These techniques, which rely upon the energy for spray dispersion being provided through the air-stream, and not, as in conventional spraying, through the liquid, are referred to as 'atomization'. Typically the droplets produced are much smaller than those produced by the usual high-volume sprayers. Because of the smaller droplet size smaller volumes of spray per hectare are needed and the use of organic solvents such as kerosene, petroleum oil, or fuel oil, in place of water, becomes economically feasible. If water is used as carrier, the concentration of active ingredient can be increased, thus giving rise to the term 'low-volume' spraying.

Not all low-volume spraying is done by aircraft, as there is some ground equipment suitable, but most of the aerial spraying done is low volume. The typical rates for ground application are commonly 100–200 l/ha, whereas for aerial application they were 15–75 l/ha. As mentioned previously, the equipment used for ground application is typically an air-blast machine, and the carrier liquids are frequently organic solvents, although water is used sometimes.

Ultra-low-volume (u.l.v.) spraying. This technique originally consisted of the production of very small droplets (c. 70 μm) by a rotary atomizer, which were carried in a light oil and blown by a fan in drift spraying. Early u.l.v. work was mainly carried out aerially from light aircraft. When applying u.l.v. sprays the pilot flies higher than in conventional spray application (5–10 m instead of 1–2 m), often flying crosswind so as to use the movement of the ambient air to distribute the spray over the crop. The higher altitude increases the spray swath to about three times that of conventional aerial spray swaths (about 30 m as compared to 10 m). However, a recent trend is for aerial u.l.v. spray applications now being used to apply the spray from heights of 2–4 m using spray swaths of 15–25 m, depending on type of aircraft and equipment used; these conditions seem to be about optimum for aircraft u.l.v. applications. Because of the small amount of liquid to be sprayed, the total time required for the operation is usually much less than half that for conventional low-volume aircraft spraying.

U.l.v. ground-spraying was introduced relatively recently, and now very diversified equipment is being used. The advantages of changing from conventional methods to u.l.v. for ground spraying are less apparent than with aerial spraying, but there are some advantages, namely in *ultra-low dosage* spraying and in cases where transport of water is a problem.

The use of u.l.v. sprays requires fine droplets to ensure adequate cover but not so fine that losses due to drift and evaporation become too great. Recent development of rotary atomizers has ensured the production of a narrow spectrum of droplet size. Several manufacturers have been modifying their ground-spraying equipment for u.l.v. application.

The real establishment of u.l.v. techniques for crop spraying was in 1963 with the good results obtained with malathion for the control of cotton insects. From that time a great many papers have been published on this topic, and a large number of pests have been successfully controlled on a wide range of crops. Further information on u.l.v. techniques can be found in Maas (1971) and Matthews (1979).

Efficiency of insecticidal spray applications

The efficiency of a sprayer (applicator) is its ability to apply a known volume of liquid, at a known and steady rate of delivery, to a designated target area, with a regular and even spread of droplets. The efficiency of a spray application, in effect, is a combination of the mechanical efficiency of the equipment and that of the operator, and is measured ultimately by the proportion of pesticide actually reaching the target insect pest (and of course applies equally to any other form of pesticide application). With many field applications of pesticides the overall efficiency is frequently very low, often less than 1% of the insecticide actually reaching the insects, and sometimes much less.

In Fig. 29 is shown a diagram from Bottrell (1979) representing the fate of an aerial discharge of insecticide, regarded as reasonably typical.

Droplet size. The main objective in crop spraying is to spread the active chemical evenly over all the plant surfaces so that a lethal dose is available for the pest to pick up. When large spray volumes are used the final coverage is often far from complete due to the coalescence of drops and run-off. Thus the pesticide should be distributed over the plant surface in spray droplets that are as small as possible to produce a complete coverage. The droplet density required in a given spray operation depends upon various factors: type of pesticide (fungicide or insecticide); mobility of pest to be controlled; mode of action of insecticide (systemic, contact, or stomach poison). Thus when a contact insecticide is sprayed against a sluggish pest, a much higher droplet density is required than for the spraying of a stomach poison for the control of a highly mobile pest. In the case of fungicides an even better coverage is required for effective control.

Generally, the smallest droplet size would be the most effective, but very small droplets do not fall freely, often being carried by wind and air currents. It appears that droplets of less than 30 μm are practically airborne – they will be carried by the surrounding air so that they will not touch large target surfaces such as cotton leaves, but they may settle on small diameter targets such as the needles of coniferous plants, hairs of caterpillars, etc. Thus, for adult mosquito control, the optimum droplet size is 5–25 μm; for tsetse in vegetation, 10–30 μm. For agricultural crop spraying the situation is complicated; for many cotton pests it was found that droplet size of 20–50 μm was best, but typically many crop areas are relatively small and situated near other crops, which means that the spray must be deposited in a much more limited target area. Consequently the optimum droplet size in many crop-spraying operations will be decided upon mainly by the necessity to avoid spray drift, and thus larger droplet sizes will be used.

In aerial crop spraying for insect control, the optimum droplet size seems to be in the range of 80–120 μm, the lower values being more suitable for the treatment of larger areas only. In ground crop spraying the u.l.v. droplet size will be in the range 60–90 μm, although with systemic insecticides the optimum droplet size may be larger.

Fig.29. The fate of an insecticide discharged by aircraft (Flint & van den Bosch, 1977, after von Rümker *et al.*, 1974, from Bottrell, 1979).

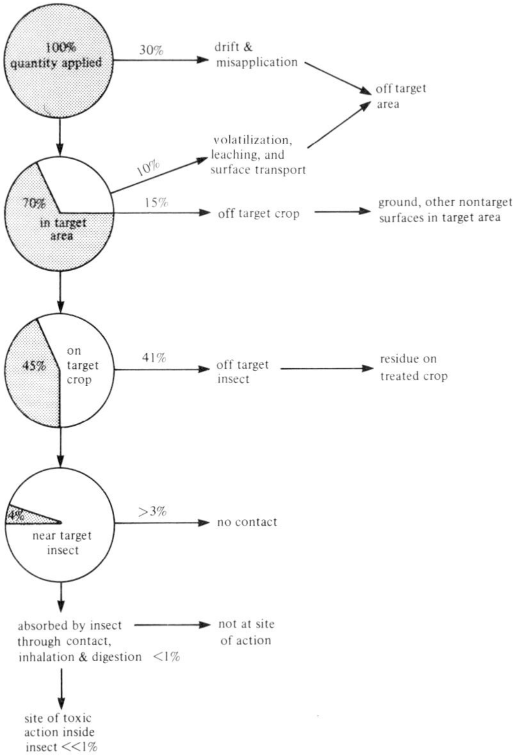

This is one aspect of spray technology that has received extensive study in recent years, and table 2.5 in Matthews (1979) lists the optimum size ranges for a selection of targets.

Target	*Droplet size* (μm)
Flying insects	10–50
Insects on foliage	30–50
Plant foliage	40–100
Soil (and drift avoidance)	250–500

Controlled droplet application (CDA). This term was coined in 1975, in discussions on herbicide application, in order to emphasize the importance of having a known and uniform size of droplet in sprays of liquid pesticides, but the basic concept had of course been known for a long time. However, awareness of the precise importance of droplet size in pesticide sprays, and methods of achieving both uniform droplet size and pattern is relatively recent (Matthews, 1977, 1979).

Solvents. Anyone with an elementary knowledge of applied mathematics will be aware that the volume of a sphere increases according to the cube of its linear measurement (radius or diameter) whereas the surface area increases according to the square. Thus smaller droplets have relatively large surface areas, which implies that the rate of evaporation of spray droplets is higher with smaller droplets. Consequently, solvents used in u.l.v. formulations must have low evaporation rates, and water should never be used. The more volatile liquids would also lead to evaporation of the solvent in the atomizer and might cause crystallization of the pesticide there. Liquid pesticides are sometimes sprayed undiluted in u.l.v. application, but more often some solvent is required.

The solvent must be non-phytotoxic, of low volatility, of high dissolving power for the pesticide, of low viscosity, and compatible with the pesticide. Not many of the solvents generally available will fit all these categories and so clearly good u.l.v. solvents are difficult to find. Xylene is too volatile, other aromatic hydrocarbons of lower volatility are often highly phytotoxic. Alcohols and ketones show higher phytotoxic effects at lower volatility, and other common solvents show the same variations in properties. Sometimes a mixture of various solvents can be used with reasonable success. In the laboratories of Phillips–Duphar have been developed several compounds called *adjuvants* which dramatically decrease the phytotoxicity of many low-volatile solvents, thus the range of solvents that can be used for u.l.v. formulations has now increased. Conventional emulsifiable concentrates usually have a low flash point, but this seldom represents a fire hazard since the concentrate is mixed with a large volume of water before spraying. However, with u.l.v. techniques the solvent must be of low volatility for with rotary atomizers with electrical systems and many rotating parts the possibility of electrical discharge is always present.

With the use of these adjuvants a wider range of low-volatile solvents can be used for u.l.v. formulations and many of these formulations are now regarded as being special ultra-low-volume (s.u.l.v.) formulations. Their characteristics are high concentration, low volatility, low phytotoxicity, with a much lower viscosity than most u.l.v. preparations, and with a flash point above 75 °C.

Spray residues. In conventional spraying the spray liquid contains a large amount of water, containing various wetting agents, dispersants and emulsifiers. The pesticide is generally present as a finely dispersed phase, solid in the case of wettable powders, or liquid when using emulsions. In u.l.v. sprays the pesticide is generally present as a true solution in an oil carrier, or sometimes the technical material is sprayed as such. Such differences in the spray make-up will have a profound effect on the behaviour of the spray droplets on the biological target.

After deposition of a spray droplet on a leaf, the droplet will assume a certain shape by spreading over the leaf surface. With conventional aqueous spray liquids this spreading depends greatly on the properties of the leaf: on hydrophilic leaves the droplet will spread to a thin film, but on hydrophobic leaves the droplets usually retain their spherical shape. The spreading of oils on most smooth leaf surfaces is much better than that of aqueous solutions. The s.u.l.v. formulations spread to a thin film on most leaves, even hydrophobic leaves, with very little run-off. However, there are times when run-off is wanted, and in these cases u.l.v. application is not the most suitable method of application.

The formation of a residue from an emulsion droplet is very complicated; evaporation of the water, breaking of the emulsion, and crystallization of the pesticide from the oil phase can occur simultaneously. Often from u.l.v. sprays the crystalline residue on the leaves is particularly coherent and very resistant to dislocation, and also the rain-fastness of u.l.v. residues seems to be better than that of conventional formulations.

Electrostatic spraying. A fairly recent innovation in crop spraying was developed by Coffee (1973), and termed electrostatic spraying. The basic principles are that the spray or dust particles are electrostatically charged and emitted from the nozzle region in a directed stream along the electrical field flux lines so that they surround the earthed target (crop plant), and the result is usually a very even cover of particles on both sides of the leaves. Since the particles are similarly charged they settle evenly on the leaves due to mutual repulsion and as there is a positive reaction on the tiny particles to the target leaves spray drift is minimized.

The method was initially developed using insecticide dusts but has since been adapted for use with low-volume sprays, and it would seem that this method of insecticide application has great potential commercially, particularly for small farmers.

The development studies on this method of pesticide application are mostly being conducted in Africa and especially on cotton crops. Since this crop is generally grown in dry areas the smallholder farmers constantly face a problem in providing water for more conventional spraying systems.

Dusting

Sometimes it is more convenient to use a dust instead of a spray: the need for water is obviated; the dust may be bought ready for use and is more easy to handle than spray concentrate; the dusting appliances are generally lighter and easier to manipulate than sprayers. For dusting, the active ingredient is diluted with a suitable finely divided 'carrier' powder, such as talc. The dust is usually applied by introduction into the air-stream of a fan or turbine blower. However, in practice, dusts are often not easy to apply; frequently the powder 'cakes', usually through absorption of atmospheric moisture, or 'balls' in the hopper (through static electrification). Also it is difficult to ensure that the dust is homogeneously mixed.

It is generally found that dusting is only practicable in the calmest weather, and that best results are obtained when the dust is applied to wet or dew-covered plants. When dusts are applied to dry foliage usually not more than 10–15% of the applied material sticks to the foliage. Thus there are not many occasions when dusting is a more suitable application method than spraying.

Fumigation

The toxicity of a gas to a pest is proportional to its concentration and to the time of exposure against that pest. Research into gas properties has shown that usually fumigation is only successful in completely closed spaces or with special precautions to lengthen the time of exposure. For stored-products fumigation, the material can either be treated in special chambers or under large gas-proof sheets. Some field crops are treated by drag sheet techniques in which the fumigant is enclosed below a light impervious sheet dragged at a rate dependent on its length behind the vaporizing appliance.

Soil can be fumigated by the injection of volatile liquids directly into the soil at frequent regular intervals. The 'DD' soil injector is used to control nematodes and other soil pests, but it is a tedious process only suitable for relatively small areas.

Smoke generators contain a blend of the pesticides and a combustible mixture which burns in a self-sustained reaction at a low temperature, so that the minimum amount of pesticide is destroyed during volatilization, and the finely dispersed pesticide is carried in the cloud of smoke.

Aerosols contain the toxicant dissolved in an inert liquid which is gaseous at ordinary temperatures but liquefiable under pressure. When the pressure is released, the solution is discharged through a fine nozzle, the solvent evaporates and the toxicant is dispersed in a very finely-divided state. Methyl chloride, at 5.5 bar, and dichlorodifluoromethane ('Freon') at 6.2 bar at ordinary temperatures are two widely used aerosol solvents.

The earlier aerosols were used against medical and household pests, and were not suitable for use on plants as the carrier was usually strongly phytotoxic; but now there are water-based aerosols (usually employing a mixture of synergized pyrethroids) available for garden use on cultivated plants.

Seed dressings

The earliest form of seed dressing was to steep the seed in liquids such as urine or wine. The object of a seed dressing is to protect the seed in the soil and also to protect the seedling for a period after germination. The dressing will form a protective zone around the seed, and the extent of the zone will depend upon whether the pesticide has any fumigant or systemic action. In the past, seed dressings have been used mainly against smuts and other diseases, but now there are many insecticides which can be successfully formulated as seed dressings against insect pests such as wireworms, chafer larvae and shoot flies, and preparations of systemic compounds will protect against aphids and other sap-sucking insects on the young plants. Seed dressings can be liquids which are adsorbed on to the seed coat, or powders which are either sufficiently adhesive to stick directly on to the seed coat or else have to be stuck to the seed with the aid of a 'sticker' (paraffin or methyl cellulose).

With the advent of precision seed drilling came the development of *pelleted seed*. The object of seed pelleting is to make an irregular-shaped, rough seed (e.g. sugar beet) into a smooth spherical shape so that it will pass easily through the drill. Also a very small seed (e.g. *Brassica* seed) is given more bulk for easier handling. A few major seed companies are now supplying an ever-increasing range of pelleted seeds. The pellet is composed of inert material (such as powdered clay, pumice or Fuller's earth), and clearly during the pelleting process it is easy to incorporate pesticides into the pellet. The process is quite expensive though, and at times of soil water shortage pelleted seed suffers from impaired germination.

Granules

A new method of pesticide formulation was that of granules; these are small solid particles and are now widely used for the treatment of seedling crops with systemic organophosphorous and carbamate insecticides. The main advantage of granular formulations is that the insecticide can be placed in such a manner that gives maximum protection to the plant, with minimal danger of large-scale soil pollution and negligible danger to the operator. This is of particular importance with highly toxic chemicals. Another major advantage of granules is that the active ingredient is less affected by the soil than would otherwise be the case. Many pesticides are strongly adsorbed in soil and rapidly become ineffective once they reach it. The rate at which the pesticide escapes from granular formulations is mainly controlled by the rate of leaching by rain water. However, the organophosphorous compounds used in granules are generally of low aqueous solubility. Other major factors controlling rate of pesticide release from granules are temperature, dosage, and size of granule. The six major organophosphorous insecticides initially formulated as granules were aldicarb, dimethoate, phorate, disulfoton, chlorfenvinphos and diazinon, used against fly maggots, beetle larvae, aphids and nematodes. Now many new and very toxic organophosphates are formulated as granules for safety reasons. Granules are sometimes applied broadcast, but more typically as row treatments at sowing by bow-wave technique or where labour permits by spot applications round the base of individual plants using hand applicators – 'Rogor' and 'Birlane' applicators are generally available for this purpose. The body of the granule is made of various inert substances; for soil application both phorate and disulfoton granules are made of Fuller's earth. Also used are coal, rice husks, corn cob grits, gypsum and other minerals. On occasions it is advantageous to make a foliar application of granules and there is a pumice formulation for this purpose; pumice is more expensive as a base but is lighter, stickier, and more effective for foliar lodging.

Encapsulation

The most recent development in pesticide formulation is the technique of micro-encapsulation. Work at Rothamsted Research Station, England, has shown that it is possible to encapsulate an insecticide in a non-volatile envelope of cross-linked gelatine in such a way that it is non-toxic by contact, but is toxic to insects ingesting it. By the addition of suitable stickers the formulation can be given considerable resistance to weathering. This type of formation would appear to be of great promise for the control of leaf-eating insects. The formulation has the advantage of being far safer to handle than other pesticide formulations where very toxic chemicals are being used, and it presents far fewer hazards to beneficial insects such as predators, parasites and pollinators. The capsules are so tiny that the formulation has the appearance of being a slightly coarse powder. However there have been reports that honey bees collected the capsules mistakenly for pollen grains, which are about the same size, and that this caused a number of larval deaths in the hives.

Where a contact kill is required, it is possible to prepare leaking capsules which will release the poison over a period of time. At Rothamsted it was found that under warm conditions a standard wettable powder of DDT lost over 90% from the target area in 35 days, while a leaking capsule formulation lost about 20% over the same period.

Baits

The use of poison baits in pest control is generally confined to the insect groups Orthoptera (crickets, cockroaches, sometimes locusts), Isoptera (termites) and Hymenoptera (ants), in addition to terrestrial molluscs and vertebrates (especially birds and rodents). These pests are typically gregarious with underground nests. Social insects in nests are particularly difficult to kill by spraying insecticides, and sometimes the nests are hard to locate. Most social insects take food back to the nest where it is shared by trophallaxis amongst other adults and the larvae. In the case of fungus-growing termites (family Termitidae) and ants (tribe Attini), the collected food material is incorporated into the fungus gardens on which the fungus grows and poisons introduced in this manner eventually get distributed throughout the colony.

The earliest poison baits relied upon inorganic stomach poisons such as salts of arsenic, lead, mercury and sodium fluoride, but these have now largely been replaced by certain organochlorine compounds.

Poison baits generally comprise four separate components (Cherrett & Lewis, 1974).

(1) *Carrier* or *matrix*: inert material which provides the structure of the bait. This may be an edible or attractive material such as meal (e.g. soyabean, groundnut) or citrus pulp, or a more inert or less attractive substance such as ground rice husks, corn cob grits, clay, vermiculite and gelatine microcapsules.

(2) *Attractants*: this may be an integral part of the carrier itself (e.g. soyabean meal, citrus pulp), or may be substances such as sugar, molasses, or vegetable oils added to the inert carrier base. A recent development in leaf-cutting ant baits is the incorporation of a trail pheromone to attract the worker ants to the baits. Obviously any chemical that will induce the insects to pick up, or eat, the bait will make this method of control more successful.

(3) *Toxicant* (*poison*): previously usually a stomach poison, but more recently contact insecticides such as aldrin, dieldrin, and heptachlor are being used.

(4) *Additives*: for specific formulation purposes, such as preservatives, materials to bind the bait together in pellets or blocks, and waterproofing agents. The physical properties of baits can be important, especially for use in the tropics with high temperatures, high humidity and torrential monsoon rains. For a broadcast bait to remain effective in the field for several weeks under wet tropical conditions it must not disintegrate in the rain. A standard ant bait is 'Mirex 450', but the compressed pellets apparently break down in heavy rain and become ineffective. Experiments by Cherrett and others have shown that leaf-cutting ant baits composed of citrus pulp with soyabean oil and aldrin can be rendered fairly waterproof by the addition of a hydrophobic surface deposit of siloxane, which prolongs the effective life of the bait under wet conditions.

Baits are often placed by hand, but clearly this is time-consuming and costly at present times, in most countries, and recent trials against leaf-cutting ants in C. America have used aerial application from low altitude with considerable success.

Another type of baiting is the use of spot sprays against adult fruit flies (Tephritidae). This technique involves the spraying of spots (i.e. squirts or blobs) of mixture, about 100 ml in volume each, scattered throughout the foliage in the orchard or plantation. The viscous liquid applied in this manner consists of protein 'solids' (hydrolysates) with added malathion as the toxicant; for full details see Drew, Hopper & Bateman (1978).

Systemic pesticides

Certain pesticides are capable of entering the plant body and being translocated to other parts of the plant through either the phloem or xylem systems. These insecticides may be applied as sprays on to the soil, sprays on to the foliage, granules on to the soil, or a foliar application of granules, or in the case of woody plants direct injection into the phloem system can be made using special injectors. Sap-feeding insects, such as aphids, are more readily killed by systemic insecticides than by those with a contact action. Parasites and predators are not affected unless they come into contact with the insecticide, that is if the plants are sprayed. Some of these insecticides are highly poisonous to man, but others are now available (such as malathion, trichlorphon and menazon) which are still effective against the insect pests but with reduced toxicity to man.

A systemic insecticide must persist in the plant body in an active form until the contaminated sap is sucked or eaten by the insect pests. There are, however, no problems of surface adhesion and weathering, although the problem of penetrating the plant has to be overcome. Consequently, these insecticides are sufficiently lipid-soluble to enable them to penetrate the plant cuticle and also soluble enough in water to be easily translocated within the plant.

They must also resist hydrolysis and enzymatic degradation for a sufficiently lengthy period of time to be effective as a pesticide. Obviously, not many insecticides possess this subtle balance between lipophilic and hydrophilic properties and, of course, non-phytotoxic at insecticidal concentrations.

The term systemic is not always used in the same sense: sometimes the substances taken up by the plant roots are not referred to under this heading.

Some pesticides which can penetrate the plant cuticle and pass through the cells are termed *translaminar* (or *penetrant*) pesticides. These can penetrate the leaf cuticle and will pass through the leaf to the other surface: hence pests on the underneath of the leaf can be killed by spraying the pesticide on the upper surface, and it can also kill leaf-miners effectively. These pesticides are not usually transmitted through either phloem or xylem systems but just diffuse through the cells.

Pesticide deposition and persistence

In temperate countries a considerable amount is known about the proportion of pesticide reaching the target area, and how long it persists there. The relationship between the amount of pesticide present and the control of the pest obtained is thus fairly well established.

The object of a pesticide spray programme is firstly to 'hit' the target, that is to place sufficient quantity of the chemical either directly on to the pest organism itself or on to the place the pest inhabits so that the pest will come into contact with enough of the poison to kill it. The second object is that enough of the pesticide remains on the foliage for sufficient duration to kill the required proportion of the pest population.

Hitting the target is often quite difficult for many pests live on the underneath surface of leaves where they are quite sheltered. Or they may live in the denser foliage in the centre of bushes, or the plants may be so closely planted that it is difficult to spray between them, or the tree too tall for the uppermost leaves to be reached. This point has already been mentioned on page 36. Then the question of persistence remains. Most spraying is done on to plant foliage and so the general target area is leaf surface. The main aspects to be considered are enumerated below.

(1) Collection of the pesticide droplets on the leaf surface (i.e. 'hitting' the target area).
  - (a) Droplet size and density: coalescence.
  - (b) Spray velocity: impaction of droplets determines collection success.
  - (c) Angle of contact between spray and leaf surface.
  - (d) Leaf surface texture (i.e. wettability); waxy or hairy leaves tend to deflect spray droplets.
  - (e) Quantity of carrier and run-off: includes droplet coalescence with subsequent run-off; deliberate run-off is sometimes the aim, either to redistribute the chemical to lower leaves or on to the soil (as for cutworm control).

(2) Weathering of the spray residue on the leaves (i.e. chemical retention).
   (a) Rain run-off: redistribution to lower leaves can be advantageous, otherwise wash-off on to the soil usually means wasted pesticide.
   (b) Degradation of the chemical (sometimes the breakdown products are still insecticidal, so effectiveness may not be reduced).
      (i) Volatilization (i.e. vaporization): this can be an advantage in an enclosed space (i.e. stored products container).
      (ii) Temperature (thermal) breakdown.
      (iii) Solar radiation; either photodegradation or reaction with other radiations of a solar origin.
      (iv) Breakdown by soil micro-organisms.

A great deal of experimental work has been done in laboratories and in small plot field experiments on these problems of deposition and persistence, as reported by Matthews (1979). But in practical terms the situation is usually very complicated, and most field applications of pesticides are made with a low level of precision, which accounts in part for the tremendous variation in results that are typical. In general, in most crop/pest situations there are so many variable factors that it is extremely difficult to take appropriate account of all of them, which is why most control programmes are rather inefficient.

Pesticide degradation by micro-organisms. A recent discovery (first noticed in the late 1970s) in the USA has led to awareness that in some soils micro-organisms are causing rapid degradation of some pesticides so that they have little effect. The precise mechanisms involved are not yet understood. The chemicals involved in this accelerated degradation are mostly herbicides (thiocarbamates whose active ingredient is eptam), but also the carbofuran insecticides sold as 'Furadan'.

Whether this phenomenon will become widespread is not at all clear, but the implications are serious especially when considering the increase in the practice of no-till and low-till farming with its concomitant use of soil pesticides.

Temperature. At Rothamsted, experiments have shown that temperatures would be reached on isolated surfaces in the tropics such that no non-systemic insecticide applied in commonly used formulations would persist for more than a few days. Under tropical conditions the majority of even the most persistent insecticides may be lost from the target surface in a few days. For example, dieldrin deposits lost 25% at 20 °C and 95% at 40 °C, in 24 hours, at the same low wind speeds (2 mph). However, when the deposit level is very low (about 0.005 $\mu g/cm^2$ on a glass surface; 0.01–0.02 $\mu g/cm^2$ on a cotton leaf) it no longer obeys these rules, but remains very firmly bound to the surface for a very long period. These data are likely to apply to insecticides applied as sprays of wettable powders, emulsions, or dusts.

Work done at Beltsville, MD, USA, in 1975 on solar degradation of pesticides in hot weather showed that, after application of dieldrin and heptachlor as sprays on short grassland (5 kg/ha) during hot sunny weather, volatilization was so great that the half-life of dieldrin was 2.7 days and for the heptachlor 1.7 days, and after 30 days there was 10% of the dieldrin left and 4% of the heptachlor.

Most agricultural crops are completely exposed to the sun and so the above result can be expected, although for some plantation crops the trees may form a canopy with shade underneath. Dieldrin applied to tree trunks and the underneath of branches for Tsetse Fly control will persist, since it is shaded from direct sunlight, and effectively kill Tsetse resting there for periods of up to four to five months.

Unfortunately, however, not a great deal of research into the persistence of most insecticides has actually been carried out in tropical countries and often spray programmes tend to be too speculative.

In temperate regions the cold winter conditions inhibit pesticide degradation and there is usually little breakdown over the winter period; this is one of the reasons why there was such dramatic chemical build-up in the soil in parts of the UK, Europe and N. America in the days when DDT and dieldrin were widely used.

Rain. Most parts of the wet tropics receive their annual rainfall torrentially and it is expected that insecticide residues on crop plant foliage will be washed off at a far greater rate than similar residues exposed to the more gentle temperate rainfall.

Tropical rain is often a problem, but in temperate regions rainfall is seldom serious. The only time it presents a problem is if rain falls immediately after a spray application to crop foliage, then the spray will be washed to the soil. The operator is expected to exercise some judgement in deciding when to apply a crop spray; unexpected showers of course sometimes necessitate an immediate additional spray. Once the spray residue is dried on the plant foliage most temperate rains do not wash off very much.

Spray additives

Spreaders (sometimes called *wetters* or *surfactants*) are substances added to the spray to reduce the surface tension of the droplets so as to facilitate contact between spray and sprayed surface. Plain water falling on a waxy leaf such as that of a *Brassica* will normally collect in large drops and will then run off, leaving the leaf surface dry. The incorporation of a spreader is now standard production practice in the manufacture of most modern pesticides. For crops with particularly waxy leaves (such as brassicas) or against pests with particularly waxy cuticles (like mealybugs, and Woolly Apple Aphid) it is necessary to add extra spreader to the spray solution. Sometimes when extensive run-off is required to enable the pesticide to penetrate to the lower part of a dense crop, this can be achieved by addition of extra spreader to the spray. From a physical point of view wetting and spreading are not quite the same, but for practical purposes they can be regarded as synonymous. Spreaders and surfactants exist in three forms: non-ionic, anionic and cationic, classified according to their ionizing properties. Surfactants are defined as surface-active components. The non-ionic detergents depend upon a balance between hydrophilic and lipophilic properties throughout the molecule for their wetting properties. The advantages of these substances include the fact that they are incapable of reacting with cations or anions present in other spray components and in hard water, and that they are not hydrolysed in either acidic or alkaline solutions. However, phytotoxicity of supplements has to be taken into account. Anionic spreaders possess a negative charge on the amphipathic ion, and typical examples include soap, sulphated alcohols and sulphonated hydrocarbons. Cationic spreaders carry a positive charge on the amphipathic ion and a negative charge on the gegenion, and examples include the quarternary ammonium and pyridinium salts. The advantage of cationic spreaders is that they cannot react with ions of heavy metals. The incompatibility of anionic and cationic additives must be borne in mind if it is necessary to add several supplements to a spray mixture.

Dispersants. Sprays must be of uniform concentration and with suspensions there is always the danger of sedimentation. By the addition of a dispersant (or protective colloid) sedimentation can be effectively delayed. The more effective colloids used as dispersants are the methyl celluloses and the sodium carboxymethyl celluloses.

Emulsifiers. These are added to emulsions to modify the properties of the interface between the disperse and continuous phases. Many of the spreaders or surfactants also function as emulsifiers (e.g. soap).

Penetrants. Oils may be added to a spray to enable it to penetrate the waxy cuticle of an insect more effectively. Some of the more effective sprays against locusts were solutions of dieldrin in light petroleum oils.

Humectants. These are substances added to a spray to delay evaporation of the water carrier, and the more commonly used compounds are glycerol and various glycols. They are more frequently used with herbicides than insecticides.

Stickers. Stickers such as methyl cellulose, gelatine, various oils and gums, are used to improve the tenacity of a spray residue on the leaves of the crop. Maximum spray retention is particularly important in the tropics where rainfall is often monsoonal and torrential. Generally a fine particle deposit is more tenacious than one of coarse particles. Spreaders usually enhance the tenacity of a deposit and its retention on the plant, although they may retain their wetting properties and thus cause the deposit to be washed off by rain or dew. Some spreaders break down on drying and form insoluble derivatives and these can greatly enhance retention of the deposit.

Lacquers. In order to achieve a slow release of a pesticide in certain locations it is possible to formulate some insecticides into lacquer, varnish or paint. The painted area then releases the insecticide slowly over a lengthy period of time. The insecticides used in this manner were mainly organochlorines and particularly DDT and dieldrin. Incorporation of insecticides into paint is of some value but the lacquers are of limited use owing to the problem of the lacquer remaining after the pesticide has dispersed. This practice is of more use against household and stored products pests than crop pests.

Waterproofing. Baits used in the tropics where they are exposed to torrential rainfall are liable to disintegrate in the rain, and hence lose their effectiveness. As mentioned on page 127, Cherrett and others have shown that leaf-cutting ant baits can be rendered sufficiently waterproof by the addition of a hydrophobic surface deposit of siloxane, without reducing the attractiveness of the bait to the foraging ants.

Synergists. These are substances which cause a particular pesticide to have an enhanced killing power. They are sometimes called *activators*. The way in which synergists act is not always fully understood, but some operate on a biochemical level inhibiting enzyme systems which would otherwise destroy the toxicant. Usually the synergist itself is not insecticidal. Piperonyl butoxide is a synergist for the pyrethrins and certain carbamates. Some pairs of organophosphorous insecticides have a mutually synergistic action (sometimes called '*potentiation*', especially in American literature). Other synergists include piprotal, propyl isome, sesamin and sesamex. Many of these are particularly effective on the pyrethrins.

Equipment for application

As already mentioned, pesticide application technology has now reached quite an advanced stage and in view of the excellent publication by Matthews (1979), and the many papers in *Pesticide Science*, *Agricultural Aviation*, *Tropical Pest Management* (formerly *PANS*), and the like, it would be presumptuous to attempt an adequate coverage of this topic in a few pages, so this section is limited to just a few of the more obvious aspects in very brief outline; for details see Matthews (1979) and also Rose (1963).

High- and low-volume spraying

As previously mentioned, the usual amount of carrier liquid (water) for high-volume spraying is quantities in excess of 600 l/ha, typically 600–1100 l/ha, and occasionally as much as 2200 l/ha.

Whilst low-volume spraying uses water volumes in the region of 50–500 l/ha, typically the rates are 15–75 l/ha for aerial application, and 100–200 l/ha for ground application. All spraying systems consist basically of a tank for holding the spray liquid, a device for applying pressure to the liquid, and a nozzle or outlet through which the liquid is forced. Thus the only basic difference between equipment for low-volume spraying and that for high-volume is the capacity of the tank. The different types of high- and low-volume sprayers are very numerous but certain types can be categorized as follows.

Hand sprayers

Compression systems

Atomizers. These consist of a simple compression cylinder with an inlet at one end for the air and an outlet at the other for the compressed air. A plunger is moved up and down the cylinder to produce the compressed air. There are seldom valves present. The outlet tube is fixed at right-angles to a fine tube leading from the liquid container, and on the compression stroke the air is forced across the open end of the feed tube and creates a vacuum which draws up the spray liquid from the tank. As the liquid is drawn up it is broken up into tiny droplets by the air-stream. Hand atomizers are useful for treating individual plants, but they are tiring to operate for long periods. On the more refined atomizers the spray is delivered continuously by means of a pressure build-up system.

Pneumatic hand sprayers. These are machines with a tank capacity varying from a half to three litres and the tank acts as a pressure chamber. An air pump is attached to the chamber and it projects inside. The outlet pipe runs from the bottom of the tank and ends in a nozzle externally. Air is pumped into the tank which compresses the liquid and forces it out of the nozzle when the release valve is opened. On release, the spray is forced out by the air pressure in a continuous fine spray. The better machines can deliver a continuous spray for up to about five minutes when fully charged with compressed air. These sprayers are most useful in glasshouses or for treatment of individual bushes under calm conditions. As with the previous type these sprayers use very fine nozzles and as such are more suitable for use with solutions or emulsions than suspensions, which tend to block the aperture.

Knapsack pneumatic sprayers. These are basically the same as pneumatic hand sprayers except that they are designed for spraying large quantities of liquid (tank capacity up to 23 l). The tank is usually carried on the operator's back, suspended on a harness with shoulder straps. The outlet pipe is extended by means of flexible tubing and terminates in some form of hand lance. The lance usually carries from one to four nozzles, and is easily carried in one hand. A hand valve on the lance base controls the flow of liquid. The air pump is operated with the sprayer on the ground, and a high pressure is normally built up, which will last for about ten minutes of operation. These sprayers are manufactured in a wide variety of models, of varying degrees of efficiency. In general they are very useful, especially for the small farmer or for pesticide trial work. They can be very effective for estate work when teams of operators are employed, and individual attention for the plant is required. Since no system of agitation is incorporated, knapsack sprayers are more suitable for use with solutions than with suspended materials. Very long lances can be obtained for orchard and plantation use.

Pump systems

Syringes. Syringes consist of a cylinder into which the spray liquid is drawn on the return stroke of the plunger, and expelled on the compression stroke. The spray is sucked in through the spray nozzle aperture, or else through a separate ballvalve-controlled inlet near the nozzle. The spray produced is drenching, and the syringe is difficult and tedious to use; but they are useful for spraying small numbers of plants. Most syringes are simple in construction, and will last for years with minimum maintenance.

Force-pump sprayers. These are sprayers with a hand-operated pump, with a lance and nozzle outlet and a feed-pipe to draw the spray liquid from a separate container. Although small in size, these sprayers (fitted with a 45 cm double-action pump) can throw a jet of spray up to a height of 12 m. These sprayers are good for spot treatments in orchards, and, provided that the solution is kept stirred, they will spray suspensions as well as solutions and emulsions. This type of sprayer is obviously tiring to use and it is quite difficult to control the rate of application, but due to the double-action hand pump the spray is continuous.

Stirrup-pump sprayers. These consist of a double-action pump suspended in a bucket. For support there is a foot stirrup reaching to the ground on the outside. A flexible outlet pipe carries the spray liquid from the pump to the spray lance which may vary in length and arrangement of nozzles. A stirrup-pump sprayer requires two operators, one to hold the lance and direct the spray, the other to stir the solution (if it is a suspension) and to work the pump. They are very useful, all-purpose machines, of robust construction which will withstand hard wear. Bush crops, buildings, and small trees can be easily sprayed with a stirrup-pump sprayer, and it is ideal for team operation. Providing the liquid is kept stirred, then quite coarse suspensions can be sprayed. A large version, mounted on wheels, with a large capacity double-action pump is available for treating larger areas.

Knapsack sprayers. These are the all-purpose, very successful, sprayers used throughout the world for spraying pesticides over smaller areas. They consist basically of a spray container which sits comfortably on the back of the operator held by shoulder straps. The pump is of double-action, built either inside or outside the spray container, and is operated by working a lever which projects alongside the operator's body. In some models the pump lever also operates an agitation paddle in the spray tank. The spray liquid is applied through a lance held in the operator's free hand; the lance is connected to the spray tank by a long flexible hose. The tank capacity is usually about 23 l. Provided that a sufficiently coarse spray nozzle is used, this sprayer can be used with any type of spray. Many of the most recent knapsack sprayers are almost entirely made of plastic, which obviates the problem of metal corrosion by the more corrosive pesticides. With a little practice the rate of spray application can be controlled quite accurately. Knapsack sprayers can be tiring to operate over a long period of time, but they are very versatile, quite robust in construction, very portable, and most useful on small farms or in teams on larger estates, or for pesticide trials.

Power-operated sprayers

Compression systems

Hand guns. Two types of compression hand-spray guns are made. One type has the spray liquid fed into a pipe through which air from a portable compressor is fed. The other type is similar to the small compressed-air spray that is worked by hand, but the air pump is replaced by an inlet from a portable compressor. Droplet size, and rate of application, can usually be carefully controlled, but the capacity of the spray tank is small and hand guns are only of value where small areas have to be covered with small amounts of spray. Suspensions may block the outlet nozzle, especially if it is of very small aperture. The advantage of these sprayers is that any type of compressor can be used, and they may also be used as paint sprayers.

Portable sprayers. Many types of small portable sprayers are manufactured; some can be carried easily by one man and others are larger and mounted on a wheeled chassis. Air is compressed by a small compressor and is forced into the spray container. The container usually holds about 45 l, is of strong welded construction and is operated at a pressure of about 7 bar. The outlet hose from the spray container may end either in single or multiple lances, or may end in a boom. Provided the spray tank is lined with an anticorrosive material, these sprayers can be used for spraying corrosive liquids, since there is no pump to be corroded. As the compressor can be used for other purposes these machines can be useful to the smaller grower, who of necessity requires versatility in his equipment.

Large mounted sprayers. These are basically similar to the smaller portable sprayers, but they have spray tanks of a much larger capacity, hence they require larger compressors. The whole machine is usually pulled by a tractor, and the outlet terminates in a spray boom of varying design, or else in a series of hand lances for the spraying of individual trees. The spray booms may cover 6–9 m or more in a single swath. The better booms can usually be adjusted for height to suit the crop being sprayed. Rates of application can be adjusted by altering the size of the jet aperture in the nozzles. The booms may be positioned vertically for fruit crop spraying. Typically these machines are usually used for low-volume pesticide application, and they may be mounted on aircraft or helicopters for low-volume aerial application.

Pump systems

Portable sprayers. The range of small portable sprayers available is now very extensive. They are all based upon the fundamental units of a power source, a pump and a spray tank. The smallest units consist of a small simple, double-action reciprocal pump harnessed to a small air-cooled engine upon a framework so that it can be conveniently carried. The largest units may include a tank complete with positive agitator, mounted upon a four-wheeled chassis. The pump outlet pipe may supply a number of hand lances, depending upon the capacity of the pump. The mounted machines may have a small boom mounted so that they can spray ground crops. These sprayers are only really suitable for small areas of orchard or plantation crops, for generally two lances is the effective limit.

Large sprayers. These have a range of tank capacity varying from about 180–1800 l and may operate at very high pressures (up to 55 bar) with a high rate of delivery. They are invariably mounted on tractors and often have booms which will deliver from 50–3500 l/ha. Some of the larger booms are built vertically so that the nozzles point upwards, to ensure that the under-leaf is sprayed as well as the top surfaces; these are used particularly for crops such as coffee and fruit trees. These sprayers may be mounted on aircraft for aerial use, usually for low-volume application.

Low-volume mist blowers. Low-volume mist blowers and fogging machines are of relatively recent development. Since the water used in high-volume spraying is purely an inert carrier, use is made in the low-volume equipment of air-streams to carry the very finely dispersed spray droplets, and so far less water need be employed. The air-blast systems consist mainly of a series of nozzles which produce a coarse spray through wide apertures which is then broken into fine droplets by a fast-moving air-stream. This basic system has many modifications, mainly in the method in which the liquid is introduced into the air-stream to be broken up. The air-stream is usually produced by a centrifugal fan of large capacity. Mist blowers vary enormously in tank size, from portable machines that can be carried by one man to large, self-mounted, tractor-drawn machines. They can be used to spray in two different methods, either as blast spraying or drift spraying. Blast spraying is usually carried out in orchards, or plantations of such crops as apples, mangoes, coffee or rubber. The air all around each tree is replaced by a mixture of air and spray droplets, and generally a good, even cover is achieved using as little as 100 l/ha. Run-off of insecticide is eliminated, and the small droplets dry out quickly so the chemical becomes firmly attached to the foliage. Drift spraying relies upon the movement of ambient air to carry the insecticide mist through the crop, and hence its use is dependent upon favourable weather conditions. The optimum droplet size is thought to be about 80–120 μm; very fine droplets either evaporate or are lost in the wind and never settle, and larger droplets settle too rapidly. Mist blowers are expensive to buy, but very efficient; up to 30 ha of orchard can be sprayed by two operators in a day. Drift on to other crops is the main danger when using this type of sprayer.

Ultra-low-volume spraying

U.l.v. application technique was developed in E. Africa in the control of Desert Locust, shortly after World War II, using solutions of DNOC and dieldrin in diesel oil. Development of u.l.v. for crop spraying started much later. The technique consists essentially of the production of very small droplets (c. 70 μm) carried in a light oil and blown by a fan in drift spraying from light aircraft.

In the early 1960s much of the development work

on u.l.v. was done by the Plant Pest Control Division of USDA in co-operation with American Cyanamid. When applying u.l.v. sprays the pilot generally flew the plane much higher (3–6 m) than in conventional spray application where the altitude is usually about 1–2 m. The higher altitude increases the spray swath to about three times that of the conventional aerial spray. Generally, with the Cyanamid u.l.v. method the spray swath was about 30 m compared to 12 m. (See also page 122.)

Electrostatic sprayers. The ICI 'Electrodyne' sprayer is a small, hand-held apparatus with a disposable combined bottle and nozzle ('Bozzle') containing the already formulated insecticide. The efficiency of the application appears to be very high, with minimal drift, but at present only the hand-held sprayer is commercially available; ICI stated that they expected it to be widely available in 1982. They hoped that tractor-mounted 'Electrodyne' sprayers might be available by 1983/4. For further details see Coffee (1973) or your local ICI representative.

Dusters

All dusters consist basically of a hopper (a container for the dust), a system of agitation to disturb the dust and a feed mechanism to pass the dust into a current of air which is carried through an outlet as a turbulent cloud.

Hand dusters are usually primitive in structure and tedious to use, but can be effective for the small farmer. They are often a crude distributing arrangement built as part of the packaging; either two concentric cylinders free to move within each other, or a cardboard piston or diaphragm for pumping the air. As the air passes through the pack it picks up a small quantity of dust and ejects it through a nozzle.

Hand pump dusters are cheap and easy to operate but do not have very much control over the amount of dust delivered. More advanced pumps have a double-action plunger which maintains a constant and even stream of air.

Bellows-type dusters generate the air-stream by the contraction and expansion of a pair of bellows. The commonest types are worn on the back in knapsack fashion. The dust hopper (containing from 3.5–7 kg of dust) is carried on the back in a metal cradle; the bellows are situated either on the back or top of the hopper and drive a stream of air through a tube into a small mixing chamber by the hopper outlet. The dust is fed into the mixing chamber by a simple agitator, and then the air and dust travel along a flexible pipe running along the side of the operator, and can be controlled easily with one hand. The outlet may vary in shape and design for different purposes. The bellows and agitator are operated by a simple up-and-down movement of a lever worked by the free hand.

Rotary hand dusters produce an air-stream by a fan driven off a hand crank through a reduction gear. They are often mounted either on the chest of the operator or on his back, using a metal frame and system of straps.

Power dusters are manufactured in a variety of forms. Traction dusters derive their motive power from the turning of the land wheels, either as a wheelbarrow type or a two-wheeled trailer type. Wheelbarrow-type dusters are suitable for use on small areas only, whereas the trailer type is usually pulled by a tractor and has a much greater capacity. Power dusters are equipped with independent engines to provide the power for their operation. The smallest types can be strapped to the chest of the operator, but the larger types become very expensive. The larger power boom dusters are effective for treatment of large areas. Some of these dusters work on the drift principle and use the movement of ambient air to distribute the powder over the crop. Aeroplane dusting is carried out in areas where the terrain and crops are suitable, and again the air-stream is used to spread the powder over the crop.

Filters and nozzles

Two components of sprayers of particular importance are filters and nozzles. All spraying machines are equipped with a series of filters to ensure that no coarse particles are permitted to pass into the feedpipe and block the nozzles. Filters are vitally important, particularly since the mixing of the spray and the filling of the tank often take place in the field; without the main tank filter the spray would frequently become contaminated with insect bodies, leaves, pieces of grass, and other detritus, which would clog the nozzles and ruin the spray programme. At various points in the pipe system additional filters may be placed and the nozzles themselves may also be fitted with gauze filters in various positions according to their design.

Nozzles are the detachable apertures which break up the pesticide liquid into spray. Many different types of nozzles are made, and each gives its own particular spray pattern. The nozzle disc controls the final shape of the spray pattern. In *cone nozzles* (which have a spray pattern in the form of a hollow or solid cone) the disc is perforated by a series of small holes. In *fan nozzles* the perforation is an elongate, horizontal slit. Discs are often made either of tungsten steel or more recently in ceramics of various sorts, these materials being less subject to abrasion and wear. Cheap nozzles made of inferior materials soon abrade and then as the apertures enlarge the rate of spray application increases. Fan nozzles are generally less subject to wear than cone nozzles, but cone nozzles give a better breakdown of the water into droplets. The size of the apertures on the disc control the rate of spray delivery, but below a certain diameter the size does not greatly affect the size of droplet produced. Some nozzles can be adjusted to obtain droplet sizes from a fine mist to a heavy drenching spray. Pesticides in suspension have to be used with particular care for nozzles with small apertures can be easily blocked by them. For many of the better makes of sprayers a range of nozzles is available so that the most suitable nozzle for a particular purpose can be used. This is of great importance in spraying field trials because at the same rate of operation (i.e. pumping and walking with a knapsack sprayer) use of the different nozzles will give spray delivery rates in either the high-volume or low-volume ranges, according to the nozzle used. Using a very fine nozzle (low-volume) to put on a high-volume spray is exhausting to the operator, and conversely with a coarse nozzle it is almost impossible to apply a low-volume spray successfully.

8 *Pesticides in current use*

As previously mentioned, there are various ecological hazards associated with the repeated use of pesticides on crops, as they are mostly potent chemical poisons, but so far as can be seen agriculture will continue to rely heavily on their use for the forseeable future in the continuing battle with insect pests. What is important is that these poisons be used judicially; essentially this means the minimum dosage, and the fewest applications, for the maximum effect. This entails a high level of understanding, both of the chemicals themselves and the basic principles of pest control.

Types of pesticides

Although in this chapter pesticides are viewed basically according to their chemical structure, it should be remembered that their mode of action may be equally important, and on this basis they could be grouped as follows.

(a) *Repellants.* These work by keeping the insects away from the host plants by the use of repellent odours. In olden days the laying of kerosene-soaked sacks between crop rows had this effect; it also works by masking the typical olfactory signals from the host plant. At present mostly used against mosquitoes and other medical pests, and less often with crop plants.

(b) *Antifeedants.* Insects will land on the crop plants but these chemicals on the foliage inhibit feeding behaviour. Products from the Neem tree are proving to be very effective in this respect (see page 67).

(c) *Fumigants.* These are mostly volatile substances (liquids, and some powders) that vaporize under ambient conditions; it includes gases of various types. The gases are mostly used in food stores and other enclosed spaces; the volatile liquids likewise, and also for soil fumigation against nematodes (eelworms) and other soil pests.

(d) *Smokes.* These are finely divided powders mixed with a combustible material. As with fumigants, only suitable for use in an enclosed space, such as stored products pests, but also important in greenhouses.

(e) *Stomach poisons.* Only effective if ingested and absorbed through the intestinal tract. Usually applied to the foliage of the plant to be eaten by the pests with biting and chewing mouthparts; also mixed with baits to encourage ingestion.

(f) *Contact poisons.* These are absorbed directly through the insect cuticle.
   (i) *Ephemeral* – short-lived, and usually applied to the foliage.
   (ii) *Residual* – persistent (remain active for a long period of time). May be applied to foliage, but more often to soil for long-term activity.

(g) *Systemic poisons.* Watered on to the soil, or applied to tree trunks, or sprayed on to the foliage; absorbed by the plant and translocated through the plant body in the vascular system. Effective especially against sap-sucking and tunnelling insects; usually applied as a spray but granules are also employed. Translaminar pesticides can be absorbed through the leaves.

Pesticide recommendations

The number of pesticides with which I have personal experience is rather limited, and so recommendations have to be made on the basis of published information, both from the latest edition of the *Pesticide Manual* (7th ed., 1983), other general publications such as *Approved*

Products – 1984 (MAFF, 1984), etc., and data sheets from the chemical companies concerned. In some sources the data available are vague and include such statements as 'generally effective against soil (sap-sucking/foliage-eating) insects . . .', or 'effective against lepidopterous pests . . .'. The practising agricultural entomologist is clearly usually concerned with a particular pest (or pest complex) on a specific crop, and such vague recommendation is of limited use. As mentioned below, the effectiveness of a particular pesticide is a somewhat variable property which really requires quite specific definition. Thus in a book of this nature, where many different countries (and continents) are concerned, it is not feasible to make very specific pesticide recommendations, so only general recommendations are made together with information as to the more appropriate methods of application. For *detailed* pesticide usage, the local Ministry of Agriculture (or equivalent) recommendations should be consulted; they are generally published annually, and may even include supplements during the year as new information becomes available.

Pesticide effectiveness

As already mentioned (page 39), the *efficiency* of a pesticide is expressed as the more or less constant proportion of the population killed by the poison (under optimum conditions) regardless of the actual number of pests involved. The percentage killed by the poison under laboratory conditions (i.e. Potter Tower) may be quite constant, assuming no resistance has developed. *Effectiveness* (*sensu stricta*) is more concerned with the number of pests surviving after treatment, or more practically the extent of crop damage occurring after treatment.

In the remainder of this book, when considering control recommendations it is usual to apply the term '*effective*' in a broad sense (*sensu lato*) which embraces both aspects mentioned above; but students should be aware of the distinction.

In the halcyon days when DDT and dieldrin were used so effectively against so many pests there was little need to query the factors affecting efficiency, since usually both were giving consistent kills of 98–99% under a wide range of conditions. Nowadays the alternative pesticides are far more variable in their effectiveness (*s.l.*) and much less predictable in terms of results.

Perusal of the pesticide literature indicates that for example some 60 chemicals are regarded as 'effective' against Lepidoptera (caterpillars), and about 50 are regarded as 'aphicides'. Clearly not all these compounds are equally effective and many will not be chosen for use in pest management schemes. In any pest population the application of a chemical poison (under controlled conditions) will only kill a certain proportion; part of the population will be killed very easily (by small doses), and at the other extreme a small number of pests will survive even a massive dose. This is generally true for all large populations of organisms studied epidemiologically. For experimental (comparative) assessment it is usual to use the rate responsible for mortality of half the trial population (hence LD_{50}). But, of course, for practical control purposes usually a kill of 95% or more is the aim, depending in part on the type of crop/produce. Unfortunately for the entomologist, on some crops there is economic (or political) pressure to aim at a kill-level of 98–99%, which in these post-organochlorine days is seldom feasible. Out of the total range of reportedly 'effective' pesticides against caterpillars or aphids, if applied under optimum conditions at the optimum dose rate (see below), it may be expected that some chemicals can give a kill of 98%, whereas others will range from 90, 80, 70, 60, or even 50 or less percentage kill. Clearly the latter chemicals will not be used in most pest management programmes, unless, for example, they are *very* cheap.

This scale of pesticide effectiveness is to be expected in respect to use under optimum conditions and at optimum dosage. On consideration of the range of environmental conditions to be expected in the field, the variation in dose rates and application methods, etc., the end-result can be quite horrendous variation in relative effectiveness. It should also be pointed out that the range of results recorded for one species of aphid could possibly be even reversed when the chemicals are used against another species of aphid.

Factors influencing pesticide effectiveness

These factors can be grouped under four different headings, as shown below.

Pest organism

Species concerned. Certain species within a group (e.g. either Lepidoptera, or more specifically Noctuidae) may show markedly less susceptibility to a particular chemical than the others in the group. This is shown strikingly in the bollworm complex (Noctuidae) on cotton in parts of Africa.

Stage of organism. Both egg and pupal instars are biologically resistant stages designed to survive under adverse environmental conditions, and many (most) pesticides are ineffective against them. Conversely, a few pesticides may be particularly effective as ovicides.

Age of organism. With caterpillars, especially Noctuidae, the older stages, particularly fourth and fifth instars, are far less susceptible to most chemical poisons, even though at this time their food consumption is at a peak; the dose that would kill a second instar larva usually has no discernible effect upon a final instar larva. This situation also prevails so far as some other insects are concerned (Diptera, Coleoptera, etc.).

Development of resistance. From the first time an insect population is subjected to insecticidal action, selection for natural resistance is operating, and after a few years many species show noticeable resistance to that chemical (and other closely related chemicals usually). As resistance develops so the effectiveness of that chemical declines.

Chemical

Target success. In laboratory testing the target insect is typically sprayed directly, and so most application uncertainties are eliminated. However, with field crop applications the target insects are frequently not reached, due to their evasion, operator carelessness, insecticidal drift, inappropriate machinery or application equipment, incorrect timing, etc. At best, most field applications have a very low efficiency rating: usually only a few percent; the most efficient recorded has been aerial spraying of locust swarms (see page 123).

Persistence. This is by definition the efficiency over a period of time. Generally the organochlorine compounds were very persistent and remained effective for months often. Some carbamates are quite long-lived, but most organophosphorous compounds are quite short-lived and their persistence is measured in weeks or even days. Persistence in the field is closely linked to the physical/climatic factors listed below, as they affect the natural rate of chemical *degradation*. The storage history of a chemical being used is of relevance, as sometimes chemicals may be stored for several years prior to sale.

Dosage. Variation in dose rate produces a corresponding variation in kill. In practical terms dosage is arrived at as a balance between cost/effectiveness and phytotoxicity. A low dosage gives only a small kill, but usually encourages resistance development; a very high dosage scarcely increases the kill above the usual level, but is likely to cause phytotoxicity.

Sometimes the 'official' dosage recommendations are not particularly clear; for example in the *Pesticide Manual* (1979) the uses for permethrin are given as 'effective against a broad range of pests. It controls leaf- and fruit-eating lepidopterous and coleopterous pests in cotton at 100–150 g a.i./ha; fruit at 25–50 g/ha; vegetables at 40–70 g/ha; vines, tobacco, and other crops at 50–200 g/ha.' For practical purposes it would be preferable that more specific local recommendations be made, and this is usually done by the local Ministry of Agriculture staff. The presentation of dosage recommendations may be complicated in that technical data are expressed as a weight/volume of active ingredient (a.i.; i.e. pure chemical) per area of application, or volume of spray, etc. Most pesticides are produced in several different concentrations and different formulations, so the a.i. for each has to be calculated prior to use. A very recent trend in some publications has been to express dosage in terms of *acid equivalent* (a.e.), which is reported to be the active ingredient expressed in terms of the parent acid! For practical purposes a simplified system of dosage recommendation may be preferable, especially for peasant farmers, when the actual amount of locally available formulation to be used can be stated.

In some rural situations, especially throughout the Third World countries, the level of understanding in the farming/smallholder community, of the basic principles of pest control, may be very low. Where understanding is minimal there is often a tendency for the farmer to apply a smaller dose than recommended to save on the costs. Where there is some understanding of pest control principles by the farmer, there is sometimes a tendency to increase the suggested dosage on the assumption that 'more must be better'. In Europe and parts of Australia there has long been an inclination on the part of farmers to apply more pesticides, more heavily, and more frequently than actually needed; also many situations have been recorded where a farmer made appropriate pesticide application but then continued the applications long after the need had been removed. Present emphasis is for better communication, greater understanding by the farming community, and a more judicious use of pesticides generally.

Formulation. There may be some differences in effectiveness according to the precise mode of formulation, and also in respect to spray additives. Occasionally in spray mixtures there may be chemical incompatibility.

Cost. Some of the newer chemicals are expensive, and occasionally prohibitively so for smallholders and peasant farmers, especially for use on their food crops.

Availability. In the more advanced countries there is usually a large selection of pesticides available, as opposed to many Third World countries where often currency exchange problems restrict levels of imports, sometimes severely. However, some of the chemicals listed in this chapter are not universally available, as in Europe sales of Japanese chemicals and purely American chemicals are very limited, and *vice versa*, although many of the more useful pesticides are quite widely manufactured now throughout the world under licence from the company with the patent.

Local approval. Because of the hazards involved in the use of these chemical poisons, most countries now have local legislation to control the importation and/or use of the more toxic or ecologically damaging chemicals. For example, few countries now permit the import or use of parathion despite its effectiveness as an insecticide, because of its high mammalian toxicity, especially its dermal toxicity. Most European countries are likewise restricting the use of the organochlorine compounds, with a view to their eventual abandonment as suitable alternatives are found (and as resistance becomes more widespread and more total), but in the USA as a result of the activities of the EPA there has for several years been a total ban on the use of the organochlorine compounds.

Toxicity to bees, livestock and fish. Some compounds may be very effective entomologically, but their particular toxicity selective to pollinating bees, fish, etc., may restrict their use in certain situations. Thus in an area adjacent to intensive freshwater fishing it would be advisable to avoid the use of endosulfan with its very high fish toxicity. Similarly in parts of Europe where apiculture is a major industry it would be advisable to restrict the use of some of the compounds more dangerous to bees.

Climate/Soil

Temperature. At low temperatures the chemical may be relatively inactive (as will the insects of course), but at high temperatures chemical degradation is accelerated. In temperate situations repeated use of organochlorines typically resulted in a soil resïdue build-up over the winter period, but a fairly rapid degradation through the warm summer. Most tropical countries never experienced the alarming accumulation of DDT or dieldrin in the soils because the higher ambient temperatures throughout most of the year degraded the chemicals quite rapidly.

Moisture. High levels of soil moisture tend to accelerate chemical breakdown; but alternatively some chemicals do not work effectively if the soil is dry.

Rainfall. Heavy or persistent rainfall tends to wash off pesticides from the upper levels of plant foliage into the soil. This tends to be more of a tropical problem rather than temperate. Heavy rainfall will also cause chemical leaching into nearby water-collection systems.

Insolation. High levels of solar radiation (especially the ultra-violet component) greatly accelerate degradation of pesticides, but this is more a problem of the tropics rather than temperate regions.

Soil types. Some soil types, especially the peat (muck) types with a high organic content, may inactivate some pesticides by surface adsorption; dosages for such soils are usually doubled relative to those for mineral soils. Soils excessively acid or alkaline may affect pesticide performance and degradation rate. The importance of soil structure (as indicated by texture) in relation to pesticide performance is now becoming widely recognized and, for the first time in 1983, the MAFF *Approved Products for Farmers and Growers* had a section on 'The definition of soil types on labels of approved products' (pages 26–7).

Crop plants

Plant size. The usual aim when applying insecticidal sprays is for foliage cover, and clearly the size of the crop plants, spacing and plant density, will affect pesticide performance. If foliage density is great, then it may not be possible to reach the lower levels with a pesticide spray; some dense plants may have a leaf area index (l.a.i.) of 3.0 or even more. Leaf size and arrangement may also be important as many pests typically lurk underneath the leaves and may not be reached by insecticidal sprays delivered vertically from above. In many orchards and plantations, tree planting is arranged so as to permit spraying machinery access between the rows with nozzles pointing upwards to deliver the spray on to the undersides of the leaves.

Crop sensitivity. Some plants are unduly sensitive to certain chemicals, as is shown particularly by the 'sulphur-shy' crops and varieties on which sulphur formulations should not be used. At very high dose rates most crops can be expected to suffer phytotoxicity; this adverse effect is often a reaction to the spray additives rather than to the pesticide itself.

Considerations in pesticide use

Efficient use of pesticides

To ensure accurate application the label should be read carefully *before* use, and the recommended doses, volumes and times of applications followed. With many of the newer, short-lived chemicals, dosage, placement and timing are critical. Attention should also be paid to maintenance of equipment, for defects such as faulty or worn nozzles can appreciably alter rates of application.

Compatibility of mixed products used in combined spray programmes is important. If no information is given on the labels consult the manufacturer or dealer. Wetting agents are sometimes used as additives to sprays for crops that are difficult to wet (e.g. brassicas, peas), but not all wetting agents are compatible with all proprietary formulations.

Taint and damage to plants are hazards to be avoided. Where chemicals are known invariably to cause damage to certain crops or varieties of plants, or if taint or off-flavours are produced in edible crops, this is mentioned under the specific chemical.

Pest resistance to pesticides. Pest resistance to chemicals is one of the major problems in agricultural entomology in all parts of the world now, and is ever increasing. Such cases arise in localized areas, or even on certain holdings, and may remain purely local for some time. Resistance does not usually develop until a particular chemical has been employed in that area for some considerable time (3–10 years). Sometimes resistance to one compound may be closely followed by resistance to other related compounds, e.g. one organochlorine compound and the others in the same group.

Information as to which major local pests may be exhibiting resistance to widely used insecticides can usually be obtained from the local Ministry or Department of Agriculture staff.

Natural control agents. It is very important that pesticide applications should not adversely affect existing natural control of the pest population. To this end, it is desirable that knowledge of the local predators and parasites of key pests be accumulated so that, whenever possible, existing levels of natural control may be maintained in addition to the artificial control measures that are being applied. Fairly obvious cases where extreme care should be taken are infestations of scale insects on trees, leaf miners, and sometimes the late stages of aphid infestations.

Storage of chemicals. Storage of chemicals is important; they should be kept in a dry place protected from extremes of temperature. Generally frost is more harmful to stored liquid formulations than is high temperature. Most formulations may be expected to be used for at least two years without loss of efficiency, but unsatisfactory storage conditions can impair stability and effectiveness.

Safe use of pesticides

All pesticides should be treated with care whether they are known to be particularly poisonous or not, and the following routine observed.

(1) Read the label, especially the safety precautions, carefully before use.
(2) Use all products as recommended on the labels and do not use persistent chemicals where there are effective, less persistent alternatives.
(3) Avoid drift on to other crops, livestock and neighbouring property and take care to prevent contamination of any water source, whether for drinking or irrigation purposes.
(4) Safely dispose of all used containers. Liquid contents must first be washed out thoroughly and the washings added to the spray tank. Packages containing powders or granules must be completely empty before disposal. Burn bags, packets and polythene containers. Puncture non-returnable metal containers (except aerosol dispensers) and bury them in a safe place. Bury glass containers or dispose of them with other refuse. On no account use empty pesticide containers for any other purpose.
(5) Return unused materials to store under lock and key.
(6) Clean any protective clothing used, and wash exposed parts of the body thoroughly when the job is completed.

Particular attention should be paid to the paragraph headed 'Caution' in the pesticide section of this chapter. It indicates:

(1) whether the chemical is particularly poisonous; if so, certain protective clothing should be worn.
(2) whether there are any special user risks involved,

even if not very toxic, such as irritation to the skin. Some protective clothing should be used with these chemicals, and their labels should be consulted for guidance.

(3) what precautions should be taken to ensure that unacceptable residues do not remain on edible crops at harvest. Where appropriate, the time at which a chemical may be applied is shown and also the minimum interval that must be observed before last treatment and harvest.

(4) whether there are risks to bees, livestock, fish or other wild life. In the case of bees, the degree of risk is given as *dangerous*, *harmful* and *safe* (see following section).

Fish are susceptible to many chemicals, and great care must be taken to prevent contamination of ponds, waterways and ditches with chemicals or used containers.

Toxicity to bees

Dangerous: highly toxic to bees working the crop or weeds, at time of treatment, toxic for 24 hours or more.

Azinphos-methyl	HCH
Bendiocarb	Heptenophos
Carbaryl	Lead arsenate
Chlorpyrifos	Methomyl
Cypermethrin	Mevinphos
Deltamethrin	Monocrotophos
Demeton-*S*-methyl sulphone	Parathion
Diazinon	Permethrin
Dichlorvos	Phosphamidon
Dieldrin	Phoxim
Dimethoate	Pirimiphos-ethyl
Endrin	Pirimiphos-methyl
Ethoate-methyl	Propoxur
Etrimfos	Quinalphos
Fenitrothion	Triazophos
Formothion	

Harmful: toxic to bees working a crop at the time of treatment, but not hazardous if applied when bees are not foraging.

Acephate	Methidathion
Bromophos	Menazon
DDT	Nicotine
DDT with malathion	Omethoate
Demephion	Oxydemeton-methyl
Demeton	Phosalone
Demeton-*S*-methyl	Sodium monochloroacetate
Endosulfan	Tetrasul
Malathion	TDE
Mecarbam	Thiometon
Mercury	Vamidothion

Safe (now referred to as 'offering minimal hazard'):

Aldicarb (granules)	Mephosfolan (soil drench)
Binapacryl	Phorate (granules)
Dicofol	Pirimicarb
Diflubenzuron	Quinomethionate
Dinocap	Rotenone
Disulfoton (granules)	Schradan
Ethiofencarb (granules)	Tetradifon

With a number of pesticides in current use there is no precise information readily available as to their toxicity to bees.

Chemicals included in the Health and Safety (Agriculture) (Poisonous Substances) Regulations (1975) (UK)

These regulations are laid down in the Statutory Instrument No. 282 (1975), and they require any user of certain chemicals in the UK to observe certain precautions in their use. This section is included here as general information to indicate which of the pesticides are most dangerous, and the type of precautions required for their safe application. However, it must be remembered that only chemicals at one time 'approved' are on this list; several of the chemicals listed below are not at present (1983/4) approved, but they were on former lists.

Part I substances

Chloropicrin Dimefox	*Full protective clothing*, e.g. rubber gloves and boots, respirator, and either an overall and rubber apron or a mackintosh when preparing the diluted chemical. The respirator may be dispensed with when the diluted chemical is applied to the soil.

Part II substances

*Aldicarb Carbofuran Disulfoton *DNOC *Endosulfan Endrin Fonofos Mephosfolan *Methomyl *Mevinphos Oxamyl Parathion Phorate Schradan *Thiofanox Thionazin	*Full protective clothing*, which includes rubber gloves and boots and, either a face shield or dust mask, and an overall and rubber apron or a mackintosh; or a hood, or a rubber coat and sou'wester, depending on the operation being performed. A respirator is required when applying aerosols or atomizing fluids in glasshouses. When the substances are used in the form of granules certain relaxations in the protective clothing requirements are allowed.

Part III substances

Amitraz *Azinphos-methyl *Chlorfenvinphos *Deltamethrin *Demephion *Demeton-*S*-methyl *Dichlorvos Ethion Heptenophos *Methidathion Nicotine *Omethoate *Oxydemeton-methyl Phosphamidon Quinalphos *Thiometon *Triazophos *Vamidothion	Rubber gloves and face shield when preparing the diluted chemical. Full protective clothing, i.e. overall, hood, rubber gloves and respirator, is needed however when applying aerosols or atomizing fluids in glasshouses.

The more-widely used insecticides and acaricides currently available for crop protection†

The data presented here are mostly derived from Martin (1972*a*), the 7th ed. by Worthing (1983), MAFF (1983) and from the pesticide data sheets provided by the various chemical companies (Bayer, Shell, ICI, Ciba-Geigy, etc.).

Insecticides in current use mostly fall into several different categories, according to their chemical structure and mode of action, but some are unfortunately intermediate in chemical structure, or else somewhat anomalous, and do not lend themselves to easy categorization. For teaching purposes it is desirable to be able to distinguish between these basic groups. However, in this book, for practical purposes the following basic types of insecticides are recognized: chlorinated hydrocarbons, substituted phenols, organophosphorous compounds, carbamates, miscellaneous compounds, natural organic compounds, organic oils, biological compounds, and insect growth regulators. Clearly some of these groups could equally well be subdivided and their arrangement here is somewhat arbitrary.

Many pesticides are sold as mixtures to be used against a pest complex.

*These pesticides are included on the *Poisons List* (1972), and the *Poisons Rules* apply general and specific provisions for the labelling, storage and sale of these chemicals (as per *Approved Products – 1983*).

†The insecticides and acaricides used mostly for protection of stored products, domestic use, and against medical and veterinary pests, are not included in this list, neither are chemicals used solely as nematicides, fungicides, etc.

In the following section only a few of the more widely used trade names for the pesticides are given, but a more extensive list of trade names is included in Appendix A.

Chlorinated hydrocarbons

These compounds are often referred to as the 'organochlorines' and collectively they are a broad-spectrum and very persistent group, which usually kill both by contact and as stomach poisons. Generally they are more effective against insects with biting mouthparts than the sap-suckers. Because of their persistence they get taken up in food chains very easily and accumulate in the body fat of the vertebrate predators at the apex of the food chains. Under normal conditions this build-up in the body fat may not affect the animal, but in times of starvation, when body fat reserves are being utilized, the amount of pesticide released into the blood may be critical or even fatal. Most countries in Europe and N. America are in the process of restricting the use of chlorinated hydrocarbons where *suitable alternatives* are available, because of the long-term contamination dangers to the environment. It can also be observed that in most of these countries the major pests have already become, or are now becoming, resistant to the organochlorine compounds, and so their period of utility is ending. In some cases, though, suitable alternatives are not readily forthcoming.

The organochlorine compounds may be subdivided into several groups: the three most commonly used groups are represented by DDT, HCH and aldrin (cyclodienes). Despite the structural differences between the subgroups they do possess several characteristics in common: they are chemically stable, have a low solubility in water, moderate solubility in organic solvents and lipids, and a low vapour pressure. The properties of stability and solubility make most of the group very persistent. They also produce similar physiological responses in the insects.

The solubility of DDT in lipids enables the poison to penetrate the insect integument quite readily (as opposed to slow penetration through animal skin). This difference in penetrative ability would account for its selective toxicity to insects. Cuticle thickness does not seem to greatly influence the susceptibility of insects to DDT, although it may penetrate more readily through the flexible intersegmental membranes. Dissolution in the epicuticular wax is apparently the essential prerequisite to toxic action. Penetration rate of the poison increases with temperature, often nearly doubling for a 20 °C rise in temperature. The precise mode of action of these poisons on the insect is not fully understood.

1. Aldrin (*Aldrin, Aldrite, Aldrex, Drinox, Toxadrin,* etc.)

Properties. A broad-spectrum, persistent, non-systemic, non-phytotoxic insecticide with high contact and stomach activity, effective against soil insects at rates of 0.5–5 kg/ha.

Aldrin is stable to heat, alkali and mild acids, but the unchlorinated ring is attacked by oxidizing agents and strong acids, and it is oxidized to dieldrin. It is compatible with most pesticides and fertilizers but is corrosive because of the slow formation of HCl on storage.

The technical product is a brown or colourless solid, practically insoluble in water but is quite soluble in mineral oils, and readily soluble in acetone, benzene and xylene.

The acute oral LD_{50} for rats is 67 mg/kg; it is absorbed through the skin.

Use. Effective against all soil insects, e.g. termites, beetle adults and larvae, fly larvae, cutworms, crickets, ants, etc.

Caution

(*a*) Harmful to fish.
(*b*) Treated seeds should not be used for human or animal consumption.
(*c*) Avoid excessive skin contact.
(*d*) Risks to wildlife are considerable on a long-term basis.

Formulations. 30% e.c.; 2.5–5% dust; m.l.; 5 and 20% granules; 20–50% w.p.; or liquid seed dressing; and as an insecticidal lacquer.

2. Chlordane (*Sydane,* etc.)

Properties. A persistent, non-systemic contact and stomach insecticide; non-phytotoxic at insecticidal concentrations. Usually a mixture of isomers (including heptachlor). The technical product is a viscous amber liquid, insoluble in water but soluble in most organic solvents.

The acute oral LD_{50} for rats is 457–590 mg/kg.

Use. Mainly used against termites, beetles (wood-boring and crop pests), and in ant baits, and sometimes for reducing the numbers of earthworms in lawns.

Caution

(*a*) Harmful to fish and livestock.
(*b*) Pre-access interval for livestock to treated areas – 2 weeks.

Formulations. Include e.c. 500 and 700 g a.i./l; dusts; granules 50 and 100 g/kg; kerosene solutions 20 and 200 g/l.

3. Chlorobenzilate (*Akar, Folbex*)

Properties. A non-systemic acaricide with little insecticidal action, particularly effective against phytophagous mites. Some phytotoxicity has been noted on pear, plum, and some apple varieties.

The technical product is a brown liquid, insoluble in water, but soluble in most organic solvents including petroleum oils. It is hydrolysed by alkali and strong acids.

The acute oral LD_{50} for rats is from 700 to 3100 mg/kg.

Use. Effective only against phytophagous mites, but will kill all active and inactive stages, including eggs.

Caution. Pre-harvest interval for edible crops is not known.

Formulations. 25% and 50% e.c.; 25% w.p.; and 'Folbex' fumigation strips for smoke generation.

4. DDT (*DDT*, many trade names)

Properties. A broad-spectrum stomach and contact poison, of high persistence, non-systemic, and non-phytotoxic except to cucurbits.

The *pp'*-isomer forms colourless crystals, practically insoluble in water, moderately soluble in petroleum oils and readily soluble in most aromatic and chlorinated solvents. The technical product is a waxy solid. DDT is dehydrochlorinated at temperatures above 50 °C, a reaction catalysed by u.v. light. In solution it is readily dehydrochlorinated by alkalis or organic bases; otherwise it is stable, being unattacked by acid and alkaline permanganate or by aqueous acids and alkalis.

The acute LD_{50} for male rats is 113 mg/kg. DDT is stored in the body fat of birds and mammals and excreted in the milk of mammals.

Use. Effective against most insects, but with little action on phytophagous mites; resistance now widespread.

Caution

(*a*) Harmful to bees, fish and livestock.
(*b*) Pre-harvest interval for edible crops – 2 weeks.
(*c*) Do not use on cucurbits or certain barley varieties, as damage may occur.

Formulations. 25% e.c.; 25% m.l.; 50% w.p.; 5% dusts; smokes. May be combined with HCH or malathion in sprays or smokes.

5. Dicofol (*Kelthane, Acarin, Mitigan*)

Properties. It is a non-systemic acaricide, with little insecticidal activity, recommended for the control of mites on a wide range of crops. Although residues in soil decrease rapidly, traces may remain for a year or more. Safe to bees.

The pure compound is a white solid, practically insoluble in water, but soluble in most aliphatic and aromatic solvents. It is hydrolysed by alkali, but is compatible with all but highly alkaline pesticides. Wettable powder formulations are sensitive to solvents and surfactants, which may affect acaricidal activity and phytotoxicity.

The acute oral LD_{50} for male rats is 809 ± 33 mg/kg.

Use. Effective only against Acarina; kills eggs and all active stages of the mites.

Caution. Pre-harvest interval for edible crops – 2–7 days.

Formulation. 18.5% and 42% e.c.; 30% dust.

6. Dieldrin (*Dieldrex, Alvit, Dilstan,* etc.)

Properties. A broad-spectrum, persistent, non-systemic, non-phytotoxic insecticide made by oxidation of aldrin. It is of high contact and stomach activity to most insects.

It is stable to light, alkali and mild acids, and is compatible with most other pesticides. Dieldrin occurs as white odourless crystals; the technical product is in light brown flakes; practically insoluble in water, slightly soluble in petroleum oils, moderately soluble in acetone, soluble in aromatic solvents.

The acute oral LD_{50} for male rats is 46 mg/kg; it can be absorbed through the skin.

Use. Effective against most insects, but resistance is now very widespread.

Caution

(*a*) Harmful to fish.
(*b*) Dangerous to bees.
(*c*) Seed dressings containing organomercury compounds can cause rashes or blisters on the skin. Those containing thiram can be irritating to skin, eyes, nose and mouth.
(*d*) Treated seed should not be used for human and animal consumption.
(*e*) Dressed seeds are dangerous to birds.

Formulations. 15% e.c.; m.l.; 50% w.p.; dry seed dressings. Some seed dressings made with thiram or organomercuric compounds.

7. Endosulfan (*Thiodan, Cyclodan,* etc.)

Properties. It is a non-systemic contact and stomach insecticide and acaricide.

The technical product is a brownish crystalline solid, practically insoluble in water, but moderately soluble in most organic solvents. It is a mixture of two isomers. It is stable to sunlight but subject to a slow hydrolysis to the alcohol and sulphur dioxide. It is compatible with non-alkaline pesticides.

The acute oral LD_{50} for rats is 110 (55–220) mg/kg.

Use. Effective against most crop mites, and some Hemiptera (aphids, capsids), caterpillars and beetles.

Caution

(*a*) This is a poisonous substance (Part II): full protective clothing should be worn.
(*b*) Extremely dangerous to fish.
(*c*) Dangerous to livestock.
(*d*) Harmful to bees.
(*e*) Pre-harvest interval for fruit – 6 weeks.
(*f*) Pre-access interval to treated areas: of unprotected persons – 1 day; animals and poultry – 3 weeks.

Formulations. 17.5% and 35% e.c.; 17.5%, 35% and 50% w.p.; 1%, 3%, 4% and 5% dusts, and 5% granules.

8. Endrin (*Endrex, Hexadrin, Mendrin*)

Properties. A very toxic, broad-spectrum, persistent insecticide and acaricide isomeric with dieldrin. It is non-systemic, and non-phytotoxic at insecticidal concentrations, but is suspected of damage to maize.

It is a white crystalline solid practically insoluble in water, sparingly soluble in alcohols and petroleum oils, moderately soluble in benzene and acetone. The technical product is a light brown powder of not less than 85% purity. It is stable to alkali and acids but strong acids or heating above 200 °C cause a rearrangement to a less insecticidal derivative. It is compatible with other pesticides.

The acute oral LD_{50} for male rats is 17.5 mg/kg.

Use. Effective against many insects and mites; used mainly on field crops.

Caution

(*a*) A very poisonous pesticide (Part II): full protective clothing should be worn.
(*b*) Dangerous to bees, fish, livestock, wild birds and animals.
(*c*) Fruit should not be sprayed after flowering.
(*d*) Minimum interval to be observed between last application and access to treated areas: of unprotected persons – 1 day; of animals and poultry – 3 weeks!

Formulations. 20% liquid formulation, or dust.

9. HCH, γ (*Lindane, HCH, BHC, Gammalin*)

Properties. Benzene hexachloride exists as five isomers in the technical form but the active ingredient is the γ-isomer. *Lindane* is required to contain not less than 99% γ-HCH. It exhibits a strong stomach poison action, persistent contact toxicity, and fumigant action, against a wide range of insects. It is non-phytotoxic at insecticidal concentrations. The technical HCH causes 'tainting' of many crops but there is less risk of this with *Lindane*.

Lindane is stable to air, light, heat and carbon dioxide; unattacked by strong acids but can be dehydrochlorinated by alkalis.

It occurs as colourless crystals and is practically insoluble in water; slightly soluble in petroleum oils; soluble in acetone, aromatic and chlorinated solvents.

The acute oral LD_{50} for rats is 88 mg/kg.

Use. Effective against many soil insects, e.g. beetle adults and larvae, fly larvae, ants, Collembola, and also against many other biting and sucking insects, e.g. aphids, psyllids, whiteflies, capsids, midges, sawflies and thrips.

Caution

(*a*) Dangerous to bees and spiders.
(*b*) Harmful to fish and livestock.
(*c*) Pre-harvest interval about two weeks on most crops.
(*d*) Treated seeds should not be used for human or animal consumption.

Formulations. Many different seed dressings, some with added organomercury compounds or with Captan or Thiram; 50% w.p. with bran as bait; dust; liquid e.c., or suspension.

10. Heptachlor (*Drinox, Heptamul, Velsicol*)

Properties. A broad-spectrum, non-systemic, stomach and contact insecticide with some fumigant action.

It is a white crystalline solid, practically insoluble in water, slightly soluble in alcohol, more so in kerosene; it is stable to light, moisture, air and moderate heat. It is compatible with most pesticides and fertilisers.

The acute oral LD_{50} for male rats is 100 mg/kg.

Use. Effective against many different insect species; used mostly against soil-inhabiting fly and beetle larvae, and in ant baits.

Formulations. As seed dressings; e.c.; w.p.; dusts and granules of various a.i. contents.

11. Mirex (*Mirex, Dechlorane*)

Properties. It is a stomach insecticide, with little contact effect, used mainly against ants.

It is a white solid of negligible volatility; insoluble in water but moderately soluble in benzene, carbon tetrachloride and xylene; unaffected by concentrated mineral acids.

Acute oral LD_{50} for male rats is 306 mg/kg.

Use. Mostly used in baits against Fire Ants, Harvester Ants and Leaf-cutting Ants.

12. Tetradifon (*Tedion, Duphar*)

Properties. A systemic acaricide toxic to the eggs and all stages of phytophagous mites except adults. At acaricidal concentrations it is non-phytotoxic.

It forms colourless crystals, almost insoluble in water, slightly soluble in alcohols and acetone, more soluble in aromatic hydrocarbons and chloroform. It is resistant to hydrolysis by acid or alkali, is compatible with other pesticides, and is non-corrosive.

The acute oral LD_{50} for rats is more than 5000 mg/kg.

Use. Effective against eggs, larvae, nymphs of phytophagous mites, but not adults. Recommended for application to top fruit, citrus, tea, cotton, grapes, vegetables, ornamentals and nursery stock.

Caution

(*a*) Do not use smokes to young cucumbers or plants that are wet as damage may occur.
(*b*) At correct dosage is harmless to beneficial insects.
(*c*) Safe to bees.

Formulations. 20% w.p.; 18% e.c.; may be combined with malathion in smoke foundations.

13. Tetrasul (*Animert V-101*)

Properties. A non-systemic acaricide, highly toxic to eggs and all stages of phytophagous mites except adults. At the correct dosage it is non-phytotoxic. As it is highly selective it does not pose a hazard to beneficial insects or to wild life.

It is a brown crystalline solid, only slightly soluble in water, moderately soluble in acetone and ether, but soluble in benzene and chloroform. Stable under normal conditions, but should be protected against prolonged exposure to sunlight; it is oxidized to its sulphone, tetradifon. It is non-corrosive, and compatible with most other pesticides.

The acute oral LD_{50} for female rats is 6810 mg/kg.

Use. Effective against eggs, larvae and nymphs of most phytophagous mites, but not adults. Recommended for use on fruit and cucurbits at the time when the winter eggs are hatching.

Caution. Harmful to bees.

Formulations. 18% e.c. and 18% w.p.

14. TDE (*DDD, Rhothane*)

Properties. It is a non-systemic contact and stomach insecticide which, not of the general high potency of DDT, is of equal and greater potency against certain insects, e.g. leaf-rollers, mosquito larvae, hornworms. It is non-phytotoxic except possibly to cucurbits.

The pure compound forms colourless crystals practically insoluble in water but soluble in most aliphatic and aromatic compounds. Its chemical properties resemble DDT, but it is more slowly hydrolysed by alkali.

The acute oral LD_{50} for rats is 3400 mg/kg.

Use. Particularly effective against leaf-rollers, mosquito larvae, hornworms; effective against many caterpillars, weevils, capsids, thrips and earwigs.

Caution

(*a*) Harmful to bees, fish and livestock.
(*b*) Pre-harvest interval for edible crops – 2 weeks.
(*c*) Pre-access interval for livestock to treated areas – 2 weeks.

Formulations. 50% w.p.; e.c. 25%; 5% and 10% dusts.

Substituted phenols

The nitrophenols are not likely to compete with the newer insecticides; they are used mainly as herbicides and for the control of powdery mildews. The dinitrophenols have mammalian toxicity so high as to restrict their usefulness in crop protection.

15. Binapacryl (*Morocide, Acaricide, Dapacryl*)

Properties. It is a non-systemic acaricide (also effective against powdery mildews) and mainly used against red spider mites. Non-phytotoxic to a wide range of apples, pears, cotton and citrus; some risks of damage to young tomatoes, grapes and roses.

It is a white crystalline powder, practically insoluble in water but soluble in most organic solvents. It is unstable in concentrated acids and dilute alkalis, suffers slight hydrolysis on long contact with water and is slowly decomposed by u.v. light. It is non-corrosive, and compatible with w.p. formulations of insecticides and non-alkaline fungicides. With organophosphorous compounds it may be phytotoxic.

Acute oral LD_{50} for rats is 120–165 mg/kg.

Use. Effective especially against red spider mites.

Caution

(*a*) Harmful to fish and livestock, but safe to bees.
(*b*) Pre-harvest interval for edible crops – 1 week.
(*c*) Pre-access interval for livestock to treated areas – 4 weeks.

Formulations. 25% and 50% w.p.; 40% e.c.; 4% dust.

16. Dinocap (*Karathane*)

Properties. A non-systemic acaricide and contact fungicide, used widely on fruit trees, vines and ornamentals.

A dark brown liquid, usually a mixture of several isomers; insoluble in water, but soluble in most organic solvents.

The acute oral LD_{50} for rats is 980–1190 mg/kg.

Use. Recommended for use against *Panonychus ulmi* on apple, and various powdery mildews.

Caution

(*a*) Phytotoxic to some plants (chrysanthemums especially).
(*b*) Irritating to eyes, skin, and respiratory systems.
(*c*) Dangerous to fish.
(*d*) Pre-harvest interval: outdoor edible crops – 1 week; edible glasshouse crops – 2 days.

Formulations. Include e.c. 500 g a.i./l; w.p. 250 g/kg.

17. DNOC in petroleum oil (*Sinox, Dinitrol*)

Properties. It is a non-systemic stomach poison and contact insecticide; ovicidal to the eggs of certain insects. It is strongly phytotoxic and its use as an insecticide is limited to dormant sprays or on waste ground, e.g. against locusts. It is also used (not usually in oil) as a contact herbicide for the control of broad-leaved weeds in cereals, and in e.c. formulations for the pre-harvest desiccation of potatoes and leguminous seed crops. It forms yellowish, odourless crystals, only sparingly soluble in water, but soluble in most organic solvents and in acetic acid. The alkali salts are water soluble. It is explosive, and is usually moistened with up to 10% water to reduce the hazard, though it is corrosive to mild steel in the presence of water.

Use. Effective against overwintering stages of aphids, capsids, psyllids, scale insects, red spider mites, and various Lepidoptera on top, bush and cane fruit. Products containing DDT also control various weevils and tortrix moths in their overwintering stages. Also used as a herbicide.

Caution

(*a*) If the concentrated substance contains more than 5% of DNOC it is a Part II substance, and protective clothing should be worn.
(*b*) Dangerous to fish.
(*c*) Very phytotoxic; use only as dormant spray or on waste ground.

Formulations. The insecticide formulation is an e.c. in petroleum oil; it may be formulated with DDT.

18. Pentachlorophenol (*Dowicide, Santophen,* etc.)

Properties. An insecticide used for termite control, a fungicide used for protection of timber from fungal rots and wood-boring insects. It is strongly phytotoxic, and is used as a pre-harvest defoliant and as a general herbicide.

It forms colourless crystals; volatile in steam; almost insoluble in water; soluble in most organic solvents. It is non-corrosive in the absence of moisture, solutions in oil cause deterioration of natural rubber but synthetic rubbers may be used in equipment and protective clothing.

The acute oral LD_{50} for rats is 210 mg/kg; it irritates mucous membranes and causes sneezing, the solid and aqueous solutions stronger than 1% cause skin irritation.

Use. Effective against termites and other wood-boring insects; use restricted by strong phytotoxicity.

Caution

(*a*) Very phytotoxic.
(*b*) Irritation to skin and mucous membranes.

Formulations. Used as such or formulated in oil. *Santobrite* and *Dowicide G* are the technical sodium salt.

Organophosphorous compounds

These were discovered and developed during the Second World War by a German research team responsible for developing nerve gases; they are amongst the most toxic substances known to man.

These compounds have phosphorus chemically bonded to the carbon atoms or organic radicals, and are effective as both contact and systemic insecticides and acaricides. Nearly 50 compounds are in current use against insects and mites. Many of these compounds are very toxic to mammals and birds and have to be handled with care. Doses may be accumulative. The systemic compounds are very effective against sap-sucking insects. All the organophosphorous compounds are relatively transient and are soon broken down to become non-toxic. In comparison with the persistent chlorinated hydrocarbons great care in timing and application is required in the use of these compounds for effective results. The mode of action in both insects and mammals appears to be inhibition of acetylcholinesterase (or some very similar enzyme). Some newer compounds are more persistent.

19. Acephate (*Orthene, Ortran, Tornado*)

Properties. A systemic insecticide of moderate persistence; residual activity, at foliar rates, lasting 10–15 days. Used against a wide range of crop pests; safe to crops.

The technical product (80–90% purity) is a colourless solid, quite stable; quite soluble in water, less so in organic solvents.

The acute oral LD_{50} for rats is 866–945 mg/kg; less for mice and many birds.

Use. Effective against many aphids, thrips, leaf miners, caterpillars and sawfly larvae.

Caution

(*a*) Harmful to bees.
(*b*) Pre-harvest interval for edible crops – 6 weeks.

Formulations. Include w.s.p. 250, 500, 700 g a.i./kg; pressurized sprays 2.5 and 10 g/l; granules.

20. Azinphos-methyl (*Gusathion, Benthion*)

Properties. A non-systemic, broad-spectrum, insecticide and acaricide of relatively long persistence, with contact and stomach action. It forms white crystals, almost insoluble in water but soluble in most organic solvents. It is unstable at temperatures above 200 °C and is rapidly hydrolysed by cold alkali and acid.

The acute oral LD_{50} for rats is 16.4 mg/kg.

Use. Effective against Lepidoptera, mites, aphids, whiteflies, leafhoppers, scales, psyllids, sawflies, thrips, grasshoppers, some fly larvae and some beetles.

Caution

(*a*) A poisonous substance (Part III): protective clothing should be worn.
(*b*) Dangerous to bees.
(*c*) Harmful to fish and livestock.
(*d*) Pre-harvest interval for edible crops – 1–4 weeks according to crop, and country.
(*e*) Pre-access interval for livestock to treated area – 2 weeks.

Formulations. 20% e.c.; 25% and 50% w.p.; 2.5% and 5% dusts; u.l.v. formulations. A formulation with demeton-*S*-methyl sulphone known as *Gusathion MS* has a more systemic action.

21. Azinphos-methyl with demeton-*S*-methyl sulphone (*Gusathion MS*)

Properties. A mixture with *Metasystox* combining the properties of both pesticides; usually the proportion is 75% to 25%. In practice the mixture acts like azinphos-methyl with a systemic action.

Use. Also effective against thrips, aphids, and fly larvae (midges).

22. Bromophos (*Brofene, Bromovur, Nexion, Pluridox*)

Properties. A broad-spectrum, contact and stomach insecticide: persists on sprayed foliage for 7–10 days. Non-phytotoxic at insecticidal concentrations, but suspected of damage to plants under glass. No systemic action.

It occurs as yellow crystals, relatively insoluble in water, but soluble in most organic solvents, particularly in tetrachloromethane, diethyl ether, and methyl benzene. It is stable in media up to pH9, non-corrosive, and compatible with all pesticides except sulphur and the organometal fungicides.

The acute oral LD_{50} for rats is 3750–7700 mg/kg.

Use. Effective against a wide range of insects on crops, at concentrations of 25–75 mg/100 l.

Caution

(*a*) Harmful to fish and bees.
(*b*) Possibly phytotoxic under glass.

Formulations. E.c. 250, 400 g a.i./l; w.p. 250 g a.i./kg; dusts 20–50 g a.i./kg; atomizing concentrate 400 g a.i./l; coarse powder 30 g a.i./kg; granules 50–100 g a.i./kg.

23. Bromopropylate (*Acarol, Neoron*)

Properties. A contact acaricide with residual action, effective against mites resistant to organophosphorous compounds.

It is a crystalline solid, only slightly soluble in water, but is readily soluble in most organic solvents.

The acute oral LD_{50} for rats was more than 5000 mg/kg.

Use. Effective against all phytophagous mites on crops; recommended for use on fruits, vegetables, cotton, flowers at 37.5–60 g a.i./100 l; on field crops at 0.5–1.0 kg/ha.

Caution

(*a*) Dangerous to fish.
(*b*) Harmful to birds.

Formulations. E.c. at 250 and 500 g a.i./l.

24. Carbophenothion (*Trithion, Garrathion, Dagadip*)

Properties. A non-systemic acaricide and insecticide, with a long residual action. It is phytotoxic at high concentrations to some plants. 50% degradation in soil occurs in 100 days or longer, depending upon the soil type.

It is a pale amber liquid; insoluble in water, but miscible with most organic solvents. Relatively stable to hydrolysis; is oxidized on the leaf surface to the phosphorothiolate; compatible with most pesticides, and is non-corrosive to mild steel.

The acute oral LD_{50} for male rats is 32.2 mg/kg.

Use. Used mainly on deciduous fruit, in combination with petroleum oil, as a dormant spray for the control of overwintering mites, aphids and scale insects; on citrus as an acaricide.

Caution. Harmful to livestock, birds and wild animals.

Formulations. E.c. 0.25, 0.5, 0.75, 1.0 kg/l; 25% w.p.; dusts of 1, 2, and 3%.

25. Chlorfenvinphos (*Birlane, Sapecron, Supona*)

Properties. A relatively short-lived insecticide, effective against soil insects, non-phytotoxic at recommended dosages. The pure compound is an amber-coloured liquid, sparingly soluble in water but miscible with acetone, ethanol, kerosene and xylene. It is stable when stored in glass or polythene vessels, but is slowly hydrolysed by water. It may corrode iron and brass on prolonged contact and the e.c. formulations are corrosive to tin plate.

Acute oral LD_{50} for rats is 10–39 mg/kg.

Use. Particularly effective against rootflies, rootworms and cutworms as soil applications. As a foliar insecticide it is recommended for the control of Colorado Beetle on potato, leafhoppers on rice, and for stem borers on maize, sugarcane and rice. The half-life in soil is normally only a few weeks.

Caution

(*a*) This is a poisonous substance (Part III): protective clothing should be worn.

(*b*) Dangerous to fish.
(*c*) Use seed dressings carefully to avoid risks to birds.
(*d*) Treated seed should not be used for human or animal consumption.
(*e*) Pre-harvest interval for edible crops – 3 weeks.

Formulations. 24% e.c.; 25% w.p.; 5% dust; 10% granules; seed dressings (liquid) 40% (+2% mercury compounds).

26. Chlorpyrifos (*Dursban, Lorsban*)

Properties. A broad-spectrum insecticide with contact, stomach and vapour action. No systemic action. At insecticidal concentrations it is non-phytotoxic. It is sufficiently volatile to make insecticidal deposits on nearby untreated surfaces. In soil it persists for 2–4 months.

It forms as white crystals; insoluble in water, but soluble in methanol and most other organic solvents. Stable under normal storage conditions. It is compatible with non-alkaline pesticides; corrosive to copper and brass.

The acute oral LD_{50} for male rats is 163 mg/kg.

Use. Effective against many soil and foliar insects, and mite pests. Also used domestically against flies, mosquitoes and household pests, as well as ectoparasites of sheep and cattle. Specifically used for control of root maggots, aphids, capsid bugs, caterpillars and red spider mites.

Caution
(*a*) Dangerous to bees, fish and shrimps.
(*b*) Pre-harvest interval for all edible crops is 2–6 weeks.

Formulations. W.p. 25%; e.c.; 0.2 and 0.4 kg/l; granules 1–10%.

27. Chlorpyrifos-methyl (*Reldan*)

Properties. An insecticide with a broad range of activity, by contact, stomach and fumigant action; no systemic activity; not persistent in soil.

It forms colourless crystals; insoluble in water, but soluble in most organic solvents. Quite stable, but hydrolysed under both acid and alkaline conditions.

The acute oral LD_{50} for rats is 1630–2140 mg/kg.

Use. Mostly used for protection of stored grains and foodstuffs, and against household pests; also mosquitoes (adults), aquatic larvae, and some foliar crop pests.

Caution
(*a*) Irritating to eyes and skin.
(*b*) Dangerous to some fish and especially to shrimps.

Formulations. Only an e.c. 240 g a.i./l.

28. Cyanofenphos (*Surecide*)

Properties. An insecticide used against a wide range of crop pests, both in the tropics and temperate regions. A colourless crystalline solid insoluble in water, but moderately soluble in aromatic solvents.

The acute oral LD_{50} for rats is 89 mg/kg.

Use. Effective against rice stem borers and gall midges, cotton bollworms, and caterpillars and other vegetable pests in temperate regions.

Caution
(*a*) A very poisonous substance, and great care is required in its use; full protective clothing is necessary.
(*b*) Very dangerous to fish.

Formulations. Include e.c. 250 g a.i./l; dust of 15 g/kg (Sumitomo Chem.Co.).

29. Demephion (*Cymetox, Pyracide*)

Properties. A systemic insecticide and acaricide effective against sap-feeding insects, and non-phytotoxic to most crops.

It is a straw-coloured liquid (a mixture of two isomers), miscible with most aromatic solvents, chlorobenzene and ketones; immiscible with most aliphatic solvents. It is generally non-corrosive and compatible with most, except strongly alkaline, pesticides.

The acute oral LD_{50} for rats is about 0.015 ml/kg; the acute dermal LD_{50} is about 0.06 ml/kg.

Use. Mainly used against aphids, on all crops. (Now generally superseded.)

Caution

(*a*) This is a poisonous substance (Part III): protective clothing should be worn.
(*b*) Harmful to bees, livestock, fish, game, wild birds and animals.
(*c*) Pre-harvest interval for edible crops – 3 weeks.
(*d*) Pre-access interval for livestock to treated areas – 2 weeks.

Formulations. 30% e.c.

30. Demeton (*Systox, Solvirex*)

Properties. A systemic insecticide and acaricide with some fumigant action, effective especially against sap-sucking insects and mites. No marked phytotoxicity has been recorded.

The technical product is a light yellow oil, hydrolysed by strong alkali, but is compatible with most non-alkaline pesticides. It is almost insoluble in water but is soluble in most organic solvents.

The acute oral LD_{50} for male rats is 30 mg/kg.

Use. Effective against sap-sucking insects and mites.

Caution

(*a*) Harmful to bees.
(*b*) Harmful to fish and livestock.
(*c*) Pre-harvest interval is not known.

Formulations. E.c. of different oil contents.

31. Demeton-*S*-methyl (*Metasystox 55, Demetox*)

Properties. A systemic and contact insecticide and acaricide, metabolized in the plant to the sulphoxide and sulphone; rapid in action; moderate persistence.

It is a colourless oil, only slightly soluble in water, soluble in most organic compounds. It is hydrolysed by alkali. The acute oral LD_{50} for rats is 65 mg/kg of the technical material, 40 mg/kg of the pure.

Use. Effective against most sap-sucking pests (aphids, leafhoppers, etc.), sawflies and red spider mites.

Caution

(*a*) This is a poisonous substance (Part III): protective clothing should be worn.
(*b*) Certain ornamentals, especially some chrysanthemums, may be damaged by sprays.
(*c*) Harmful to bees, fish, livestock, game, wild birds and animals.
(*d*) Pre-harvest interval for edible crops – 2–3 weeks.
(*e*) Pre-access interval for livestock to treated areas – 2 weeks.

Formulations. 25 and 50% e.c. with emulsifier chosen to reduce dermal hazards.

32. Diazinon (*Basudin, Diazitol, DBD, Neocid,* etc.)

Properties. A non-systemic insecticide with some acaricidal action, used mainly against flies, both in agriculture and veterinary practice. At higher dosages it may be phytotoxic.

It is a colourless oil almost insoluble in water but is miscible with ethanol, acetone, xylene, and is soluble in petroleum oils. It decomposes above 120 °C and is susceptible to oxidation; stable in alkaline media, but is slowly hydrolysed by water and dilute acids. It is compatible with most pesticides but should not be compounded with copper fungicides.

The acute oral LD_{50} for male rats is 108 mg/kg.

Use. Especially effective against flies and their larvae (e.g. Anthomyiidae on vegetables and Carrot Fly), also used against mites, thrips, springtails, glasshouse pests, and some bugs (aphids, capsids, etc.).

Caution

(*a*) Dangerous to bees.
(*b*) Harmful to fish, livestock, game, wild birds and animals.
(*c*) Overdosage may lead to phytotoxicity on some crops.
(*d*) Pre-harvest interval for edible crops – usually 2 weeks.

(*e*) Pre-access interval for livestock to treated areas – 2 weeks.
(*f*) Should no longer be used on rice in S.E. Asia.

Formulations. Aerosol solutions; 25% e.c.; 40 and 25% w.p.; 4% dust; 5% granules.

33. Dichlorvos (*Vapona, Nogos, Oko, Mafu, Dedevap, Nuvan*)

Properties. A short-lived, wide-spectrum, contact and stomach insecticide with fumigant and penetrant action, non-phytotoxic. It is used as a household and public health fumigant, especially against mosquitoes and other Diptera, in addition to crop protection uses.

It is little soluble in water but is miscible with most organic solvents and aerosol propellants. It is a colourless to amber liquid, stable to heat, but is hydrolysed; corrosive to iron and mild steel but non-corrosive to stainless steel and aluminium.

The acute oral LD_{50} for male rats is 80 mg/kg.

Use. Especially effective against flies; often used for glasshouse fumigation – kills most glasshouse pests. Also used on outdoor fruit and vegetables where rapid kill is required close to harvest. Will kill sap-sucking and leaf-mining insects also.

Caution
(*a*) A poisonous substance (Part III) requiring protective clothing to be worn.
(*b*) Dangerous to bees.
(*c*) Pre-harvest interval – 1 day.
(*d*) Pre-access interval to treated areas – 12 hours.

Formulations. 50 and 100% e.c.; 'Vapona Pest Strip'; 0.4 to 1.0% aerosols; 0.5% granules.

34. Dimefox (*Terra-systam, Hanane*)

Properties. A systemic insecticide and acaricide of very high toxicity, used mainly for soil treatment of hops against aphids and red spider mites; non-phytotoxic at insecticidal concentrations.

It is a colourless liquid, miscible with water and most organic solvents. It is resistant to hydrolysis by alkalis but is hydrolysed by acids, slowly oxidized by vigorous oxidizing agents, rapidly by chlorine. Hence, for decontamination, treat with acids followed by bleaching powder. It is compatible with other pesticides, but the technical product slowly attacks metals.

The acute oral LD_{50} for rats is 1–2 mg/kg; the acute dermal LD_{50} for rats is 5 mg/kg; the hazards of vapour toxicity are high.

Use. Effective against sap-sucking insects (aphids) and mites, but toxicity hazards are high. (Now generally superseded).

Caution
(*a*) This is a very poisonous substance (Part I): full protective clothing must be worn.
(*b*) Dangerous to fish, livestock, game, wild birds and animals.
(*c*) Pre-harvest interval for picking hops – 4 weeks.
(*d*) Pre-access interval to treated areas, of unprotected persons – 1 day; of livestock and poultry – 4 weeks.

Formulations. *Terra-systam* is a 50% w.v. solution.

35. Dimethoate (*Rogor, Roxion, Cygon, Dantox*)

Properties. A systemic and contact insecticide and acaricide, used mainly against fruit flies (Olive Fly and Cherry Fly) and aphids.

The pure compound is a white solid, only slightly soluble in water, soluble in most organic solvents except saturated hydrocarbons such as hexane. It is stable in aqueous solution and to sunlight, but is readily hydrolysed by aqueous alkali. It is incompatible with alkaline pesticides.

The acute oral LD_{50} for rats is 250–65 mg/kg.

Use. Effective against aphids, psyllids, some flies, sawflies, Woolly Aphid and red spider mites. Mainly used against fruit flies, Olive Fly and Cherry Fly, and aphids.

Caution
(*a*) Dangerous to bees.
(*b*) Harmful to fish, livestock, game, wild birds and animals.

(*c*) Do not use on chrysanthemums, hops or on ornamental *Prunus* spp.
(*d*) Pre-harvest interval for edible crops – 1 week.
(*e*) Pre-access interval for livestock to treated areas – 1 week.

Formulations. 20 and 40% e.c.; 20% w.p.; 5% granules.

36. Disulfoton (*Disyston, Murvin 50, Parsolin*)

Properties. A systemic insecticide and acaricide used mainly as a seed dressing or granules to protect seedlings from insect attack. It is metabolized in the plant to the sulphoxide and sulphone. It is a colourless oil with a characteristic odour, only slightly soluble in water, but readily soluble in most organic solvents. It is relatively stable to hydrolysis at pH below 8.0.

The acute oral LD_{50} for male rats is 12.5 mg/kg.

Use. Effective against aphids on vegetables and fruit, and Carrot Fly, leafhoppers on rice, vegetables, cotton, some flies, some leaf miners, some beetles.

Caution

(*a*) This is a poisonous substance (Part II): full protective clothing should be worn.
(*b*) Dangerous to fish.
(*c*) Pre-harvest interval for edible crops – 6 weeks.

Formulations. *Disyston*, 50% impregnated on activated carbon; also 5 and 10% granules, based on Fuller's earth (FE) or pumice (P).

37. Ethion (*Embathion, Nialate, Hylemox, Rhodocide*)

Properties. It is a non-systemic insecticide and acaricide, used mainly in combination with petroleum oils on dormant fruit as an ovicide and scalecide. It is non-phytotoxic. It is a pale-coloured liquid very slightly soluble in most organic solvents including kerosene and petroleum oils. It is slowly oxidized in air and is subject to hydrolysis by both acids and alkalis.

The acute oral LD_{50} for rats is 208 mg/kg (for the pure substance) and 96 mg/kg (technical grade).

Use. Effective against eggs and dormant stages of pests (scales, leafhoppers, red spider mites, Heteroptera) on fruit trees. Some use against anthomyiid fly maggots on cereals and vegetables (in the soil).

Caution. This is a Part III poisonous substance: full protective clothing should be worn. It is phytotoxic to some varieties of apple.

Formulations. 25% w.p.; e.c.; 4% dust, 50% seed dressings.

38. Ethoate-methyl (*Fitios*)

Properties. It is a systemic insecticide and acaricide with contact action, particularly effective against fruit flies. The pure compound is a white cyrstalline solid, almost insoluble in water, but soluble in benzene, chloroform, acetone, ethanol. It is stable in aqueous solution but hydrolysed by alkali.

The acute LD_{50} for male rats is 340 mg/kg; non-irritant.

Use. Effective especially against Olive Fly (60 g a.i./l) and fruit flies (50 g/100 l); recommended for control of aphids, and red spider mites on fruit, arable and vegetable crops at rates of 70–170 g a.i./450–1125 l/ha.

Caution

(*a*) Harmful to fish, livestock, game, wild birds and animals.
(*b*) Dangerous to bees.
(*c*) Pre-harvest interval for arable crops – 1 week.
(*d*) Pre-access interval for animals to treated areas – 1 week.

Formulations. 20, 40% e.c.; 25% w.p.; 5% dust and 5% granules.

39. Etrimfos (*Ekamet*)

Properties. A broad-spectrum contact insecticide, non-systemic, used against a broad range of pests on many different crops; a moderate residual activity of 7–14 days.

It is a colourless oil, insoluble in water, but soluble in most organic solvents.

The acute oral LD_{50} for rats is 1800 mg/kg.

Use. Effective against many Lepidoptera, Diptera, and Coleoptera, and to a variable extent Hemiptera, on fruit trees, citrus, olive, grapevine, potato, vegetables, maize, lucerne and paddy rice (including granules against Pyralidae in paddy rice).

Caution

(*a*) Dangerous to fish and bees.
(*b*) Harmful to livestock, game, wild birds and animals.
(*c*) Pre-harvest interval for edible crops – 3 days.
(*d*) Pre-access interval for livestock – 1 week.

Formulations. Include e.c. 520 g a.i./l (= 500 g/kg); granules 50 g/kg.

40. Fenitrothion (*Accothion, Folithion, Sumithion*)

Properties. It is a contact and stomach insecticide, particularly effective against rice stem borers, but has a wide spectrum of activity; also a selective acaricide but of low ovicidal activity. It is a brownish-yellow liquid, practically insoluble in water, but soluble in most organic solvents, and is hydrolysed by alkali. Of moderate persistence.

The acute oral LD_{50} for rats is 250–500 mg/kg.

Use. Effective against lepidopterous larvae (rice stem borers especially), aphids, whiteflies, scales, mealybugs, capsids, psyllids, some fly larvae, some beetles, locusts, thrips and sawflies.

Caution

(*a*) Dangerous to bees.
(*b*) Harmful to fish, livestock, game, wild birds and animals.
(*c*) Pre-harvest interval for edible crops – 2–3 weeks.
(*d*) Pre-access interval for livestock to treated areas – 1 week.

Formulations. 50% e.c.; 40 and 15% w.p.; 5, 3, and 2% dusts.

41. Fenthion (*Baytex, Lebaycid, Queleton,* etc.)

Properties. It is a contact and stomach insecticide with a useful penetrant action, which, by virtue of low volatility and stability to hydrolysis, is of high persistence. It is a colourless liquid, practically insoluble in water, but readily soluble in most organic solvents. It is stable at temperatures up to 210 °C and is resistant to light and to alkaline hydrolysis.

The acute oral LD_{50} for male rats is 215 mg/kg. It is of greater toxicity to dogs and birds, and is used for the control of weaver birds in Africa.

Use. Effective against fruit flies, many caterpillars, leafhoppers and plant bugs, aphids, thrips, mites, sawflies, some beetles; also weaver birds.

Caution

(*a*) Harmful to birds and wildlife.
(*b*) Pre-harvest interval for edible crops – 7–42 days according to country and crop.

Formulations. 50, 40, 25%, w.p.; 60% fogging concentrate; 50% e.c.; 3% dust, *Queleton* for use against weaver birds.

42. Fonofos (*Dyfonate*)

Properties. It is an insecticide particularly suitable for the control of soil maggots and other soil insects. It has caused some damage to seeds when placed in their proximity. Persistence in soil is moderate, of the order of eight weeks.

It is a pale yellow liquid; practically insoluble in water, but miscible with most organic solvents such as kerosene, xylene, etc. Stable under normal conditions.

The acute oral LD_{50} for male rats is 8–17 mg/kg.

Use. Effective against most soil pests, such as root maggots, soil caterpillars, wireworms and other beetle larvae, crickets, symphylids.

Caution

(*a*) This is a very poisonous substance (Part II), and full protective clothing should be worn.
(*b*) Dangerous to fish.

(*c*) Pre-harvest interval for edible crops – 6 weeks.

Formulations. 5 to 10% granules.

43. Formothion (*Anthio, Aflix*)

Properties. It is a contact and systemic insecticide and acaricide effective against sap-sucking insects and mites. In plants, it is metabolized to dimethoate. In loamy soil the half-life is 14 days.

It is a yellow viscous oil or crystalline mass, slightly soluble in water, miscible with alcohols, chloroform, ether, ketones, and benzene. It is stable in non-polar solvents, but is hydrolysed by alkali and incompatible with alkaline pesticides.

The acute oral LD_{50} for male rats is 375–535 mg/kg.

Use. Effective against many sap-sucking insects and mites, especially aphids and red spider mites, and some fly larvae.

Caution

(*a*) Dangerous to bees.
(*b*) Harmful to fish and livestock.
(*c*) Pre-harvest interval for edible crops – 1 week.
(*d*) Access of animals to treated areas – 1 week.

Formulations. 25% w/v e.c. is the usual formulation.

44. Heptenophos (*Hostaquick, Ragadan*)

Properties. A translocatable insecticide with rapid initial action and short residual effect. It penetrates plant tissues and is quickly translocated in all directions. It is used against sap-sucking insects, and some Diptera, and ectoparasites of domesticated animals.

A pale, brown liquid, scarcely soluble in water, but quite soluble in most organic solvents.

The acute oral LD_{50} for rats is 96–121 mg/kg.

Use. Effective against many sap-sucking insects, but especially aphids.

Caution

(*a*) This is a poisonous substance (Part III): protective clothing should be worn.
(*b*) Pre-harvest interval for edible crops is 1 day.
(*c*) Dangerous to bees.
(*d*) Harmful to fish.

Formulations. For agricultural use only as an e.c. 500 g a.i./l.

45. Iodofenphos (*Nuvanol N, Elocril, Alfacron*)

Properties. A non-systemic, contact and stomach insecticide and acaricide, used mainly for control of stored product pests. For crop protection purposes persistence is one to two weeks.

It forms colourless crystals, almost insoluble in water, slightly soluble in benzene and acetone. It is stable in neutral media but unstable in strong acids and alkalis.

The acute oral LD_{50} for rats is 2100 mg/kg.

Use. Effective against various pests of stored products and public hygiene, and for crop protection against a wide range of coleopterous, dipterous, and lepidopterous pests.

Caution

(*a*) Toxic to bees.
(*b*) Pre-harvest interval for edible crops – 7–14 days.

Formulations. 50% w.p.; 20% e.c.; and 5% powder.

46. Malathion (*Malathion, Malastan, Malathexo*)

Properties. It is a wide-spectrum, non-systemic insecticide and acaricide, of brief to moderate persistence and low mammalian toxicity. It is generally non-phytotoxic, but may damage cucurbits under glasshouse conditions, and various flower species.

It is a colourless or pale brown liquid, of slight solubility in water, miscible with most organic solvents though not in petroleum oils. Hydrolysis is rapid at pH above 7.0 and below 5.0; it is incompatible with alkaline pesticides and corrosive to iron.

The acute oral LD_{50} for rats is 2800 mg/kg.

Use. Effective against aphids, thrips, leafhoppers, spider mites, mealybugs, scales, various beetles, caterpillars and flies.

Caution

(*a*) Harmful to bees and fish.

(*b*) To avoid possible taint to edible crops allow 4 days from application to harvest (7 days for crops for processing).

(*c*) Pre-harvest interval – 1 day.

Formulations. As e.c. from 25 to 86% (many of 60%), 25 and 50% w.p.; dusts of 4%, and as atomizing concentrates (95%) for u.l.v. applications.

47. Mecarbam (*Murfotox, Pestan, Afos*)

Properties. An insecticide and acaricide with slight systemic properties, used for control of Hemiptera and Diptera. At recommended rates it persists in soil for 4–6 weeks.

It is a pale brown oil, almost insoluble in water, slightly soluble in aliphatic hydrocarbons, miscible with alcohols, aromatic hydrocarbons, ketones and esters. It is subject to hydrolysis; is compatible with all but highly alkaline pesticides; slowly attacks metals.

The acute oral LD_{50} for rats is 36 mg/kg.

Use. Effective against scale insects and other Hemiptera, Olive Fly and other fruit flies, leafhoppers and stem flies of rice, and rootfly maggots on vegetables.

Caution

(*a*) Pre-harvest interval for edible crops – 2 weeks.

(*b*) Pre-access interval to treated areas – 2 weeks.

(*c*) Harmful to bees, fish, livestock and game.

Formulations. E.c. 68, 40%; w.p. 25%; dusts; *Murfotox oil* 5% in petroleum oil; 5% granules.

48. Menazon (*Saphizon, Saphicol, Sayfos, Aphex*)

Properties. It is a systemic insecticide, used mainly against aphids. It is regarded as non-phytotoxic.

It forms colourless insoluble crystals, stable up to 35°C; is weakly basic, and is compatible with all but strongly alkaline pesticides, but may be decomposed by the reactive surfaces of some 'inert' fillers.

The acute oral LD_{50} for female rats is 1950 mg/kg.

Use. Most frequently used against aphids, as a seed dressing, also as a drench and a root dip. Also used for Woolly Aphid on apple.

Caution

(*a*) Harmful to bees and livestock.

(*b*) Pre-harvest interval for edible crops – 3 weeks.

Formulations. Seed dressings as 50 and 80%, 70% w.p.

49. Mephosfolan (*Cytro-lane*)

Properties. An insecticide with contact, stomach, and systemic action following root or foliar absorption.

An amber liquid, insoluble in water, but soluble in many organic solvents; stable in water under neutral conditions, but hydrolysed by acid or alkali.

The acute oral LD_{50} for rats is 4–9 mg/kg.

Use. Effective against aphids, whiteflies, many caterpillars (bollworms, stem borers, *Spodoptera* spp.) and mites, on most of the major crops.

Caution

(*a*) A very poisonous substance (Part II substance; *Poisons Rules* apply in UK): full protective clothing must be worn.

(*b*) Dangerous to fish, game and wildlife.

(*c*) Harmful to livestock.

(*d*) Pre-harvest interval for treated hops – 4 weeks (in the UK).

(*e*) Pre-access interval for livestock – 10 days.

Formulations. E.c. 250, 750 g a.i./l; liquid; granules 20–100 g/kg.

50. Methidathion (*Supracide, Ultracide*)

Properties. It is a non-systemic insecticide, with some acaricidal activity, and capable of foliar penetration. Non-phytotoxic to all plants tested; rapidly metabolized and excreted by plants and animals.

It forms colourless crystals, of slight solubility in water, readily soluble in acetone, benzene, and methanol. It is stable in neutral and weakly acid media, but much less stable in alkali. It is compatible with many fungicides and acaricides.

Acute oral LD_{50} for rats is 25–48 mg/kg.

Use. Of promise against lepidopterous larvae; foliar penetration enables it to be used against leaf-rollers. Used on a wide variety of crops against leaf-eating and sucking insects and mites, especially against scale insects.

Caution

(*a*) This is a poisonous substance (Part III), and protective clothing should be worn.
(*b*) Harmful to bees and livestock.
(*c*) Dangerous to fish.
(*d*) Pre-harvest interval for edible crops – 3 weeks.
(*e*) Pre-access interval for livestock to treated areas – 2 weeks.

Formulations. 40% e.c. and w.p.; and 20% e.c. and w.p.

51. Mevinphos (*Phosdrin, Menite, Phosfene*)

Properties. It is a contact and systemic insecticide and acaricide of short persistence. Although non-persistent, its high initial kill provides a relatively long period before build-up recurs. It is non-phytotoxic.

The technical product is a pale yellow liquid, miscible with water, alcohols, ketones, chlorinated and aromatic hydrocarbons, but only slightly soluble in aliphatic hydrocarbons. Stable at ordinary temperatures, but hydrolysed in aqueous solution, and rapidly decomposed by alkalis, hence incompatible with alkaline fertilizers and pesticides. It is corrosive to cast iron, mild and some stainless steels, and brass; relatively non-corrosive to copper, nickel and aluminium; non-corrosive to glass, and many plastics, but passes slowly through thin films of polyethylene.

Acute oral LD_{50} for rats is 3.7–12 mg/kg.

Use. Effective against sap-feeding insects (aphids, etc.) at 140–280 g/ha, mites and beetles at 210–350 g/ha, caterpillars at 280–560 g/ha; and some fly larvae. Especially useful for giving rapid kill close to harvest.

Caution

(*a*) This is a poisonous substance (Part II); full protective clothing should be worn.
(*b*) Dangerous to livestock, bees, fish, game and wild animals.
(*c*) Pre-harvest interval for edible crops – 3 days.
(*d*) Pre-access interval to treated areas – 1 day.

Formulations. Being water soluble, formulation is unnecessary, but e.c. of 5, 10, 18, 24, 48 and 50% technical are available, together with dusts and w.p.'s.

52. Monocrotophos (*Nuvacron, Azodrin, Monocron*)

Properties. A fast-acting insecticide with systemic stomach and contact action, used against a wide range of pests on a variety of crops; persistence of 1–2 weeks. It has caused phytotoxicity under cool conditions to some apples, cherries, and sorghum varieties, and is incompatible with alkaline pesticides.

It is a crystalline solid, miscible with water, soluble in acetone, and ethanol, sparingly soluble in xylene but almost insoluble in kerosene and diesel oils. Unstable in low molecular weight alcohols and glycols. Corrosive to iron, steel, and brass, but does not attack glass, aluminium and stainless steel.

Acute oral LD_{50} for rats is 13 mg/kg.

Use. Effective against a wide range of pests including mites, bugs, leaf-eating beetles, leaf miners and caterpillars.

Caution

(*a*) Dangerous to fish and livestock, bees and birds.
(*b*) Pre-harvest intervals is 3–30 days, according to crop and country; a very toxic chemical.

Formulations. Water miscible concentrates contain 200–600 g a.i./l.

53. Naled (*Dibrom, Ortho, Bromex*)

Properties. A non-systemic, contact and stomach insecticide and acaricide with some fumigant action, used mainly under glass and in mushroom houses.

The technical product is a yellow liquid, insoluble in water, slightly soluble in aliphatic solvents, and readily soluble in aromatic solvents. It is stable under anhydrous conditions, but rapidly hydrolysed in water (90–100% in 48 hours at room temperature), and by alkali; stable in glass containers, but in the presence of metals and reducing agents, rapidly loses bromine and reverts to dichlorvos.

Acute LD_{50} for rats is 430 mg/kg.

Use. Mainly against glasshouse and mushroom pests, but also in fruit fly baits.

Caution

(*a*) Pre-harvest interval – 24 hours.
(*b*) It is phytotoxic to cucurbits.

Formulations. 4% dust and e.c. 1 kg/l.

54. Omethoate (*Folimat*)

Properties. A systemic insecticide and acaricide with a broad range of action and little phytotoxicity, except to some peach varieties.

It is a colourless oily liquid, readily soluble in water, acetone, ethanol and many hydrocarbons; insoluble in light petroleum; hydrolysed by alkali.

Acute oral LD_{50} for male rats is about 50 mg/kg.

Use. Effective against a wide range of insects, particularly caterpillars, and Homoptera; also Orthoptera, thrips, some beetles, and phytophagous mites.

Caution

(*a*) This is a Part III substance.
(*b*) Harmful to bees.
(*c*) Damaging to various varieties of peach.
(*d*) Pre-harvest interval for edible crops – 21–28 days.

Formulations. Include e.c. and granules with a range of a.i. contents.

55. Oxydemeton-methyl (*Metasystox-R*)

Properties. A systemic and contact insecticide and acaricide, used against sap-sucking insects and mites, with a fast kill and moderate persistence.

It is a clear brown liquid, miscible with water and soluble in most organic solvents except light petroleum. It is hydrolysed by alkali.

The acute oral LD_{50} for male rats is 65 mg/kg.

Use. Effective against aphids, and red spider mites on most crops, also leafhoppers, whiteflies, psyllids, thrips, some flies, sawflies. Only a limited effect on *Brassica* aphids.

Caution

(*a*) This is a poisonous substance (Part III), and protective clothing should be worn.
(*b*) Harmful to bees, livestock, fish, game and wild animals.
(*c*) Pre-harvest interval for edible crops – 2–3 weeks.
(*d*) Pre-access interval for livestock to treated areas – 2 weeks.

Formulations. As e.c. of various a.i. contents (25, 50%).

56. Oxydisulfoton (*Disyston-S*)

Properties. A systemic insecticide and acaricide particularly suitable for seed treatment against virus vectors.

It is a pale-coloured liquid, slightly soluble in water, readily soluble in most organic solvents.

The acute oral LD_{50} for rats is about 3.5 mg/kg.

Use. Particularly effective against sap-sucking insects and mites.

Caution. This is a very poisonous substance, and full protective clothing should be worn.

Formulations. As seed dressings of various a.i. contents; also as e.c. and granules.

57. Parathion (*Folidol, Bladan, Fosfex, Fosferno, Thiophos*)

Properties. A non-systemic, contact and stomach insecticide and acaricide, with some fumigant action. It is non-phytotoxic except to some ornamentals, and under certain weather conditions, to pears and some apple varieties; hazardous to operators; now banned in many countries.

It is a pale yellow liquid, scarcely soluble in water, slightly soluble in petroleum oils, but miscible with most organic solvents. In alkaline solution it rapidly hydrolyses; on heating it isomerizes.

The acute oral LD_{50} for male rats is 13 mg/kg; for females 3.6 mg/kg; acute dermal LD_{50} respectively 21 and 6.8 mg/kg.

Use. Effective against most Homoptera, Diptera, springtails, mites, millipedes and some nematodes. Prolonged use may result in extensive destruction of predators and parasites.

Caution

(*a*) This is a very poisonous substance (Part II), and full protective clothing should be worn; easily absorbed through the skin.
(*b*) Dangerous to bees, fish, livestock, game, wild birds and animals.
(*c*) Pre-harvest interval for edible crops – 4 weeks.
(*d*) Pre-access interval for livestock to treated areas – 10 days.

Formulations. To w.p. and e.c. of various a.i. contents; also to dusts, smokes and aerosols.

58. Parathion-methyl (*Dalf, Metacide, Folidol-M, Nitrox-80*)

Properties. A non-systemic, contact and stomach insecticide, with some fumigant action, and a range of action similar to that of parathion but of lower mammalian toxicity. It is non-phytotoxic.

The technical product is a brown liquid of about 80% purity, scarcely soluble in water, slightly soluble in light petroleum and mineral oils, but soluble in most other organic solvents. It is hydrolysed by alkali at a faster rate than parathion, and readily isomerizes on heating. Compatible with most other pesticides.

The acute oral LD_{50} for male rats is 14 mg/kg; for female rats 24 mg/kg. It is hazardous to wildlife but is of brief persistence.

Use. As for parathion, but not effective against Acarina.

Caution

(*a*) This is a poisonous substance, so protective clothing should be worn.
(*b*) Hazardous to wildlife.

Formulations. To e.c. and dusts of various a.i. contents, *Nitrox 80* is 80% solution in an aromatic petroleum solvent.

59. Phenisobromolate (*Acarol, Neoron*)

Properties. A contact acaricide with residual activity; of promise for use on many crops.

It is a crystalline solid, insoluble in water, but readily soluble in most organic solvents; stable in neutral media.

The acute oral LD_{50} for rats is 5000 mg/kg.

Use. Of promise for use against mites on pome and stone fruits, citrus, hops, cotton, beans, cucurbits, tomatoes, strawberries and ornamentals.

Formulations. As 500 and 250 g/l e.c.

60. Phenthoate (*Elsan, Cidial, Papthion, Tanone*)

Properties. A non-systemic insecticide and acaricide with contact and stomach action. It may be phytotoxic to some peach, fig and grape varieties, and may discolour some red-skinned apple varieties.

It is a crystalline solid, almost insoluble in water, but miscible with most organic solvents.

The acute oral LD_{50} for rats is 250–300 mg/kg.

Use. Effective against caterpillars, aphids, jassids, mites, and is also used for the protection of stored grain.

Formulations. 50% technical; 5% in mineral oil; 40% w.p.; 2% granules; 85% pure compound.

61. Phorate (*Thimet, Rampart, Granutox, Timet*)

Properties. A persistent systemic insecticide used in granular and e.c. formulations for the protection of seedlings from sap-feeding and soil insects; some fumigant action. In temperate climates effective soil persistence of 15–20 weeks is expected.

It is a clear liquid, only slightly soluble in water, but miscible with carbon tetrachloride, dioxane, xylene and vegetable oils. It is hydrolysed by alkalis and in the presence of moisture.

The acute oral LD_{50} for male rats is 3.5 mg/kg; for females 1.6 mg/kg.

Use. Effective against aphids, wireworms, various fly maggots (Frit, Carrot), capsids, leafhoppers, various weevils.

Caution

(*a*) This is a very poisonous substance (Part II), and full protective clothing should be worn.
(*b*) Dangerous to fish and livestock.
(*c*) Pre-harvest interval for edible crops – 6 weeks.
(*d*) Pre-access interval for livestock to treated areas – 6 weeks.

Formulations. As e.c. of various a.i. content; 5, 10 and 15% granules.

62. Phosalone (*Zolone, Embacide, Rubitox*)

Properties. A non-systemic insecticide and acaricide, used on deciduous tree fruits, field and market garden crops, against a wide range of pests. It persists on plants for about two weeks before being hydrolysed.

It forms colourless crystals; insoluble in water and light petroleum, but soluble in acetone, benzene, chloroform, ethanol, methanol, toluene and xylene. It is stable under normal storage conditions, non-corrosive, and compatible with most other pesticides.

The acute oral LD_{50} for male rats is 150 mg/kg.

Use. Effective against a wide spectrum of pests: caterpillars on fruit, cotton bollworms, fruit fly maggots, aphids, psyllids, jassids, thrips, various weevils and red spider mites.

Caution

(*a*) Harmful to bees, fish and livestock.
(*b*) Pre-harvest interval for edible crops – 3 weeks.
(*c*) Pre-access interval for livestock to treated areas – 4 weeks.

Formulations. 30, 33, 35% e.c.; 30% w.p.; 2.5 and 4% dusts. Various e.c. formulations under heading *Zolone DT* with DDT for use on cotton.

63. Phosfolan (*Cyalane, Cyolon*)

Properties. A systemic insecticide of short persistence in soil, and readily metabolized in plants and animals.

A pale yellow solid, soluble in water and most organic solvents (except hexane); the aqueous solution is stable under neutral conditions, but hydrolysed by acids and alkalis.

The acute oral LD_{50} for rats is 8.9 mg/kg; high dermal absorbability.

Use. Effective against a broad range of sucking insects (thrips, whiteflies, jassids), mites, and caterpillars (especially *Prodenia* and *Spodoptera* spp., and Cabbage Looper).

Caution

(*a*) A very poisonous substance; full protective clothing must be worn (very high dermal toxicity).
(*b*) Dangerous to livestock and wildlife.

Formulations. Include e.c. 250 g a.i./l; and granules 20–50 g/kg.

64. Phosmet (*Imidan, Appa, Prolate, Germisan*)

Properties. A non-systemic acaricide and insecticide, used at concentrations safe for a variety of predators of mites and thus useful for integrated control programmes.

It is a white crystalline solid, with an offensive odour, scarcely soluble in water, but more than 10% soluble in acetone, dichloromethane and xylene.

The acute oral LD_{50} for male rats is 230 mg/kg; it is readily degraded both in laboratory animals and in the environment.

Use. Used mainly against phytophagous mites; when used as recommended should not affect mite predators.

Formulations. 20 and 30% e.c. and 50% w.p. Storage above 45 °C may lead to decomposition.

65. Phosphamidon (*Dimecron, Dicron, Famfos*)

Properties. A systemic insecticide and acaricide, rapidly absorbed by the plant, but only a little contact action.

Non-tainting, and non-phytotoxic except to some cherry varieties and sorghum varieties related to Red Swazi. No fumigant action.

It is a pale yellow oil, miscible with water, and readily soluble in most organic solvents except saturated hydrocarbons. It is stable in neutral and acid media but is hydrolysed by alkali. Compatible with all but highly alkaline pesticides. It corrodes iron, tin plate and aluminium, and is packed in polyethylene containers.

The acute oral LD_{50} for rats is 28.3 mg/kg. The half-life in plants is about 2 days.

Use. Effective against sap-feeding insects and leaf-eating ones; particularly aphids, bugs, many caterpillars (but not Noctuidae), rice stem borers, thrips, Colorado Beetle, and other beetles, grasshoppers, sawflies, fly larvae, and phytophagous mites.

Caution

(*a*) This is a poisonous substance (Part III): protective clothing should be worn.
(*b*) Dangerous to bees.
(*c*) Harmful to livestock.
(*d*) Pre-harvest interval for edible crops – 3 weeks.
(*e*) Pre-access interval for livestock to treated areas – 2 weeks.

Formulations. *Dimecron 20*: 20 kg/100 l in isopropanol, the water content rigidly controlled to delay hydrolysis; similarly for *Dimecron 50* and *100*; 50% w.p.

66. Phoxim (*Baythion, Valexon, Volaton*)

Properties. An insecticide of brief persistence and no systemic action, with low mammalian toxicity, effective against a broad range of insects; most useful against soil insects and stored products pests.

It is a yellow liquid virtually insoluble in water, slightly soluble in light petroleum, soluble in alcohols, ketones, and aromatic hydrocarbons. Stable to water and acid media, but unstable to alkali. Believed to be compatible with most pesticides of a non-alkaline nature.

The acute oral LD_{50} for rats is more than 2000 mg/kg.

Use. Effective against a broad range of insects, especially stored products pests and insects affecting man. Also successfully used against soil pests (dipterous maggots, rootworms and wireworms); u.l.v. applications against grasshoppers.

Caution

(*a*) Dangerous to bees by both contact and vapour effect.
(*b*) Harmful to fish.

Formulations. 50% e.c.; 5% granules; and a concentrate for u.l.v. application. Other experimental formulations are under test. *Baythion* is the trade name for use against pests of man and stored products; *Valexon* for agricultural use.

67. Pirimiphos-ethyl (*Primicid, Fernex, Primotec*)

Properties. It is a broad-spectrum insecticide, particularly against soil-inhabiting Diptera and Coleoptera. No phytotoxicity has been recorded using recommended rates; high rates of seed dressing have resulted in seedling abnormalities though. Stable for 5 days at 80 °C. In the pure state it is a pale straw-coloured liquid; insoluble in water but miscible with most organic solvents; corrosive to iron and unprotected tin plate.

Acute oral LD_{50} for rats is 140–200 mg/kg.

Use. Effective particularly against soil-inhabiting Diptera and Coleoptera; effective as foliage spray at conventional rates against species of Lepidoptera, Coleoptera, Homoptera and Tetranychidae.

Caution

(*a*) Dangerous to bees.
(*b*) Hazardous to birds and wild animals.

Formulations. A 20% s.d.; 250 g/l, 500 g/l e.c.; 5 and 10% granules with 5 and 10% thiram added.

68. Pirimiphos-methyl (*Actellic, Blex, Actellifog*)

Properties. It is a fast-acting, broad-spectrum insecticide of limited persistence, with both contact and fumigant action; non-phytotoxic. It penetrates leaf tissue to the extent that an insect on one side of a leaf is killed by chemical applied to the other side, and there is also a slight systemic action.

It is a straw-coloured liquid, insoluble in water, but soluble in most organic solvents; decomposed by strong acids and alkalis; does not corrode brass, stainless steel, nylon or aluminium.

Acute oral LD_{50} for female rats is about 800 mg/kg (low mammalian toxicity).

Use. Effective against species of Lepidoptera, Coleoptera, aphids, Tetranychidae, and many other crop pests (Homoptera, Heteroptera, Thysanoptera, Diptera, Orthoptera); used against stored products pests.

Caution

(*a*) Dangerous to bees.
(*b*) Harmful to fish.
(*c*) Pre-harvest interval – 3–7 days.

Formulations. 25 and 50% e.c.; 5 and 10% granules, and 100 g/l and 500 g/l u.l.v. formulations.

69. Profenofos (*Curacron*)

Properties. A broad-spectrum, non-systemic insecticide, with both contact and stomach action, used against many different crop pests.

A pale yellow liquid, barely soluble in water, but miscible with most organic solvents. Fairly stable under neutral and slightly acidic conditions.

The acute oral LD_{50} for rats is 358 mg/kg.

Use. Effective against many pests of cotton and vegetables, both insects and mites. Usual rates are 250–500 g a.i./ha for sucking insects and mites, and 400–1200 g a.i./ha for biting and chewing insects.

Formulations. E.c. 500 g a.i./1, 400 g a.i./l; u.l.v. 250 g a.i./l; granules 50 g a.i./kg; and mixtures with chlordimeform.

70. Prothoate (*Fac, Fostin, Oleofac, Telefos*)

Properties. An acaricide and insecticide, with systemic action, used mainly against phytophagous mites and some sap-sucking insects.

It is a colourless crystalline solid, virtually insoluble in water but miscible with most organic solvents. It is stable in neutral, moderately acid and slightly alkaline media, but is rapidly decomposed in strong alkali.

The acute oral LD_{50} for male rats is 8 mg/kg.

Use. Effective for protection of fruit, citrus and vegetable crops from tetranychid and some eriophyid mites, and some insects, notably aphids, Tingidae, Psyllidae, and Thysanoptera.

Caution

(*a*) This is a very poisonous substance, and full protective clothing should be worn.
(*b*) Dangerous to fish, livestock and game.

Formulations. 20% tech., 40% tech., and 3% tech.; 5% granules.

71. Quinalphos (*Savall*)

Properties. A contact and stomach insecticide and acaricide, with good penetrative properties; in plants degradation occurs within a few days.

A colourless crystalline solid, insoluble in water, but soluble in most organic solvents. Generally stable under ambient conditions, but susceptible to hydrolysis, especially under alkaline conditions.

The acute oral LD_{50} for rats is 62–137 mg/kg.

Use. Effective against scales, and against caterpillars on fruit trees, cotton, and vegetables; also Celery Fly, leatherjackets in cereals, late generation Carrot Fly, and the pest complex on rice.

Caution

(*a*) A poisonous substance (Part III; *Poisons Rules* apply in UK): protective clothing should be worn.
(*b*) Irritating to the eyes.
(*c*) Dangerous to bees, fish, game, wild birds and animals.
(*d*) Harmful to livestock.
(*e*) Pre-harvest interval for edible crops – 1–3 weeks.

Formulations. Include e.c. 250 g a.i./kg, 200 g/l; dust 15 g/kg; u.l.v. formulation 300 g/kg; granules 50 g/kg.

72. Quinomethionate (*Morestan, Erade, Forstan*)

Properties. A selective, non-systemic, acaricide; and a fungicide specific to powdery mildews.

It forms yellow crystals, practically insoluble in water and sparingly soluble in organic solvents. In chemical properties it is closely related to thioquinox, but is more stable to oxidation.

The acute oral LD_{50} for rats is 2500–3000 mg/kg.

Use. Controls red spider and other phytophagous mites on a wide range of crops; and powdery mildews.

Caution

(*a*) Safe to bees.
(*b*) Certain blackcurrant varieties may be damaged by sprays.
(*c*) Pre-harvest interval for edible crops – 3–28 days according to crops and country.

Formulations. 25% w.p., and also formulated as smokes (20%), and 2% dust.

73. Schradan (*Systam, Pestox 3*)

Properties. A systemic insecticide and acaricide with little contact effect; effective against sap-feeding insects and mites; non-phytotoxic at insecticidal concentrations.

It is a brown viscous liquid, miscible with water and most organic solvents; slightly soluble in petroleum oils, and readily extracted from aqueous solution by chloroform. Stable to water and alkali, but hydrolysed under acid conditions.

The acute oral LD_{50} for male rats is 9.1 mg/kg.

Use. Used mainly against aphids and red spider mites on a range of crops.

Caution

(*a*) This is a very poisonous substance (Part II), and full protective clothing should be worn.
(*b*) Dangerous to fish, livestock, game, wild birds and animals.
(*c*) Safe to bees.
(*d*) Pre-harvest interval for edible crops – 4–6 weeks.
(*e*) Pre-access interval for livestock to treated areas – 4 weeks.

Formulations. 30% aqueous solution; also anhydrous, 75–80% or 60% with anhydrous surfactant.

74. TEPP (*Nifos T, Vapotone, Bladen, Fosvex*)

Properties. A non-systemic aphicide and acaricide of brief persistence; very high mammalian toxicity.

It is a colourless hygroscopic liquid, miscible with water and most organic solvents, but only slightly soluble in petroleum oils; is very rapidly hydrolysed by water, and is corrosive to most metals. It is rapidly metabolized in the animal body.

The acute oral LD_{50} for rats is 1.2 mg/kg; the acute dermal LD_{50} is 2.4 mg/kg.

Use. Used only against aphids and phytophagous mites.

Caution

(*a*) An extremely poisonous pesticide and full protective clothing must be worn. High mammalian dermal toxicity.
(*b*) Dangerous to livestock, game, wild birds and animals.

Formulations. For agricultural purposes TEPP refers to a mixture of polyphosphates containing at least 40% tetraethyl pyrophosphate. As an aerosol, a solution in methyl chloride is used.

75. Terbufos (*Counter*)

Properties. An insecticide with strong initial, and with residual activity against soil insects.

A pale yellow liquid, almost insoluble in water but soluble in many organic solvents. It decomposes on prolonged heating at high temperatures, at low pH, and with strong alkalis.

The acute oral LD_{50} for rats is 1.6–4.5 mg/kg, and the acute dermal LD_{50} for rats is from 1.0–7.4 mg/kg.

Use. Effective against many different types of soil insects.

Caution

(*a*) This is a very poisonous substance, and full protective clothing must be worn.
(*b*) This poison can be absorbed through the skin.
(*c*) Generally dangerous to game and wildlife.

Formulation. Because of its toxicity it is only formulated as granules, from 20–150 g a.i./kg.

76. Tetrachlorvinphos (*Gardona, Rabon, Ravap*)

Properties. A selective insecticide used against lepidopterous and dipterous pests on the aerial parts of crops; very low mammalian toxicity; some translaminar action.

It is a white crystalline solid, scarcely soluble in water, but soluble in chloroform (40% w/w), methyl chloride (40%); less so in xylene (15%) and acetone (20%). It is temperature-stable, but slowly hydrolysed by water, particularly under alkaline conditions.

The acute oral LD_{50} for rats is 4000–5000 g/kg.

Use. Effective against lepidopterous and dipterous pests of fruit, rice, vegetables, cotton and maize. With certain exceptions, it does not show high activity against Hemiptera, and because of rapid breakdown is not effective in the soil. Shows promise against pests of stored products.

Formulations. 240 g/l e.c.; 50 and 70% w.p.; and 5% granules.

77. Thiometon (*Ekatin, Intrathion*)

Properties. A systemic insecticide and acaricide suitable for control of sucking insects and mites in orchards, vineyards, hop gardens, and on beet. At 0.02% a.i. systemic effects persist for 2–3 weeks.

It is a colourless oil, fairly soluble in water, slightly soluble in light petroleum oils but soluble in most other organic solvents.

The acute oral LD_{50} for rats is 120–30 mg/kg.

Use. Mainly used for aphid control (and also mites) on fruit, potatoes, beet, cereals, and vegetables.

Caution

(*a*) This is a poisonous substance (Part III), and protective clothing should be worn.
(*b*) Harmful to bees, fish, livestock, game, wild birds and animals.
(*c*) Pre-harvest interval for edible crops – 3 weeks.
(*d*) Pre-access interval for livestock to treated areas – 2 weeks.

Formulations. 20% e.c., coloured blue; and a dry spray dust.

78. Thionazin (*Nemafos, Zinophos, Nemasol*)

Properties. A soil insecticide and nematicide, of relatively brief persistence.

It is a brown liquid, fairly soluble in water, miscible with most organic solvents; readily hydrolysed by alkali.

The acute oral LD_{50} for rats is 12 mg/kg, with the dermal LD_{50} being 11 mg/kg.

Use. Soil application is effective against symphylids and Cabbage Root Fly (drench); granules recommended for use on cotton, cucurbits, groundnuts, brassicas and tomato.

Caution

(*a*) This is a very poisonous substance (Part II), and full protective clothing should be worn.
(*b*) Dangerous to fish and livestock.
(*c*) Pre-access interval for livestock to treated areas – 8 weeks.

Formulations. 25 and 46% e.c.; 5 and 10% granules.

79. Thioquinox (*Eradex, Eraditon*)

Properties. A non-systemic acaricide effective against eggs; and a fungicide specific against powdery mildews.

It is a brown powder, practically insoluble in water and most organic solvents, but is slightly soluble in ethanol and acetone. Stable to temperature and light; resistant to hydrolysis, but susceptible to oxidation without any reduction of biological activity.

The acute oral LD_{50} for rats is 3400 mg/kg; some dermal irritation caused to some operators.

Use. Specific against eggs of phytophagous mites; and the powdery mildews.

Formulations. 50% w.p.

80. Triazophos (*Hostathion*)

Properties. A broad-spectrum insecticide and acaricide with some nematicidal properties; used for either foliar or soil application. It can penetrate plant tissues generally but has no systemic action.

A pale brown liquid, with little solubility in water, but soluble in most organic solvents.

The acute oral LD_{50} for rats is 82 mg/kg.

Use. As a foliar spray against aphids on fruit, at 75–125 g a.i./100 l, and on cereals at 320–600 g a.i. (40% e.c.)/ha. Incorporated into the soil, it has been used to control wireworms and some cutworms at 1–2 kg a.i./ha.

Caution

(*a*) This is a poisonous substance (Part III), so full protective clothing should be worn.
(*b*) Dangerous to bees, game and all wildlife.

Formulations. E.c. 400 g a.i./l; u.l.v. concentrates at 250 and 400 g a.i./l; w.p. 300 g a.i./kg; granules 20 and 50 g a.i./kg.

81. Trichloronate (*Agritox, Agrisil, Phytosol*)

Properties. A persistent (in soil) non-systemic insecticide recommended for control of root maggots, wireworms, and other soil insects; acts as contact and stomach poison. Recently shown to be effective against Stem Eelworm. Foliar uses are under test. Soil applications have given up to five months residual control.

A brown liquid, practically insoluble in water, but soluble in acetone, ethanol, aromatic solvents, kerosene, and chlorinated hydrocarbons. It is hydrolysed by alkali.

The acute oral LD_{50} for rats is 16–37 mg/kg. The acute dermal LD_{50} for rats is 135–340 mg/kg.

Use. Effective against root maggots, wireworms and other soil and grassland insects. Also gives control of some nematodes. Very persistent in soil. Probably effective against cutworms, termites, and Collembola.

Caution

(*a*) A very poisonous substance, so full protective clothing must be worn. Very easily absorbed through mammalian skin.
(*b*) Use in some countries prohibited because of high toxicity and excessive persistence.
(*c*) Pre-harvest interval for edible crops varies – 4 weeks (Norway), 8 weeks (Holland).
(*d*) Dangerous to fish, livestock, game, wild birds and animals.

Formulations. 50% e.c. and granules of various a.i. content (2.5, 7.5%); also as a 20% seed dressing powder.

82. Trichlorphon (*Dipterex, Anthon, Chlorofos*)

Properties. A contact and stomach insecticide with penetrant action, recommended for use against flies, some bugs, some beetles, lepidopterous larvae, and ectoparasites of domestic animals. Its activity is attributed to its metabolic conversion to dichlorvos. Moderate persistence.

It is a white crystalline powder, quite soluble in water, insoluble in petroleum oils, poorly soluble in carbon tetrachloride and diethyl ether; soluble in benzene, ethanol, and most chlorinated hydrocarbons. It is stable at room temperature but is decomposed by water at higher temperatures (and acid media) to form dichlorvos.

The acute oral LD_{50} for male rats is 630 mg/kg.

Use. Effective against most lepidopterous larvae, flies and fly maggots, some Homoptera, many Heteroptera, some beetles. Used for household and veterinary pests, under different trade names.

Caution

(*a*) Harmful to fish.
(*b*) Pre-harvest interval for edible crops – 2–14 days, according to the country.

Formulations. 50% w.p.; 50 and 80% soluble powders; 50% e.c.; 5% dust; and 2 and 5% granules.

83. Vamidothion (*Kilval, Vation, Vamidoate*)

Properties. A systemic insecticide and acaricide of high persistence and particular value against Woolly Aphid. It is metabolized in the plant to the sulphoxide which has similar biological activity but greater persistence.

It forms a colourless waxy solid, very soluble in water (4 g/ml) and most organic solvents, but almost insoluble in light petroleum and cyclohexane. Undergoes slight decomposition at room temperature; non-corrosive, and compatible with most pesticides.

The acute oral LD_{50} for male rats is about 100 mg/kg. The toxicity of the sulphoxide is reduced by about half.

Use. Especially effective against Woolly Aphid; used for control of sap-feeding insects and mites on fruit, rice, cotton, and many other crops.

Caution

(*a*) This is a poisonous substance (Part III), and protective clothing should be worn.
(*b*) Harmful to bees, fish, livestock, game, wild birds and animals.
(*c*) Pre-harvest interval for edible crops – 4–6 weeks.

Formulations. Solution containing 40% w/v.

Carbamates and related compounds

The successful development of the organophosphates as pesticides directed attention to other compounds which act as anticholinesterases. This led to the discovery of carbaryl with its broad spectrum of activity. It appears that the action of the carbamates on both insects and mammals is an inhibition of the cholinesterase.

84. Aldicarb (*Temik*)

Properties. A systemic, soil-applied, insecticide, acaricide and nematicide; used against a wide range of leaf-eating and sap-sucking insects, stem and cyst nematodes, and certain mites. Usually applied to seed furrows, or as bands, or broadcast on to the soil; moisture is required to release the active chemical from the granules, so rainfall or irrigation should follow application. Uptake by the plant roots is rapid; protection is given for up to 80 days.

It is formed as colourless crystals, only slightly soluble in water, but soluble in most organic solvents. Generally stable, except to concentrated alkali; non-flammable, and non-corrosive.

The acute oral LD_{50} for male rats is 0.9 mg/kg, and the acute dermal LD_{50} for male rabbits is 5 mg/kg.

Use. Effective against a wide range of insects, mites and nematodes, especially aphids, whiteflies and leafminers. Mealybugs, scale insects, capsid bugs, spider mites and tarsonemid mites are also controlled. Both stem and cyst nematodes are killed. Reduces virus damage to many crops by controlling the insect and nematode vectors.

Caution

(*a*) This is a very poisonous substance (Part II), so full protective clothing must be worn. Easily absorbed through mammalian skin.
(*b*) Dangerous to fish, game and wildlife.
(*c*) Pre-harvest interval (for tomatoes in UK) – 6 weeks.

Formulations. Because of its toxicity and handling hazards, only formulated as granules (50, 100, and 150 g a.i./kg) for soil application.

85. Bendiocarb (*Garvox*)

Properties. A contact and stomach-acting insecticide, used against a wide range of pests on crops, stored products, and household/domestic pests; short-lived.

A colourless solid, insoluble in water and scarcely soluble in most organic solvents; stable to hydrolysis at pH 5, but at pH 7 decomposition is quite rapid (4 days at 25 °C).

The acute oral LD_{50} for most mammals is 34–64 mg/kg; dermal toxicity moderate.

Use. Effective against Collembola, symphylids, Lepidoptera, and Coleoptera, especially in soil; also used against Frit Fly, and many turf pests (wireworms, corn rootworms, etc.) in N. America. Mainly used on sugar beet, maize and turf; experimentally on cotton, rice, tobacco, potatoes, and many cereals. Also used in households and food stores, against flies, mosquitoes, cockroaches, wasps, ants, fleas, etc.

Caution

(*a*) A very toxic chemical; full protective clothing must be worn.
(*b*) Very dangerous to bees and fish.
(*c*) Dangerous to livestock and wildlife.

Formulations. Include w.p. 800 g a.i./kg; dust 10 g/kg; granules 30 and 100 g/kg; and a paint-on bait for fly control.

86. Bufencarb (*Bux*)

Properties.A non-persistent carbamate insecticide, effective against a wide range of soil and foliage insects. Degradation in soil is fairly rapid, with no expected seasonal accumulation following use of granules.

The technical product is a yellow solid, of low melting point; almost insoluble in water, very soluble in methanol and 1,2-dimethylbenzene, but less so in aliphatic hydrocarbons such as hexane. Stable in neutral or acid solutions, but the rate of hydrolysis increases with either a rise in temperature or increase in pH.

The acute oral LD_{50} for rats is 87 mg/kg; acute dermal LD_{50} for rabbits is 680 mg/kg.

Use. Active at rates of 0.5–2.0 kg a.i./ha against a range of soil insects, such as corn rootworm, rice water weevil, root mealybugs; and foliage insects such as rice leafhoppers and planthoppers; and graminaceous stem borers.

Caution

(*a*) This is a very poisonous substance, and full protective clothing must be worn.
(*b*) This poison can be absorbed through the skin.
(*c*) Dangerous to fish.

Formulations. E.c. 240 g a.i./l and 360 g a.i./l; dusts 20 and 40 g a.i./kg; granules 100 g a.i./kg.

87. Carbaryl (*Sevin, Carbaryl 85, Murvin, Septon*)

Properties. A contact insecticide with slight systemic properties and broad-spectrum activity. No evidence of phytotoxicity at recommended rates. Has growth regulatory properties and may be used for fruit thinning (apples). Also used for killing earthworms in turf. Persistence more like an organochlorine compound than an organophosphate.

It is a white crystalline solid, barely soluble in water, but soluble in most polar organic solvents (such as dimethyl sulphoxide). It is stable to light, heat and hydrolysis under normal conditions, and is non-corrosive. Compatible with most other pesticides, except those strongly alkaline, such as lime-sulphur or Bordeaux mixture, which cause hydrolysis.

The acute oral LD_{50} for male rats is 850 mg/kg.

Use. Effective against many insect pests, especially capterpillars, midges, beetles, Orthoptera, capsids and other bugs. Generally more effective against chewing insects than sap-suckers.

Caution

(*a*) Dangerous to bees.
(*b*) Harmful to fish.
(*c*) Pre-harvest interval for edible crops – 1 week.

Formulations. 50 and 85% w.p.; 5 and 10% dusts.

88. Carbofuran (*Furadan, Curaterr, Yaltox*)

Properties. A broad-spectrum systemic insecticide, acaricide and nematicide, used against both foliage feeding insects, mites and soil pests. In plants its half-life is less than five days; in soil the half-life is 30–60 days.

It is a white odourless, crystalline solid, fairly soluble in water and acetone; non-inflammable, but unstable in alkaline media.

The acute oral LD_{50} for rats is 8–14 mg/kg.

Use. Effective against various sap-feeding insects and mites, especially leafhoppers on rice, and also root-eating caterpillars and beetle larvae on cereals.

Caution

(*a*) This is a poisonous substance (Part II), and protective clothing should be worn.

(*b*) Harmful to fish.

Formulations. 75% w.p.; 4% flowable paste; 2, 3 and 10% granules.

89. Ethiofencarb (*Croneton*)

Properties. A systemic insecticide, formulated for both soil and foliar application; specific effect against aphids.

A yellow oil, slightly soluble in water.

The acute oral LD_{50} for rats is 411–99 mg/kg; very toxic to fish.

Use. Effectiveness confined to aphids; in the UK recommended for use on potato.

Caution

(*a*) Harmful to fish.

Formulations. Both e.c. 500 g a.i./kg; and granules 100 g/kg.

90. Methiocarb (*Mesurol, Draza, Baysol*)

Properties. A non-systemic insecticide and acaricide, with a broad spectrum of activity; persistent; a powerful molluscicide.

It is a white crystalline powder, practically insoluble in water, but soluble in most organic solvents. Hydrolysed by alkali.

The acute oral LD_{50} for male rats is 100 mg/kg.

Use. Mainly used for control of slugs and snails, can control cutworms and various beetles.

Caution

(*a*) Harmful to fish.

(*b*) Pre-harvest interval for edible crops – 7 days.

(*c*) Pre-access interval for livestock to treated areas – 7 days.

Formulations. 50 and 70% w.p.; *Draza*, 4% granules.

91. Methomyl (*Lannate, Halvard, Nudrin*)

Properties. A broad-spectrum, systemic and contact insecticide and acaricide. It is also effective as a nematicide.

It is a white crystalline solid, slightly soluble in water, soluble in acetone, ethanol, methanol; the aqueous solution is non-corrosive. It is stable in solid form and aqueous solution under normal conditions, but decomposes in moist soil.

The acute oral LD_{50} for male rats is 17 mg/kg. Not a skin irritant.

Use. As a soil treatment it has given systemic control of certain insects and nematodes. It controls, as a foliar spray, many insects such as aphids, Colorado Beetle, leafrollers and many other caterpillars, and red spider mites.

Caution

(*a*) This is a poisonous substance, and protective clothing should be worn.

(*b*) Pre-harvest interval for edible crops – about 3 weeks.

Formulations. 90% w.p.; 25% w.p.

92. Oxamyl (*Vydate*)

Properties. A carbamate insecticide and nematicide with both contact and systemic action; moderate residual effect. Will control nematodes with both soil and foliar application.

A white crystalline solid, stable in solid form and most solutions, but decomposes to innocuous materials in natural waters and soil. Aeration, sunlight, alkalinity and higher temperatures increase the rate of decomposition. Fairly soluble in water; more soluble in ethanol, propanone and methanol. Aqueous solutions are non-corrosive.

The acute oral LD_{50} for male rats is 5.4 mg/kg.

Use. Effective against aphids, thrips, leaf beetles, flea beetles, leaf-miners, and mites, as foliar sprays at rates of 0.2–1.0 kg/a.i./ha; and nematodes by both foliar and soil application.

Caution

(*a*) A very poisonous substance (Part II): full protective clothing must be worn.
(*b*) Dangerous to fish, game and wildlife.

Formulations. Water soluble liquid at 240 g a.i./l; granules 50 and 100 g a.i./kg.

93. Pirimicarb (*Pirimor, Aphox, Fernos*)

Properties. A selective insecticide of use against Diptera and aphids, being effective against organophosphorous-resistant aphid strains. It is fast-acting, and has fumigant and translaminar properties; it is taken up by the roots and translocated in the xylem vessels. Non-acaricidal.

It is a colourless solid, virtually insoluble in water, but soluble in most organic solvents. Forms well-defined crystalline salts with acids; these salts are water soluble.

The acute oral LD_{50} for rats is 147 mg/kg.

Use. Effective against organophosphorous-resistant strains of aphids, and also against dipterous maggots.

Formulations. 50% w.p. and 5% granules.

94. Promecarb (*Carbamult, Minacide*)

Properties. A non-systemic contact insecticide, useful against coleopterous pests, Lepidoptera and Diptera.

It is a colourless crystalline solid, slightly soluble in water, soluble in acetone and ethylene dichloride; hydrolysed by alkali.

The acute oral LD_{50} for rats is 70–100 mg/kg.

Use. Effective against Coleoptera, Lepidoptera, and fruit leaf-miners.

Caution. This is a poisonous substance, and protective clothing should be worn.

Formulations. 25% e.c.; 37.5 and 50% w.p.; 5% dust.

95. Propoxur (*Baygon, Blattanex, Unden, Suncide*)

Properties. A non-systemic insecticide, with rapid knockdown, used against Hemiptera, flies, millipedes, ants and other household and public health pests. Non-phytotoxic.

It is a white crystalline powder, only slightly soluble in water, but soluble in most organic solvents; unstable in highly alkaline media.

The acute oral LD_{50} for rats is about 100 mg/kg.

Use. Effective against jassids, aphids, other bugs, flies, millipedes, termites, ants and other household pests, ticks and mites, mosquitoes.

Caution

(*a*) This is a poisonous substance, and protective clothing should be worn.
(*b*) Dangerous to bees.

Formulations. E.c.; w.p.; dusts; granules; baits; and pressurized sprays of different a.i. concentrations; some sprays with added dichlorvos.

96. Thiofanox (*Decamox*)

Properties. A systemic insecticide and acaricide, used as a seed or soil treatment against root and foliar-feeding pests.

A colourless solid, soluble in water and in most organic solvents; stable under normal storage conditions, and quite stable to hydrolysis.

The acute oral LD_{50} for rats is 8 mg/kg; and dermal LD_{50} for rabbits 39 mg/kg.

Use. Controls aphids, leafhoppers and capsids on beet and potato; also Beet Leaf Miner, flea beetles, *Atomaria* and Colorado Beetle; applied as granules in-furrow.

Caution

(*a*) A very toxic chemical (Part II substance; *Poisons Rules* apply in UK); full protective clothing must be worn; can be absorbed through the skin.
(*b*) Dangerous to bees, fish, game, wildlife and livestock.

Formulations. Mostly as granules, 50, 100, and 150 g a.i./kg; and a seed treatment 429 g/l.

Miscellaneous compounds and fumigants

These are organic fumigants and inorganic salts which posses toxic qualities to insect pests. The gas and fumigants act through the insect respiratory system, but the inorganic salts act usually as stomach poisons. Many of these compounds are some of the earliest pesticides used in agriculture.

97. Aluminium phosphide (*Phostoxin*)

Properties. The phosphine liberated is highly insecticidal and a potent mammalian poison, used only for fumigation of stored products.

Aluminium phosphide is a yellow crystalline solid, stable when dry, but reacts with moist air liberating phosphine. Phosphine is highly toxic, and spontaneously inflammable in air. The residue is harmless.

Use. Fumigation of stored products and containers only.

Caution. This is a very potent poison – fumigation takes from 3–10 days and should only be undertaken by trained personnel.

Formulations. *Phostoxin* evolves a non-inflammable mixture of phosphine, ammonia and carbon dioxide; manufactured as 3 g tablets or pellets (0.6 g).

98. Amitraz (*Mitac, Taktic, Triatox, Azaform, Baam*)

Properties. An acaricide effective against a wide range of phytophagous mites; all stages are susceptible. It has some insecticidal properties, and is effective against some Hemiptera and eggs of Lepidoptera. Relatively non-toxic to predacious insects and bees.

It is insoluble in water, but soluble in methyl benzene and propanone. At acid pH it is unstable, and will slowly deteriorate under moist conditions; generally compatible with most commonly used pesticides.

The acute oral LD_{50} for rats is 800 mg/kg.

Use. Effective against phytophagous mites, especially red spider mites, at concentrations from 20–50 g a.i./100 l, depending upon the species. Also effective against eggs of *Heliothis* spp., and used against ticks and mites on cattle and sheep.

Caution

(*a*) This is a poisonous substance (Part III): protective clothing should be worn.
(*b*) Pre-harvest interval for fruit is 2 weeks.
(*c*) Harmful to fish.

Formulations. E.c. 200 g a.i./l; d.p. 250 and 500 g a.i./kg.

99. Carbon disulphide (*Weevil-Tox*)

Properties. A fumigant with insecticidal properties, also used for soil treatment against nematodes and soil insects.

A yellow mobile liquid with boiling point 46 °C; the vapour is 2.6 times as dense as air; very inflammable and dangerous. The vapour is highly poisonous, producing giddiness and vomiting in 30 min at 6.8 g/m^3; used on stored grain with added carbon tetrachloride (to reduce fire hazard).

Use. Used for fumigation of nursery stock, stored grains, and for soil treatment to kill insects and nematodes.

Caution

(*a*) Very poisonous, and great care is needed in its use.
(*b*) Very inflammable and dangerous to use.

Formulations. The pure liquid is used for fumigation; soil applications use either the liquid, or emulsion solutions in liquid.

100. Copper acetoarsenite (*Paris Green*)

Properties. A stomach poison used as baits, very phytotoxic. Introduced about 1867 for the control of Colorado Beetle.

It is a green powder of low solubility in water; in the presence of water and carbon dioxide it is readily decomposed to give water-soluble and phytotoxic arsenical compounds.

Acute oral LD_{50} for rats is 22 mg/kg; a violent poison to man when ingested.

Use. Effective only as baits because of high phytotoxicity.

Caution. A very poisonous substance which should be handled with great care; a persistent and accumulative poison.

Formulations. Most specifications require a content of at least 35% arsenious oxide, 20% cupric oxide, 10% acetic acid; and not more than 1.5% arsenious oxide in a water-soluble form.

101. Cryolite (*Sodium aluminium fluoride*)

Properties. A naturally occurring mineral, about 98% pure, with insecticidal action, both contact and stomach; non-phytotoxic at insecticidal concentrations.

A crystalline solid, but synthetically produced as an amorphous powder; both are insoluble in water, but soluble in dilute alkalis; incompatible with strongly alkaline pesticides.

Acute toxicity to mammals is low.

Use. Applied as a suspension, against a wide range of insects.

Formulations. A suspension of 2 g/l in dilute alkali.

102. Cyhexatin (*Tricyclohexyltin hydroxide*) (*Plictran*)

Properties. An acaricide effective against the motile stages of a wide range of phytophagous mites. No phytotoxicity reported on deciduous fruit or on glasshouse crops, but some spotting on citrus fruit has been observed.

It is a white crystalline powder, insoluble in water, poorly soluble in most organic solvents, but 22% w/v soluble in chloroform. Stable to ambient temperatures, and in neutral and alkaline media. Degraded on exposure to u.v. light in thin layers.

The acute oral LD_{50} for rats is 540 mg/kg.

Use. Effective against motile stages of phytophagous mites, but not the eggs.

Caution

(*a*) Virtually harmless to bees.
(*b*) Some phytotoxicity to citrus fruit reported.

Formulations. 50% a.i. w.p.; and 25% a.i. w.p. (available only in Europe).

103. Ethylene dibromide (= Dibromoethane) (*Bromofume, Dowfume–W*)

Properties. An insecticidal fumigant used against stored products pests and for treatment of fruit and vegetables, and for soil treatment against certain insects and nematodes; very phytotoxic.

It is a colourless liquid, insoluble in water, but soluble in ethanol, ether, and most organic solvents; stable and non-inflammable.

The acute oral LD_{50} for male rats is 146 mg/kg, dermal application will cause severe burning.

Use. Used against stored products pests; for ripening treatment of fruit and vegetables; and for soil treatment against certain insects and nematodes. In the USA it is sprayed on imported fruit (and sometimes fruit for export), to kill fruit flies, but recently (1984) its use for this purpose has been banned by the EPA following demonstration of carcinogenic properties.

Caution

(*a*) This is a dangerous substance, possibly carcinogenic; full protective clothing must be worn.
(*b*) If used for soil treatment, planting must be delayed for 8–14 days because of its phytotoxicity.

Formulations. *Dowfume W–85* (1.44 kg a.i./l) in an inert solvent, for soil application; for mill use a solution in carbon tetrachloride.

104. Lead arsenate (*Gypsine, Soprabel*)

Properties. A non-systemic stomach insecticide, with little contact action. Non-phytotoxic, though the addition of components causing the production of water-soluble arsenical compounds may lead to leaf damage.

Formulated as either paste or powder: $PbHAsO_4$. It is poisonous to mammals when ingested, 10–15 mg/kg being fatal.

Use. Used against various caterpillars, sawfly larvae, and tipulid larvae. Does not give complete control of caterpillars, but is selective and does not harm predators of red spider mites.

Caution

(*a*) Dangerous to bees.

(*b*) Harmful to fish and livestock.
(*c*) Pre-harvest interval for edible crops – 6 weeks.
(*d*) Pre-access interval for livestock to treated areas – 6 weeks; 3 weeks in wet weather.

Formulations. As wettable powders or pastes.

105. Mercurous chloride (*Calomel, Cyclostan*)

Properties. A general poison, but being phytotoxic its use is generally limited to soil application, mainly for the control of root maggots. It is also used as a fungicide and for the control of club-root of brassicas. Its biological activity is attributable to its reduction to metallic mercury.

It is a white powder, insoluble in water, but soluble in ethanol and most organic solvents. In the presence of water it is slowly dissociated to mercury and mercuric chloride, a reaction hastened by alkalis.

The acute oral LD_{50} for rats is 210 mg/kg.

Use. Slightly effective against soil-inhabiting root maggots of vegetables. Use generally restricted also by the high cost of the chemical.

Caution. All mercury compounds are general poisons.

Formulations. The pure compound is used as a seed dressing; as a 4% dust on a non-alkaline carrier.

106. Methyl bromide (= Dibromomethane) (*Bromogas, Embafume*)

Properties. A general poison with high insecticidal and some acaricidal properties, used for space fumigation and for the fumigation of plants and plant products in stores. It is a soil fumigant, used for the control of nematodes, fungi and weeds.

It is a colourless gas, scarcely soluble in water, soluble in most organic solvents; stable, non-corrosive and non-inflammable.

It is highly toxic to man; in many countries its use is restricted to trained personnel.

Use. Effective against stored products pests and soil pests.

Caution. A very poisonous gas; use should be restricted to trained personnel.

Formulations. Packed as a liquid in glass ampoules (up to 50 ml), in metal cans and cylinders for direct use. Chloropicrin is sometimes added, up to 2%, as a warning gas.

107. Sulphur (Lime-sulphur) (*Cosan, Hexasul*)

Properties. A non-systemic direct and protective fungicide and acaricide. Generally non-phytotoxic, except to certain varieties known as 'sulphur-shy'.

It is a yellow solid, existing in allotropic forms; practically insoluble in water, slightly soluble in ethanol and ether. Slowly hydrolysed by water; compatible with most other pesticides, except petroleum oils.

Use. Used mainly against phytophagous mites, especially against 'big-bud' mites on blackcurrants. Also effective against powdery mildews, and apple scab.

Caution

(*a*) Some fruit varieties are 'sulphur-shy'; the manufacturer's instructions on labels of products as to susceptible varieties should be followed.
(*b*) To avoid possible taint, do not use on fruit for processing.

Formulations. Dusts; w.p.; as finely ground 'colloidal' suspensions or more usually as lime-sulphur.

Natural organic compounds and pyrethroids

Many plants produce toxic compounds, usually thought to act as deterrents to grazing by herbivores, and some of these are selectively poisonous to insects and have proved to be successful as insecticides (see also page 67). Nicotine and rotenone are two of the best-known examples. The flowers of *Pyrethrum* were found to contain insecticide and, when analysed, yielded a mixture of various pyrethrins. Extracts of the dried flowers (sold as 'Pyrethrum') were so useful as insecticide that some of the pyrethrins were eventually synthesized. But they are not at all persistent, rapidly degraded by both heat and sunlight, and although they have a spectacular knockdown effect their killing effect is limited. The broad spectrum of activity, and low vertebrate toxicity, together with rapid knockdown and lack of resistance, made these compounds very useful for certain types of insect pest control. This led to a search for synthetic chemical homologues and analogues possessing better insecticidal qualities, and now after some years a range of pyrethroids has been developed.

The new compounds, some of which are not closely related to the pyrethrins, are collectively termed the 'pyrethroids', and sometimes lumped together with the juvenile hormones (JH) and insect growth regulators (IGR) as the 'third-generation insecticides'. The pyrethroids have the same broad spectrum of activity and a tendency not to induce resistance, but their longevity and stability have been greatly enhanced, and their insect-killing ability many times improved over the naturally occurring pyrethrins. However, their increased toxicity makes most of them dangerous to bees and other pollinating insects, and dangerous to the natural predators and parasites; so this limits their usefulness a little, especially in crop/pest situations where natural control is important. The pyrethroids are very effective insecticides, albeit rather toxic, and their use is becoming more widespread, especially in situations where pests have developed resistance to some of the widely used organochlorines, organophosphates, and also now carbamates. In the New World the pyrethroids are now widely employed for cotton protection and on top fruit. But it does appear that some resistance is developing to some of the earlier (most used) pyrethroids in some situations.

Most of the pyrethroids were initially used for protection of livestock and in domestic situations (public health), and their crop use came later. Now for crop protection purposes it appears that, in addition to their broad-spectrum activity, some pyrethroids show particular effectiveness against certain groups of insects, for example according to Shell Co. alphamethrin is very effective against beetles and fenvalerate against aphids. The pyrethroids not included in this chapter are those that at the time of writing were almost entirely used for livestock protection and in domestic situations.

108. Bioallethrin (*D-Trans, Esbiol*)

Properties. A powerful contact insecticide, by nature a synthetic pyrethroid, with a rapid 'knockdown' effect. Metabolic detoxification is delayed by addition of synergists such as piperonyl butoxide. It is much more persistent than the natural pyrethrins. It occurs as two very similar isomers. Insoluble in water, but miscible with most organic solvents.

The acute oral LD_{50} for male rats is 500–780 mg/kg.

Use. Mostly used against household pests at present, especially Diptera.

Formulations. Mostly in combination with synergists and other insecticides in kerosene as fly sprays; also as aerosols, 0.1–0.6% a.i.; and impregnated dusts.

109. Cypermethrin (*Cymbush, Ripcord*)

Properties. A synthetic pyrethroid, broad-spectrum in action, and effective against a wide range of insect pests by contact and stomach action; persistent; non-systemic; no recorded phytotoxicity.

A thick brown semi-solid substance, liquid at 60 °C; stable; insoluble in water but soluble in most organic solvents; more stable in acid than alkaline media; some decomposition in sunlight.

Acute oral LD_{50} for rats is 303–4123 mg/kg, depending on carrier used; for mice 138 mg/kg; a slight skin irritant and eye irritant.

Use. Effective against leaf- and fruit-eating Lepidoptera and Coleoptera, as well as Hemiptera, cutworms in litter and soil, biting flies and animal ectoparasites.

Caution

(*a*) Dangerous to bees, and to fish.
(*b*) Slight skin and eye irritant.

Formulations. E.c. 25–400 g a.i./l; u.l.v. 10–75 g/l, etc.

110. Deltamethrin (*Decis*)

Properties. A very potent insecticide, with contact and stomach action against a wide range of pests; many more times toxic to insects than the pyrethrins (30–1000 ×).

The technical product is a colourless crystalline powder, insoluble in water, but soluble in most organic solvents; stable on exposure to both air and sunlight.

The acute oral LD_{50} for rats is about 135 mg/kg.

Use. Effective against a wide range of insects, especially aphids, mealybugs, scales, psyllids, whiteflies, and caterpillars on fruit trees.

Caution

(*a*) A poisonous substance, recommended in the UK for inclusion in Part III category, and under the *Poisons Rules*.
(*b*) Irritating to eyes and skin.
(*c*) Dangerous to bees.
(*d*) Very dangerous to fish.

Formulations. Include e.c. 25 g a.i./l; w.p. 25 and 50 g/kg; dusts 0.5 and 1.0 g/kg; u.l.v. concentrate 4, 5, and 10 g/l; granules 0.5 and 1.0 g/kg.

111. Fenvalerate (*Belmark, Pydrin, Sumicidin*)

Properties. A synthetic pyrethroid, with highly active contact and stomach action against a wide range of pests, including strains resistant to organochlorine, organophosphorous and carbamate insecticides; persistent and with good foliar retention.

The technical product is a viscous brown liquid, largely insoluble in water but soluble in most organic solvents; stable to heat and sunlight; more stable in acid than alkaline media (optimum stability at pH 4).

Acute oral LD_{50} for rats is 450 mg/kg.

Use. Effective against most insect pests; widely used against Lepidoptera and Coleoptera foliage and fruit pests, as well as Hemiptera (said by Shell Co. to be especially effective against aphids); not recommended against mites. Used in public health and to control livestock ectoparasites, including ticks.

Caution. Dangerous to bees and to fish.

Formulations. Include 25–300 g a.i./l e.c.; 25–75 g/l u.l.v.; 100 g/l s.c.; also as mixtures with fenitrothion.

112. Nicotine (*Nicofume*)

Properties. A non-persistent non-systemic, contact insecticide with some ovicidal properties. Can be used as a fumigant in closed spaces; water-insoluble salts (the so-called 'fixed' nicotines) have been used as stomach insecticides. Nicotine is prepared from tobacco (*Nicotiana tabacum*) by steam distillation or solvent extraction.

A colourless liquid, darkening on exposure to air, miscible with water below 60 °C (forming a hydrate); miscible with ethanol, ether, and readily soluble in most organic solvents.

The acute oral LD_{50} for rats is 50–60 mg/kg.

Use. Effective against aphids, capsids, leaf-miners, thrips on a wide range of horticultural crops; sawflies and Woolly Aphid on apple. Also used for fumigation of glasshouses.

Caution

(*a*) Harmful to bees.
(*b*) Dangerous to fish, livestock, game, wild birds and animals (Part III poison).
(*c*) Pre-harvest interval for outdoor edible crops – 2 days.
(*d*) Pre-access interval for livestock to treated areas – 12 hours.

Formulations. Marketed as the 95% alkaloid, or as nicotine sulphate (40% alkaloids); also as 3–5% dusts. For fumigation nicotine 'shreds' are burnt, or the liquid nicotine is applied to a heated metal surface.

113. Permethrin (*Ambush, Kalfil, Talcord*)

Properties. A synthetic pyrethroid insecticide, with contact action against a wide range of insect pests; persistent; non-systemic.

The technical product is a brown liquid, insoluble in water but soluble in most organic solvents; stable to heat, but some photochemical degradation; more stable in acid than alkaline media.

Acute oral LD_{50} varies with *cis/trans* ratio, carrier used, and conditions for use; for rats 430–4000 mg/kg.

Use. Effective against foliage and fruit-eating Lepidoptera, Coleoptera, Hemiptera, and others, as well as animal ectoparasites. In many uses now generally superseded by cypermethrin.

Caution

(*a*) Dangerous to bees.
(*b*) Extremely dangerous to fish.

Formulations. E.c. 100–500 g a.i./l; solution of 50 g/l; u.l.v. concentrates 50 and 100 g/l; etc.

114. Pyrethrins (*Pyrethrum*)

Properties. The pyrethrins is a general term usually including the cinerins as well as the pyrethrins extracted from flowers of *Pyrethrum cinerariaefolium*. They are powerful contact insecticides; non-systemic, causing a rapid paralysis or 'knockdown'. They are unstable to sunlight and are rapidly hydrolysed by alkalis with a loss of insecticidal properties. Their metabolic detoxification may be delayed by the addition of synergists such as piperonyl butoxide, sesamin, etc.

The acute oral LD_{50} for rats is about 200 mg/kg.

Use. Effective against flies, and household pests, usually with added synergist.

Formulations. As dusts with added non-alkaline carrier; as aerosols the extract is dissolved in a volatile solvent such as methyl chloride or dichlorodifluoromethane, usually with added synergist.

115. Resmethrin (*Chryson, For-Syn, Synthrin*)

Properties. A synthetic pyrethroid, with powerful contact action against a wide range of insects. Plant toxicity is low. However it is not synergised to any extent by pyrethrin synergists. Toxicity to normal houseflies is about 20 times that of natural pyrethrins.

It occurs as a mixture of isomers; a colourless waxy solid; insoluble in water, but soluble in most organic solvents. More stable than pyrethrins, but is decomposed quite rapidly on exposure to air and light.

The acute oral LD_{50} for rats is about 2000 mg/kg.

Use. Effective against a wide range of insect pests, both household and agricultural.

Formulations. May be formulated with or without other pyrethroids and pyrethrin synergists in aerosol concentrates; also as water-based sprays; e.c.; w.p.; and concentrate for u.l.v. application.

The mixture of isomers is usually 20–30% of cis-isomer, and 80–70% of the trans-isomer, the former known sometimes as Cismethrin, and the latter as Bioresmethrin.

116. Rotenone (*Derris, Cube, Rotacide*)

Properties. This is the name given to the main insecticidal compound of certain *Derris* spp., and *Lonchocarpus* spp.; known for many years to be effective as a fish poison and an insecticide. It is a selective non-systemic insecticide with some acaricidal properties; non-phytotoxic; but of low persistence in spray or dust residues. Insoluble in water but soluble in polar organic solvents; readily oxidized in presence of light and alkali to less insecticidal products.

The acute oral LD_{50} for rats is 132–1500 mg/kg; it is very toxic to pigs.

Use. Effective against aphids, caterpillars, thrips, some beetles, and red spider mites.

Caution

(*a*) Dangerous to fish.
(*b*) Toxic to pigs.
(*c*) Pre-harvest interval for edible crops – 1 day.

Formulations. Usually as dusts of the ground root with a non-alkaline carrier; dusts may be stabilized by addition of a small quantity of a strong acid such as phosphoric acid.

Organic oils

Also known as hydrocarbon oils, they comprise the oils distilled from crude mineral oils (petroleum or mineral oils) or from coal tars (tar oils). These oils are complex chemically, of such a nature as to defy separation into individual compounds. They are all strongly phytotoxic and use is generally restricted to dormant season washes, when they are destructive to insect and mite eggs.

117. Petroleum oils (*Volck*, etc.)

Properties. Also known as mineral oils; refined grades are known as white oils. The use of kerosene as an insecticide probably dates from about the time of its introduction as an illuminant. Oils of higher distillation range came into use about 1922. They consist largely of aliphatic hydrocarbons, both saturated and unsaturated. They are produced by distillation and refinement of crude mineral oils. The ones used as pesticides generally distil above 310°C (636°F), namely 64–79%, 40–49%, and 10–25% respectively; density rarely exceeds 0.92 at 15°C (60°F). Viscosity and density vary according to the geographical area (oil-field) from which the crude oil came.

Use. Effective against certain insects such as mealybugs, scales and thrips, and against red spider mites; they are ovicidal. Their use is limited by their phytotoxicity; a semi-refined oil can be used as a dormant ovicide; for foliage use, a refined oil of narrow viscosity range is required. Relatively harmless to mammals.

Caution. Generally very phytotoxic; can be used as herbicide (especially TVO).

Formulations. Available either alone or in mixture with DNOC or DNOC and DDT.

118. Tar oils

Properties. Produced by distillation of tars resulting from the high-temperature carbonization of coal and coke oven tars. Although used for wood preservation since 1890, the introduction of the formulated products known as tar oil washes for crop protection dates from about 1920. These oils are brown to black liquids, of density 1.05–1.11; insoluble in water but soluble in organic solvents. They consist mainly of aromatic hydrocarbons but contain 'phenols' and 'tar acids'; they are highly phytotoxic.

Use. Effective for the control of the eggs of many insect species, particularly of aphids, Lepidoptera, psyllids and other bugs. Because of their phytotoxicity use is restricted to the dormant season. An effective fungicide.

Caution

(*a*) Dangerous to fish.
(*b*) Irritating to skin, eyes, nose and mouth.
(*c*) Strongly phytotoxic; use only as dormant washes.

Formulations. As miscible winter washes, and stock emulsion winter washes. The most insecticidally active tar oil is creosote.

Biological compounds

A number of polyhedral and granular viruses, and bacteria, and a few fungi, are commonly recorded as causing diseases in insect populations. On several occasions when part of a pest population has been found to be naturally infected by a pathogen it has been possible to make up a spray solution from a suspension of macerated diseased insect bodies and to spray this solution over uninfected parts of the population, often with spectacular success. Generally, the only insect groups that regularly suffer from viral or bacterial epizootics are Lepidoptera and Coleoptera.

The only biological compounds at present readily available commercially for use against insect pests are a spore suspension of *Bacillus thuringiensis*, for use against lepidopterous larvae only, and *Heliothis* Nuclear Polyhedrosis Virus. Various strains of *Bacillus popilliae* cause Japanese Beetle Milky Disease, and commercial formulations are now available in the USA.

119. *Bacillus thuringiensis* Berliner (*BTB-183, Thuricide, Biotrol-BTB*)

Properties. This biological compound is a suspension of spores in an inert powder, specific in action to lepidopterous larvae, on which it acts as a stomach poison only. There are no phytotoxic effects, and it is innocuous to both mammals, other animals, and insect predators and parasites. It is a stable product which has been kept at room temperature for over six years without any detectable loss in potency. However, since it is a viable biological compound it has to be protected from extremes of heat and light, and corrosive fumes. This bacterium, in addition to forming resistant spores, also produces a crystal of thermolabile endotoxin called a parasporal body. It is this parasporal body which is the insecticidal agent in any preparation of *B. thuringiensis*. The crystalline toxin is insoluble in water and is thus deposited intact upon the leaf surface, and after ingestion by the caterpillar acts as a stomach poison. It has no contact action whatsoever. Some caterpillars are killed after feeding within a few hours; other less susceptible species are not killed directly but suffer a gut paralysis which stops them feeding, and they die within a few days. A few important pest species of Lepidoptera do not appear to be susceptible to this pathogen.

Use. This is effective only against the larvae of Lepidoptera, and as already mentioned is not equally effective against all caterpillars.

Caution. There appear to be no real restrictions against the use of this insecticide at all. However, it is incompatible with certain insecticides which appreciably alter the pH.

Formulations. It is formulated as a wettable powder containing 25 billion viable spores per gram of product (2.5%), and 97.5% of inert ingredients.

120. Codling Moth GV

Properties and Use. A granulosis virus specifically pathogenic to Codling Moth; used in fruit orchards.

At present only available in the USA.

121. *Heliothis* PHV (*Elcar, Viron H*)

Properties. A nuclear polyhedrosis virus (*Baculovirus heliothis*) specific to the genus *Heliothis*; it acts after ingestion and the virus multiplies inside the insect haemocoel, killing the caterpillar. Death of young larvae occurs in about 2 days, older larvae take 4–9 days to die, depending on the dose. Storage at above 25°C will deactivate the virus; for prolonged storage the product may be frozen. There is no evidence that resistance can be developed to this virus. No phytotoxicity has been recorded.

Use. Effective against the caterpillars of *Heliothis* only: 5 pest species in total.

Formulations. As a w.p. with 0.4% viral bodies, in an inert powder; to be applied in 20–200 l water per ha (Sandoz, Inc.).

(Several other polyhedrosis viruses, for example *Spodoptera* PHV, are now commercially available in the USA.)

122. *Trichoplusia* PHV (*Viron T*)

Properties. At present only available in the USA.

123. *Verticillium lecanii* (*Vertalec, Mycotal*)

Properties and Use. An entomophagous fungus specific to certain Homoptera, and developed for use in glasshouse situations with a controlled and limited environment. In the UK two formulations are commercially available; '*Vertalec*' for use against aphids, and '*Mycotal*' against Glasshouse Whitefly. These products are really only suitable for glasshouse use for the required conditions are very high humidity (more than 85% RH) and a temperature not falling below 15°C. Death of infected insects will take 10–14 days. Effective against strains resistant to chemical insecticides.

Formulation. As a w.p. for high-volume spraying in water.

Insect growth regulators (IGR) and juvenile hormones (JH)

In the search for the effective, specific pesticides, which have minimal disruptive effects on the environment and do not encourage resistance build-up in the pests, insect hormones have shown great promise. They fall into two distinct categories. The juvenile hormones (JH), if applied to full-grown larvae (usually caterpillars), disturb the process of metamorphosis to the extent that the insect dies as a deformed pupa/adult. These hormones are normally species-specific, and so it is only economic to contemplate commercial production of JH for very important species of pests, such as Cotton Boll Weevil and Pink Bollworm. Insect growth regulators (IGR) often act as pesticides by interfering with cuticle formation at the time of ecdysis, killing the moulting larva. These chemicals are not as specific as JHs, and will be effective against members of the same group, e.g. caterpillars or fly maggots (larvae).

124. Diflubenzuron (*Dimilin*)

Properties. An insect growth regulator, produced in 1974, which operates by ingestion, and interference with the deposition of new cuticle at ecdysis. A species-dependent ovicidal (contact) action has been shown for *Spodoptera* as well as prevention of egg eclosion after uptake by females. No systemic action, so it is not effective against sap-sucking insects; in soil it is rapidly degraded.

The technical product is pale brown crystals, insoluble in water and apolar solvents, but it is soluble in polar solvents.

The acute oral LD_{50} for rats was more than 4640 mg/kg.

Use. Effective against a broad range of leaf-eating insects and some mites, at rates of 1.5–30 g a.i./100 l. Also used against fly and mosquito larvae.

Formulation. A water dispersible powder of 250 g a.i./kg.

125. Methoprene (*Altosid, Manta, Kabat*)

Properties. An insecticide of the insect growth regulator group, of short field persistence being rapidly degraded by micro-organisms and sunlight. No recorded damage to non-target organisms, especially predators.

A pale amber liquid, insoluble in water, but soluble in most organic solvents. Completely non-toxic to laboratory tested mammals.

Use. Most effective against Diptera (larvae); developed initially for destruction of mosquito larvae, now used for systemic treatment of cattle against hornfly, and being tested against various stored products pests. Treated larvae develop apparently normally into pupae, but the pupae die without giving rise to adults.

Formulation. Liquid 10% (8 g a.i./l.).

126. Kinoprene (*Enstar*)

Properties. A chemical analogue of a naturally occurring insect growth regulator, specific to the Homoptera; it inhibits development, causes sterility, and kills developing eggs. The compound is stable, but short lived, being rapidly degraded. It is soluble in organic solvents; a pale amber liquid.

The acute oral LD_{50} for rats is 4900 mg/kg; harmless to mammals.

Use. Effective only against Homoptera; aphids, whiteflies, scales and mealybugs are said to be controlled.

Formulation. An e.c., *Enstar* 5E, is a 500 g/l concentrate containing 65.3% a.i., to be applied as a h.v. spray, at present only recommended for glasshouse use, on ornamentals in the USA.

Pesticides/pest chart

In the event of wanting to kill insects which are not dealt with here as major pests, and so do not have specific control recommendations, it can be borne in mind that various pesticides are generally toxic to groups of closely related insects. If no recommendation is available then the table may be of assistance in deciding which chemical to use. However, it must be stressed that such extrapolation can be dangerous, for although some chemicals are specific to certain groups of insects, many notable exceptions occur. For example, carbaryl is a pesticide generally effective against caterpillars in most parts of the world, but in Malawi, on the cotton crop, it is effective against the Red Bollworm (*Diparopsis castenea*) but has no toxic effect upon *Heliothis* caterpillars, and DDT had to be used against the latter pests; against a mixed infestation a spray mixture was employed.

The relative effectiveness of different pesticides has already been discussed on pages 39 and 138, and this must be borne in mind when using the chart below.

The subdivision of the pests into fifteen main categories, as listed on the chart heading, is an attempt to be as specific as possible within the confines of available space. Clearly the lumping of all Hymenoptera together is far from satisfactory as there are major differences in response to insecticides shown by sawfly larvae, wasps and ants. The Hemiptera are a problem group in this respect as a few chemicals will kill most, if not all, of the plant and animal bugs, but most insecticides will kill some bugs, but not many; one or two pesticides only kill aphids!

The pesticides regarded as 'generally effective' are marked with a cross (x); the presence of a dot does not necessarily mean that the chemical is ineffective against those animals (although usually it does), sometimes it indicates that there was a lack of information published.

It must be remembered that this table does not take into account any local development of resistance, nor any local government restrictions on pesticide usage; local advice should always be sought for specific recommendations.

This table is intended as a *guide* only.

CHEMICAL	Orthoptera	Isoptera	Homoptera	Aphids	Mealybugs	Scales	Heteroptera	Thysanoptera	Lepidoptera	Diptera	Coleoptera	Hymenoptera	Acarina	Nematoda	Myriapoda	Miscellaneous	*Persistence*	*Mode of action*
	PESTS AGAINST WHICH GENERALLY EFFECTIVE																	
Chlorinated hydrocarbons																		
1. Aldrin	x	x	x	.	x	.	.	x	x	x	x	x	.	.	.	(Soil Pests)	Long	C,S
2. Chlordane	.	x	.	.	.	.	x	.	.	.	x	x	.	.	.	Earthworms	Long	C,S
3. Chlorobenzilate	.	.	.	.	.	.	.	.	.	.	.	.	x	.	.	.	—	C
4. DDT	x	x	x	x	x	x	x	x	x	x	x	x	.	.	.	.	Long	C,S
5. Dicofol	.	.	.	.	.	.	.	.	.	.	.	.	x	.	.	.	Long	C
6. Dieldrin	x	x	x	x	x	x	x	x	x	x	x	x	.	.	.	Spiders	Long	C,S
7. Endosulfan	.	.	x	x	.	.	x	.	x	x	x	.	x	.	.	.	Long	C,S
8. Endrin	x	x	x	x	x	x	x	x	x	x	x	x	x	.	.	.	Long	C,S
9. HCH, γ	x	.	x	x	.	.	x	x	x	x	x	x	.	.	x	Spiders	M–L	C,S,F
10. Heptachlor	.	.	.	.	.	.	.	x	.	x	x	x	.	.	.	(Soil Pests)	—	C,S,F
11. Mirex	.	.	.	.	.	.	.	.	.	.	.	x	.	.	.	.	—	S
12. Tetradifon	.	.	.	.	.	.	.	.	.	.	.	.	x	.	.	.	Long	Syst.
13. Tetrasul	.	.	.	.	.	.	.	.	.	.	.	.	x	.	.	.	—	C
14. TDE	.	.	.	.	.	.	x	x	x	x	x	.	.	.	.	.	—	C,S
Substituted Phenols																		
15. Binapacryl	.	.	.	.	.	.	.	.	.	.	.	.	x	.	.	Fungi	Mod.	C
16. Dinocap	.	.	.	.	.	.	.	.	.	.	.	.	x	.	.	Fungi	Mod.	C
17. DNOC*	x	.	x	x	x	x	x	.	x	.	.	.	x	.	.	(Eggs)	—	C
18. Pentachlorophenol*	.	x	.	.	.	.	.	.	.	.	x	.	.	.	.	Fungi	—	
Organophosphorous compounds																		
19. Acephate	.	.	x	x	.	.	.	x	x	x	.	x	.	.	.	.	Mod.	Syst.
20. Azinphos-methyl	x	.	x	x	x	x	x	x	x	x	x	x	x	.	.	.	Long	C,S
21. Azinphos-methyl + d-*S*-ms	x	.	x	x	x	x	x	x	x	x	x	x	x	.	.	.	—	C,S,Syst
22. Bromophos	.	.	x	.	.	.	x	.	x	x	x	.	x	.	.	.	Mod.	C,S
23. Bromopropylate	.	.	.	.	.	.	.	.	.	.	.	.	x	.	.	.	Long	C
24. Carbophenothion	.	.	x	x	x	x	.	x	.	x	.	.	x	.	.	.	Long	C,S
25. Chlorfenvinphos	.	.	x	.	.	x	.	.	x	x	x	.	.	.	.	.	Mod.	C,S
26. Chlorpyrifos	.	.	x	x	.	.	x	.	x	x	x	x	x	.	.	V,H	Long	C,S,F
27. Chlorpyrifos-methyl	.	.	.	.	.	.	.	.	x	x	x	.	x	.	.	SP,H	Mod.	C,S,F

CHEMICAL	Orthoptera	Isoptera	Homoptera	Aphids	Mealybugs	Scales	Heteroptera	Thysanoptera	Lepidoptera	Diptera	Coleoptera	Hymenoptera	Acarina	Nematoda	Myriapoda	Miscellaneous	*Persistence*	*Mode of action*
28. Cyanofenphos	.	.	.	.	.	.	.	.	x	x	.	.	.	.	.	H	—	C,S
29. Demephion	.	.	.	x	.	.	.	.	.	x	.	.	x	.	.	.	Mod.	Syst.
30. Demeton	.	.	x	x	x	x	x	x	.	.	.	.	x	.	.	.	—	Syst.,F
31. Demeton-*S*-methyl	.	.	x	x	x	x	x	x	.	x	.	x	x	.	.	.	Mod.	C,Syst.
32. Diazinon	.	.	x	x	.	x	x	x	x	x	.	x	x	.	.	V	Mod.	C,S
33. Dichlorvos	x	x	x	x	x	x	x	x	x	x	x	x	x	.	.	PH,V,H	Short	C,S,P
†34. Dimefox	.	.	x	x	.	.	.	.	.	.	.	.	x	.	.	.	Mod.	Syst.
35. Dimethoate	.	.	x	x	x	x	.	x	.	x	.	x	x	.	.	.	Mod.	C,Syst
36. Disulfoton	.	.	x	x	.	.	.	.	.	x	x	.	x	.	.	.	Mod.	Syst.
37. Ethion	.	.	x	x	.	x	x	x	.	x	.	.	x	.	.	.	Mod.	C,S
38. Ethoate-methyl	.	.	.	x	.	.	.	.	.	x	.	.	x	.	.	.	Mod.	Syst,C
39. Etrimfos	.	.	x	.	.	.	.	.	x	x	x	.	.	.	.	.	Mod.	C
40. Fenitrothion	x	.	x	x	x	x	x	x	x	x	x	x	x	.	.	.	Mod.	C,S
41. Fenthion	.	.	x	x	.	.	x	x	x	x	x	x	x	.	.	Birds	Mod.	C,S,P
42. Fonofos	x	.	.	.	.	.	.	.	x	x	x	.	.	.	x	(Soil Pests)	Mod.	C
43. Formothion	.	.	x	x	.	.	.	.	x	x	.	.	x	.	.	.	Mod.	C,Syst.
44. Heptenophos	.	.	x	x	.	.	x	.	.	x	.	.	x	.	.	V	Short	S,P
45. Iodofenphos	.	.	.	.	.	.	.	.	x	x	x	.	.	.	.	PH,SP	Mod.	C,S
46. Malathion	.	x	x	x	x	x	x	x	x	x	x	x	x	.	.	.	S–M	C,S
47. Mecarbam	.	.	x	x	x	x	x	.	.	x	.	.	x	.	.	.	Long	C,Syst.
48. Menazon	.	.	.	x	.	.	.	.	.	.	.	.	.	.	.	.	Mod.	Syst.
49. Mephosfolan	.	.	x	x	.	.	.	.	x	.	.	.	x	.	.	.	Mod.	C,S,Syst.
50. Methidathion	.	.	x	x	x	x	x	.	x	.	.	.	x	.	.	.	Mod.	C,P
51. Mevinphos	.	.	x	x	.	.	.	.	x	x	x	.	x	.	.	.	Short	C,Syst.
52. Monocrotophos	.	.	x	.	.	.	x	.	x	x	x	.	x	.	.	.	Mod.	Syst.,C
53. Naled	.	.	x	.	.	.	.	x	.	x	.	.	x	.	.	.	Short	C,S,F
54. Omethoate	x	.	x	x	x	x	.	x	x	.	x	.	x	.	.	.	Mod.	Syst.
55. Oxydemeton-methyl	.	.	x	x	.	.	.	x	.	x	.	x	x	.	.	.	Mod.	Syst.,C
56. Oxydisulfoton	.	.	x	x	.	.	.	.	.	.	.	.	x	.	.	.	—	Syst.
†57. Parathion	.	.	x	x	x	x	x	x	x	x	.	.	x	x	x	.	Mod.	C,S,F
58. Parathion-methyl	.	.	x	x	x	x	x	x	x	x	x	.	.	x	x	.	—	C,S,F
59. Phenisobromolate	.	.	.	.	.	.	.	.	.	.	.	.	x	.	.	.	Long	C
60. Phenthoate	.	.	x	x	.	.	x	.	x	.	x	.	x	.	.	SP	—	C,S
61. Phorate	.	.	x	x	.	.	x	x	x	x	x	.	.	.	.	.	Long	Syst.
62. Phosalone	.	.	x	x	.	.	.	x	x	x	x	.	x	.	.	.	Mod.	C
63. Phosfolan	.	.	x	x	.	.	x	x	x	.	.	.	x	.	.	.	Short	Syst.
64. Phosmet	.	.	.	.	.	.	.	.	x	.	.	.	x	.	.	.	—	C
65. Phosphamidon	x	.	x	x	.	x	x	x	x	x	x	x	x	.	.	.	Short	Syst.
66. Phoxim	x	.	.	.	.	.	.	.	x	x	x	.	x	.	.	SP,H	Short	C

CHEMICAL	Orthoptera	Isoptera	Homoptera	Aphids	Mealybugs	Scales	Heteroptera	Thysanoptera	Lepidoptera	Diptera	Coleoptera	Hymenoptera	Acarina	Nematoda	Myriapoda	Miscellaneous	*Persistence*	*Mode of action*
67. Pirimiphos-ethyl	.	.	x	.	.	.	.	.	x	x	x	.	x	.	.	.	Mod.	C
68. Pirimiphos-methyl	x	.	x	x	x	x	x	x	x	x	x	x	x	.	.	SP	Mod.	C,F,P
69. Profenofos	.	.	x	.	.	.	x	.	x	.	x	.	x	.	.	.	—	C,S
†70. Prothoate	.	.	x	.	.	.	x	x	.	.	.	.	x	.	.	.	—	Syst.
71. Quinalphos	.	.	.	.	.	x	.	.	x	x	.	.	x	.	.	.	Short	C,S,P
72. Quinomethionate	.	.	.	.	.	.	.	.	.	.	.	.	x	.	.	Fungi	Mod.	C
73. Schradan	.	.	.	x	.	.	.	.	.	.	.	.	x	.	.	.	Long	Syst.
†74. TEPP	.	.	.	x	.	.	.	.	.	.	.	.	x	.	.	.	Short	C
†75. Terbufos	.	.	.	.	.	.	.	.	.	x	x	.	.	.	x	(Soil Pests)	Mod.	C
76. Tetrachlorvinphos	.	.	x	.	.	.	.	.	x	x	x	x	.	.	.	.	Short	C,S,P
77. Thiometon	.	.	.	x	.	.	.	.	.	.	.	.	x	.	.	.	Mod.	Syst.
78. Thionazin	.	.	.	.	.	.	.	.	.	x	.	.	x	x	x	.	S–M	C
79. Thioquinox	.	.	.	.	.	.	.	.	.	.	.	.	x	.	.	Fungi	—	C
80. Triazophos	.	.	.	x	.	.	x	.	x	x	x	.	x	x	.	.	Mod.	C.P.
81. Trichloronate	.	x	.	.	.	.	.	.	x	x	x	.	.	x	.	.	Long	C,S
82. Trichlorphon	.	.	x	.	.	.	x	x	x	x	x	.	.	.	.	V,H	Short	C,S,P
83. Vamidothion	.	.	x	x	.	.	.	.	.	.	.	.	x	.	.	.	Long	Syst.
Carbamates and related compounds																		
†84. Aldicarb	.	.	x	x	x	x	x	x	.	x	x	.	x	x	.	.	Long	Syst.
85. Bendiocarb	.	.	.	.	.	.	.	.	x	x	x	x	.	.	x	SP & V	Short	C,S
86. Bufencarb	.	.	x	.	x	.	.	.	x	.	x	.	.	.	.	.	Short	C(?)
87. Carbaryl	x	.	x	.	.	x	x	x	x	x	x	x	.	.	.	Earthworms	Long	C,Syst.
88. Carbofuran	.	.	x	x	.	.	.	.	x	x	x	.	x	x	x	.	Mod.	Syst.
89. Ethiofencarb	.	.	.	x	.	.	.	.	.	.	.	.	.	.	.	.	—	Syst
90. Methiocarb	.	.	.	.	.	.	.	.	x	.	x	.	x	.	.	Mollusca	Mod.	C,S
91. Methomyl	.	.	.	x	.	.	.	.	x	.	x	.	x	x	.	.	Mod.	Syst,C
†92. Oxamyl	.	.	x	x	.	.	.	x	.	x	x	.	x	x	.	.	Mod.	C,Syst.
93. Pirimicarb	.	.	.	x	.	.	.	.	.	x	.	.	.	.	.	.	Short	Syst,P,F
94. Promecarb	.	.	.	.	.	.	.	.	x	x	x	.	.	.	.	.	—	C
95. Propoxur	.	x	x	x	.	.	x	.	.	x	.	x	x	.	x	H,V	Mod.	C
†96. Thiofanox	.	.	x	x	.	.	x	.	.	.	x	.	x	.	.	.	Mod.	Syst.
Miscellaneous compounds and fumigants																		
† 97. Aluminium phosphide	.	.	.	.	.	.	.	.	x	.	x	.	x	.	.	SP	—	F
98. Amitraz	.	.	x	.	.	.	.	.	Eggs	.	.	.	x	.	.	V	Mod.	C
† 99. Carbon disulphide	.	.	x	.	.	.	.	.	x	x	x	.	x	x	.	SP	—	F
†100. Copper acetoarsenite	.	.	.	.	.	.	.	.	.	.	x	.	.	.	.	.	Mod.	S
101. Cryolite	x	.	.	.	.	.	.	.	x	.	x	.	.	.	.	.	Long	C,S
102. Cyhexatin	.	.	.	.	.	.	.	.	.	.	.	.	x	.	.	.	Long	C

CHEMICAL	Orthoptera	Isoptera	Homoptera	Aphids	Mealybugs	Scales	Heteroptera	Thysanoptera	Lepidoptera	Diptera	Coleoptera	Hymenoptera	Acarina	Nematoda	Myriapoda	Miscellaneous	*Persistence*	*Mode of action*
†103. Ethylene dibromide	.	.	.	.	.	.	.	.	x	x	x	.	x	x	.	SP	—	F
†104. Lead arsenate	.	.	.	.	.	.	.	.	x	x	.	x	.	.	.	.	Long	S
105. Mercurous chloride	.	.	.	.	.	.	.	.	.	x	.	.	.	.	.	Fungi	Long	S,C
†106. Methyl bromide	.	x	.	.	.	.	.	.	x	x	x	x	x	x	.	Fungi & SP	—	F
107. Sulphur (Lime-sulphur)	.	.	.	.	.	x	.	x	.	.	.	.	x	.	.	Fungi	Long	C
Natural organic compounds and pyrethroids																		
108. Bioallethrin	.	.	.	.	.	.	.	.	.	x	.	.	.	.	.	H	Short	C
109. Cypermethrin	.	.	x	.	.	.	x	.	x	x	x	.	.	.	.	PH & V	Long	C,S
110. Deltamethrin	.	.	x	x	x	x	.	x	x	x	.	.	.	.	.	H	Mod.	C,S
111. Fenvalerate	.	.	x	x	.	.	.	.	x	x	x	.	.	.	.	PH & V	Long	C,S
112. Nicotine	.	.	.	x	.	.	x	x	.	x	.	x	.	.	.	.	Short	C
113. Permethrin	.	.	x	.	.	.	x	.	x	.	x	.	.	.	.	H	Long	C
114. Pyrethrins (Pyrethrum)	.	.	x	.	.	.	x	x	x	x	.	x	x	.	.	.	Short	C
115. Resmethrin	.	.	.	.	.	.	.	.	.	x	.	.	.	.	.	H	Mod.	C
116. Rotenone	.	.	.	x	.	.	.	x	x	.	x	x	x	.	.	.	Mod.	C
Organic oils																		
†117. Petroleum oils*	.	.	.	.	x	x	.	x	.	.	.	.	x	.	.	.	Long	C
†118. Tar oils*	.	.	x	x	x	x	x	.	x	.	x	.	x	.	.	Fungi	Long	C
Biological compounds																		
119. *Bacillus thuringiensis*	.	.	.	.	.	.	.	.	x	.	.	.	.	.	.	.	Mod	S
120. Codling Moth GV	.	.	.	.	.	.	.	.	x	.	.	.	.	.	.	.	—	S
121. *Heliothis* PHV	.	.	.	.	.	.	.	.	x	.	.	.	.	.	.	.	—	S
122. *Trichoplusia* PHV	.	.	.	.	.	.	.	.	x	.	.	.	.	.	.	.	—	S
123. *Verticillium lecanii*	.	.	x	x	.	.	.	.	.	.	.	.	.	.	.	.	—	C
Insect growth regulators																		
124. Diflubenzuron	.	.	.	.	.	.	.	.	x	x	.	x	x	.	.	PH	Short	S
125. Methoprene	.	.	.	.	.	.	.	.	.	x	.	x	.	.	.	V,PH	Short	S
126. Kinoprene	.	.	x	x	x	x	.	.	.	.	.	.	.	.	.	.	Short	S

† Indicates general high level of toxicity, but only used against certain pests
H Used mainly against Household pests
SP Used mainly against Stored Products pests
PH Used mainly for Public Health
V Used against Veterinary pests also (mostly ectoparasites)

* Herbicide
Persistence:
Long (> 2 weeks)
Moderate (1–2 weeks)
Short (< 1 week)

Mode of action:
C – contact action
S – stomach poison
Syst. – systemic
F – fumigant
P – penetrant

9 *Major temperate crop pests*

Descriptions, biology and control measures

The orders and families of pests in this chapter are now rearranged according to the 10th edition of Imms textbook *Imms' general textbook of entomology* (Richards & Davies, 1977), as this classification appears to be widely accepted. Within each family the genera and species are arranged alphabetically. The only exceptions are the very large families Scarabaeidae and Chrysomelidae within the Coleoptera, as they have very well-defined subfamilies with important biological differences between the different subfamilies; within these two families the genera are separated into their respective subfamilies.

The scientific name of each pest used in this section is the one in general current usage at the time of writing, or the one given in the second edition (revised) of Kloet & Hincks *A check list of British insects* (parts 1–5; 1964–78). I have not attempted to be completely up-to-date with reference to insect pest names, as this is basically not feasible; the object of using a name is that the insect concerned be easily identified in respect to literature sources, etc. Similarly, no attempt has been made to incorporate synonymy in this chapter, with exception for a few very recent and important name changes.

The distribution maps are mostly based on the maps produced by the CIE, *Distribution maps of pests: Series A (Agriculture)* (at present numbers 1–457), together with information that appears to be reliable from the many sources listed in the bibliography. In some cases the symbol ● has been used on a map to indicate an area where it can reasonably be supposed the pest occurs, but to date no precise records are available. For further information on pest distributions see page 20.

Once again it must be stressed that the chemical control recommendátions given in this chapter are tentative and intended more as a guide than as definite recommendations. This is, in part, because inevitably chemical control recommendations are out of date by the time the book is published, and also there are no generally agreed international recommendations; a chemical approved in one country may be banned in another, as most countries have different approval criteria. This is especially apparent now with the blanket ban on the use of organochlorines in the USA through the agency of the EPA, whereas in most countries the rationale was that these chemicals should be withdrawn from use as soon as suitable alternatives were discovered. The question of *effectiveness* of chemical insecticides is raised on page 138. Also the space available here does not permit recommendations to be made in detail. This is basically a textbook for use in teaching about crop pests and their control. For details of practical control measures local advice from the Ministry of Agriculture must be sought.

Order

Collembola

(springtails)

Family
Sminthuridae

Sminthurus spp.
(esp. *S. viridis* (L.))

Common name. Garden Springtails (Lucerne 'Flea')
Family. Sminthuridae
Hosts. (main). Alfalfa, clovers.
(alternative). Lettuce, cucumber, cereals, root crops, grass, many seedlings.
Damage. Leaves, cotyledons and stems of seedlings are nibbled and holes made; leaves may become skeletonized in heavy attacks. Damage usually restricted to seedlings.
Pest status. Springtails are sporadic pests, on seedlings of a wide range of crop plants, but populations can become very large and then significant damage may be done. Destruction of precision-drilled root crop seedlings (e.g. sugar beet) can be serious.
Life history. Eggs are laid throughout the year, usually in batches of 20–60. They are pale yellow, globular and 0.3 mm in diameter. Hatching takes 4–6 days usually.

Nymphs are globular, greenish, with no distinction between thorax and abdomen. They differ from the adults only in size. Development takes 4–10 weeks.

Adults are about 2–3 mm in length; antennae are quite long and elbowed. The springing organ is well developed and the insect jumps readily.

There are several generations annually as breeding is more or less continuous, depending on environmental conditions; being tiny and soft-bodied they are very sensitive to humidity levels. In Australia the hot dry summer is spent in the egg stage; low humidity delays egg-hatching.

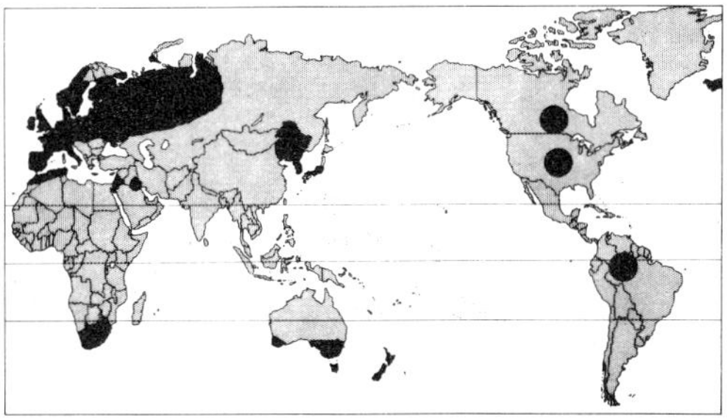

Distribution. A widespread genus, found throughout Europe, parts of Asia, N. Africa, S. Africa, Australia, New Zealand, Canada, and parts of the USA and S. America. *S. viridis* distribution is shown on CIE map no. A.63.

Quite a serious pest in Australia, on pastures and vegetables. Together with *Bourletiella* spp., these round-bodied forms are soil surface-dwellers, but the small elongate, white-bodied *Onychiurus* spp. are subterranean soil-dwellers which feed mostly on the plant roots and seedling stems, and sometimes attack the first leaves. These other two genera are also cosmopolitan and polyphagous in feeding habits.
Control. When populations are large, control measures may be required, and insecticides that have been found to be effective include dieldrin and aldrin as either dusts or sprays, and sprays of malathion or parathion. Springtails are normally preyed upon by predacious mites and natural control is often quite effective.

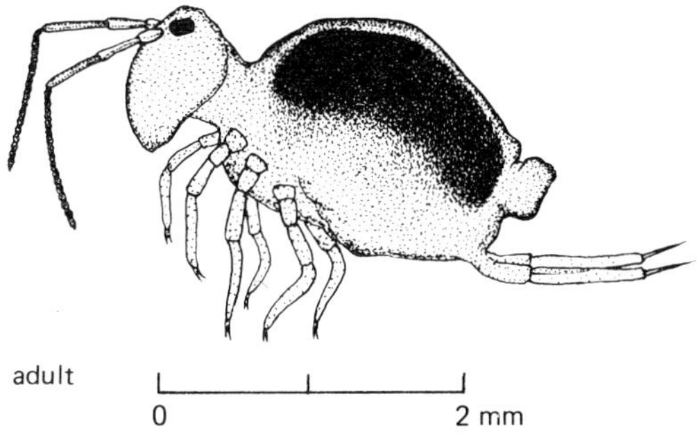

Order

Orthoptera

(crickets, grasshoppers, etc.)

Families
Gryllidae (crickets)
Gryllotalpidae (mole crickets)
Acrididae (short-horned grasshoppers)

Acheta spp.

Common name. Field Crickets
Family. Gryllidae
Hosts. Polyphagous pests that feed at night on many different garden and field crops; mostly damaging herbaceous plants.
Damage. All parts of the plant body may be damaged, and sometimes the bark of woody stems gnawed; entire seedlings may be eaten. Crickets often cut seedlings and leave them for a day to wilt before dragging them down into the nest the following night.
Pest status. Sporadically serious pests on many different crops, especially at the seedling stage, but infestations usually localized.
Life history. These species do not usually make a nest as such, but adults may make tunnels in the soil or else inhabit cracks in the soil.

Eggs are laid in the soil after rains, in batches of up to 30, by the female using her long ovipositor. Each female lays about 2000 eggs. Hatching takes 10–20 days or more, according to temperature.

Nymphal development through about 12 instars usually takes 40–60 days.

Adults are large, fat-bodied insects 2–3 cm in length, black or brown in colour, with long cerci, and the females have a long ovipositor. Adults usually live for 2–3 months; they are omnivorous in diet and eat other insects.

In the tropical regions there may be four generations per year, but in California two or three generations are usual and in cooler regions these species are univoltine, overwintering in the egg stage.

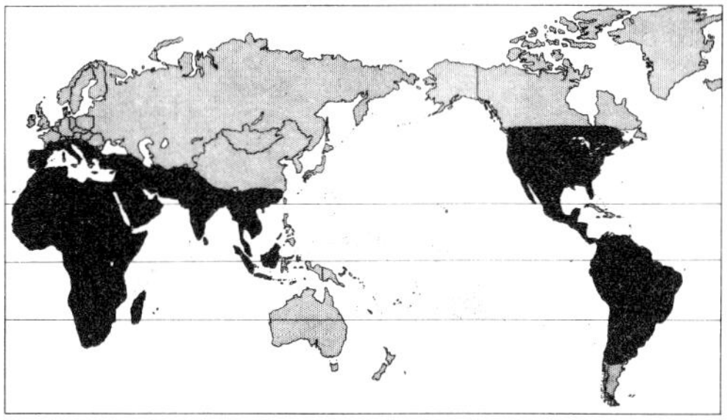

Distribution. Several species of *Acheta* are known:
A. assimilis F. – found throughout USA and in S. America. In Canada the species recorded as this is apparently *Gryllus pennsylvanicus* (Burm.).
A. bimaculata (De Geer) – (Two-spotted Cricket) Africa, southern Europe, and parts of Asia.
A. testaceus Wlk. – found in S. China and parts of S.E. Asia.

Other species of 'field' crickets include *Teleogryllus* spp. in Asia and Australia, and the *Brachytrupes* spp. of Africa and Asia.

Many species of *Gryllus* are also referred to as Field Crickets, and very closely resemble species of *Acheta*.

The migratory Mormon Cricket (*Anabrus simplex* Hald.), a grassland pest of Canada and the USA is actually a tettigoniid grasshopper.
Control. Field infestations may be controlled using poison baits, and the other methods recommended for grasshoppers (page 194), and when they occur as domestic pests in houses chlordane sprays have been effective.

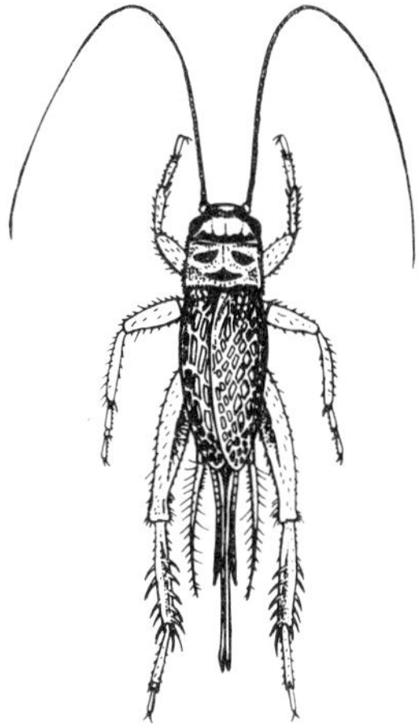

Gryllotalpa spp.

(*G. gryllotalpa* (L.))
(*G. africana* Pal.)
(*G. hexadactyla* Perty)

Common name. Mole Crickets
Family. Gryllotalpidae
Hosts (main). Potato

(alternative). Garden vegetables, grasses, and many herbaceous crops.

Damage. Roots, tubers, and underground stems are eaten by these subterranean pests; ground fruits such as strawberry may be eaten. Small seedlings may be entirely eaten, by the tunnelling nymphs and adults.

Pest status. Widespread and polyphagous pests, only occasionally serious. In the UK *G. gryllotalpa* has become so scarce that it is now a protected species under recent conservation legislature.

Life history. The female constructs a breeding burrow and lays the eggs in the terminal chamber some 10–15 cm beneath the soil surface; she usually makes three or more breeding chambers and lays in total about 100 eggs. Hatching takes 2–3 weeks.

First instar nymphs remain in the breeding chamber and are fed by the mother. Later instars remain in burrows during the day and forage for food on the soil surface at night. There are 9–11 nymphal instars, taking about 10 months for development.

Adults are about 2–3 cm in length, brown in colour, with a velvety surface; wings are folded and appear short (leaving part of the abdomen uncovered), but the insects fly well. Adults live for several months, and at times have nocturnal dispersal flights. Mostly they feed underground in an extensive tunnel system, but sometimes they feed on the surface at night.

There is only one generation per year, and overwintering occurs as large nymphs or young adults.

Distribution. *G. africana* Pal. – (African Mole Cricket) Africa, Asia up to Japan, Korea and the USSR, and Australasia (CIE map no. A.293).

G. gryllotalpa (L.) – (European Mole Cricket) Europe, Asia, N. Africa, and USA.

G. hexadactyla Perty – (American Mole Cricket) USA.

In the USA other mole crickets are placed in the genus *Scapteriscus* and several species are widespread and abundant, but they resemble *Gryllotalpa* in appearance.

Control. When control is required it has been usual to use chlordane dust broadcast on the soil surface, or else sprays of aldrin, dieldrin or heptachlor. Poison baits of bran with chlordane wettable powder added have also been effective, but several applications have usually been needed, at about 10-day intervals.

It should be remembered that in the UK *G. gryllotalpa* is a protected species and is sufficiently rare now that it is unlikely to be encountered.

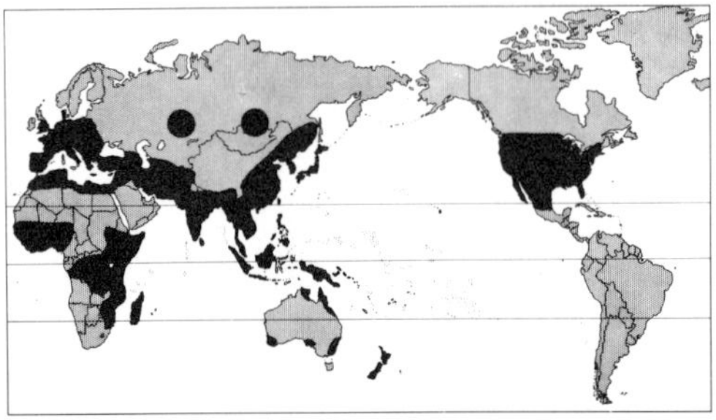

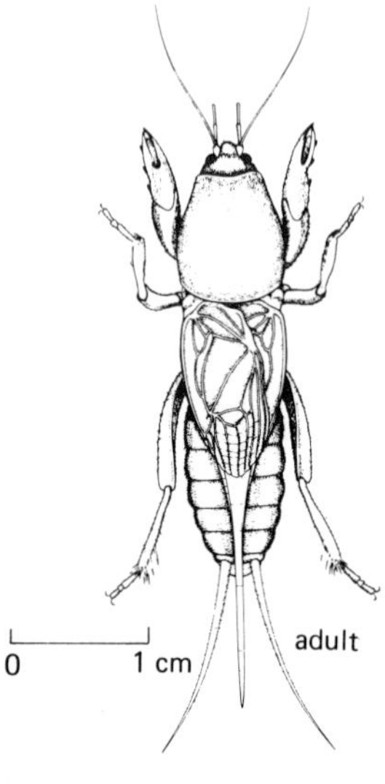

Grasshoppers and locusts

Family. Acrididae
Hosts (main). Grasses and cereals (wheat, barley).
(alternative). Alfalfa, clovers, cotton, buckwheat, tobacco, flax, tomato, potato, and other crops are occasionally attacked.
Damage. Grazing by both adults and nymphs causes loss of leaf lamina; severe attacks by large populations can result in complete defoliation.
Pest status. In places where the natural climatic climax vegetation of grassland has been ploughed for agriculture, or where the grassland persists for pasture, grasshoppers (Acrididae) are often the dominant insect fauna, and may be serious pests, especially where the grasslands (prairies, steppes, etc.) are used for cereal growing. Only occasionally are populations sufficiently large for the species to be a serious pest, but the various species are widespread and common.
Life history. Eggs are laid in pods in the soil. In temperate regions oviposition occurs in the late summer and autumn and the eggs overwinter, hatching in the spring. Egg-pods may contain from 15 80 eggs, each female laying several pods over a period of 2–3 months.

There are usually six or seven nymphal instars. During the later stages food may become scarce (oviposition sites are often concentrated in small locations) and nymphal migrations are common. Nymphal development usually takes 40–60 days.

The adults are mostly large insects, up to 50 mm in length, and sometimes 70 mm, either brown or green in basic coloration. Several species are regularly migratory in habits and may be serious pests. The Acrididae are basically grassland insects and feed predominantly on Gramineae, but some are quite polyphagous and will feed on many dicotyledonous plants. The species that are short-winged and flightless are often good climbers and may defoliate trees up to 4 m in height.

The true locusts are species that are dimorphic with solitary and gregarious phases that differ in coloration and in behaviour – only the gregarious phase being migratory. Not all common names are accurate in this respect.

There is usually only one generation per year; overwintering occurs in the egg stage in temperate regions.

Grasshopper and locust pests

The main tropical species of classical notoriety are as follows:

Locusta migratoria L. – (Migratory Locust) a pest of cereals and pastures and some other crops; occurs as three distinct subspecies, the African, Asiatic and the Oriental Migratory Locusts. The Asiatic subspecies extends as far as Australia.
Nomadacris septemfasciata (Serv.) – (Red Locust) only Africa; on Gramineae and other plants.
Schistocerca gregaria (Forsk.) – (Desert Locust) occurs throughout Africa to India; polyphagous but with a preference for Gramineae; devastatingly serious pest in the past.

Two other more tropical genera of importance are *Oxya*, the Small Rice Grasshoppers of Asia (CIE map no. A.295), and *Zonocerus*, the Elegant Grasshoppers of Africa, which are quite serious, polyphagous pests (CIE map no. A.322).

Temperate pests species include:

Austracris guttulosa (Wlk.) – (Spur-throated Locust) Australia; polyphagous.
Austroicetes cruciata (Saussure) – (Small Plague Grasshopper) Australia; grasses and wheat attacked.
Camnula pellucida (Scudder) – (Clearwing Grasshopper) Canada and USA; cereals and some other crops attacked.
Chortoicetes terminifera (Wlk.) – (Australian Plague Locust) feeds on pastures, field crops and vegetables (polyphagous); recent plague in 1984 throughout NSW and into Victoria and Queensland.
Dociostaurus maroccanus (Thnb.) – (Mediterranean Locust) Mediterranean Region and Middle East; on cotton and cereals (CIE map no. A.321).
Gastrimargus marmoratus – (Marbled Grasshopper) China, Japan and S.E. Asia.

Gastrimargus musicus (F.) – (Yellow-winged Locust) Australia; cereals and sugarcane attacked.
Melanoplus spp. – about 10 species occur in Canada and USA; several are important pests; some migratory; polyphagous on cereals, vegetables and field crops.
Phaulacridium vittatum (Sjostedt) – (Wingless Grasshopper) Australia; polyphagous.
Schistocerca americana (Drury) – (American Grasshopper) USA; Gramineae.
Schistocerca spp. – (Locusts) Canada and USA; alfalfa, clovers, Gramineae.

Control of grasshoppers (Acrididae)

Many species of predators and parasites feed on Acrididae and are of importance in the natural control of populations. Important predators include man, many birds, lizards, toads, frogs, some small mammals, some wasps, meloid beetles. Parasites include various Diptera and Hymenoptera mostly. The Meloidae egg predators/parasites are particularly important in controlling many pest populations.

Chemical control is used in several different aspects:

(1) Poison baits – bran mixed with aldrin, HCH, dieldrin powders.
(2) Swarming locusts sprayed aerially with DNOC, dieldrin, HCH, parathion.
(3) Ground spraying or dusting with dieldrin, HCH, parathion, malathion or carbaryl.

When large swarms, or scattered large populations, are to be controlled then clearly aerial spraying or the use of large power-driven sprayers or dusters is necessary.

The most important locust species are dealt with in the Anti-Locust Research Centre (1966) *Locust handbook*, but a more exhaustive treatment is given in the COPR (1982) book *The locust and grasshopper agricultural manual* where several hundred species are considered.

Order

Dermaptera

(earwigs)

Family
Forficulidae

Forficula auricularia L.

Common name. Common Earwig
Family. Forficulidae
Hosts (main). Cultivated flowers (chrysanthemum, carnation, etc.).

(alternative). Many flowers and some soft-stemmed crops; peach, plum and other soft fruits.

Damage. Petals of flowers are chewed, causing an unsightly appearance; young flower buds sometimes excavated, causing blindness; young leaves may also be eaten, and holes in soft fruits.

Pest status. Not usually a serious pest, but very widespread and common; often found in cavities in damaged fruits, and distorted leaves, but are seldom the causal agent.

Life history. Eggs are laid during the winter in a cell in the soil; some 50–100 eggs are laid, and the female stays with the eggs until they hatch in February/March.

The tiny white nymphs are fed by the attendant female on regurgitated food until they mature and disperse in May/June. Nymphal development takes about 10 weeks.

The female lays a second batch of eggs about May/June; these hatch in late June or early July and develop into adults by mid-September.

The adults are about 20–25 mm long, with characteristic short elytra and stout terminal pincers. They are nocturnal in habits and feed at night; although they are winged they seldom fly. In diet they are omnivorous and eat some aphids and other small insects.

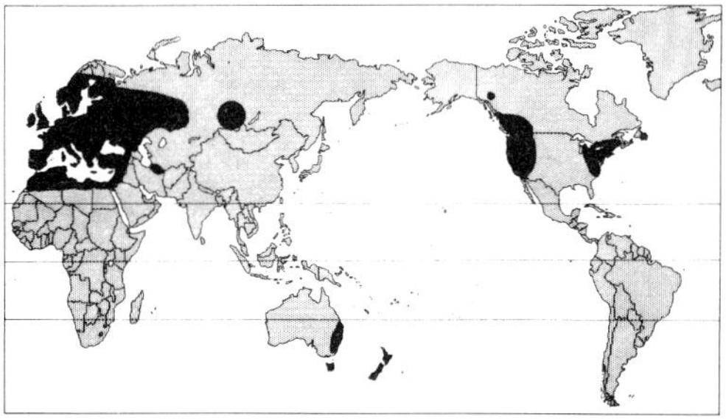

Typically there are two generations annually; the second-generation adults overwinter in the soil and litter.

Distribution. Widely distributed throughout Europe, and introduced into the USA and Canada, as well as Australia and New Zealand (CIE map no. A.79). Other very similar species are to be found throughout the warmer parts of the world where they cause similar damage to some plants.

Euborellia bores young pods of groundnut in India, and also attacks cotton bolls, sorghum and cabbage. *Nala lividipes* (Duf.) (also Labiduridae) attacks some seedlings (maize, sorghum, beetroot) in Australia.

Control. Removal of plant debris and rubbish prevents populations from building up; in gardens earwigs can be trapped under inverted flowerpots stuffed with straw.

If chemical control is required then HCH, malathion and trichlorphon are effective. In glasshouses parathion or HCH can be used as smokes.

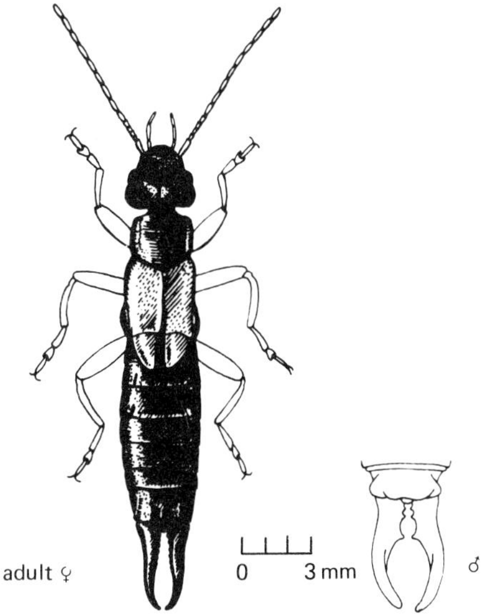

Order

Homoptera

(plant bugs)

Families
Cicadellidae (leafhoppers)
Delphacidae (planthoppers)
Cercopidae (spittlebugs and froghoppers)
Psyllidae (jumping plant lice)
Aleyrodidae (whiteflies)
Aphididae (aphids; plant lice)
Pemphigidae (woolly aphids)
Phylloxeridae (phylloxeras)
Pseudococcidae (mealybugs)
Margarodidae (fluted scales)
Coccidae (soft scales)
Diaspididae (hard or armoured scales)

Cicadella aurata (L.)
(= *Eupteryx aurata* (L.))

Common name. Potato Leafhopper
Family. Cicadellidae
Hosts (main). Potato
(alternative). Not known.
Damage. Sap-sucking causes yellowing, wilting and foliage distortion; leaves often have reddish discoloration along the midribs; not a virus vector.
Pest status. A widespread but not usually serious pest of potato in the UK.
Life history. The slender white eggs are laid into the plant foliage tissues, usually on the stems and leaf veins; eggs hatch after several days.

Nymphs are white, and gradually (through four instars) assume adult coloration, becoming more greenish. Nymphal development takes 3–5 weeks.

Adults are slender, pale green with dark markings, wedge-shaped and quite large (7 mm long), with well-developed hind legs.

There are usually several generations per year.
Distribution. Widely distributed throughout the UK, and parts of Europe.

In most parts of the world there are some 4–6 species of leafhoppers to be found reguarly on local potato crops. *Cicadella viridis* L. is the Green Leafhopper found on hazelnut in southern Europe, and in Japan. *C. spectra* Dist., is the White Rice Leafhopper of Japan.

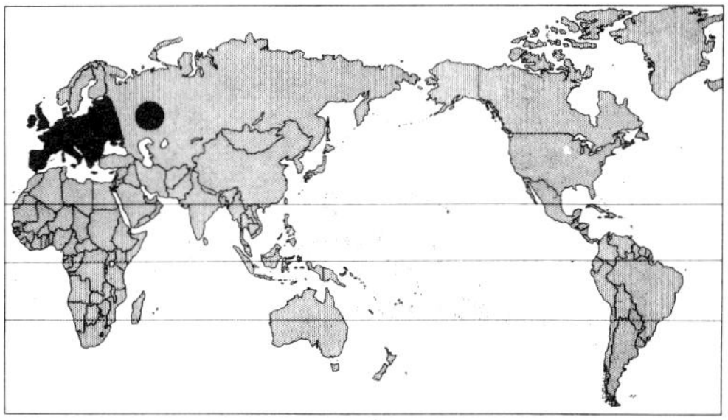

Control. Usually specific control measures are seldom taken against this pest as the sprays applied to kill aphids usually keep leafhoppers in check.

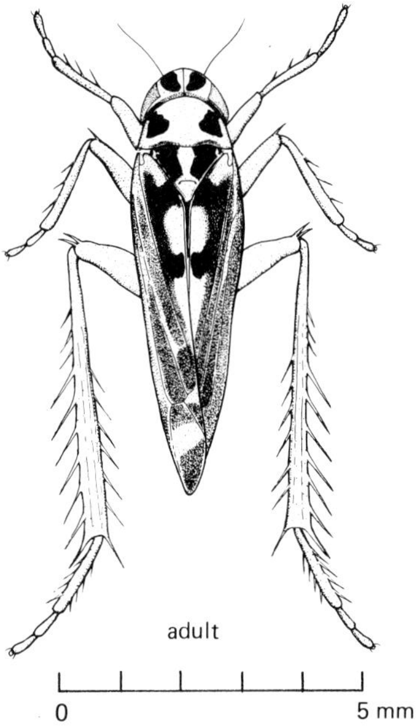

Circulifer spp.

(*C. tenellus* (Baker))
(*C. opacipennis* (Leth.))

Common name. Beet Leafhoppers
Family. Cicadellidae
Hosts (main). Sugar beet, spinach.
(alternative). Cucurbits, tomato, beans, many ornamentals, weeds and wild plants.

Damage. Direct damage includes the curling, stunting and distortion of foliage, caused by sap-sucking; both species are vectors of beet 'curly top' disease.

Pest status. The disease carried by these bugs is quite serious, and pesticide application is usually required in areas of regular occurrence.

Life history. The adult bugs overwinter, usually in beds of herbaceous weeds.

Eggs are laid in the plant tissues after the start of spring growth, in both stems and leaves; each female lays 300–400 eggs. Hatching requires from 5–40 days, according to temperature, in the USA.

There are five nymphal instars, and the time of development from egg to adult takes 1–2 months.

Adults are small, 3 mm in length, pale green or brownish, often with dark markings but the colour pattern is quite variable.

Generations usually overlap; in the USA there are three annually in the north and five or more in California. The first generation is on spring weeds, mostly mustards, and the later generations are on both weeds and crop plants.

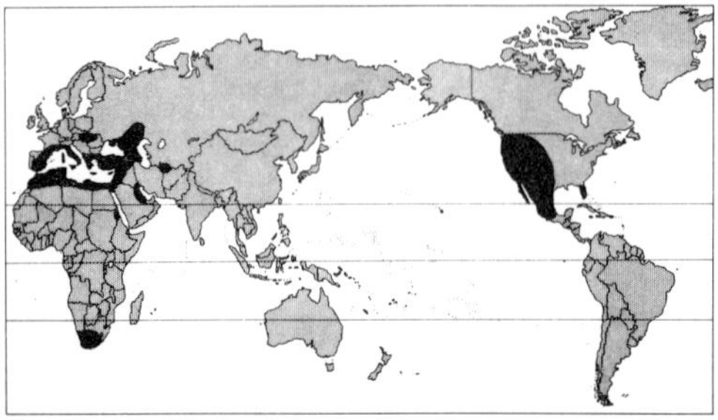

Distribution. *C. tenellus* occurs in Spain, Sicily, Israel, North Africa, Sudan, S.W. Africa, S. Africa, Hawaii, USA, Mexico and the West Indies (CIE map no. A.133).

C. opacipennis is Old World in the Mediterranean Region and Middle East (CIE map no. A.134).

Control. Early planting, and the controlling of weeds in the vicinity of crops will help to keep populations of leafhoppers in check. The replacing of weeds by perennial grasses around the fields deters the pests. But prevention of 'curly-top' disease is difficult because of continuous reinfestation of the crop, though chemical application will minimize the disease effects.

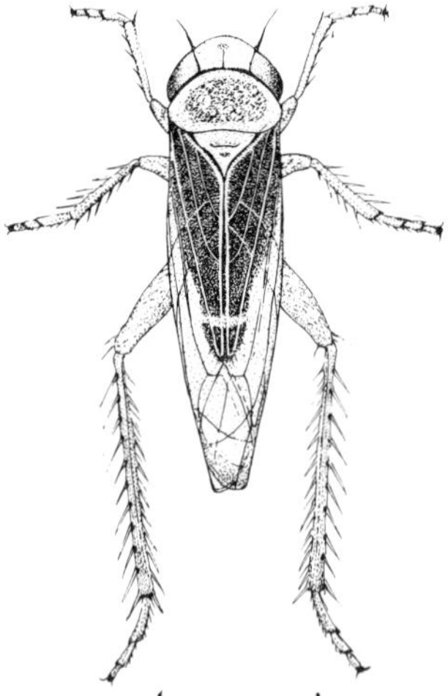

adult

Empoasca spp.

Common name. Green Leafhoppers (Potato Leafhopper, etc.)
Family. Cicadellidae
Hosts (main). Potato, apple.

(alternative). Beans, tomato, castor, cotton, egg-plant, clovers, rhubarb, etc.; recorded from more than 200 host plants.

Damage. Feeding causes foliage curling, stunting and dwarfing, accompanied by 'hopperburn' (yellowing or browning of foliage) caused by saliva injection into the phloem system. These species do not appear to transmit diseases.

Pest status. The many species of *Empoasca* are found worldwide on a vast range of host plants, and many are quite important pests.

Life history. In temperate regions the adult leafhoppers overwinter; in the tropics breeding is continuous.

Eggs are embedded in the leaf veins or the leaf petioles; greenish in colour, banana-shaped and 0.8 mm long. Hatching requires 6–10 days or more, according to temperature.

Nymphs are yellowish-green, frog-like, and live on the underneath of leaves. There are five nymphal instars, reaching up to 2 mm in length.

Adults are 2–3 mm in length, pale green, narrow-bodied and wedge-shaped, with long wings; very agile and active. Females lay from 60–100 eggs usually.

In the USA *E. fabae* is quite strongly migratory, and usually does not overwinter north of the Gulf States; the spring migration takes adults to northern USA and Canada.

There are 2–3 generations per season in Canada, but in the south breeding is continuous with about ten generations annually.

Distribution. The genus is quite cosmopolitan, but appears to be absent from Australasia. Many species are known as crop pests; some are temperate, some tropical, a few are cosmopolitan in either the New World or the Old World, and some are restricted to Africa. A few of the best-known species are as follows:

E. fabae (Harris) – (Potato Leafhopper) Canada, USA, C. & S. America (CIE map no. A28).
E. flavescens (F.) – (Green Leafhopper) Europe, Asia; also recorded in Africa, Canada and USA (CIE map no. A.326).
E. fascialis (Jacobi) – (Cotton Jassid) Africa, south of the Sahara (CIE map no. A250).
E. lybica (de Berg) – (Cotton Jassid) Mediterranean, Eastern Africa (CIE map no. A.223).

Four other species of *Empoasca* are recorded on potato in USA; several others are found on fruit trees in both the New World and Old World.

Control. Cotton with pubescent foliage is unattractive to jassids, but on some crops insecticidal treatments are required; for the list of effective chemicals see page 201.

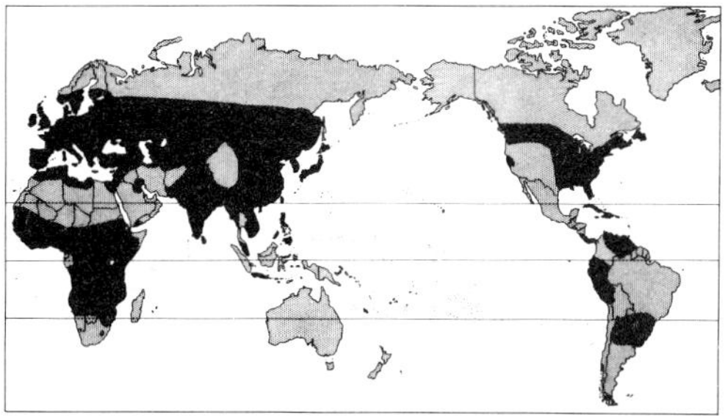

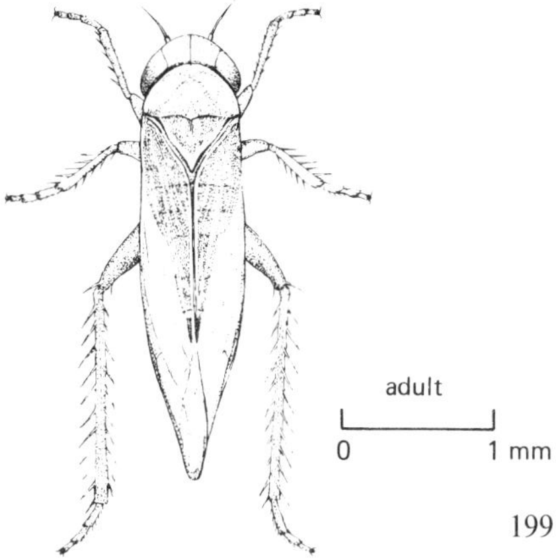

Typhlocyba spp.

(= *Edwardsiana* part)

Common name. Fruit Tree Leafhoppers
Family. Cicadellidae
Hosts (main). Apple, plum.

(alternative). Blackberry, gooseberry, loganberry, peach, rose, potato, strawberry, hop, currants, hazelnut, raspberry, many ornamentals and wild plants, mostly trees and woody shrubs.

Damage. Feeding damage consists of mottling and distortion of foliage, leaf-curl, and stunting of both foliage and fruit. In dry seasons severe attacks may cause defoliation. These species do not seem to transmit pathogens.

Pest status. Common and widespread pests, frequently encountered, but not often very serious, although sometimes control measures are required. Many species are known.

Life history. Overwintering takes place in the egg stage.

Eggs are laid in the young stem tissues, under the bark of trees; summer eggs are usually laid in leaf tissue. Overwintering eggs hatch in April and May and nymphal development takes 3–6 weeks.

First-generation adults appear in June/July; second generation in August/September; adults may live as long as two months. The adults are white, sometimes with dark spots on the thorax, about 3 mm in length.

There are two generations per year, so far as is known, for all species in most areas.

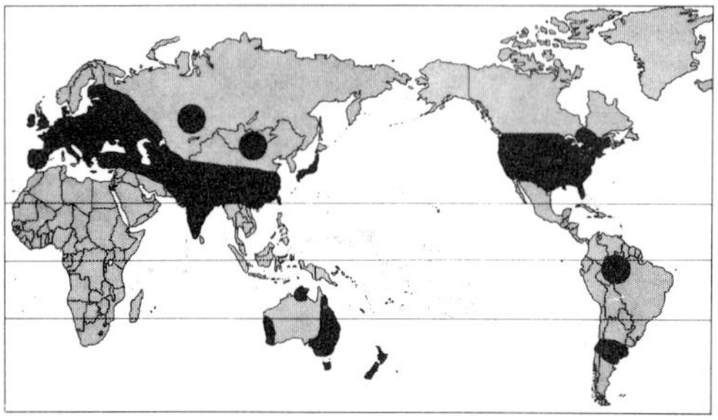

Distribution. For many years the genus *Typhlocyba* was regarded, *sensu lato*, as containing many species in different subgenera; the adults were mostly quite similar in appearance, whitish in colour (some with dark markings), and found throughout the Holarctic Region on a wide range of trees and shrubs. But in a recent taxonomic revision the subgenera were reinstated as genera, and now only a few species remain in *Typhlocyba* (*s.s.*). The more important pest species in this group are:
Edwardsiana crataegi (Doug.) – (Apple Leafhopper) on fruit trees (Rosaceae); Europe, Asia, Australasia, N. and S. America (CIE map no. A.432).
Edwardsiana rosae (L.) – (Rose Leafhopper) polyphagous on fruit trees and bushes; Europe, Japan, USA.
Typhlocyba pomaria (McAtee) – (White Apple Leafhopper) on many fruit trees; N. America.
Typhlocyba quercus (F.) – (Fruit Tree Leafhopper) polyphagous on fruit trees, nuts, strawberry, etc.; Europe.
Most of these species are only identifiable after dissection of the male genitalia, and so to most field entomologists many species are indistinguishable; however *T. quercus* has characteristic red spots on the elytra.

Control. When control measures are required, then the insecticides mentioned on page 201 may be used.

0 3 mm adult

Leafhopper pests (Homoptera; Cicadellidae (=Jassidae))

This is a large group of bugs, second in abundance only to the aphids, with some 8500 species, worldwide in distribution with several large cosmopolitan genera, recorded from a wide range of host plants. Most species are restricted in their choice of hosts, but some are polyphagous. Identification of leafhoppers is difficult as the main taxonomic character is the male aedeagus: thus field recognition is only feasible to generic level. As an indication of the abundance of this group in the UK some 37 species have been recorded on top and soft fruit (including hop).

In the tropics breeding is often continuous, with up to 10 generations per year, but in temperate regions the grassland species are generally bivoltine, and the arboreal species univoltine, although up to five generations per year have been recorded in some species. In temperate regions overwintering occurs in any stage, but most frequently as the adult or the egg.

Most are phloem-feeders, but the Typhlocybinae feed from the mesophyll layer of leaves. Crop damage may be direct, caused by feeding, or by transmission of viruses and mycoplasms causing diseases; some have toxic saliva and cause 'hopperburn'.

Eggs, up to 300 per female, are laid under the epidermis of the host plant, singly, or in rows, or clusters, using the short sharp ovipositor.

Dispersal is by young adults in flight, and some species regularly undergo considerable migration, both in Europe and North America and also in the Far East (China and Japan).

Most populations are checked by high levels of natural control through predators and parasites, but some species do regularly require pesticide application.

Control of leafhoppers (jassids)

Many species, especially in the tropics, have now established resistance to a number of the organophosphorous insecticides, and so care has to be taken when selecting candidate chemicals. For some species it appears that host-plant breeding for resistance is most successful, as exemplified by *Empoasca* and the 'hairy' strains of cotton.

Pesticides that have been recorded as successful for control of leafhoppers include DDT, azinphos-methyl, carbaryl, dimethoate, demeton-*S*-methyl, endosulfan, formothion, malathion, phorate and oxydemeton-methyl, but the most frequently recommended chemical is malathion.

Several foliage sprays are generally recommended at 1–2 week intervals, taking care to reach the undersurface of the leaves where the leafhoppers live.

Leafhopper pest species

In addition to the four genera/species mentioned in the text, other temperate leafhoppers of importance as crop pests include:

Amrasca terraereginae (Paoli) – on sunflower in Australia.

Aphrodes spp. – (Strawberry Leafhoppers) Europe and Canada; virus vectors.

Austroasca viridigrisea (Paoli) – (Vegetable Jassid) Australia; on alfalfa, clovers and potato.

Cicadella aurata (L.) – (Potato Leafhopper) UK and Europe.

Cicadella viridis L. – (Green (Hazelnut) Leafhopper) S. Europe, Japan.

Cicadella spectra Dist. – (White Rice Leafhopper) Japan.

Dalbulus maidis D. & W. – (Corn Leafhopper) (on maize); USA, C. and S. America.

Edwardsiana rosae (L.) – (Rose Leafhopper) Europe, Japan, USA; polyphagous.

Erythroneura mori Mats. – (Blood Spot Leafhopper) China and Japan; on persimmon and mulberry.

Erythroneura spp. – (Striped Apple Leafhoppers) five species recorded on apple in the USA.

Erythroneura spp. – (Grape Leafhoppers) China, Canada, USA.

Eurypteryx stellulata Burm. – (Cherry Leafhopper) Europe.

Euscelis obsoletus – (Apple Leafhopper) Europe.
Euscelis spp. – on strawberry; Europe.
Graminella spp. – (Cereal (Grass) Leafhoppers) cereals and grasses; USA.
Ledra aurita (L.) – on hazelnut; S. Europe (not UK).
Macropsis trimaculata (Fitch) – on plum; USA.
Macropsis spp. – on *Rubus* spp.; cause stunt disease; UK.
Macrosteles fascifrons (Stål) – (Six-spotted Leafhopper) on lettuce, celery and carrot it is a serious pest in Canada and USA as a virus vector; also feeds on potato, oats, barley, flax, sunflower, rye, parsnip, strawberry and clovers, and transmits viruses on these crops. Regularly a migrant species into Canada from the USA.
Nephotettix spp. – (Green Leafhoppers) on rice and grasses; Africa and Asia; most species tropical in distribution (CIE maps nos. A.286 and 287).
Zygina pallidifrons (J.Ed.) – (Glasshouse Leafhopper) on rose, tomato, cucurbits and other glasshouse crops; Europe.
Zygina spp. – five species recorded on fruit trees in the UK, and several in Japan.

Laodelphax striatella (Fall.)

Common name. Small Brown Planthopper
Family. Delphacidae
Hosts (main). Maize, oats, wheat.
(alternative). Rice, barley, millets, sugarcane, and some grasses.
Damage. Some direct damage is caused by sap-sucking, seldom serious, but indirect damage by virus transmission may be serious.
Pest status. A major pest of cereals; a vector of Northern Cereal Mosaic Virus and Oat Rosette Virus, as well as several rice diseases. Recently a serious pest of rice in Japan owing to its acquisition of resistance to some insecticides.
Life history. Eggs are laid in masses on the leaves or stems; each female laying some 60–260 eggs. Hatching takes 5–15 days.

The nymphs are generally smaller than other Delphacidae; the fourth-instar nymphs hibernate in a state of diapause, for overwintering, but there are usually 6–7 generations per year first.

Adults are macropterous or brachypterous; the male about 3.5 mm long and the female 2.0 mm long.

The acquisition period for Rice Dwarf virus and for Black-streaked Dwarf is 15 and 30 minutes respectively, and the incubation period for symptoms is 5–10 days and 7–12 days.
Distribution. A palaearctic species occurring from Europe through northern Asia to Japan, Korea and China, as well as N. Africa, Sumatra, Philippines and China (CIE map no. A.201).

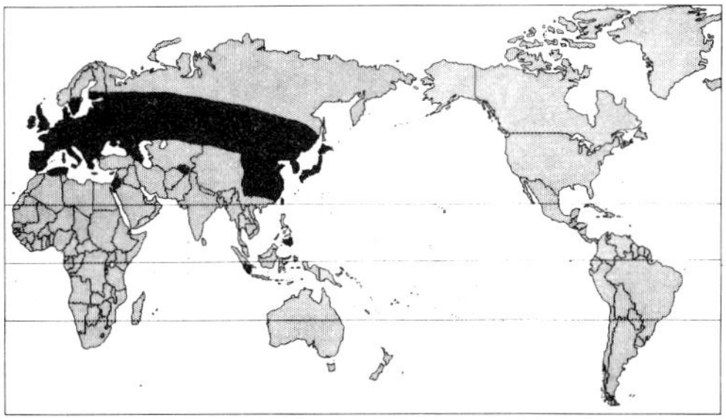

Other temperate species of Delphacidae of some importance include:
Peregrinus maidis (Ashmead) – (Corn Planthopper) on maize, sugarcane, sorghum; cosmopolitan, but more tropical (CIE map no. A.317).
Perkinsiella saccharicida Kirk. – (Sugarcane Planthopper) sugarcane, maize, etc.; S. America, Hawaii, S. Africa, Madagascar, China, S.E. Asia, Australia (CIE map no. A.150).
Sogatella furcifera (Horv.) – (White-backed Planthopper) on rice and some grasses; India, S.E. Asia, Australia, China, Korea, Japan (CIE map no. A.200).
Sogatodes oryzicola (Muir) – (Rice Delphacid) on rice in the USA.
Nilaparvata lugens (Stål) – (Brown Rice Planthopper) only on *Oryza*; India, S.E. Asia, Australia, China, Korea and Japan (CIE map no. A.199): really a tropical species as it only occurs in Japan and Korea as a migrant annually.
Control. As for the leafhoppers (Cicadellidae), but resistance to malathion is widespread.

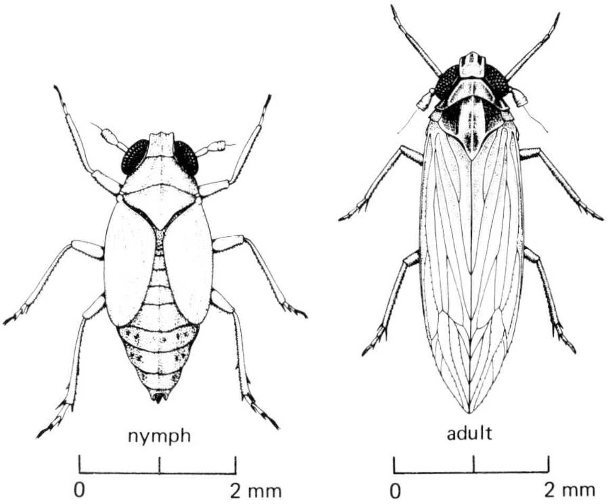

Philaenus spp.
Cercopis spp.

Common name. Spittlebugs (Froghoppers)
Family. Cercopidae
Hosts (main). Sugar beet, strawberry, pear, apple, clovers.
(alternative). Currants, cherry, plum, hazelnut, blackberry, hop; many other plants both cultivated and wild (dock, willow-herbs, and grasses).
Damage. Nymphs in their conspicuous spittle masses suck sap and cause some leaf distortion and loss of sap. Adults may feed on pear fruitlets causing corky blisters and sometimes deformed fruits; on leaves small necrotic spots.
Pest status. Seldom of any economic importance, but several species are very widespread and abundant; the nymphs in their spittle masses are virtually indistinguishable but *Cercopis* are on roots.
Life history. Eggs are laid in batches of 1–30, cemented together, on plant stems; overwintering occurs in the egg stage; hatching occurs in April.

The young nymphs find a site on the host plant and start feeding and producing the spittle mass which surrounds the soft delicate body. Spittle is an excretion from the alimentary canal (mostly water) mixed with another secretion from the Malpighian tubules. There are five nymphal instars and development takes 1–3 months. Most spittle masses contain several small nymphs, but usually only one large one. Nymphs of *Cercopis* are usually on roots rather than foliage.

Adults usually appear in early June. *Cercopis* species are usually a bright red and black in colour, but *Philaenus* are brown with certain mottled patterns. In size they are some 6–12 mm long; when disturbed they jump readily (hence the name 'froghoppers') and also fly. The adults live for several months and feed continually. Eggs are laid in September, and egg-laying usually continues until the adults are killed by cold weather.
Distribution. Europe, parts of Asia, Japan, USA and Canada. The two most abundant species are probably *Philaenus spumarius* (L.), called in the UK the Common Spittlebug and in USA the Meadow Spittlebug, and *Cercopis vulnerata* Ger. the (UK) Red and Black Froghopper.
Control. Control is seldom needed in Europe, but in the USA and Canada infestation of clover crops is often heavy enough to warrant insecticide use. Then the sprays recommended are HCH, carbaryl, diazinon, endosulfan and malathion. Sometimes a surfactant is needed to break down the spittle mass.

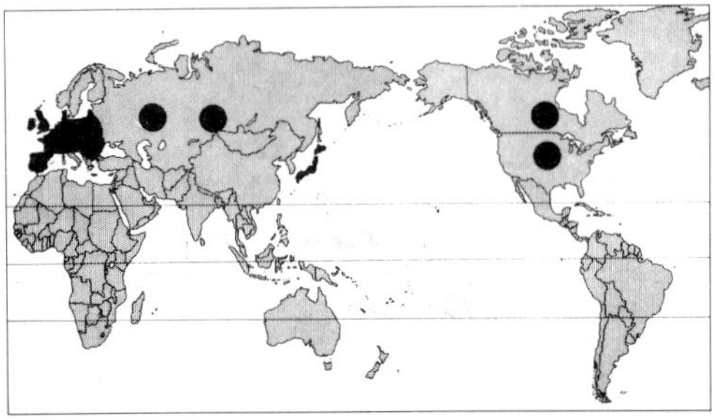

Nymph of *Philaenus* on strawberry leaf

Psylla mali Schmid.

Common name. Apple Sucker (Psyllid)
Family. Psyllidae
Hosts (main). Apple; most varieties.
(alternative). Not known; probably monophagous.
Damage. Nymphs suck sap in the spring from flower trusses and leaf buds; heavy infestations cause brown coloration of petals (like frost damage), and even death of flower buds. Adult feeding damage is negligible.
Pest status. Formerly a pest of some importance on apple, but now routine spraying with organophosphorous compounds against aphids and caterpillars controls it; seldom seen now outside gardens or unsprayed orchards.
Life history. Apple Psyllid overwinters in the egg stage.

Eggs are laid on fruit spurs, along leaf scars, at the base of leaf buds, or scattered along the twigs.

Hatching occurs at about bud-burst. Nymphs are bright green with red eyes, and distinctively flattened; later instars have conspicuous lateral wing buds; development takes 4–6 weeks, through five nymphal instars. Some waxy filaments are produced.

Adults are about 3 mm in length, and stay on the apple tree to feed and live throughout summer and early autumn. Sexual dimorphism is distinct as the male genitalia and female ovipositor are very conspicuous. Egg-laying is from end of August to end of September, when adults die.

There is only one generation per year.

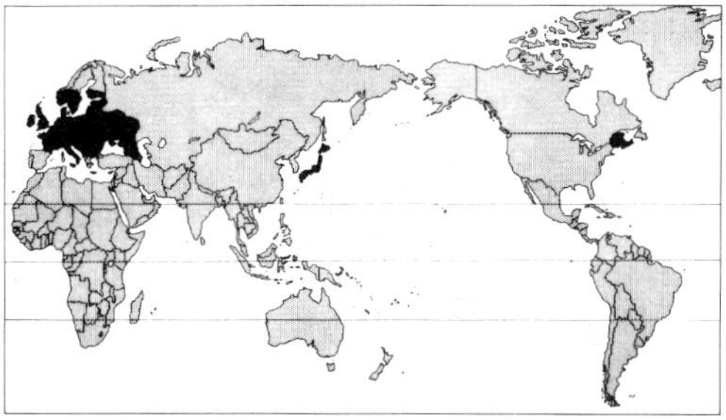

Distribution. Occurs throughout Europe, up to southern Finland, down to Italy, Turkey, western USSR, Japan, and eastern Canada (New Brunswick, Nova Scotia) (CIE map no. A.154).
Control. In most commercial orchards this insect is killed in the egg stage by tar winter washes, and the nymphs killed by routine organophosphorous sprays in the spring to control aphids and caterpillars. Effective insecticides include γ-HCH, azinphos-methyl, dimethoate, chlorpyrifos, cypermethrin, demeton-*S*-methyl, dichlorvos, fenitrothion, malathion and nicotine. DNOC and tar oil are effective as dormant winter sprays against the eggs.

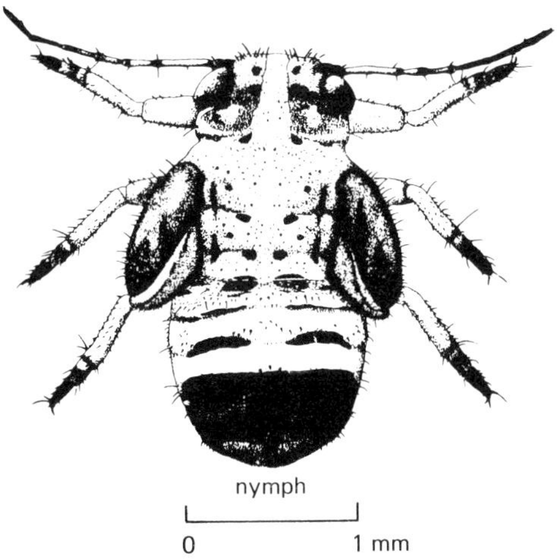

Psylla pyricola (Förs.)
(= *P. simulans* Förs.)

Common name. Pear Sucker (Psyllid)
Family. Psyllidae
Hosts (main). Pear; most varieties.
(alternative). Not known; probably monophagous.
Damage. Nymphs feed on developing leaf and blossom trusses in the spring, and on leaves in the summer; feeding causes necrosis of phloem sieve tubes, thought to be due to toxic compounds in the saliva; blossoms may be killed. Sooty moulds are usually associated with infestations of this insect.
Pest status. A pest of increasing importance over recent years, and large populations may be found in many commercial pear orchards. Autumn infestations may damage the developing fruit buds (causing crop losses the following year), and may cause premature defoliation in September/October. In Europe it transmits the causal organism of 'Pear Decline' disease.
Life history. Pear Psyllids overwinter as winged adults; some on the pear trees, some elsewhere. Eggs are laid on shoots and spurs between late February and mid-April. Hatching occurs from bud-burst until the end of flowering.

First-instar nymphs are orange in colour, with red eyes; later instars have conspicuous lateral wing buds and flattened bodies.

First-summer generation adult females lay eggs along the leaf upper midrib. The summer nymphs live on the underside of leaves with drops of honey-dew. Honey-dew production is usually copious, and sooty moulds are often present.

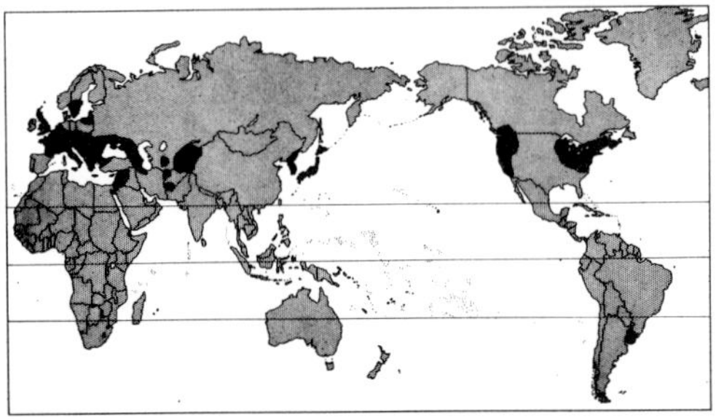

Three generations per year is usual, with adult population peaks in early June, late July, and mid-October.

Four species of Psyllidae may be recorded on Pear, but this is usually the only economic pest species.
Distribution. Widely distributed throughout Europe (not Ireland), parts of the Middle East, USSR, Afghanistan, Korea, Japan, parts of Canada and the USA, and S. America (Uruguay) (CIE map no. A.156).
Control. In many years predators and parasites keep this pest in check. Resistance to organophosphorous compounds is widespread in some regions. Pesticide use should be very discriminative.

Pesticides generally effective include γ-HCH, amitraz, deltamethrin, permethrin, and the organophosphorous compounds azinphos-methyl, chlorpyrifos, demeton-*S*-methyl, dichlorvos, dimethoate, fenitrothion, malathion, oxydemeton-methyl and vamidothion; also diflubenzuron. Tar oil, and DNOC as dormant sprays are not as effective as only the adults on the pear trees are killed.

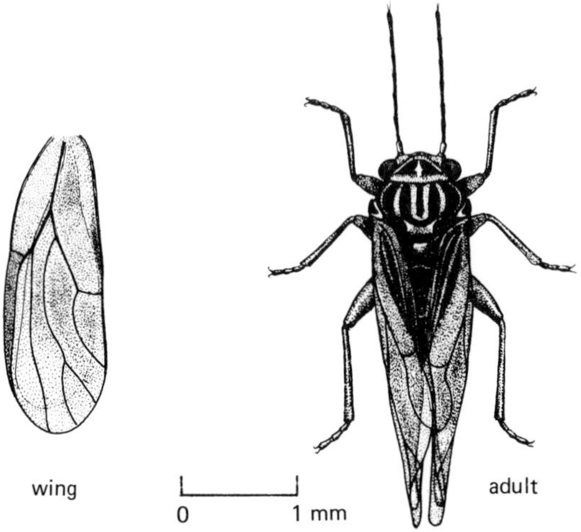

Psyllid pests (Homoptera; Psyllidae)

Psyllids, sometimes called 'suckers', may either be regarded as constituting a single large family (*sensu stricta*) of about 2000 species, or else about eight closely related families (*sensu lato*). Both adults and nymphs feed by stylet-insertion, usually into the phloem, on the foliage of Dicotyledones mostly. Most species are narrowly host-specific, being either monophagous or oligophagous. They transmit a number of viruses, bacteria and mycoplasms which cause plant diseases. Many species are gall-forming, presumably through cecidogenic chemicals in the saliva. Most galls are formed by feeding nymphs but some adults cause plant distortion by feeding, or by oviposition. Many nymphs produce terminal waxy filaments, sometimes in very large quantities, and most produce honey-dew.

Certain plant families appear to be favoured as hosts, and these include the Malvaceae, Moraceae, Lauraceae, and Sterculiaceae. The hosts include almost equally trees, shrubs, and herbaceous plants. The group is equally well represented in both the temperate and the tropical regions. About a dozen species are crop pests of some importance, and others occur on ornamentals and forest trees. The majority of pest species are free-living and feed on actively growing plant tissues, but most produce some host growth distortion, if not an actual gall.

In the tropics breeding is continuous and up to ten generations per year may occur. Temperate species are mostly univoltine, and most overwinter in the egg stage, but a few have up to three or four generations annually and some overwinter as adults.

Most psyllid populations are kept in check by high levels of natural parasitism and predation, but sometimes chemical control is needed, and in places resistance to various organophosphorous compounds is established.

Temperate species of some importance as crop pests include:

Paratrioza cockerelli Sulc. – (Potato Psyllid of N. America) causes leaf-rolling and foliage distortion; on potato and tomato.

Psylla pruni (Scopoli) – on *Prunus*; throughout Europe, UK, Georgia, Caucasus and Irkutsk.

Psylla pyri (L.) – (Pear Psyllid) restricted to pear as host; found in Europe (not UK) from S. Fennoscandia to Italy and the USSR (Crimea) (CIE map no. A.155).

Psylla pyrisuga Förster – also on pear in Europe, UK, Russia and Japan.

Trioza alacris Flor – (Bay Leaf-roll Psyllid) causes red galls by the rolling of the leaf edge on bay; occurs in both UK and Western Europe.

Trioza apicalis Förster – (Carrot Psyllid) this European species also occurs on potato and parsnip, and causes leaf-curl, the heavier infestations causing most foliage distortion.

Trioza brassicae Vasil'ev – (Brassica Psyllid) recorded from *Brassica* spp. in Russia.

Trioza nigricornis Förster – a truly polyphagous species recorded from 24 plant species in 11 families, although some records are dubious. Distribution in Europe (not UK) from Sweden to Spain, Turkey and the USSR.

Trioza trigonica Hodkinson – probably from carrot, in Europe and the Mediterranean Region.

Trioza tremblayi Wagner – Italy; from onions and some other hosts, probably quite polyphagous.

Trioza diospyri (Ashmead) – (Persimmon Psylla) on persimmon; USA.

Aleyrodes proletella (L.)

(= *A. brassicae* Wlk.)

Common name. Brassica Whitefly
Family. Aleyrodidae
Hosts (main). Cabbage, Brussels sprout, cauliflower, broccoli, and kale.

(alternative). Swede, turnip, mustards, wild Cruciferae; many Compositae, and other plants.

Damage. Sap-sucking can cause foliage discoloration and withering, and general loss of plant vitality; some honeydew production.

Pest status. An occasional and local pest in the southern parts of the UK, often quite abundant and very conspicuous; seldom requires control measures.

Life history. Eggs are elongate-oval, and laid upright in a semicircle group on the underneath of *Brassica* leaves; initially pale and translucent, becoming darker; egg-laying takes place from mid-May until September. Hatching takes about 12 days.

Nymphs are scale-like, and covered with wax, white in colour with two yellow spots on the abdomen. On the dorsal surface of the last abdominal segment is the vasiform orifice characteristic of the group. Nymphal development takes about ten days.

The fourth instar is called the 'pupa', and is thicker, immobile, pale in colour with red eyes. Pupation takes about four days.

Adults are tiny moth-like bugs, about 1.5 mm long, with dark head and thorax and yellow abdomen covered by a conspicuous white waxy layer; the forewings have a faint dark bar. The adults fly readily if disturbed.

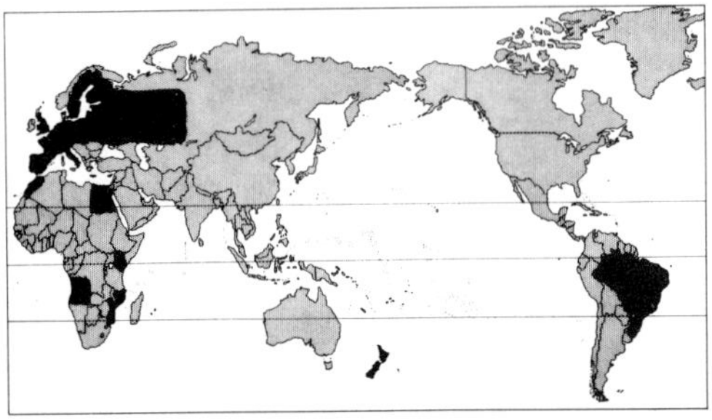

There are usually 4–5 generations per year, and the adults overwinter under the leaves of *Brassica* crops. In the summer the life-cycle takes about a month.

Distribution. Recorded from Europe, USSR, North and Eastern Africa, New Zealand, and Brazil.

Several other species of *Aleyrodes* are pests of some crops and several ornamentals in parts of Asia, Canada and the USA; some species are quite polyphagous; the most notable being:

Aleyrodes lonicerae Wlk. – (Strawberry Whitefly) polyphagous, but most on Rosaceae and Labiatae; Europe, Near East, West Asia and China.

Aleyrodes spiraeoides Quaint. – (Iris Whitefly) polyphagous on ornamentals and weeds; Canada and USA.

Control. It is seldom necessary to use chemical control measures against this pest, but malathion would probably be effective against adults (not to be used on Brussels sprouts).

Adults and nymphs of *Aleyrodes lonicerae* on underside of *Oxalis* leaf

0 ⎯⎯ 1 cm

Bemisia tabaci (Genn.)

Common name. Tobacco Whitefly (Cotton Whitefly)
Family. Aleyrodidae
Hosts (main). Tobacco, tomato, cotton, sweet potato, cassava, legumes.

(alternative). Totally polyphagous; recorded from more than 200 hosts in 63 different plant families.

Damage. Sap-sucking causes little damage directly, but infestations are quite conspicuous; a very serious virus vector in the tropics on the crops listed above.

Pest status. A sporadically serious pest of tomato and tobacco in warmer regions transmitting virus diseases.

Life history. The egg is about 0.2 mm long, pear-shaped, and stands upright on the underneath of the leaf: they are white when laid but turn brown; hatching takes about seven days. Each female lays about 200 eggs.

The nymphs move only a short distance before settling down and feeding; once settled they do not move again. All nymphs are flattened, scale-like, oval, somewhat spiny, and covered with wax.

The fourth instar, or 'pupa' is about 0.7 mm, and the red eyes of the adult can be seen. Total nymphal development takes 2–4 weeks according to temperature.

The adult is a minute whitefly, 1 mm in length, white with a waxy bloom. In the warmer parts of the world breeding may be more or less continuous with many generations per year.

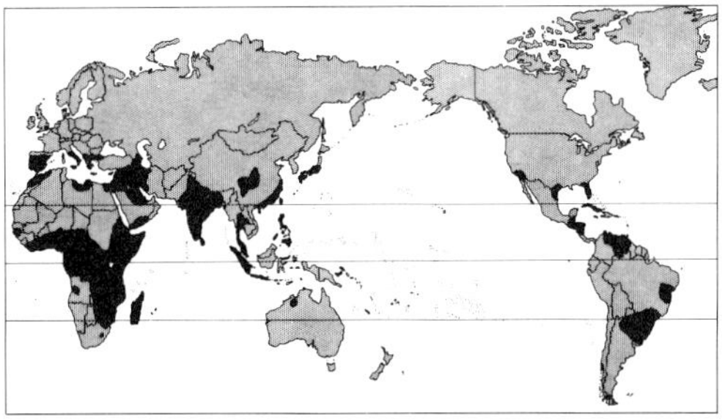

Distribution. Cosmopolitan throughout the warmer parts of the world, occurring as far north as southern Europe (once in UK), Japan, and southern USA (CIE map no. A.284).

Control. When control is necessary, sprays of DDT, dimethoate, permethrin, or pirimiphos-methyl should be effective. Some chemicals (such as DDT) only kill the adults, but the latter two should kill both nymphs and adults if sprayed on the underneath leaf surfaces.

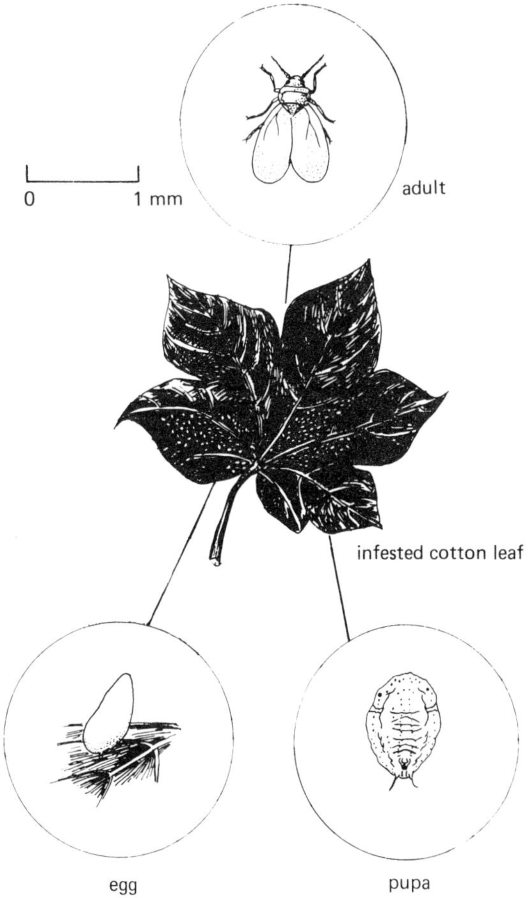

Trialeurodes vaporariorum (Westw.)

Common name. Glasshouse Whitefly
Family. Aleyrodidae
Hosts (main). Tomato, cucumber, tobacco, potato, many ornamentals.

(alternative). Totally polyphagous; recorded from more than 300 hosts in 82 families.

Damage. Sap-sucking causes wilting of leaves; in severe cases leaves may die; copious honey-dew is produced and sooty moulds may be a problem. It does not appear to be a virus vector.

Pest status. A serious pest in temperate glasshouses, and of field crops where the summers are warm enough. Yield losses can be heavy.

Life history. Eggs are laid in groups on the underneath of leaves; yellow when laid, becoming black after about nine days (at 21 °C). Each female lays 200–250 eggs, about eight per day during a life-span of 3–6 weeks.

Nymphs are flattened, scale-like, pale green, and spiny in appearance. Total development takes from 18 days at 21 °C. The fourth instar is the 'pupal' stage.

Adults are tiny (1 mm long) moth-like bugs, snowy-white due to a covering of fine wax; they fly readily if disturbed. They live mostly on the undersurface of the host plant leaves.

The life-cycle can be completed in about 27 days at 21 °C. In temperate regions this is a glasshouse pest, but will survive on crops and other plants outside during a warm summer, and sometimes populations outside survive mild winters. It is regularly a pest of field crops in the summer in Canada and the USA.

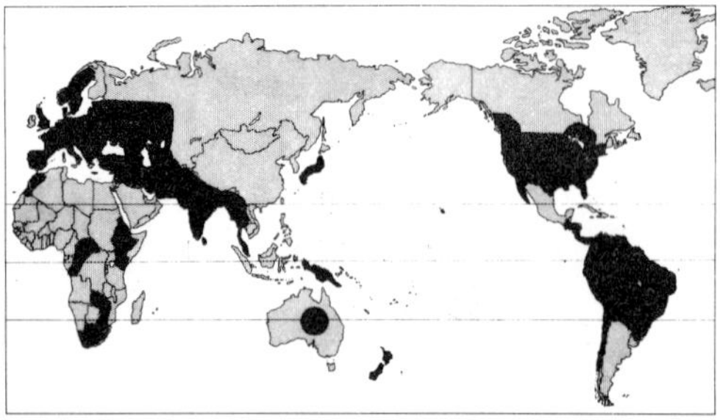

Distribution. Found throughout Europe, western Asia, India, Japan, Australasia, parts of Africa, Canada, USA, Hawaii, Central and South America.

Several other New World species are known:
Trialeurodes floridensis (Q.) – Avocado Whitefly.
T. packardi (M.) – Strawberry Whitefly; polyphagous.
T. vittata (Quaintance) – Grape Whitefly.

Control. In glasshouses this pest is now usually controlled biologically through the introduction of the parasite *Encarsia formosa* Gahan (Hymenoptera: Aphelinidae). The parasites may be purchased from several suppliers in each country, and if they are introduced when the whitefly population is still quite small then successful control is usually achieved. For further details see page 110.

For outdoor use the chemicals recommended are HCH, permethrin, dichlorvos, dimethoate, malathion, or pirimiphos-methyl, and aldicarb granules. Some of these chemicals may also be used for fumigation and treatment of infested glasshouses.

A recent development is the use of a fungus (*Verticillium lecanii*) for controlling glasshouse whiteflies; this is reputed to be very successful.

Leaf of *Cineraria* infested underneath with adults and nymphs

0 2 cm

Acyrthosiphon pisum Harris

(= *Macrosiphum pisi* (Kalt.))

Common name. Pea Aphid
Family. Aphididae
Hosts (main). Pea.

(alternative). Only other Leguminosae, including *Phaseolus*, *Vicia*, *Vigna*, *Lotus*, *Medicago*, *Trifolium*, etc.

Damage. Leaves become cupped and there is general foliage distortion with pods malformed; plants may be stunted, and severe loss of yield may result. Young shoots become clustered with soft pink or green aphids, which readily drop to the ground if disturbed.

Pest status. A sporadically serious pest of peas wherever they are grown; usually only serious on pea, but occurring widely on other Leguminosae. More than 25 different virus diseases are transmitted by this aphid on Leguminosae.

Life history. Eggs are laid low on the haulm, usually on *Trifolium* and *Medicago*, where they overwinter. Hatching occurs in February/March, and migrating forms are produced in mid-May when they move to pea crops and other legumes. Here they breed from June to September. They seldom appear on the pea crop before flowering. Each parthenogenetic female usually produces 4–12 young per day and populations on peas build up rapidly. Oviparous females are produced about October and sexual eggs laid on the winter hosts. In mild winters adult aphids will survive, but in severe winters in Europe only eggs survive.

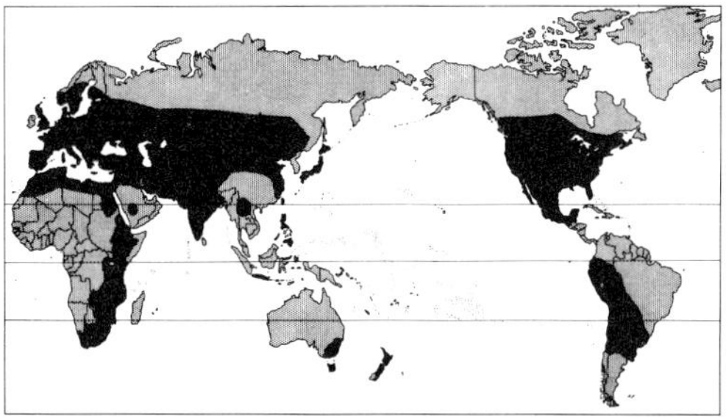

Males are seldom recorded, but both apterous and winged males are sometimes seen. Adults are large aphids (about 3 mm long), green, yellow or pink in colour, with long antennae and siphunculi; cauda is long and pointed; legs long and slender.

One generation probably only takes seven days for complete development. The 'species' exists as a complex of strains and subspecies with different host preferences, in both Europe and North America.

Distribution. Cosmopolitan, but more temperate, extending into the highland areas of Africa, India and Java (CIE map no. A.23).

Many different species of *Acyrthosiphon* occur on a wide range of cultivated and wild host plants.

Control. The usual aphicides are effective against this pest when required (see page 229). In the UK the economic threshold is regarded as 5–10% of the growing tips infested with aphids.

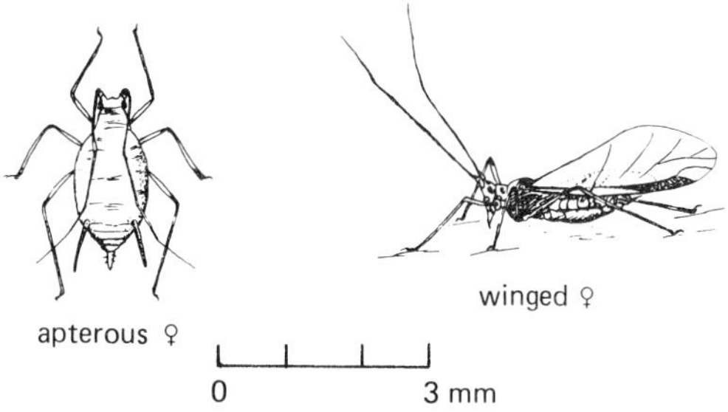

infested pea shoot

Aphis craccivora Koch

(= *A. leguminosae* Th., etc.)

Common name. Groundnut Aphid

Family. Aphididae

Hosts (main). Groundnut, and many other Leguminosae.

(alternative). Polyphagous on many cultivated and wild plants.

Damage. Some host plant wilting, especially in hot weather, and on alfalfa foliage may be distorted or killed, and seed pods distorted. Vector of several viruses.

Pest status. A serious pest on groundnut owing to transmission of Rosette Virus; on other crops a sporadic but widespread and common pest. About 14 virus diseases are spread by this insect.

Life history. The adults are small black aphids, rather like Black Bean Aphid in appearance; some 1.5–2.0 mm in length; siphunculi and cauda black; antenna about two-thirds as long as body.

Nymphs are wingless, rounded in body shape, and dark-coloured. They appear in the crop soon after germination; the adults have usually overwintered on nearby leguminous plants. The woody Leguminosae (broom, laburnum, etc.) are probably the primary hosts.

Rosette Virus is transmitted in a persistent manner; acquisition period is about four hours, and the infective period lasts for more than ten days, going through the moult. The virus is transmitted by all active stages, but the nymphs are more effective than the apterae.

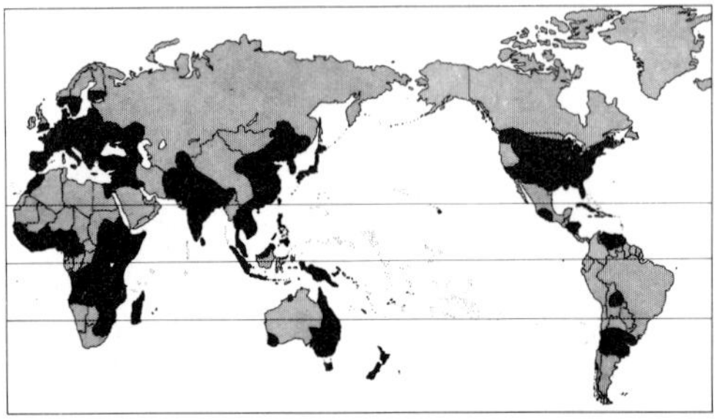

Distribution. Completely cosmopolitan with a more or less continuous distribution throughout both temperate and tropical regions. (CIE map no. A.99).

Control. Cultural control on groundnut can be effective through early planting, and close-spacing. For chemical control menazon seed-dressing is said to give control for about five weeks. Spray recommendations include dimethoate weekly or menazon fortnightly.

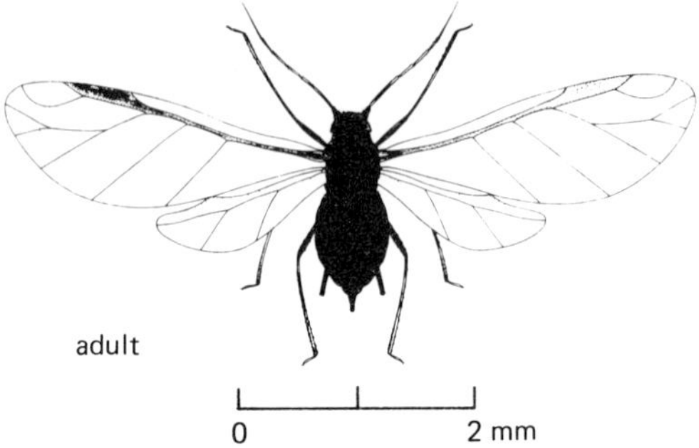

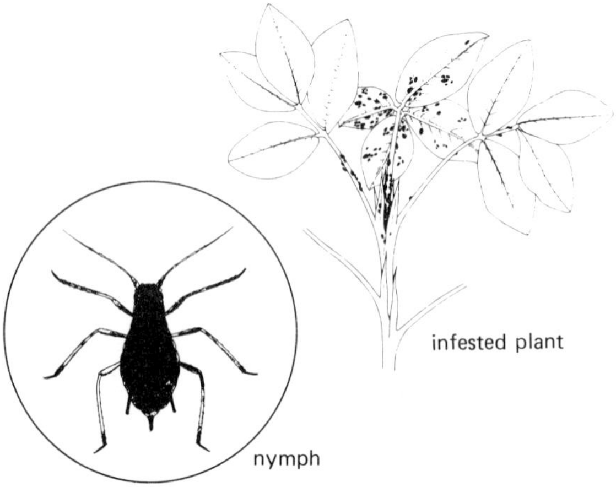

Aphis fabae Scopoli

Common name. Black Bean Aphid
Family. Aphididae
Hosts (main). Beans (*Vicia*, *Phaseolus*, *Vigna* and *Glycine* spp.), pea, vetch.

(alternative). Sugar beet, spinach, mangel; many other cultivated plants and weeds, especially Chenopodiaceae, also rhubarb, poppy, thistles.

Damage. Dense colonies of small black, soft-bodied aphids, round the stems and under young leaves. Most damage is direct by loss of sap in young tissues, followed by loss of seed yield (field beans) or root yield (beet); excessive excretion of honey-dew, sooty moulds present.

Pest status. A serious pest of field beans and in some places of sugar beet; heavy infestations invariably cause serious yield reductions. Recorded vector for about 30 virus diseases; a widespread and very abundant pest species which occurs as a number of distinct races.

Life history. In Europe overwintering eggs are laid on the bark of Spindle Tree (*Euonymus europaeus*) and Snowball Tree (*Viburnum opulus*).

Eggs hatch over the period February/April. The emerging fundatrices feed on the young leaves and reproduce parthenogenetically. The young are wingless when mature, but the next generation are almost entirely winged females, which in late May and early June fly on to the herbaceous host plants, where breeding is continuous for the summer period, with a population peak about July.

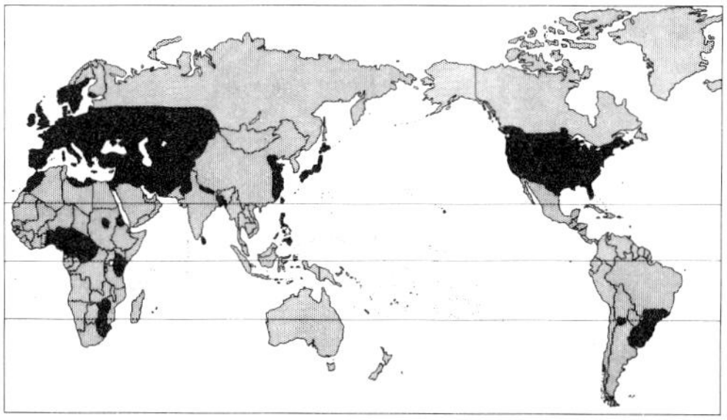

In the autumn winged females and males are produced. They seek out the primary host plants where they mate and the females lay their eggs in bud axils and cracks in the bark. The eggs are green when laid, later shiny black.

Characteristic features include the dark (greenish) body, short antennae, white waxy spots on the abdomen on nymphs, and black tapering siphunculi $1\frac{1}{4}$–$1\frac{3}{4}$ times length of cauda.

In the summer one generation can develop in 7–8 days, but under cooler conditions 18 days may be required. In the UK each female produces 90–120 offspring.

Distribution. Cosmopolitan, but more temperate, and very abundant; not so common in the tropics (CIE map no. A.174).

Control. Early sowing of root crops helps to minimize crop damage. In the UK this pest tends to be a problem in alternate years. Monitoring of winter egg populations on primary hosts can give indications as to both size of subsequent pest populations and the timing of the dispersal flights. Insecticidal use has to be carefully planned so as not to harm pollinating insects, especially bees. In northern situations, where the winged aphids arrive later, it is sometimes only necessary to spray headlands.

As a preventative measure foliar application of disulfoton or phorate granules are recommended; for eradicative treatment a spray of pirimicarb is the least harmful to predacious ladybirds and pollinating bees. The economic threshold is 5% or more field bean plants infested on SW headland, on spring-sown crops. Autumn-sown crops are generally not at risk.

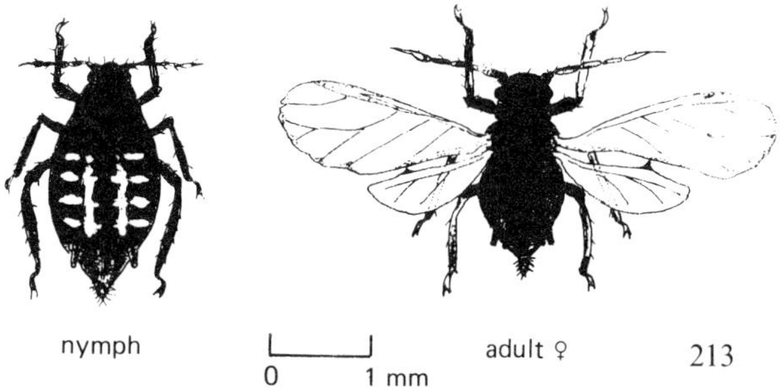

Aphis gossypii Glover

Common name. Melon/Cotton Aphid
Family. Aphididae
Hosts (main). Cotton (other Malvaceae) and Cucurbitaceae.

(alternative). Polyphagous, on many legumes, and many other plants; on glasshouse crops in more northern regions.

Damage. Leaves are cupped or otherwise distorted, with clusters of soft dark aphids on young shoots or under leaves; usually honey-dew and sooty moulds present.

Pest status. A widespread pest of sporadic importance; outbreaks on young plants are common in spells of dry weather which clear up rapidly in the rains. Recorded as a vector of some 44 virus diseases.

Life history. Only female adults are found; winged or wingless; small or medium-sized (1–2 mm long); greenish-black in colour; eyes red; siphunculi and cauda black; antennae rather short (about half body length). Adults live for 2–3 weeks and produce 2–9 offspring per day. Females mature in 4–20 days.

In temperate regions overwintering eggs have been found on *Rhamnus* and Bignoniaceae; it has been thought that this indicates there may be two or more similar polyphagous species being confused under the name of *A. gossypii*.

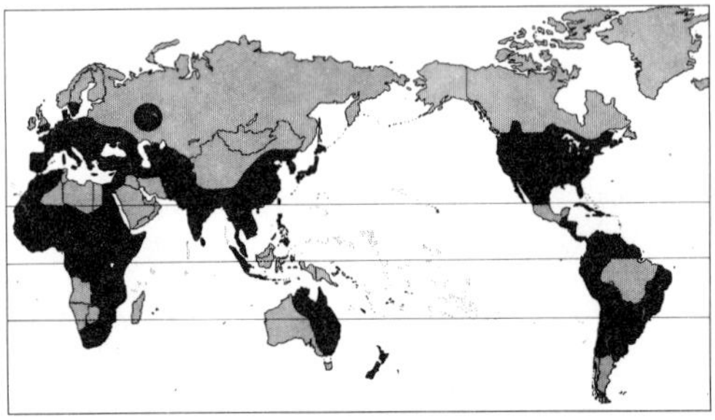

Distribution. A cosmopolitan species that appears to prefer warmer regions. There may in fact be more than one species of aphid under this name (CIE map no. A.18).

Control. Control measures are not often required for this pest on most crops, but on cotton the pest is more serious.

Early sowing of cotton crops helps to reduce infestation levels, and some varieties are more susceptible to aphid than others.

For chemical control it should be noted that some populations have shown resistance to some of the usual pesticides.

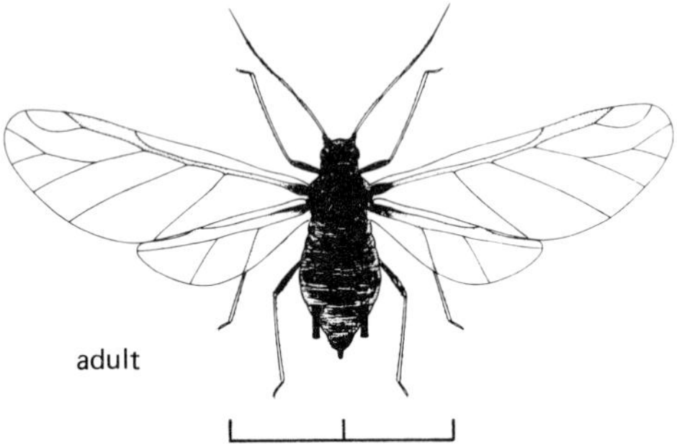

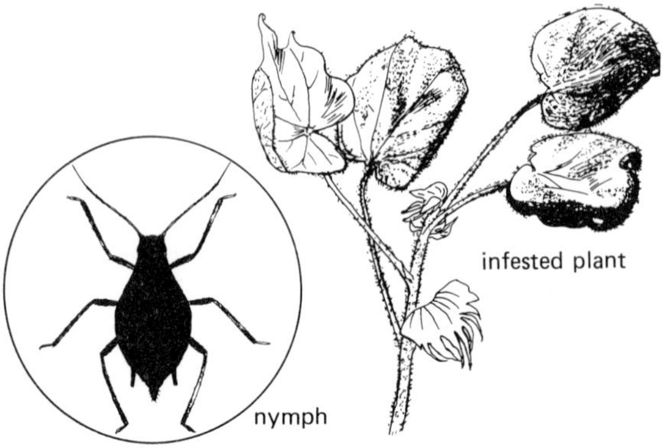

Aphis pomi Deg.

Common name. Green Apple Aphid
Family. Aphididae
Hosts (main). Apple
(alternative). Pear, quince, rowan and hawthorn (Rosaceae).

Damage. Infestation of young shoots; leaves may be curled, and sooty moulds on the upper leaf surfaces. Heavy summer infestations may actually kill the shoot tip, and tree growth may be seriously impaired.

Pest status. A pest on young apple trees and nursery stock, with heavy infestations in late spring and summer; seldom a pest on mature trees.

Life history. Eggs overwinter on the young wood of the host trees, they are often laid in very large numbers on the young twigs. Hatching occurs over a period of 2–3 weeks, from mid-March to early April, according to temperature, to coincide with bud-burst (flowering).

Adults are smallish green aphids with dark siphunculi; early populations are all female, and they reproduce viviparously and parthenogenetically. The first winged adults are seen in June, and in the period June/July infestations spread to new trees. At this time commercial orchards become infested from nearby trees.

The colonies are usually attended by a small dark ant species.

Distribution. Occurs in Europe, Mediterranean Region, into the Middle East, and N. America (CIE map no. A.87).

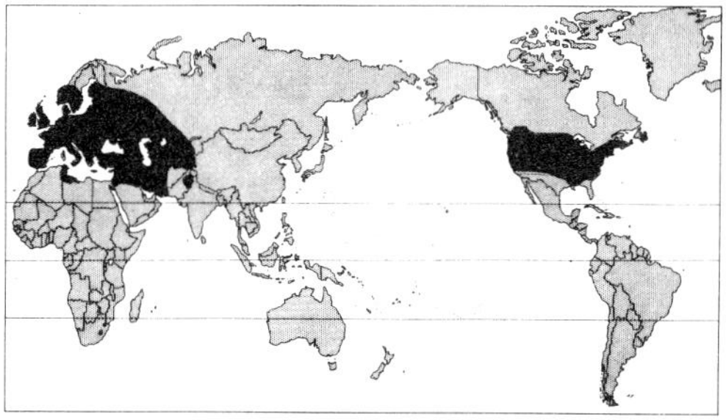

Formerly thought to occur also in the Far East, Southern Africa and Australasia, but it is now assumed that these records were from misidentifications.

Control. On established trees most infestations will be controlled by the natural predators and parasites, but on young apple trees it may be necessary to apply pesticides. Usually when deciding on pesticide application it is necessary to consider the total apple tree pest spectrum, especially the aphid complex, in order both to keep the number of treatments down to a minimum and to cause the least damage to predators, parasites and pollinators.

Infested shoot of apple (scale: 0 – 2 cm)

Aulacorthum solani (Kalt.)

(= *Macrosiphum solani* (Kalt.))

Common name. Glasshouse–Potato Aphid (Foxglove Aphid)
Family. Aphididae
Hosts (main). Potato, and other Solanaceae.
(alternative). Sugar beet, beans, etc.; polyphagous; primary host is foxglove.
Damage. Leaves cupped or otherwise distorted, and quite yellow. Clusters of small pale green aphids on young shoots and underneath young leaves; honey-dew and sooty moulds usually present. Often found, together with Bulb and Potato Aphid, on sprouting tubers in the chitting boxes; the sprouts are often killed. Damages many greenhouse crops throughout the year.
Pest status. A pest of potato in both Europe and N. America, sometimes serious. Polyphagous in diet, it is a vector for more than 30 different virus diseases.
Life history. It may breed continuously in greenhouses, but field populations usually overwinter on foxglove and hawkweeds (Canada) as eggs; adults may overwinter on sprouting potatoes in store.

The adult is quite a large aphid, 2–3 mm in length; shiny yellow-green or brownish in colour, characterized by having the front of head emarginate, nymphs with green or orange patch round the base of the siphunculi; siphunculi long and slender, tapering, and with blackish tips; adults often have abdomen with a pattern of bars and dark spots.

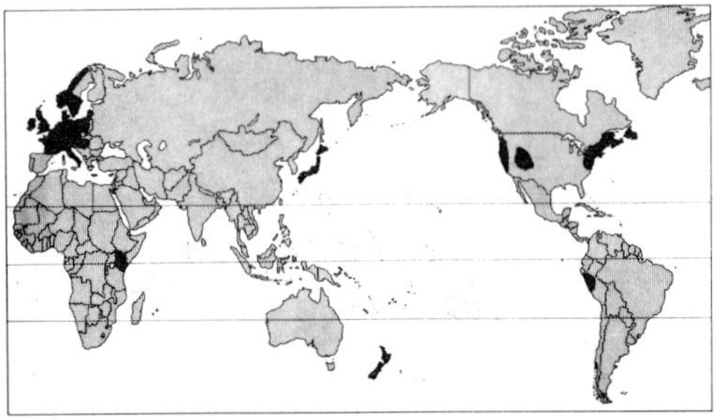

One generation takes about two weeks under favourable conditions. Population peak on field crops is in July, with migration to other hosts in September/ October for egg-laying.
Distribution. Found in Europe, Japan, New Zealand, Kenya, Peru, USA and Canada (CIE map no. A.86).
Aulacorthum circumflexum (Buckt.) is the Mottled Arum Aphid, found throughout Europe and N. America on various Liliaceae.
Control. In most temperate potato-growing areas some 5–6 species of aphids may be regularly found infesting field crops; several other species are more important pests than this species, and generally the aphid complex requires pesticide applications, especially to reduce the incidence of virus diseases. For the recommended chemicals see page 229.

The economic threshold for potato aphids in the UK is the presence of an average of 3–5 aphids per true leaf in a sample of 30 each of top, middle and lower leaves taken across the field.

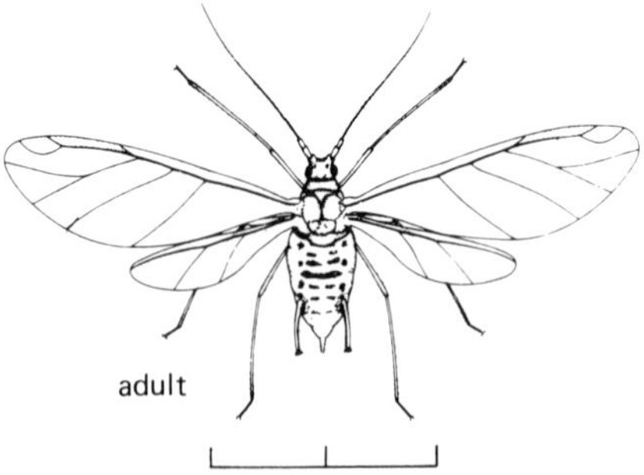

Brachycaudus helichrysi (Kalt.)

Common name. Leaf-curling Plum Aphid
Family. Aphididae
Hosts (main). Plum and damson (primary hosts).
(alternative). Clovers, chrysanthemum, aster and other plants.
Damage. Leaf distortion caused by feeding colonies under the leaves; leaves are often killed and young shoots may be killed.
Pest status. A serious and regular pest of plum and damson, both in the nursery and on fruiting trees.
Life history. Eggs are laid on the spurs and shoots of the plum and damson trees; overwintering takes place here. Hatching occurs very early in the spring, often before bud-burst, and the nymphs feed from the bases of the fruit buds. As the buds open the young leaves and flowers are colonized, and then later the aphids move on to the leaves of the new shoots. Migration of winged adults starts during May, but so long as young growth continues aphid breeding also continues on the plum trees.

The first-generation adult females are usually reddish-brown, but both adults and nymphs of later generations are yellowish-green. Large populations usually continue on the primary hosts until July.
Distribution. Recorded from Europe, Israel, India, Japan, and Australia; possibly widely distributed throughout the whole of Asia.

In India recorded from almond, apricot, citrus and peach, as well as plum.

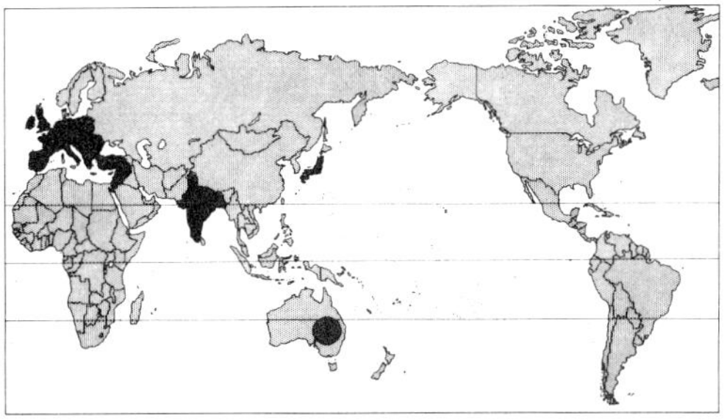

Brachycaudus persicaecola (Boisd.) is the Black Peach Aphid of the USA, on peach and plum; it overwinters as wingless forms on the tree roots, attended by ants, and then most move up on to the young foliage.
Control. A winter application of tar oil is said to be very effective in killing the eggs of this species.

If the winter wash is omitted then spring sprays should be applied from bud-breaking to the bud-burst stage.

Infested plum shoots showing distorted shoot (above) and curled leaves

Brevicoryne brassicae (L.)

Common name. Cabbage Aphid
Family. Aphididae
Hosts (main). Cabbage and other *Brassica* spp.
(alternative). Most of the Cruciferae; also celery.
Damage. Masses of soft, mealy-grey aphids are found in clusters on leaves, stems and flowers, in discrete colonies. Plants are weakened, wilted, turn yellowish and may die if weather conditions are dry; seedlings very susceptible.
Pest status. A serious pest of brassicas, affecting both yield of crop and crop quality. Moulted skins, honey-dew and sooty moulds reduce market value of the crop. A vector of Cauliflower Mosaic Virus and 22 other viruses attacking Cruciferae.
Life history. Eggs overwinter on both cultivated brassicas and wild Cruciferae, but in warmer regions adults survive. Eggs hatch over March/April, and the young nymphs are greenish-yellow and covered with a grey mealy wax. Apterous adults are also mealy in appearance. The first adults are wingless and viviparous, but winged females are produced from June to August and they migrate to the new season's crop plants. Winged adults are greenish with short antennae (shorter than body), pairs of black segmental patches on the abdomen, a black basal wing vein, and short barrel-shaped siphunculi; body length 2.0–2.5 mm.

Later populations tend to infest flower heads and stalks rather than leaves. Egg-laying starts in September and ceases in December usually.

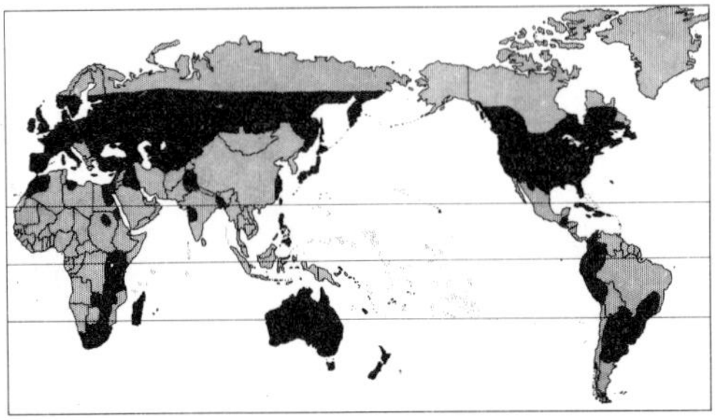

Distribution. Originally Palaearctic or Holarctic in origin, but now almost completely cosmopolitan, although mostly confined to higher altitudes in the tropics (CIE map no. A.37).
Control. Cultural control measures include destruction of crop residues, and removal of cruciferous weeds in the vicinity of the crops.

Weather conditions and natural predation and parasitism greatly influence the rate of population build-up and decline.

Pesticide recommendations are now restricted to systemic compounds, because of the problems of actually reaching the target site under the leaves and protected by body wax. Both granules and sprays are recommended (disulfoton, phorate, demephion, dimethoate, formothion, demeton-*S*-methyl, etc.). Sprays are most successful if applied when the aphid colonies are small.

Cryptomyzus ribis (L.)

Common name. Red Currant Blister Aphid
Family. Aphididae
Hosts (main). Red currant, white currant, flowering currant.

(alternative). Occasionally on black currant; secondary host is hedge woundwort.

Damage. Characteristic purple blisters on young leaves; young shoots may sometimes be destroyed. On black currant the blisters are yellowish-green in colour.

Pest status. Often a serious pest on red currant; infestation is spectacular and quite characteristic; the species is common in both Europe and the USA.

Life history. Eggs overwinter on the stems of the currants, and hatch as the buds are breaking. The first two generations are wingless, but subsequent generations include some winged females that disperse on to other hosts (hedge woundwort). After several generations on hedge woundwort winged males and females fly back on to the currants where the females lay the sexual eggs for overwintering.

The adult aphids are creamy-white in colour with knobbed hairs on their bodies. Honey-dew production is quite prolific and sooty moulds may be a problem on the foliage of the bushes.

Distribution. Recorded from Europe and the USA.
Cryptomyzus galeopsidis (Kalt.) is the Black Currant Aphid of Europe, to be found on all currants, also fragile and creamy-white, but producing no leaf deformation, and the summer hosts are dead nettle and hemp nettle.

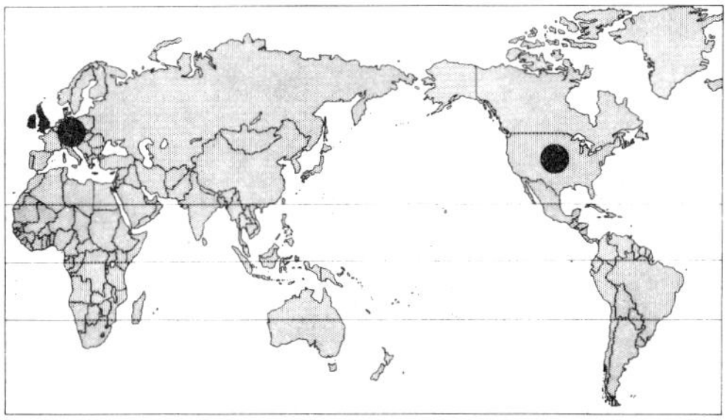

Control. As with the other fruit tree/bush aphids, chemical control is either a winter wash to kill the eggs, or a spray at bud-burst to kill the young nymphs; choice of insecticidal spray usually being determined by consideration of total local pest spectrum.

Infested leaves showing characteristic 'blistering'

Hyalopterus pruni (Geoff.)

Common name. Mealy Plum Aphid
Family. Aphididae
Hosts (main). Plum, damson, peach.

(alternative). Apricot, grapevine, almond; reeds and aquatic Gramineae are the secondary hosts.

Damage. Very heavy infestations are usual with this species, with dense colonies under the leaves of young growth, causing yellowing and premature defoliation; honey-dew production is excessive and sooty moulds usually abundant. Infestations are very messy, but leaves are not usually curled by this species.

Pest status. A slightly serious pest on fruit trees, very conspicuous and characteristic, and a widespread species.

Life history. Eggs overwinter on the trees, and hatch in late April or early May. Only apterous females are produced until late June when the first alates appear; the winged females migrate on to reeds and other aquatic grasses.

Infestations on plum trees usually become obvious about June or early July. The aphids are green in colour with a coating of white powdery wax.

In Israel there is a parthenogenetic cycle that overwinters on the reeds, as well as the sexual cycle on the fruit trees.

Honey-dew production is copious and the entire foliage becomes sticky and may be covered with sooty moulds, as well as flies and wasps and other insects that come to feed on the sugar.

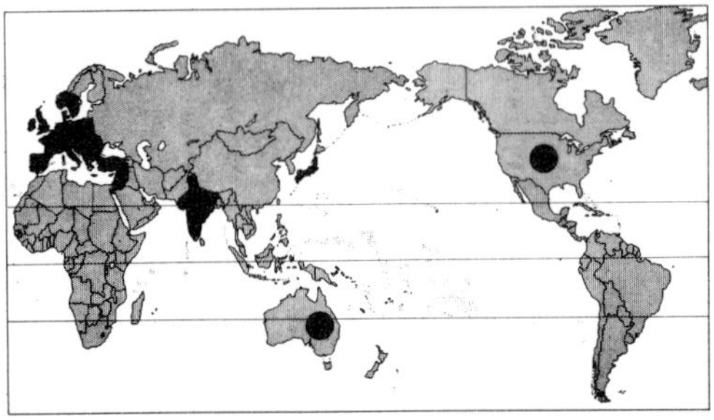

Distribution. Europe, Israel, India, Australia, Japan and the USA, are recorded localities for this species; probably quite cosmopolitan.

The closely related *Hyalopterus arundinis* F. is the Large-tailed Peach Aphid of China, recorded from peach, apricot and plum.

Control. A winter wash of tar oil usually gives the best control of this pest, as with the other plum aphids.

Plum leaf (ventral view) showing infestation of aphid nymphs

Lipaphis erysimi (Kalt.)

(= *Hyadaphis erysimi* (Kalt.))

(= *Rhopalosiphum pseudobrassicae* (Davis))

Common name. Turnip Aphid

Family. Aphididae

Hosts (main). *Brassica* spp. other Cruciferae, both crops and weeds.

(alternative). Recorded from Leguminosae, Solanaceae, opium poppy, barley, *Chrysanthemum*, and *Ranunculus scleratus*. In Europe mostly on cruciferous weeds.

Damage. Feeding causes leaves to curl, and to form small pockets in which colonies of green aphids live on the undersides of leaves; leaves wilt, turn yellow; seedlings may die; older plants become stunted and also may die; often much honey-dew.

Pest status. A serious pest of Cruciferae in parts of Asia and the New World; direct damage is often severe; a vector of Turnip Mosaic Virus and 14 other viruses; total crop failures are known; generally on weeds in Europe rather than crops.

Life history. Usually confused with Cabbage Aphid, but not so mealy in appearance. The biology is very similar.

In the tropics this species is parthenogenetic and males are never found. The young are produced viviparously, each female producing 100–200 nymphs.

In temperate regions males are found; the species overwinters as eggs on cruciferous weeds. The four nymphal instars can develop in 6–12 days under warm conditions.

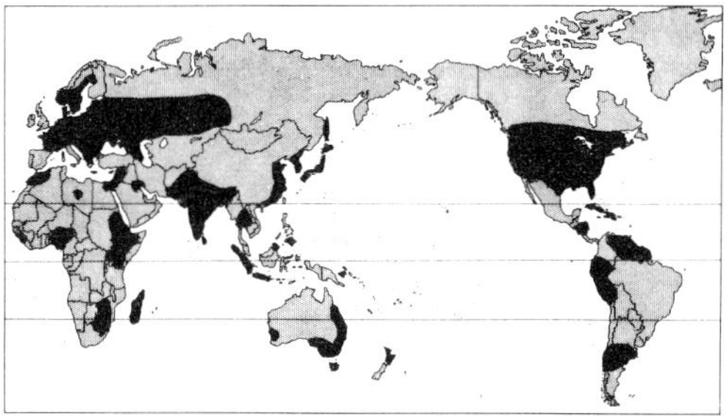

Adults are olive green, with dark markings and head; siphunculi and antennae are quite short; body length about 1.9 mm; adults live for 14–30 days, and may be either alate or apterous.

In cooler regions there are usually 6–8 generations annually, but in southern USA as many as 45 generations have been recorded, as a result of more or less continuous breeding.

Distribution. Thought to be Palaearctic in origin, now quite cosmopolitan in distribution, and is well represented throughout the tropics (CIE map no. A.203).

Control. The usual aphicides (see page 000) are effective against this insect, but control of Turnip Mosaic Virus is difficult; seed beds are sometimes protected with fine screen netting. Control regularly needed in China, Canada, and the USA.

Under field conditions the level of natural predation and parasitism is high, but in the production of Chinese Cabbage, Chinese Flowering Cabbage, etc., where the growing period is very short there is seldom time for natural control to make any appreciable effect on the aphid population on the crop.

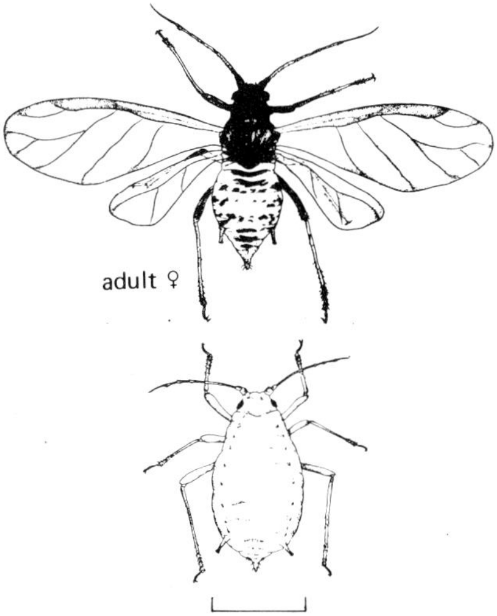

Myzus persicae (Sulz.)

(many synonyms)

Common name. Peach–Potato Aphid (Green Peach Aphid)
Family. Aphididae
Hosts (main). Peach (primary host).

(alternative). Potato, tomato, tobacco, beet, cereals, vegetables; about 110 plants in 35 different families are recorded; truly polyphagous.

Damage. Direct damage is usually leaf-curl on young foliage, with aphid clusters underneath. These symptoms often followed by virus disease symptoms.

Pest status. A very important pest on many crops in many parts of the world, doing damage both directly by feeding and by virus transmission. This is probably the best-known and most intensely studied aphid in the world. Recorded transmitting more than 100 virus diseases of plants in some 30 different families.

Life history. In temperate regions the eggs overwinter on the twigs of the primary host (peach), or else parthenogenetic females overwinter on secondary hosts. The success of the latter depends upon the severity of the winter. In the tropics there is continuous parthenogenetic viviparous breeding with no alternation of generations (and hosts) and males are never recorded. Overwintering eggs hatch from late January to April.

There are usually several generations on the peach trees, then about mid-May winged females are produced and they disperse to secondary hosts. On the secondary hosts continuous breeding produces a mixture of winged and apterous females; the winged forms dispersing within the crop and between crops, producing new colonies and spreading viruses. Unlike many other aphids this species does not produce dense colonies, individuals are more spread over the plant. At the end of the summer winged parthenogenetic females disperse back to the primary host, where they produce a population of wingless oviparous females that mate with winged males of the same generation that come directly from secondary hosts.

The adult is small to medium-sized (1.25–2.5 mm long); green with a dark thorax and dark mark on abdomen; antennae two-thirds as long as body; siphunculi slightly swollen with dark tips, quite long; front of head emarginate.

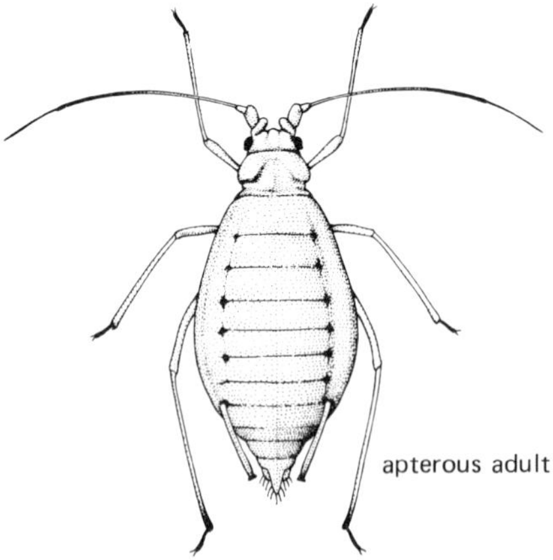

apterous adult ♀

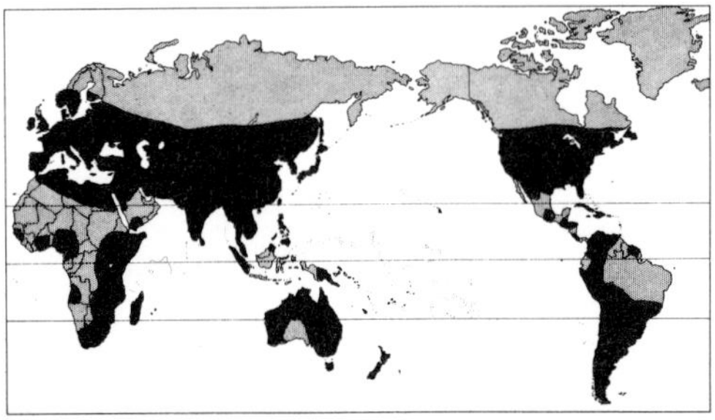

Distribution. A completely cosmopolitan species with a worldwide distribution (CIE map no. A.45).

Many physiological races of *M. persicae* have been recorded, showing no morphological differences, but distinct host feeding preferences.

Control. The use of chemicals to prevent virus spread by controlling the aphid vectors is generally not very successful. Thus, there is need for a careful integrated approach to control, paying special attention to predators, parasites, alternate hosts and crop manipulation.

For suggested aphicides see the control section on page 229, also page 216 for control of potato aphids in the UK. In the UK the treatment threshold on sugar beet is more than one aphid per 4 plants (average) before the 20-leaf stage.

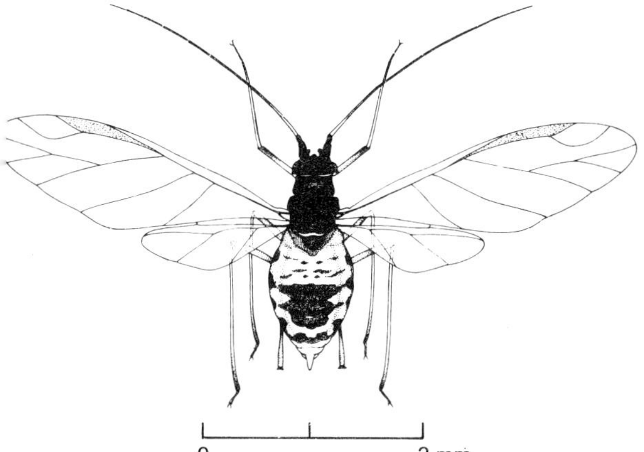

alate adult ♀

Rhopalosiphum maidis (Fitch)

(= *Aphis maidis* Fitch)

Common name. Corn Leaf Aphid (Cereal Leaf Aphid)
Family. Aphididae
Hosts (main). Maize, barley.

(alternative). Wheat, sorghum, millets, sugarcane, rice, tomato, tobacco, manila hemp, and other crops, some grasses, some weeds.

Damage. Leaves and inflorescence covered with colonies of dark bluish-green aphids, with a slight waxy covering. Leaves may become mottled and distorted, and new growth dwarfed. Inflorescence may be so damaged as to be sterile. Honey-dew production is prolific.

Pest status. A serious pest of maize (corn) and sometimes barley; most serious on young plants. A vector of about 10 virus diseases.

Life history. It is not known whether this aphid overwinters as eggs in temperate regions or whether it migrates annually up from more southern areas.

Reproduction is almost entirely parthenogenetic in most parts of the world, but males are reported to be more abundant in Korea, indicating a probable oriental origin for this species.

Adults are apterous or winged, about 2 mm long, with characteristically short antennae and short siphunculi with a dark purple area around the base.

In warm regions the life-cycle takes about eight days, and there may be up to 50 generations per year (Texas).

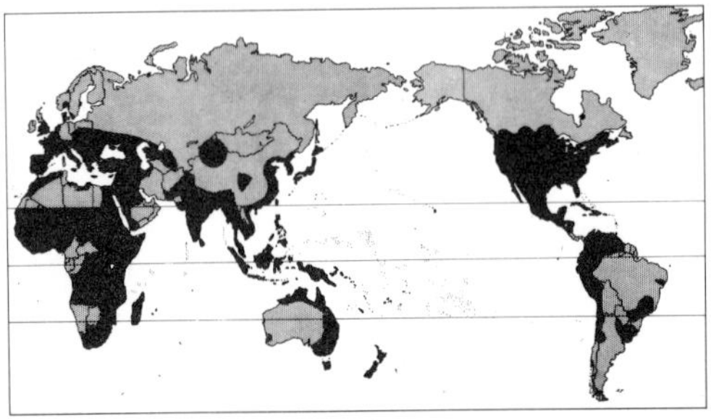

Distribution. Almost completely cosmopolitan in distribution, found throughout the tropics and temperate regions of the world (CIE map no. A.67).

Rhopalosiphum fitchii (Sanderson) is the second most important grain aphid pest in the USA, and also occurs on apple; it is the Apple–Grain Aphid.

Control. Burning seed crop stubbles soon after harvest effects a measure of cultural control.

If the plants are growing vigorously the aphids are usually kept under control by the predators and parasites.

For chemical control recommendations see page 229.

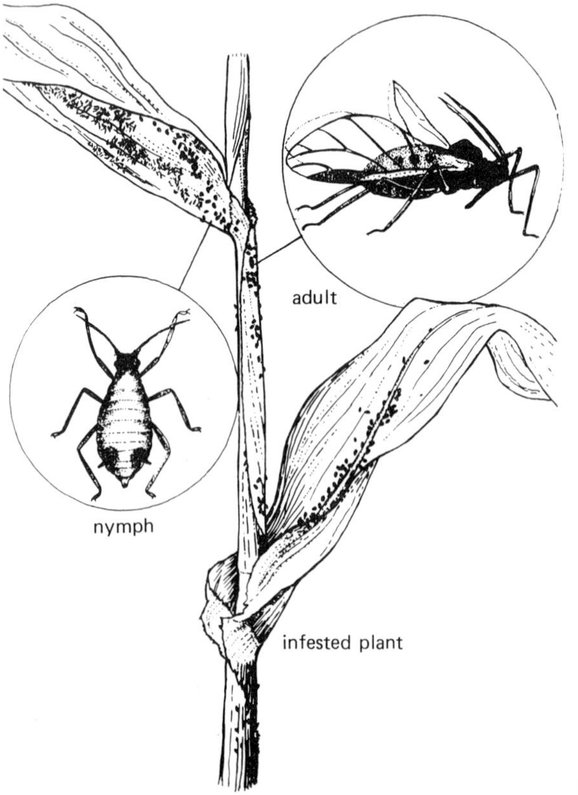

Schizaphis graminum (Rondani)

(= *Toxoptera graminum* (Rond.))

Common name. Wheat Aphid (Greenbug)
Family. Aphididae
Hosts (main). Wheat
(alternative). Oats, barley, rye, maize, sugarcane, and many grasses.
Damage. Feeding causes a yellowing of the leaves. Usually present in very large colonies on the leaves, stems and inflorescence. Often much grain deformation. Vector of several virus diseases.
Pest status. The most damaging aphid on small grains in both Canada and the USA, as well as in South Africa. A migrant species in North America; often occurs in epidemic proportions when weather conditions are suitable.
Life history. In warm climates breeding is continuous, but in cooler regions eggs are laid and adults hibernate on the host plants; in colder regions eggs are the only overwintering stage. In Canada this species apparently cannot survive the winter and it is an annual migrant from the USA.

The species is characterized by being green in colour, medium-sized (1.5–2 mm); siphunculi of moderate length, with a black apex, and held close to the body; both nymphs and adults have a dark green mid-dorsal stripe.

This species can still breed at temperatures as low as 4 °C, whereas the predators and parasites that often check the populations require temperatures of 16 °C or more.

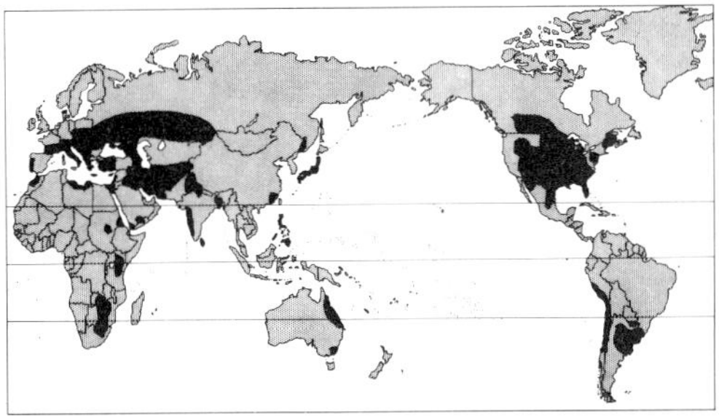

Distribution. Almost cosmopolitan in distribution, but not occurring in the UK or northern Europe; present records are somewhat scattered (CIE map no. A.173).
Control. Natural control of this species is often very important and successful, and, combined with cultural methods such as destruction of volunteer cereals, it may only require pesticide application occasionally.

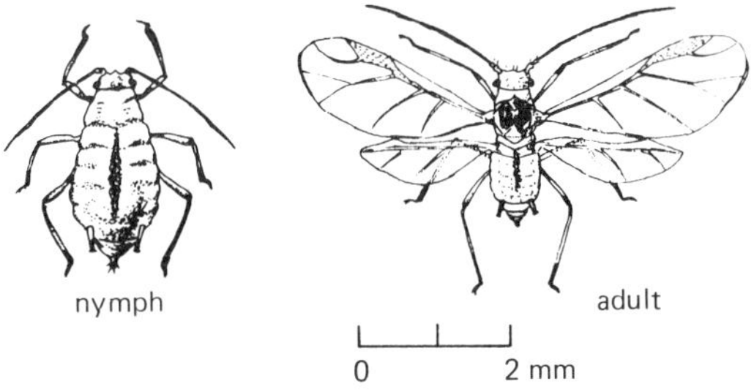

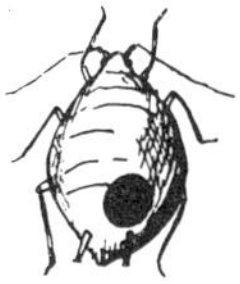

mummified nymph

nymph being parasitized

Toxoptera aurantii (Boy.)

Common name. Black Citrus Aphid
Family. Aphididae
Hosts (main). Citrus

(alternative). Apple, pear, tea, cocoa, coffee, *Ficus* spp., asparagus, *Cola*, mango, lychee, cinchona; more than 120 hosts known; truly polyphagous.

Damage. Distortion of young leaves, with clusters of black aphids under young leaves and on flush growth; usually also sooty moulds growing on the honey-dew.

Pest status. A universal pest on *Citrus*, but widespread and polyphagous, and a greenhouse pest in cooler temperature regions.

Life history. Only females are known, and they produce 5–7 living young per day, up to a total of about 50 per female. The young are dark brown in colour.

Adults are shiny black in colour and may be winged or apterous, from 1.2–1.8 mm in body length, with relatively short antennae.

Life-cycle takes only six days at 25°C but is increased to 20 days at 15°C, with a similar effect seen at high temperatures; at above 30°C aphid populations decline sharply. Infestations are usually attended by ants.

Distribution. Cosmopolitan throughout the warmer parts of the world including the Mediterranean Region, Florida and California, and in greenhouses in Europe, UK and farther north in the USA (CIE map no. A.131).

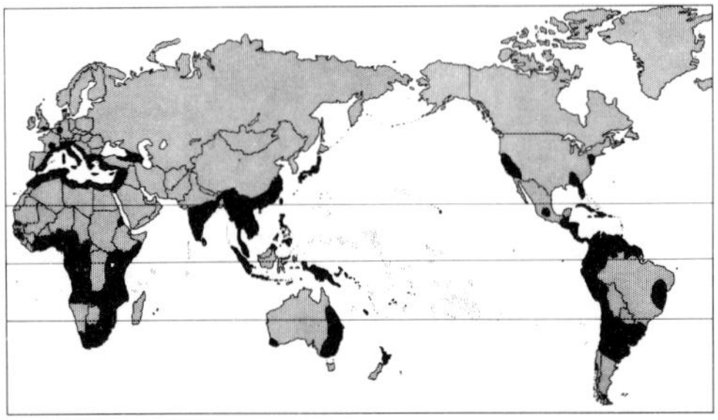

Control. Natural control can be assisted by sticky-banding or spray-banding trees to keep ants away; a large number of predators and parasites normally attack this aphid.

If sprays need to be used, see the recommendations on page 229.

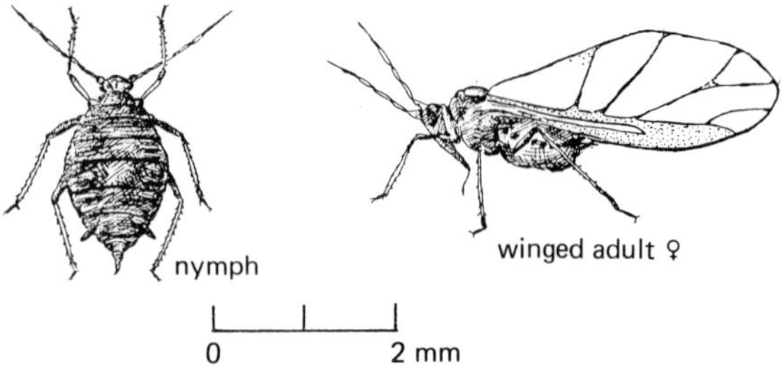

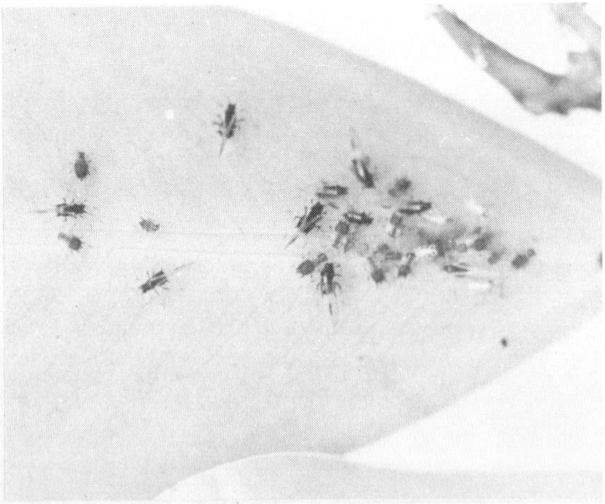

Leaf of *Ficus* with aphid infestation

Aphid pests (Homoptera; Aphididae)

This is not the largest family of insects, there being about 3500 species described, but it is certainly one of the most abundant groups, and is clearly the most abundant of the Homoptera and characteristic of the northern temperate regions. It is found equally well represented on woody and herbaceous plants, and many species alternate between woody hosts in the winter and herbaceous plants in the summer. Almost every species of higher plant is attacked by at least one species of aphid. Some aphids are quite host-specific and will only feed on one genus of plants; at the other extreme are polyphagous species recorded from dozens of different families! On some of the more important crops there may be found up to a dozen different species of aphids, at different levels of abundance. Identification in the field is difficult for many species, as generally the taxonomic characters are esoteric and microscopic; but some species may be recognized on sight (using a hand lens), with practice, by a combination of colour, size, host plant (and location) and obvious anatomical characters. Many distribution and host records made in the past are now discounted owing to misidentification and to confusion over names. Generally most crops can be expected to act as host for half-a-dozen or even more aphid species, and so field infestations have to be identified with extreme care. In the UK in 1977 two keys were published for the field identification of aphids on field crops (brassicas, potato and sugar beet) and cereals, by MAFF, and these do permit field identifications with some measure of certainty; other such keys are clearly required for all the major crops.

The life-cycles are spectacular in that they involve polymorphism, parthenogenesis, viviparity, and alternation of generations on different host plants. A typical life-cycle would be represented as follows. The winter is spent as eggs laid the previous autumn by sexual females. In the spring the eggs hatch into apterous parthenogenetic viviparous females; these produce a new generation of similar forms, but with a few winged females. Several similar generations are produced during the summer and winged viviparous females become more common. By mid-summer the winged females are very abundant and swarms are produced for dispersal purposes in numbers sufficient to darken the sky! Towards the end of summer the progeny of the dispersed females that landed on appropriate host plants, together with those of the apterous forms which remained on the original plant, give rise to sexual females and to males. The new adults mate and the oviparous females lay eggs which will overwinter on the food plant. Many species are truly migratory and the winter is spent on woody trees and the summer on herbaceous plants.

In the tropics the situation is simpler in that males are rare or totally absent and alternation of hosts not practised.

A few species make galls on the host plant, but these are mostly members of the closely related family Pemphigidae (sometimes regarded as a subfamily of Aphididae).

Some of the polyphagous and cosmopolitan 'species' such as *Myzus persicae* occur as distinct physiological races with different host preferences, but no observable differences in anatomy or morphology.

Most aphids are phloem feeders but some use the xylem system of the host plants also. Honey-dew excretion is general, but some species are more prolific in their production. Aphid saliva does in some cases seem to be toxic and direct feeding damage usually consists of leaf-curling and wilting in severe attacks, with some growth stunting but young soft shoots may be killed. Many species are virus vectors and thus major crop pests; for example *Myzus persicae* is recorded to transmit more than 100 virus diseases of plants in about 30 different families including most of the more important (non-cereal) crops.

Aphid populations are preyed upon very heavily by many insect predators, especially Coccinellidae and Syrphidae, and many parasitic Hymenoptera, and all management programmes should endeavour not to upset the natural control usually present.

Aphid pest species

In addition to the species already specifically mentioned, the following are temperate pest species of some importance:

Amphorophora rubi (Kalt.) – (Blackberry Aphid) on raspberry and blackberry; Europe.

Aphis grossulariae Kalt. – (Gooseberry Aphid) on gooseberry; Europe.

Aphis idaei v.d.G. – (Raspberry Aphid) only on raspberry and loganberry; Europe.

Aphis nasturtii (Kalt.) – (Buckthorn–Potato Aphid) on potato, overwinters on buckthorn; Europe, Canada, USA.

Aphis schneideri (Borner) – (Permanent Currant Aphid) only on currants; Europe.

Aphis spiraecola Patch – (Spiraea Aphid) on citrus, stone fruit and other fruits; cosmopolitan (CIE map no. A.256).

Cavariella aegopodii (Scopoli) – (Willow–Carrot Aphid); on carrot, celery, parsnip, parsley, fennel, overwinters on willows, virus vector; Europe, Canada, USA.

Cavariella konoi Tak. – (Celery Aphid) on celery, carrot, parsnip; Canada, USA.

Cavariella pastinaceae (L.) – (Parsnip Aphid) on parsnip, celery, carrot; Europe, Canada.

Chaetosiphon fragaefolii (Ckll.) – (Strawberry Aphid) occurs only on cultivated strawberries, virus vector; Europe and USA.

Chromaphis juglandicola (Kalt.) – (Walnut Aphid) on walnut; Europe, Asia, USA.

Cryptomyzus galeopsidis (Kalt.) – (Blackcurrant Aphid) on all currants, spends summer on dead nettles; Europe.

Dysaphis devecta (Wlk.) – (Rosy Leaf-curling Aphid) only on apple; Europe.

Dysaphis mali (Ferr.) – (Rosy Apple Aphid) only on apple; USA.

Dysaphis plantaginea (Pass.) – (Rosy Apple Aphid) only on apple; Europe.

Dysaphis pyri (Fonsc.) – (Pear–Bedstraw Aphid) pear, summers on bedstraws; Europe.

Dysaphis tulipae (B.de F.) – (Tulip Bulb Aphid) tulip bulbs, other bulbs and corms; Europe, USA.

Forda spp. – (Root Aphids) on wheat roots and grasses; Canada.

Hypermyzus lactucae (L.) – (Currant–Sowthistle Aphid) on currants, summers on sowthistle; Europe, USA.

Hysteroneura setariae (Thom.) – (Rusty Plum Aphid) on plum, peach, other fruit trees, also grasses; mostly USA, but widely distributed (CIE map no. A.255).

Macrosiphum albifrons Essig – (Lupin Aphid) native to the USA (California), now introduced into the UK and killing lupin plants (1984); a very spectacular new pest to the UK causing serious damage.

Macrosiphum avenae (F.) – (Grain Aphid) cereals only, virus vector; Europe, Western Asia, Canada and USA (CIE map no. A.204).

Macrosiphum euphorbiae (Thos.) – (Potato Aphid) primary hosts are Rosaceae, on summer hosts quite polyphagous (potato, beans, tobacco, etc.), virus vector; Europe, Africa, India, Japan, China, Australia, New Zealand, N. and S. America (CIE map no. A.44).

Megoura viciae Buckton – (Vetch Aphid) on vetches, peas, broad bean and other Leguminosae (only); cosmopolitan.

Melanocallis caryaefoliae (Davis) – (Black Pecan Aphid) on walnut; USA.

Metopolophium dirhodum (Wlk.) – (Rose–Grain Aphid) overwinters on Rosaceae, summers on wheat, oats, other Gramineae, and potato; cosmopolitan.

Metopolophium festucae (Theob.) – (Grass Aphid) only on Gramineae, mostly on cultivated grasses; Europe.

Myzocallis coryli (Goet.) – (Filbert Aphid) on hazelnut; USA.

Myzus ascalonicus Doncaster – (Shallot Aphid) on onions, lettuce, beet, tobacco, many vegetables and field crops (polyphagous), virus vector; Europe, USA, Canada (CIE map no. A.113).

Myzus cerasi (F.) – (Cherry Aphid) only on cherries; Europe, USA.

Myzus ornatus (Laing) – (Violet Aphid) more polyphagous probably than *M. persicae* which it resembles, also virus vector; cosmopolitan (CIE map no. A.264).

Nasonovia ribisnigri (Mosley) – (Lettuce Aphid) on lettuce and chicory, several weeds, overwinters on currants and gooseberry; Europe.
Ovatus crataegarius (Wlk.) – (Mint Aphid) USA.
Rhopalosiphoninus latysiphon Davids. – (Bulb and Potato Aphid) on potatoes in store, and stolons, also celery, overwinters on wheat and grasses, most serious on glasshouse bulb crops; Europe, Canada.
Rhopalosiphoninus staphyleae (Koch) – (Mangold Aphid) polyphagous but only important on clamped mangolds; Europe.
Rhopalosiphum insertum (Wlk.) – (Apple–Grass Aphid) primary hosts are apple, pear, rowan or hawthorn, migrates to oats and grasses; Europe.
Rhopalosiphum rufiabdominalis (Sasaki) – (Rice Root Aphid) on rice, other cereals, sugarcane, *Prunus*, potato; cosmopolitan but more tropical (CIE map no. A.289).
Rhopalosiphum padi (L.) – (Oat–Bird Cherry Aphid) primary host is bird cherry, secondary mostly Gramineae, wheat, oats, rice, other cereals, also apple, virus vector; cosmopolitan but more temperate (CIE map no. A.288).
Roepkea bakeri (Cowan) – (Clover Aphid) on clovers; Canada.
Phorodon humuli (Schrank) – (Damson–Hop) Aphid; on hop and plums; Europe, Japan, Canada, USA.
Sitobion avenae (F.) – (Grain Aphid (many synonyms)) on wheat, barley, oats, and other Gramineae, virus vector; Europe, USA.
Sitobion fragariae (Wlk.) – (Blackberry–Cereal Aphid) overwinters on *Rubus* and migrates on to cereals and grasses in summer; Europe.
Therioaphis maculata (Buckton) – (Spotted Alfalfa Aphid) on alfalfa and some other forage legumes; Mediterranean Region, Middle East, India, and USA (CIE map no. A.126).
Therioaphis riehmi (Born.) – (Sweet Clover Aphid) on clovers; Canada.
Therioaphis trifolii (Monell) – (Clover Aphid) on clovers; USA and Canada.

Control of aphids

Direct damage to crop plants by feeding aphids is mainly the removal of nutrients and water from the tissues which is insignificant in small infestations but causes loss of vigour and wilting if populations are large. It does appear that some species have toxic substances in the saliva and these species cause foliage distortion and stunting of plants, and young shoots of plum, apple, etc., may be killed. Young plants are generally more susceptible to damage, and growth may be seriously impaired, whereas older or mature plants tolerate quite large aphid infestations without showing any deleterious effects. Grain aphids on young cereal ears may cause blindness or shrunken kernels; on nearly ripe grain the effect is negligible. Foliage discoloration is quite common, but the most spectacular is that caused on apple leaves by Rosy Leaf-curling Aphid. Honey-dew production may be quite serious, partly by allowing sooty moulds to develop, and in cereal ears the stickiness makes harvesting difficult.

Aphids are vectors of most plant viruses in temperate regions and thus their indirect damage is often very serious. Since only one or two infested aphids may innoculate a crop, prevention of diseases by insecticide application is extremely difficult; plant breeding for resistance is the most successful method of combatting virus diseases.

In a few cases the destruction of alternative host plants has been a method of controlling an aphid species, but this is a drastic procedure and seldom likely to be successful. Sometimes the spraying of alternative hosts with insecticides has been profitable, and the monitoring of aphid populations on the alternative hosts has been used to predict both size of populations and timing of immigration on to field crops (Way *et al*, 1981).

It is difficult to protect a field crop from aphid infestation, but is quite feasible to protect seed beds under a gauze shelter for the crops that are planted out as young seedlings (e.g. brassicas, tobacco, etc.). Experiments have shown that surrounding seed beds with reflective strips, or blue-coloured fabric, has a protective action in that this colour deters aphids from settling (alternatively they are attracted to yellow).

Natural control by predators and parasites is very important, particularly so on orchard crops and perennial plants, and so pesticide use must be very judicious. It has been recorded many times that careless use of some insecticides actually exacerbated aphid pest situations on some crops. This is one of the reasons why granular formulations of aphicides are popular.

Economic thresholds have been worked out for a number of important aphid pests in the UK, and include the following:

Apple:	*Rhopalosiphum insertum*	– if 50% trusses have aphids at bud-burst.
	Dysaphis plantaginea	– any aphids found at bud-burst.
	(Then spray an aphicide before flowering.)	
Wheat:	*Sitobion avenae*	– if five aphids per ear found at start of flowering, and weather conditions fine.
	Metopolophium dirhodum	– 30+ aphids found per flag leaf.

Other thresholds are known; for further details the local ADAS (MAFF) staff should be consulted.

Many aphid infestations are attended by ants whose presence clearly deters many of the natural enemies; natural control can be greatly enhanced by either sticky bands or spray-banding the host tree trunk to deny access to the ants; often such a measure is all that is required to control an aphid population in the field.

Being small and soft-bodied, aphids are very sensitive to environmental conditions, and most populations are severely depleted by a spell of bad weather; they are easily washed off the plants by heavy rain, which is presumably why most infestations occur on the underneath of leaves. Many gardeners still prefer to remove aphid infestations by the use of soapy water.

When insecticides have to be used, the following are all standard recommendations:

(1) Winter washes on fruit trees with tar oils or DNOC to kill overwintering eggs; care must be taken as these oils are very phytotoxic.
(2) Seed dressings with systemic insecticides (menazon, etc.) to protect seedlings in the early stages of growth when they may be most susceptible.
(3) Granules applied bow-wave at time of drilling, or applied later to the young plants (aldicarb, disulfoton, phorate, etc.).
(4) Foliar or soil application of systemic insecticides, such as dimethoate, demeton-*S*-methyl, disulfoton, formothion, menazon, mevinphos, thiometon, etc., or foliar application of suitably formulated granules for foliar lodging. Insecticides with translaminar properties may be very effective.
(5) Foliar application of contact insecticides such as HCH, malathion, deltamethrin, permethrin, pirimicarb, etc.; sprays must reach the underneath of the leaves where many aphids live in order to be effective.
(6) Against glasshouse aphids there is now a microbial aphicide available, reputed to be very effective. The fungus is *Verticillium lecanii*, but is only effective under very moist conditions (more than 85% RH and temperature higher than 15 °C) and so is only suitable for use in protected cultivation.

The precise insecticide most suitable for a particular species should be decided after consultation with local Ministry of Agriculture recommendations, as resistance to many chemicals is widespread, some crops may be sensitive to particular pesticides, and not all of the many aphicides are equally effective.

Eriosoma lanigerum (Hausmann)

Common name. Woolly Apple Aphid
Family. Pemphigidae
Hosts (main). Apple
(alternative). Pear, quince, ornamental species of *Malus*, cotoneaster, pyracantha, *Crataegus* and *Sorbus* spp. In the USA also on elm.
Damage. Conspicuous dense colonies of white woolly (waxy) aphids on tree branches spoil the appearance of nursery stock and form characteristic corky galls. Direct damage is mainly loss of sap, but the galls often split and provide entry points for fungal diseases.
Pest status. A common and widespread pest of apple throughout the world, but seldom very serious; reported to be particularly damaging in Egypt and Kenya.
Life history. The entire life-cycle is spent on the host tree, and breeding is by successive generations of wingless, viviparous, parthenogenetic females. Overwintering young nymphs hide in crevices and under loose bark; they are waxless and inconspicuous.

Winged adult females are produced in small numbers in July and later in September, and these each lay one single egg which either does not hatch or else the nymph dies. In the USA the winged forms migrate to elm trees, but this does not occur in Europe.

In S. Africa, America and Australia a form occurs which lives on the tree roots.

The adult is 1.5–3 mm, covered with a fine bloom of wax, has no external siphunculi and a short, rounded cauda.

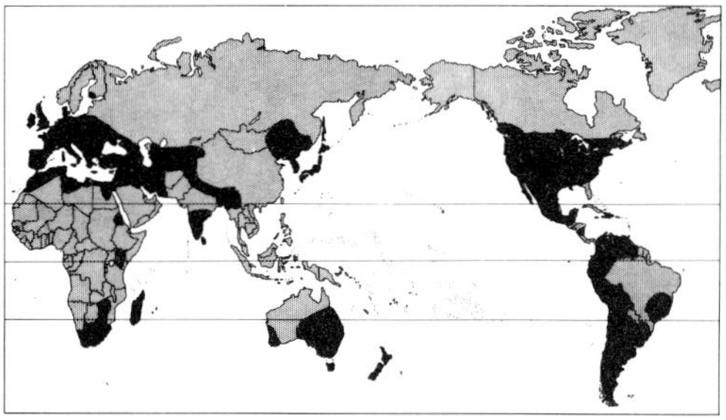

Main infestations occur on pruning scars, or where the bark is damaged or cracked; later infestations spread to the young growth.
Distribution. This insect clearly originated in the eastern states of the USA, but is now almost completely cosmopolitan except in the hotter parts of the tropics (CIE map no. A.17).

In the USA there is the closely related *Eriosoma pyricola* B. & D. (Woolly Pear Aphid), on a similar range of host trees.

The other species of *Eriosoma* in Europe, Asia, and N. America (and several other closely related genera) are pests of forest trees such as elms and oaks.
Control. In Kenya and parts of the USA and Europe (including parts of the UK) this pest is kept under control by the parasite *Aphelinus mali* (Aphelinidae).

When chemical control is necessary, winter washes of tar oils or DNOC usually keep the pest population in check; for later control measures the systemic insecticides are most effective (see page 230). Spot sprays have to be drenching, and added wetters are often required in order to penetrate the waxy covering of the insects. Vamidothion is especially effective against Woolly Aphid.

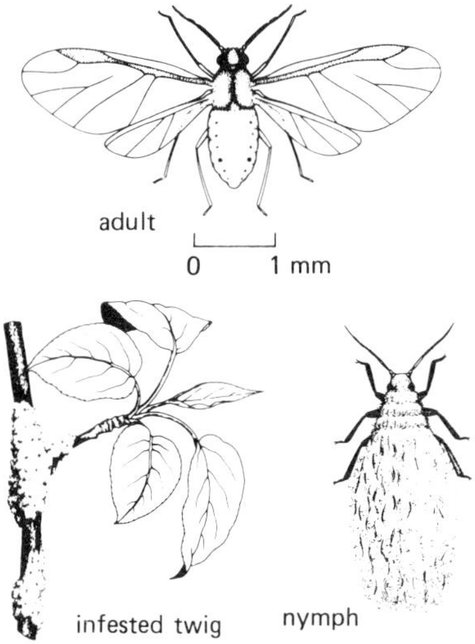

Pemphigus bursarius (L.)

(= *Aphis bursarius* L.)

Common name. Lettuce Root (Woolly) Aphid
Family. Pemphigidae
Hosts (main). Primary host is *Populus* spp.
(alternative). Secondary host is lettuce, and weeds in the Compositae.
Damage. This species infests the roots of lettuce, and heavy infestations cause the plant to wilt and turn the foliage yellow. There is no virus transmission recorded.
Pest status. A regular and sometimes serious pest of lettuce, especially when the soil is dry.
Life history. The winter is spent as eggs in cracks in the bark on several different species of poplar; hatching occurs in March/April in the UK and the nymphs cause hollow flask-shaped galls in the leaf petioles. During June winged aphids are produced, and as the galls open and dry the winged females fly off to infest the secondary hosts on their root systems. Viviparous parthenogenetic reproduction continues on the lettuce roots for several wingless generations, and then in September winged adults are produced which make the return migration back to the poplar trees where the overwintering eggs are laid.

Some of the wingless forms usually remain on the lettuce (and weed) roots and they often survive over winter, even in the absence of lettuce plants, and will colonise lettuce planted in the same ground the following spring.

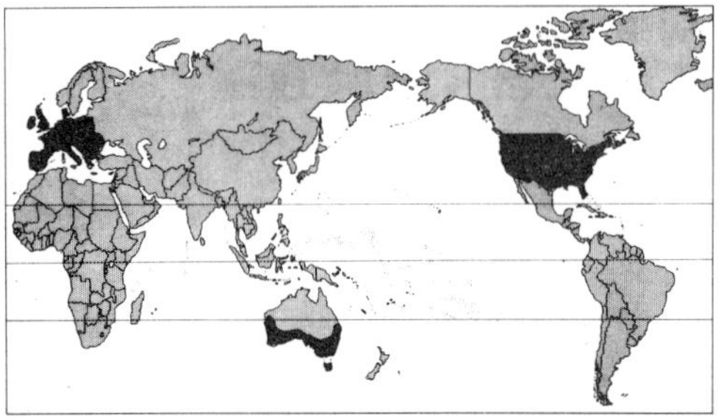

There is copious wax produced and infested roots of lettuce are white and woolly in appearance. Adult aphids are greenish in colour, covered with white wax, about 2 mm long; and without protruding siphunculi.
Distribution. Widespread throughout Europe, Australia and the USA.

In Canada there are three species of some importance:
Pemphigus populivenae Fitch – (Sugar Beet Root Aphid) polyphagous.
Pemphigus populicaulis Fitch – (Poplar Leaf-petiole Gall Aphid) polyphagous.
Pemphigus populitransversus Riley – (Poplar Petiole Gall Aphid) on brassica foliage.
Control. Crop residues should be destroyed and the soil well ploughed to destroy any overwintering soil populations. Destruction of composite weeds such as dandelion, sowthistles, and the like help to reduce populations, but the poplars that are primary hosts are usually grown as ornamentals or for windbreaks, so removal of the primary hosts is not a feasible idea.

Irrigation alleviates the symptoms of attack by root aphid. Certain lettuce varieties show high resistance to this pest.

Established infestations are seldom destroyed, but preventative treatment using diazinon granules or spray can be quite successful.

Petiole leaf galls on poplar

Viteus vitifoliae (Fitch)

(= *Phylloxera vastatrix* Planch)
(= *Daktulosphaira vitifoliae* (Fitch))

Common name. Grape Phylloxera (Vine 'Louse')
Family. Phylloxeridae
Hosts (main). Grapevine
(alternative). Other species of *Vitis* in the USA.
Damage. Root-feeding causes callus tissue on older roots and nodules on fine roots, leading to death of roots; leaf feeding causes globular galls on the leaves.
Pest status. Previously a serious pest in vineyards in Europe. In eastern USA the plants tolerate pest populations. Now seldom serious after widespread use of American rootstocks for grafting purposes. At the turn of the century this pest almost destroyed the European wine-production industry.
Life history. In the USA this insect has an autoecious migration; it hibernates on the roots, and then develops galls on the leaves, although the radicolous cycle continues through oviparous parthenogenesis. In Europe where no foliage of American vines is available, the entire life-cycle is confined to the roots only (anholocyclic). When alate forms appear and give rise to the sexual generation the latter die; but if sucker growth survives, then the sexual adults may give rise to the leaf-gall generation.

The American vines have roots that tolerate callus formation by special compensatory growth, and so the practice of grafting European vines on to American rootstocks was established.

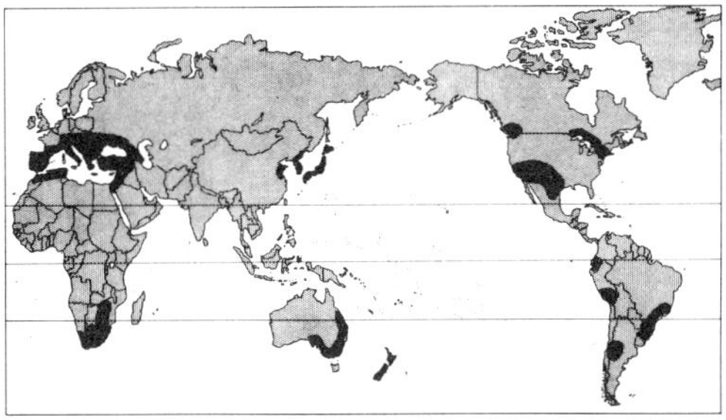

Distribution. Native to eastern USA this pest has been introduced into California, Europe, and is now spread into S. Africa, Australia, New Zealand, Japan and Korea, and parts of S. America (CIE map no. A.339).
Control. Since the widespread use of resistant American rootstocks, or hybrids, this pest has become minor in status, and is more of academic interest now. In parts of southern France where water is plentiful Phylloxera infestations were controlled by flooding the vineyards after harvest for several weeks. In these regions the local vines were often not grafted on to American rootstocks. In 1984 there was a severe outbreak of Phylloxera on these vines in the Narbonnais region, but it was restricted to this region.

Pseudococcus longispinus (T.T.)

(=*P. adonidum* (L.)

Common name. Long-tailed Mealybug
Family. Pseudococcidae
Hosts (main). Citrus, apple, pear, grapevine, fig, coffee, cocoa, sugarcane, coconut and other palms.

(alternative). Polyphagous on many crop plants and ornamentals, and house plants.

Damage. Groups of waxy mealybugs congregate on shoots, leaves or fruits; heavy infestations stunt or even kill young plants; some deformation is usual; sooty moulds usually present on the accumulated honey-dew.

Pest status. Usually not too serious on any one crop, but several species of 'long-tailed mealybugs' are very wide-spread and common on many plants.

Life history. Each adult female lays about 100–200 eggs in a single mass, then dies. The young nymphs disperse to find feeding sites. The first two instars are identical, but the males have a quiescent third instar in a small cocoon and then a fourth instar with large wing buds. Males take 30–34 days to develop from the egg stage. Female nymphs only have three instars before becoming adult; as they grow larger they gradually produce more wax.

The adult female is distinctively long-tailed in that there is a pair of long caudal waxy tassels. Several species of mealybug have females with long posterior 'tails' so this character is not specific.

Ants are usually associated with infestations of this species.

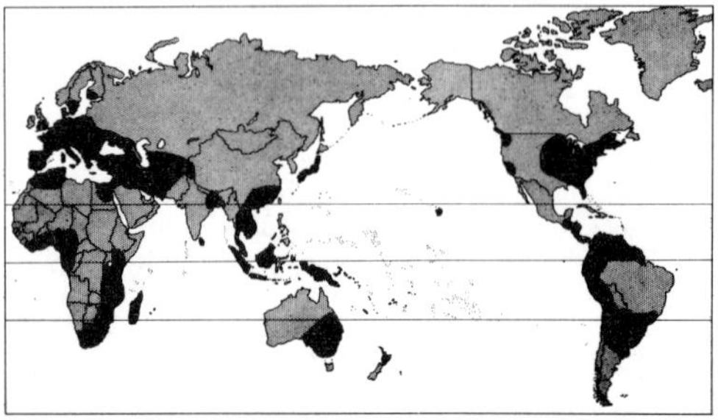

Distribution. Very widely distributed; cosmopolitan in the warmer parts of the world; widely found in greenhouses in Europe, UK and the USA (CIE map no. A.93).

Pseudococcus citriculus Green is another 'long-tailed mealybug', quite polyphagous in diet but as yet only known from China.

Pseudococcus maritimus (Ehrh.) is the Grape Mealybug; quite polyphagous; from USA, but not particularly 'long-tailed'.

Control. Often not needed; but see page 241.

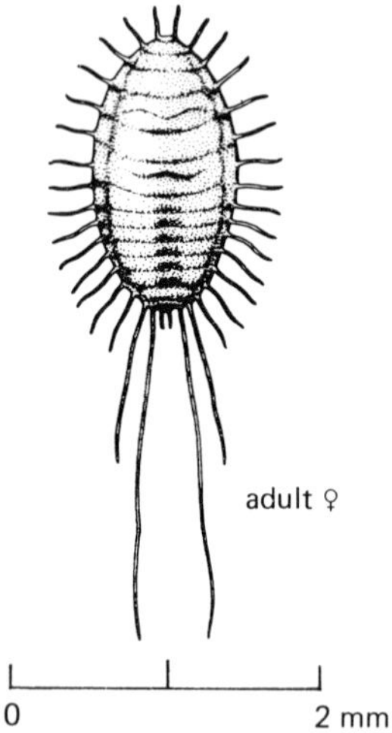

Icerya purchasi Maskell

(= *Pericerya purchasi* Mask.)

Common name. Cottony Cushion Scale (Fluted Scale)
Family. Margarodidae
Hosts (main). Citrus, mango, guava, rose.
(alternative). Polyphagous; on many woody plants.
Damage. Leaves and twigs infested with clusters of large, fluted white scales; infested leaves often turn yellow and fall prematurely; shoots, branches, and entire young trees may be killed. Copious quantities of honey-dew produced.
Pest status. A widespread, polyphagous pest, important on several different crops.
Life history. The adults are all females, quite large (c. 3.5 mm), stout, with a brown body covered with wax, and a large, posterior, white, fluted egg-sac containing about 100 eggs. Hatching takes a few days to two months according to climate.

The three nymphal stages are shiny, reddish-coloured under the wax; they sit mostly along the midrib under the leaves. Fully grown scales are mostly found on the twigs and branches.

The adult females are hermaphrodite and self-fertilizing; though males have occasionally been recorded.
Distribution. Cosmopolitan throughout the warmer parts of the world (CIE map no. A.51); this pest originated in Australia, and was introduced into California in 1868 and is now occurring in all *Citrus*-growing areas; first recorded in Israel in 1911.

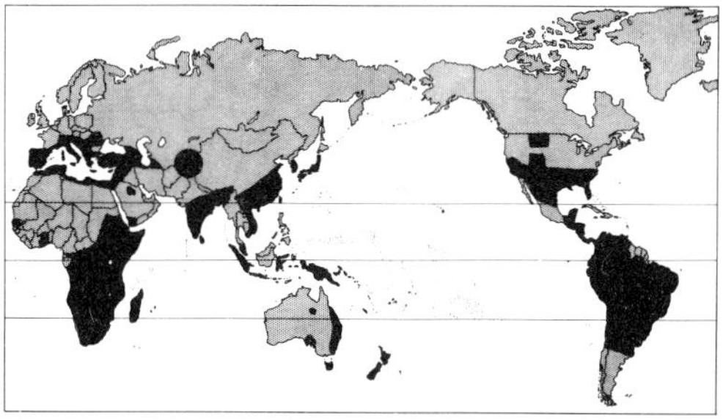

Icerya aegyptica (Dgl.) – (Egyptian Fluted Scale) polyphagous; found in tropical Asia, and parts of Africa and Australia (CIE map no. A.221).
Icerya seychellarum (Westw.) – (Seychelles Fluted Scale) also polyphagous; found in tropical Asia, up to Japan, and in parts of Africa (CIE map no. A.52).
Control. In many areas successfully controlled by the ladybird *Rodolia cardinalis* imported from Australia.

If chemical control is required then malathion and white oil has proved effective; otherwise see page 241.

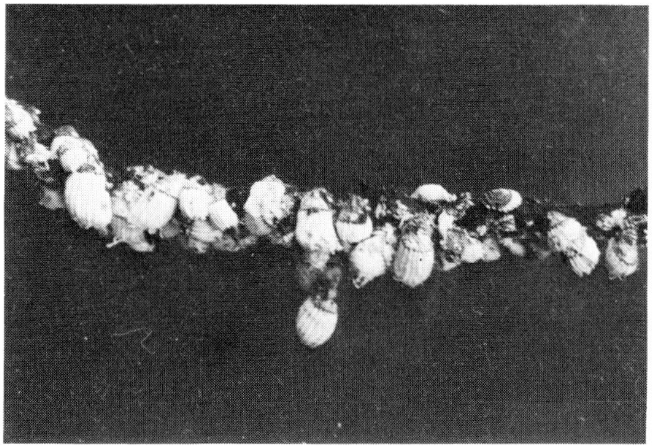

Adults on twigs and nymphs on leaves of *Cassia*

Parthenolecanium corni (Bouché)

(= *Eulecanium corni* (Bch.))

Common name. Plum Scale (Brown Scale)
Family. Coccidae
Hosts (main). Plum, almond, apple, cherry, grapevine, peach.

(alternative). Walnut, elm, acacia, blackberry, hazelnut, currants, gooseberry, raspberry; also on broom, cotoneaster; in total recorded from more than 300 hosts.

Damage. Twigs infested with smallish brown scales, often with woolly egg masses under the posterior part of the scale; young nymphs on leaves and flush growth: few obvious signs of damage, but infested leaves may fall prematurely.

Pest status. Only abundant in gardens and orchards that are neglected; usually killed by regular winter washes with tar oils.

Life history. The females lay 200–300 eggs each, in one mass underneath the convex brown scale, lifting up the posterior end of the scale; as oviposition finishes the female dies, but the dead dried scale remains *in situ* for months or even years.

Hatching occurs June/July, and the nymphs move up on to the fresh green growth where they feed. In the first two instars the nymphs wander intermittently; in August they moult into the second instar. On deciduous hosts the second instar nymphs (about 0.5 mm) move on to twigs and branches where they settle to overwinter, and at this time they change from green to brown in colour. In the spring they resume feeding and growth takes place, and they moult into adults in April.

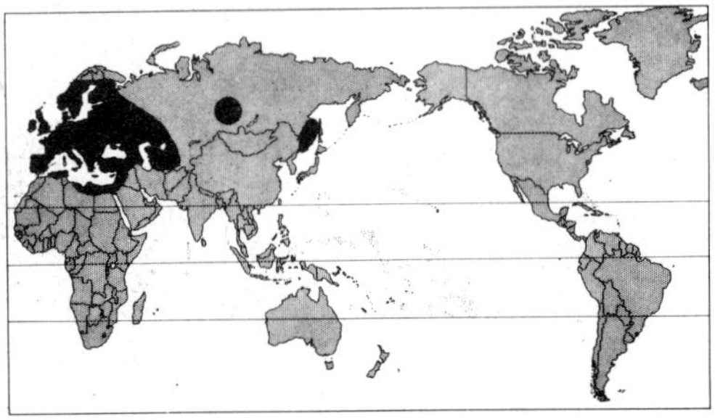

Some populations are bisexual and males are produced at this time. The females grow rapidly and become quite convex, and during May/June each female lays her eggs in the white egg 'case' and then dies. Adult females are about 3 mm long.

There is only one generation per year in the UK.

Distribution. Found throughout Europe, to Turkey, N. Africa, eastern and western USSR (CIE map no. A.394). Recorded by Davidson & Peairs (1966) as occurring throughout the USA and S. Canada.

Parthenolecanium persicae (F.) – (Peach Scale) recorded from peach, grapevine, apple, quince, guava, lemon; Europe, Middle East, Australasia, Canada, USA, S. America (CIE map no. A.395).

Several other species are found in the USA and southern Canada.

Control. In commercial orchards regular use of winter washes and spring sprays keep Plum Scale suppressed. If an extra spray is needed for this pest it is recommended that the first of two sprays, a fortnight apart, be applied in early July (UK), taking care to wet thoroughly all the bark and the underside of the leaves with diazinon, fenitrothion or malathion as a high-volume aqueous spray.

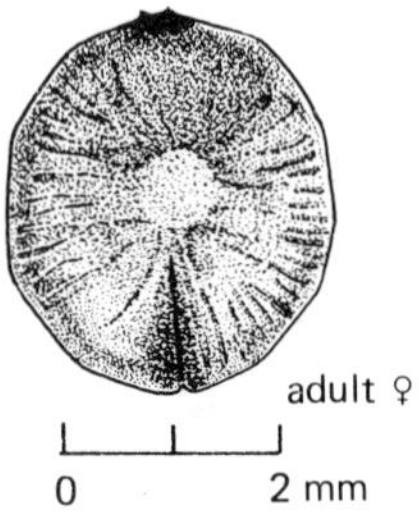

Saissetia oleae (Oliver)

Common name. Black Scale (Olive Scale)
Family. Coccidae
Hosts (main). Citrus, olive, oleander, *Ficus*.
(alternative). A wide range of trees, shrubs and grasses; polyphagous.
Damage. Large black scales on the twigs and shoots; with heavy infestations shoots and leaves may wither; often extensive sooty mould; fruit development may be impaired.
Pest status. A widespread and polyphagous pest, with conspicuous infestations; usually not serious.
Life history. Each female scale lays about 2000 eggs (1000–4000), over a period of 2–4 weeks; hatching takes 15–20 days.

The crawlers start feeding within a few hours, preferably under the leaves and on shoot tips, but adults prefer shoots and twigs. Nymphal development takes 2–3 months, but if conditions become unfavourable the nymphs can go into diapause.

Winged males are rarely found and reproduction is normally parthenogenetic. The life-cycle under favourable conditions takes 3–4 months. Adult females are blackish, strongly convex and 2–3 mm in length.
Distribution. Cosmopolitan in distribution, but absent from much of Africa; mostly a subtropical species (CIE map no. A.24); the more northern records refer to greenhouse infestations.

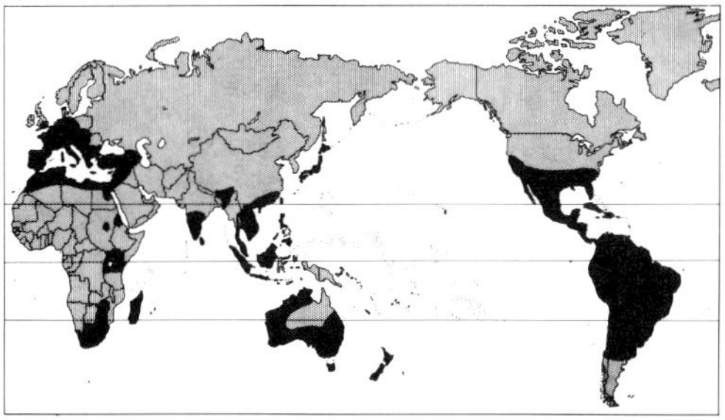

The records from Africa south of the Sahara may in fact belong to a very closely related species, *Saissetia privigna* De Lotto.

The closely related *Saissetia coffeae* (Wlk.), (Helmet Scale) is an almost completely cosmopolitan species found throughout the warmer parts of the world, on a wide range of crops and wild plants (CIE map no. A.318).
Control. If control is required, see page 241.

infested olive twig

Scales on twig of *Ficus microcarpa*

Lepidosaphes ulmi (L.)

Common name. Mussel Scale (Oystershell Scale)
Family. Diaspididae
Hosts (main). Apple, pear.

(alternative). Deciduous tree and bush fruits; also on hawthorn, cotoneaster, sloe, heather, and other trees and shrubs.

Damage. Heavy infestations occur when branches and twigs are heavily encrusted with scales; there is a general debilitating effect on growth and development.

Pest status. A regular pest of fruit trees; both widespread and polyphagous; sometimes quite serious.

Life history. Crawlers hatch in late May and early June, and wander for 3–4 days before settling and starting to feed; soon after settling they moult into the second instar nymphs. After one more moult the adult stage is reached in late July.

The adult females lay eggs in August, under the scale, and the female then dies. Overwintering takes place in the egg stage; up to 80 eggs under one dead female scale. Most scales are found on the twigs and smaller branches.

Adults are elongate and shaped like a mussel shell, 3.5 mm long; greyish-brown in colour. Both sexual and parthenogenetic races occur; in the former race winged males are found in August.

Only one generation per year in the UK.

Distribution. A cosmopolitan species with worldwide distribution in cooler regions, but not occurring in S. Africa (CIE map no. A.85).

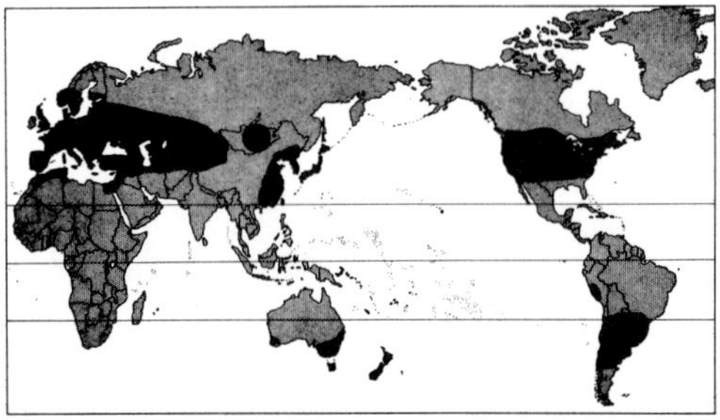

Control. *Aphytis mytilaspidis* (LeB) (Hym., Aphelinidae) is an external parasite of Mussel Scale that is quite abundant and widespread.

Not usually a serious pest in commercial orchards because of the regular use of routine sprays and winter washes. Should a summer spray be needed, then the recommendation is two sprays a fortnight apart starting in early June (or early May under glass) in the UK, using diazinon, fenitrothion or malathion (also see page 241).

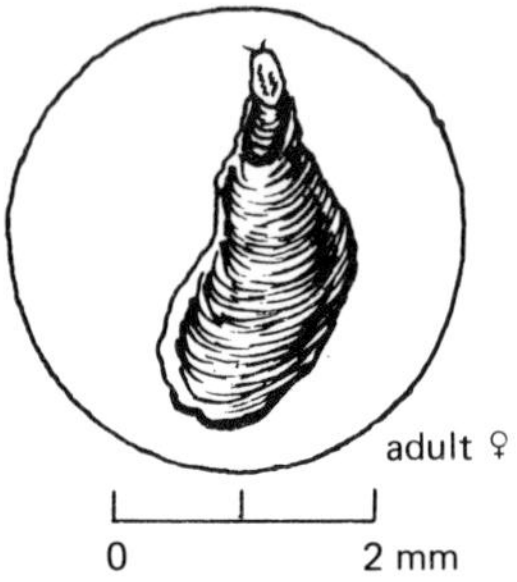

Infested twig of plum

Quadraspidiotus perniciosus (Comstock)

(= *Aspidiotus perniciosus* Comst.)

Common name. San José Scale
Family. Diaspididae
Hosts (main). Apple, peach, pear, plum, currants.
(alternative). Most deciduous fruit trees and shrubs; wide range of other trees and shrubs; polyphagous on woody plants; more than 700 host plants recorded.
Damage. Tiny circular scales can be seen on the bark of lightly infested trees, or on the fruits; in a heavy infestation the bark may be completely covered with overlapping scales; bark cracks, exudes sap, whole tree may die.
Pest status. A very serious pest of deciduous fruit trees and bushes; infested trees may die, and fruit is unsalable. Entire orchards debilitated and even destroyed.
Life history. One of the few viviparous hard scales, each female producing 100–400 nymphs. The crawlers are minute (0.2 mm), yellow and very active. Overwintering occurs in the nymphal stage.

The adult female is subcircular, flattened but with a raised central nipple, grey in colour, 1–2 mm in diameter, and completely covered by the circular yellow scale. Male scales are oval, about 1 × 0.5 mm.

Total life-cycle takes 18–20 weeks in California, and in the warmer parts of the world there may be 4–5 generations per year.
Distribution. Essentially a subtropical species, but extending throughout Japan and into Canada; probably summer temperatures are the key factor to its distribution. In Europe found in Switzerland and Germany, but not yet recorded in the UK (CIE map no. A.7).

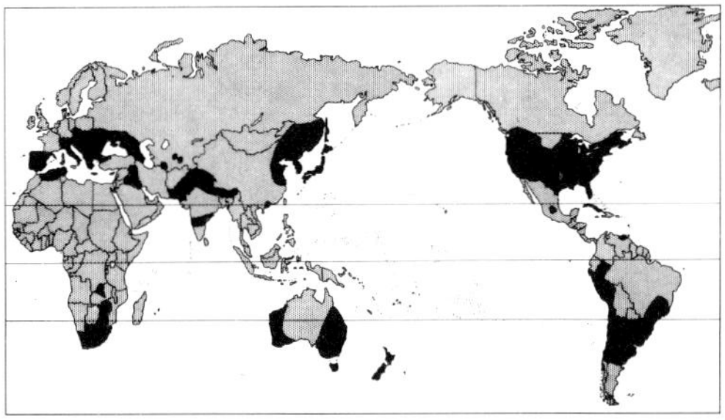

Quadraspidiotus juglansregiae (Comstock) – (Walnut Scale) USA.
Quadraspidiotus ostraeformis (Curt.) – (Oyster Scale) on apple, peach, pear, apricot, plum, cherry, nectarine, currants, and birch; Europe, USA, Canada.
Quadraspidiotus pyri (Licht.) – (Pear Scale) on pear, apple, peach, poplar and ash; Europe.
Control. As an extremely important fruit pest there is legislation in most countries against this species being accidentally introduced and established; there are also international phytosanitary regulations. For chemical control see page 241.

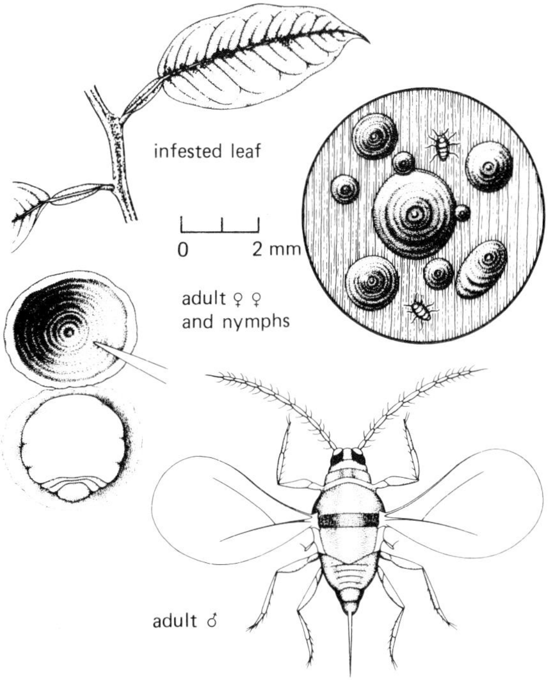

Mealybugs and scale insects (Homoptera; Coccoidea)

This superfamily contains the mealybugs and scale insects (*sensu lato*); a large group which is most important in the warmer parts of the world, but is widespread in N. America and in greenhouses in northern Europe. Generally, it may be said that the aphids are the main temperate bug pests and the scale insects (and mealybugs) the main tropical bug pests of agriculture on a world basis.

The hosts are mostly woody plants (trees and shrubs), but some species attack herbaceous plants and Gramineae. The crops most heavily attacked by both mealybugs and scale insects are undoubtedly citrus and coffee, and the various palms (Palmae). Some species are quite host-specific and are found only on *Citrus* for example, but others are polyphagous and on many different woody plants.

Field determination to genus is difficult as the main taxonomic character is chaetotaxy, so only tentative field identifications are feasible, especially as many trees will be host to several different, closely related mealybugs and scales.

The life-cycle is unusual in that the females are degenerate, apterous, obscurely segmented, with atrophied appendages and a scale-like body with either a waxy or powdery coating. Adult males are small, usually two-winged (hindwings lost), and with long antennae; in some species males have never been recorded. Females are oviparous or viviparous and the newly hatched nymphs usually shelter under the dead female 'scale' for a while. Eggs are laid under the 'scale' of the female, and by the time oviposition has ceased the female is dead and shrunken, but the eggs and the newly hatched young are protected physically. The minute first-instar nymphs are termed 'crawlers' and are the dispersive stage, having well-developed legs and being very active, and small enough to be carried on air currents.

Mealybugs retain a measure of mobility, even as adults, but the scales settle as nymphs on either twigs, fruit or leaves, and once their mouthparts ('stylet') are inserted into the host phloem system most species are immobile. In a number of species it is usual for young nymphs to feed on leaves and then to migrate to their permanent site on twigs. Some are typically found on the fruits of the host plant.

Many Coccoidea are honey-dew excretors which means that their infestations are usually attended by ants and associated with sooty mould infestations.

Predators and parasites are important in controlling many populations, especially the predacious Coccinellidae and parasitic Hymenoptera, and care has to be taken in any management programme not to upset the usually high level of natural control being exerted.

Important pest species of Coccoidea

In addition to the species already specifically mentioned, the following are pest species that occur in the cooler parts of the world and are of some importance as agricultural pests.

Pseudococcidae (Mealybugs)

Dysmicoccus boninsis (Kuw.) – (Grey Sugarcane Mealybug) on sugarcane and Gramineae; New World, Japan, Egypt, Australia (CIE map no. A.116).

Dysmicoccus brevipes (Ckll.) – (Pineapple Mealybug) polyphagous; cosmopolitan in tropics and sub-tropics (CIE map no. A.50).

Planococcus citri (Risso) – (Citrus Mealybug) polyphagous; cosmopolitan in the warmer parts of the world, and in greenhouses in Europe and N. America (CIE map no. A.43).

Phenacoccus aceris (Sigm.) – (Apple Mealybug) on apple; USA, Europe and now recorded in the UK.

Margarodidae (Fluted scales)

This small group is mostly tropical in distribution, but *Icerya purchasi* is included in this book as it does occur in warmer temperate regions.

Drosicha spp. – (Giant 'Mealybugs') polyphagous; in tropical Asia.

Coccidae (Soft scales)

Ceroplastes rubens Mask. – (Pink Waxy Scale) polyphagous; India, China, Japan, Australia, E. Africa (CIE map no. A.118).

Chloropulvinaria psidii (Mask.) – (Guava Mealy Scale) polyphagous; cosmopolitan in warmer regions (CIE map no. A.59).

Coccus hesperidum L. – (Soft Brown Scale) polyphagous on trees and shrubs; cosmopolitan in warmer climates (CIE map no. A.92).

Coccus viridis (Green) – (Soft Green Scale) polyphagous on trees and shrubs; pantropical (CIE map no. A.305).

Coccus spp. – (Soft Scales) several species on various hosts in the USA and Asia.

Eulecanium tiliae (L.) – (Hazelnut Scale) on hazelnut, apple, pear, etc.; Europe.

Gascardia destructor (Newst.) – (White Waxy Scale) polyphagous on trees; Australasia, Florida, and southern Africa (CIE map no. A.117).

Mesolecanium nigrofasciatum Pergande – (Terrapin Scale) polyphagous on woody hosts; Canada and USA.

Pulvinaria ribesiae Sigm. – (Woolly Currant Scale) on currants, gooseberry and rowan; Europe.

Pulvinaria vitis (L.) – (Woolly Vine Scale) on grapevine, peach, apricot; Europe.

Pulvinaria spp. – (Cottony Scales) on fruit and other trees; Europe, USA and Canada.

Diaspididae (Armoured scales)

Aonidiella aurantii (Maskell) – (California Red Scale) polyphagous on a wide range of hosts; cosmopolitan in warmer regions (CIE map no. A.2).

Aonidiella spp. – on many different host plants; USA.

Aspidiotus hederae Bouché – (Oleander Scale) on olive, apple, mango, palms, citrus, oleander (polyphagous); cosmopolitan (CIE map no. A.268).

Aspidiotus destructor Sign. – (Coconut Scale) polyphagous; pantropical (CIE map no. A.218).

Chionaspis furfura (Fitch) – (Scurfy Scale) polyphagous on fruit trees; USA and Canada.

Chrysomphalus aonidum (L.) – (Purple Scale (Florida Red Scale)) polyphagous on both monocotyledons and dicotyledons; pantropical (CIE map no. A.4).

Chrysomphalus dictyospermi (Morgan) – (Spanish Red Scale) polyphagous on fruit trees and shrubs; cosmopolitan in warmer regions (CIE map no. A.3).

Pseudaulacaspis pentagona (Targ.) – (White Peach Scale) polyphagous on fruit trees; cosmopolitan in warmer regions (CIE map no. A.58).

Control of mealybugs and scale insects
(Coccoidea)

Mealybugs and their close allies have the body surface covered with wax, often as a thick flocculent layer, which affords considerable protection against contact insecticides as the body is virtually unwettable. Similarly the scales have an expanded dorsal shield under which the body is sheltered. Thus the mature females and the larger nymphs are difficult to kill with contact insecticides, even when extra wetters are added to the sprays. Systemic insecticides can be used, but success is generally minimal. The only stage really vulnerable is the first-instar 'crawler', and this stage persists for only a few days.

Natural control is usually very important in most orchards, and in the past careless use of broad-spectrum insecticides, such as DDT, led to many pest population resurgences and outbreaks, especially of aphids, scale insects and spider mites. These population outbreaks were due entirely to the accidental destruction of the natural predators and parasites.

Most orchard crops are now carefully managed with an IPM approach, and the different aspects to be considered are mentioned below.

(1) **Cultural control**

(a) Phytosanitation – heavily infested fruit trees are often best left alone to permit the parasites to emerge.

(b) Clean planting material – all new planting material should be pest-free (use oils, fumigation, washing or hot-water treatment).

(2) **Biological control**

(a) Natural control – allow some trees to remain unsprayed on alternate years, to permit parasite populations to survive and build up. Use of sticky bands or spray banding to keep ants out of the tree allows the parasites to be more effective.

(b) Biological control – supplementation of existing natural control by introducing predators or parasites; especially important to fill any gaps in the local spectrum of predators and parasites.

(3) **Pesticides**

(a) Winter (dormant) washes with tar oils or DNOC, on deciduous trees, against overwintering scales and eggs; generally very successful.

(b) Spring sprays of petroleum or white oils against young nymphs, but care must be taken for these oils are basically phytotoxic.

(c) Contact insecticides against young nymphs, using malathion, diazinon or fenitrothion – usually two sprays at 14-day intervals are required.

Order

Heteroptera

(animal and plant bugs)

Families
Miridae (capsids)
Lygaeidae (lygaeid bugs)
Pentatomidae (shield or stink bugs)

Calocoris norvegicus (Gmel.)

(= *C. bipunctatus* F.)

Common name. Potato Capsid
Family. Miridae
Hosts (main). Potato, brassicas.
(alternative). Hawthorn, rose, nettle, and many wild plants.
Damage. Feeding punctures cause necrotic spots and eventually holes in young leaves. In severe attacks the toxic saliva may kill shoots and distort the plant.
Pest status. A common and widespread pest in the UK and Europe, and also Canada, but usually not too serious.
Life history. Eggs are yellow, about 2 mm long, shaped like a curved cylinder, embedded in the stem of the overwintering woody host (hawthorn, dog-rose), in bark crevices.

Nymphs are small green bugs resembling the adults but wingless. They feed on the overwintering host after hatching in April, and eventually migrate via hedgerow weeds on to the summer hosts of nettle and potato crops; they are common on potatoes from about late May onwards. There is a summer generation found on either nettles or potato crops.

The adult is quite a large bug, 7 mm long, green in colour, with two spots on the pronotum, black in colour; antennae are long, and eyes prominent.

There are usually two generations per year; eggs of the second generation are laid in September in the stems of the overwintering woody hosts, usually in bark crevices.

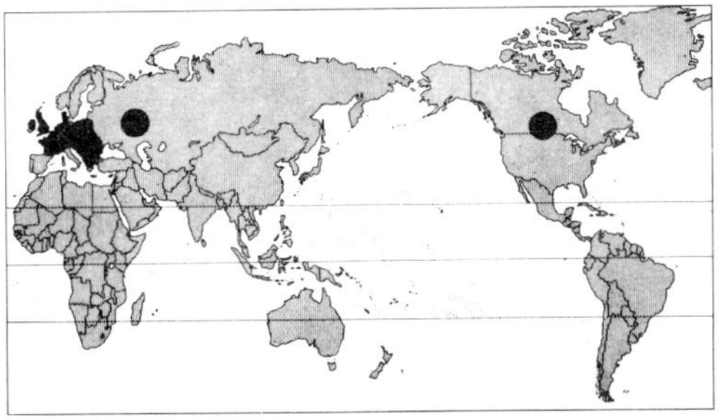

Distribution. Found throughout the UK, most of Europe, and in Canada. The precise extent of its distribution is not known at present.
Control. Control is not often required, but the chemicals that have been effective in the past are DDT and γ-HCH. Phorate or fenitrothion are generally now recommended for potato capsids.

Damage to potato crops is often restricted to headlands near hedgerows.

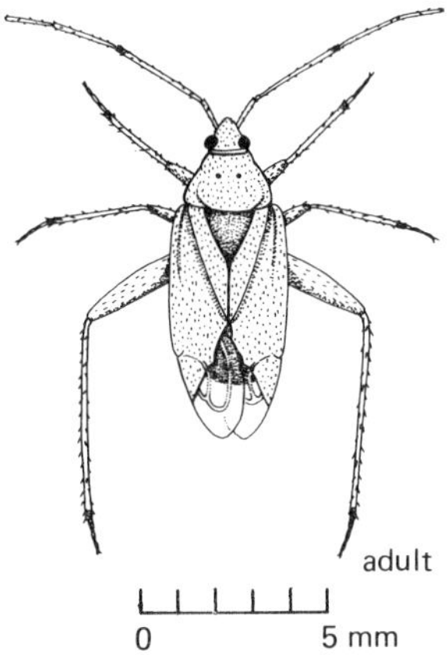

Lygocoris pabulinus (L.)

(= *Lygus pabulinus* (L.))

Common name. Common Green Capsid
Family. Miridae
Hosts (main). Potato, sugar beet, beans, apple.
(alternative). Swede, other fruit, field and vegetable crops (polyphagous); also hawthorn and dog-rose.
Damage. Toxic saliva injected when feeding causes necrotic spots on foliage and fruits, leading to leaf tattering and fruit calluses; in severe attacks shoots may be killed.
Pest status. Sometimes a serious pest on fruit crops, for apples and pears with callused spots may be unsalable; on field crops damage is seldom serious; but the pest is widespread and very common in the UK and Europe.
Life history. Overwintering occurs in the egg stage; eggs are laid in crevices in the bark of the overwinter host plants (trees); but there are two generations per year. Egg-hatching occurs from mid-April to mid-May, and is completed by petal-fall. Young nymphs feed on young leaves, either at shoot tips or else the blossom trusses. Older nymphs feed on fruitlets and young shoots. Most fruit tree infestation, and field crop infestation, ends with the appearance of the adults of the first generation in June. The winged adults are active and disperse freely on to other crops and weeds (mostly herbaceous plants) where the second generation develops. Second-generation adults appear in mid-August and they return to the woody hosts (fruit trees and bushes, etc.) where the overwintering eggs are laid.

The adult is a bright green, oval-shaped bug, about 6 mm in length; the forewings are slightly curved laterally.
Distribution. A very widespread and common insect throughout the UK, and in Europe.

Three species of *Lygocoris* are agricultural pests in Japan; there are four other species to be found in the UK.
Control. On field crops control measures are seldom warranted, but on fruit damage may be serious and so pesticides may have to be used. When capsids move into the orchard from neighbouring shelter then a barrier spray of 20–30 m along the nearest headland may be sufficient, and the entire orchard need not be sprayed.

There is no really effective ovicide and so spraying aims at killing the young nymphs. On apple and pear DDT may be used until flowering, otherwise sprays should not be applied until petal-fall and then the recommended chemicals are chlorpyrifos, dimethoate, fenitrothion, formothion, malathion or triazophos.

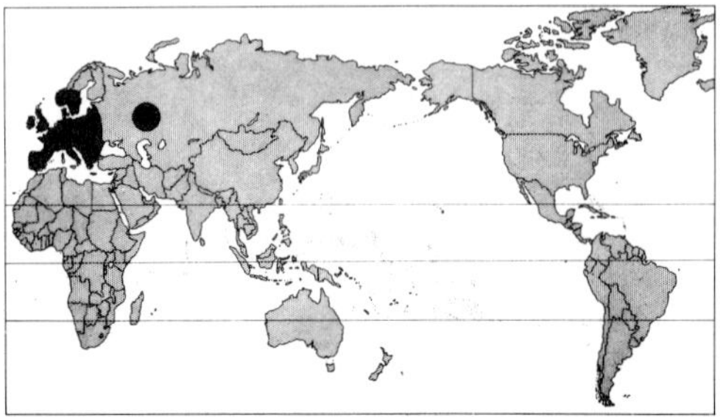

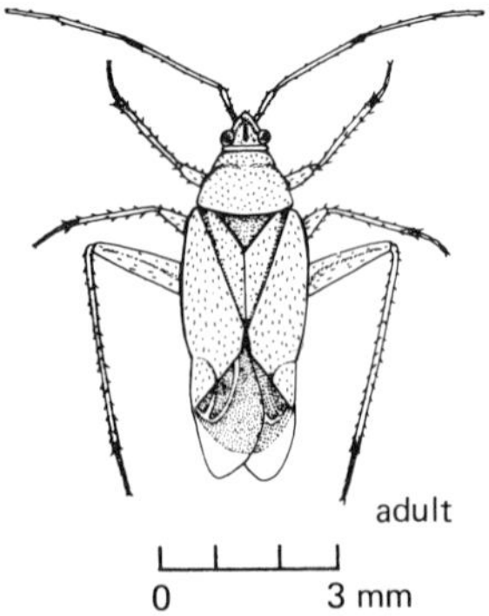

Lygus spp.

(*L. lineolaris* (Beauv.))
(*L. pratensis* (L.))
(*L. rugulipennis* Popp.) etc.

Common name. Tarnished Plant Bugs
Family. Miridae
Hosts (main). Potato, sugar beet, celery, peach, apple, alfalfa, pear, plum, quince.

(alternative). A wide range of field, fruit and vegetable crops; polyphagous.

Damage. Heavy feeding causes deformed fruits, kills buds, flower buds, shoots, and reduces seed yield on clovers and alfalfa. On leaves the toxic saliva causes speckling and tattering.

Pest status. The many species of *Lygus* are very important crop pests; they cause damage that is often serious on a vast range of cultivated plants. A number of other species are abundant on weeds and wild plants, as the genus is a large one.

Life history. Overwintering takes place in the adult stage, they hibernate in foliage, often in conifers or in leaf litter. They emerge from hibernation in March or April. Oviposition takes place in May, eggs being laid in unopened flower buds and stems of the host plants.

Nymphs are greenish with two black spots on each thoracic dorsum; the six nymphal instars take a month or more to complete, and the first adults of the new generation appear in July. The cycle is then repeated and the second-generation adults appear at the end of September. In southern USA breeding is continuous and there may be five or more generations annually.

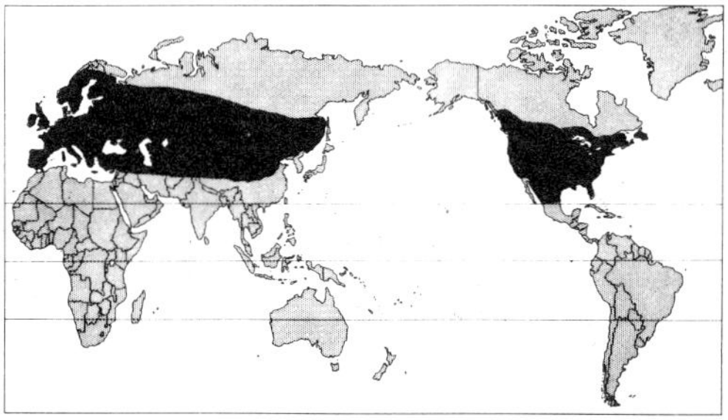

The adult is a smallish bug, 6 mm in length, rounded/oval in shape, brownish/greenish in colour with darker markings, and usually with the rusty appearance that gives them the name of 'tarnished' plant bugs. There is some variation in the dark markings of adults which makes specific identification difficult.

Distribution.
Lygus lineolaris (= *L. oblineatus* (Say)) – Canada, USA, and Mexico (CIE map no. A.38).
Lygus pratensis – Europe and Asia (more northern parts) (CIE map no. A.39).
Lygus rugulipennis – British Isles and Europe.
Lygus disponsi Linn. is the Japanese Tarnished Plant Bug.

At least 12 other species of *Lygus* are crop pests in Canada and the USA, and several other species occur in the UK and Europe on wild hosts and weeds.

Control. In the past DDT sprays were effective against these pests; present recommendations include the use of phorate granules at or before planting field crops, and nicotine or dimethoate sprays on foliage. Generally the recommendation varies according to the crop concerned and the country.

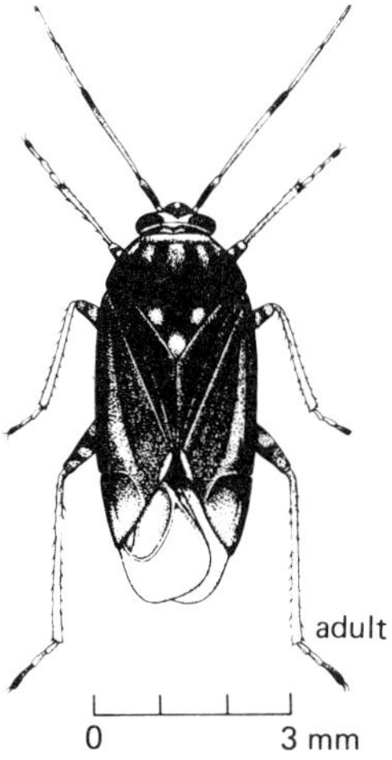

Orthops campestris (L.)

Common name. Stack Bug (Carrot Plant Bug)
Family. Miridae
Hosts (main). Carrot, celery, parsnip, parsley, dill.
(alternative). Other (wild) Umbelliferae; also recorded from dahlia.
Damage. Adults and nymphs suck sap from young leaves, flower buds and unripe fruits; only damaging really to seed crops.
Pest status. Occasionally serious damage done to seed crops, otherwise a common and widespread pest, frequently encountered.
Life history. Overwinters as the adult bug, hibernating presumably in either foliage or leaf litter; recorded in Sweden overwintering as eggs.

Eggs are laid in the spring in the buds and young leaves of the host plant; hatching requires only a few days.

The nymphs are small and pale green; in the UK nymphal development takes about two months but in Canada only 4–5 weeks are needed.

The adult is a pretty little bug, pale green with dark brown markings, and about 3.5–4 mm long.

In Canada there are 2–3 generations per year, but in the UK this species is univoltine.
Distribution. Throughout Europe, as far north as Sweden, Canada and the USA.

Five other species of *Orthops* occur in the British Isles on a wide range of different host plants.

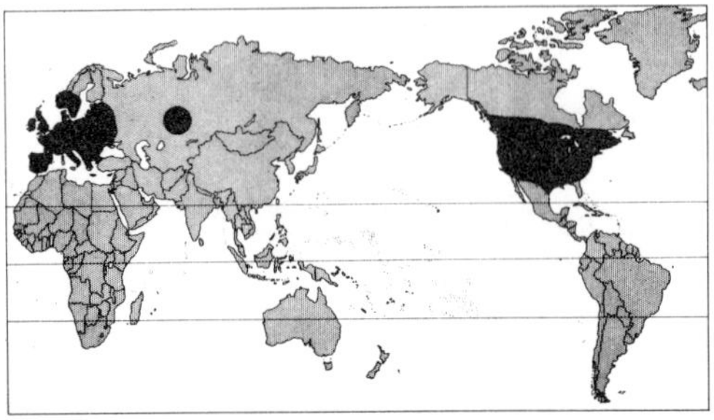

Control. This pest seldom needs actual control measures, but on a seed crop it might be necessary to apply insecticides, in which case the chemicals used for other capsids should be effective. However, in the UK MAFF recommendations for capsid control do differ for the different species.

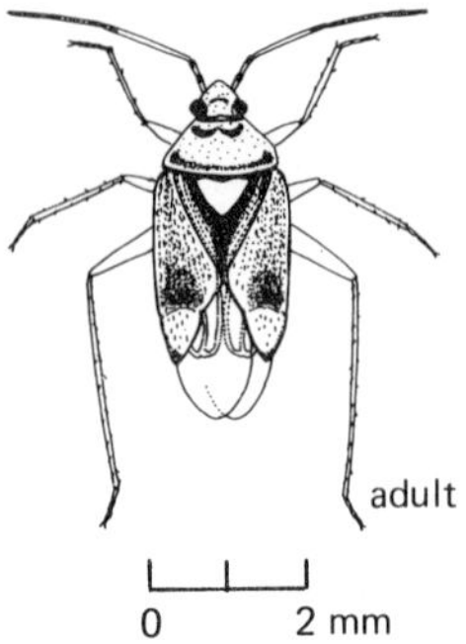

Plesiocoris rugicollis (Fall.)

Common name. Apple Capsid
Family. Miridae
Hosts (main). Apple, currants, and gooseberry.
(alternative). Willows, bog myrtle.
Damage. Feeding bugs puncture young leaves, shoots and fruit; leaves are distorted with necrotic spots and tattered holes; fruits with calloused patches and sometimes distorted. Commercial value of fruits severely reduced.
Pest status. Only recorded as an apple pest in the UK since about 1914, when it rapidly became a serious pest, but it was virtually eradicated by the widespread use of DDT from 1946 onwards. Now more of a garden pest, and it may be locally very abundant.
Life history. Overwintering occurs in the egg stage; eggs are laid in crevices in the bark of young branches. Hatching takes place from mid-April to early May.

Young nymphs feed first on the rosette leaves surrounding the blossom trusses; later they feed on fruitlets and young foliage.

Adults are found from early June to late July, and they live for 4–6 weeks. Eggs are laid in June and July; then active life ceases. It is thought that on the wild hosts there might be a second generation. The adults are small dark capsids, green and brown in colour, from 5.5–7 mm in length; the head, pronotum and scutellum is green, sometimes margined with yellow. They are most frequently found sheltering in unopened leaves and at the base of young shoots (see illustration).
Distribution. Found throughout the British Isles and parts of Europe.
Control. Should control be needed then the chemicals likely to be effective include DDT, carbaryl, chlorpyrifos or fenitrothion; sprays should be applied at the green cluster stage (flower buds evident but not opening yet, leaves unfolding).

capsid damage to apple

Adult bugs in apple shoot

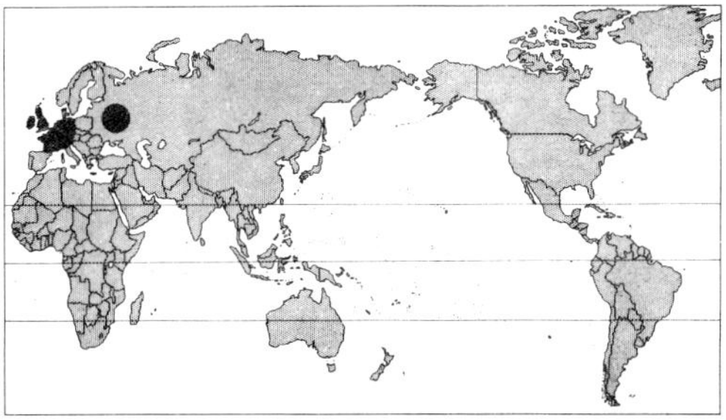

Blissus leucopterus (Say)

Common name. Chinch Bug
Family. Lygaeidae
Hosts (main). Maize, sorghums.
(alternative). Wheat, millets, rice, barley, rye, oats and many grasses.

Damage. Sap-sucking causes growth retardation, and sometimes plant death; heavy infestations result in major crop losses.

Pest status. A serious pest of cereals in the USA and Canada, together with several other closely related species.

Life history. The adult bugs hibernate in hedgerows and grass tussocks. In the spring they emerge and the females lay eggs, each female laying several hundred eggs (about 20 eggs per day) behind the lower leaf sheaths, or in the soil by the plant stems. Hatching takes 1–3 weeks.

Nymphal development passes through five instars. On wheat, barley and rye the plants ripen and become less succulent while the bugs are still nymphs; so the nymphs crawl to adjacent maize or sorghum fields (which have a longer development period) where damage may be serious.

Adults reach maturity at the end of June and fly to maize and sorghum fields to oviposit; this starts the second generation. Second-generation adults appear in late summer or early autumn (fall) and as the weather cools they pass into a state of hibernation in grass tussocks. The adults are black and white, with whitish wings, and small in size (5 mm long).

In the extreme south of the USA there may be three generations per year.

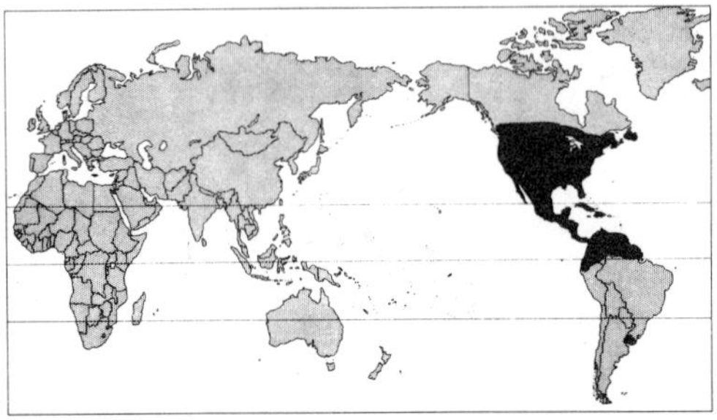

Distribution. Southern parts of Canada, throughout the USA and parts of C. and S. America including the W. Indies (CIE map no. A.333).

Blissus insularis Barber – (Southern Chinch Bug of the USA) mostly on grasses.
Blissus occiduus Barber – (Western Chinch Bug) Canada and USA; serious on wheat, barley and various grasses.
Blissus leucopterus hirtus Mont. – (Hairy Chinch Bug) predominately a turf pest, in the USA.
Blissus pallipes Distant – (Long Wheat Grain Stink Bug (Chinch Bug)) on wheat and barley in China.

Control. Weather is the key factor to Chinch Bug populations; large populations develop in hot dry weather; heavy rain and low temperatures increase egg and nymphal mortality greatly.

An egg parasite (*Eumicrosoma benefica* Gahan – Scelionidae) is common. Cultural control measures include the growing of resistant varieties, and the avoidance of adjacent plantings of maize and small grains. The use of physical barriers is no longer recommended, but insecticidal spray barriers may effectively kill the migrating nymphs. Chemicals generally effective include carbaryl, chlorpyrifos, diazinon, ethion fenthion and fenitrothion.

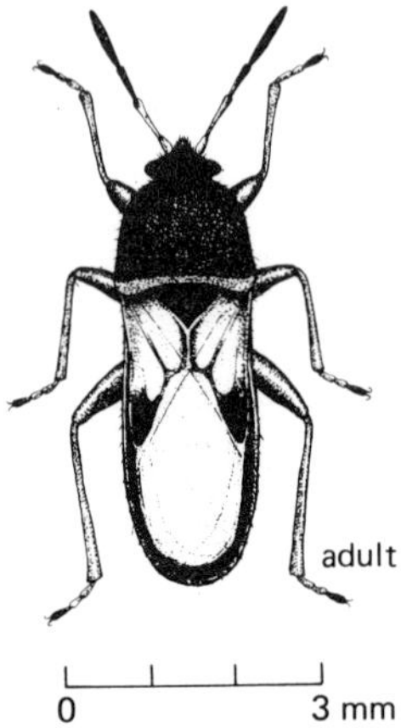

Nysius spp.

(*N. ericae* (Schill.), etc.)

Common name. False Chinch Bugs (Seed Bugs)
Family. Lygaeidae
Hosts (main). Brassicas, beets, potato.
(alternative). Polyphagous on many field, vegetable and fruit crops; some species more restricted to a few host plants; also on many weeds.
Damage. Sap-sucking causes wilting and necrosis of tissues; heavily attacked seedpods suffer yield losses and damaged seeds are nonviable.
Pest status. Quite serious pests on the main host crops in several different parts of the world; widespread and polyphagous mostly, the genus is an important pest.
Life history. Adults overwinter amongst plant remnants and in leaf litter, and become active in the spring, when females lay eggs on the foliage of low-growing plants (often weeds) or on the soil. Egg hatching takes about 4–8 days.

The nymphs are reddish and grow through five moults before becoming adult after about 3–4 weeks.

The adult bugs are small (about 3–5 mm long), and grey-brown in colour, with darker markings. Summer adults live for about four weeks, but autumn adults overwinter in hibernation.

In Australia and southern USA the life-cycle takes about four weeks (summer) and there are usually 4–5 generations per year.

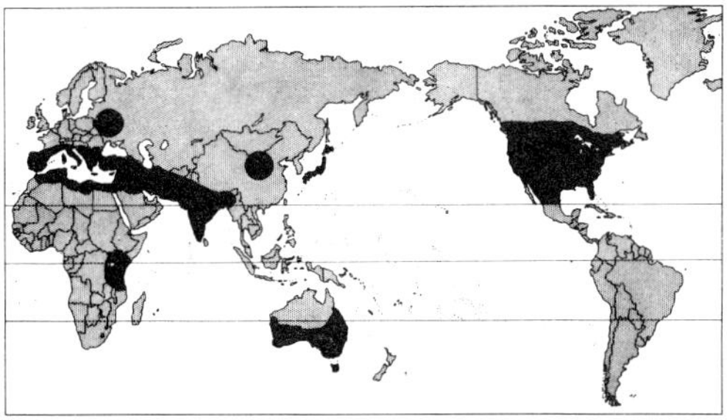

Distribution. The main species recorded worldwide are as follows:
Nysius caladoniae Distant – (Caledonia Seed Bug) USA.
Nysius clevelandensis Evans – (Grey Cluster Bug) polyphagous; Australia.
Nysius cymoides Spin. – on brassicas mostly; Mediterranean Region.
Nysius ericae (Schill.) – (False Chinch Bug) polyphagous; Canada and USA.
Nysius inconspicuous Distant – on sesame; India.
Nysius plebejus Distant – polyphagous; Japan.
Nysius raphanus Howard – (False Chinch Bug) polyphagous; USA
Nysius sp. – on pyrethrum; East Africa.
Nysius spp. – (Seed Bugs) recorded from many hosts; USA, two species occur on wild plants in the UK.
Nysius turneri Evans – (Invermay Bug) Australia.
Nysius vinitor Bergroth – (Rutherglen Bug) polyphagous; Australia.
Control. When damage warrants control measures, the recommended insecticides are the same as those for Chinch Bug, and in addition endosulfan, malathion and parathion as sprays.

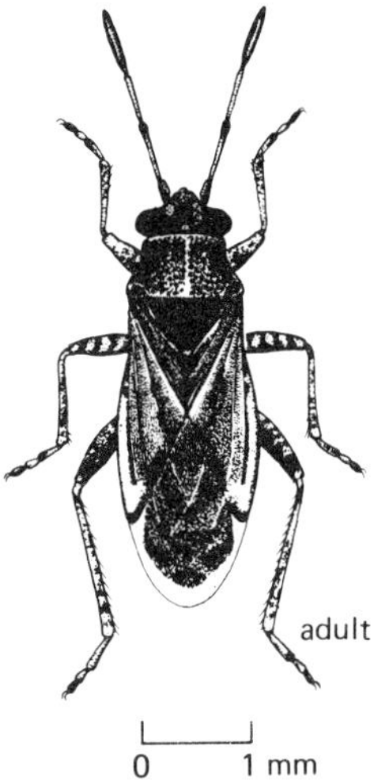

Eurygaster spp.

(*E. austriaca* (Schr.))
(*E. integriceps* Put.)

Common name. Wheat Shield Bug; Sunn Pest, Senn Bug
Family. Pentatomidae (Scutelleridae)
Hosts (main). Wheat, barley.
(alternative). Rye, maize, and other Gramineae.
Damage. The main damage is sap-sucking from young grains at the milky stage; damaged ears remain empty.
Pest status. At the moment mostly an historical pest, the last serious outbreak ended in the early 1960s, and, except for Iran and Turkey, no large-scale control measures have been needed. In 1959/60 Algeria lost an estimated 583 000 tonnes of wheat (200 ha infested), and these typical loss figures show that the Sunn Pest was in many areas the main limiting factor in the cultivation of cereals. In practice field infestations of *Eurygaster* are assessed together with *Aelia* species.
Life history. Eggs are laid in small clusters on the plant foliage, usually leaf blades. The adult is a largish rounded shield bug, with greatly expanded scutellum which extends right to the posterior end of the abdomen; body coloration is yellow-brown, with two black and white spots at the base of the scutellum, body length 12–13 mm.

The adults hibernate in vegetation and litter on the hillsides. Populations are (were) often very large, with 10–60 bugs per square metre of hibernation site. The mountain hibernation sites are often 10–20 km from the fields infested. The preferred habitat for these bugs is in semi-arid areas with a mild winter. The spring migration to the wheat field may take as long as a month. The new generation of adults appears in the crops about June; there are usually 1–3 generations per year according to climatic conditions. Field populations often reach a density of 10–50 bugs per square metre.
Distribution.
Eurygaster austriaca occurs in southern Europe, Algeria, Morocco, to Iran, Turkey, and the USSR (CIE map no. A.361).
E. integriceps occurs in Greece, Asia Minor, Pakistan and the USSR (CIE map no. A.40).
E. koreana Wagner is found in Korea and Japan.

In the Middle East and S.E. Europe an associated pest on wheat is the pentatomid genus *Aelia* (*A. melanota* Fieb., *A. furcula* Fieb., *A. rostrata*, *A. acuminata* L., *A. germari* Kust.).
Control. The egg-parasite *Trissolcus* (Hym., Scelionidae) is of some importance in the natural control of these bugs.

When populations were (are) very large then chemical control is the only practical solution; the following chemicals have been effective when sprayed on to the crops as the hibernating adults arrive: DDT, fenthion, parathion, methomyl; sometimes more effective as mixtures.

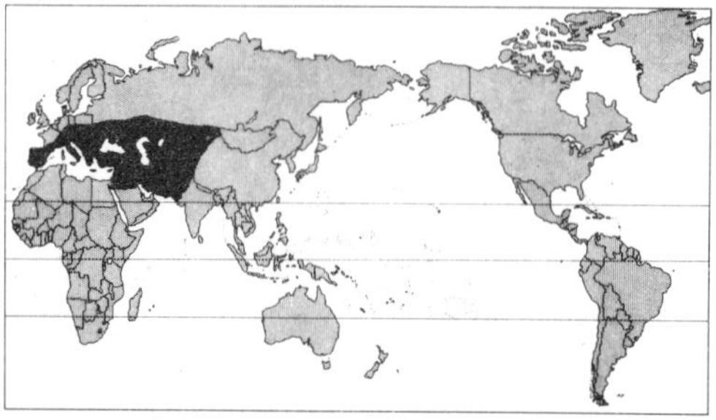

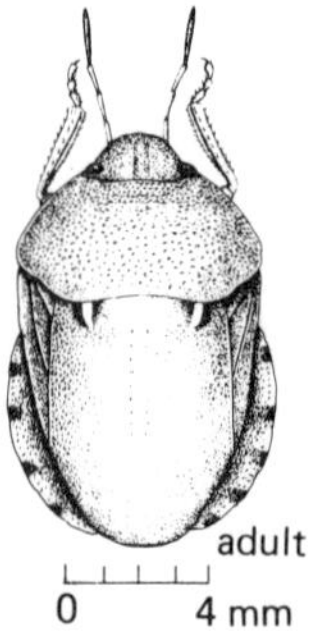

adult of *E. integriceps*

Nezara viridula (L.)

Common name. Green Stink Bug (Green Vegetable Bug)
Family. Pentatomidae
Hosts (main). Vegetables and legumes.
(alternative). Many other crops and ornamentals; truly polyphagous, recorded from more than 100 different plant hosts.
Damage. Toxic saliva injected at feeding causes death of tissues and necrosis; fruits and leaves become distorted and spotted; young fruits may be shed.
Pest status. A common and worldwide pest, sometimes quite serious, but to be found regularly on a very wide range of crops in the warmer parts of the world.
Life history. In cooler regions the adults hibernate over the winter period, amongst vegetation. They emerge in the spring, mate, and the females lay their eggs. Eggs are barrel-shaped, large (1.2 × 0.75 mm), and laid in clusters under the leaves, usually 50–60 per batch (10–130 recorded); each female lays some 100–300 eggs.

There are five nymphal instars, the first remains clustered by the egg-raft, but later ones disperse and feed from the soft parts of the plant, preferably from developing fruits and seeds. The nymphs are orange and brown in colour.

Development is generally slow; from egg to adult taking 6–10 weeks; with a pre-oviposition period of about 2–3 weeks. The threshold temperature for development is 12°C. In the eastern Mediterranean there are three generations per year.

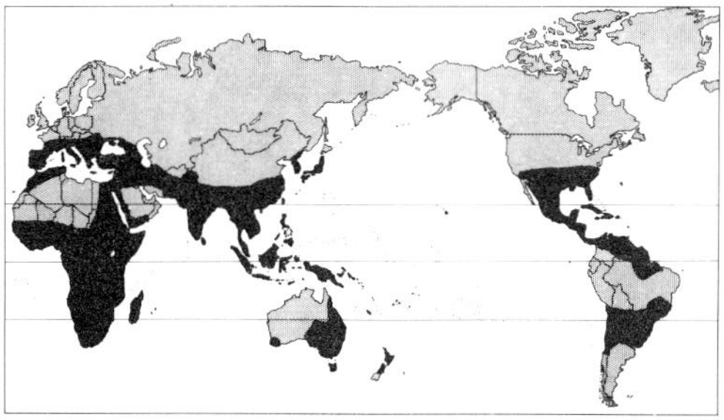

The adult is a large green bug, about 15–18 mm long and 8–10 mm broad. Some individuals have a yellow edge to the pronotum, and some are brownish rather than green in colour, but these are not common.
Distribution. Completely cosmopolitan in the warmer parts of the world, from southern Europe and Japan, to South Africa and Australasia (CIE map no. A.27).
Control. *Teleonemus basalis* (Woll.) (Hym., Scelionidae) is an important egg-parasite, and has been introduced into Australia in an attempt to control this pest.

Chemical control is not often required, and is in fact difficult to achieve; chemicals used with some success include DDT, HCH, phorate, malathion, dimethoate, and phosphamidon.

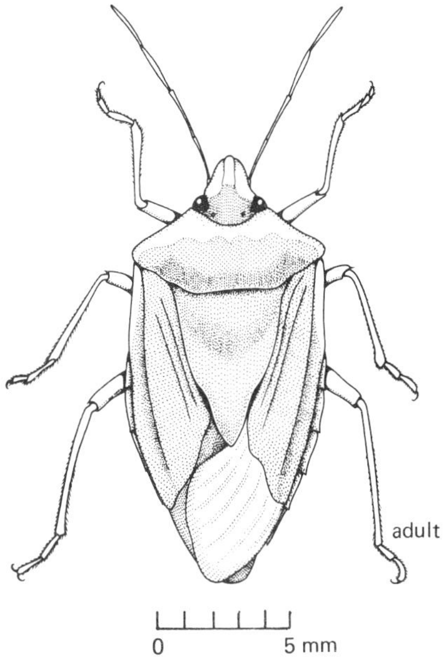

Order

Thysanoptera

(thrips)

Family
Thripidae (thrips)

Frankliniella spp.

Common name. Flower Thrips
Family. Thripidae
Hosts (main). Leguminosae, cotton, barley, oats, wheat.
(alternative). Polyphagous, on coffee, tomato, tobacco, potato, capsicums, sweet potato and many other crops and ornamentals.
Damage. Direct damage consists of leaf and flower distortion (usually following bud infestation); seedling growth may be retarded and yields may be reduced. Mature plants are generally little affected by thrips. Some species are important virus vectors; Tomato Spotted Wilt Disease is particularly serious on several crops.
Pest status. Several species are quite important pests in many parts of the world. Many species are regularly found in flowers of crop plants but appear to do little actual damage. Some of the virus diseases are serious and have severe effects on crop yields.
Life history. Eggs are inserted in the tissues of the host plant, usually in the buds, flowers or unfurled leaves.

Nymphs are pale-coloured and wingless, and found inside or under the curled leaves. There are three nymphal instars; and pupation takes place in the soil.

Adults are pale brown, dark brown or black, with paler bands across the abdominal segments; body length 1.0–1.5 mm.

In the warmer parts of the world the life-cycle takes some 2–5 weeks.

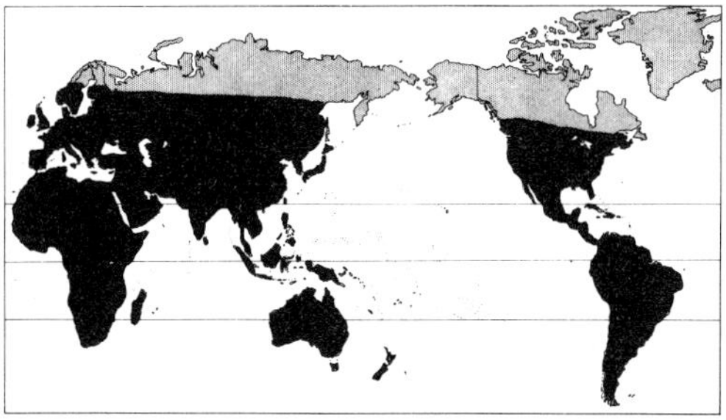

Distribution. Several of the more important species of *Frankliniella* include:
Frankliniella schulzei (Trybom) – Cotton Bud Thrips of Africa.
Frankliniella tritici (Fitch) – (Flower Thrips) on alfalfa, oats, beans, asparagus, etc.; Canada and USA.
Frankliniella fusca (Hinds) – (Tobacco Thrips) on tobacco, cotton, tomato, and cucumbers in greenhouses; Canada and USA (also a virus vector).
Frankliniella occidentalis (Pergande) – (Western Flower Thrips) on several crops; (also a virus vector).
Frankliniella intonsa Trybom – (Flower Thrips) many crop plants; Japan.

Several other species are of regular occurrence on ornamentals, and are minor pests on many crops; about ten different species of *Frankliniella* are recorded on cotton worldwide. The genus is completely worldwide in distribution, but the different species tend to be on different hosts in different countries.
Control. For control measures see page 258.

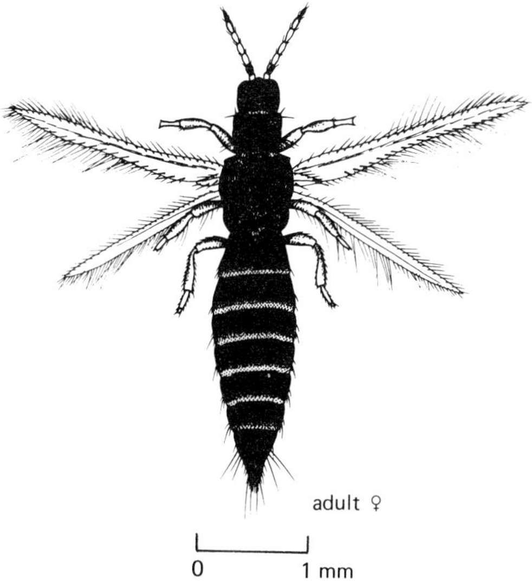

Heliothrips haemorrhoidalis (Bouché)

Common name. Greenhouse Thrips (Black Tea Thrips)
Family. Thripidae
Hosts (main). Tea in warmer regions; greenhouse crops in temperate parts.

(alternative). A very wide range of crop plants; truly polyphagous.

Damage. Feeding causes silvery patches on the underside of leaves, covered with black spots; severe attacks destroy leaves and may even kill the plants.

Pest status. Seldom a severe pest on crops in the field, but may be very damaging on greenhouse crops, and losses of yield may be serious.

Life history. Eggs are bean-shaped, 0.3 mm long, pushed into leaf tissue by the female and the wound covered by a drop of excreta. Each female lays some 25 eggs over a 7-week period.

Nymphs are pale-whitish in colour, with red eyes, and they typically carry a drop of excreta on the tip of the upturned abdomen. The excreta drops are deposited at intervals, causing the small black spots on the leaf surface. The nymphs usually congregate on a damaged leaf, then change into a 'prepupal' stage with short wing pads; after a day or two they moult again into the 'pupal' stage, which has longer wing pads; both stages are yellow with red eyes.

The adults are all females, which reproduce parthenogenetically; each is dark brown or black, 1.5 mm in length, legs, antennae and wings are whitish.

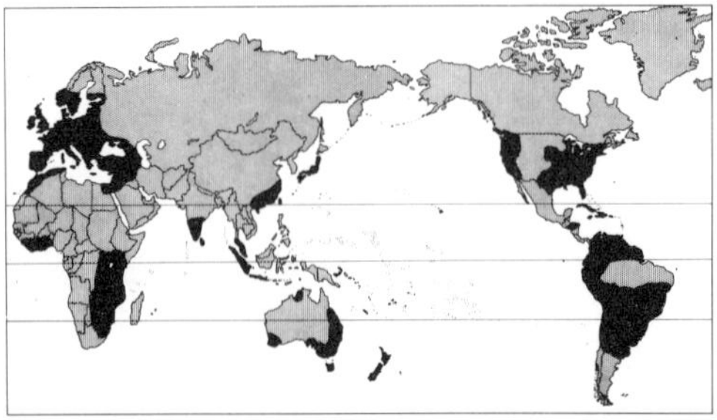

The life-cycle takes from eight weeks at 19 °C to 12 weeks at 15 °C. This species is unusual in that pupation takes place on the plant and not in soil.

Distribution. This species is cosmopolitan in distribution, but more frequently in greenhouses in temperate regions; however records from Asia are sparse (CIE map no. A.135).

Control. If sprays are used against this pest they should be full-cover and directed as far as possible at the underside of the leaves.

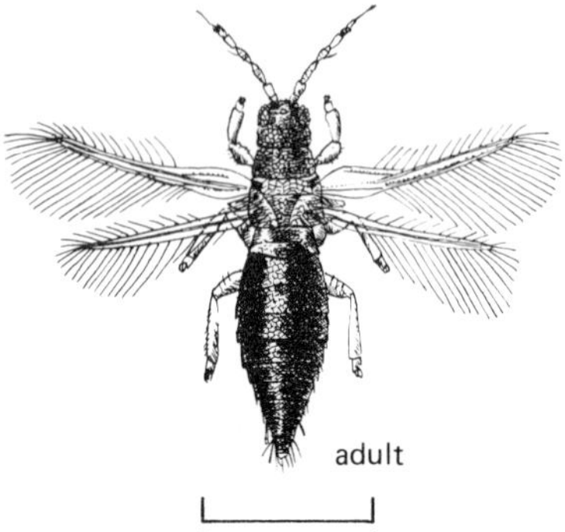

adult

0 1 mm

eggs

nymph with drop of excreta

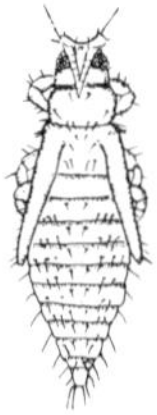

pupa

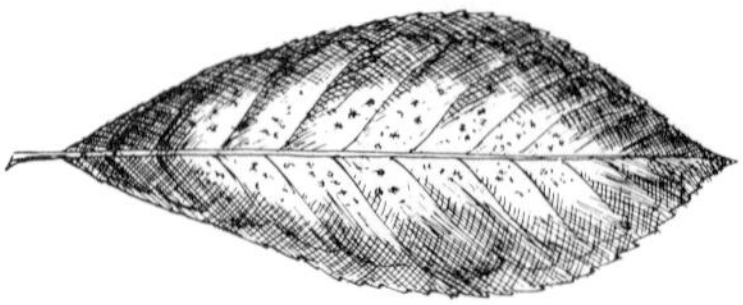

damaged leaf

Kakothrips robustus (Uzel)

(= *Frankliniella robusta*)

Common name. Pea Thrips
Family. Thripidae
Hosts (main). Pea and broad bean.
(alternative). Other beans.
Damage. Feeding of nymphs and adults causes silvering and distortion of young pods, and flowers fail to develop; heavy attacks may stunt the plants, and severe early attacks destroy the flowers so no pods are produced.
Pest status. Mostly a pest of pea and broad bean in gardens and market gardens, but sometimes damaging to field peas.
Life history. Eggs are laid in the tissues of the stamen sheaths, flower parts or young pods; hatching takes about nine days.

Nymphs are yellow-orange and feed initially in the flowers and then later on the surface of the young pods; development takes about three weeks, then the large nymphs descend to the ground and pupation takes place in the soil. The nymphs in the soil overwinter, and in May turn into 'prepupae' and then pupae, and adults emerge in May/June.

The adults are shiny black in colour, about 1.7 mm long, and really are indistinguishable from *Frankliniella* and other adults. The population peak is about mid-June, but thrips are to be found on the peas until the end of July. There is only one generation per year.

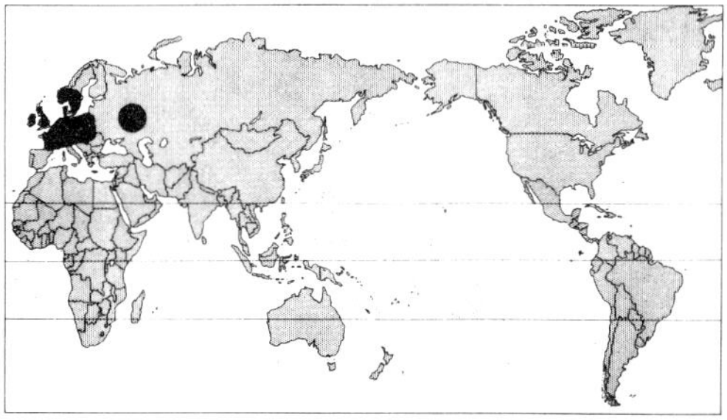

Distribution. Widespread throughout the British Isles and much of Europe.
Control. The insecticides generally recommended against this pest are sprays of dimethoate or fenitrothion, and also malathion.

Formerly the chemical most successfully used was DDT.

Damaged pea pod with thrips nymphs

Thrips tabaci Lind.

Common name. Onion Thrips
Family. Thripidae
Hosts (main). Onions and leek.

(alternative). Tobacco, tomato, cotton, pineapple, pea, beet, brassicas, and many other crops; polyphagous, with more than 300 species of plants recorded.

Damage. Leaves of attacked plants silvered and flecked, sometimes wilting. Heavy attacks may destroy seedlings; occasionally entire crops may die. An important virus vector.

Pest status. A polyphagous pest on many crops throughout the world; vector of virus diseases of tobacco, tomato, pineapple, and some other crops. Generally a serious pest in N. America.

Life history. Eggs laid in notches in epidermis of leaves and stems of young plants; hatching takes 4–10 days.

Nymphs are white or yellow and the three instars (two moults) take only about five days to develop, then they enter the soil to pupate, which takes 4–7 days usually.

Adults are small, yellowish-brown in colour, with dark transverse bands across the body; they are only about 1 mm in length.

One generation develops in as short a time as 2–3 weeks; in the tropics there may be ten or more generations per year, but in the UK there are generally only two.

Distribution. A completely cosmopolitan species, extending as far north as Scandinavia, Finland and Canada (CIE map no. A.20).

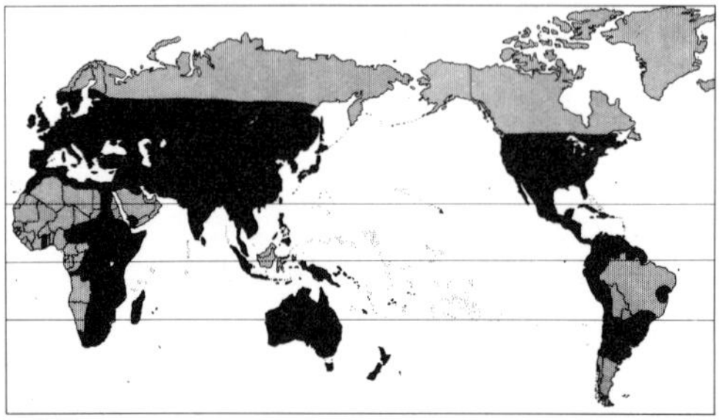

Other important species of *Thrips* include the following:

Thrips angusticeps Uzel – (Cabbage Thrips) on brassicas, sugarbeet, peas, apple and pear flowers; UK and Europe.

Thrips flavus Schrank – (Honeysuckle Thrips) also on *Rubus*; Europe, Japan.

Thrips atratus Haliday – (Carnation Thrips) Europe, including the UK.

Thrips linarius Uzel – (Flax Thrips) on flax; Europe (not UK) and USSR.

Thrips major Uzel – (Rose Thrips) from many flowers, especially Rosaceae; UK and most of Europe.

Thrips nigropilosus Uzel – (Chrysanthemum Thrips) on chrysanthemum, flax, oats, wheat; Europe, Japan, New Zealand, Canada, USA (CIE map no. A.416).

Four other species of *Thrips* are recorded as crop pests in Japan.

Control. Formerly DDT sprays were used to control this pest; DDT is sometimes still recommended, and other chemicals recommended include malathion, fenitrothion, phosfolan, and some of the other chemicals listed on page 258.

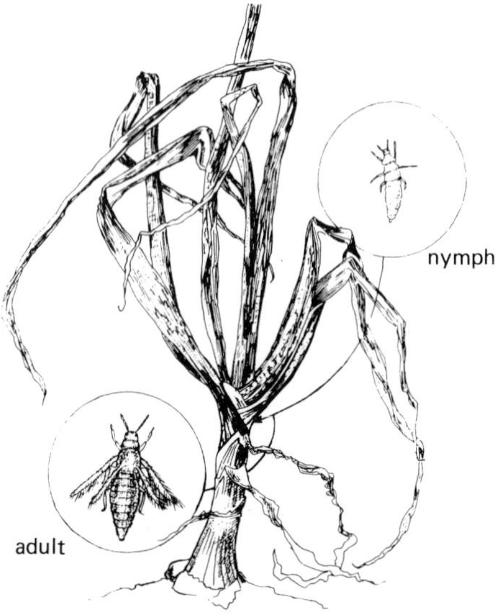

infested onion plant

Thrips (order Thysanoptera)

A large worldwide group with at least 5000 species, well represented in the tropics but most abundant in temperate regions where their extensive summer dispersal flights earn them the name of 'thunder-flies'.

Many species are phytophagous and feed on the leaves of their host plants, but some infest flowers and affect fruits, and a few make galls. A few are predacious and eat other insects, and many are fungus-feeders and to be found in leaf litter. Some of the phytophagous species are polyphagous as well as cosmopolitan, but others are more restricted in their diets and distributions. Some species are important as vectors of virus diseases, especially some of those that occur in glasshouses in Europe and N. America.

Adults are generally black or brown in colour, and most are quite similar in appearance, so field identification is very difficult in most cases. Reliance on host plant data usually gives a good indication as to identity. The nymphs are often red, orange, or yellow in colour in contrast to the drab adults. Pupation almost always takes place in the soil; *Heliothrips haemorrhoidalis* is unusual in that pupation takes place on the host plant.

Pest species occur in the two largest families. The Thripidae are almost all sap-feeding and there are over 160 genera. Eggs are inserted into the host plant tissues using the saw-like ovipositor. Most of the pest species belong to this family. The Phlaeothripidae contains over 300 genera and a few are pests by their actions of leaf-rolling and leaf-distortion. In some species the adults guard the egg-mass on the leaf surface, and also remain with the nymphs after hatching.

Important pest species of thrips

In addition to the species already mentioned in this chapter, the following are species of some importance as crop pests, and most of them are temperate in distribution.

Thripidae

Anaphothrips obscurus (Mull.) – (Grain Thrips) on cereals; Canada, USA and Japan.

Aptinothrips rufus (Gmel.) –
Aptinothrips stylifer (Trybom) – } on oats, barley and wheat; in Europe, Canada and USA.

Aptinothrips spp. – (grass thrips) on grasses and cereals; Europe, Canada, USA.

Baliothrips biformis (Bagnall) – (Rice Thrips) on rice; tropical Asia, Taiwan, and Japan (also UK, Romania, Brazil) (CIE map no. A.215).

Baliothrips minutus (van Dev.) – (Sugarcane Thrips) USA.

Caliothrips spp. – on cotton and many other hosts in the tropics.

Hercinothrips bicinctus (Bagnall) – (Banana Thrips) on bananas in tropics and greenhouse crops in UK, Europe and USA.

Hercinothrips femoralis (Reuter) – (Banded Greenhouse (Sugarbeet) Thrips) polyphagous on field crops and ornamentals in greenhouses; cosmopolitan (CIE map no. A.402).

Limothrips cerealium (Halstead) – (Grain Thrips) wheat, maize, rye, oats, barley, grasses and citrus; cosmopolitan (CIE map no. A.245).

Limothrips denticornis (Halstead) – (Barley Thrips) on cereals; Canada.

Scirtothrips aurantii Faure – (Citrus (Tea) Thrips) polyphagous; from Egypt and Africa (CIE map no. A.137).

Selenothrips rubrocinctus (Giard) – (Red-banded Thrips) polyphagous; pantropical (CIE map no. A.136).

Taeniothrips inconsequens (Uzel) – (Pear Thrips) in flowers of pear, plum, apple; Europe and USA.

Taeniothrips simplex (Morison) – (Gladiolus Thrips) Europe; on Liliaceae.

Taeniothrips sjostedti (Trybom) – (Bean Flower Thrips) polyphagous in flowers; Malta and Africa.

Phlaeothripidae

Gigantothrips elegans – (Giant Fig Thrips) on leaves of *Ficus*; Asia.

Gynaikothrips ficorum – (Banyan Leaf-rolling Thrips) on *Ficus retusa* (California) and *F. microcarpa* (tropical Asia).
Gynaikothrips kuwani – (Cuban Laurel Thrips) on several plants; Asia and USA.
Haplothrips leucanthemi Schrank – (Clover Thrips) Canada.
Haplothrips tritici – (Wheat Leaf-rolling Thrips) on wheat; Europe.
Hoplandothrips spp. – Coffee Leaf-rolling Thrips, etc.; Africa
Liothrips oleae – (Olive Thrips) Mediterranean Region.
Liothrips spp. – many ornamentals; USA, Asia.

Control of thrips

Parasitic Hymenoptera are of some importance, especially some species of Eulophidae.

Ploughing after crop harvest will help to kill pupae in the soil and adults hiding in leaf litter.

Early planting helps to avoid some seedling infestations. Chemical application as a means of virus control is generally unsuccessful. For chemical control of the thrips themselves the following treatments are recommended:

(1) Soil application of DDT or HCH, or granules of aldicarb or phorate.
(2) Foliar sprays of DDT, carbaryl, fenitrothion, or dimethoate.
(3) In glasshouses use smokes or fogs of DDT, HCH, parathion, nicotine or pirimiphos-methyl; or sprays of diazinon, dichlorvos or resmethrin.
(4) A recent trend in the UK, as part of the general biological control programmes in commercial glasshouses, is to spray a special formulation of deltamethrin and polybutene ('*Thripstick*') on to the floor covering to kill the thrips as they fall to the ground to pupate.

An extensive list of insecticides, dosages and methods of application, effective against thrips (but somewhat out of date now), is given as appendix 5 in Lewis (1973), pages 304–6.

Order

Coleoptera

(beetles)

Families
Carabidae (ground beetles)
Silphidae (carrion beetles)
Scarabaeidae (chafers; white grubs)
  Cetoniinae (rose chafers)
  Melolonthinae (cockchafers)
  Rutelinae (flower or June beetles)
Buprestidae (jewel beetles; flat-headed borers)
Elateridae (click beetles; wireworms)
Bostrychidae (black borers)
Nitidulidae (sap beetles)
Cucujidae (flat bark beetles)
Cryptophagidae (cryptophagid beetles)
Byturidae (fruitworm beetles)
Coccinellidae (ladybird beetles)
Tenebrionidae (flour beetles)
Meloidae (blister beetles)
Cerambycidae (longhorn beetles)
Bruchidae (bruchid beetles)
Chrysomelidae (leaf beetles)
  Cassidinae (tortoise beetles)
  Chrysomelinae
  Criocerinae
  Eumolpinae
  Galerucinae
  Halticinae (flea beetles)
Apionidae
Brenthidae (brenthid weevils)
Curculionidae (weevils proper)
Scolytidae (bark beetles)

Harpalus rufipes (Deg.)

(= *Ophonus rufipes* Deg.)

Common name. Strawberry Seed Beetle
Family. Carabidae
Hosts (main). Strawberry
(alternative). Weeds, such as fat-hen, etc.

Damage. Adult beetles bite the seeds from the fruits, spoiling the appearance and market value of the fruits; some seeds are eaten on the fruit which then becomes holed. Similar damage is done by feeding linnets.

Pest status. A sporadically serious pest of strawberry in parts of the UK and Europe, in that the quality of the fruit is reduced. Several different species of Carabidae are involved as strawberry pests; see below.

Life history. Eggs are laid in summer in the soil, amongst weeds; each female lays 10–20 eggs. The larvae are omnivorous and feed on other insects and seeds of several weeds, including fat-hen. The larvae overwinter in the soil and leaf litter, and pupation takes place the following spring.

The adult is a largish black beetle, typically carabid in appearance, some 12–18 mm in length; the body colour is dull black with a golden-grey pubescence, and the legs are reddish. A very active beetle that runs rapidly when disturbed: it is nocturnal in habits and hides during the day. Some adults survive and hibernate over winter as well as the large larvae.

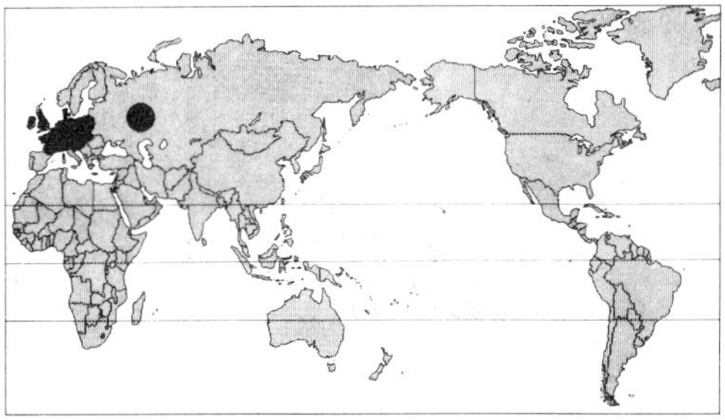

Distribution. In the UK and Europe there are actually about six species of strawberry ground beetles:

Pterostichus cupreus (L.)
Pterostichus madidus (F.)
Pterostichus melanarius (Illiger)
Abax parallelepipedus (P. & M.)
Nebria brevicollis (F.)
} All formerly in the genus *Feronia*.

Superficially all these beetles look quite similar, and their leg colour is rather variable, from reddish to black.

These species tend to eat the fruit rather than the seeds and their damage resembles that done by slugs, but they are seldom serious pests.

Control. Slug pellets (methiocarb) broadcast over the rows are used against these pests, as also is a bait made of malathion in crushed oats.

In some regions this is one of a group of carabids that are imported general insect predators in agricultural crops, and sometimes this species feeds heavily on cereal aphids.

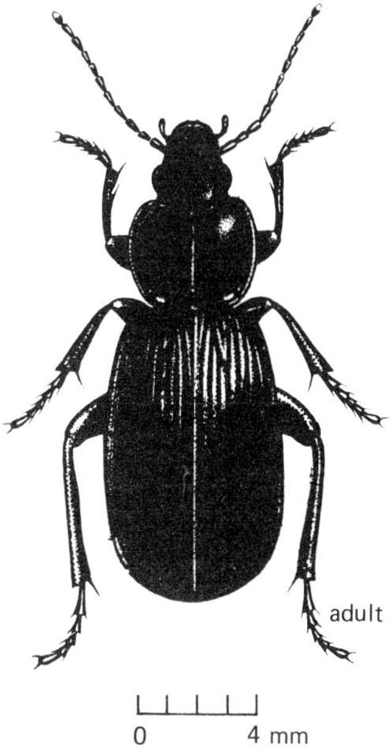

Aclypea opaca (L.)

Common name. Beet Carrion Beetle
Family. Silphidae
Hosts (main). Sugar beet, mangolds.
(alternative). Other Chenopodiaceae, and wheat, barley, oats, cabbage, swede, potato.
Damage. Adults and larvae feed on the leaves of beet seedlings; larvae also eat roots in the soil. The foliage is smeared with black excrement. Heavy attacks can result in extensive defoliation.
Pest status. In the UK damage is seldom serious, but this is a major pest on the Continent, and damage may be extensive.
Life history. Eggs are laid in the soil from April to June; hatching takes 5–9 days. The larvae feed actively for about three weeks; they are flattened in shape with lateral expansions of the dorsal tergites, and they resemble woodlice, some 10–12 mm long.

Pupation takes place in an earthen cell in the soil, and requires about 14 days.

Adults are largish black beetles, about 10 mm in length, oval and black with ridged elytra. Adults and larvae are often found feeding together on the leaves of the host plants. Adults hibernate in litter or under stones during the winter, becoming active again the following spring.

It is thought there is only one generation per year.

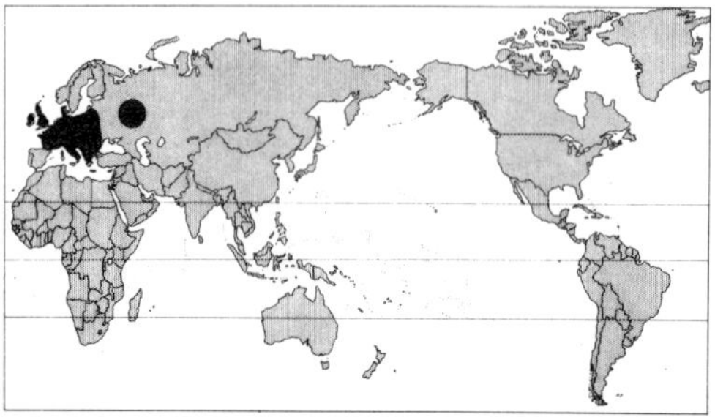

Distribution. Found throughout the British Isles and most of Europe.
Control. When infestations and damage warrant control measures, sprays of DDT, parathion, or HCH have been effective.

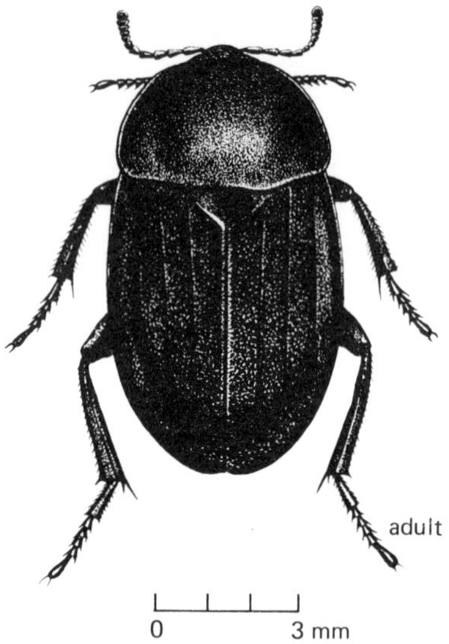

Cetonia aurata (L.)

Common name. Rose Chafer
Family. Scarabaeidae (Cetoniinae)
Hosts (main). Rose flowers damaged by adult beetles.
(alternative). Larvae in soil feed on grass roots; also roots of strawberry and raspberry.
Damage. Adults eat the flowers of rose; other members of this subfamily eat other flowers, of both crop plants and ornamentals. Larval damage to grassland is usually slight as populations are small; not a common species.
Pest status. Not usually a serious pest; adult damage is unsightly, and bare patches in grassland caused by the larvae tend to be obvious but uncommon.
Life history. Globular white eggs are laid in the soil in early summer.

The larvae are typical chafer grubs (see *Melolontha* spp.) but the reddish bristles on the body, and a brown patch on the sides of the prothorax, distinguish this species from *Melolontha*; larval development takes 2–3 years, and full-grown larvae measure 25–36 mm. Pupation takes place in late spring, and the adults appear in late May and early June.

Adults are brilliant metallic golden-green, and very attractive insects, 14–20 mm in length, and to be seen flying over grassland or around roses in bright sunlight; the underside of the body is reddish and the legs are black; the antennae have only three plates in the club.

One generation takes 2–3 years.

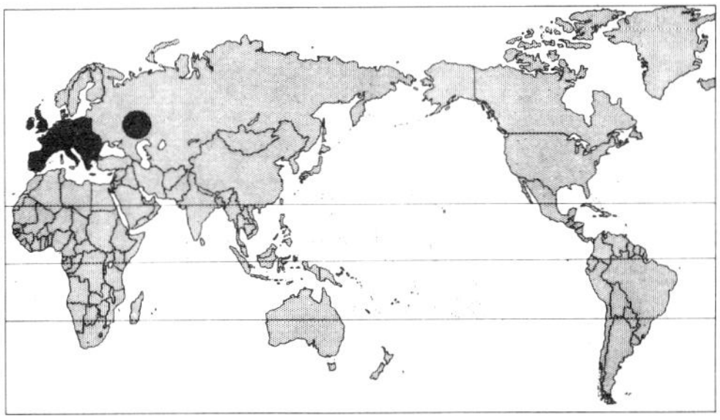

Distribution. Occurs throughout the UK and Europe. Other genera of similar appearance occur throughout the rest of the world.

A second species occurs in the UK: *Cetonia cuprea* F.

Control. For control measures see page 269.

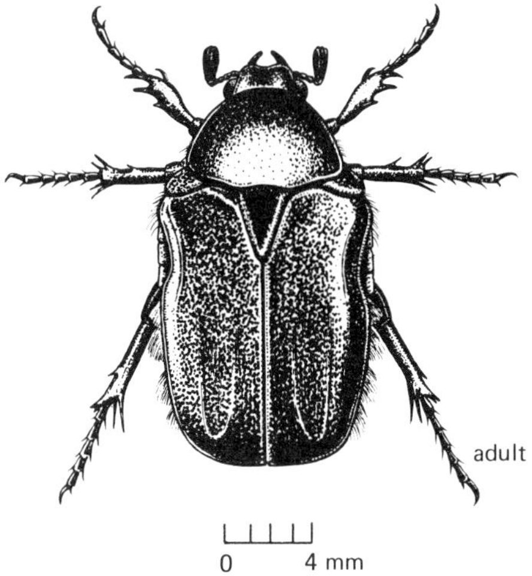

Amphimallon spp.

(*A. majalis* (Raz.))
(*A. solstitialis* (L.))

Common name. European Chafer; Summer Chafer
Family. Scarabaeidae (Melolonthinae)
Hosts (main). Pastures and nursery stock.
(alternative). Roots of cereals, legumes, also conifers.
Damage. Larvae in soil feed voraciously on roots of many crop plants; plants may be completely killed.
Pest status. Not important in the UK but serious pests on the Continent and in Canada, where infestations may be heavy (in 1962 in Ontario up to 450 larvae per square metre) and damage to crops extensive.
Life history. Eggs are laid in the soil during June/July, and require several weeks to hatch.

The young larvae feed for a while in the autumn, and then become inactive for the winter; feeding resumes in the spring and they usually become fully grown by the second autumn, when they hibernate again. Fully grown larvae are rather small for chafer grubs, only 24–30 mm long.

Pupation takes place in an earthen cell in the soil during the second spring, and adults emerge in June.

The adult is a rather small (16 mm long), reddish-brown chafer, with a darker head and thorax, distinctly 'hairy', and antennae with only three lamellae. They are long-lived and feed on foliage of trees usually.

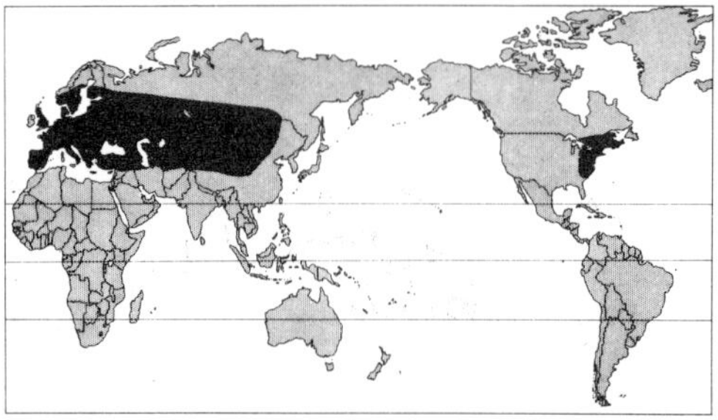

Under suitably mild climatic conditions the life-cycle may be completed in one year, but in the UK it requires two years, and should food be scarce it can require three years for complete development.
Distribution.
Amphimallon majalis is found in Europe (not UK) and Canada (Ontario) and eastern USA (CIE map no. A.371). *Amphimallon solstitialis* occurs throughout Europe and Asia (CIE map no. A.391).
Control. See page 269 for control measures.

A. solstitialis

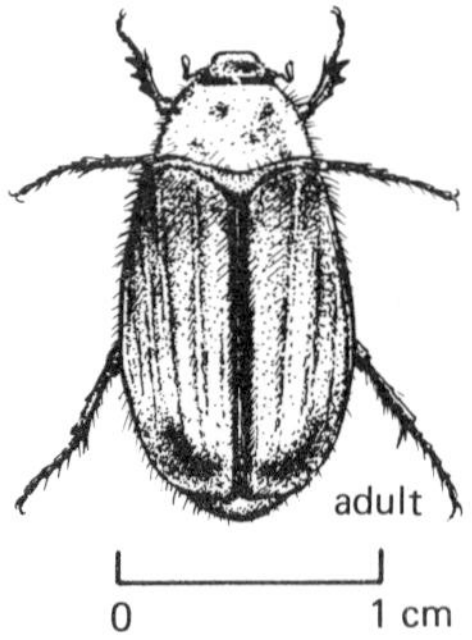

Melolontha spp.

(*M. hippocastani* F.)
(*M. melolontha* (L.))
(*M. japonica* Burm.)

Common name. Cockchafers
Family. Scarabaeidae (Melolonthinae)
Hosts (main). Adults damage apple fruitlets, and leaves of some trees.

(alternative). Larvae in soil eat roots of many different plants; polyphagous.

Damage. Adult damage to apple fruitlets consists of small bites out of a series of adjacent fruits; larval damage to the roots of seedlings may destroy the plants; damage to small plants such as strawberry may be serious, especially if insect populations are large.

Pest status. These are common and widespread insects, sometimes to be found in very large numbers when they are serious pests; obviously they are more damaging to young or small plants. Adult damage to apple fruits is often serious.

Life history. Eggs are white and globular, and laid in soil in batches of 12–30, at a depth of up to 10 cm, under close plant cover. Each female lays about 60–70 eggs. Hatching takes 3–5 weeks in the summer.

Larvae are white-bodied, plump, and typically scarabaeiform in shape. Generally, it is thought they do not move about much in the soil, mostly just a vertical movement according to seasonal conditions; so large populations develop where the females laid their eggs.

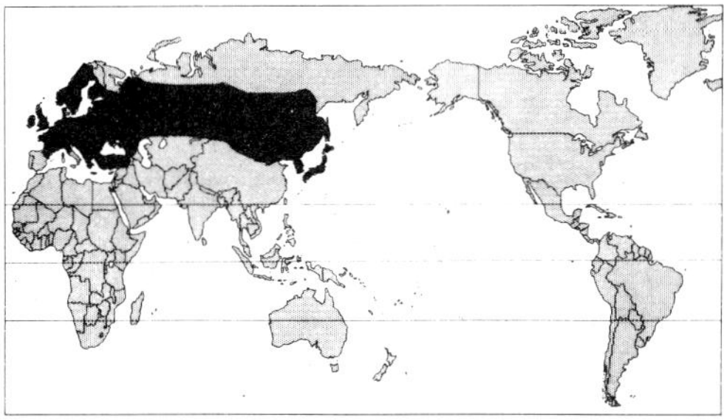

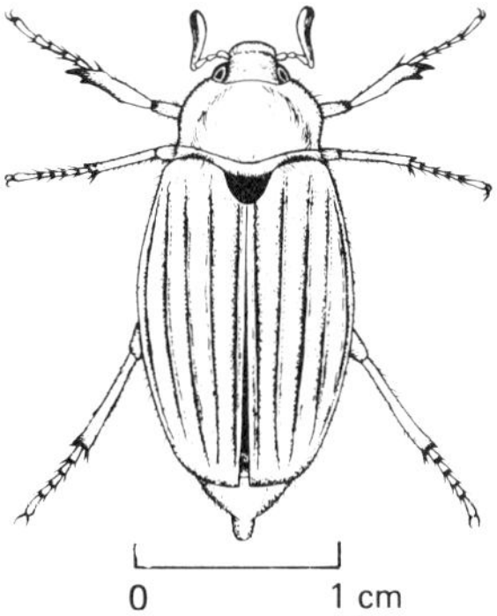

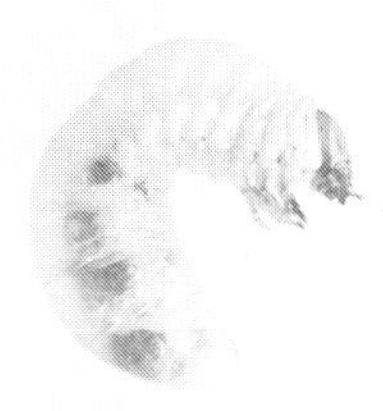

larva

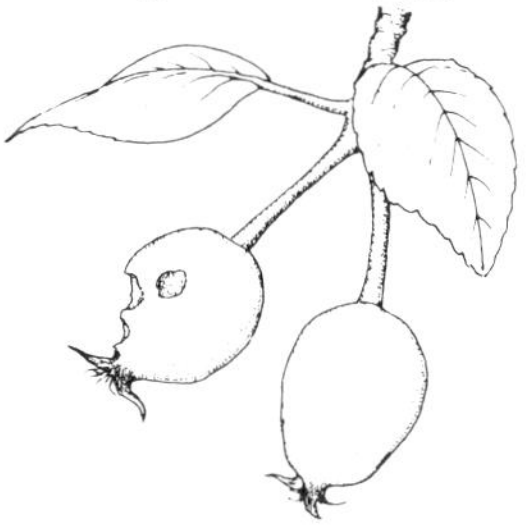

Damage to young apples and leaves by adult chafers

Larval development is slow; during the first year the small larvae cause little damage, but during the second and third years their root feeding seriously affects tree seedlings and herbaceous plants, including Gramineae. Fully grown larvae reach a length of 44 mm, after a developmental period of 2–3 years. In the autumn of the second, third or even fourth year, the mature larvae burrow quite deep into the soil and pupate in an earthen cell. Often adults are formed by the end of the autumn, but they remain inside the cell over winter and emerge from the soil in the spring (May and early June).

The adult is a large brown beetle, about 25 mm long, with a dark head and thorax and reddish-brown ridged elytra. It is nocturnal and flies at night to feed on tree foliage and sometimes apple fruitlets. Mating takes place after emergence and then there is a pre-oviposition period in the female of about three weeks.

This genus is actually associated with deciduous woodland and is not ecologically a grassland inhabitant, although it can be found in young sand-dunes. Most damage is done to crops grown in the vicinity of woodlands.

In the UK the species generally requires three years for complete development.

Distribution. *Melolontha hippocastani* occurs in northern UK, northern Europe and Asia, almost to Japan (CIE map no. A.194).

Melolontha melolontha is recorded from throughout the UK and from Europe (CIE map no. A.193).

Melolontha japonica is recorded from Japan.

In the Old World the most important temperate genus of cockchafer is *Melolontha* (supported by *Amphimallon*); the New World equivalent is the very large genus *Phyllophaga* which occurs as at least a couple of dozen pest species from northern Canada, throughout the USA, and C. America, down into S. America. Morphologically they closely resemble the two genera here illustrated.

Control. For control measures see page 269.

Phyllopertha horticola (L.)

Common name. Garden Chafer
Family. Scarabaeidae (Rutelinae)
Hosts (main). Adults damage flowers and opening buds; also fruitlets of apple and pear.

(alternative). Larvae in soil eat roots of cereals, nursery stock, grass, etc.

Damage. Essentially a grassland species; adults eat flowers and young leaves of fruit trees, shrubs and herbaceous plants; larvae eat roots of grasses, cereals, and other plants growing in ploughed-up grassland. Adults may eat holes in fruitlets of apple and pear, as does *Melolontha*.

Pest status. This is a very abundant species in the UK, but not too serious a pest. The closely related *P. nazarena* is quite a serious pest of cereal crops in the eastern Mediterranean Region.

Life history. Adults emerge in late May/June, often in large swarms, and can be seen flying on warm days; the males fly looking for females. After mating, the female burrows back into the soil and lays her eggs. Thus infested pasture becomes reinfested each summer. Eggs are laid at a depth of 3–5 cm in the soil, and hatch in 25–40 days. The egg-laying period is from 1–23 days.

The larvae feed on grass roots until late autumn; when fully grown the larvae are only about 12 mm long, and pale brown in colour. In late autumn the larvae burrow deeper into the soil and construct an earthen cell for hibernation and pupation. The different larval instars take approximately three weeks, four weeks and 8–10 weeks. Large larvae pass into diapause about the end of October and hibernate deep in the soil until pupation at the end of April/May.

The adults are small chafers, only about 9 mm long, with a metallic green head and thorax and reddish-brown elytra.

There is only one generation per year.

Distribution. *Phyllopertha horticola* is found throughout the UK and most of Europe.

Phyllopertha nazarena Mars, is the Nazarene Chafer of the eastern Mediterranean.

Phyllopertha pubicollis Waterhouse is a common pest species found in China and eastern Asia.

Control. For control measures see page 269.

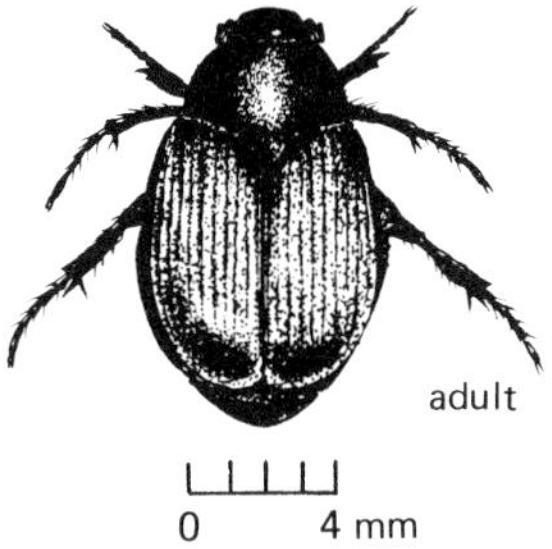

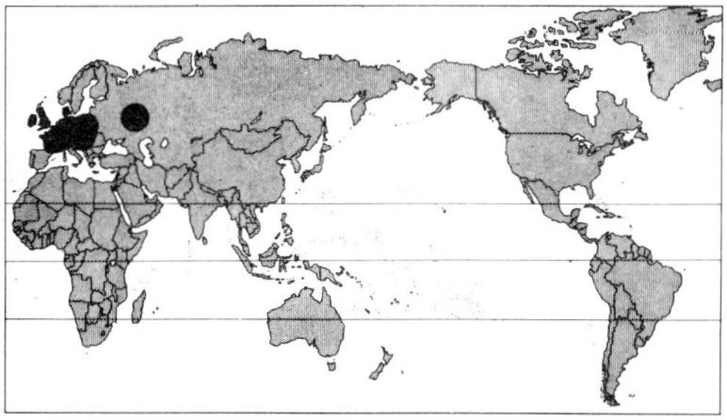

Popillia japonica Newman

Common name. Japanese Beetle
Family. Scarabaeidae (Rutelinae)
Hosts (main). Apple, cherry, plum, raspberry, strawberry, clovers, maize, rose.

(alternative). Many trees, shrubs, herbs; polyphagous; more than 250 plants recorded as hosts.

Damage. Adults eat foliage, flowers and fruits; on flowers and leaves holes are eaten (between the veins on leaves); only ripe fruits are eaten by the adults. Larvae in the soil eat roots of grasses, and also damage a wide range of plants.

Pest status. Adults are serious foliage (and flower) pests on a wide range of cultivated plants, and the larvae are serious pasture pests.

Life history. Eggs are laid in the soil (several centimetres deep), preferably in pasture, and at above 18 °C development takes place; hatching usually requires about 14 days. Oviposition takes place in the autumn.

The young larvae start feeding, but as the weather cools they burrow deeper into the soil, until on the approach of winter they cease feeding and build their hibernation cells. They require temperatures (soil) of above 18 °C for development but can survive cold northern winters if the ground is protected by a cover of snow. The hibernation cell is about 5–15 cm deep. In late spring, as the ground warms, the larvae resume activity and feeding and tunnel towards the surface. They feed in the surface layers of the soil. Fully grown larvae are whitish with a brown head, typically scarabaeiform, about 25 mm long. The shape of the anal slit and surrounding bristles distinguishes the larvae from other local chafers.

Pupation takes place in another earthen cell, about 10 cm deep; there is a period of prepupal quiescence lasting about ten days, then the pupal period proper for 8–20 days.

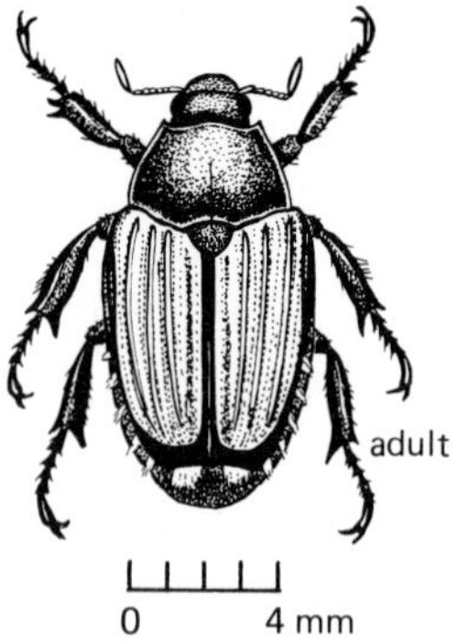

Popillia beetle on flower of *Melastoma* in China

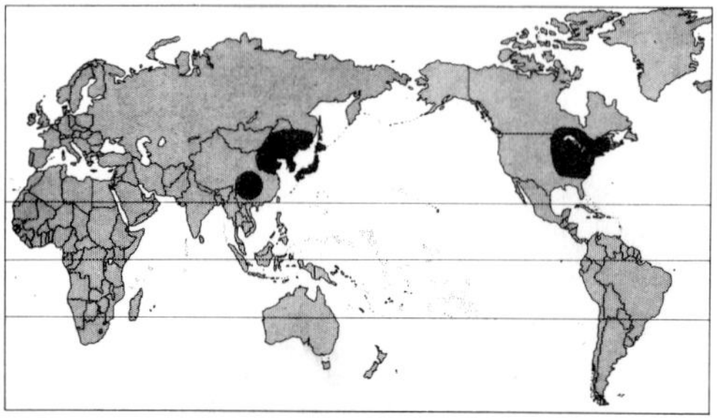

Adults emerge in June, July and August, and fly in warm sunny weather, often in very large numbers; their feeding activity is closely correlated to sunny conditions. The adult is a very pretty beetle, about 11 mm long, with bright metallic green head, thorax and legs, and reddish-brown elytra. It closely resembles *Phyllopertha horticola*, but differs in having green (not brown) legs and tufts of white bristles along the abdomen edges.

Distribution. Japanese Beetle is endemic to Japan, Korea and China, and was accidentally introduced into eastern USA in 1916 where it has spread widely and has become a very serious pest. At present it has not spread into Europe (CIE map no. A.16).

Other species of *Popillia* are endemic in China and its environs, also in India and parts of Africa.

Control. For control measures see page 269.

Chafers and chafer (white) grubs
(Coleoptera: Scarabaeidae)

This large family, with more than 19000 species, falls into several distinct groups, here regarded as subfamilies but sometimes given status as separate families. The adults are rather similar in appearance, but the larvae which are the major pests are almost indistinguishable from each other, though easily recognized as a group; they are fleshy grubs with a swollen abdomen, usually adopting a **C**-shaped position, with a well-developed head capsule and large jaws, and thoracic legs. This typical larval shape is known as 'scarabaeiform'. Some larvae live in rotting wood or rotting vegetation or animal dung, but the majority live in soil and eat plant roots and these pest species are referred to as 'white grubs' or 'chafer grubs'. The feeding larvae generally live at depths of 2–10 cm in the soil, but hibernation may take place as deep as 110 cm in Canada where the winter is very cold. Pasture populations of larvae are regularly more than 30 per square metre, and up to 900 per square metre are recorded in Canada. Some species are restricted in their diet to roots of Gramineae, but many are polyphagous; many crops are damaged as well as pastures and grass leys; damage to qualify turf such as golf courses, playing fields and lawns may be very costly. In the tropics most damage is done to sugarcane and cereal crops; in temperate regions quality turf and root crops may be the most severely damaged, although many crops are attacked. Damage to root crops consists of holes eaten into the tuber or root (see page 99), usually rather wide and shallow holes. Crops most at risk from chafer grubs are those which are grown in newly ploughed grassland, and both wireworms and leatherjackets are to be expected also.

Subfamily Cetoninae (rose chafers)

Largely a tropical group (2600 species); these beetles have a brightly coloured, flattened body, and the elytra are characteristically emarginated laterally. In flight the elytra are only raised slightly and the wings project through these emarginations. The beetles are usually diurnal in habit and fly to flowers (usually on trees and shrubs) to feed on nectar and pollen with their weak mouthparts. They will also feed on soft over-ripe fruits, but because their very weak mouthparts are almost nonfunctional the adults are not pests themselves.

Pest species are found in the subtropical Asiatic genus *Protaetia* and the temperate *Cetonia*.

Subfamily Coprinae (=Scarabaeinae) (dung beetles)

These black rounded beetles are not crop pests, although common on pastures, and neither are their coprophilous larvae.

Subfamily Dynastinae (rhinoceros and elephant beetles)

These are tropical rain forest beetles, many large in size, all black in colour, and nocturnal in habits. The larvae generally live in rotting vegetation where they feed saprophagously, and the adults do damage by eating the foliage of trees and palms, though a few attack Gramineae. The main pests are the many species of *Oryctes* (rhinoceros beetles) that damage the crowns of palms (Palmae), and the giant *Xylotrupes* feeds on the foliage of some tropical trees. *Heteronychus* includes the cereal beetles of Africa and Australia that are also pests of sugarcane.

Subfamily Melolonthinae (cockchafers)

Usually dull brown beetles with fat rounded bodies; elytra held vertically in flight; nocturnal, and fly to lights at night; claws of hind legs equal in size and immovable. The larvae are serious crop pests in the soil. The adults have strong biting mouthparts and eat leaves of trees and shrubs and bite pieces out of young fruits such as apples.

Some important tropical pest species are in the genera:

Dermolepida – (Sugarcane Beetles; Greybacks) Australia.
Lepidiota – (Sugarcane Whitegrubs) tropical Asia and Australasia.
Leucopholis – (Chafer (White) Grubs) Philippines.
Schizonycha – (Chafer Grubs) polyphagous; throughout Africa.

Some of the more important temperate pest species are in the genera *Amphimallon* and *Melolontha*, as well as the following:

Holotrichia – (Chafer Grubs) China and parts of Asia.
Hoplia philanthus – (Fuessly) – (Welsh Chafer) Wales and western England.
Phyllophaga anxia Le. – (Common June Beetle) polyphagous; Canada.
Phyllophaga fusca (Froel.) – (Northern June Beetle) Canada.
Phyllophaga spp. – (June Beetles, etc.) at least 20 species are pests throughout Canada, USA, C. and S. America; adults eat leaves of trees and field crops, and larvae polyphagous.
Serica brunnea (L.) – (Brown Chafer) polyphagous, but basically a woodland species, not grassland; UK and northern Europe.
Serica orientalis Mots. – (Oriental Brown Chafer); China.
Serica spp. – (Brown Chafers) polyphagous; India and parts of Asia.

Subfamily Rutelinae (flower beetles; June (May) beetles)

The common English name refers to the time of flying adults (to lights), and so there is variation in different parts of the world: in Australia the several important species of *Anoplognathus* are referred to as Christmas beetles. The body is smooth, oval and shiny; occasionally bright metallic; the hind legs have thickened tibiae and long movable claws of unequal length. Adults are mostly nocturnal and will fly to lights at night; they have well-developed mandibles and feed on foliage of trees and shrubs as well as ornamental herbaceous plants, eating both leaves and flowers. The genus *Popillia* is active during daylight hours: beetles can be found both feeding and flying by day.

Common pests include *Phyllopertha* and *Popillia*, as well as the following:

Adoretus spp. – (Flower Beetles) adults and larvae polyphagous; pantropical in the Old World and now established in the USA.
Anomala spp. – (Chafers, etc.) polyphagous; found throughout Africa, Asia (both tropical and temperate), and in the USA (16 pest species recorded in Japan).
Anoplognathus spp. – (Christmas Beetles) polyphagous; Australia.
Lachnosterna spp. – (White Grubs) polyphagous root pests; Canada, USA, West Indies.

Control of chafers and chafer grubs

Adults are generally difficult to control because of their mobility and nocturnal activity, but *Popillia* adults feed in daylight and can be sprayed whilst on the foliage of the plants. Some population control of adults has been achieved in places using ultra-violet light traps at night.

Larvae are the main pests, so most control measures are directed against them. Several different approaches have been used in the past, with some success, and these aspects need to be borne in mind in any future IPM programme.

(1) Natural control – predators of some importance include wasps in the families Scoliidae and Tiphidae (Digger Wasps), some beetles (Carabidae, Histeridae), many different species of birds, moles and other Insectivora, and in North America skunks may be serious predators. Deliberate introduction of predatory insects to control chafer grubs has met with mixed success. Parasites include several species of Tachinidae, several nematodes, a bacterial disease called 'milky' disease, and the fungus *Cordyceps*.

(2) Cultural control – deep ploughing will expose larvae and pupae to both predators (such as bird flocks following a plough) and adverse weather conditions. Crop rotation may be used to avoid the more susceptible crops while there are large numbers of larvae still in the ground. The avoidance of planting susceptible crops in newly ploughed grassland. Rototilling the soil will physically kill some of the larvae and pupae.

(3) Chemical control – to date the only really successful insecticides for killing chafer grubs has been the organochlorines, aldrin, DDT, dieldrin, HCH (BHC), heptachlor and chlordane, but resistance to these chemicals has long been established in both USA and some parts of Europe.

In the USA the organophosphorous compound isofenfos has given good results at 2.2 g a.i./ha for control of chafer grubs on golf course turf, and it was expected to be registered for use on corn and turf-grass in the USA in 1981/82.

Agrilus spp.

Common name. Bark Borers (Jewel Beetles)
Family. Buprestidae
Hosts (main). Citrus, apple, pear, peach, walnut, almond, hazelnut, rose.

(alternative). All stone fruits; most other fruit trees, chestnut, oak, poplar, and many other trees, also raspberry and currants.

Damage. The larvae tunnel under the bark of woody plants, sometimes on the tree trunks, sometimes on twigs and thin branches which may be girdled and die. It is thought that they mostly attack trees already weakened or stressed in some way.

Pest status. Individually each species is only occasionally serious, but the genus taken worldwide is abundant, attacks a very wide range of plants, and is quite serious.

Life history. Eggs are laid in crevices in the bark, and hatch after 2–3 weeks; the young larvae make long winding tunnels in the inner bark; in a tree trunk the tunnel may eventually reach a metre in length; in twigs and branches girdling often results. Mature larvae have the characteristic flattened body with large thorax (especially prothorax) and well-developed mandibles, that earns the group their name of 'flat-headed borers'. When burrowing, these larvae do not have frass-holes (as do Cerambycidae) but they pack the frass tightly into the tunnel behind them (see *Chrysobothris* page 271). In the warmer parts of the world larval development takes less than one year for these small species, but in the cooler temperate regions development takes two years. Fully grown larvae measure 12–25 mm, according to species.

Pupation takes place in the gallery, often quite deep in the heartwood or in the centre of the twig; often occurring in the spring or early summer.

Adults emerge between May and September, but with a peak in June/July according to species and geographical location: they are small bronze or greenish beetles, 7–20 mm in length according to species; they may live for several weeks and feed on the tree foliage.

The life-cycle usually takes two years to complete, but the Citrus Bark Borers and the Cane Borer of the USA develop in one year.

Distribution. Worldwide there is a total of 700 species of *Agrilus* spread throughout Europe, Asia and N. America, and parts of Africa, although some species are only recorded from wild hosts. There are none recorded in Australia.

Agrilus rubicola Ab. is the Currant (Cane) Borer of USA and Canada.

Control. Good husbandry reduces the likelihood of attack, as these pests appear to prefer host trees under physiological or water stress. Pruning of infested twigs and branches helps to reduce pest populations.

On *Citrus* dimethoate as a foliar spray has been reported to be effective. For contact insecticidal use it is necessary to time sprays to catch the emerged adults before they have time to oviposit; DDT and dieldrin have been effective.

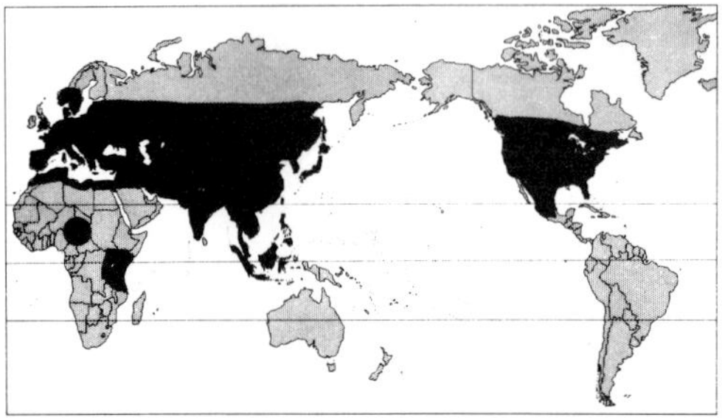

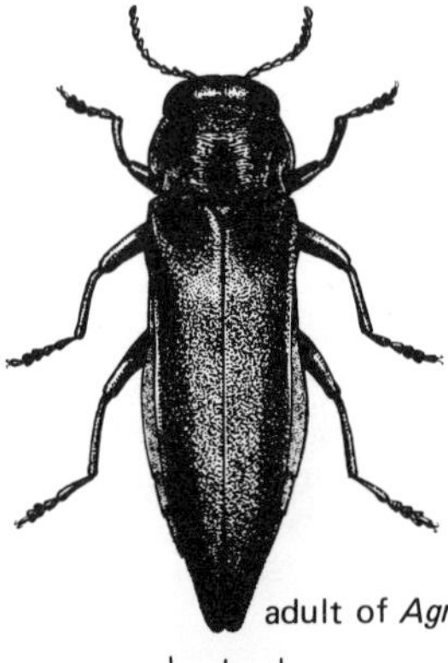

adult of *Agrilus viridis*

0 2 mm

Chrysobothris spp.

Common name. Flat-headed Borers
Family. Buprestidae
Hosts (main). Apple, rose.
(alternative). Most deciduous trees, both fruit and forest; many woody shrubs; polyphagous.
Damage. Larvae tunnel under the bark, in the inner phloem; smaller trees may be girdled; transplanted trees may be severely weakened. Adults damage the foliage to some extent by leaf-eating.
Pest status. Several species (see below) are serious pests of fruit and other trees (oaks), especially in N. America, but the genus is cosmopolitan and polyphagous.
Life history. Eggs are laid in crevices in the bark. When the larvae emerge they start burrowing through the bark at the site where the eggs were laid. Larval tunnelling is mostly in the region of the inner phloem. In a healthy tree the larvae may be killed by the heavy sap flow. The irregular tunnel is filled with tightly packed powdery frass. The mature larva, at about 25 mm, in either the spring or autumn, bores deeply into the heartwood to make the pupation cavity where it usually hibernates for a while. Pupation takes place in the spring and adults emerge over the period May to July, or even later. There is usually one generation per year, but sometimes two years are required for the life-cycle to be completed.

The adults are also small, some 12–15 mm in body length, dark bronze or greenish in colour, often with some pale spots on the elytra. The adults feed on the leaves of the host tree, and do cause a little damage.

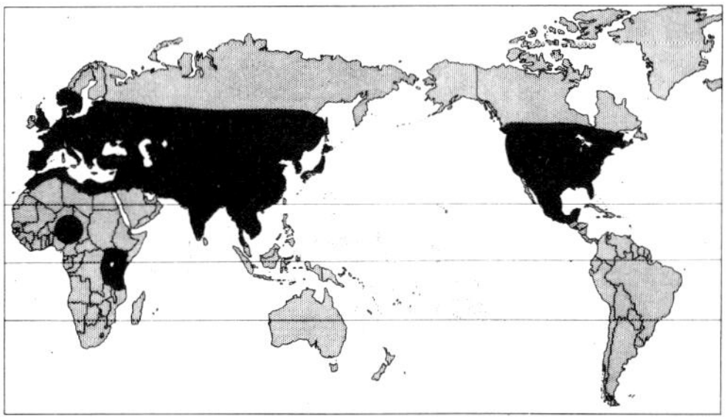

Distribution. A total of 300 species of *Chrysobothris* is reported worldwide, throughout Europe, Asia, N. America, Mexico, and also in E. Africa. One of the most important pest species is *C. femorata* (Oliver), the American Flat-headed Appletree Borer; the Pacific Flat-headed Borer (*C. mali* Horn) is equally serious along the west coast of Canada and the USA.

Other important temperate and subtropical Buprestidae include:
Capnodis spp. – on almond, peach, apple, and most stone fruits and many ornamental trees: subtropical Asia (Israel, USSR, India, etc.).
Sphenoptera spp. – more than 300 species recorded, from a wide range of woody trees and plants: from India, Pakistan, Afghanistan.
Trachys spp. – leaf miners in Japan recorded from *Prunus* spp. and beans.
Control. For control measures see *Agrilus* spp.

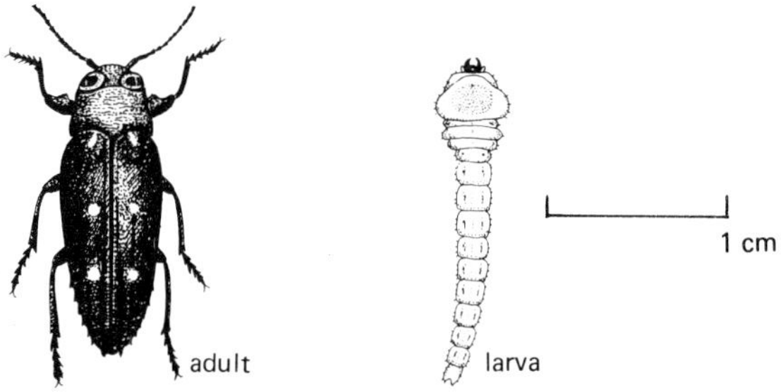

Larval tunnel in wood, with most of the packed frass removed

Agriotes spp.

Common name. Wireworms (Click Beetles)
Family. Elateridae
Hosts (main). Cereals, potato, grasses.
(alternative). Sugar beet, other root crops, sugarcane; most other field and garden crops; polyphagous.
Damage. Larvae in the soil eat the roots of plants, and tunnel tubers and root crops; also bore into stems of young cereals and other seedlings.
Pest status. Worldwide these wireworms are quite serious polyphagous pests, especially so in recently ploughed grassland. An easily recognized group, all similar in appearance and doing similar damage.
Life history. Eggs are laid singly or in small clusters in the soil, in May and June; bare soil is generally avoided and weedy ground preferred; hatching requires about one month. Each female lays about 100–300 eggs.

The larvae grow slowly but soon attain their elongate cylindrical shape that is so characteristic, and their shiny golden brown colour; fully grown larvae reach 15–25 mm in length after some 2–5 years development. Very young larvae feed on soil humus, but as they grow larger they feed more and more on plant tissues. They are very sensitive to soil moisture levels and may be most serious in irrigated fields. Permanent pasture is the natural home of these insects. Larvae reach maturity usually in July/August, when they burrow deeper into the soil and make a pupation chamber some 10–25 cm below the surface. Pupation takes 3–4 weeks; the young adults typically remain in the chambers and hibernate over winter; adults emerge in the spring in April–June.

The adults of *Agriotes* and many other genera are small, elongate beetles, brown or blackish in colour, 10–20 mm in length, and they all bear a strong resemblance to each other. A few species of *Ctenicera* (*Corymbites*) in the USA are brightly metallic, as are some of the large tropical species to be found in Africa and Asia.

In the UK the typical life-cycle takes 4–5 years, but in warmer parts it may be only 2–3 years.
Distribution. The genus *Agriotes* occurs throughout Europe, Asia and N. America. Other very similar genera are found in other parts of the world. These include:
Ctenicera (= *Corymbites*) – in pastures; Europe and USA.
Athous – (Garden Wireworms); UK and Europe.
Ectinus – on wheat in Asia, including Japan.
Melanotus – sugarcane, sweet potato, etc.; Asia and USA.
Limonius – sugar beet; USA.
Lacon – (Tropical Wireworms); pantropical.
Control. Prompt ploughing after harvest will reduce wireworm populations somewhat. Avoidance of sowing susceptible crops has a similar effect. For crops at risk, seed dressings are recommended, followed by a broadcast treatment of HCH or aldrin; granules of phorate or fonofos are also effective.

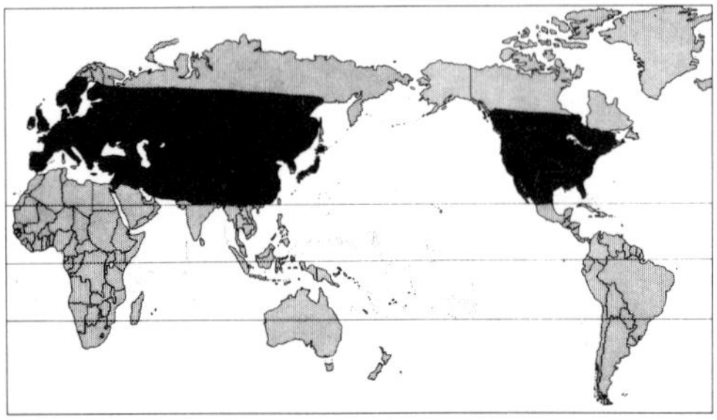

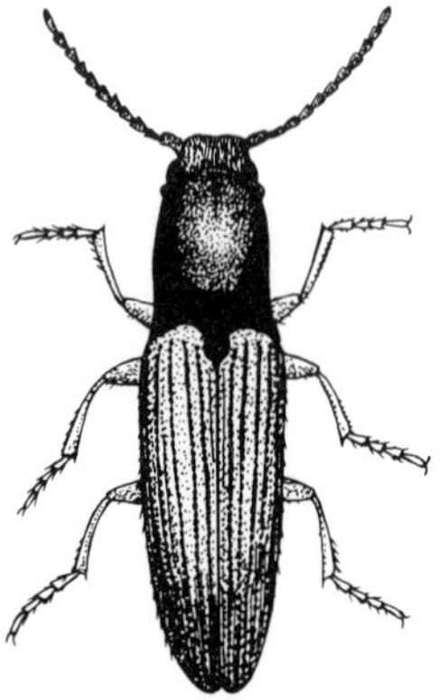

adult click beetle

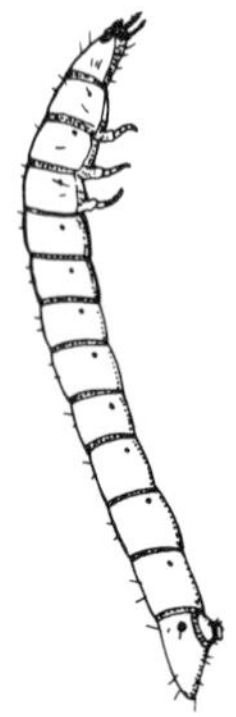

wireworm

0 5 mm

Trogoderma granarium Everts.

Common name. Khapra Beetle
Family. Dermestidae
Hosts (main). Stored cereals.
(alternative). Stored groundnuts, pulses, spices, and cereal products.

Damage. The larvae bore out stored grains and seeds; the species is fairly polyphagous; in the hot humid tropics development is very rapid and large populations build up quickly. In Europe it is mainly found in maltings and heated food stores.

Pest status. In the tropics and sub-tropics this cosmopolitan pest is possibly the most destructive pest of stored grains. In the few countries where it is not yet recorded (E. and S. Africa) there is legislation designed to prevent its accidental establishment.

Life history. Eggs are laid in the stored produce, and the hatching larvae bore into the grains. The larvae are elongate, dark and very 'hairy', and some of the hairs are spear-headed. Larvae develop at different rates, some coming to maturity in two weeks whereas others take months or even a year or more; they are able to enter facultative diapause if food is scarce, or conditions are too cold or too hot or wet, and they may persist in this condition for many months. They have been recorded living for eight years in the larval stage, even when food was available. The larvae often live gregariously, with no tendency towards cannibalism as shown by *Tribolium*.

The adult is a small, oval, dark beetle, 2–4 mm in length; they are wingless, and neither do they feed, living for about 14 days; they are only dispersed by man.

Under optimum conditions the life-cycle can be completed in about three weeks (37°C and 25% RH).

Distribution. Almost cosmopolitan in the warmer parts of the world, and in maltings and heated food stores in Europe and N. America. It prefers a hot dry climate.

This is the only truly phytophagous species in the family Dermestidae. Its close relative *Trogoderma versicolor* will also feed and thrive on cereals, but is more usually found feeding on fur and dead insects.

Control. Because of its speed of development, small size, and powers of larval survival, this extremely important pest of stored cereals and grains is very difficult to control if once established in a food or grain store. Repeated fumigation with methyl bromide, etc., is probably the most effective method, but pirimiphos-methyl is reputed to be effective.

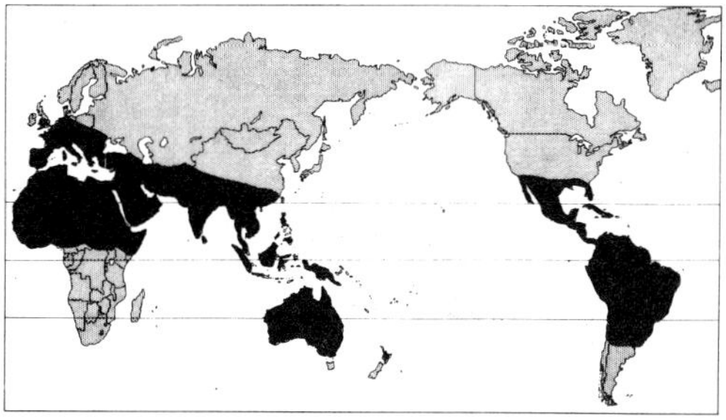

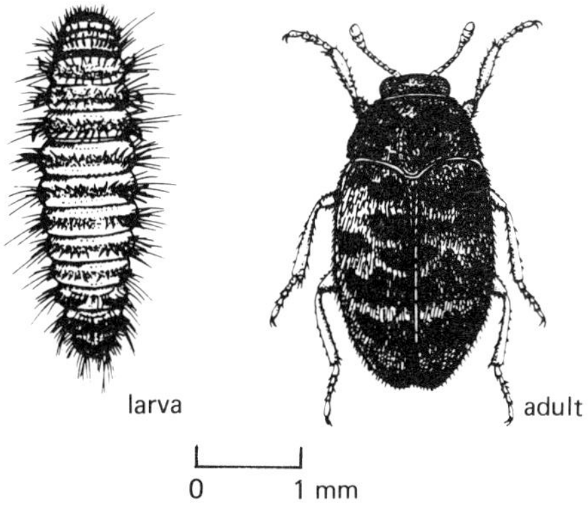

Stored products beetles (and other pests)

As listed on page 595, the more important pests of stored products are the few moths in the Pyralidae, some mites (Acarina), and a miscellany of beetles that belong to several unrelated families. The different types of damage inflicted by these different insects are mentioned on page 100. It should be emphasized that in any study of post-harvest crop losses that the effects of rodents (the synanthropic rats and mice) in many rural situations are extremely important, as is granivorous bird damage on the ripening crop in the field, for all the small-grained cereals.

So far as the insects and mites are concerned, in the damage done, and in various aspects of infestation, they are able to be considered collectively. Certainly the methods of controlling insect populations, and of minimizing damage to the stored products, are equally applicable to the moths and their larvae as to the beetles.

Stored produce infestation control

In any produce infestation control programme the approach should ideally be for several phases simultaneously, the major aspects being:

(1) **sound buildings**: if there are holes or cracks in the walls, and if doors do not fit properly, or windows are broken, then access for rats, mice, and insects is made easy. Also fumigation will not be effective if the building cannot be properly sealed.

(2) **use of containers**: on a large scale this would include silos that are specially designed for bulk storage of grain or pulses, and that hold many tonnes of produce, and could also include the underground silos used for long-term grain storage in many countries as part of the national strategic (famine) reserves. Some of these silos, which are usually air-tight, are hermetically sealed, sometimes with a toxic or inert gas pumped into the top prior to sealing (carbon dioxide or nitrogen). Even if air-filled, such silos are effective in that after sealing the small amount of air present is soon depleted of oxygen by the insects present which then soon suffocate.

Shipment of some bulk produce is now being effected through the use of freight containers; well-made containers should be virtually air-tight and produce may be fumigated after being packed. However, in practice, many of the containers being used for international transport of agricultural produce are damaged slightly and no longer gas-tight, so attempts at fumigation are often less than successful.

The use of clean sacks for bagged produce is of prime importance, but will be referred to in (6).

(3) **store hygiene**: this refers to general cleanliness, removal of debris, old sacks, spilled grain, etc., which should be burned or otherwise destroyed. The cleaning of storage premises using industrial vacuum cleaners will also remove insect and mite eggs from the crevices; ordinary sweeping will usually leave mites on the floor surface.

(4) **clean produce**: all produce should be inspected prior to admittance into a store to ensure that only uninfested material is brought in. Some pests, for example *Sitophilus* weevils on maize and bruchids on pulses, etc., are naturally found in small numbers in field infestations, and care has to be taken to ensure that the harvested crop is not already field-infested. Infested produce should be treated in a special gas-tight fumigation chamber prior to admittance into a store.

(5) **drying and cooling**: at the time of harvest most grain and pulse crops are somewhat moist and require drying prior to storage. In more simple situations traditional air-drying is practised, but this is fraught with problems owing to the vagaries of the weather, and a recent trend is for bulk cleaning and drying in regional stores. Usually the grain is first screened to remove trash and then dried. The longer the storage time anticipated then the lower the moisture content of the grain should be. Wheat

should never be stored at all at moisture levels exceeding 16%; at levels of 14% or less most insects show a marked reduction in rate of development. At moisture levels of above 13% *Aspergillus* and *Penicillium* readily develop. With most (but not all) stored products pests it is well known that at low relative humidities their rate of development drops significantly.

Cooling the produce to a safe level of about 15 °C also slows down the rate of insect development greatly, and for some serious sub-tropical pests development will cease altogether at these temperatures. The temperature threshold for development for *Cryptolestes ferrugineus* is 20 °C; for *Lasioderma serricorne* it is 19 °C; for *Rhizopertha dominica* it is 18 °C; for *Oryzaephilus* it is 17.5 °C; but for *Sitophilus granarius* this temperature is 13 °C.

(6) **pesticide treatment**: this is done in several different ways:

(a) *treatment of buildings, containers and sacks*: the empty buildings need fumigation or spraying to kill eggs and larvae (and adults) that may be hiding in crevices and cracks. Sacks likewise need regular cleansing treatment. In some species pupation takes place in a cocoon attached to a solid substrate such as a crevice wall or sack. Piles of sacks may be easily fumigated in sealed bins using a mixture of ethylene dichloride and carbon tetrachloride, poured over as a liquid.

(b) *fumigation*: provided that the granary or the storage bins are reasonably gas-tight, on-farm fumigation of grains may be carried out using the following; a 1:1 mixture of ethylene dichloride and carbon tetrachloride, or 'Phostoxin' tablets that release phosphine gas on contact with air moisture. Methyl bromide is very effective, but because of its toxicity it may only be applied by a registered operator, and its use is generally restricted to large commercial stores. Most silos are reasonably gas-tight and fumigants can be added after the silo is filled and prior to sealing. Bulk bag fumigation can be carried out under fumigation sheets, but the sheet edges have to be adequately sealed for effective treatment. Most large storage premises have a fumigation chamber, preferably large enough for a loaded lorry to be driven in, for disinfestation purposes. Successful fumigation depends in part on the toxicity of the gas employed, its concentration, and the duration of exposure. Penetration of bulk grain, or stacks of bags, may be quite a slow process, and so usually about three days duration is required. For more details about fumigation see Monro (1980).

(c) *pesticide admixture*: in buildings known to be regularly infested, or in situations (countries) of known high risk, the addition of insecticides as either spray or dusts to the grain as it passes into storage may be worthwhile. The chemicals usually recommended nowadays include malathion (both as e.c. and dust), pirimiphos-methyl (both as e.c. and dust) and fenitrothion. These are also the chemicals recommended for the general spraying or treatment of buildings and storage premises, and they usually kill beetles, moths and mites. Pesticide admixture to stored grain is now banned in the USA and some other countries.

Rhizopertha dominica (F.)

Common name. Lesser Grain Borer
Family. Bostrychidae
Hosts (main). Stored cereals.
(alternative). Dried cassava, flours, cereal products, etc.
Damage. Both adults and larvae feed on the stored grains, usually from the outside, and in a rather haphazard manner; they are both primary pests and can attack rice grains more readily than *Sitophilus*. Adults are quite long-lived.
Pest status. A serious pest of stored grains throughout the warmer parts of the world; particularly serious in Australia. In World War I wheat from Australia sent abroad was heavily infested with *R. dominica* which then became widely established in the USA and many other countries, causing serious post-harvest losses in a range of different cereals.
Life history. Eggs are laid amongst the cereal grains. The larvae are scarabaeiform and have thoracic legs, so are much more mobile than weevil larvae. The larvae eat into the grains from the outside in a haphazard manner. Pupation takes place within an eaten grain.

The adult is a tiny dark beetle, 2–3 mm in length, with a typical bostrychid cylindrical body, round prothorax and deflexed head, as well as conspicuous body sculpturing. The other members of the family, such as *Apate* (black borers), are tree borers, and it is the adult stage that does the boring not the larvae. The adults are long-lived and feed quite voraciously.

The life-cycle takes four weeks at about 34 °C.
Distribution. Originally from S. America, it is now cosmopolitan throughout the warmer parts of the world. It has not established itself in the cooler parts of western Europe though.

The Greater Grain Borer is *Prostephanus truncatus*, native to C. America where it does little damage generally, has recently been accidentally spread into Tanzania where it has revealed a devastating pest potential for both stored maize and dried cassava tubers. It is now spreading through Africa, and causing great concern.
Control. The usual fumigation practices employed for stored products pests are effective against this pest. Threshold temperature for development is about 18 °C. Pirimiphos-methyl is reported to be particularly effective against this pest; HCH, carbaryl and the pyrethroids have also been used successfully.

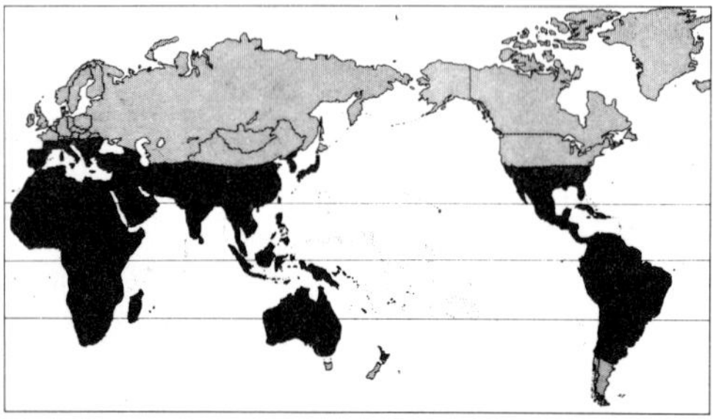

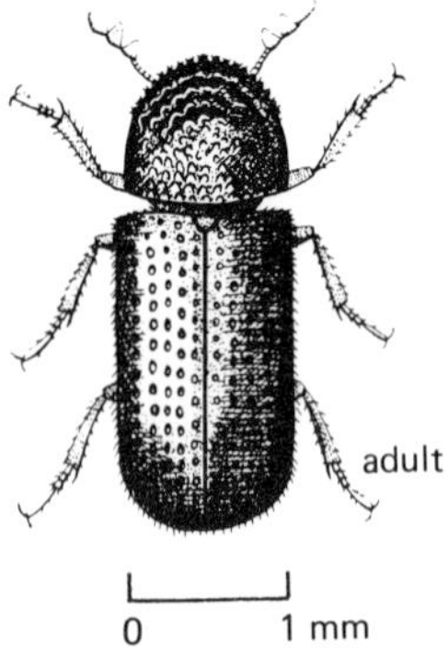

Meligethes spp.

Common name. Blossom Beetles
Family. Nitidulidae
Hosts (main). Rape and mustards.

(alternative). Seed crops of cabbage, swede, turnip, and other Cruciferae; flower heads of many other plants.

Damage. Adults eat holes into unopened flower buds for oviposition sites; they also eat the flowers. In large populations flower damage is heavy and seed production severely reduced, but in small beetle populations they may be beneficial by effecting pollination of flowers. Larval damage is generally insignificant.

Pest status. A serious pest of cruciferous crops grown for seed in the UK and Europe.

Life history. Eggs are laid inside unopened flower buds, two or three per bud, each female laying from 20–200 eggs according to species. Hatching requires 3–10 days.

The larvae live inside the flower buds and feed on the pollen and nectaries. The three larval instars require 2–3 days, 7–11 days, and 9–11 days respectively for development. Fully grown larvae leave the plant and descend to the soil, where they pupate in earthen cells; pupation takes 10–16 days.

The adults are tiny, black shiny beetles, 1.5–2.0 mm long, with stout legs and clubbed antennae. The adults overwinter in hibernation, and emerge from the leaf litter in the spring, usually May or early June. They are active beetles and fly readily; on warm days in midsummer numbers in flight may be very large. Feeding adults hole and destroy flowers; holes in unopened buds are made for oviposition.

There may be either just one generation per year, on rape for example, but if suitable crops (or plants) are available there may be a second generation.

Distribution. The two main pest species in the UK are:

Meligethes aeneus (F.) } Bronzed Blossom Beetles
Meligethes viridescens (F.) }

However, in the UK there is a total of 33 species of *Meligethes* recorded! This genus is common throughout the UK and western Europe.

Control. Autumn-sown crops of rape, drilled through wheat stubble, are less damaged by blossom beetles as flowering may be slightly before the main emergence of the beetles in early June.

On winter rape a spray of insecticide should be applied before flowering when more than 15–20 beetles per plant are seen.

On spring rape two sprays may be required, one at early bud stage if more than three beetles per plant are present, and the second at yellow-bud stage if the crop becomes reinfested. The pesticides found to be effective include HCH, DDT, dieldrin, azinphos-methyl, malathion, endosulfan and phosalone.

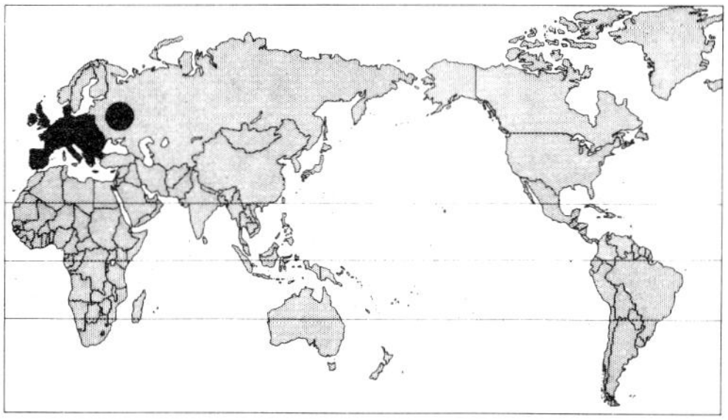

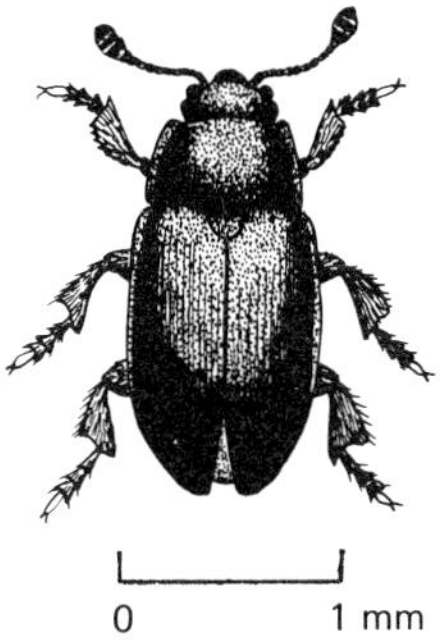

adult of *Meligethes aeneus*

Hole in rape flowerbud

Cryptolestes ferrugineus (Steph.)

(= *Laemophloeus ferrugineus* Steph.)

Common name. Rust-red Grain Beetle
Family. Cucujidae
Hosts (main). Stored cereals.
(alternative). Other stored produce, flours and cereal products.
Damage. Feeding larvae bore into the kernel of the seed and feed mainly on the germ, thus particularly affecting the germination of the grain.
Pest status. A serious pest of stored cereals, especially destructive to seed grain as germ damage impairs germination. As a general stored products pest this is not too serious.
Life history. Eggs are laid amongst the stored grains, and the larvae are primary pests in that they can bore into intact seeds and grains. Once inside the grain the larva feeds mostly on the seed germ. The larvae are thin-bodied, quite elongate, with well-developed thoracic legs and antennae, and pronounced 'tail-horns'.

Pupation takes place within the damaged cereal grains.

The adult is a tiny flattened beetle, about 2 mm in body length, with elongate, thin, clubless antennae; body colour is a rusty-red; and the beetles walk with a strange swaying gait.
Distribution. This species is found mostly in on-farm stores and general food stores, widely dispersed throughout the warmer parts of the world, and in heated stores in cooler temperate regions.

In flour mills there is often to be found the closely related species *Cryptolestes turcicus* (Grouvelle), in both Europe and Asia.
Cryptolestes pusillus (Sch.) (Flat Grain Beetle) is widespread in Asia and N. America.
Control. The usual fumigation and pesticide treatments given to stored grain kill this pest species; see also page 274.

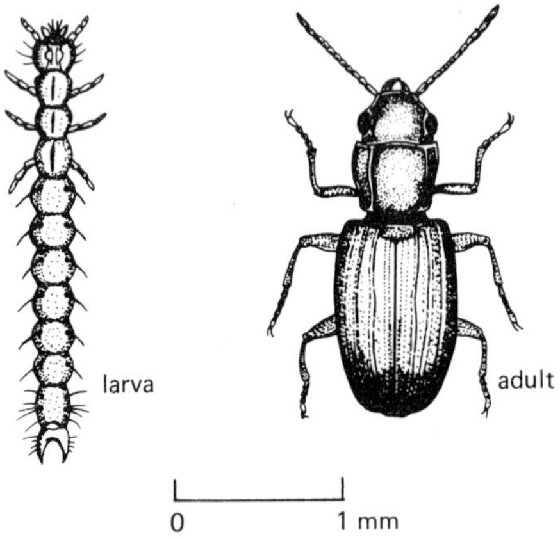

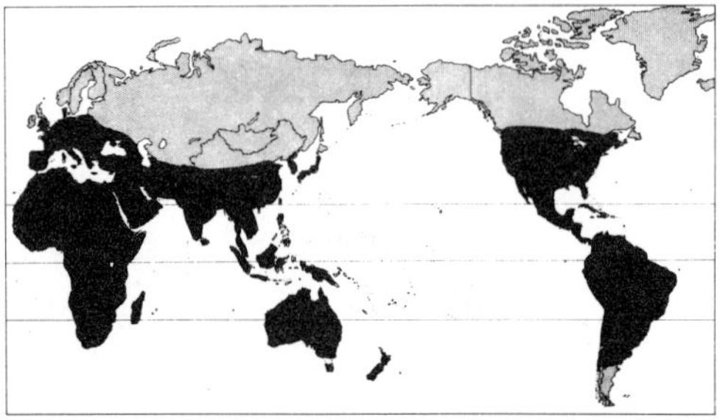

Oryzaephilus spp.

(*O. mercator* (Fauvel))
(*O. surinamensis* (L.))

Common name. Grain Beetles (Merchant and Saw-toothed)
Family. Cucujidae (Silvaniidae)
Hosts (main). Stored grains (damaged).
(alternative). Other stored foodstuffs.
Damage. Their diet consists of fragments of plant and animal material and debris; they are general feeders, secondary on stored grains following damage by the more destructive primary pests such as *Sitophilus* and *Ephestia* spp.
Pest status. These are secondary pests, but polyphagous and very widespread, tiny and easily unnoticed in small numbers, with considerable fecundity and capable of rapid development under warm and humid conditions.
Life history. Eggs are laid loosely in the grain, some 6–10 per day, up to a total of about 400. Egg development takes 4–12 days; optimum temperature appears to be 30 °C.

The tiny white, flattened, active larvae attack broken and damaged grains where they prefer to feed on the germ; larval development takes 12–20 days; conditions of high humidity (60–90% RH) are preferred.

Pupation takes place within the damaged grain, and requires 5–15 days. The adults are small (3 mm), narrow, flattened beetles, reddish in colour, with distinctive serrations along the sides of the thorax, and three longitudinal ridges. Adults also feed and are long-lived; they are recorded living for ten months or more (two years); in cool situations the adults overwinter in cracks and crevices.

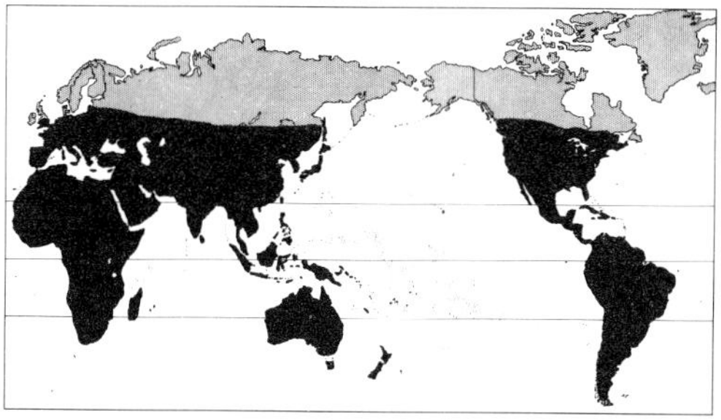

Development occurs at temperatures between 18 °C and 38 °C, and the time required varies from 80–20 days; it is fastest in grain with a moisture content exceeding 14%.
Distribution. These two species are virtually cosmopolitan in distribution in food stores and warehouses. *O. surinamensis* is distinguished by the shape of the head behind the eye, the temple being flat and nearly equal in length to the vertical eye diameter. In *O. mercator* the temple is drawn out into a short rounded point; the male genitalia are also different.
Control. These species do not breed or develop at temperatures of less than 18 °C, so they are less serious in temperate food stores than in the tropics, and may often be controlled by cooling the grain. *O. mercator* is the more temperature-sensitive of the two species.

For general control measures see page 274.

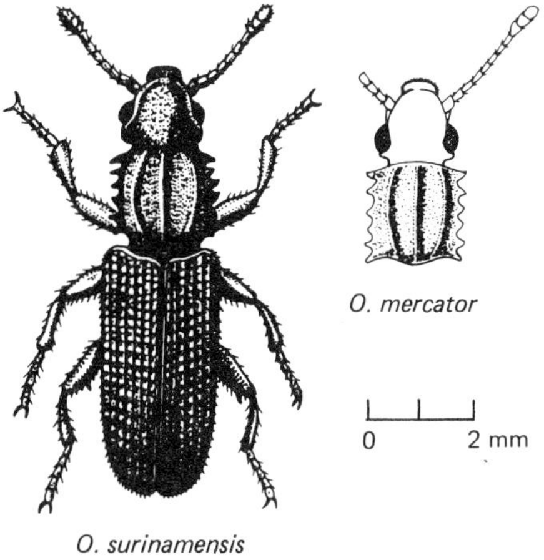

Atomaria linearis Steph.

Common name. Pygmy Mangold Beetle
Family. Crytophagidae
Hosts (main). Sugar beet, red beet, mangolds.
(alternative). Spinach, and various weeds in the Chenopodiaceae.
Damage. Feeding adults eat pieces of the hypocotyl and roots, and hole the cotyledons and heart leaves; the underground damage to the hypocotyl is most serious. Damaged seedlings may wither and die if the beetle population is large; sometimes total crop failures are recorded. It does not appear to transmit viruses.
Pest status. Occasionally a very serious pest on seedling beet, especially if sugar beet are sown after beet or mangolds, in many parts of the UK.
Life history. Adults overwinter in the soil of old beet fields; in the spring they aggregate in large numbers on fine days and dispersal flights occur from mid-April to June according to the weather. First flights occur when the air temperature approaches 20 °C. Dispersal flights are made to the new beet fields, and beetles attack the tiny seedlings.

Eggs are laid in the soil around the beet seedlings, from April onwards. The larvae feed on the fine roots of the beet plants but apparently cause no real damage; when fully grown they are about 3 mm long and grey in colour. In about three weeks the first adults emerge from the soil; these mate and lay eggs giving rise to the second generation. However, the egg-laying period is protracted and the generations overlap, so that new beetles continue to emerge through the summer and autumn, many of which spend the winter hibernating in the soil of the beet field. At times, shortly after singling, there have been field populations of 7 500 000 per ha recorded. The adults are tiny beetles, about 1.5 mm long, brown to black in colour, with quite long and clubbed antennae.

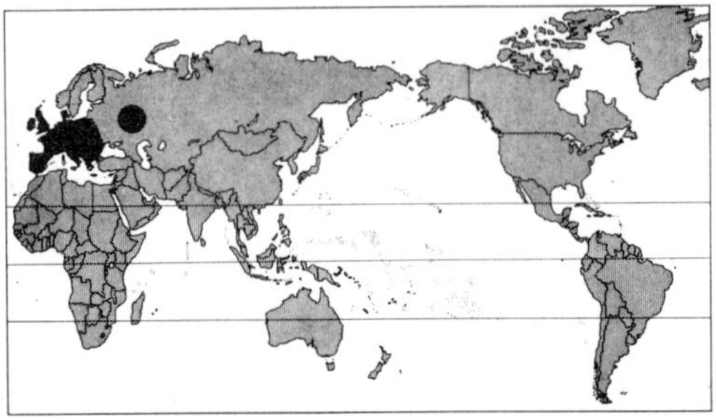

Distribution. Recorded from the British Isles and parts of Europe. In the UK there is a total of 34 species of *Atomaria* recorded.
Control. Crop rotation is the main method of control for this pest. It has generally not been too serious since 1935 in the UK when a clause was inserted into sugar beet factory contracts forbidding the growing of beet after beet or mangolds, a restriction designed to control beet cyst eelworm.

Should insecticides be required then the usual recommendation is for HCH applied to the seedbed before drilling, or as a foliar spray. Granules of aldicarb or oxamyl applied to the seed furrow are also effective.

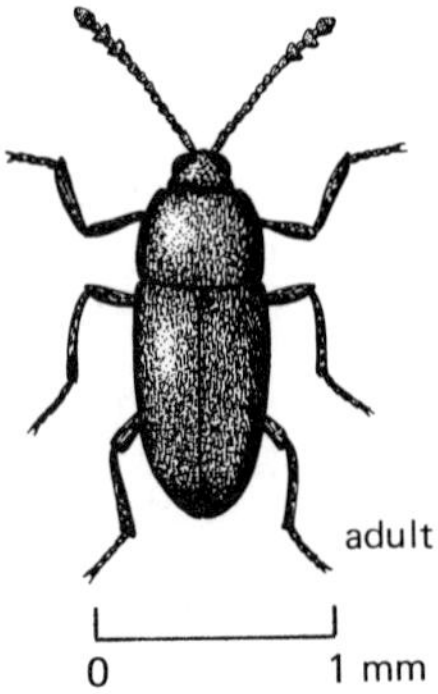

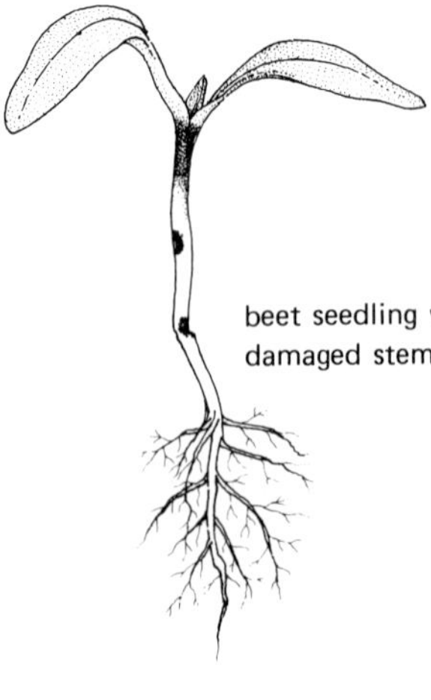

Byturus spp.

(*B. tomentosus* (Deg.))
(*B. bakeri* Barber)
(*B. rubi* Barber)

Common name. Raspberry Beetles (Fruitworms)
Family. Byturidae
Hosts (main). Raspberry
(alternative). Loganberry and blackberry.

Damage. The larvae feed in the ripening fruits; sometimes they damage buds and shoot tips; one larva may damage several fruits. Adults feed on open flowers but may damage the buds of *Rubus*.

Pest status. A serious pest of raspberry; if unchecked whole crops may be rendered unmarketable.

Life history. Adults emerge from the soil in late April to early June. They may be found feeding in open flowers of hawthorn and apple, but later they move to the canes and feed on young buds. Mating begins in June, and in June/July eggs are laid in the flowers of raspberry and loganberry; on blackberry the eggs are laid much later. Hatching takes 10–12 days.

The larvae are yellowish with a brown head. First-instar larvae feed on the surface of the young fruit, but soon move to the calyx and feed on the basal drupelets; as the fruit starts to ripen they burrow into the plug and continue to feed on the inner surface of the drupelets. Sometimes young larvae feed in the unopened buds on the tips of young canes. Mature larvae are about 8 mm long; they descend to the soil and make a small pupal cell in the

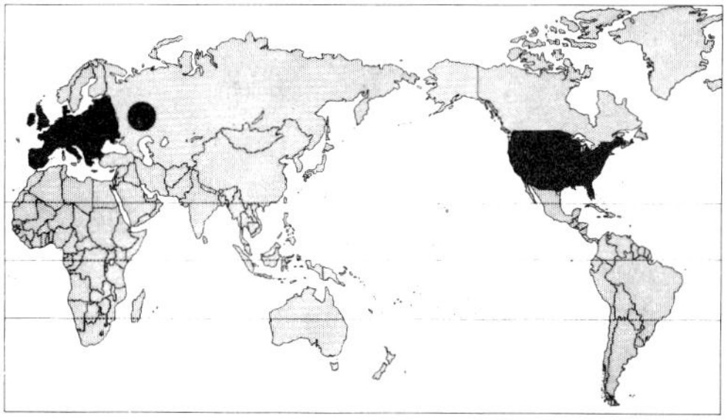

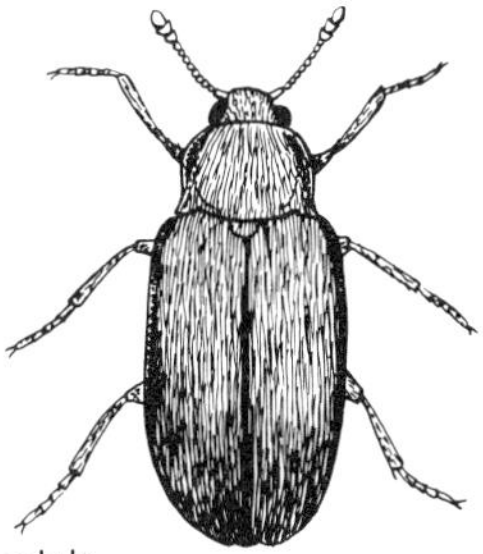

soil, at a depth varying from 2–30 cm. After about a month in the cell pupation takes place; pupation takes 4–5 weeks. The young adults remain in the pupal cell from early autumn through the winter, and emerge the following April and May.

The adults are small brownish beetles, about 4 mm in length; on emergence they are usually yellowish-brown but they become greyer with age.

Distribution. *Byturus tomentosus* is recorded throughout the UK and Europe, as also is *B. ochraceus* (Scriba).

In the USA are found the species *B. bakeri* Barber, the Western Raspberry Fruitworm, and *B. rubi* Barber, the Eastern Raspberry Fruitworm.

Control. By the use of carefully timed insecticidal sprays the larvae of these beetles can be killed; the recommendation is high-volume sprays of azinphos-methyl, derris, fenitrothion or malathion. On raspberry a single spray when the first pink fruit are seen is usually sufficient. On loganberry an extra spray is recommended at the time when about 80% of the blossom is over (about two weeks earlier). On blackberry a single spray when the first flowers open should kill the adults before eggs are laid.

Adult beetle on raspberry flower (M.A.F.F.)

Epilachna spp.

Common name. Melon Beetles (etc.)
Family. Coccinellidae
Hosts (main). Cultivated Cucurbitaceae and Solanaceae; in USA beans (*Phaseolus* spp.).

(alternative). Maize, sorghum, finger millet, rice, wheat, cotton, sesame, lettuce; in USA soybean, cowpea, and solanaceous weeds.

Damage. Both adults and larvae eat leaves and damage fruits. Adults tend to hole leaves, but larval feeding often leaves the veins intact; heavy infestations may totally destroy the leaf laminas. Stems may be gnawed and fruits holed.

Pest status. Quite serious on many crops in southern Europe, Asia and Africa; the Mexican Bean Beetle is a serious pest of beans and other legumes in N. and C. America.

Life history. The eggs are yellow, elongate, and laid in clusters, usually on the underside of leaves; each egg is placed vertically. Average oviposition is 12 batches per female, with 20–30 eggs per batch (up to 50). Incubation takes 4–5 days.

The larvae are pale-yellowish, and covered with delicate spines; when feeding they eat the lamina between veins, initially making rows of small windows but later eating the entire lamina between veins. Fully developed larvae are yellow in colour and broad, with a dark head, and branched spines; body length about 6–7 mm; larval development takes 16–28 days through four instars.

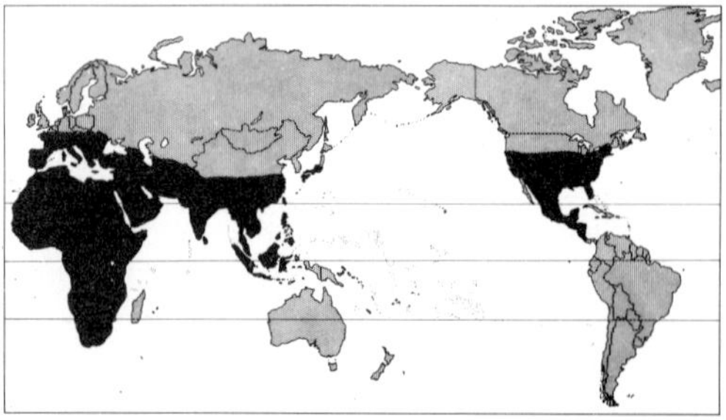

Pupation takes place on the leaf surface, and requires about seven days. The adults are small, oval (6–8 mm long) beetles, varying in colour from reddish to yellow-brown, according to both species and age; and black spots on the elytra. They look like typical 'ladybirds', but are the only members of this family to be truly phytophagous; they are strong fliers.

The life-cycle takes from 33–5 days in Africa and southern USA, where there may be 4–5 generations per year; further north there is only one generation annually.

Distribution. There are several important species of *Epilachna* and the genus is cosmopolitan, but not present in S. America or Australia.

E. chrysomelina (F.) – (the 12-spotted Melon Beetle) southern Europe, Middle East and Africa (CIE map no. A.409).

E. varivestis Muls. – (Mexican Bean Beetle) southern Canada, USA and Mexico (CIE map no. A.46).

E. sparsa (Hbst.) – India, S.E. Asia, China, Japan.

Other species occur throughout Africa, the warmer parts of Asia, the USA and C. America.

Control. Should insecticides be required, the following chemicals have been effective, especially if applied to the undersides of the leaves against the very young larvae: HCH, carbaryl, diazinon, methomyl, parathion, parathion-methyl, rotenone and fenitrothion.

Adult on leaves of *Solanum*

0 1 cm

eggs

damage larvae

Tenebrio molitor L.

Common name. Yellow Mealworm Beetle
Family. Tenebrionidae
Hosts (main). Stored grains, flours, etc.

(alternative). Virtually all stored plant products, some animal material and dead insects.

Damage. The larvae can damage intact grains, as well as eating flours and meals, and general plant and animal debris; polyphagous and somewhat omnivorous.

Pest status. Not often a serious pest as development is slow and populations are seldom very large. Often reared as food for insectivorous birds, reptiles and amphibians.

Life history. Eggs are laid singly or in groups, loosely in the produce. Hatching takes 10–12 days. Each female lays about 500 eggs.

The larvae are shiny brown, cylindrical in shape, and hard-bodied. They are the mealworms reared as animal food. In many warmer countries there are soil-dwelling tenebrionid larvae that eat roots (*Gonocephalum* spp., etc.) and they are collectively referred to as 'false wireworms'. Mature larvae reach 30 mm in length. At 25 °C larval development takes nine months. Pupation takes about three weeks.

Adults are large, oblong, black beetles, 12–16 mm in length, with distinctive striations along the elytra. They are long lived, being active for about three months.

There is only one generation per year; at low temperatures (18 °C or lower) larval development may take as long as 1.5 years.

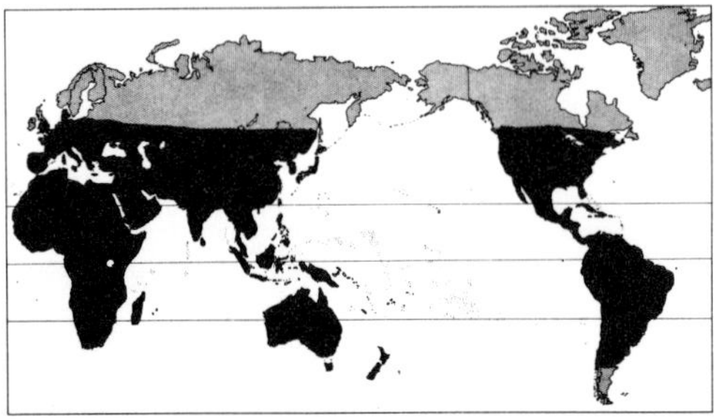

Distribution. A cosmopolitan species found in small numbers in most parts of the world.

Tenebrio obscurus F. is the Dark Mealworm Beetle, more or less sympatric in distribution.

Control. Serious infestations are seldom encountered, because of the long development period required; normal store hygiene generally keeps populations in check.

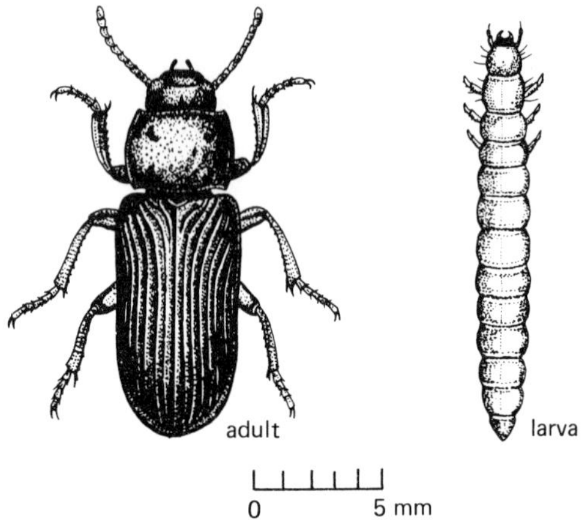

Tribolium spp.

(*T. castaneum* (Herbst))
(*T. confusum* (J.V.))

Common name. Flour beetles (Red and Confused)
Family. Tenebrionidae
Hosts (main). Stored cereals, flours and meals.
(alternative). Many types of stored foodstuffs.
Damage. Both larvae and adults feed on damaged grains and fragments; populations are often very large and damage may be extensive. Infestations are indicated by the presence of the small, brown, adult beetles in the grain/produce; the larvae usually escape notice.
Pest status. These are serious pests in stored flours, meals and processed cereals; also serious secondary pests in stored grains throughout most of the world.
Life history. The eggs are small, white, sticky and cylindrical, and laid scattered in the produce. Hatching takes 5–12 days. Each female lays about 400–500 eggs.

Larvae resemble small white mealworms, with a brown head and two upturned brown pointed structures on the last abdominal segment. In stored grain the larvae live inside the grains and are responsible for most of the damage. At 25 °C larval development takes about 25 days; at lower temperatures several months may be required.

Pupation takes place inside the damaged grains usually, and takes 10–17 days. The brown pupa still bears the characteristic tail-horns of the larva.

The adults are small, flattened, oblong, reddish-brown beetles, 3–4 mm in length, with head and thorax densely punctured and elytra ridged. They are long-lived, and may survive for 1–2 years, and can exist for a month without food. The adults are active and fly readily, especially in late afternoon. In dense populations the adults are cannibalistic, and will eat eggs and young larvae; populations of *Tribolium* thus tend to be well-dispersed throughout the stored grains.

At temperatures about 30 °C the life-cycle may be as short as 35 days.
Distribution. These species are cosmopolitan throughout the warmer parts of the world, but *T. confusum* tends to be more temperate.
Tribolium destructor Uytt. is the Dark Flour Beetle; generally more tropical in distribution than the other two species.
Control. Intact grain, uninfested by primary pests, is generally safe from attack by *Tribolium*. However, infestation by these species is very common and often heavy, and sometimes control will be required. See page 274.

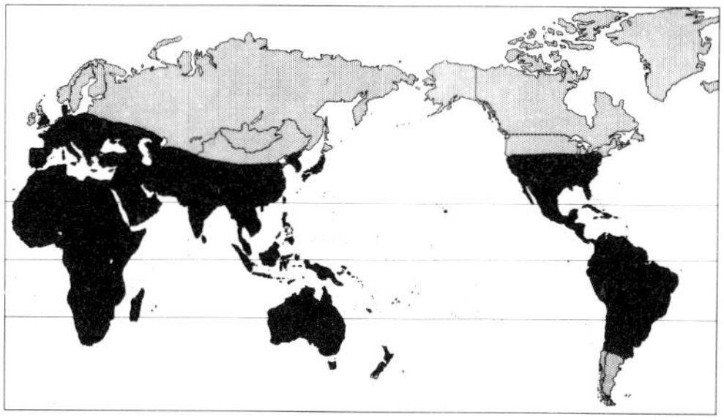

Epicauta spp.

Common name. 'Black' Blister Beetles
Family. Meloidae
Hosts (main). Pulse crops and forage legumes.
(alternative). Potato, beet, rape, cabbage, celery, onion, cucurbits, tomato, capsicums, eggplant, many garden flowers and ornamentals.
Damage. Adult beetles cluster at shoot tips and eat young leaves and sometimes flowers; irregular holes are made in the leaves; heavy infestations result in defoliation; lesser infestations give the plants a ragged appearance.
Pest status. Quite serious defoliating pests on a wide range of crops, in most parts of the world. The larvae are, however, beneficial as predators of grasshopper egg-pods in the soil, and they may be used for biological control purposes.
Life history. Eggs are laid in holes in the soil, in clusters of 100 or more; the triungulin larvae hatch after 5–10 days. Females lay large numbers of eggs (2000–10 000). The triungulin larvae are very active and burrow, seeking egg-pods of grasshoppers (Acrididae) and some other Orthoptera. After locating an egg-pod, the triungulin metamorphose into the sluggish, grublike, almost legless larvae that eat the eggs of the grasshoppers. There are finally six larval instars; in the USA the larvae over-winter in a pseudopupal stage. Pupation takes place in the soil in the early spring.

The adults are elongate, black-bodied, blister beetles, usually between 12 and 20 mm in length; some are totally black in body colour, some black with a reddish head capsule, some with white stripes or spots along the body. Stripes and spots are generally caused by white scale-like setae.

Little is known of the biology of many of these species.
Distribution. This genus of blister beetle is probably best represented in the New World where the most species are recorded, but some are found in Africa and Asia. The more important species are probably as follows:
Epicauta aethiops (Lat.) – (Grey Blister Beetle) N. Africa.
E. albovittata (Gestro) – (Striped Blister Beetle) E. Africa.
E. maculata (Say) – (Spotted Blister Beetle) Canada and USA.
E. pennsylvatica (DeG.) – (Black Blister Beetle) Canada and USA.
E. gorhami Mars. – (Striped Blister Beetle) China.
E. rufipes (Ill.) – (Red-headed Blister Beetle) Indonesia.
E. tibialis Waterhouse – (Black Blister Beetle) China.
E. vittata F. – (Striped Blister Beetle) Canada, USA, and S. America.

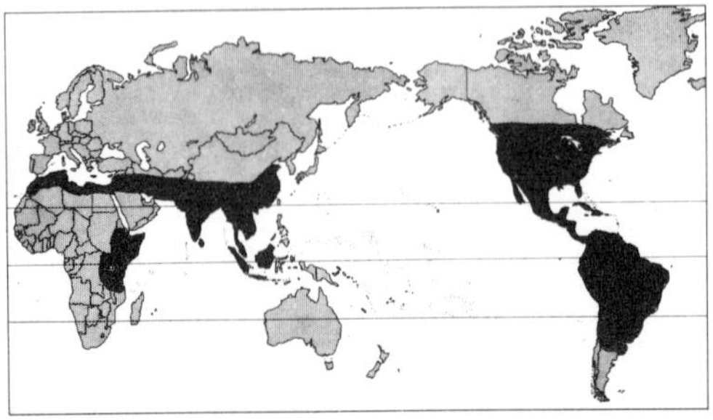

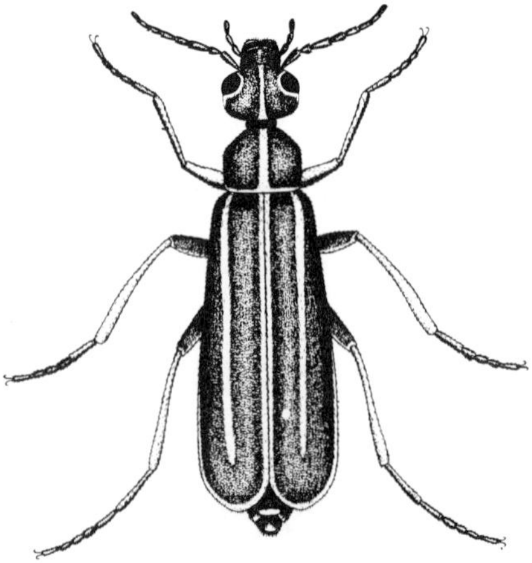

Epicauta albovittata (Africa)

0 5 mm

There are listed ten species of *Epicauta* as pests in Canada, and eight in the USA. Other important genera of blister beetles are the cosmopolitan *Lytta*, the pantropical *Mylabris* (Banded Blister Beetles), and in Australia the largest genus is *Zonitis*; each of these genera contain many species.

Control. These insects are important in the natural control of grasshoppers and locusts (Acrididae) as the larvae are egg-predators.

The adults are difficult to kill with chemicals because these blister beetles have high levels of natural resistance to many chemicals (DDT, HCH), but at high dose rates DDT, parathion, carbaryl and endosulfan have been effective.

Mylabris (Banded Blister Beetle) (China)

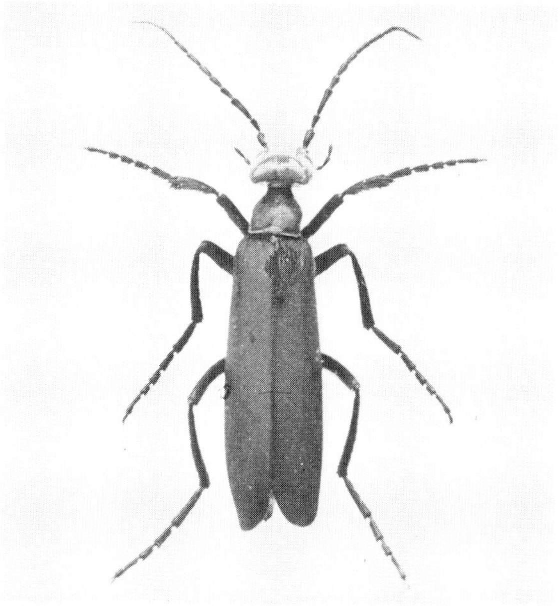

E. tibialis (China)

Mylabris feeding on a pigeon pea inflorescence

Apriona spp.

Common name. (Mulberry/Apple) Longhorn Beetles
Family. Cerambycidae
Hosts (main). Mulberry, apple.
(alternative). Fig, jackfruit, peach; wild species of *Ficus* and other Moraceae.

Damage. The larvae tunnel in the branches and tree trunk, just under the bark, and sometimes into the heartwood; in small branches they bore down the centre. Frass expulsion holes are made from which sap often exudes. Heavily attacked trees may die, especially as stressed trees are preferred as hosts.

Pest status. Widespread pests in Asia, causing spectacular damage; not often serious pests; frequently stressed or sickly trees appear to be favoured as hosts.

Life history. Eggs are laid in crevices in the bark, and hatch after about 7–10 days. The larvae burrow in the sapwood under the bark, making frass holes at intervals. Larval development under warm conditions takes 9–10 months, but under cooler conditions it is thought that nearly two years may be required. Pupation takes place at the end of the larval tunnel which is blocked by wood fragments.

The adults emerge in the spring; they are large, dark grey beetles, 4–6 cm in length, with conspicuous blackish tubercles at the bases of the elytra. The males are distinctly smaller than females, but have longer antennae. Both adults feed on tree bark or foliage, and they may live for several months.

In warmer regions the life-cycle is probably univoltine, but farther north it is likely that two years are required for complete development.

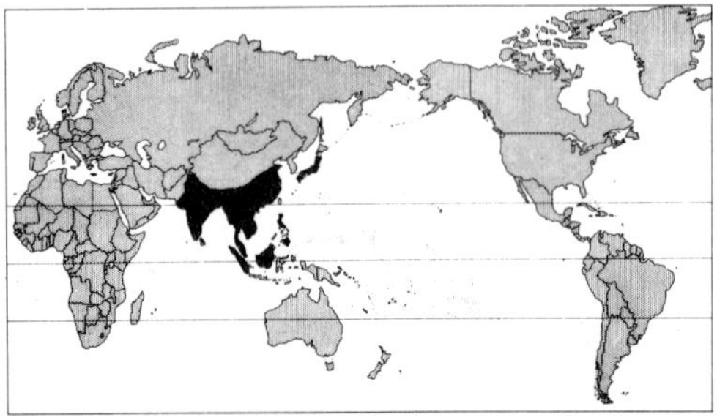

Distribution. The main species concerned as pests are as follows:
Apriona cinerea Chevr. – (Apple Stem Borer) India.
A. germari (Hope) – (Mulberry Longhorn Beetle) India, throughout S.E. Asia to S. China.
A. japonica Thom. – (Mulberry Borer) Japan.
A. rugicollis Chevr. – also from mulberry; India.

Control. When control is required it has been found that spraying the tree trunk with dieldrin or parathion will kill eggs and young larvae. With established larvae injecting the tunnel with dieldrin/kerosene has been effective.

After tunnel treatment it is recommended that all old frass be removed from under the tree, so that any new frass expelled is immediately obvious and the chemical treatment can be repeated.

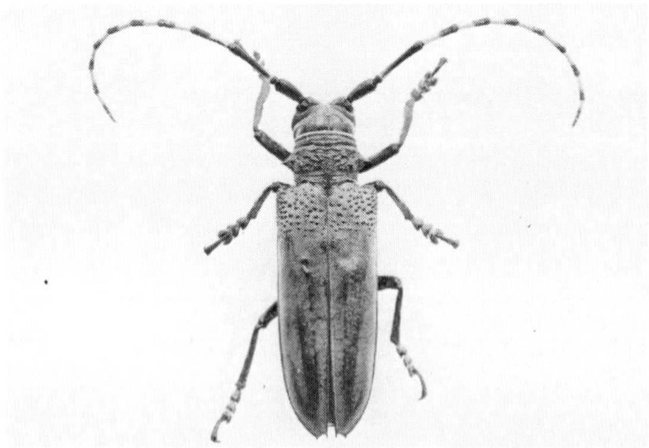

Adult *A. germari*

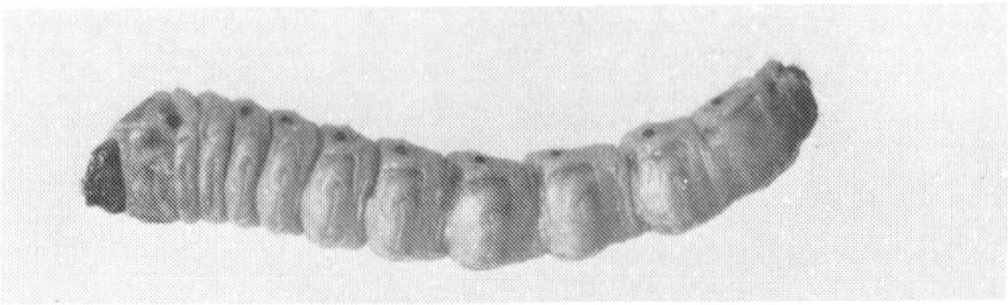

larva

Batocera spp.

Common name. Longhorn Beetles
Family. Cerambycidae
Hosts (main). Mango, fig, apple, walnut.
(alternative). Jackfruit, guava, pomegranate, rubber and papaya; wild species of *Ficus*, various ornamental and wild trees. In India 30 different host plants recorded.
Damage. The boring larvae tunnel through the sapwood under the bark of trees, both in branches and the trunk. Tree vigour is affected, terminal foliage may die, and in heavily infested trees death of the entire tree may result.
Pest status. Widespread pests that are frequently encountered in the warmer parts of Asia in both fruit trees and ornamentals, but seldom serious pests.
Life history. Eggs are laid singly in the bark of the tree, in small cuts made by the female biting; on average about 250 eggs are laid per female (maximum about 350).

On hatching the larvae tunnel straight into the wood, feeding in the vascular tissues of the tree, and eventually making a long irregular tunnel in the sapwood, extruding frass at intervals. There are usually some tunnels deep into the heartwood into which the larvae hide if disturbed. Full-grown larvae measure up to 8–10 cm long. Larval development takes 3–6 months in the hottest tropics but may be two years or more in northern China and Japan. Pupation takes place in the tunnel system, usually just under the bark, and the tunnel is packed and closed off with frass.

The adults emerge in the spring, and are magnificent insects, the largest species have a longitudinal span of nearly 20 cm. They are greyish brown in colour with conspicuous white spots on the elytra and spots on the prothorax either white or red (in life). The smallest species (*B. rubus*) has a body length of 3–4 cm, but *B. rufomaculata* females measure up to 7 cm. Adults live for 2–6 months.

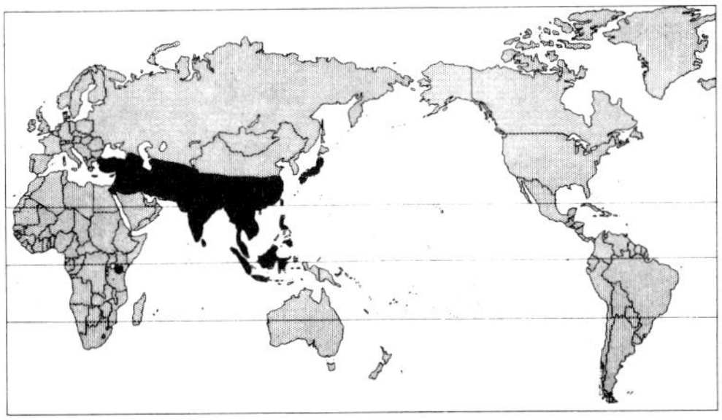

Distribution.
Batocera rubus (L.) – (White-spotted Longhorn) from fig, mango, jackfruit, etc.; from India, through S.E. Asia to southern China.
B. rufomaculata (De Geer) – (Red-spotted Longhorn Beetle) from fig, mango, guava, jackfruit, pomegranate, apple, rubber, and walnut; W., S. and E. Asia.
B. horsfieldi Hope – (Square-spotted Longicorn Beetle) from walnut; India and China. Three other species are recorded from mango in India.

In Japan is found *Batocera lineolata* Chevr., the White-striped Longicorn Beetle. Three other species are recorded from Indonesia.
Control. When control is actually required, the recommendations made for *Apriona* species (page 288) should be effective.

Larva in branch

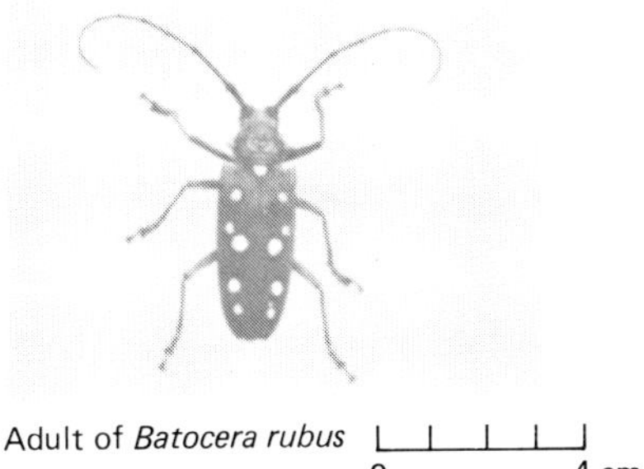

Adult of *Batocera rubus*

Acanthoscelides obtectus (Say)

Common name. Bean Bruchid
Family. Bruchidae
Hosts (main). Beans, mostly *Phaseolus* species.
(alternative). Other pulse crops (Leguminosae).
Damage. The larvae bore into the bean seeds and eat the cotyledons; there may be several larvae inside one seed. Seeds with mature larvae or pupae inside have small windows; emergence holes are about 2 mm in diameter.
Pest status. A serious pest of beans throughout the warmer parts of the world; infestations often start in the field and are then carried into produce stores where breeding continues.
Life history. The dirty white eggs are laid by the female on the young pods in the crop. Each female lays 40–60 eggs (200 maximum recorded); hatching takes 3–9 days. The larvae bore into the pods and enter young seeds; there may be several larvae in one seed. Larval development through four instars takes from 12 days to six months according to prevailing conditions. The larvae are legless, thick-bodied, white and curved.

Pupation takes place within the bored seed, behind a thin window; pupation takes 8–25 days; the young adults emerge by pushing out the 'window' on the seed.

The adults are 2–3 mm in length, brownish-black with pale patches on the elytra, and the characteristic emarginated eyes are obvious.

The life-cycle takes 4–6 weeks at 28 °C and 70% RH; in southern USA and eastern Mediterranean Region there are typically six generations annually; usually one or two in the field and the remainder in storage. Field generations occur as far north as Canada and Europe.

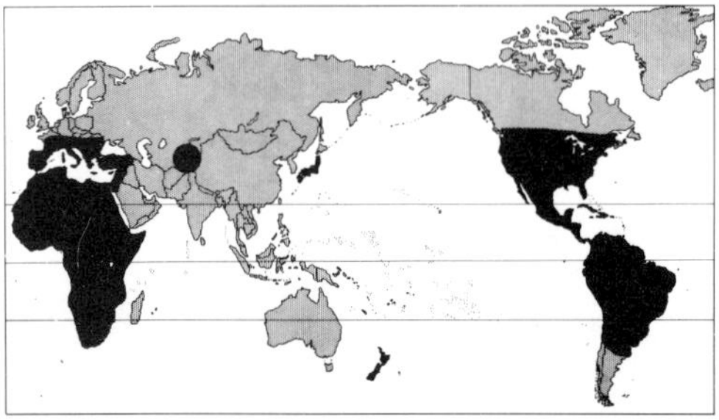

Distribution. Cosmopolitan in both tropical and temperate regions, including Europe, Africa, Japan, New Zealand, S. and C. America, USA and Canada. Its status throughout the remainder of Asia is not known at present.
Control. Field infestations may be minimized by careful crop hygiene; in particular the destruction of crop residues after harvest. For insecticide recommendations see Pea Beetle on page 291, and for storage treatments see page 274.

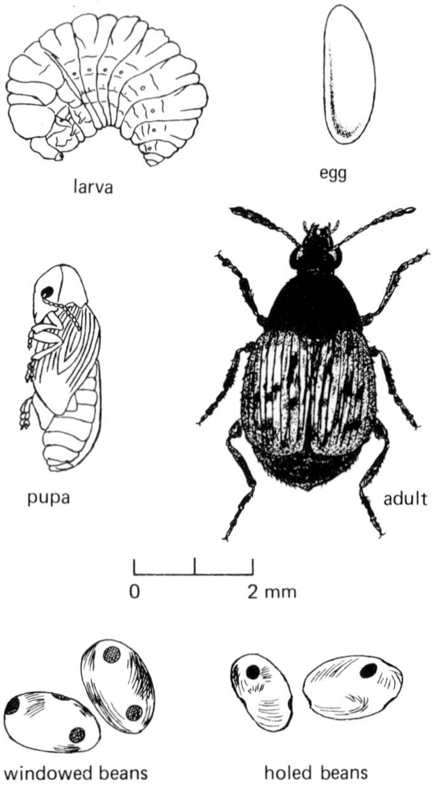

Bruchus spp.

(*B. pisorum* (L.))
(*B. rufimanus* Boh.)

Common name. Pea Beetle; Bean Beetle
Family. Bruchidae
Hosts (main). Pea, broad (field) bean (seeds).
(alternative). *Phaseolus* spp., other pulses, and forage legumes.
Damage. These bruchids only develop in the seeds of growing crops; they do not attack dried seeds or stored pulses; the larvae bore inside the developing seeds where they eat the endosperm of the cotyledons and often the germinal region also; the latter is more usual with Pea Beetle.
Pest status. Quite serious pests of leguminous crops in the field, in temperate regions: crop losses of 20–35% are not uncommon.
Life history. The adults overwinter by hibernating in sheltered places. In late spring the females lay eggs on the developing pea and bean pods. The flattened oval eggs are laid on the side of the young pods; each female lays up to 500 eggs; hatching takes about 7–18 days.

The larvae bore into the pod and then into a developing seed; the entry hole in the pod closes. There may be several larvae inside one bean seed. Larval development takes usually 2–3 months.

Pupation takes place in the hollowed-out seed just under the surface, and takes 10–15 days.

The young adults usually remain inside the pupal 'cell' for some time before emerging, and sometimes emergence takes place after the crop has been put into storage. These beetles cannot develop in dried seeds, however, and so there is no infestation of the stored seeds. Some adults hibernate inside the seeds, especially in crop remnants left in the field, but others emerge and hibernate in litter. The adult beetles are 3–4 mm long, and have somewhat variable markings.

There is only one generation per year.

Distribution.
Bruchus pisorum (Pea Beetle) is cosmopolitan in all but the coldest regions; including Europe, Asia, Australia, USA and Canada (probably native to USA).
B. rufimanus (Bean Beetle) is likewise recorded from Europe, throughout Asia, Australia and the USA (probably native to Europe/Asia).
B. ervi Fröl. (Mediterranean Pulse Beetle) this attacks mostly lentil in the Mediterranean Region.
B. chinensis L. – (Chinese Pulse Beetle) recorded from China on stored pulses.
B. brachialis Fohr. – (Vetch Bruchid) recorded from Canada and the USA.

Control. Crop hygiene, especially destruction of plant remains after harvest, helps to minimize infestation. Uninfested seed only should be sown; infested seed should be fumigated. Field infestations have generally been controlled with the use of DDT, rotenone, malathion and parathion, applied as sprays, or sometimes as dusts.

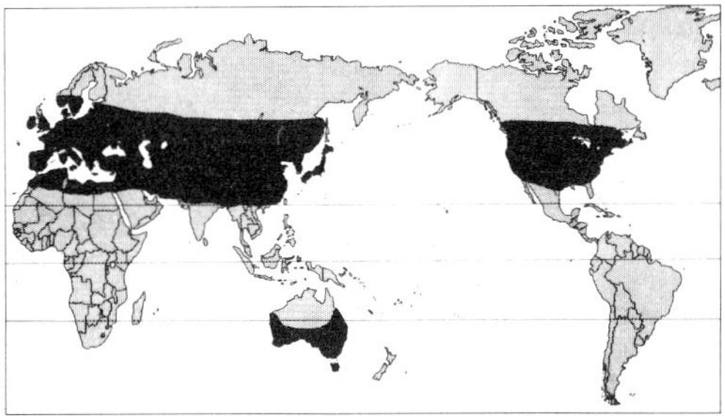

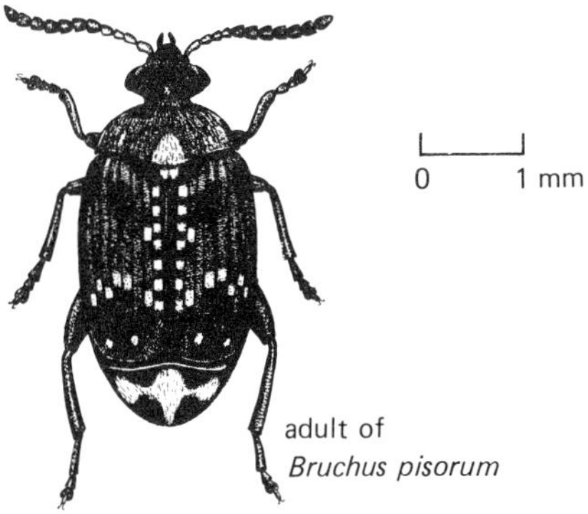

adult of
Bruchus pisorum

Cassida spp.

Common name. Tortoise Beetles
Family. Chrysomelidae (Cassidinae)
Hosts (main). Sugar beet, beets, spinach.
(alternative). Artichoke, safflower; wild Chenopodiaceae.

Damage. Adults and larvae eat holes in the leaf lamina and cotyledons; young plants may be killed.

Pest status. In the UK these are more of academic interest, but in Europe they are quite serious pests of beet. The group is more renowned as pests of sweet potato.

Life history. Eggs are laid from May onwards, singly or in small groups, on the foliage of beet seedlings; hatching takes about 12 days. About 300 eggs are laid per female.

The larvae are pale green, slug-like, with small fleshy spines; the tail is forked and held upright over the back, usually with the cast exuviae adhering; larvae of this genus usually cover their backs with excrement (as do many other Chrysomelidae), presumably for protection, either against predators or the danger of desiccation. Larval development takes about two months, when the larvae are about 6 mm long.

The pupa is less spiny than the larva and is fixed to the leaf surface. Adults are characteristic in appearance with their oval dorsal shield; these species vary in size, 3–7 mm long. Body colour is greenish or brown with varying dorsal colour patterns faintly imposed.

In the UK *C. nobilis* is univoltine, but the other two species have two generations annually. The adults hibernate overwinter.

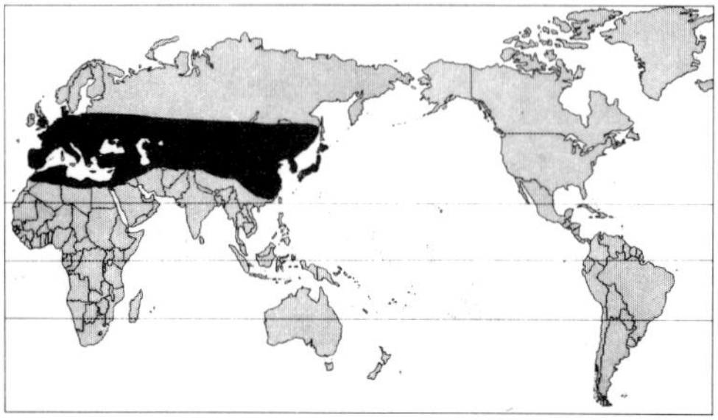

Distribution. The three main species in Europe (and UK) are:

Cassida nebulosa (L.) – (Beet Tortoise Beetle) Europe, Asia to Japan and China.

C. nobilis (L.) – recorded from parts of Europe.

C. vittata de Villers – (Beet Tortoise Beetle) Europe, Mediterranean Region.

In the UK there are another nine species of *Cassida*, mostly recorded feeding on thistles and other Compositae, but including *C. viridis* on mints (*Mentha* spp.)

In Asia and Africa the dominant tortoise beetle is *Aspidomorpha*, found mostly on Convolvulaceae (sweet potato, morning glory, etc.) as about two dozen different species. In N. America there are several different species in different genera, mostly recorded from sweet potato and other Convolvulaceae.

Control. Often damage is only slight, but on occasions control is required and the insecticides found to be effective are DDT, dieldrin, and HCH as sprays, also rotenone, carbaryl and malathion (USA).

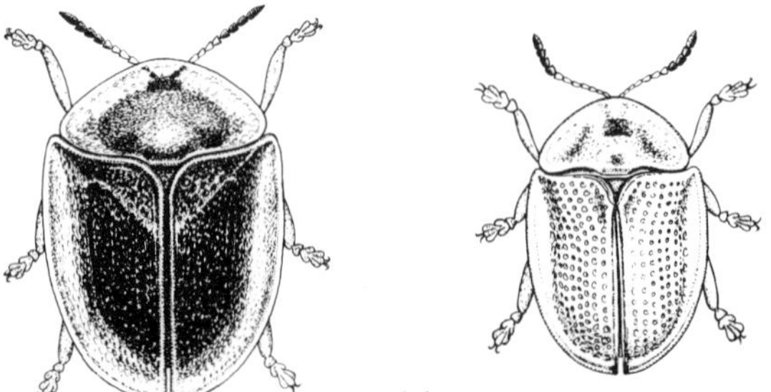

adults
C. nebulosa 0 2 mm *C. nobilis*

Leptinotarsa decemlineata (Say)

Common name. Colorado Beetle
Family. Chrysomelidae (Chrysomelinae)
Hosts (main). Potato

(alternative). Eggplant, tomato, and various other (only some) cultivated and wild Solanaceae.

Damage. Both adults and larvae feed on the leaves, and a heavy infestation results in complete defoliation and a severely reduced yield; a 50% loss is usual. A characteristic black and messy excrement is left on the leaves.

Pest status. A very serious pest of potato, and other solanaceous crops, in Europe and the USA. In the UK, in 1933, was passed the Colorado Beetle Order as a legislative measure of control against accidental introductions.

Life history. Eggs are orange-yellow, laid in batches, usually underneath the leaves. Each female lays 500–800 eggs throughout the summer period (up to 2500 eggs are recorded); hatching requires 4–6 days.

Larvae are brown initially, later becoming bright pink with two rows of black spots laterally; they generally stay on the same plant, and eat the foliage. Larval development takes 20–4 days. Fully grown larvae (about 12 mm long) burrow into the soil to pupate inside an earthen cell; pupation takes about 5–16 days; then the first-generation adults emerge, usually in July. Second-generation adults remain in the soil and overwinter in hibernation, at a depth of 10–25 cm.

The adults are rounded beetles with alternating black and yellow stripes along the elytra; 10–12 mm in length. The spring dispersal flights are usually 2–5 km.

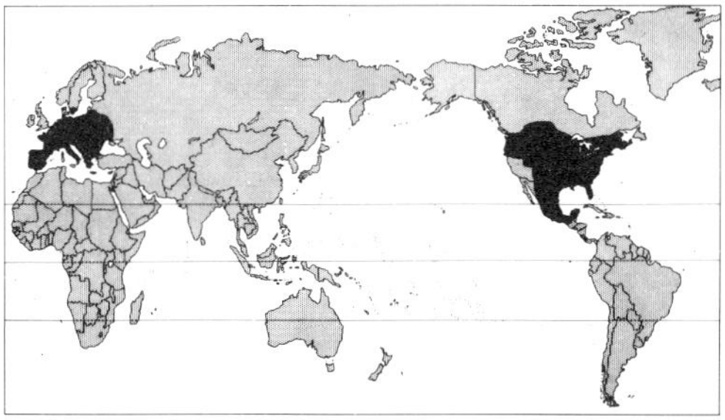

Distribution. Endemic to semi-desert regions in the western USA, on wild Solanaceae, it spread and became more numerous with widespread potato cultivation. Now recorded from throughout the USA, southern Canada, Mexico and Costa Rica, and in 1922 became established in France; now spread throughout most of Europe, but to date kept out of the UK (CIE map nos. A.6 and 139).

Control. Control is best effected by low-volume sprays of azinphos-methyl, carbaryl, DDT, or phosphamidon on the foliage of the plants, in addition to injecting the soil with D-D or metham-sodium to kill pupae and resting adults.

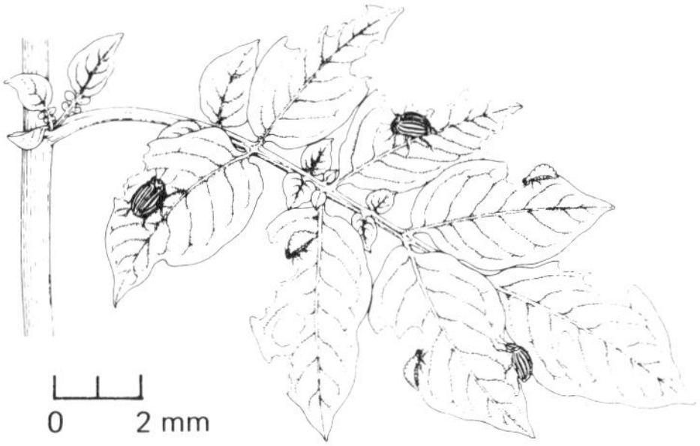

eggs

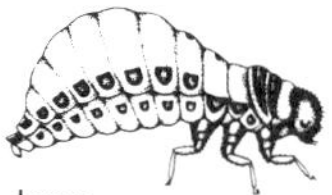
larva

pupa

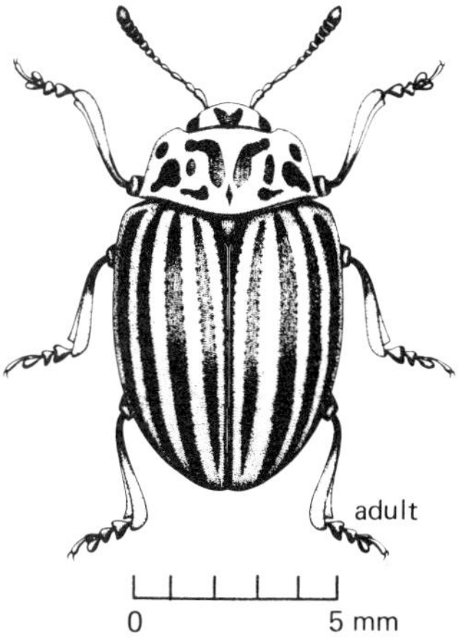

adult

Phaedon spp.

Common name. Mustard Beetles
Family. Chrysomelidae (Chrysomelinae)
Hosts (main). Mustards

(alternative). Rape, turnip, watercress, late cauliflower, and cress and swede grown for seed.

Damage. Both adults and larvae eat the foliage and 'bark' the stems, and eat developing seed pods; most serious on seed crops. Heavy infestations may completely defoliate the crop.

Pest status. A serious pest on mustard crops; less damaging on the others as most plants recover from initial damage.

Life history. Adult beetles overwinter by hibernating in crop stubble, leaf litter and hedgerows. In the spring the adults fly in search of young mustard crops (rough leaf stage) where they feed on the foliage. Eggs are laid on the undersides of the leaves, partly embedded in small holes made by the female's mandibles, in May and June. Each female lays some 300–400 eggs over a three-week period. Hatching takes 8–12 days.

Larvae are yellowish with black spots and dark legs; they feed on the leaves making shot-holes in the lamina, for 15–25 days, during which time they moult three times. Mature larvae are about 6 mm long; they drop to the ground and burrow into the soil to make a pupal cell 2–5 cm underground. Pupation takes about 10–12 days. Adults emerge in late June and July and start the second generation. The adults are small rounded beetles, shiny metallic blue in colour, 3–4 mm long, with lines of punctures along the elytra. The life-cycle takes 35–45 days. Of the two overlapping generations the first does little damage, but the second generation in August/September may be very damaging.

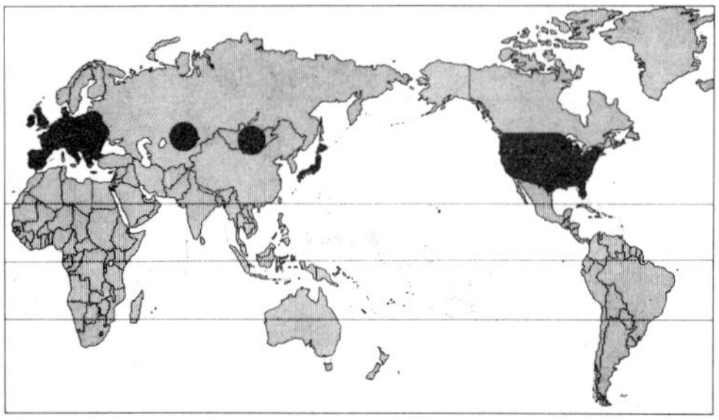

Distribution. The three most important species in Europe are *Phaedon amoraciae* (L.), *P. cochleariae* (F.), and *P. tumidulus* (Gm). In the USA *Phaedon viridus* (Mel.) is the Watercress Leaf Beetle, and in Japan *P. brassicae* Baly is the Brassica Leaf Beetle.

Control. Present recommendations include spraying or dusting with DDT or HCH. Crop hygiene, particularly destruction of stubble and crop residues, will deplete the overwintering population; early harvesting, and careful crop rotation will reduce damage levels to some extent.

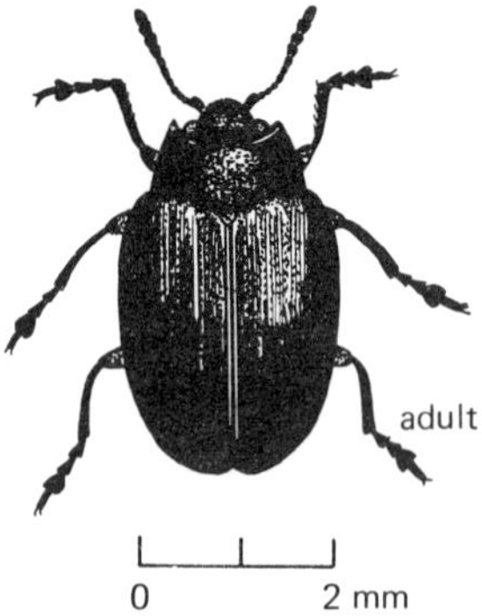

Crioceris asparagi (L.) and **Crioceris** spp.

Common name. Asparagus Beetle(s)
Family. Chrysomelidae (Criocerinae)
Hosts (main). Asparagus
(alternative). None recorded, but presumably wild species of *Asparagus*.

Damage. Adults and larvae eat the foliage, and contaminate spears with messy black excrement; heavy infestations often defoliate the crop. Larvae of *C. duodecimpunctata* are different in that they feed on or in the berries.

Pest status. These are serious pests of cultivated asparagus throughout Europe, Asia and N. America. The adults feed extensively, but actually do little damage.

Life history. The adults hibernate overwinter, in crevices, under stones or in the soil. Eggs are elongate, flask-shaped, and laid in conspicuous rows (usually 3–7) along the needle-like leaves; they measure 2 mm in length. Hatching takes 5–7 days in late June and July. Each female lays about 250 eggs.

The larvae are fleshy and soft-bodied, and pass through three instars in about 2–3 weeks. Mature larvae are greenish-grey in colour, with head and legs shiny black; 7 mm in length.

Pupation takes place in a spun cocoon in the soil, about 2–5 cm deep, taking about 10–20 days to develop.

Adults emerge about August, and start the second generation. They are elongate little beetles with black head and the elytra marked with either pale and dark bands (*C. asparagi*) or pale with dark spots (*C. duodecimpunctata*, etc.); body length 6–7 mm.

There are usually two or three generations per year.

Distribution. *Crioceris asparagi* is apparently native to Europe but now occurs in Canada and the USA.
C. duodecimpunctata (L.) – (Spotted Asparagus Beetle) also endemic to Europe and now widespread throughout Canada and the USA.

In Japan are found the species *C. orientalis* Jacoby and *C. quatuordecimpunctata* Scopoli, but the extent of their distributions is not known at present.

In W. Africa *C. nigropunctata* is recorded from asparagus foliage, and in Kenya *C. viridissima* is found also on the foliage.

Control. Insecticides used successfully against these beetles include DDT, HCH, carbaryl, rotenone, and malathion, as foliage sprays.

Adult beetles on asparagus shoots

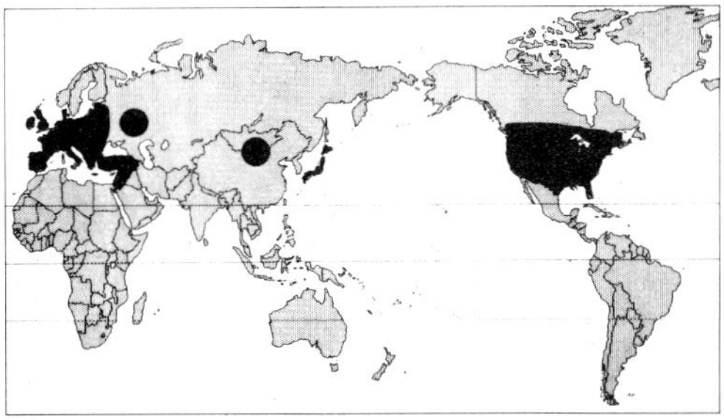

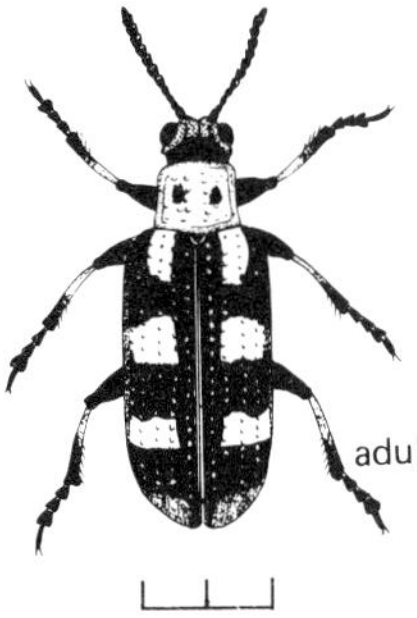

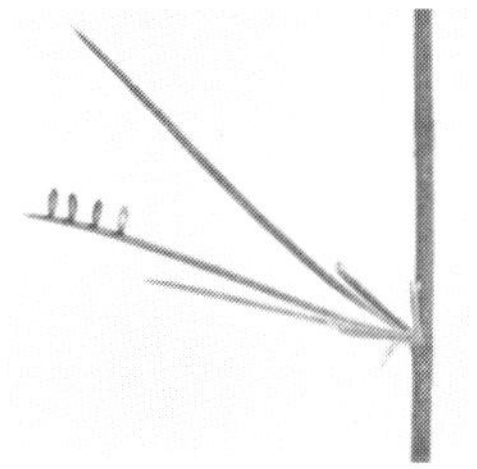

Eggs on a leaf

Oulema melanopa (L.)

(= *Lema melanopa* L.)

Common name. Cereal Leaf Beetle
Family. Chrysomelidae (Criocerinae)
Hosts (main). Wheat, barley, oats.
(alternative). Various grasses.

Damage. Adults and larvae eat longitudinal strips from the leaf lamina (rather similar to slug damage). Virus vector on cocksfoot grass in Europe.

Pest status. Only sporadically serious, but quite widespread and common on both cereals and grasses.

Life history. Eggs are laid along the leaf midribs on the upper surface, from May to July; they are covered with a secretion that hardens on exposure to air. Hatching takes about two weeks.

The larvae feed on the upper surface of the leaf, skeletonizing the lamina in longitudinal strips. The larva is cruciform, with dark head and legs, with the body yellow. The body is usually covered with dried excrement so the larva generally resembles a shiny blob on the leaf surface. There are four larval instars; development takes 3–4 weeks. Mature larvae fall to the ground and pupate in an earthen cell some 7–8 cm deep in the soil.

Adults emerge in the late summer and feed on grasses until September, when they hibernate in leaf litter and cracks in the soil. They are easily recognized, for the head is black, thorax pale brown and have shiny blue elytra; body length is 4–5 mm; they fly readily if disturbed. In the spring the adults emerge again and may be seen in April; they mate towards the end of May, and egg-laying starts a week later. The life-cycle takes 40–50 days, but it is univoltine.

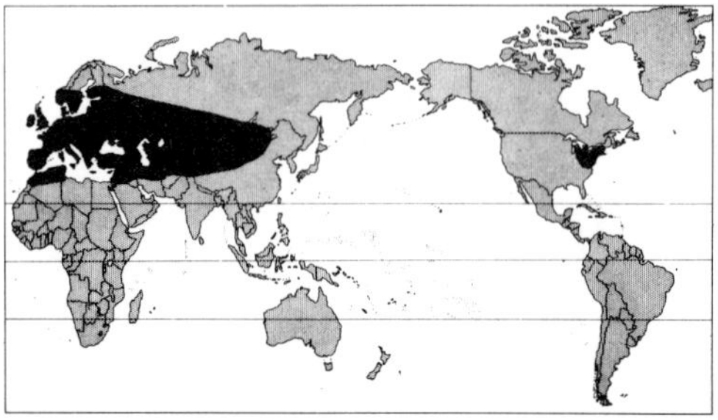

Distribution. This species occurs throughout Europe and most of temperate Asia, N. Africa, and now USA and Canada (CIE map no. A.260).

Other important species include:
Oulema oryzae Kuw. – (Rice Leaf Beetle) China, Japan.
Oulema bilineata – a pest of tobacco in S. Africa.
Oulema trilineata O. – (Three-lined Potato Beetle) USA.
Oulema erichsoni Mats. – (Wheat Leaf Beetle) Japan.
Oulema dilutipes Fair. – also from Japan.
Oulema spp. on eggplant in Asia and the USA.

Control. Damage is seldom serious enough in Europe to warrant control measures being taken, but in N. America it is a more serious pest.

Early sowing of cereals will minimize effects on the growing crops, and pubescent varieties of wheat are less attractive as hosts.

Insecticides that have proved successful on these pests include DDT, HCH and phosphamidon.

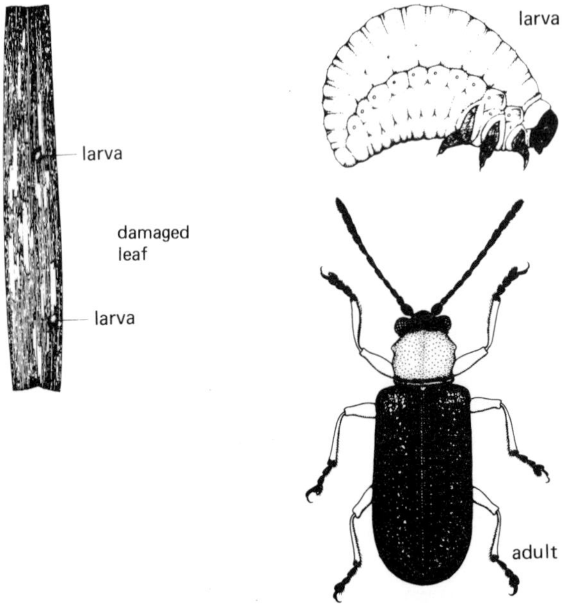

Colaspis brunnea (F.)

(= *C. flavida* Say.)

Common name. Grape Colaspis
Family. Chrysomelidae (Eumolpinae)
Hosts (main). Pulse crops and other legumes; grapevine and strawberry (roots).
(alternative). Rice
Damage. The larvae feed on the roots of leguminous plants, especially lespedeza clover, and will attack roots of rice seedlings (and seeds) if this crop is planted after legumes. The effect on the rice crop is to reduce plant stand and reduce tillering. Adults damage the foliage of many different crop plants.
Pest status. Occasionally a serious pest in southern USA when rice is grown after a legume crop.
Life history. Eggs are laid in the soil around the roots of leguminous plants. The larvae spend up to seven months in earthen cells at about subsoil level; if rice is planted after the legumes the larvae feed on the seeds, seedlings and roots. Pupation occurs in the soil, and takes 3–7 days to complete. Adults are small, brownish leaf beetles, about 5 mm in length.

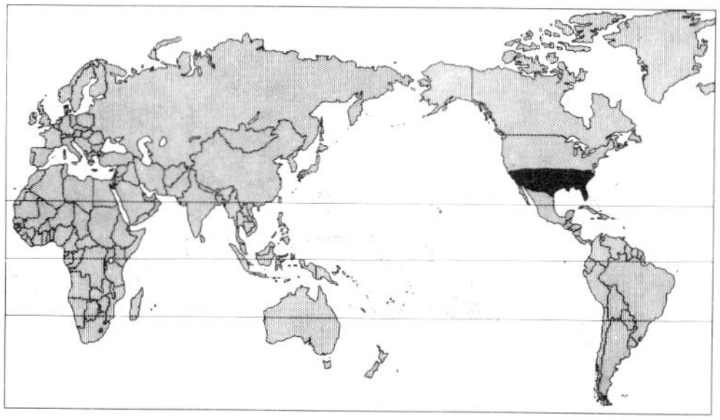

Distribution. This species is only recorded from southern USA; there are in addition several other species of *Colaspis* to be found in the USA, as well as the Banana Fruit-scarring Beetle (*C. hypochlora* Lefevre) in Mexico, C. America, and northern S. America.
Control. Usually controlled by the use of seed dressings of aldrin or heptachlor, and phorate granules have also been successfully used for controlling this pest.

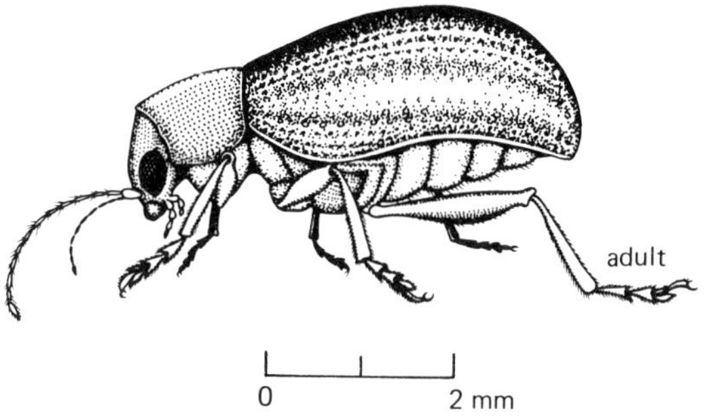

Diabrotica spp.

(*D. balteata* LeConte)
(*D. longicornis* S. & L.)
(*D. undecimpunctata* Mann.)

Common name. Banded Cucumber Beetle
Northern Corn Rootworm
Spotted Cucumber Beetle
Family. Chrysomelidae (Galerucinae)
Hosts (main). Cucurbits, maize.
(alternative). Sunflower, tobacco, groundnut, and many other crops.

Damage. The soil-dwelling larvae eat the roots of maize (corn), hence their names of 'rootworms' and they burrow into the stem often killing the seedling; the adults eat foliage, and the silks of maize, and several species show preference for flowers and foliage of Cucurbitaceae.

Pest status. Collectively this group is important as they are quite serious pests of both maize and cucurbits. Some species transmit Cucumber Mosaic Virus, and bacterial wilts of both maize and cucumbers.

Life history. Eggs are laid in the soil; they may overwinter in northern locations.

The larvae are small, wrinkled, yellowish-white with a brown head capsule. After a total developmental time of about 30 days they grow to 10–18 mm in length. Pupation takes place in an earthen cell in the soil.

The adult beetles are quite small, only 5–6 mm in length, but distinctively marked with spots and bands across or along the elytra; basic body coloration is pale greenish-yellow.

The life-cycle takes some 50 days under warm conditions; in the southern States there may be as many as six generations per year (*D. balteata*), but in Canada the other two species may be univoltine.

D. longicornis appears to be restricted to maize, whereas the other species tend to be more general feeders. It might, in point of fact, be more satisfactory to consider each species separately as there are some fundamental differences in their biology and distributions within the New World.

Distribution. These are essentially New World species and recorded distribution extends from Canada, throughout the USA, C. and S. America.

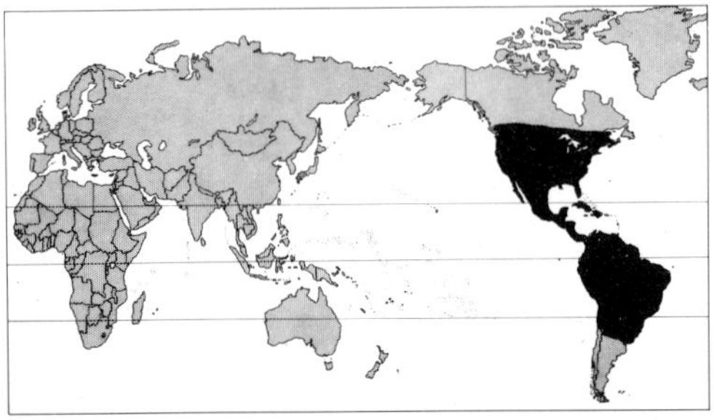

D. undecimpunctata

Other important species include:

Diabrotica v. virginifera LeConte – Western Corn Rootworm.

Diabrotica virginifera zeae K. & S. – Mexican Corn Rootworm.

A closely related group of cucumber pests are placed in the genus *Acalymma*, specifically *Acalymma vittata* (F.) – Striped Cucumber Beetle; Canada and USA; and *A. trivittatum* (Mann.) – Western Striped Cucumber Beetle of the USA. Both species closely resemble *Diabrotica* spp. but are distinctively striped longitudinally.

Control. In areas at risk, the avoidance of winter crops such as alfalfa, which encourages the pests in the spring, will reduce populations.

Most species are now showing resistance to the previously used organochlorines and if control is really required then soil application of granules of diazinon, phorate or parathion are generally effective.

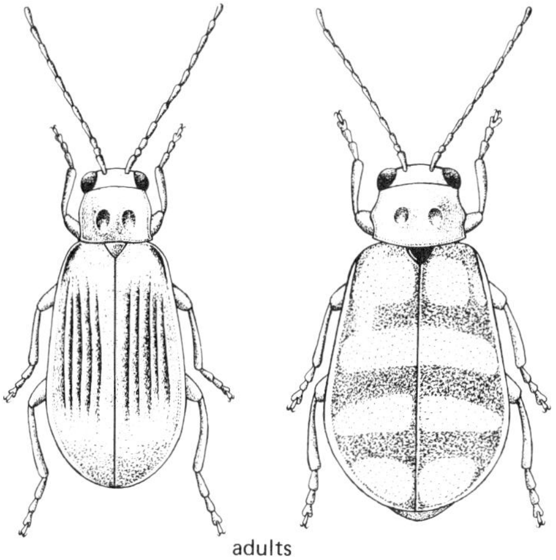

adults

D. longicornis *D. balteata*

Chaetocnema concinna (Marsh.)

Common name. Mangold Flea Beetle
Family. Chrysomelidae (Halticinae)
Hosts (main). Mangolds, sugar beet.
(alternative). *Rumex*, *Polygonum*, and other Chenopodiaceae.
Damage. Adults eat small shot-holes in the cotyledons and first leaves; if the weather is hot and the beetle population large then seedlings may be destroyed. Larvae in soil eat the roots but cause no discernible damage.
Pest status. Occasionally a serious pest of beets in the UK and Europe, but outbreaks are irregular; damage levels are associated with weather, most serious under dry conditions; under dry cool conditions growth is slowed, under dry hot conditions wilting is most likely.
Life history. Eggs are laid in the soil around the plants in April and May; hatching takes 15–30 days.

The larvae feed on the seedling roots for about 4–6 weeks, by which time they reach 4–6 mm in body length; they then pupate in the soil.

Adults are tiny flea beetles, uniform metallic bronze-green in colour, 1.5–2.5 mm in length, with deeply punctured elytra. They emerge in July and feed on the crop throughout the summer, causing no damage to the plants which are by then well grown. In the autumn the adult beetles disperse to hedgerows, woods and the like where they hibernate in the leaf litter or grass tussocks. They emerge in the first warm days of the following spring, and generally disperse when air temperatures approach 21 °C; thus damage is most likely to be severe in a warm dry spring.

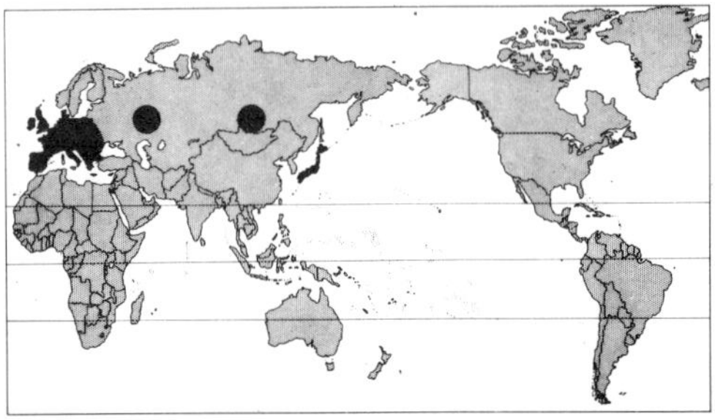

Distribution. At present only recorded from the UK, western Europe, and Japan, with the closely related (?) species *C. tibialis* Illiger in Bulgaria. In Japan another five species are recorded, including *C. cylindrica* Baly, the Barley Flea Beetle. In the UK a total of nine species of *Chaetocnema* are known.

In Canada *C. pulicaria* Melsh. is the Corn Flea Beetle which, together with *C. ectypa* Horn (Desert Corn Flea Beetle) causes serious damage to maize crops both in Canada and the USA; they are both vectors of Corn Wilt bacteria. In the USA *C. confinis* Crotch is the Sweet Potato Flea Beetle; *C. denticulata* (Illiger), the Toothed Flea Beetle, is another pest species on maize. In Australia *C. australica* Baly is the Couch Flea Beetle and does not appear to be a serious crop pest.
Control. For control measures see page 305.

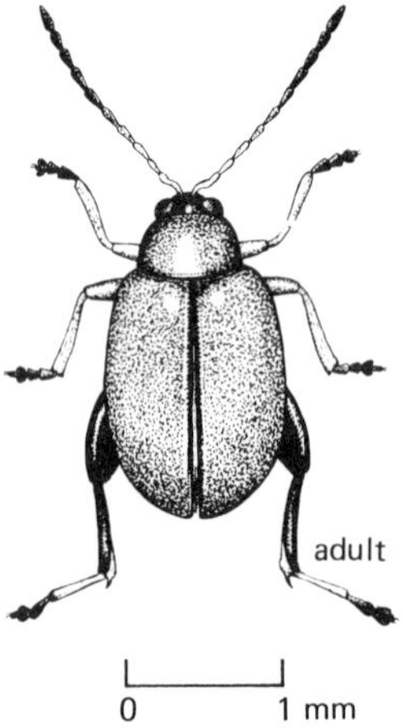

Crepidodera ferruginea (Scop.)

Common name. Wheat Flea Beetle
Family. Chrysomelidae (Halticinae)
Hosts (main). Wheat (and grasses).

(alternative). Thistles (*Cirsium* etc.) and grasses are the natural hosts for this species.

Damage. The larvae bore into the central shoot of wheat seedlings, causing dead-hearts and patchiness in the crop. Typically a 'ley pest'. Adults feed on thistles. A small neat hole in the base of the shoot is made by the attacking larva.

Pest status. Sporadic attacks recorded on wheat throughout the UK and Europe; generally unimportant and more of academic interest than economic.

Life history. Eggs are laid on the soil surface by the grass or wheat plants; hatching takes 3–4 weeks. The larvae feed in the base of the seedling plant; they are creamy-white with a black head capsule and posterior plate. Larvae normally feed inside the grass shoots during the winter and spring.

Pupation takes place in the spring (May) and requires about four weeks. Adults emerge over a period from June to mid-September, and egg-laying follows shortly. Adults are rusty-reddish in colour with yellow legs; body length is 3.0–3.5 mm.

Infestations of wheat crops usually occurs when a ley is ploughed in late summer and then sown with winter wheat: the larvae disperse from the rotting turf and enter the young wheat shoots.

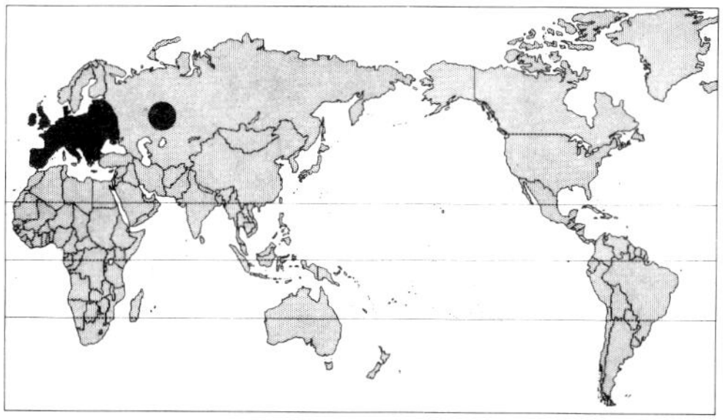

Distribution. Recorded only from the UK and Europe.

Control. Control may be achieved by leaving a gap of at least four weeks between the ploughing of grassland and the drilling of a following crop of winter wheat.

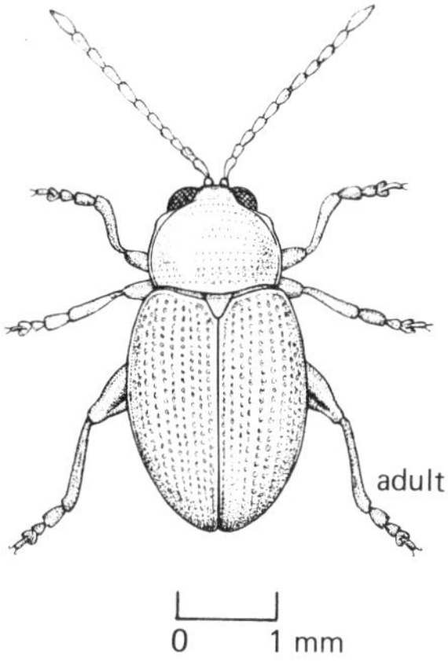

Phyllotreta spp.

Common name. Cabbage Flea Beetles
Family. Chrysomelidae (Halticinae)
Hosts (main). *Brassica* spp. and other Cruciferae.
(alternative). Some species are more polyphagous and also feed on cereals, and some on cotton.
Damage. Adults feed on cotyledons and young leaves, making shot-hole damage; seedlings may be destroyed. Larvae live in the soil (mostly) and feed on the roots of the host plants but do little damage. Some adults transmit mosaic viruses.
Pest status. A very widespread and abundant group of species worldwide, occasionally causing economic damage; otherwise regular minor pests.
Life history. Eggs are laid on the soil next to the host plants, in batches either small (3–4) or large (20–30), according to the species concerned. Hatching takes 1–2 weeks. Oviposition generally takes place in Europe from April to July.

The larvae either burrow into the soil, to feed on the roots, or in the case of *P. nemorum* they bore into stem or leaf tissue and feed in the mesophyll where they make small blister mines. Most larvae are mature by July when they measure 5–6 mm in length; their general body colour is whitish or pale yellow, with very short legs, dark head, and dark skeletal plates on both thorax and abdomen.

Pupation for all species (including *P. nemorum*) takes place in an earthen cell in the soil; and takes 2–3 weeks.

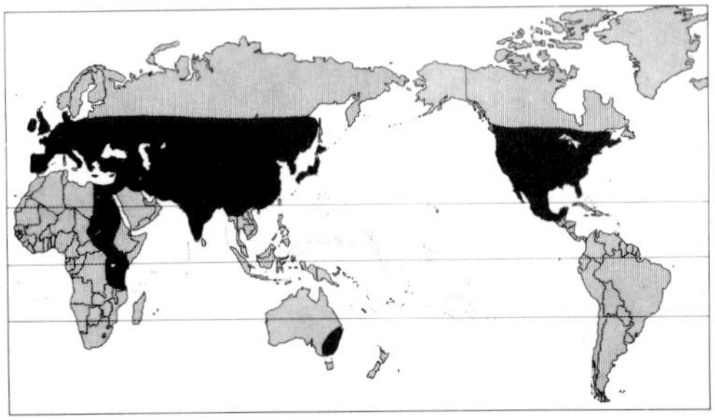

Young adults emerge in Europe in July and August; they range in size from less than 2 mm up to 3.5 mm in length. There are two predominant colour patterns: one group resembles *P. nemorum* and is black with two broad yellow bands, the other group is unicolorous shiny black, often with a metallic sheen, although leg coloration may vary according to species. The adults overwinter in leaf litter or crevices in the soil. The life-cycle takes 5–6 weeks usually; some species in some locations are univoltine, whereas others may typically have 2–3 generations per year. The adult beetles emerge from hibernation when the warm weather comes in the spring, over a period from April to June.
Distribution. The genus *Phyllotreta* is abundant throughout Europe, Asia, north-eastern Africa, S.E. Australia, and throughout N. America.

The more important species are as follows:
Phyllotreta aerae All. – (Radish Flea Beetle) on crucifers; Europe and Canada.
P. albionica (Lec.) – (Cabbage Flea Beetle) on crucifers; Canada.
P. armoricae (Koch) – (Horseradish Flea Beetle) on crucifers; Canada and USA.
P. atra (F.) – (Small Black Flea Beetle) on crucifers; Europe.
P. cheiranthei Weise – on crucifers and cotton; Egypt, E. Africa, Sri Lanka.
P. consobrina (Curtis) – (Turnip Flea Beetle) on crucifers; UK (rare in Europe).
P. cruciferae (Goeze) – (Cabbage (Crucifer) Flea Beetle) on crucifers; Europe, Asia, USSR, Canada and USA.
P. diademata Foudras – (Crown Flea Beetle) Europe.
P. nemorum (L.) – (Large Striped Flea Beetle) on crucifers; Europe, Asia, USSR, and S.E. Australia.
P. nigripes (F.) – (Turnip Flea Beetle) on crucifers; Europe.
P. pusilla Horn – (Western Black Flea Beetle) on crucifers and other plants; USA.
P. ramosa (Crotch) – (Western Striped Flea Beetle) on crucifers and some ornamentals; USA.

P. robusta Lec. – (Garden Flea Beetle) crucifers and some garden plants; Canada.
P. striolata (F.) – (Striped Flea Beetle) on crucifers; China, Japan, Canada and USA.
P. undulata Kuts. – (Small Striped Flea Beetle) on crucifers; Europe.
P. vittula Redt. – (Barley Flea Beetle) attacks barley, also other temperate cereals, as well as Cruciferae; Europe.

Most temperate countries have a large number of *Phyllotreta* species on their faunal checklists; the UK is fairly typical and has 14 species, the majority of which are agricultural pests.

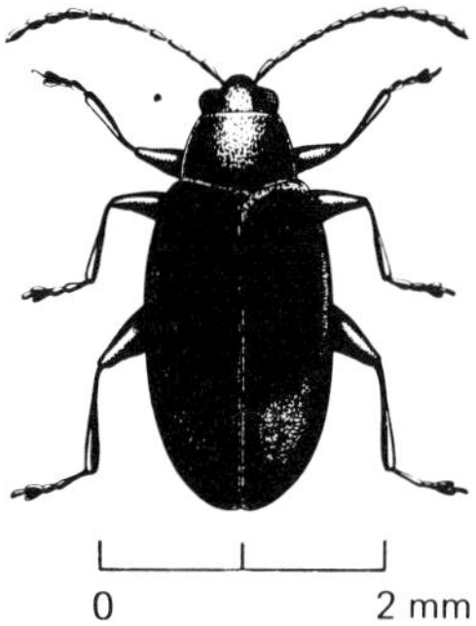

P. cruciferae

adults

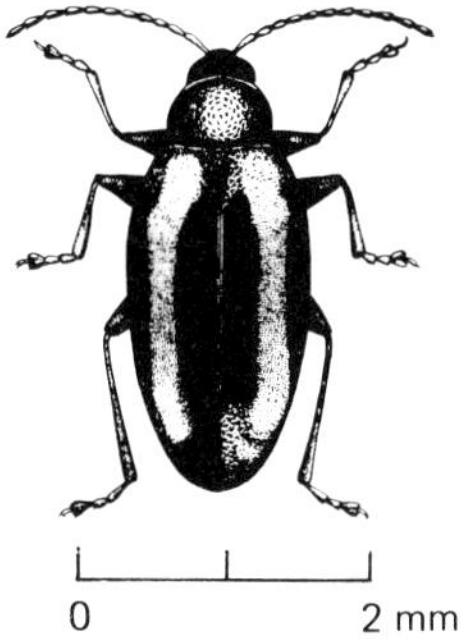

P. nemorum

Damage to leaves of radish by *P. striolata*

Flea beetles (Coleoptera; Chrysomelidae; Halticinae)

This group contains a very large number of mostly tiny dark leaf beetles all with similar oval, rounded bodies, longish antennae, and stout hind femora with which they make their flea-like prodigious leaps. They feed on the foliage of a wide range of plants, from grasses and herbs, to shrubs and trees. Typical feeding damage consists of small shot-holes in the leaf lamina (see illustrations on pages 85 and 303) and seedling stems may be gnawed. Generally the adults overwinter by hibernating in leaf litter and plant debris, thus infestations may be very heavy in the early spring and seedling damage may be very serious. Entire crops may be destroyed whilst still at the cotyledon stage.

The larvae are elongate and cylindrical in shape, like tiny wireworms, and are usually to be found in the soil where they feed on plant roots and generally are of little consequence economically. However there is some diversity of larval biology; the larvae of *Prodagricomela nigricollis* mine leaves of *Citrus* in China (Hill, 1983; page 466) in a most conspicuous manner; *Phyllotreta nemorum* larvae mine the leaves and stems of *Brassica* spp., making small blister mines that later dry up and are generally quite inconspicuous. Larvae of *Epitrix cucumeris* and *E. tuberis* make tunnel mines on the surface of potato tubers in Canada and the USA. *Psylliodes chrysocephala* has larvae that make galls in the stems of overwintering brassicas. In some species of *Altica* the larvae live on the leaf surface alongside the adults (as do many other Chrysomelidae) where they hole the lamina of the leaf and cover the surface with a messy black excrement (Hill, 1982; page 130).

Important pest species of flea beetles

In addition to the species included in this chapter, there are many other species of economic significance throughout the world. Field identification of flea beetles is often very difficult, even in the cases where the adults are brightly coloured or patterned, because of the very large number of species that occur in most of the genera. Groups of some economic importance include:

Altica spp. (= *Haltica*) – a large, worldwide genus of small bluish beetles, with several pest species in Asia, at least five pest species in the USA, and seven species mostly from crop plants in the UK.

Aphthona euphorbiae (Schrank) – (Large Flax Flea Beetle) recorded only from flax; throughout Europe and the UK.

Aphthona spp. – many are recorded from wild herbaceous plants in Europe, but some species prefer tree hosts in eastern Europe and Asia; eight species recorded from the UK.

Argopistes spp. – (Citrus Flea Beetles) from China and Japan; also one species recorded from privet in China.

Disonycha spp. – on spinach and other vegetables; in Canada and the USA.

Epitrix cucumeris (Harris) – (Potato Flea Beetle) on potato, tobacco, eggplant, cucurbits, and other crops; Canada and the USA.

E. fuscula Crotch – (Eggplant Flea Beetle) on Solanaceae; USA.

E. hirtipennis (Melsh.) – (Tobacco Flea Beetle) on tobacco and tomato; Canada and USA.

E. subcrinata LeConte – (Western Potato Flea Beetle) mostly on potato; Canada and USA.

E. tuberis Gentner – (Tuber Flea Beetle) polyphagous on potato and many vegetable and other crops; Canada and USA.

Epitrix spp. – other pest species are known in N. America, mostly preferring Solanaceae as hosts; in Europe several species feed on wild Solanaceae.

Longitarsus parvulus (Payk.) – (Flax Flea Beetle) apparently only recorded from flax; found throughout Europe; in the UK a total of 41 species of *Longitarsus* are recorded from wild herbaceous hosts in several different families (Compositae, Labiatae, Leguminosae, etc.).

Prodagricomela nigricollis Chen – (Citrus Flea Beetle/Leaf Miner) only recorded to date from *Citrus* spp. in S. China.

Systena blanda Melsh. – (Pale Striped Flea Beetle)

polyphagous on many crops; Canada and USA.
S. elongata (F.) – (Elongate Flea Beetle) USA.
S. frontalis (F.) – (Red-headed Flea Beetle) on clovers and many other crops; Canada.
Systena spp. – at least three other species are polyphagous crop pests throughout Canada and USA.

Control of flea beetles

Because of the adults hibernating over winter, it is mostly the seedling stage of the crop that has to be protected; damage to trees and established shrubs and herbs is seldom serious as compensatory growth will maintain an adequate leaf area index.

Cultural methods. Use of good seedbeds (moist, fine tilth, fertilized, etc.) will help young plants to grow quickly through the susceptible seedling stage.

Chemical methods. Insecticides may be used in several different ways in order to combat flea beetles, as follows:

(1) Seed treatment – dressings, etc., of HCH; followed, if required, by later treatment of the growing crop (sprays of HCH, carbaryl, diazinon).
(2) Dusting or spraying of the crop after infestation is evident, with DDT, HCH, carbaryl or diazinon.
(3) Soil application of granules at sowing – primarily for control of other soil-dwelling pests (e.g. Cabbage Root Fly; beet seedling pests); carbofuran is recommended for this purpose.

Psylliodes affinis (Payk.)

Common name. Potato Flea Beetle
Family. Chrysomelidae (Halticinae)
Hosts (main). Potato, tomato.

(alternative). Preferred hosts include *Solanum nigrum* (Black Nightshade) and *S. dulcamara* (Bittersweet).

Damage. Adults hole the leaves of young plants in the spring, and large plants in August (second generation); larvae in the soil eat roots. The adults in the spring may often be found on wild Solanaceae rather than 'on the crops.

Pest status. Occasionally attacks have been severe in both the UK and Europe, mostly on potato crops growing on heavy land. Usually a minor pest on these two crops.

Life history. The hibernating adults emerge and become active in the spring, and the females lay eggs, singly or in groups, in the soil at the base of the host plant. Hatching takes about 8–10 days.

The larvae, white-bodied with a darker head, feed on the plant roots in the soil doing little noticeable damage, and mature in about four weeks at a length of 5–6 mm. Pupation takes place in the soil, also requiring about four weeks.

Adults are quite large (2–3 mm long) and brown in colour; femora dark with the rest of the legs pale. They emerge in August/September and may be found on potato crops in large numbers when feeding damage becomes very obvious but injury is negligible. In the autumn the adults disperse to hedgerows where they hibernate until the following spring.

There is only one generation per year.

Distribution. *P. affinis* is widespread throughout the UK and parts of Europe, other species occur in other parts of the world, also on Solanaceae, but some on other crops as well.

Out of the 14 different species of *Psylliodes* recorded in the UK, several are pests, including *P. attenuata* (Koch), the Hop Flea Beetle.

P. elliptica All. attacks fodder beet, wheat and barley in Israel.

P. punctulata Melsh. is the Hop Flea Beetle of Canada and USA; a polyphagous pest recorded from crops in the Solanaceae, Chenopodiaceae, Cruciferae, as well as hop, rhubarb, clovers and various weeds.

Several other species are pest of Cruciferae and Solanaceae in Japan.

Control. For control information see page 305.

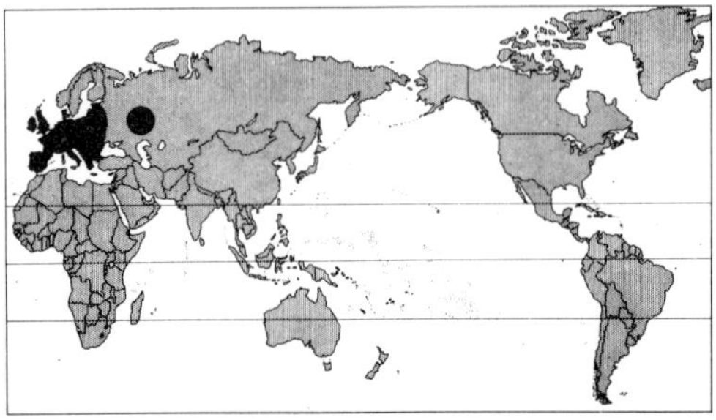

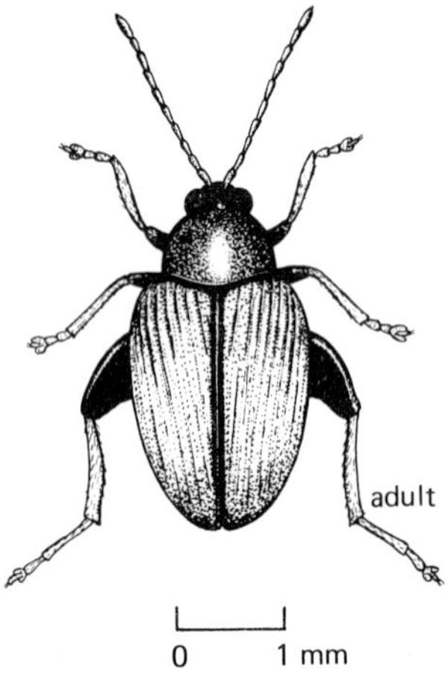

Psylliodes chrysocephala (L.)

Common name. Cabbage Stem Flea Beetle
Family. Chrysomelidae (Halticinae)
Hosts (main). *Brassica* species.
(alternative). Other members of the Cruciferae.
Damage. Larvae tunnel inside stems and leaf-veins; some stems are completely hollowed; severely attacked plants may die. Damage is most common in autumn and winter on young plants. Adults are rarely found in the field, and cause little damage.
Pest status. A locally serious pest in parts of the UK and Europe, and now Canada, especially on crops of *Brassica* spp. planted out in the autumn.
Life history. Adults hibernate over winter, and emerge in the spring, when the females lay their eggs; each female may lay up to 1000 eggs, singly, in the soil or on cruciferous plants; hatching requires about 20 days. Some females may live for as long as 1.5 years and they oviposit twice, in the late summer/autumn and then again in the spring.

The larvae bore into the host plant, usually at the base of a petiole, from where they tunnel either into the stem or into the leaf-vein. The larvae grow rapidly to a size of about 8 mm, when they leave the plant to pupate in the soil at the end of April or in May.

Adults emerge in June. They are large flea beetles, 4–5 mm in length, shiny and green or blue (colour is rather variable), with reddish legs. Most of July is spent in aestivation, and egg-laying starts in August and continues until the winter. Some females die after laying their eggs, but others continue feeding for a while and then hibernate for the winter; they will lay the rest of their eggs in the spring. Some females are recorded living for 1.5 years.

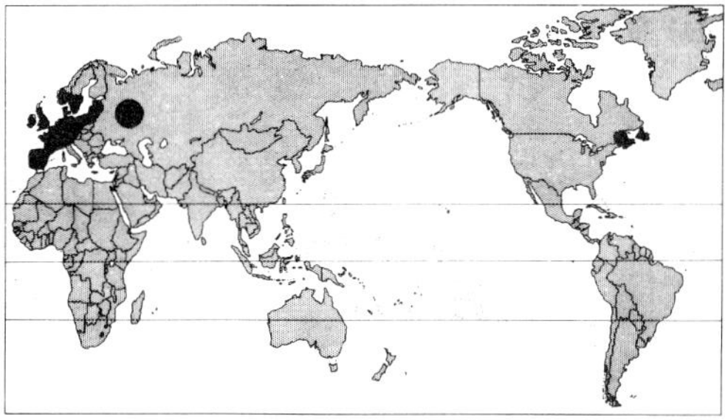

The eggs hatch, and the young larvae attack the autumn/winter-sown or planted-out spring cabbage, kale, and rape. In these hosts the larvae feed and develop over the winter, usually leaving the plants in February to pupate in the soil.

The generations generally overlap considerably and all stages may be found at any one time.
Distribution. Recorded throughout the UK and Europe, often more abundant in coastal regions; now recorded from Canada (Newfoundland).

In Canada the closely related *P. punctulata* Melsh. (Hop Flea Beetle) is a polyphagous pest recorded from many Cruciferae, but the larvae eat the roots in the soil. *P. punctifrons* Baly is the Cabbage Flea Beetle of Japan.
Control. Brassicas sown *in situ* by drilling withstand attack by this pest far better than do transplanted ones; and well-grown, healthy plants tolerate some attack with no signs of injury.

For chemical recommendations see page 305.

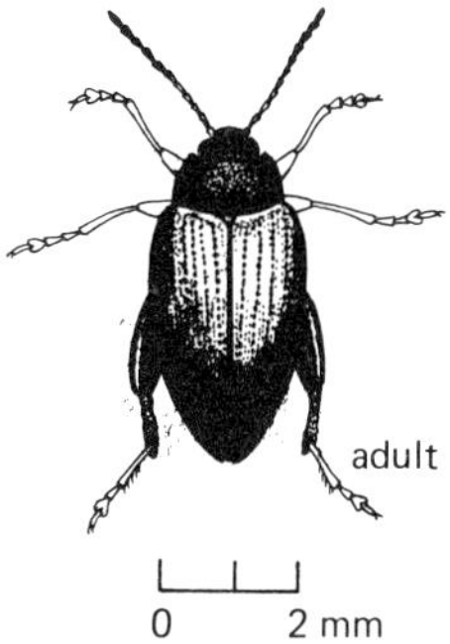

Apion spp.

Common name. (Clover) Seed Weevils, etc.
Family. Apionidae
Hosts (main). Flowers of Leguminosae, especially clovers.
(alternative). Other species attack cotton, lettuce, cabbage, carrot, cashew, jute, citrus and other crops.
Damage. The larvae develop and feed within the young ovules or pods and eat the seeds; sometimes half the seeds in one flowerhead may be destroyed; feeding adults slightly damage the foliage, but this is generally unimportant. A few species have larvae that make tunnel mines in the leaves, or in plant stems.
Pest status. Not very important as pests, although occasional severe infestations do occur with some regularity; but widespread and abundant, and frequently encountered.
Life history. Eggs are deposited singly in green flowerheads, before they open in May; each female only lays a small number of eggs (5–20); hatching takes 4–8 days. Usually each female lays several eggs per flowerhead, and the oviposition period lasts for about one month.

The tiny white larvae burrow into the florets and feed on the developing ovules; a single larva may damage 6–10 ovules. Larval development proceeds through three instars in about 18 days, and at about 2 mm length it pupates inside the flowerhead, usually inside a single ovule. Pupation takes about six days.

The young adults are tiny beetles with a characteristic globular body and elongate tapering snout, with small non-geniculate antennae; body length is only 1.5–3.0 mm; coloration is typically black and shiny, but some species are reddish and a few brightly metallic blue or green. Identification to species is very difficult and can only be done by taxonomists. The adults seldom fly, and stay on the host plant feeding on the foliage; sometimes damage to buds may be serious. Egg-laying proceeds and the second generation commences. At the end of the summer adult beetles disperse to hedgerows and woods where they hibernate over winter.

In Europe there are generally two generations each year.
Distribution. The genus *Apion* is completely cosmopolitan and found on all the continents in the world. In some countries there are many species, most of which are tiny, shiny black, pear-shaped weevils: the UK has 83 species recorded! The majority are recorded from the Leguminales as hosts.
Control. Careful management of clover crops can reduce infestation levels, especially early harvest of forage crops.

Chemical control has, to date, relied on high-volume sprays of DDT or HCH, given as two sprays, the first when buds are forming and the second as the buds open.

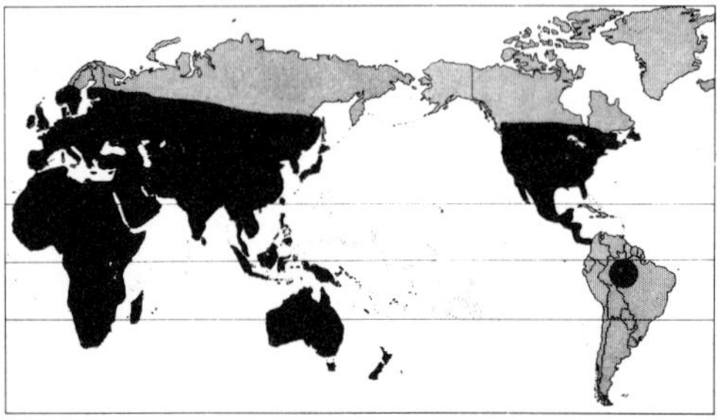

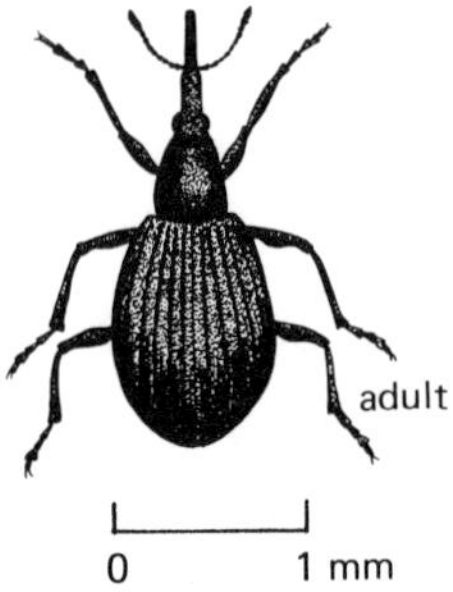

Araecerus fasciculatus (Deg.)

Common name. Coffee Bean Weevil; Nutmeg Weevil
Family. Anthribidae
Hosts (main). Nutmegs, coffee, cocoa beans.
(alternative). Seeds of many plants, both ripe in the field and in stores.
Damage. The larvae develop inside the dried seeds, usually staying inside the same seed for the whole larval period, and pupation takes place there.
Pest status. A serious pest of stored seeds in many parts of the tropics, and of regular occurrence in temperate food stores, but in the UK fails to survive the winter.
Life history. Eggs are laid singly on the ripening or ripe seeds, both in the field and in storage.

The larvae are typical weevils, being fat-bodied, white, legless but with a dark brown head capsule and well-developed jaws. The larvae stay inside the seeds for their period of development and then pupate there.

The adult resembles a small brown bruchid, but without emarginate eyes; body length about 3 mm; they are active fliers and disperse readily. In Hong Kong, where they live mainly on nasturtium seeds and wild hosts, adult beetles are frequently seen on kitchen windows attempting to gain access to domestic premises.

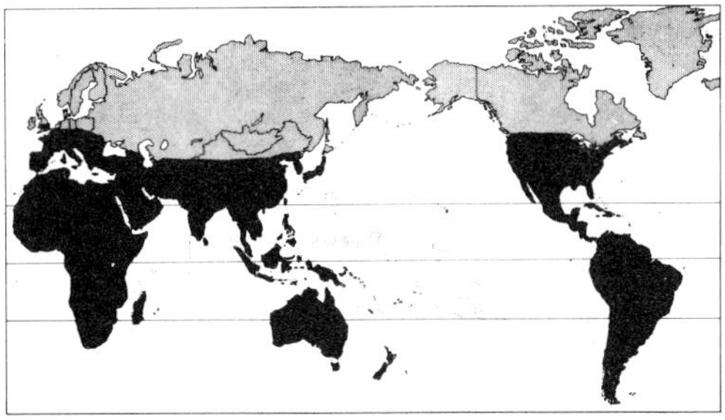

Distribution. Cosmopolitan throughout the warmer parts of the world, and in heated stores in temperate regions; also found in unheated stores in UK and Europe, but here the insects die over winter due to the cold.
Control. Generally not required as field infestations of crops are usually light, and warehouse infestations are controlled by the regular fumigations.

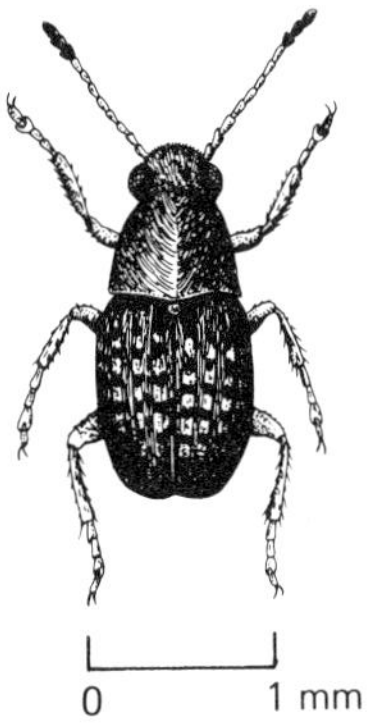

Anthonomus pomorum (L.)

Common name. Apple Blossom Weevil
Family. Curculionidae
Hosts (main). Apple
(alternative). Pear, quince.

Damage. Adults pierce unopened blossom buds to feed on the developing anthers; larvae feed inside the unopened ('capped') buds and destroy the flowers; one larva per bud is usual.

Pest status. Formerly a very serious orchard pest but, since the use of DDT, has declined to occasional pest status; often more common in orchards near to woodland.

Life history. Adults hibernate over winter, and may be found on the trees in the spring on warm sunny days, usually on the flower buds; usually from mid-March to mid-April, or bud-breaking to green cluster stage. They feed mainly on the blossom buds by piercing the bud scales that still ensheath the fruit buds.

Eggs are laid from bud-burst until green cluster, laid individually amongst the anthers, one per bud; each female lays 40–50 eggs.

The larvae feed first on the anthers and later on the bases of the petals so that they remain unopened. The petals eventually turn brown and show the characteristic 'capped' appearance. Larval development takes 3–4 weeks.

Pupation takes place within the capped bud, and takes some 16–20 days. The young adult escapes by gnawing a hole through the base of the dead petals. The adult is a small weevil, about 5 mm in length, with a long rostrum; body colour is dark blackish-brown, but with a pale band in the shape of a **V** across its back. The adults feed on the leaves making small holes in the lamina. After about three weeks the adults seek sheltered situations in which they hibernate until the following spring. Adult activity ceases about the end of June.

There is only one generation per year.

Distribution. This species is recorded from Europe and Japan.

Other species of importance include:
Anthonomus grandis Boh. – Cotton Boll Weevil of Mexico and the USA, and South America (CIE map no. A.12).
Anthonomus musculus Say – (Cranberry Weevil) USA.

Other species also occur on fruit crops and are dealt with on the following pages.

Control. For control measures see page 312.

This pest is usually heavily parasitized by the ichneumon wasp *Scambus pomorum* (Ratz.), and parasitism levels of 25% have been recorded in several UK orchards.

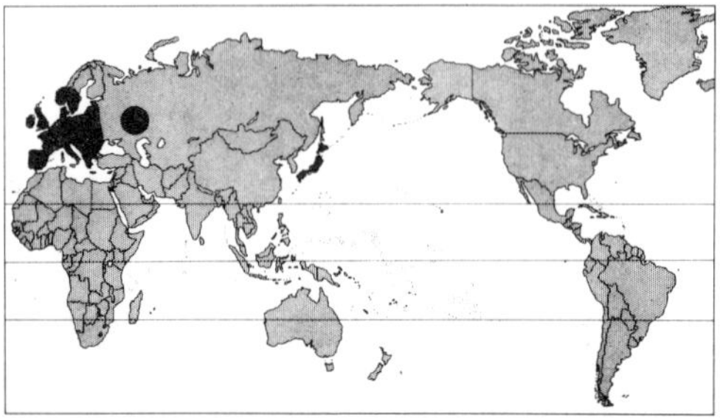

'Capped' blossom with larva inside

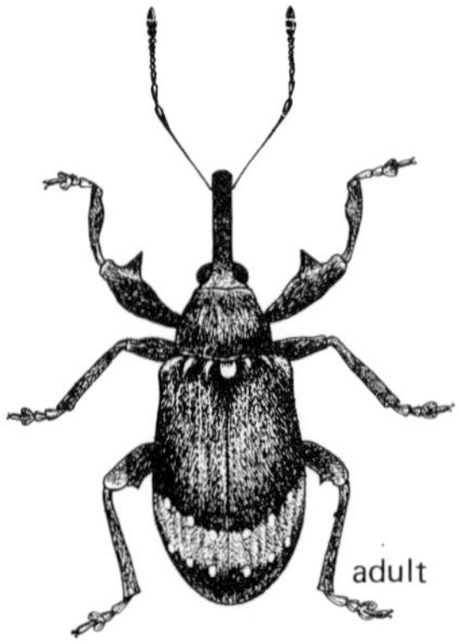

adult

0 1 mm

Weevils (Coleoptera; Curculionidae)

A very large group of beetles, containing more than 60 000 species, worldwide in occurrence, equally well distributed throughout the tropics and temperate regions, and many are pests of cultivated crops and forest trees. There is some diversity of life-styles that makes generalization a little difficult, but there is an overall similarity.

All the larvae show great morphological similarity, being fat-bodied, sluggish, legless, whitish in colour with a brown well-sclerotized head capsule, and well-developed mandibles. They are most frequently to be found in the soil where they bite and chew fine roots and sometimes the root damage is sufficient to warrant their categorization as pests. Some larvae burrow into the root nodules of Leguminosae (*Sitona*), some bore into the tap root, some bore into herbaceous stems and rhizomes (banana weevils; coconut weevils; Sisal Weevil). A few larvae are aquatic, and some are leaf-miners in dicotyledonous plants. Others make globular galls in herbaceous stems (Cabbage Stem Weevil), or in root crops (Turnip Gall Weevil), or corms (*Otiorhynchus*). Many species develop inside seeds and grains (Hazelnut Weevil; Cabbage Seed Weevil; *Sitophilus*) and a few inside unopened flower buds (blossom weevils), and others inside developing fruits (Cotton Boll Weevil; Mango Seed Weevil). The pantropical palm weevils (*Rhynchophorus*) have larvae that bore the crown of palms, destroying the growing point, and may eventually bore right through the trunk. Another rather specialized case is the Tea Root Weevil in Africa, whose larvae tunnel the roots of woody shrubs (tea, coffee, etc.) and often ring-bark them, killing the entire bush. Several genera are timber borers and they tunnel the sapwood of crops such as cashew (Cashew Weevil) in Africa; there are many species in *Pinus* and other conifers, and some in deciduous trees (oak, alder, birch, etc.), and these are forestry pests of some consequence.

Adult weevils may be roughly divided into two groups: 'broad-nosed' weevils and 'snouted' weevils with a long rostrum and terminal mouthparts. The 'broad-nosed' weevils are foliage eaters; some make regular peripheral feeding notches in the leaf margin (*Systates*; *Sitona*), others tend to make holes in the lamina of the leaf, or else leave ragged-edged feeding scars on the leaf edges (*Phyllobius*; *Otiorhychus*).

The 'snouted' weevils usually (in the case of the females) combine the act of feeding with the preparation of an oviposition site. The long rostrum is used to make a deep excavation into the host plant tissues, the female feeds whilst excavating the hole, then she turns round and lays an egg there. The larva then develops *in situ*, in the shoot, stem, fruit, seed, etc. This is shown particularly well in the case of the palm weevils (*Rhynchophorus*), and the Chinese Bamboo Weevil (Hill, 1983; p.477). Species of *Alcidodes* are known as stem-girdling weevils in that the adults eat the stem bark, usually girdling the stem just above ground level.

Pupation usually takes place in the soil, except for the deep plant tissue borers which remain inside the stem/fruit/seed to pupate. Thus infested fruits/seeds show a characteristic emergence hole through which the young adult has departed (see Hazelnut Weevil; page 318).

Important pest species of weevils

In addition to the species already included in this chapter, there are other important pest species that require mention.

Alcidodes spp. – (Stem-girdling Weevils) several hosts; Africa and India.

Baris spp. – (Melon Weevils, etc.) larvae bore fruits of Cucurbitaceae in the Mediterranean Region; also pineapple in S. America.

Conotrachelus nenuphar Herbst – (Plum Curculio) on top fruits; USA and Canada.

Conotrachelus spp. – (Quince/Butternut/Black Walnut Curculios) USA.

Cosmopolites sordidus (Germ.) – (Banana Weevil) pantropical (CIE map no. A.41).

Curculio sayi – (Small Chestnut Weevil) on Sweet Chestnut; USA.

Cyrtotrachelus longimanus – (Bamboo Weevil) China.
Diocalandra spp. – (Coconut Weevils) Old World tropics (CIE maps nos. A.248 and 249).
Listroderes costirostris (Klug) – (Vegetable Weevil) polyphagous; S. America and USA.
Listronotus oregonensis (LeC.) – (Carrot Weevil) USA.
Lissorhoptrus spp. – (Rice Water Weevils) USA, S. America, and now Japan.
Lixus spp. – (Beet/Cabbage Weevils) Mediterranean Region and Asia.
Lixus concavus Say – (Rhubarb Weevil) USA.
Odoiporus longicollis (Oliv.) – (Banana Stem Weevil) tropical Asia and the Pacific Islands.
Magdalis spp. – (Bark Weevils) larvae bore under tree bark, adults hole leaves; on many fruit trees, elms, etc.; Europe, USA.
Myllocerus spp. – (Grey Weevils) polyphagous; India and Asia.
Polydrusus spp. – (Leaf Weevils) on fruit and nut trees, larvae in the soil; Europe.
Rhynchophorus (Palm Weevils) pantropical (CIE maps nos. A.258 and 259).
Pissodes spp. – (Conifer Weevils) Asia, Canada, USA.
Sphenophorus spp. – (Billbugs) mostly on cereals; Canada and USA.
Sternochetus spp. – (Mango (Seed) Weevils) Old World tropics (CIE map no. A.180).
Systates spp. – (Systates Weevils) throughout Africa.
Tachypterellus spp. – (Apple/Cherry Curculios) USA.
Trichobaris spp. – (Potato Stalk Borers) Canada.
Tychius spp. – (Clover Seed Weevils) Canada.

Control of weevils

Adults. Generally persistent contact or stomach poisons have to be used, both as foliar sprays and on the mulch underneath, for the control of adults. The successful insecticides have been DDT, HCH, aldrin, dieldrin, as well as carbaryl, chlorpyrifos, fenitrothion and toxaphene

Larvae. Deep-boring larvae have been killed in many hosts by injecting the infested tunnel galleries with demeton-*S*-methyl, paradichlorobenzene, carbaryl or oxydemeton-methyl.

Larvae in the soil have been successfully killed by the use of seed dressings, sprays, dusts, and granules of the insecticides aldrin, dieldrin, DDT, chlorpyrifos, carbaryl and fenitrothion.

Anthonomus piri Kollar

(= *Anthonomus cinctus* Redt.)

Common name. Apple Bud Weevil
Family. Curculionidae
Hosts (main). Apple, pear.
(alternative). Not known at present.
Damage. The larvae feed inside fruit buds, which remain unopened in the spring, and eventually die. Adults bite buds and leaf petioles; may cause leaf-fall.
Pest status. Not a common pest in the UK but regarded as serious on pear throughout much of Europe.
Life history. The adult beetles emerge from overwintering hibernation in January, and during February the females lay eggs, singly, into the fruit buds; the adults presumably then die.

The eggs hatch into larvae also in February, and each larva feeds inside the unopened bud reducing it to a hollow shell. Pupation takes place within the hollowed bud and takes about six weeks.

Adults emerge in May and feed on the tree foliage until about mid-June, when they seek hibernation sites where they remain until January of the following year. The adult beetles are 5–6 mm in length; reddish-brown in colour with a broad straight white band across the elytra.

Adults bite small holes in leaf petioles and buds, and may cause leaf-fall and bud death.

There is only one generation per year.

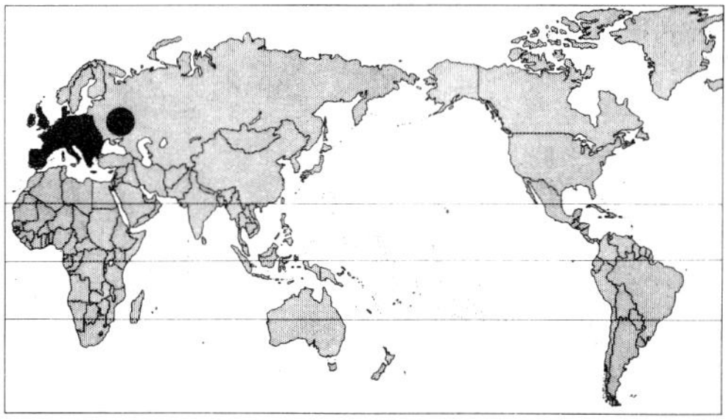

Distribution. This species is recorded throughout the UK and Europe. A total of 13 species of *Anthonomus* is recorded from the UK, and the genus is equally well represented in N. America, and also in Japan.
Control. In the UK control measures are seldom warranted, but the species is regarded as a more serious pest in Europe; for control measures see page 312.

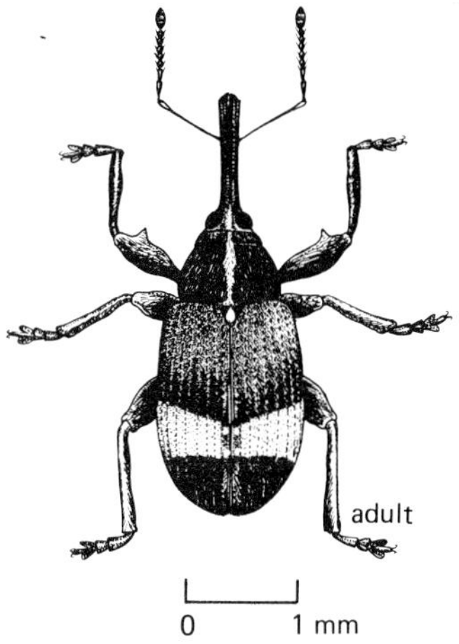

Anthonomus spp.

(*A. rubi* (Hb.))
(*A. signatus* Say)

Common name. Strawberry Blossom Weevils (UK; USA)
Family. Curculionidae
Hosts (main). Strawberry
(alternative). Raspberry, blackberry, dewberry, wild strawberry, and various wild Rosaceae.
Damage. Eggs are laid inside unopened flower buds and the petiole is then partially severed to prevent further development of the unopened flower. Sometimes the buds are completely severed and fall to the ground.
Pest status. These are regular pests of strawberry and other cane fruits, and sometimes serious damage is done, both in Europe and the USA.
Life history. The adults hibernate in sheltered locations and appear in the strawberry fields in April and May. The female lays eggs singly into unopened flower buds through deep feeding/oviposition punctures; she then bites through the bud petiole either partially or completely, which prevents further development of the unopened bud; egg-laying takes place mostly in June.

Both larval and pupal development takes place within the unopened flower bud.

Young adults emerge about the second week of July, through a circular hole in the side of the bud. They feed on the foliage for a few weeks and then proceed to their winter hibernation quarters. The adult is a small, long-nosed, black weevil, 2–3 mm in body length. There is no pale transverse band across the elytra in either species.

There is only one generation per year.
Distribution. *Anthonomus rubi* is the British and European species; and *A. signatus* is the N. American counterpart with apparently identical biology.
Control. Some varieties of strawberry are apparently more susceptible to infestation and damage than others; and sometimes early varieties suffer more injury than later ones.

For control measures see page 312.

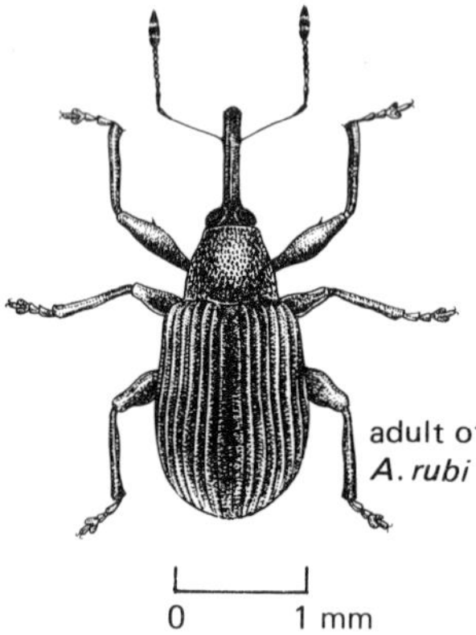

adult of *A. rubi*

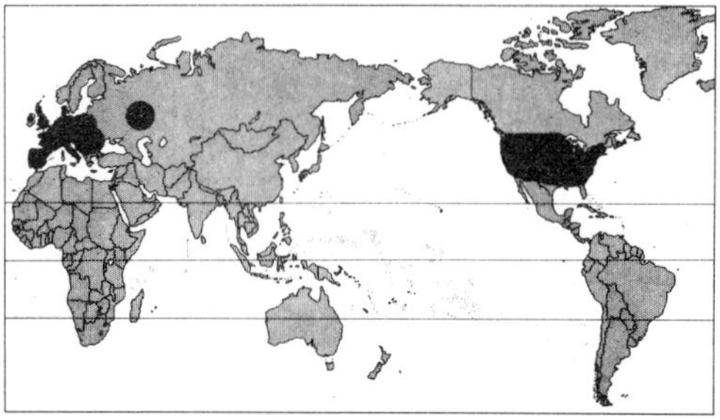

Ceutorhynchus assimilis (Payk.)

Common name. Cabbage Seed Weevil
Family. Curculionidae
Hosts (main). Rape
(alternative). Cabbage, and other cruciferous seed crops; and wild Cruciferae.

Damage. The larvae develop inside the seed pod and feed upon the developing seeds; one infestation often destroying all the seeds in a single pod. One larva per pod is most usual, with about 25% of the seeds destroyed.

Pest status. A serious pest in the UK with the recent increase in cultivation of rape as an oil seed. The oviposition holes in the pod are also used by Turnip Seed Midge (*Dasyneura brassicae*) as entry holes.

Life history. The adult weevils overwinter by hibernating in sheltered places. They emerge in May and June and infest winter rape during the flowering period. Eggs are laid from mid-May in holes made in the pods by the feeding females. Hatching takes 8–10 days.

The larvae feed on the developing seeds inside the pod, totally eating some and damaging others. After 4–5 weeks the larvae are mature and about 4 mm long; they bore through the pod wall and make an emergence hole through which they depart. Pupation takes place in the soil; pupation takes about 12 days.

The second generation starts with young adults that disperse to attack other crops through July and August. They then disperse again to hibernate in hedgerows and leaf litter. The adult is a small dark weevil, only 2–3 mm long, with typical geniculate antennae just more than halfway down the elongate snout.

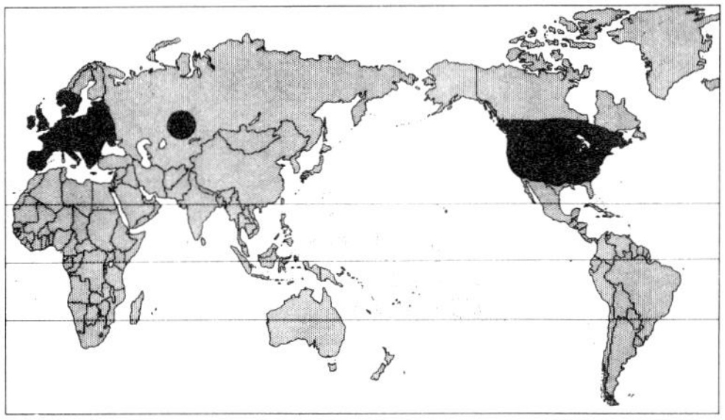

Distribution. Recorded from the UK and Europe, and also Canada and the USA.

Control. In Canada this pest is parasitised by at least ten species, some of which were introduced in 1949 in a deliberate attempt at biological control.

In the UK when adult populations reach an average of one weevil per plant during the flowering period then it is thought that control measures are probably justified. Because of the danger to bees insecticide sprays should *never* be applied during flowering, and a post-flowering spray is recommended, using either phosalone or triazophos. A pre-flowering spray against adults can be made using azinphos-methyl, endosulfan, HCH or malathion.

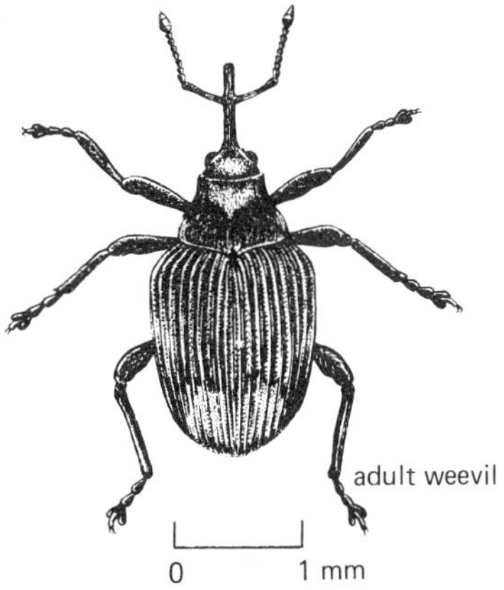

Adult on rape flowers

Emergence hole in rape seed pod

Ceutorhynchus pleurostigma (Marsh.)

Common name. Turnip Gall Weevil
Family. Curculionidae
Hosts (main). Turnip
(alternative). Cabbage, and other cruciferous crops; also charlock and wild Cruciferae.
Damage. Larvae make round marble-like galls on the roots of Cruciferae; infestations are typically patchy throughout a crop; severely attacked young plants may die, especially under dry conditions. Damage symptoms are sometimes confused with those shown by club root.
Pest status. Quite a serious pest in the UK and Europe, for seedlings may be killed, and galled root crops are rejected as vegetables.
Life history. Eggs are laid singly in the root of the host plant, often several eggs in one root; most eggs are laid in the period August/September. Hatching takes 5–17 days.

The feeding of the larva results in the formation of a globular gall on the root. During the autumn the larvae are small, and development ceases over the cold winter period. In the spring the larvae recommence feeding, and reach maturity at about 4 mm body length in March and April, when they leave the galls to pupate in the soil. Pupation takes about 12 days.

Adults emerge generally in May and feed on the foliage of the crucifers; there may sometimes be a second generation. The adult is a small dark weevil about 3 mm long, with a rounded body, antennae situated near the tip of the snout, and a spine on each fore-femur.

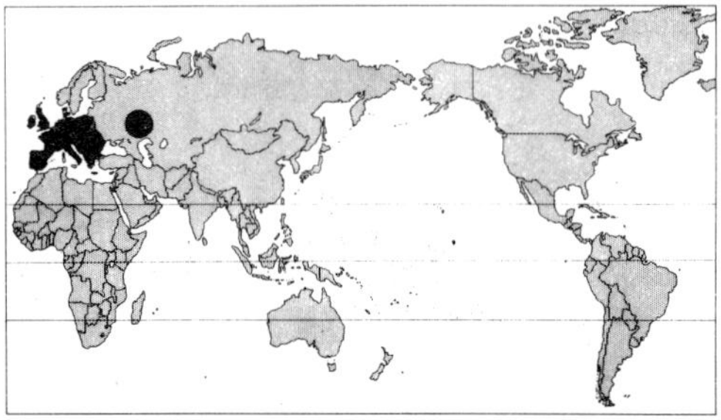

The adults are long-lived, and generally the life-cycle stages are quite variable in duration, so that the overall life-cycle is not synchronized and both adults and arvae may be found throughout most of the year.
Distribution. This species is, to date, only recorded from the UK and western Europe.
Control. Cultural control measures include removal and drying of brassica stubble (ploughing of crop remains encourages this pest), crop rotation, and also rejection of galled seedlings at transplanting.

Insecticides may be used as a seed dressing of HCH, also as a spray at planting-out, or a band application of dust over young seedlings; dieldrin has also been successful.

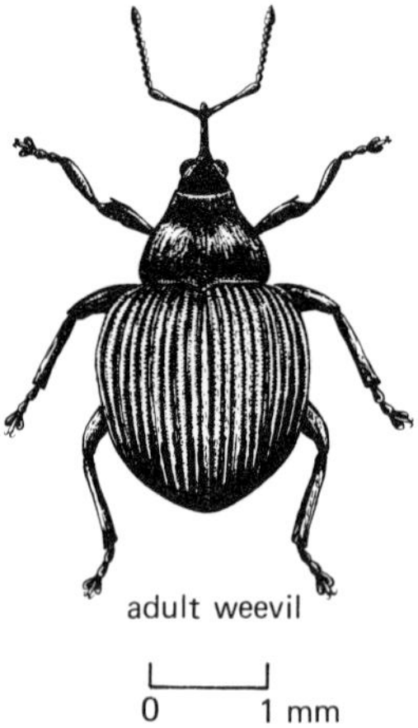

adult weevil

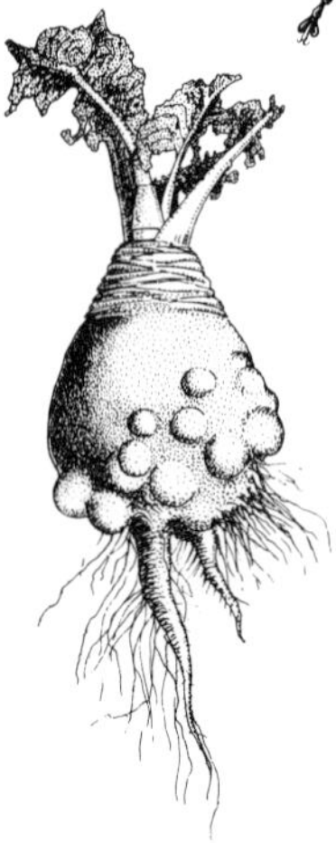
infested turnip

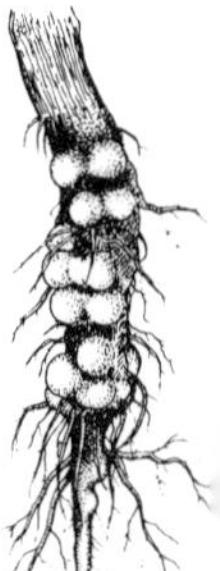
damaged cabbag[e] stem

Ceutorhynchus quadridens (Panz.)

Common name. Cabbage Stem Weevil
Family. Curculionidae
Hosts (main). Cabbage
(alternative). Brussels sprout, broccoli, swede and mustard.
Damage. The larvae inhabit stems and petioles of cruciferous seedlings, which become soft and spongy, and stems are easily broken.
Pest status. This can be a very damaging pest, but it is not a common species and infestations tend to be sporadic.
Life history. Adults hibernate overwinter and become active again in the spring. Females bite holes in the stems of cruciferous plants, usually seedlings, and then lay eggs, usually 2–3 eggs per hole. Hatching takes about 5–7 days.

The larvae tunnel inside the stem; there are usually several larvae in each stem, often some 10–20 per stem. Larval development takes from 3–6 weeks; fully-grown larvae measure 4–6 mm long, and leave the stems to pupate in the soil. Pupation takes about three weeks.

The young adults emerge from the soil and feed on the foliage of the cruciferous plants, either crops or weeds. At the advent of winter they seek hibernation sites in which to overwinter. Adults closely resemble *C. assimilis* but have a whitish patch at the base of the elytra, and the tarsi and antennae are reddish in colour; body length is 3–4 mm.

There is only one generation per year.

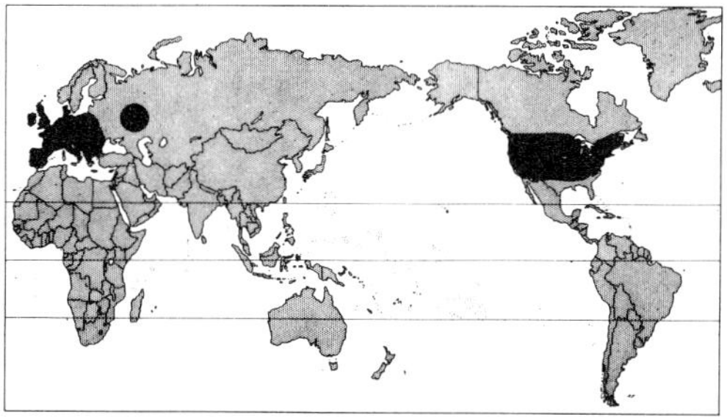

Distribution. Recorded from UK and Europe, and also the USA.

Identification of *Ceutorhynchus* species is difficult as this is a very large European genus of weevils, and most of the adults resemble each other quite closely; in the UK 52 species are recognized, many of which feed on Cruciferae, although some appear to prefer Compositae and others Labiatae.
Control. Seed dressings of HCH have been very effective at controlling this pest; also the use of HCH as a band spray. Carbofuran used to control Cabbage Root Fly will also protect the seedling from Stem Weevil.

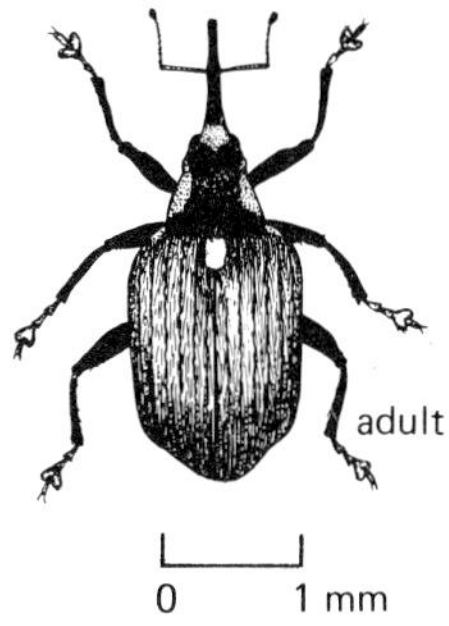

Curculio spp.

(*C. nucum* L.; = *Balaninus nucum* (L.))
(*C. neocorylus* Gibson)
(*C. occidentalis* (Casey))

Common name. Hazelnut/Filbert Weevils (UK; USA)
Family. Curculionidae
Hosts (main). Hazelnut; both cultivated and wild plants.
(alternative). Not known; possibly monophagous on *Corylus* spp.
Damage. The larvae destroy the kernel of the nut by feeding inside; the mature larva enlarges the oviposition hole in order to depart the nut for pupation in the soil under the bush.
Pest status. Pests of very regular occurrence on hazelnut, but not often causing serious damage.
Life history. The mature larvae overwinter in earthen cells in the soil; pupation takes place in the spring and adults emerge in May; they fly on warm sunny days. The female weevil bores a hole in the side of the young nut with her snout and then deposits a single egg inside. Hatching occurs in June and the larva feeds on the developing kernel, which it usually destroys completely. When fully mature at the end of July or in August, the larva escapes from the nut by enlarging the oviposition hole made by the female, and falls to the ground. After burrowing into the soil the larva makes an earthen cell 5–10 cm deep in the ground in which it proceeds to hibernate for the winter. Pupation takes place in April, and the adult weevils emerge generally in May.

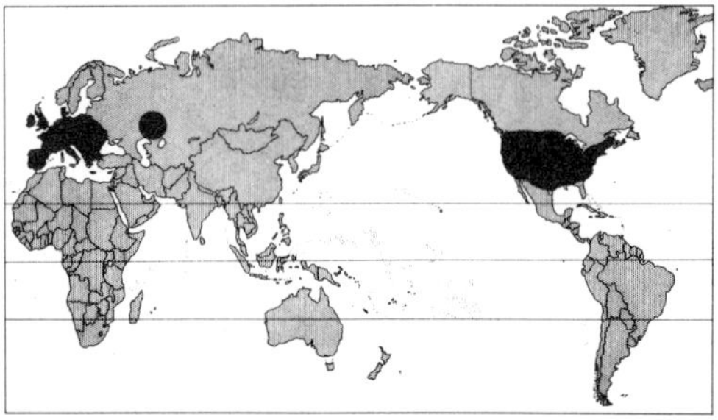

The adults are large brown weevils, some 4–7 mm in length; underneath the brown hair/scales the body is actually blackish.

There is only one generation per year.

Distribution.
Curculio nucum is found throughout the UK and Europe. *C. neocorylus* and *C. occidentalis* are both N. American species found in the USA.

In the UK the other known species of *Curculio* are smaller and make galls on trees (oak, birch, blackthorn, etc.) or inhabit and share hymenopterous galls on these hosts.

C. sayi is the Small Chestnut Weevil of the USA.

Control. There appear to be some differences in host preference in that some species of *Corylus* are more likely to be attacked by a particular species of *Curculio* than others: filberts appear to be more damaged than the others.

For control see page 312.

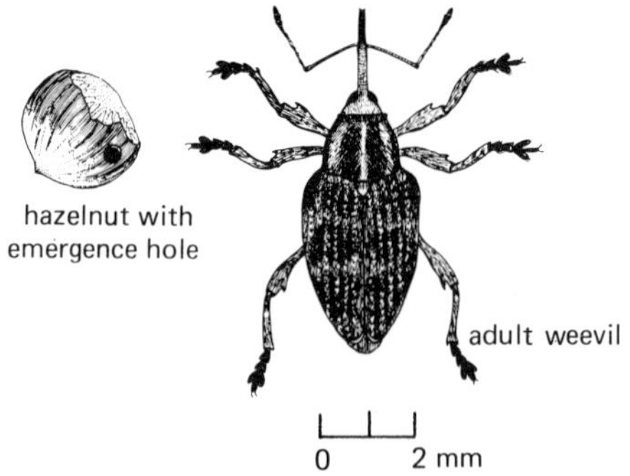

Graphognathus spp.

Common name. White-fringed Weevils
Family. Curculionidae
Hosts (main). Cotton, pulses, sweet potato.
(alternative). Totally polyphagous, recorded from 380 species of host plant.
Damage. The larvae are the major pests, living in the soil they feed on the plant roots; adults feed on the foliage and do occasionally defoliate the plants.
Pest status. The larvae are quite serious polyphagous soil pests, and in the USA quarantines have been established to prevent their spread; the adults are really only minor pests.
Life history. The females are parthenogenetic and flightless, and lay their eggs on the soil surface 5–25 days after emergence; each female can lay up to 1500 eggs, often in groups of up to 60; egg-laying continues for two months. Incubation takes 11–30 days.

The larvae live in the soil and feed on plant roots, and overwinter as larvae, although some eggs may also overwinter in sheltered locations.

Pupation takes place in the soil in an earthen cell in the following spring. Adults (all females) emerge in May (USA) and being wingless do not disperse far, so heavy infestations may build up locally. The adults are typical broad-nosed weevils, about 12 mm long, dark grey in colour with conspicuous white lateral edges to the elytra. Adults are quite long-lived and survive for 3–4 months.

There is only one generation per year.

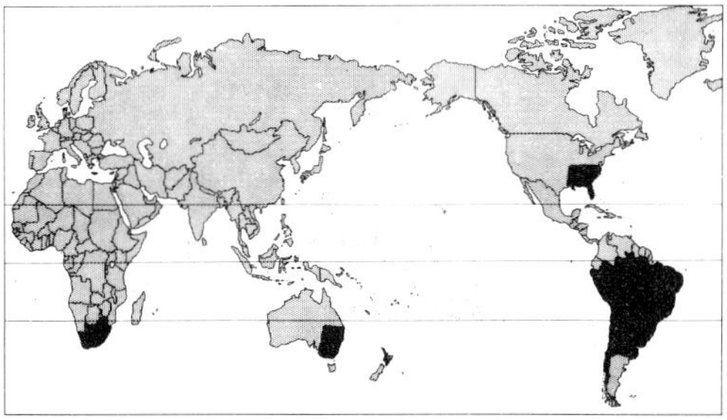

Distribution. This genus is S. American in origin and was first recorded in Florida in 1936. It has since spread throughout southeastern USA, and also New Zealand, Australia, and S. Africa (1941).

The species concerned are *Graphognathus leucoloma* (Boh.) which occurs as five distinct races (*dubious*, *fecundis*, *imitator*, *pilosus* and *striatus*) (CIE map no. A.179); *G. minor* (Buch.) and *G. peregrinus* (Buch.).
Control. In the USA quarantine measures are employed to prevent further spread of these pests; often using fumigation of plant materials.

Cultural control measures include planting of oats or other small grains in infested ground, and rotations that incorporate legumes only once in 3–4 years. Discing and harrowing generally reduces the numbers of larvae and pupae in the soil.

Insecticides found to be effective include DDT, aldrin, dieldrin, heptachlor, and chlordane, incorporated into the top 10 cm of the soil or sprayed on the foliage. Present chemical recommendations include soil fumigation with D-D, or methyl bromide and carbon disulphide.

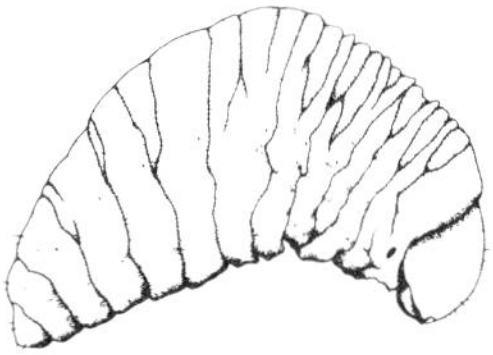

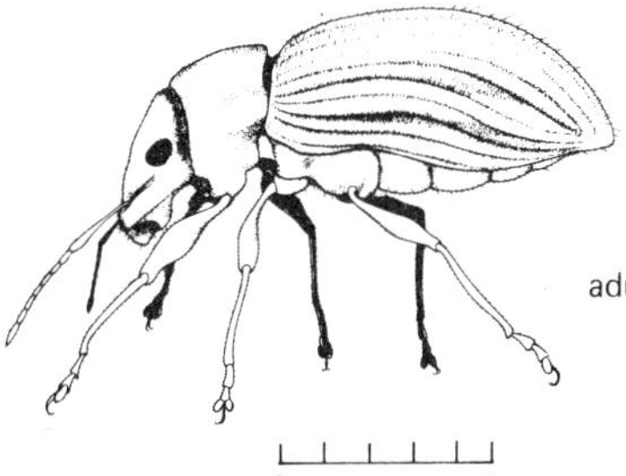

Hypera spp.
(= *Phytonomus* spp.)

Common name. Alfalfa (Clover Leaf) Weevils
Family. Curculionidae
Hosts (main). Alfalfa, clovers, trefoils.
(alternative). Other Leguminosae.
Damage. Larvae feed exposed on the tips of foliage at night, eating buds and stem tips, and also the leaf lamina. Flower bud destruction is important in seed crops. Adults make small feeding holes in the leaves.
Pest status. Serious pests of forage legumes in temperate regions; injury usually most severe in dry seasons.
Life history. Eggs are laid in the spring in holes eaten in the stems or buds by the females; from 1–40 eggs are laid per cavity, and each female lays from 200–800 eggs. Some species lay only 200–300 eggs and these are usually laid singly. Incubation takes 2–3 weeks.

The hatching larvae feed *in situ* inside the stem for 3–4 days, then they move to the shoot tips where they feed concealed in the folded leaflets and buds. Later they tend to feed on the opened leaves and may eat all the lamina except for the main veins. Larval development takes 20–60 days, when a size of 2–3 mm is reached; body colour is generally greenish.

Pupation takes place in a spun cocoon either in the foliage or else in the soil; development takes about three weeks.

Young adults emerge and then feed on the foliage for a while before hibernating in a sheltered location. In the spring the adults become active in March and April (in Europe) and feed on the foliage for a time before laying eggs. The adult is a small weevil, some 3–6 mm in length, greyish-brown in colour with distinctive longitudinal striping; the snout is rather short.

There is only one generation per year.

The different species do tend to have a slightly different biology and habits.
Distribution. The main species concerned are as follows:
Hypera nigrirostris (F.) – (Clover Leaf Weevil) Europe, Japan, Canada, USA.
H. postica (Gyll.) – (Alfalfa Weevil) Europe, Asia, N. Africa, Canada and USA (CIE map nos. A.304 and 456).
H. punctata (F.) – (Trefoil Leaf Weevil) Europe, Canada, USA.
H. meles (F.) – (Clover Head Weevil) Canada and USA.

In the UK 16 species of *Hypera* are recorded, most from Leguminosae.
Control. In Canada these weevils are quite heavily parasitized and also attacked by a fungus that appears to be of importance in regulating numbers.

The most successful insecticide has been DDT as a heavy spray about 10–14 days after the first cut and before the flower buds appear.

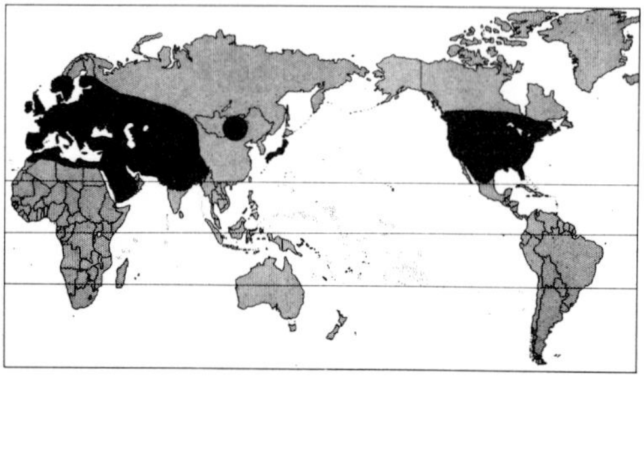

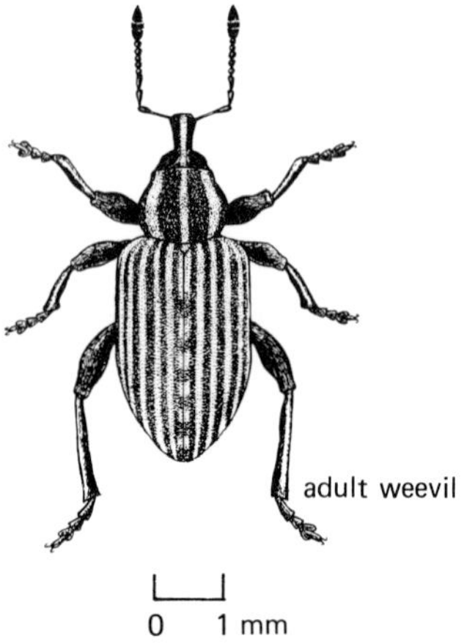

Otiorhynchus clavipes (Bonsdorff)

Common name. Red-legged Weevil (Plum Weevil)
Family. Curculionidae
Hosts. Raspberry, strawberry, plum, apricot, peach, nectarine, gooseberry; and larvae also on various pot plants and glasshouse crops
Damage. Both adults and larvae are pests; adults damage buds, grafts, and young shoots, as well as leaves; larvae in soil eat the roots of many plants, including strawberry in beds and glasshouse plants.
Pest status. Quite serious pests of many fruit crops and ornamentals throughout Europe.
Life history. Adults emerge in April/May during warm weather; they feed on buds, stalks and leaves of the host plants, and start egg-laying in late May or June. Each female lays up to 500 eggs in litter or cracks in the soil, usually in batches; laying takes place over a period of several weeks. Many eggs may be sterile. Hatching takes 8–24 days.

The larvae burrow into the soil under the plants; they feed on the host plant roots all summer and autumn and overwinter as larvae, becoming fully grown the following spring when they measure about 14 mm in length.

Pupation takes place in the earthen cell, some 15–20 cm deep in the soil. The earliest adults emerge in April/May, but those from the last-laid eggs do not emerge until the end of the summer. Developmental time from egg to adult takes from 12–18 months. Late-hatching eggs give larvae that do not pupate until the following autumn and the adults may stay in the pupal cells or hibernate in leaf litter; these are the first adults to become active in the spring.

The adults are shiny black in colour, with reddish legs, and about 12 mm long. They are wingless, and nocturnal in habits so only feed at night.
Distribution. This species appears to be recorded only from the UK and Europe.
Control. For control measures see page 322.

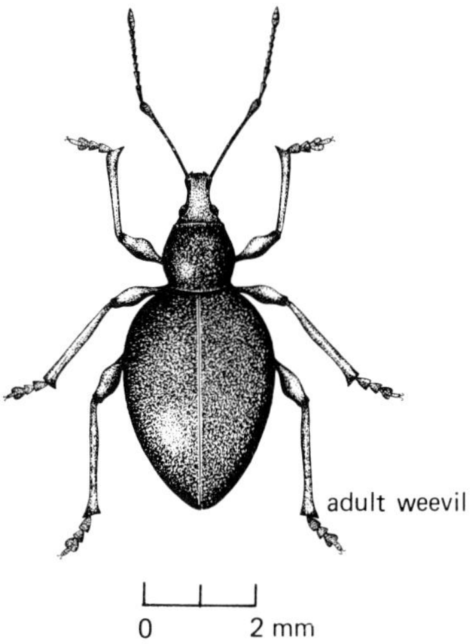

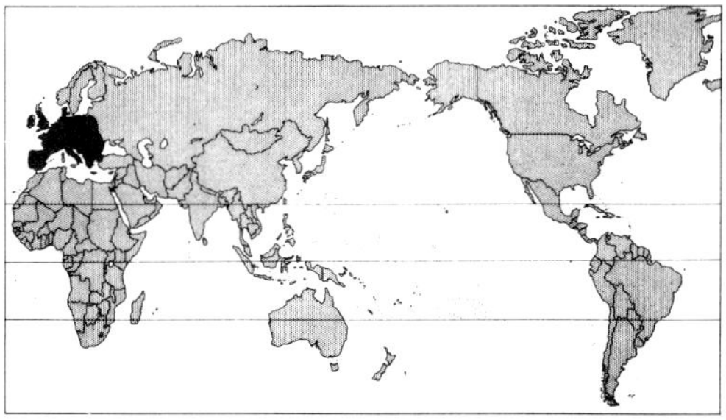

Otiorhynchus (wingless) weevils
(Coleoptera; Curculionidae)

This is a large genus of medium- to large-sized weevils, well-represented throughout the temperate regions. In the UK alone there are 19 different species recorded. In the USA they were formerly called *Brachyrhinus* but this name is now rejected.

Adults differ considerably in size, coloration, and surface adornment; but they all have a rounded body with a shortish rostrum ('snout' or 'nose'), and very long antennal scapes, and some have the fore-femora toothed ventrally. They are nocturnal in habits and feed on the plant foliage, usually making a characteristic leaf-edge notching. Most of the species are wingless, and a few have the elytra actually fused together and immobile; thus their powers of dispersal are limited. Often dispersal is effected through the accidental transportation of larvae and/or pupae in soil with pot plants. In some species it is the adult weevil that is the main pest, destroying shoots and buds, eating leaves and sometimes bark; whilst the larvae in the soil do insignificant damage to the roots. *O. sulcatus* is unusual in that only females are recorded and they reproduce parthenogenetically.

Larvae are all soil-dwelling, white, apodous, with a brown head capsule and large mandibles. In some species the larvae are the main pests, destroying plants such as strawberry, cyclamen, primulas, and alpine herbs and shrubs, whereas the adult damage is usually slight, although gnawed leaves lose a little of their aesthetic appearance. In glasshouses root damage to a wide range of plants is quite common.

Major pest species

Otiorhynchus clavipes Bonsdorff – (Plum (Red-legged) Weevil) Europe.
O. cribricollis Gyllenhal – (Apple Weevil) S. Europe, N. Africa, W. USA (CIE map no. A.423).
O. ligustici (L.) – (Alfalfa Snout Weevil) Europe and USA.
O. ovatus (L.) – ('Strawberry Root Weevil') Australia, Canada, USA, Europe.
O. rugifrons (Gyllenhal) – ('Strawberry Root Weevil') Europe.
O. rugosostriatus (Goeze) – ('Strawberry Root Weevil') Australia, Europe, USA.
O. singularis (L.) – (Clay-coloured Weevil) Europe, USA.
O. sulcatus (F.) – (Black Vine Weevil) Europe, Australia, New Zealand, Canada and USA (CIE map no. A.331).

Control of wingless weevils (*Otiorhynchus* spp.)

It is generally agreed that control measures applied once damage symptoms are evident are invariably too late, and the usual practice is to aim at eradication of the local population and to prevent spread to adjacent crops.

Cultural methods. Adequate crop rotation and crop spacing, as well as correct spacing of rows and plants along rows, will minimize damage, because of the limited powers of dispersal possessed by these insects. Ploughing and cultivation of old beds will destroy larvae and pupae, and clean ditches decrease the extent of adult movement between adjacent fields.

Chemical methods.
Larvae. On strawberry, if an attack is noticed during flowering or fruiting, it is generally too late to protect that crop, and measures taken are usually aimed at eradicating the population to prevent subsequent damage. The strawberry crown should be drenched with DDT, carbaryl, or chlorpyrifos solution; this should kill the young larvae after hatching, before they burrow deeper into the soil. γHCH can be worked into the soil in infested land prior to planting. Larvae in pots may be killed by mixing aldrin or DDT dust with the potting compost.
Adults. Generally adults are difficult to kill in the open. Formerly DDT was quite effective; carbaryl or chlorpyrifos are used on open crops and usually give some measure of success. In infested glasshouses adults are more easily killed using parathion smokes.

NB It should be remembered that in many countries there are restrictions regarding the use of DDT and parathion on edible and garden crops.

Otiorhynchus spp.

(*O. ovatus* (L.))
(*O. rugosostriatus* (Goeze))
(*O. rugifrons* (Gyll.))

Common name. Strawberry Root Weevils
Family. Curculionidae
Hosts (main). Strawberry
(alternative). Many other field, garden and glasshouse crops, and wild plants.
Damage. The larvae eat the roots of the plants; large larvae cut through the main roots and often burrow into the base of the crown; attacked plants wilt and often die.
Pest status. Four species of *Otiorhynchus* are common on strawberry in Europe, and several other species also attack this host; collectively they are seriously damaging to this crop. They tend to be more serious pests in lighter soils.
Life history. Adults are seen in the fields from the middle of April onwards; they feed on leaves (making edge notches) at night and hide in the litter during the day. Most adults actually emerge in June and July, and most eggs are laid from mid-August to September. Larval damage is first noticed in the autumn, and becomes worse each succeeding year.

Life-history details are much the same as for *O. clavipes* (page 321).

Larvae overwinter and resume feeding in the spring prior to pupation. The adults are brown or blackish in colour, 5–10 mm long, with very heavy sculpturing on prothorax and elytra. Adult feeding damage is usually leaf-notching and seldom of importance, except as an indication of infestation prior to obvious larval damage.
Distribution. These species are widespread throughout Europe, Australia, the USA and parts of Canada. *O. ovatus* is recorded attacking brassica plants and potato stems in Canada; and is also alleged to bite people if handled.

In the UK another strawberry root weevil is *Sciaphilus asperatus* (Bonsd.).
Control. For control measures see page 322.

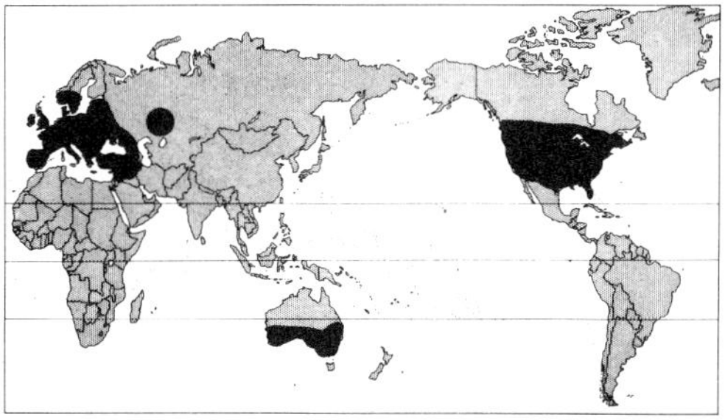

O. ovatus

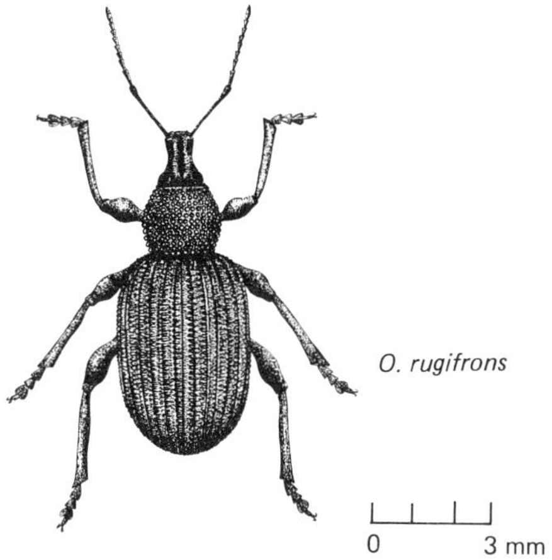

O. rugifrons

Otiorhynchus singularis (L.)

Common name. Clay-coloured Weevil (Raspberry Weevil)
Family. Curculionidae
Hosts (main). Apple, raspberry.
(alternative). Other top and soft fruits, and hop; deciduous woodland and hedges.
Damage. The adults feed on buds and grafts in the early spring, later on the flush growth and gnaw the bark. On raspberry they eat fruit blossoms and young fruit buds, and also the rind (bark) of the young spawn (new shoots); young stems are often ring-barked and die. Larvae in the soil are seldom pests.
Pest status. The adult weevil is quite a serious pest of top and soft fruits, throughout Europe and the USA.
Life history. Adults emerge from hibernation as early as February in the UK, feeding first on unopened buds. This species breeds naturally in deciduous woodland and hedgerows and invades adjacent fruit plantations and orchards. This is one of the smaller species; adults measure about 6–8 mm in body length, coloured brown with small paler markings on the elytra, often obscured by adhering soil.

Eggs are laid during the summer months, and the larvae feed on plant roots (seldom causing any noticeable injury) throughout the autumn and winter, and pupate usually in early February. Pupation only takes 1–2 weeks, even in February.

There is only one generation per year.

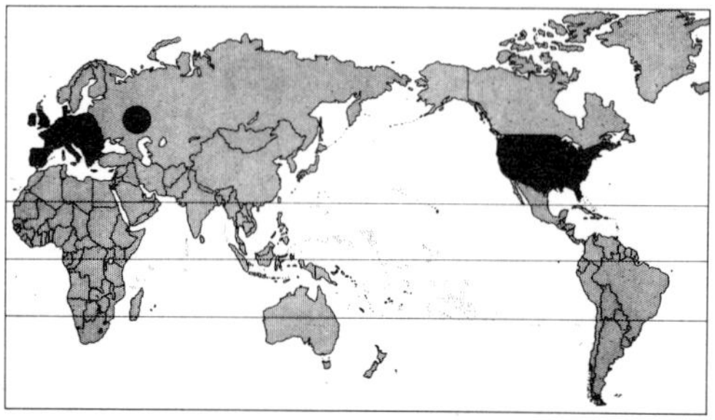

Distribution. This species is recorded from Europe and the USA only, to date.
Control. For control measures see page 322.

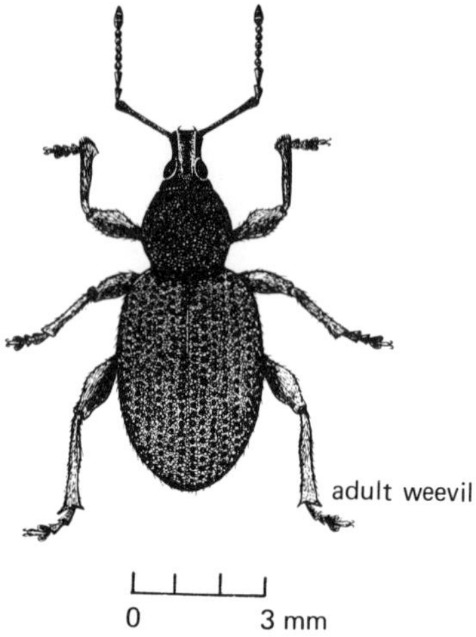

Otiorhynchus sulcatus (F.)

Common name. Black Vine Weevil
Family. Curculionidae
Hosts (main). Strawberry, grapevine, *Cyclamen*.
(alternative). Raspberry, and many different glasshouse crops (roots).
Damage. Larvae in the soil eat the roots of many cultivated and wild plants; often found in open habitats as well as being the commonest species in glasshouses. Adult leaf damage is negligible.
Pest status. The larvae of this species are serious glasshouse and pot plant pests, as well as being damaging to various field crops.
Life history. In the open the adults emerge from May onwards, after overwintering as larvae. Adult feeding damage consists of the usual leaf edge-notching, and is seldom important. In heated glasshouses in the UK adult emergence is typically in the autumn. In the following spring the adult females (males are unknown) lay their parthenogenetic eggs in the pots or beds containing suitable host plants. The oviposition period is very protracted, and may last through the spring and summer periods. Larvae from spring-laid eggs may be damaging by the autumn to pot plants, with severe damage done by late autumn, prior to pupation in December. Larvae from late-laid eggs will not become adult until the following year. Generally the life-cycle takes a year for completion, and all stages of this weevil may occur at the same time, particularly in glasshouses.

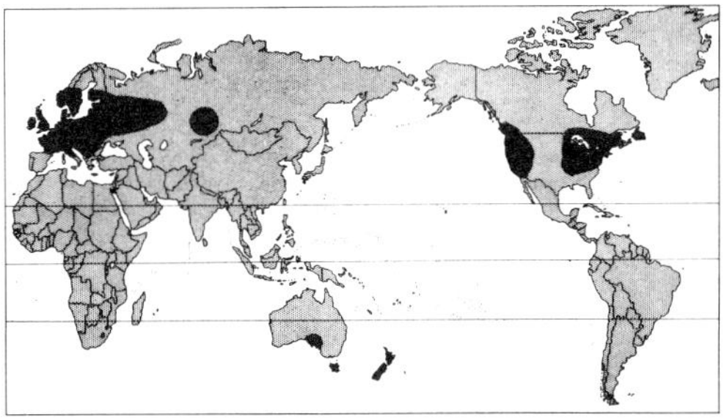

The adult female is about 9 mm long, black in colour, with a rough sculpturing on the elytra; the body is minutely studded with short yellow bristles.
Distribution. Found throughout Europe, USSR, Australia, Tasmania, New Zealand, St. Helena, Canada and the USA (CIE map no. A.331).
Control. For control measures see page 322.

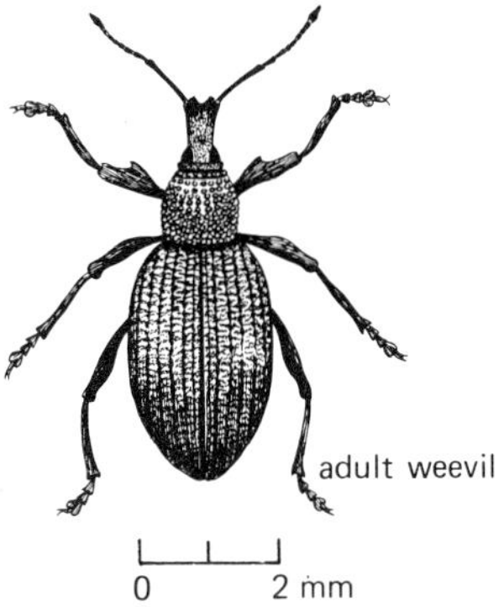

Philopedon plagiatus (Schall.)

Common name. Sand Weevil
Family. Curculionidae
Hosts (main). Sugar beet
(alternative). Wild Chenopodiaceae, especially on sandy soils and near the coasts.

Damage. Feeding adults notch leaves of sugar beet; on seedlings they destroy the cotyledons and often bite through the soft stem, killing the seedling. Larvae in the soil feed mostly on roots of chenopod weeds.

Pest status. Only a pest on sugar beet, and only important on light sandy soils, especially in coastal regions of the UK and northern Europe.

Life history. This appears to be a sand dune dwelling insect in its natural habitat, where there are many pioneer plants belonging to the Chenopodiaceae, but it has spread inland into coastal sugar beet growing regions.

Eggs are laid in sandy soils in May and June. The larvae feed underground on the roots of many different wild plants; it is thought they feed for about 18 months before they pupate at the end of the second year. The young adults overwinter in the soil and become active in April when they start to feed, and then lay their eggs until about mid-June when they die.

The adult is a stout little brown weevil, usually distinctively striped and 4–7 mm in length. The striped effect is given by alternating lines of pale and dark setae, which are flattened and scale-like.

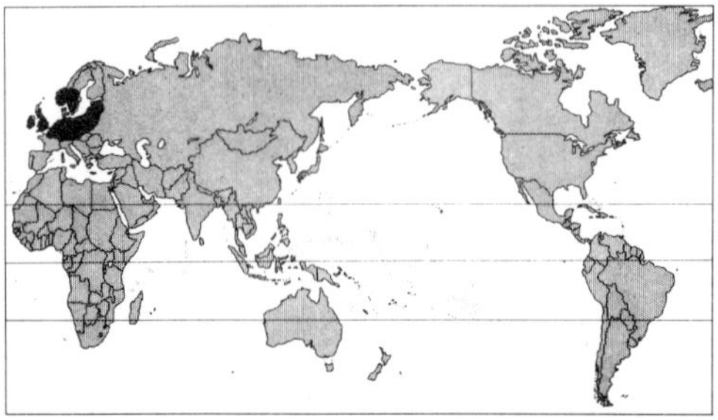

Distribution. To date, only recorded from the UK and northern Europe.

Control. Directly damage is observed in beet fields it is recommended that application of insecticide dusts or sprays be made. Formerly control was achieved with DDT or HCH, or trichlorphon.

Treatments with carbofuran granules on light soils, in the seed furrow, or thiofanox on heavier soils, when applied to protect against other pests, will also kill Sand Weevil.

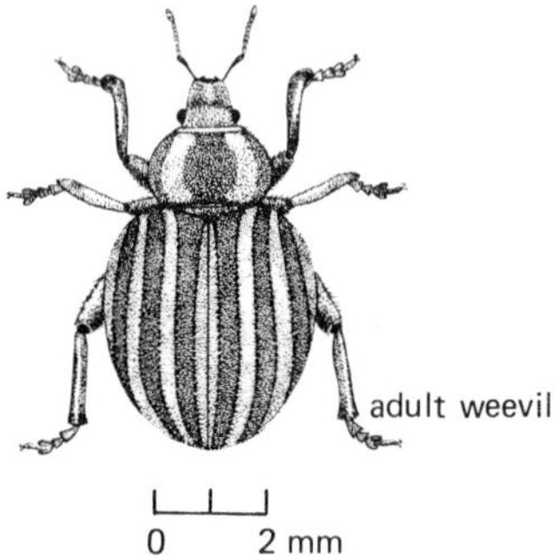

Phyllobius spp.

Common name. Common Leaf Weevils
Family. Curculionidae
Hosts (main). Apple, pear, hazelnut etc.
(alternative). Hawthorn, roses, nettles, elm, and other wild plants; larvae feed on grass roots mostly.
Damage. Adults are widespread, often very common, and leaf damage is frequently encountered; larvae eat grass roots and sometimes cause bare patches in grassland.
Pest status. Adults are rarely of any consequence on fruit and tree crops, but larval damage to grass swards is sometimes quite serious.
Life history. Eggs are laid in the soil, usually at a depth of 2–3 cm, in the spring. Details of the life-cycle of these species are not known, except that the larvae feed on the roots of grasses and are typical weevil larvae in all respects. Pupation takes place within the soil under the sward. The larvae overwinter in the soil, and pupate in the spring in an earthen cell.

Young adults are to be seen on foliage mostly in June when the signs of their feeding activity are numerous. The adults are rather elongate small weevils, about 5–6 mm in body length, and several species are a shiny golden/green colour, but others are more brownish; body coloration is generally somewhat variable.

There is only one generation per year.
Distribution. The main species concerned are as follows:
Phyllobius argentatus (L.) – Silver-green Leaf Weevil.
P. maculicornis Germar – Green Leaf Weevil.

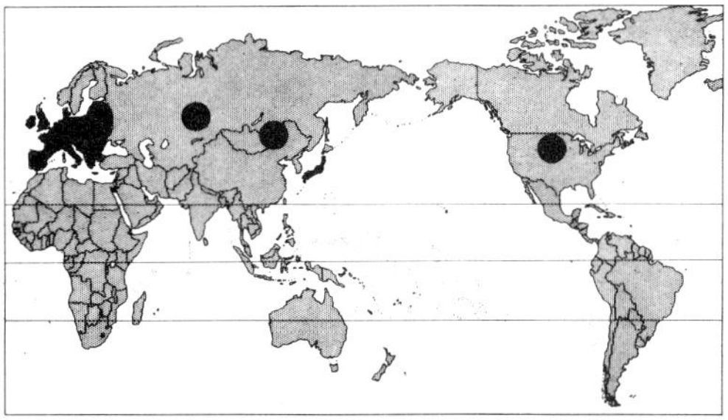

P. oblongus (I.) – Brown Leaf Weevil.
P. pyri (L.) – Common Leaf Weevil.

These species are all recorded from the UK and Europe; four other species are recorded as pests in Japan, and one in the USA.

In the UK there are 10 different species recorded.
Control. Adult damage to leaves is of regular occurrence but seldom worthwhile attempting to control; DDT and HCH have given good control in the past as have carbaryl, chlorpyrifos and fenitrothion.

Larvae in pastures and grassland are sometimes seriously damaging to the sward, especially of ryegrass and fescues; chemical control is generally not successful, but rolling of the sward several times will kill some of the larvae in the soil and prevent further damage.

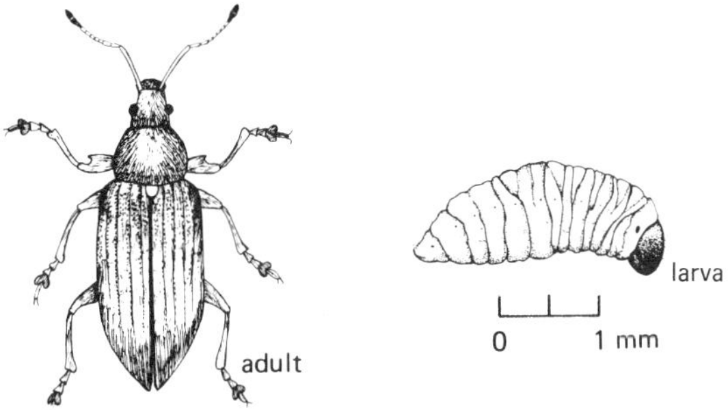

Hazelnut leaves eaten by adults

Rhynchites (Caenorhinus) aequatus (L.)

Common name. Apple Fruit Rhynchites
Family. Curculionidae (Attelabidae)
Hosts (main). Apple
(alternative). Plum (hawthorn is the natural host).

Damage. Adults make tiny feeding/oviposition punctures into young fruitlets, from petal-fall until July; the fruit becomes distorted and blemished. Not every hole contains an egg. A single fruitlet may have several hundred holes.

Pest status. Quite a regular pest in many parts of the UK and Europe, and similar weevil pests occur on these crops in other parts of the world.

Life history. The adult weevils emerge in the spring and may be seen on apple trees from pink bud onwards. During the period from petal-fall (May) until July the female weevils feed on the young fruitlets and lay eggs in some of the punctures. The eggs hatch into small larvae which feed on the flesh of the fruit. As the larvae are so small in size they do not prevent the fruit from continuing to develop; many of the oviposition holes close up as the fruit grows, but the fruit remains distorted and blemished.

The adult is a tiny weevil, about 4–5 mm in length (including snout) and a distinctive reddish-brown in colour with the body quite hairy.

There is only one generation per year.

Some of the similar fruit pest species in N. America have two generations each year and the second-generation adults hibernate over winter.

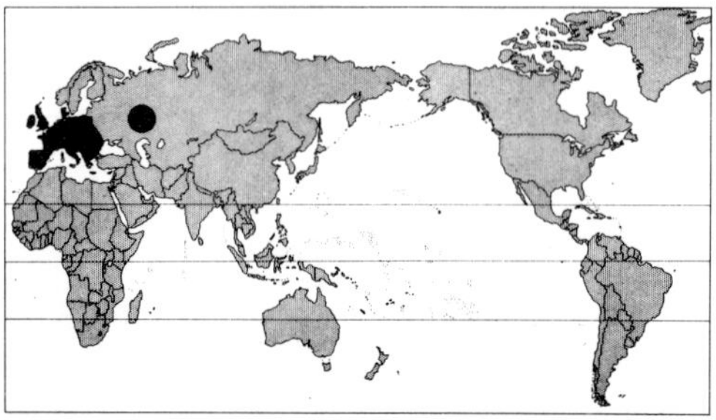

Distribution. This species is recorded from the UK and Europe. Similar species occur in the New World, as mentioned on page 330.

Control. For control measures see page 312; this pest has been particularly susceptible to both DDT and HCH applied as a spray at pink bud stage.

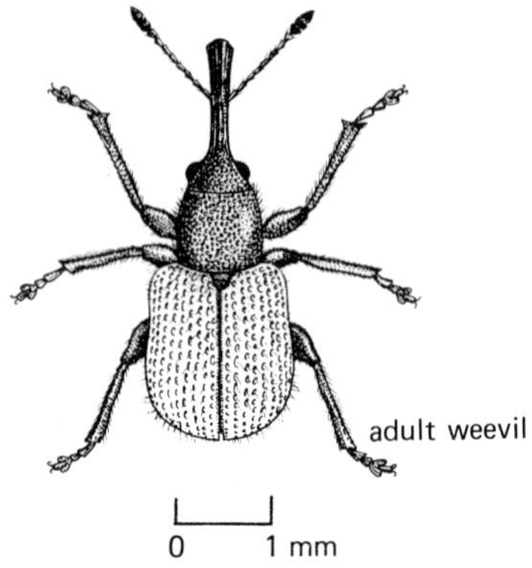

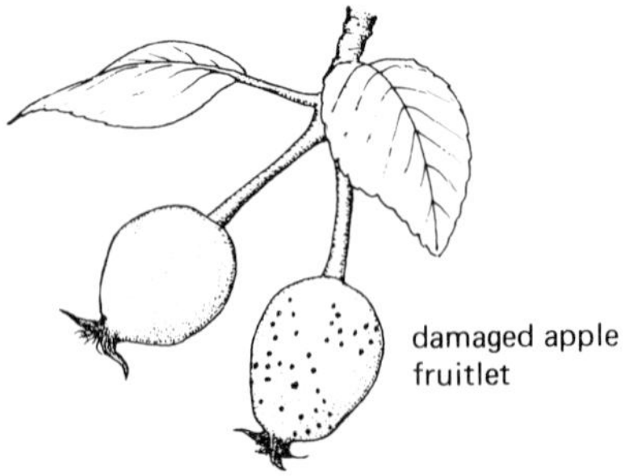

Rhynchites (Caenorhinus) germanicus Herbst.

Common name. Strawberry Rhynchites
Family. Curculionidae (Attelabidae)
Hosts (main). Strawberry
(alternative). Blackberry, loganberry and raspberry.
Damage. The adult weevils feed on folded leaflets in April and then females feed on the petioles and blossom trusses, making small cavities in which eggs are laid; after laying eggs the female makes a ring of punctures on the stalk below the eggs; the blossom truss (or leaf) wilts and then dries out and finally dies.
Pest status. Quite a serious pest of strawberry in parts of the UK and Europe. On the *Rubus* hosts damage may also be serious as the tips of new growth are destroyed.
Life history. Adults overwinter in the soil and start to emerge in March/April, when they feed on the folded young leaves. In April the females start laying eggs; they bite holes in the side of the truss stalks and then lay up to four eggs per stem inside the holes; then the female makes a ring of small punctures in the stalk below the oviposition site. Later the truss or leaves wither and die. Eggs are laid from middle of April to August, but in a favourable season most are laid in May.

The larvae feed inside the dried-up part of the stem which had been cut off by the ovipositing female. When fully fed the larvae leave the stem and burrow into the soil to pupate in an earthen cocoon. After a few weeks in the cocoon pupation is completed and the young adults are formed, but they remain in the soil until the following spring.

The adult beetles are tiny, 2–3 mm long, bright blue-green in colour and with a long slender snout. In body shape it closely resembles the other species of *Rhynchites* illustrated in this book.
Distribution. To date, this species is only recorded from the UK and Europe; the precise extent of its distribution is not known as yet.
Control. For chemical control, DDT as a high-volume spray, or as a dust, has been very effective; as an alternative chemical carbaryl, chlorpyrifos or fenitrothion should be effective.

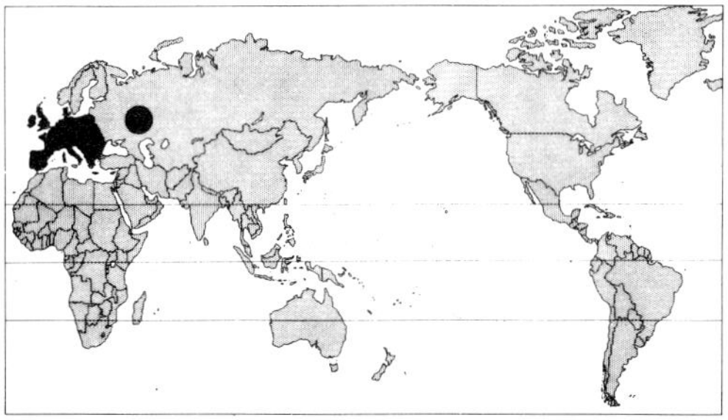

North American fruit curculios

There is a group of small weevils, rather squarish in body shape, but with a long narrow snout, that feed on the fruits of native N. American trees and the injury they inflict on the fruits is much like that done by the Apple Fruit Rhynchites (*Rhynchites aequatus*) in Europe (page 328). These native New World weevils are as follows:

Conotrachelus nenuphar (Herbst) – (Plum Curculio) on plum, apple, pear, quince and related hosts (Rosaceae).

Conotrachelus crataegi Walsh – (Quince Curculio) on quince and pear.

Conotrachelus juglandis LeConte – (Butternut Curculio) on butternut.

Conotrachelus retentus (Say) – (Black Walnut Curculio) on walnuts and butternut.

Tachypterellus consors cerasi List – (Cherry Curculio) on wild and cultivated cherries.

Tachypterellus quadrigibbus (Say) – (Apple Curculio) on apple, pear, crab apple and hawthorn.

This group of fruit weevils, together with the *Rhynchites* species, typically cause two types of damage: they are either twig-cutters or else lay eggs in young fruits and the larvae develop inside the fruits.

Rhynchites (R.) caeruleus (Degeer)

Common name. Apple Twig Cutter (Weevil)
Family. Curculionidae (Attelabidae)
Hosts (main). Apple
(alternative). Pear, plum, and oaks.

Damage. The adult weevil cuts off shoots after having laid an egg in the terminal part of the shoot; if completely cut through, the shoot falls to the ground, otherwise the cut part dangles distally. The damage is usually done in early June.

Pest status. A serious pest of apple and other fruit trees in that each oviposition site destroys a terminal growing point. Damage is often more serious on young trees and small bush trees, but established standards are also damaged.

Life history. The adults emerge from the earthen pupation cells in the spring and are to be found on the apple trees in May when they feed on the young foliage. In early June, or when 10–20 cm of new growth is formed, the female lays a single oval yellow egg in the shoot tissues which are still quite soft; the egg is usually laid 10–15 cm from the shoot tip. Having laid the egg, the female weevil then bites through the shoot just below where the egg is situated, and the shoot often falls to the ground.

The larva feeds inside the cut shoot, on the ground, and hollows out the pith. When fully grown the larva leaves the shoot and burrows into the ground to make an earthen pupation cell. The pupa overwinters in a dormant state, and the adult emerges in the following spring.

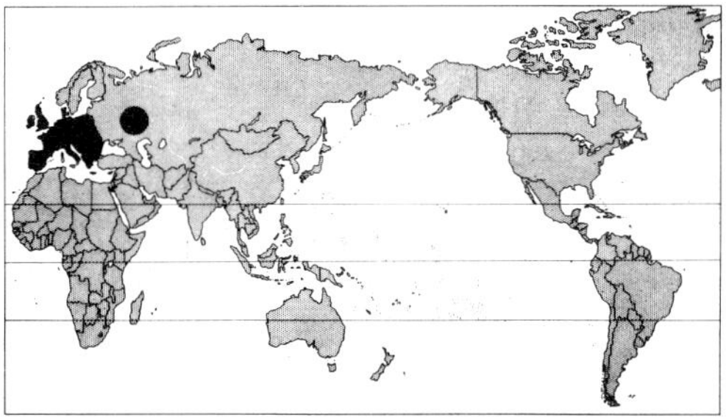

The adults are small, bright blue weevils, about 4 mm long, and rather hairy. It is thought that the natural hosts for these insects are the several species of oak native to western Europe, but with the advent of widespread apple cultivation the insects have adapted to these new hosts.

Distribution. This species is recorded from the UK and western Europe; the actual extent of distribution is not known at present.

Control. For many years the standard control recommendation has been high- or low-volume sprays of DDT, applied at pink bud stage. Should alternatives be required, it is thought probable that the following insecticides would be effective: azinphos-methyl, carbaryl, chlorpyrifos, fenitrothion or phosalone.

To avoid damage to pollinating bees crops in flower should not be sprayed (that is trees with open blossoms).

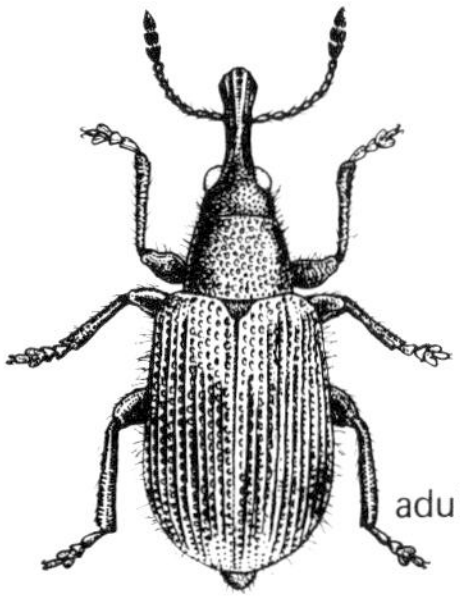

Sitona spp.

Common name. Clover Weevils; Pea and Bean Weevils
Family. Curculionidae
Hosts (main). Clovers, *Vicia* beans, pea.
(alternative). Other cultivated and wild Leguminosae.
Damage. Larvae in the soil eat out the root nodules, and may cause appreciable crop losses; adults notch leaf edges and occasionally cause a crop loss, but are known as virus vectors. Young seedlings may be defoliated.
Pest status. Formerly regarded as being unimportant pests, they are now thought of as quite serious, particularly on *Vicia* beans; adult damage is generally unimportant.
Life history. Adults overwinter in leaf litter and hedgerows; they fly into the bean crops (and clovers) during the first warm sunny days in the spring. Damage symptoms generally are most severe along the field margins (headlands) when the crop plants are about 5–8 cm tall. First adults are usually seen in early May. Each female lays up to 1400 eggs on the soil around the seedlings, but many eggs do not hatch. Incubation takes some 14–21 days.

The larvae burrow into the soil and feed on the root nodules, hollowing out the nodules and feeding inside. Larval development takes 6–7 weeks. Mature larvae are about 5 mm long and pupate inside an earthen cell at a depth of about 5 cm. Pupation takes about 2–3 weeks.

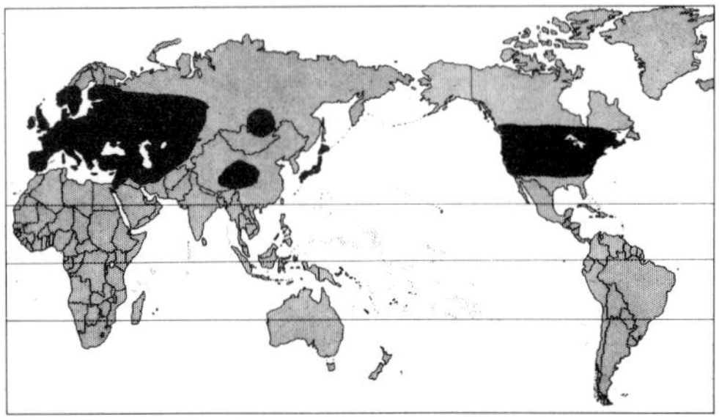

Young adults emerge, usually to find that the host crop has either been harvested or else there is little fresh growth, and so move away from the field seeking fresh growth; this takes place in late July and early August in the UK. The adult weevils are all similar in shape, and usually with pale and dark longitudinal banding, usually 3–6 mm in length; there are slight differences in size, colour pattern, shape and host preferences, between the different species of *Sitona*, but specific identification is very difficult. *S. hispidulus* has a different biology in that egg-laying starts in the autumn (until the spring) and so both adults and eggs overwinter; this is only a pest on clovers.

There is only one generation per year.
Distribution. This is a very large genus, found predominantly in Europe and Asia: there are 20 species recorded in the UK, 12 in Israel, and in Japan three species are serious pests. In the New World there are three important European pest species in the USA and at least five species in Canada. They are all confined to the Leguminoseae or Leguminales as hosts.

The main pest species are as follows:
Sitona cylindricollis (Fahraeus) – (Sweetclover Weevil) Europe (rare in UK), Asia (including USSR, China, Tibet), Canada and USA (CIE map no. A.372).
S. hispidulus (F.) – (Common Clover Weevil) Europe, western Asia, Japan, Canada and USA (CIE map no. A.437).
S. humeralis Stephens – (Clover Weevil) Europe.
S. lepidus Gyllenhal – (Clover Weevil) Europe.
S. lineatus (L.) – (Pea and Bean Weevil) Europe, Israel, Japan, Canada and USA.
S. macularius (Marsh.) – (Broom and Clover Weevil) Europe.
S. puncticollis St. – (Clover Weevil) Europe.
S. sulcifrons (Thunb.) – (Common Clover Weevil) Europe.

Three other species are of importance in Israel, and there are other important species both in Japan and Canada.

Control

Cultural methods. Crops grown on a good tilth generally suffer less damage than crops on poor cloddy soils. Late sown crops are also less damaged, but sowing late is seldom feasible.

Chemical methods.

Adults. Dusts and sprays of DDT dieldrin, and HCH have been effective at killing adult weevils on young crops, but there may be residue problems; an alternative insecticide is permethrin, but repeated sprays may be required if more weevils fly into the crop. Other recommended chemicals include fenitrothion and triazophos, as foliar sprays.

Larvae. Soil treatment using granules of carbofuran or phorate, applied in the seed furrow at the time of sowing kill both ovipositing female weevils and young larvae, but are not persistent enough for control on autumn-sown beans.

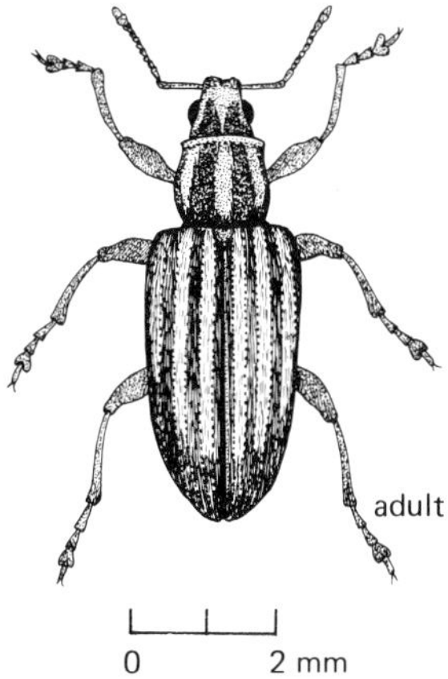

adult *Sitona* weevil

Field beans with notched leaves

Sitophilus granarius (L.)

Common name. Grain Weevil
Family. Curculionidae
Hosts (main). Stored wheat.
(alternative). Other grains in storage, flours, and some foodstuffs.

Damage. The larvae feed on and inside wheat grains, which are gradually hollowed out and destroyed; grain stored in bulk is more heavily infested than grain in sacks. Infestation causes heating and moisture condensation.

Pest status. A serious, temperate, primary pest of stored grains, of worldwide importance: infestations are confined to grain and produce stores.

Life history. Eggs are laid inserted into grains; each female lays about 200 eggs, usually a few per day. Females live for about nine months, but do not lay eggs either above 35 °C or below 13 °C, so females that emerge in the autumn do not lay eggs until the following spring. The larvae remain inside the grain for their whole developmental period and then pupate there. The emerging young adult makes a small circular hole on leaving the grain. Full-grown larvae are about 5 mm in length, and the adults about 3–4 mm. All species of *Sitophilus* are very similar in appearance, but *S. granarius* are wingless and hence cannot fly; the elytra tend to be more ridged and the prothorax punctures more oval and less circular. At 2 °C adults survive for 40 days, and at 5 °C more than 100 days; mating will not take place at less than 12 °C, although activity starts at 10 °C. In bulk grain stores in the UK temperatures seldom fall below 5 °C, so this species is well adapted for survival in a temperate climate. In the UK there are usually 3–4 generations per year. Development does not proceed in grain with a moisture content less than 9.5%.

Distribution. *S. granarius* is found throughout the cooler parts of the world in grain and produce stores, and to some extent also in the warmer parts: usually in the tropics it is replaced by the two tropical species *S. oryzae* and *S. zeamais*.

Control. For control measures see the general recommendations for stored products beetles, page 274.

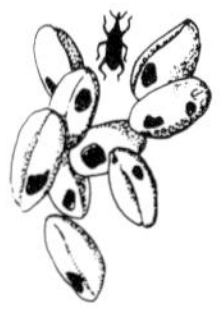

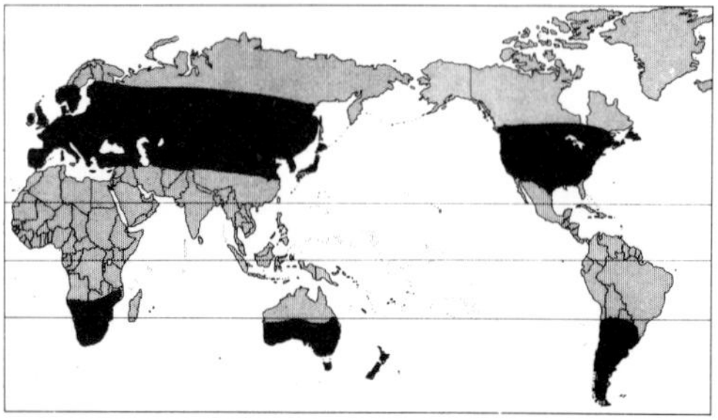

Sitophilus spp.

(*S. oryzae* (L.))
(*S. zeamais* Motsch.)

Common name. Rice and Maize Weevils
Family. Curculionidae
Hosts (main). Rice and maize grains in store.
(alternative). Other stored grains and foodstuffs.
Damage. Larvae develop inside the stored grains, which become hollowed out.
Pest status. These are very serious primary pests of all stored grains in the warmer parts of the world. On maize, infestations often start in the field and are later carried into the stores; adults are winged and fly readily.
Life history. Eggs are deposited inside the grains, in small holes chewed by the female; each female may lay 300–400 eggs over a period of several weeks. The larvae remain inside the grains and pupate when they reach a body length of about 4 mm. The adults emerge through the characteristic small circular holes in the grains and are active insects, flying readily and often infesting ripe crops in the field. They are small dark weevils, about 3.5–4.0 mm in length, often with a shiny appearance, and many specimens have four, more or less clearly defined, large reddish patches (spots) on the elytra, which distinguish them from the more temperate *S. granarius*, as well as slightly different sculpturing. These two species are extremely difficult to separate, and usually reliance is placed on male genitalia. For most practical purposes they may be conveniently considered together. Adults generally survive for up to five months.

The life-cycle takes 5–8 weeks, according to temperature, and in warmer regions breeding is continuous, with 6–8 generations per year. Development ceases below 17 °C; optimum conditions are temperatures of 27–31 °C and humidities above 60% RH.
Distribution. These two species are essentially tropical in distribution, but they do survive and thrive in temperate regions (although only inside grain stores) and in cooler regions they may die over the winter period. Their distributions are generally worldwide throughout the warmer parts of the world, but more abundant in the actual tropics.
Control. For control measures see the recommendations for stored produce beetles, on page 274.

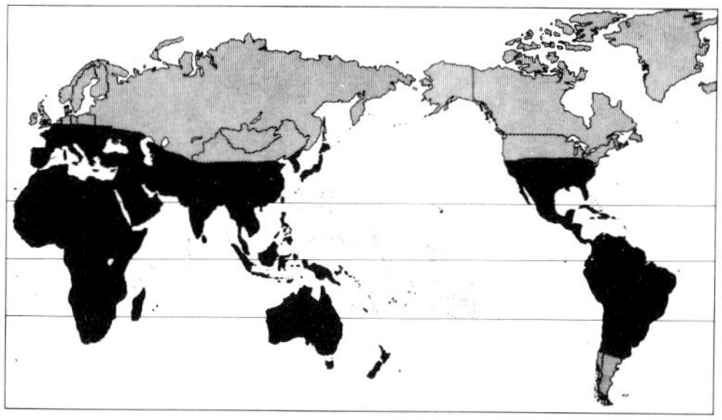

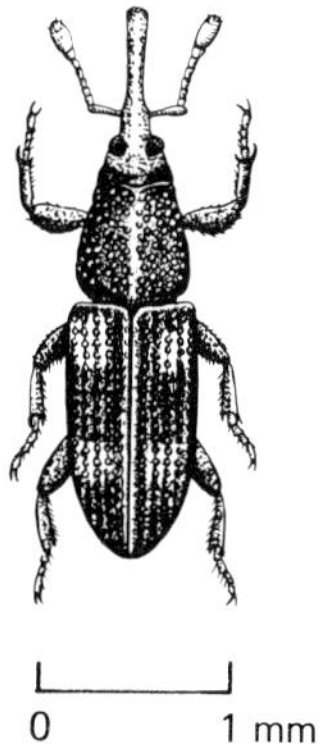

adult weevil of
S. zeamais

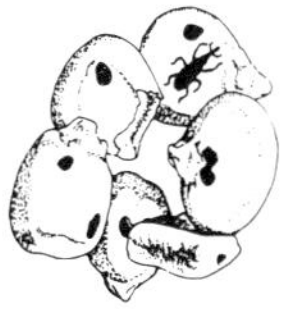
infested maize

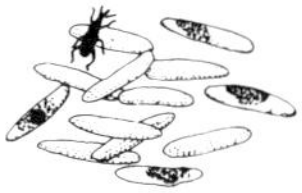
infested rice

Tanymecus spp.

(*T. dilaticollis* Gylh.)
(*T. palliatus* F.)

Common name. Southern Grey Weevil; Grey (Beet) Weevil
Family. Curculionidae
Hosts (main). Sugar beet
(alternative). Maize, sorghum, wheat, oats, barley, sunflower, peach, almond, apple.
Damage. The adult weevils eat the foliage of the host plant; the larvae in the soil eat the roots, but seldom do serious injury.
Pest status. These are quite serious leaf-eating pests on a wide range of crop plants throughout much of southern Asia.
Life history. Life-history details are not known at present, apart from their being typical weevils with soil-dwelling larvae that pupate in the soil.

The adult is actually black-bodied, but covered with grey scales, so the overall appearance is dark grey; body length is 8–11 mm.
Distribution. *Tanymecus dilaticollis* is found throughout eastern Europe and southwest Asia, including part of the USSR (CIE map no. A.357).
T. palliatus is recorded from the UK, but not as a pest here. It is a pest of beet in China; if in fact these records are of the same species it would indicate that it might occur right across Asia from the UK to China.

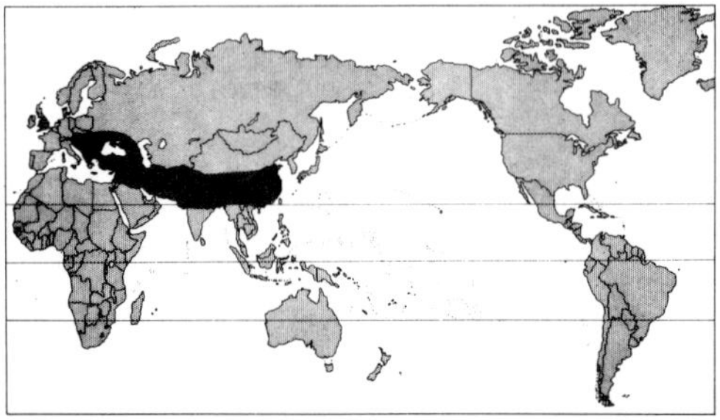

In India *T. circumdatus* Wiedemann is a regular pest of apple, and *T. hispidus* Marshall attacks ber (*Zizyphus*: jujube).

As crop pests the *Tanymecus* weevils are in many respects similar to the many species of *Myllocerus* found in India and other parts of southern Asia.
Control. In India these weevils (adults) have been controlled using sprays of diazinon and fenitrothion, as well as HCH dusts.

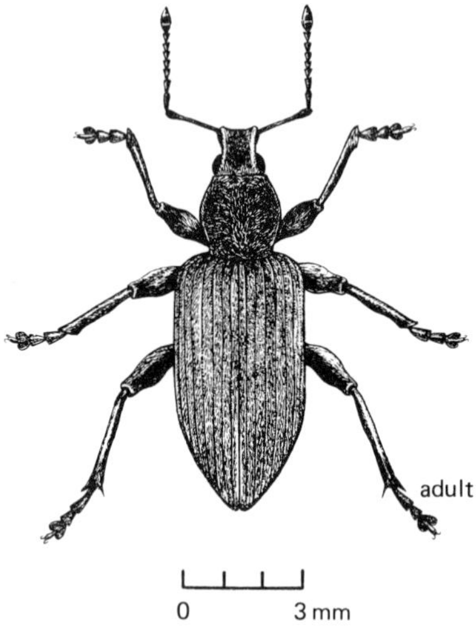

Scolytus rugulosus (Muller)

Common name. Fruit Bark Beetle
Family. Scolytidae
Hosts (main). Apple, plum, damson.
(alternative). Other stone- and pome-fruit trees.
Damage. Adult beetles tunnel under the bark to make breeding galleries that are infected with fungi to produce food both for the female beetle and the developing larvae.
Pest status. Of sporadic importance as orchard pests, but occasionally serious damage is done.
Life history. *Scolytus* is one of the monogamous groups, there being almost equal numbers of males and females that disperse together. The female beetle makes the tunnel and mating often takes place outside the tunnel. After initial mating the male constructs a nuptial chamber (where further mating takes place) within the gallery made by the female. The only other activity the male indulges in is to push out the wood fragments produced by the female tunnelling. This species makes a short straight tunnel with short lateral egg-galleries. The larvae hatch in the short lateral tunnels and as they develop they tunnel further and eventually enlarge the short tunnel distally to make a small pupation chamber. There are two generations per year in the UK.

The adult is a tiny beetle about 2.0–2.5 mm in length, dark reddish-brown and shiny, but the hind part of the elytra usually more conspicuously reddish; legs are rust-coloured.

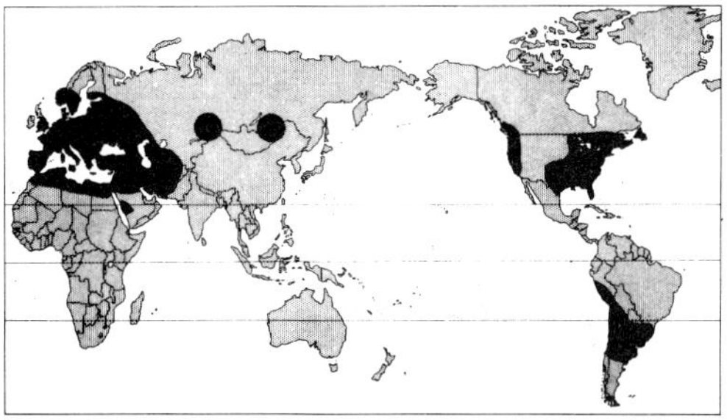

Distribution. This species is recorded throughout Europe, N. Africa, Middle East, N. and S. America (CIE map no. A.392).

Other important species of *Scolytus* include:
Scolytus amygdali (Guer.) – (Almond Bark Beetle) Mediterranean and Middle East.
S. mali (Bechstein) – (Large Fruit Bark Beetle) Europe, Canada and USA.
S. multistriatus (Marsham) – (Smaller Elm Bark Beetle) Europe, Canada and USA (CIE map no. A.347).
S. quadrispinosus Say – (Hickory Bark Beetle) USA.
S. ratzburgi Janson – (Birch Bark Beetle) Europe.
S. scolytus (L.) – (Larger Elm Bark Beetle) Europe, USSR, Near East (CIE map no. A.348).

There are several other species of *Scolytus* important as forestry pests in Canada and the USA, and several others in Japan.
Control. For information on control of bark beetles see page 339.

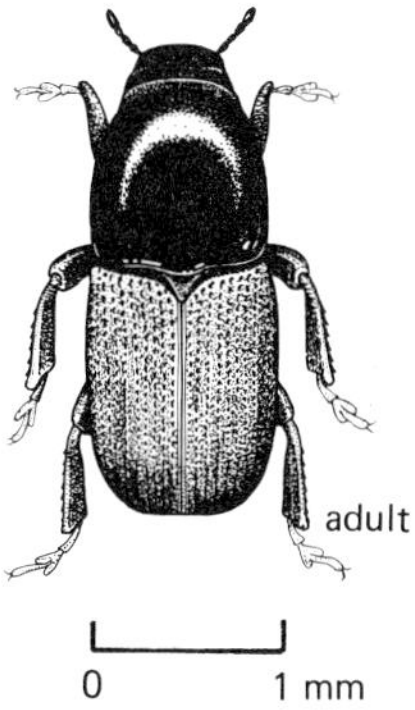

Bark and ambrosia beetles (Coleoptera; Scolytidae)

These beetles constitute the subfamilies Scolytinae and Platypodinae of the family Scolytidae (although some authorities regard them all as Curculionidae, *sensu lato*). Because of their basic biological differences it is preferable to regard them here as forming a separate family – the Scolytidae.

The adults are small, dark-coloured beetles, cylindrical in shape, with a deflexed head adapted for burrowing in wood and other plant tissues. In terms of burrowing beetles they are rather unusual in that it is the adults who tunnel in the trees and not the larvae. The adults burrow under the bark into sapwood (generally) to make breeding galleries; each gallery being tunnelled by one female. Some species are monogamous (*Scolytus*), others bigamous (*Ips*), and some entirely polygamous (*Xyleborus*) (termed spanandrous) and have a sex ratio of 1:10–50. The male is often smaller than the female, short-lived, flightless, and does not leave the breeding gallery; males are produced from unfertilized eggs and thus are haploid. The adult beetle finds a suitable host tree (or plant) and tunnels under the bark and builds an extensive breeding gallery and inoculates the tunnel lining of frass and faeces with a special fungus (ambrosia fungus). The fungus mycelium spreads over the tunnel walls and is used as food both by the maturing female and the developing larvae. The structure of the tunnel varies with the different genera (and species) of beetle concerned. The female beetle in many species has special body cavities (called mycangia) in which fungal spores or conidia of the 'ambrosia' fungus are carried and kept alive. Some of these fungi belong to the genus *Ambrosiella*. In some species the mycangia are cavities behind the mandibles, in others thoracic tubes, others carry the fungus in the crop.

The different feeding life-styles can be summarized into five basic types (see Beaver, 1977), as follows:

(1) phloeophagy – feeding on the phloem/cambium layer; a very common type, mostly associated with trees in temperate regions.
(2) herbiphagy – feeding on the tissues of soft herbs and woody twigs; an uncommon type; e.g. *Hylastinus*, the Clover Root Borer.
(3) xylophagy – these are the true wood feeders and are quite rare.
(4) spermophagy – feeding on fruits and seeds, and sometimes also leaf-stalks in the tropics; a rare type confined mostly to the tropics; e.g. *Hypothenemus hampei* (Coffee Berry Borer) and *Hypothenemus* spp.; also *Coccotrypes*, the Date Stone Borer.
(5) xylomycetophagy – these species feed on the ambrosia fungus cultivated within a breeding gallery tunnel system either in wood (under the bark) or twigs; a very common type, most abundant in the tropics but quite plentiful in warmer temperate regions; e.g. *Scolytus* and *Xyleborus*.

It is thought that the ambrosia feeders are scarce in the cool temperate regions because the fungus finds conditions too cool for proper development.

Bark beetles are characterized by having very effective dispersal and host-finding mechanisms. They usually have to find a new host for each generation, as most hosts do not survive long enough for more than one generation to develop. Most species attack trees, and most of these seem to prefer sickly or dying trees; very few species attack healthy trees, although some of the twig-borers apparently attack healthy twigs, and they also attack and destroy stressed tree seedlings (*Xylosandrus* spp.). Most species are host-selective as to species, and size-selective, usually only attacking trees of a particular size range. Some species are basically forestry pests and prefer newly felled timber as hosts. The group is actually of more importance as forestry pests than as agricultural pests, and the most heavily attacked group of trees are the gymnosperms. Flight activity is controlled by light, temperature and wind; on emergence the females are photopositive, but this later disappears. Distances of dispersal are quite prodigious, being generally 1–30 km per day for several days. Usually a large number of individuals attack the same tree simultaneously; aggregation pheromones are used to facilitate this, but not for most spanandrous species apparently.

The ambrosia beetles all exist in an ectosymbiotic relationship with the fungi, the mycelium of which is an efficient extractor of nutrients from the wood surrounding the gallery. Each beetle usually has one specific fungus, but in addition is usually associated with several others, and also some yeasts and bacteria. The fungal development is essential for both larval development and for maturation of the female's ovaries, as she does not feed prior to dispersal. Fungal development is closely controlled by humidity and temperature; it ceases when the moisture level of the wood drops below about 40–50%. As well as transporting ambrosia fungi, some species also carry pathogenic fungi which often cause the death of the tree. Thus some *Xyleborus* transmit pathogenic fungi to cocoa and mango trees in the tropics, and *Scolytus scolytus* and *S. multistriatus* transmit the causal organism of Dutch Elm Disease (*Ceratostomella ulmi*) in Europe.

Bark beetle pests

In this chapter *Scolytus* and *Xyleborus* are looked at in some detail, but other important genera and species include the following:

Blastophagus (= *Tomicus*) spp. – attack pines, spruces, firs; temperate regions.

Coccotrypes dactyliperda F. – (Date Stone Borer) Mediterranean Region.

Conophthorus spp. – (Pine Cone Beetles) USA and Canada.

Dendroctonus spp. – abundant in pine forests; C. America and USA.

Hylastinus obscurus Marsham – (Clover Root Borer) UK, Europe, Canada, USA.

Hylesinus oleiperda F. – (Olive Bark Beetle) Mediterranean Region.

Hypoborus ficus Erichs. – (Fig Bark Beetle) Med.

Hypothenemus hampei (Ferr.) – (Coffee Berry Borer) Africa, S.E. Asia, and now S. America (CIE map no. A.170).

Ips spp. – attack pines, spruces, firs, larch, etc.; northern Europe, Asia, Canada and USA.

Xylosandrus compactus (Eichh.) – (Black Twig Borer) polyphagous; pantropical (CIE map no. A.244); this species attacks healthy trees.

Xylosandrus morigerus (Bldf.) – (Brown Coffee Borer) polyphagous; S.E. Asia (CIE map no. A.292); a primary pest on some plants, secondary on others.

The most abundant genus in the Platypodinae is *Platypus*; these species are mostly tropical in distribution and most prefer as hosts freshly felled timber, and so are of no agricultural importance.

Control of bark beetles

These beetles are difficult to control because of the nature of their infestations, but the most vulnerable phase in their life-history is the attack/dispersal phase of the adults.

Cultural methods. These consist mainly in trying to keep the trees healthy. Also included is the removal and destruction of infested twigs, branches and trees, and the prompt removal of felled timber. In timber production the drying of felled timber is effective in that the fungus is soon killed. Pruning of crops such as tea and coffee causes some problems in that the pruning sites are often vulnerable to attack both by Scolytidae and sometimes termites.

In some forest situations the trapping of adult beetles is having a definite regulatory effect on pest populations; traps used include sticky traps and ultraviolet light traps, but the most effective are those baited with either aggregation or sex pheromones.

Chemical methods. These fall under several headings. Pheromone traps using aggregation pheromones are quite effective. Use of fungicides to destroy both the ambrosia fungus and the pathogenic fungi can be effective, but is expensive as it requires individual trees being injected using vascular injectors.

Against agricultural pests there are, at present, no good standard recommendations. Some success has been achieved using dieldrin as a post-pruning application, or a dieldrin/Bordeaux mixture, or HCH plus heavy sticker solution. Application of tar oil to the tree trunks in the spring will drive out adults from the gallery usually, but application has to be done with great care because of its phytotoxicity to the tree foliage.

Dead elm tree showing *Scolytus* breeding galleries under the bark

Scolytus breeding gallery

Xyleborus dispar (F.) & spp.

(= *Anisandrus dispar* F.)

Common name. Shot-hole Borer (Ambrosia Beetle) (Fruit Tree Ambrosia Beetles)

Family. Scolytidae

Hosts (main). Plum, damson, apple, pear, hazelnut, chestnut, etc.

(alternative). Many other deciduous trees (oaks, beeches, elms, rowan, holly, etc.).

Damage. The adult beetles tunnel right into the heartwood of trunks and larger branches, and then innoculate the breeding gallery with ambrosia fungus.

Pest status. A localized but quite important pest; but the genus *Xyleborus* occurs worldwide as a large number of important pest species on a wide range of hosts.

Life history. This is a spanandrous (polygamous) species with usually a sex ratio of one male to about 50 females, and the male is physically much smaller than the female. The female makes the breeding gallery by herself, unaccompanied by the male who remains behind in the gallery where he was bred. The main gallery is bored deep into the heartwood as a single tunnel with small lateral egg-chambers where the larvae develop and feed on the ambrosia fungus. Most larvae hatch in May and June, but generally both larvae and adults may be found in the galleries at all times of the year; adults are generally most abundant over the period January to June.

The adults are black in colour with antennae and legs yellow; the male is 2.0–2.5 mm long with short elytra, the female is 3.0–3.5 mm with elytra twice as long as the prothorax, and the male has a generally rounder and less cylindrical body.

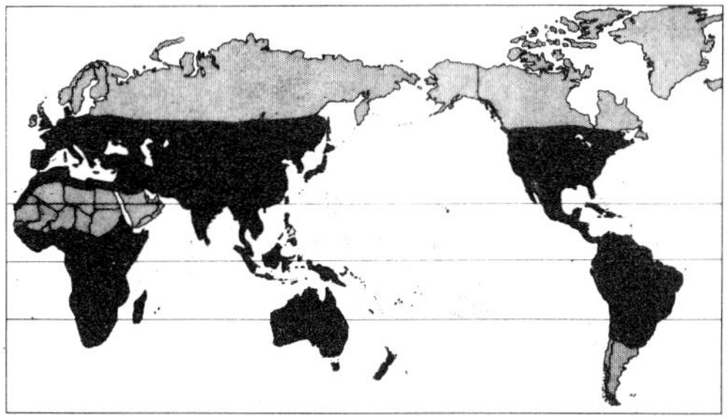

There is usually one generation per year.

The various species of *Xyleborus* are similar in appearance and biology.

Distribution. There is a large number of important pest species to be found worldwide on many different trees and woody shrubs, including:

Xyleborus dispar (F.) – (Shot-hole Borer) UK and Europe.

Xyleborus fornicatus (Eichh.) – (Tea Shot-hole Borer) polyphagous; Madagascar, India, S.E. Asia, Papua New Guinea (CIE map no. A.319); a primary pest.

X. ferrugineus (F.) – polyphagoûs; pantropical (CIE map no. A.277).

X. perforans (Woll.) – (Coconut Shot-hole Borer) polyphagous; pantropical (CIE map no. A.320).

X. saxeseni (Ratz.) – (Fruit Tree Wood Ambrosia Beetle) Europe.

In India there are nine other species of importance as fruit tree pests. In Japan a total of 15 species of *Xyleborus* are agricultural pests including:

X. apicalis Blandford – Apple Ambrosia Beetle.

X. atratus Eichhoff – Mulberry Ambrosia Beetle.

X. semiopacus Eichhoff – Apple Ambrosia Beetle.

X. sobrinus Eichhoff – Citrus Ambrosia Beetle.

Control. For control measures see page 339.

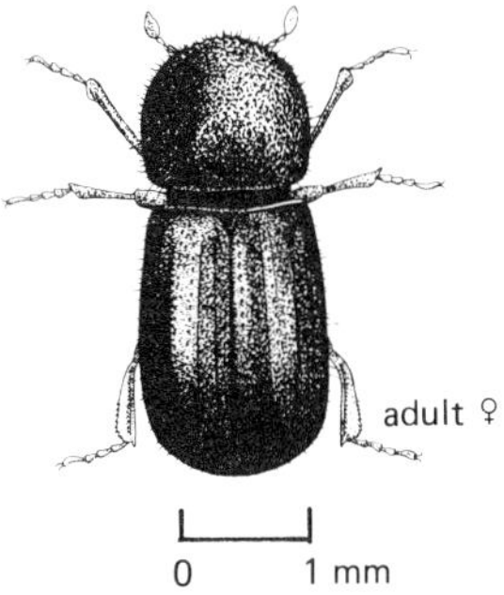

Order

Diptera

(true flies)

Families

Tipulidae (crane flies; leatherjackets)
Cecidomyiidae (gall midges)
Syrphidae (hover flies)
Sciaridae (fungus flies)
Tephritidae (fruit flies)
Agromyzidae (leaf miners)
Psilidae
Ephydridae (marsh flies)
Chloropidae ('gout' flies)
Anthomyiidae (root flies)

Tipula spp.

Common name. Leatherjackets (Common Craneflies)
Family. Tipulidae
Hosts (main). Grass, both as pasture and quality turf.
(alternative). Cereals, root crops, vegetables; polyphagous root-eaters.
Damage. The larvae in the soil eat plant roots, and tunnel tubers and root crops; seedlings may be destroyed; on damp nights they feed on the surface.
Pest status. Basically, these are grassland species and most important as agricultural pests on crops grown in newly ploughed grassland. Generally quite important pests; in Canada populations of about 1000 larvae per square metre of rangeland have been recorded.
Life history. Eggs are laid in the soil, each female laying about 300 eggs, and require about 14 days to hatch. The tiny larvae are very susceptible to drought and sunlight, when they die of desiccation. The larvae are fat legless maggots, that telescope their body when resting. They feed on plant roots throughout the autumn, winter and spring. In the summer the dark greyish larvae are fully grown at about 4 cm body length; they pupate just under the soil surface. The mobile pupae push their way to the soil surface until the front part is projecting above the surface; then the young adults emerge. Most adults emerge in August/September.

The adults are large grey-brown flies, with body length of 18–25 mm, but very long legs; the genitalia of the

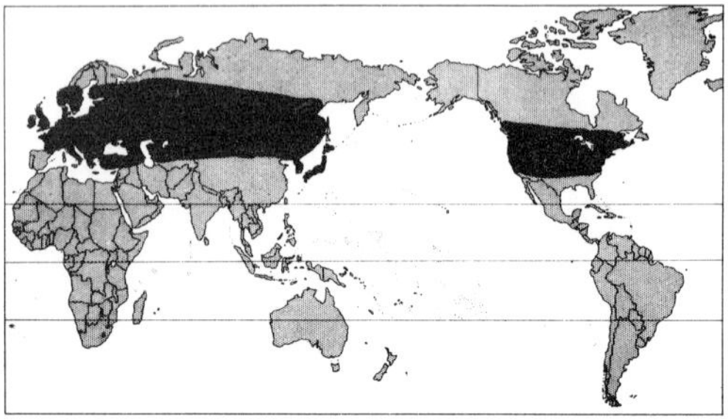

males are exerted and conspicuous, and the female has a short ovipositor.

There is only one generation per year.

Distribution. This genus is abundant throughout Europe, Asia, Canada and the USA, represented by about a dozen different species; the most important species are as follows:

Tipula oleracea L. – (Common Crane Fly/Leatherjacket) Europe.

T. paludosa Meigen – (European (Marsh) Crane Fly/Leatherjacket) Europe, Asia, Canada and USA (CIE map no. A.370).

T. simplex Doane – (Range Crane Fly) USA.

T. aino Alex. – (Rice Crane Fly) Japan.

An equally abundant close relative, but usually less damaging to crops, is the group of Spotted Crane Flies, (*Nephrotoma* spp.), also Holarctic in distribution. *Limonia* spp. are often referred to as 'Small Yellow Crane Flies' and are widespread in Palaearctic Region.

Control. Leatherjackets are preyed on quite heavily by birds such as starlings and plovers, and ploughing will expose the larvae to other bird predators. A virus disease is also quite abundant, but seldom affects more than a small proportion of any population. Rolling the soil will kill some larvae.

Insecticidal control is recommended if populations are large; either baits using bran with DDT, HCH, or fenitrothion; or methiocarb (slug) pellets; or sprays of aldrin, DDT, HCH, chlorpyrifos, fenitrothion, quinalphos, or triazophos; chlorpyrifos is reported to be particularly effective.

The UK economic threshold is a total of 10–15 larvae in ten separate 30 cm length of drill (cereals, especially winter wheat) examined in a diagonal across the field (12 or 18 cm row spacing).

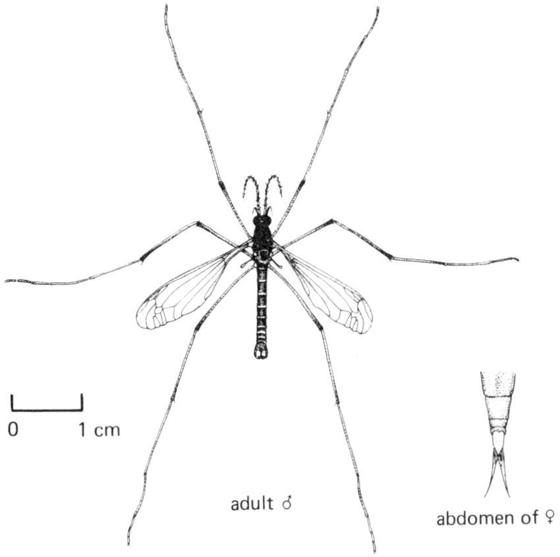

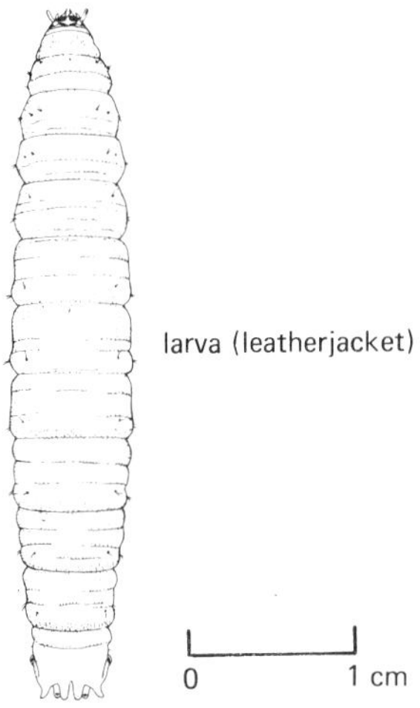

Contarinia pisi (Winn.)

Common name. Pea Midge
Family. Cecidomyiidae
Hosts (main). Peas
(alternative). Beans of some species.

Damage. Larvae deform flowers, cause a shoot 'nettlehead', and infest pods, sometimes causing them to be deformed. Early shoot destruction causes loss of yield, but late attacks on peas for freezing may be beneficial in stopping plant growth and aiding simultaneous ripening of the pods.

Pest status. A pest of regular severity throughout Europe; occasionally beneficial on crops for freezing.

Life history. Adults emerge from the cocoons in the soil during June, depending upon the prevailing temperatures. The female midge lays her eggs inside the shoots and flower buds, and sometimes inside the young pods, using her delicate ovipositor; usually in batches of 20–30 per head/flower. Hatching takes about four days.

The larvae are whitish, gregarious, and when disturbed are able to jump considerable distances. They feed in the same manner as many fly maggots, by scraping the plant tissues with their mouthhooks and sucking up the exuding sap. Feeding on young tissues causes growth deformation; some flower buds never open, the terminal shoot telescopes into a 'nettle-head' and further growth virtually ceases. After about ten days the larvae are mature, 2–3 mm long, and they descend to the soil to pupate. They may pupate immediately (inside a silken cocoon) and in 11–14 days give rise to a second generation of adults that attack the same or adjacent pea crops, or else the larvae remain inside the cocoons and overwinter. Thus the larvae from the large first generation, together with those from the smaller second generation, overwinter in their cocoons in the soil, and usually pupate early the following summer. A few may remain as larvae in the soil for several years.

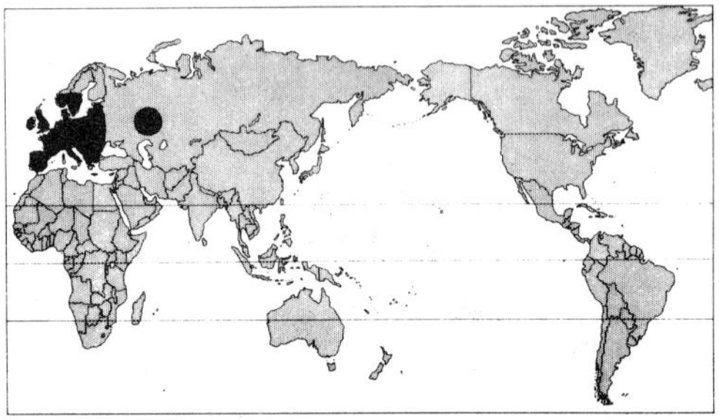

Adults are tiny, delicate midges, grey in body colour, and 2–3 mm long; they only live for about four days. Second-generation adults appear in July and August.

There are usually two generations per year.

Distribution. This pest is recorded throughout the UK and Europe, but the extent of its occurrence through Asia is not known.

Control. Susceptible crops should be very carefully (re timing) sprayed with the insecticides recommended on page 346.

Pea pod with midge larvae inside

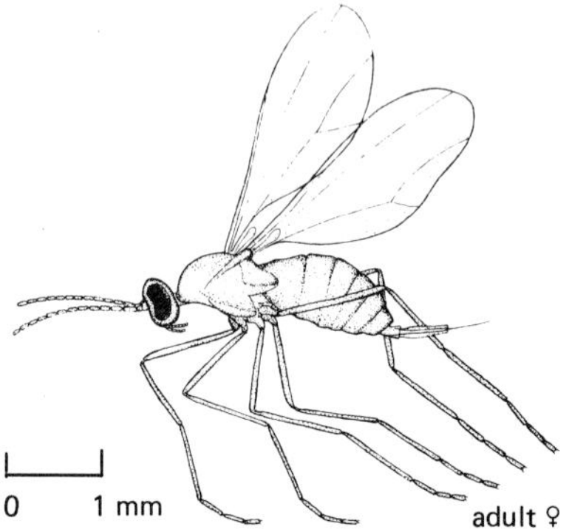

Gall midges (Diptera; Cecidomyiidae)

A very large family of minute, fragile flies; they are characterized by having reduced wing venation and long moniliform antennae with whorls of bristles. The larvae are peripneustic, with a reduced head, and usually a distinct sternal spatula. Most species are phytophagous and virtually every species of flowering plant is attacked by at least one species of gall midge. Many midge species are damaging to cultivated plants, and these are reviewed in the eight-volume monograph by Barnes (1946–69), but only a few are pests of any consequence (see below). For the field entomologist most midge species cannot be easily identified (if at all!) as both adults and larvae show striking morphological similarity throughout the family. However, most species are host-specific and larval damage (especially galls) is often distinctive. Thus by knowing the identity of the host plant, and recognizing the type of damage done, it is usually possible to arrive at the identity of the midge concerned. The larvae are often brightly coloured, orange, red or yellow (although most are white), which aids in their recognition.

Larvae show greatly diversified habits, and have been classified as follows:

(1) Zoöphagous species: they mostly prey on Homoptera and on mites; a few prey on other Diptera (both larvae and pupae), and a very few are parasites of aphids and other small Homoptera.
(2) Saprophagous species: found in decaying vegetable matter, fungi and also in the excrement of lepidopterous caterpillars, and dung.
(3) Phytophagous species; these are in turn further subdivided:
   (a) plant-feeders making no galls (cereal and grass midges – in flowerheads).
   (b) plant gall inquilines: living inside galls induced by some Hymenoptera, Coleoptera or Diptera (Tephritidae, and other Cecidiomyiidae).
   (c) true gall-formers (cecidogenous species):
      (i) leaf and leaflet semi-galls (*Dasineura* spp. on legumes, rose, violet, etc.).
      (ii) leaf galls proper (Longan Gall Midge, etc.).
      (iii) shoot and bud galls (Pea Midge; Hawthorn Shoot Midge, etc.).
      (iv) seed and fruit galls (Sorghum Midge; Pea Midge; Pear Midge, etc.).
      (v) stem galls (*Asphondylia morindae* on *Aporusa chinensis*, etc.).
      (vi) root galls.

Gall midges of importance

The three main genera of pest species are as follows; for an indication as to their abundance it may be noted that in the UK there are listed the following number of species for each genus – *Asphondylia* (15 spp.), *Contarinia* (72 spp.), and *Dasineura* (136 spp.). The two groups of plants that are probably the most attacked by gall midges are the Leguminosae and Gramineae. The main pest species of note include, as follows:

Asphondylia sesami Felt – (Sesame Gall Midge) E. Africa and India.
A. websteri Felt – (Alfalfa Gall Midge) USA.
Contarinia johnsoni Felt – (Grape Blossom Midge) USA.
C. mali Barnes – (Apple Blossom Midge) Japan.
C. merceri Barnes – (Cocksfoot/Foxtail Midge) Europe.
C. medicaginis Kieffer – (Lucerne Flower Midge) Europe.
C. nasturtii (Kieffer) – (Swede Midge) Europe.
C. pisi (Winn.) – (Pea Midge) (see page 344).
C. pyrivora (Riley) – (Pear Midge) Europe, USA.
C. humuli (Theobald) – (Hop Strig Midge) Europe.
C. rubicola Kieffer – (Blackberry Flower Midge) Europe.
C. sorghicola (Coq.) – (Sorghum Midge) Africa, Japan, Australia, USA, S. America (CIE map no. A.72).
C. tritici (Kirby) – (Yellow Wheat Blossom Midge) Europe, Asia, Japan (CIE map no. A.182).
C. vaccinii Felt – (Blueberry Tip Midge) USA.
Dasineura affinis (Kieffer) – (Violet Leaf Midge) Europe.
D. brassicae (Winnertz) – (Brassica Pod Midge) Europe.
D. coffeae Barnes – (Coffee Flower Midge) Africa.
D. crataegi (Winnertz) – (Hawthorn Button-top Midge) Europe.

D. leguminicola (Lintner) – (Clover Seed Midge) Europe, Canada, USA.
D. mali (Kieffer) – (Apple Leaf Midge) Europe, Japan.
D. plicatrix (Loew) – (Blackberry Leaf Midge) Europe.
D. pyri (Bouché) – (Pear Leaf Midge) Europe, New Zealand.
D. rhodophaga (Coq) – (Rose Midge) USA.
D. ribicola (Kieffer) – (Gooseberry Leaf Midge) Europe.
D. rosarum (Hardy) – (Rose Leaf Midge) Europe.
D. tetensi (Rub.) – (Black Currant Leaf Midge) Europe.
D. trifolii (Loew) – (Clover Leaf Midge) Europe, USA.
D. viciae (Kietter) – (Vetch Leaf Midge) Europe.
Geromyia pennisiti (Felt) – (Millet Grain Midge) Africa and India.
Haplodiplosis marginata (vR) – (Saddle Gall Midge) on cereals; Europe (page 347).
Japiella medicaginis (Rubs.) – (Lucerne Leaf Midge) Europe, USA.
Mayetiola spp. – (Hessian Fly and Stem Midges) cosmopolitan (page 348).
Mycophila etc. spp. – (Mushroom Midge complex) cosmopolitan.
Neolasioptera murtfeldtiana (Felt) – (Sunflower Seed Midge) USA.
Orseolia oryzae (W-M) – (Rice Stem Gall Midge) Africa, India, S.E. Asia (CIE map no. A.171).
Prodiplosis citrulli Felt – (Cucurbit Midge) USA.
Resseliella occuliperda (Rüb.) – (Red Bud Borer (of rose)) Europe.
R. soya Monzen – (Soybean Stem Midge) Japan.
R. theobaldi (Barnes) – (Raspberry Cane Midge) Europe.
Rhopalomyia chrysanthemi (Ahl.) – (Chrysanthemum Gall Midge) USA, Japan.
Sitodiplosis mossellana Gehin – (Orange Wheat Blossom Midge) Europe, Asia, N. America (CIE map no. A.183).

Control of gall midges

It is a little difficult to generalize about control of midges owing to the diversity of life-styles; larvae living internally inside plant tissues are difficult to destroy, but the more exposed foliage feeders are vulnerable to both predation and to insecticides.

Cultural control. Varietal resistance is being used quite successfully to combat some species. Other aspects include early or late planting/sowing; stubble and crop residue destruction; crop rotation can also be effective in minimizing damage.

Biological control. Many hymenopterous parasites and several predacious beetles and bugs (Heteroptera) are important natural enemies of many pest species; some pests are regularly kept in check by natural enemies.

Chemical control. Insecticide sprays are used to kill both the delicate little adults in the crop, and the larvae *in situ* on or in the plants. For accurate spray timing a regular infestation/crop monitoring is required, using traps for adults (wind, suction, sticky, etc.), and foliage inspection for larvae and damage symptoms. The insecticides generally used in the past included DDT, endrin, carbaryl, parathion-methyl, and phosalone. Present recommendations include sprays of azinphos-methyl, demeton-*S*-methyl sulphone, demeton-*S*-methyl, dimethoate, fenitrothion and triazophos. Some of the different species require the use of insecticides with different properties. Usually several sprays are required to give adequate temporal coverage.

Haplodiplosis marginata (von Roser)

(= *H. equestris* (Wagner))

Common name. Saddle Gall Midge
Family. Cecidomyiidae
Hosts (main). Barley, wheat.
(alternative). Oats, rye.

Damage. Blood-red larvae feed in depressions on the stem under the leaf sheath of the first and second internodes; the stem weakens and often breaks and the head falls. Heavy infestations invariably result in considerable loss in yield.

Pest status. Of some importance as a pest in the UK, in eastern regions, but a regularly serious pest in northern Continental Europe. Most damaging on winter wheat and spring barley.

Life history. Adult midges emerge from the soil in late May to early June, usually during warm wet weather. The bright red eggs are laid in short chains on the leaves, on either side. Hatching takes 1–2 weeks.

The larvae move into the leaf sheaths and feed on the stem surface under the sheath; the larvae feed here for several weeks causing a saddle-shaped depression in the stem; at maturity they are blood-red and about 4–5 mm long. Fully grown larvae enter the soil where they remain for the winter before spinning a cocoon and pupating in the spring. Some larvae may remain free in the soil for two winters before they pupate. Pupation only takes a few days.

The adults are larger than most, being 3–5 mm long, females the larger; both sexes are red in colour. They fly weakly and do not disperse far.

There is only one generation per year.

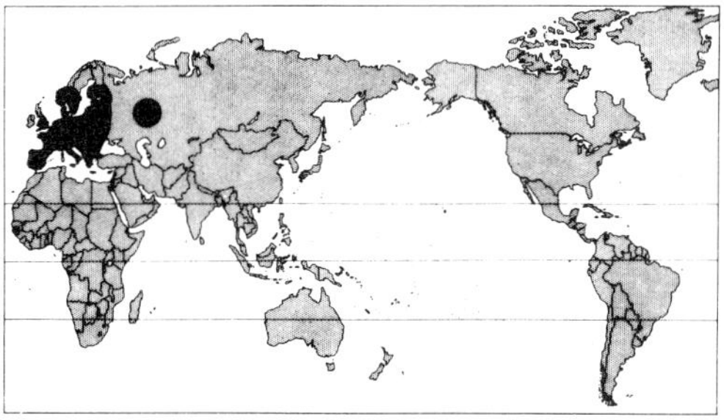

Distribution. This is basically a European species, first recorded in the UK in 1889 in Lincolnshire; found throughout most of Europe.

Control. Cultural methods of control include adequate crop rotation, early sowing and good weed control will reduce levels of infestation.

When insecticides are required, formerly DDT and parathion were used quite successfully, but the more recent recommendation is usually sprays of fenitrothion.

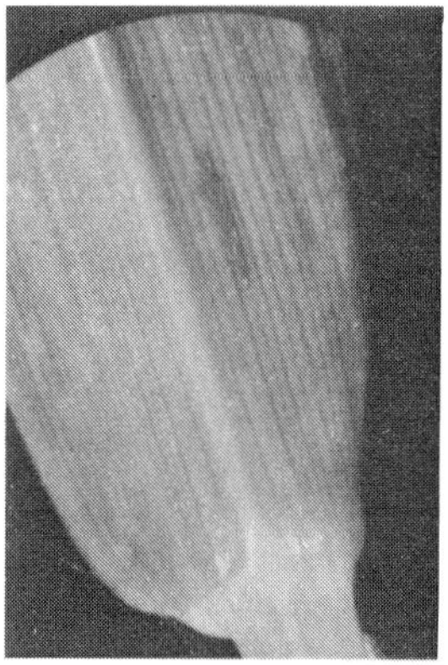

Eggs on barley leaf

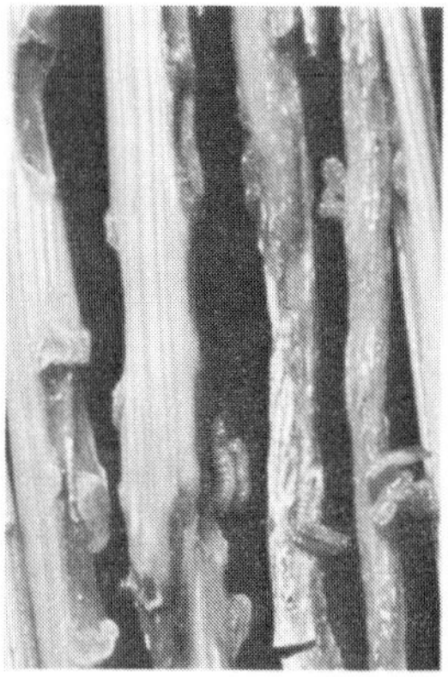

Galls and larvae on wheat stems

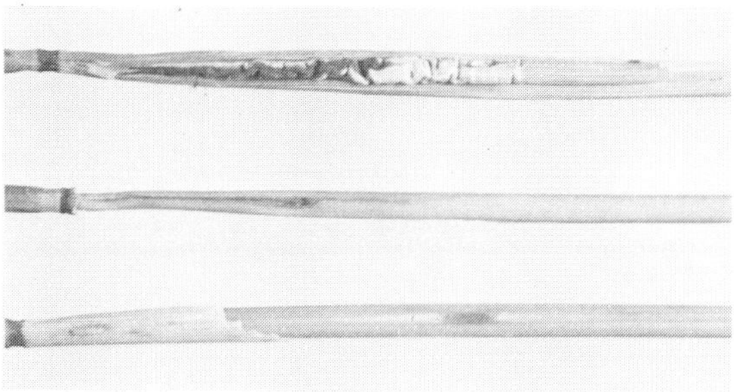

Galled wheat stems (MAFF)

Mayetiola destructor (Say)

Common name. Hessian Fly
Family. Cecidomyiidae
Hosts (main). Wheat
(alternative). Barley, rye, couch grass, and other grasses.

Damage. In early summer stems of winter wheat or spring barley lodge, and white larvae can be found at the point of stem breakage; sometimes stems do not break but larval feeding causes 'whiteheads' and poor quality grain.

Pest status. A serious pest of wheat in N. America and parts of Europe; seldom serious in the UK.

Life history. Adult midges emerge in May and June and the females lay their eggs on the leaf blades; each female lays up to 300 eggs. Hatching requires 5–7 days. The larvae crawl beneath the leaf sheaths to feed just above a node. After about three weeks the larvae are fully grown, about 4 mm long, and they moult, forming the puparium with a brown skin ('flax seed') in which the prepupa lies. In the spring generation pupation takes about 12–20 days. Adults of the second generation emerge in September and they use either grasses or volunteer wheat plants for oviposition. Second-generation prepupae overwinter inside the puparia, and actually pupate in the late spring, but some prepupae may stay within the puparium in the soil for up to four years. In N. America the autumn generation of larvae are very damaging to winter wheat seedlings; in the UK crop damage is only done by the first-generation larvae.

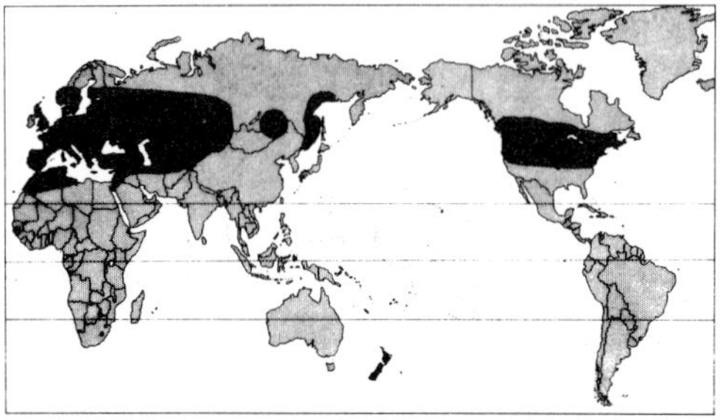

The adult is a small midge (4 mm long), with black thorax and brown abdomen; females live for about six days.

There are usually two generations per year, except in the extreme north.

Distribution. There are several species of *Mayetiola* important as cereal pests, and all are very similar but show some differences in host preferences; they include:
M. avenae (Marchal) – Oat Stem Midge.
M. mimeuri Mesnil; *M. secalis* Bollow; and *M. secalina* (Loew).
Distribution includes Europe, N. Africa, most of Asia, New Zealand, Canada and the USA (CIE map no. A.57).

Control. *Cultural control*. In N. America much breeding research has produced many resistant varieties that are tolerant of this pest. Also early sowing reduces attack levels, and stubble burning kills diapausing larvae/prepupae. In most countries levels of natural parasitism are usually quite high. In the USA the usual practice is to sow wheat on so-called 'fly-free' dates; this approach is used because the adult midges are so short-lived. Generally all control methods used for this pest are cultural; insecticide use is not recommended.

Eumerus spp.

(*E. strigatus* Fall.)
(*E. tuberculatus* Rond.) etc.

Common name. Small Narcissus Flies (Onion Bulb Fly; Lesser Bulb Fly)
Family. Syrphidae
Hosts (main). Narcissus bulbs, snowdrops, onions.
(alternative). Secondary pests in potato, carrot, and cabbages.
Damage. Larvae are found inside bulbs, and occasionally in damaged root crops, where they feed and hollow out the bulb; larvae are usually gregarious and many will be found inside the same bulb.
Pest status. Generally *Eumerus* spp. are regarded as secondary pests and appear to attack only damaged bulbs and roots, but they are quite common and widespread.
Life history. Eggs are laid during May in groups of 10–20 on or near the bulbs. The emerging maggots penetrate the host bulb or root at the point of damage, and work through the root or bulb eating continuously, until usually the bulb is quite hollow. Pupation takes place in the soil. Some larvae may not pupate until the following spring.

Adults emerge in July (in the UK) and the second generation starts; the larvae feed during the autumn and winter, and pupate the following spring. Adults emerge at the end of April and in early May.

In Israel *E. amoenus* has a phenology adapted to its wild hosts and larval development ceases (due to aestivation) when the bulbs enter summer dormancy.

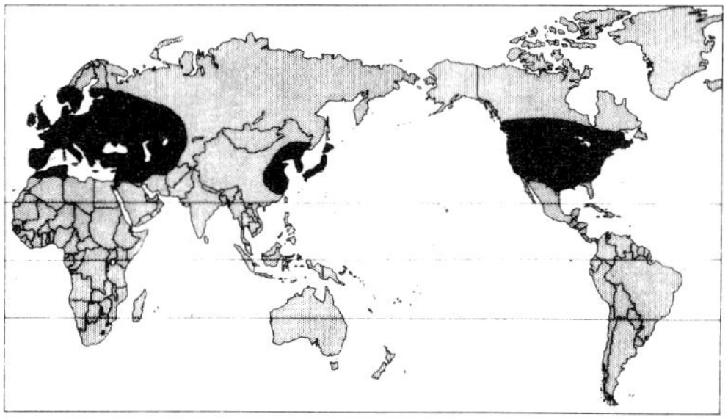

Adults are small syrphids, about 6 mm in body length, shiny and blackish with white markings.

In the UK there are two generations per year; in Israel probably three.
Distribution. The two main species are found throughout Europe, much of Asia (including Japan), Canada and the USA.

In the Mediterranean Region the common species is *Eumerus amoenus* Loew, the Mediterranean Lesser Bulb Fly, and in the Orient there are also the pest species *E. chinensis* S. & E. and *E. okinawensis* Shiraki.

E. figurans Walker is the Ginger Maggot of the USA.
Control. Generally regarded as secondary pests that only attack damaged bulbs and roots, so infestation by these insects should be examined carefully to determine the cause of primary damage.

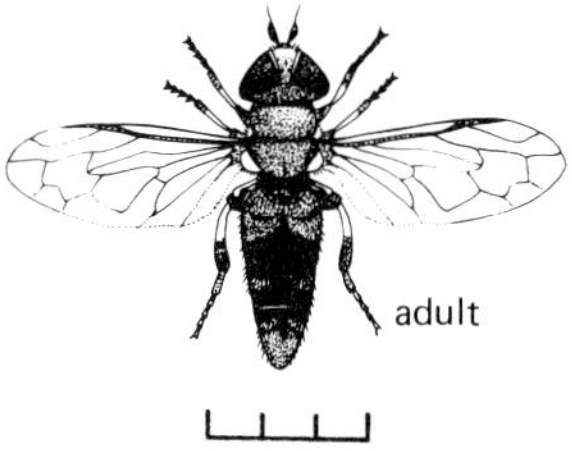

Merodon equestris (F.)

Common name. Large Narcissus Fly
Family. Syrphidae
Hosts (main). *Narcissus* bulbs.
(alternative). Bulbs of the Liliaceae and Amaryllidaceae.

Damage. A single large maggot eats out the interior of the attacked bulb; small bulbs are destroyed, large bulbs produce flowerless distorted growth.

Pest status. A serious primary pest of bulb crops, but heavy infestations are not very common, except in some localized areas.

Life history. Adults are flying from the end of April to the end of June in the UK, being especially active on warm sunny days. Each female lays about 40 eggs, one per bulb; the female places the egg as close to the bulb in the soil as possible, often crawling down the hole left by the withering leaves. On hatching the larva crawls down the side of the bulb and penetrates through the base plate. It gradually moves up to the growing point surrounded by the fleshy scale leaves. Eventually the larva has eaten out the centre of the bulb and made a large cavity. Fully grown larvae measure up to 18 mm; they leave the bulbs in about March the following year to pupate in the soil. Pupation takes 5–6 weeks.

The adult is a large fat-bodied, furry fly about 13 mm in length, with a variable body coloration, but usually blackish banded yellow, grey or orange.

There is only one generation per year.

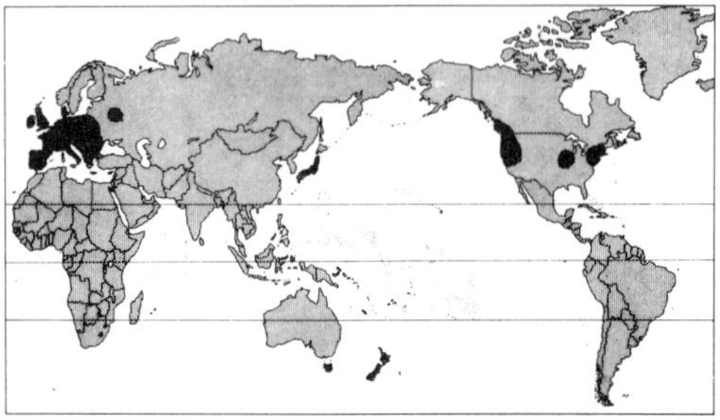

Distribution. Found throughout much of Europe, also in Japan, Tasmania, New Zealand, Canada and the USA (CIE map no. A.120).

Merodon geniculata Strobl is the Mediterranean Narcissus Bulb Fly; recorded from Algeria, Morocco, Israel, Cyprus and Turkey.

Control. In areas at risk from this pest preventative measures are recommended, and these include both a high standard of crop cultivation and the treatment of bulbs with a persistent insecticide prior to planting, usually aldrin. Treatment of the growing crop has seldom been successful. At harvest time hot-water dipping will kill any larvae inside the bulbs; also a cold soak in HCH solution. All bulb treatments, however, need to be done with considerable precision and care to avoid damage to the bulb, and local recommendations should be followed carefully.

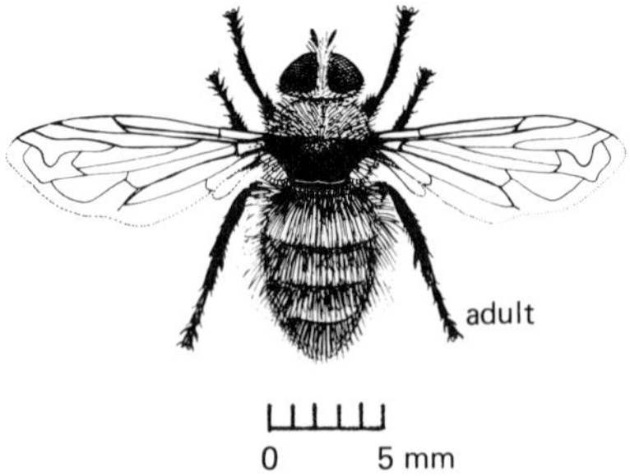

Sciara/Megascelia spp. (etc.)

Common name. Mushroom Fly complex
Family. Sciaridae/Phoridae
Hosts (main). Mushrooms (*Agaricus* spp.)
(alternative). Many pot plants and garden plants.
Damage. Larvae in the soil are partly saprophagous, but to some extent will also attack healthy plants; some species less saprophytic than others. The mushroom flies feed on the fungal mycelium and also bore the fruiting body, both stalk and cap.
Pest status. A rather heterogeneous group of more primitive flies whose larvae live in decaying vegetation and humus-rich soil; serious pests of mushrooms.
Life history. Eggs are laid in the soil; the females apparently being attracted by the smell of rotting vegetable matter or fertilizer. Most females lay 100–300 eggs.

The larvae are rather elongate little maggots, white in colour, finally measuring some 4–6 mm in length; some species (Sciaridae) have a characteristic black head capsule, but the phorid fly larvae lack the black head.

The adults are small dark flies, mostly between 3–5 mm in body length. Most species are quite short-lived and die after about a week of activity.

Under warm conditions (about 20 °C) the life-cycles may be completed in 4–5 weeks, but under cooler (field) conditions probably 6–8 weeks is required. Under warm conditions (protected cultivation) breeding may be continuous, but the larger 'wild' species in Europe are univoltine.

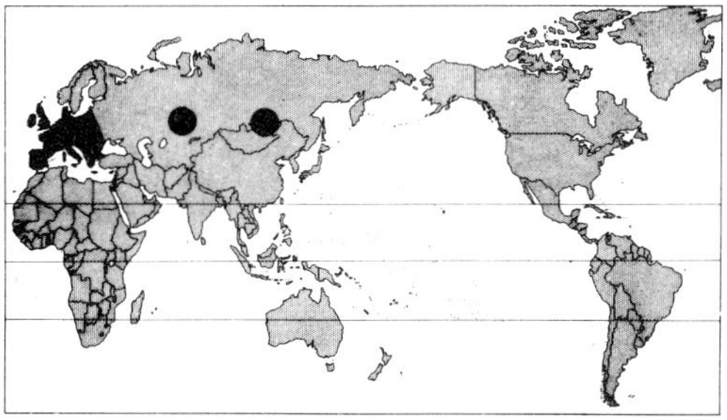

Distribution. The main groups of flies concerned are as follows:

Sciara spp., *Bradysia* spp., *Lycoriella* spp. – Sciaridae: Some species attack mushrooms and others are more important as pests of pot plants and glasshouse crops.

Megascelia spp., especially *M. nigra* – Phoridae (Scuttle Flies)

The limits of distribution of this group of flies are not known at present.

Control. Under conditions of protected cultivation soil sterilization would clearly be effective, but often this practice is not feasible. Admixture of insecticides with the compost is usually recommended, using pirimiphos-ethyl, or thionazin; also malathion, HCH, DDT, diazinon, malathion and chlorfenvinphos may be used, some as drenches, some as granules, fogs or smokes. Dichlorvos is quite effective for killing the adults in an enclosed space.

In a field situation it is not usual to attempt to control these larvae. Some are preyed upon by predacious nematodes and also some fungi.

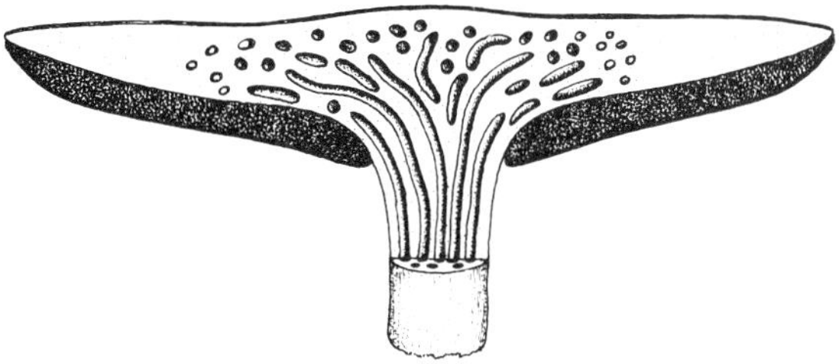

mushroom with maggot tunnels

Ceratitis capitata (Wied.)

Common name. Medfly (Mediterranean Fruit Fly)
Family. Tephritidae
Hosts (main). Fruits of peach, citrus, plum.
(alternative). Many subtropical fruits, including *Ficus*, *Solanum*, cocoa, coffee, mango, guava, etc.
Damage. Eggs are laid inside the fruits and the maggots bore through the fruit while feeding; often associated with fungal and bacterial rots; severely attacked fruits often fall.
Pest status. A very serious pest of many subtropical and deciduous fruits. Many countries have legislation to control accidental introduction of this pest.
Life history. Eggs are laid in groups, under the skin of the fruit, by the female's protrusible ovipositor; each female lays 200–500 eggs; incubation takes 2–3 days.

The maggots bore through the pulp of the fruit as they feed and develop; they are white and typically muscoid in appearance. Typically 10–12 maggots per fruit, but up to 100 have been recorded. The three larval instars develop in only 10–14 days under warm conditions.

Pupation takes place in the soil under the tree, in a longish brown puparium; usually the fruit has fallen by the time the maggots leave to pupate. Pupation takes about 14 days.

The adult fly is brightly decorative, with red/blue iridescent eyes and the body blackish with yellow and white markings; length 5–6 mm. Males have characteristic triangular expansions at the end of the antennal arista. Female flies become sexually mature after 4–5 days, and the first eggs are laid about eight days after emergence. Adults require sugary foods, and with food may live for 5–6 months.

The life-cycle takes 30–40 days under warm conditions, and there may be 8–10 generations per year.
Distribution. Essentially a subtropical species; recorded throughout southern Europe, Near East, Africa, S.W. Australia, Hawaii, C. and S. America (CIE map no. A.1). This pest has several times been accidentally introduced into the USA (Florida, California, Texas) but each time the population has been eradicated.
Control. See the following section on control of fruit flies.

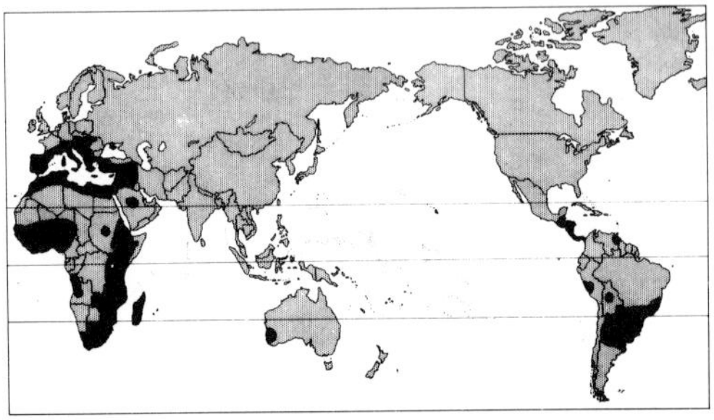

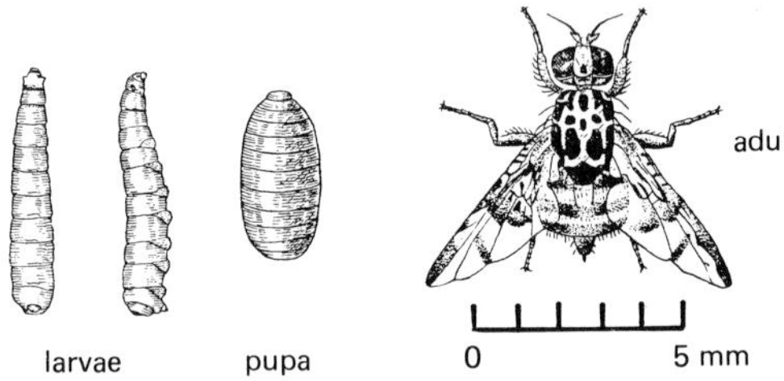

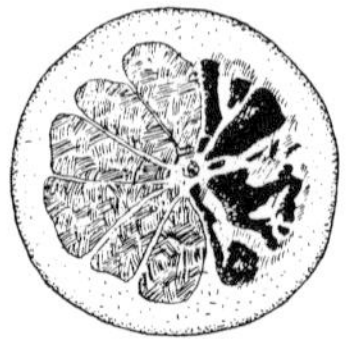

egg-laying female and egg cavity with larval tunnel

Fruit flies (Diptera; Tephritidae)

A large family of more than 1500 species of moderately sized flies. Most are brightly coloured and have mottled wings, and most females have a short horny ovipositor, although a few are quite elongate. The larvae are white, typical maggots in appearance, develop inside a wide range of fruits and are serious pests on a wide range of crops. The group is worldwide in distribution but with most species in the tropics; some species are essentially subtropical and some clearly temperate in distribution. The more serious crop pests are to be found in the warmer countries.

The larvae (maggots) are all phytophagous and may be grouped according to their biology. Most species live inside fleshy fruits, and these include the economic pest species mentioned later. Others live inside the flowerheads of Compositae (*Tephritis*, *Urophora* spp.); some are leaf miners or stem miners (*Philophylla*, *Euriba* spp., etc.); and some species of *Urophora* are gall makers on various parts of the plant body.

In many warmer parts of the world fruits such as peach and *Citrus* (and mango in the tropics) are often found to be infested by a complex of fruit fly larvae, involving several different species, and sometimes different genera. Usually each individual fruit may only be attacked by a single species, but the crop (on the whole tree) may be infested by several different species. Young larvae are difficult to identify, but the mature (third instar) larvae may usually be identified by the shape of the posterior spiracular plate; but for certain identification it may be necessary to rear adults. Similarly the fruits of Cucurbitaceae are attacked by many different Tephritidae. Some of these fruit flies have allopatric distributions and so are somewhat restricted geographically, but others are sympatric and generally infest similar ranges of host plants. With the fruits here mentioned, in most warmer parts of the world it is to be expected that there will be several different species of Tephritidae infesting local crops.

Control of fruit flies

These flies are serious pests because of their abundance and the fact that eggs are laid inside the developing fruits and the larvae (maggots) develop internally, and are thus almost completely safe from insecticides. Thus control measures have to be taken against the adults, which are both highly motile and long-lived. The adults feed on sugary foods and so can be poisoned using baits.

The present range of techniques employed against fruit flies include:

(a) bagging of ripening fruits for protection (sometimes including leaving of some exposed fruits as decoys for ovipositing females).
(b) collection and destruction of all infested and fallen fruits.
(c) sterilization and release (SIRM) of male flies (by exposure either to chemical sterilants or irradiation).
(d) male annihilation using traps baited with sex attractants (pheromones, methyl eugenol, or the commercial preparation 'Cu-lure').
(e) destruction of female flies (and males) using spot-baits of protein hydrolysates with added insecticide (malathion or naled). Further details are available in Drew, Hopper & Bateman (1978).

Until very recently, in the USA, cargoes of fruits both being imported and exported were fumigated with ethylene dibromide, to kill any fruit flies in the fruit, but since it was shown that this chemical can be carcinogenic, its use has been discontinued in the USA and a search for a suitable alternative is under way.

Major fruit fly pests

The more important species of fruit flies regarded as economic pests are listed below:

Anastrepha fraterculus (Wied.) – C. and S. America (CIE map no. A.88).
Anastrepha ludens (Lw.) – (Mexican Fruit Fly) C. America (CIE map no. A.89).
Anastrepha mombinpraeoptans Sein – (West Indian Fruit Fly) West Indies, C. and S. America (CIE map no. A.90).
Ceratitis capitata (Wied.) – (Medfly) (see page 352).
Ceratitis catoirii G.-M – Mauritius and Reunion only (CIE map no. A.226).
Ceratitis coffeae (Bezzi) – (Coffee Fruit Fly) E. Africa.
Ceratitis cosyra (Wlk.) – (Mango Fruit Fly) Africa south of the Sahara.
Ceratitis rosa Karsch – (Natal Fruit Fly) Africa (CIE map no. A.153).
Dacus ciliatus Lw. – (Lesser Pumpkin Fly) Africa and India (CIE map no. A.323).
Dacus cucumis (French) – (Cucumber Fly) Australia.
Dacus cucurbitae Coq. – (Melon Fly) tropical and subtropical Asia, E. Africa, Japan (CIE map no. A.64).
Dacus depressus Shiraki – (Pumpkin Fruit Fly) Japan.
Dacus dorsalis (Hend.) – (Oriental Fruit Fly) tropical and subtropical Asia (CIE map no. A.109).
Dacus musae (Tryon) – (Banana Fruit Fly) Australia (Queensland), Papua New Guinea.
Dacus oleae (Gmel.) – (Olive Fruit Fly) Mediterranean and S. Africa (CIE map no. A.74)
Dacus tryoni (Frogg.) – (Queensland Fruit Fly) E. Australia (CIE map no. A.110).
Dacus tsuneonis Miyake – (Japanese Fruit Fly) China and Japan.
Dacus zonatus (Saund.) – (Peach Fruit Fly) India (CIE map no. A.125).
Euriba zoe (Meig.) – (Chrysanthemum Blotch Miner) Europe.
Paroxyna misella (Lw.) – (Chrysanthemum Stem Fly) Europe.
Pardalaspis cyanescens Bezzi – (Solanum Fruit Fly) Madagascar (CIE map no. A.140).
Pardalaspis quinaria Bezzi – (Rhodesian Fruit Fly) Africa (CIE map no. A.161).
Platyparea poeciloptera (Schrank) – (Asparagus Fly) Continental Europe.
Philophylla heraclei (L.) – (Celery Fly) (see page 355).
Rhacochleana japonica Ito – (Japanese Cherry Fruit Fly) Japan.
Rhagoletis cerasi (L.) – European Cherry Fruit Fly (see page 356).
Rhagoletis cingulata (Lw.) – (Cherry Fruit Fly) N. America (CIE map no. A.159).
Rhagoletis completa Cress. – (Walnut Husk Fly) USA (CIE map no. A.337).
Rhagoletis fausta (O.S.) – (Black Cherry Fruit Fly) USA, Canada (CIE map no. A.160).
Rhagoletis indifferens Curran – (Western Cherry Fruit Fly) USA.
Rhagoletis mendax Curran – (Blueberry Maggot) USA.
Rhagoletis pomonella (Walsh) – (Apple Maggot) N. America (see page 357).
Staurella camelliae Ito – (Camellia Fruit Fly) Japan.
Strauzia longipennis (Wied.) – (Sunflower Maggot) USA.
Toxotrypana curvicauda Gerst. – (Papaya Fruit Fly) USA, India.
Trupanea amoena von Frau. – (Lettuce Fruit Fly) Japan, Europe.
Trypeta trifasciata Shiraki – (Chrysanthemum Fruit Fly) Japan.
Zonosemata electra (Say) – (Pepper (Sweet) Maggot) Canada, USA.

Philophylla heraclei (L.)

(= *Euleia heraclei* (L.))

Common name. Celery Fly
Family. Tephritidae
Hosts (main). Celery, parsnip.
(alternative). Some other Compositae (lettuce, etc.) and Umbelliferae, including parsley.

Damage. The maggots make blotch mines in the leaves, starting as small blisters. Heavily infested leaves shrivel and die; heavily infested plants are small and stunted; celery petioles remain green and taste bitter, parsnip roots are small.

Pest status. A regularly serious pest of these crops, with leaf attacks of 90% recorded; attacks occur in the UK from April to October, or later.

Life history. Eggs are laid by young adults in May, inserted singly under the epidermis on the underside of leaves; each female lays about 100 eggs. Hatching takes 6–14 days.

The maggots burrow within the leaf tissue making a blotch mine; the three larval instars take 14–19 days. Mature larvae are about 8 mm long.

Pupation takes place either within the leaf mine or in soil underneath the plant, and requires 3–4 weeks. The yellowish puparium is 3–5 mm long, and wrinkled.

Second-generation adults emerge over the period July–September and usually the second-generation pupae overwinter, but sometimes there may be a third generation.

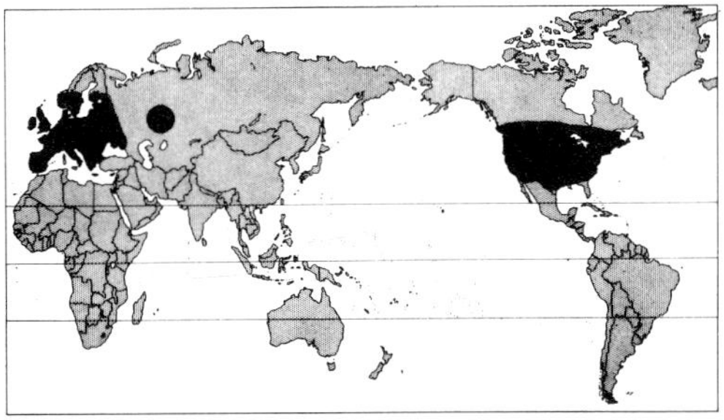

The adult is a small brown fly, some 5 mm long with wingspan 10 mm, with green eyes and mottled iridescent wings; legs are yellow. Adults emerge over the period April–June in the following spring.

There are 2–3 generations per year.

Distribution. Found throughout Europe, Canada and the USA.

Control. For this leaf-mining fruit fly the usual controls are as follows:

(a) avoidance of growing susceptible crops on or near ground infested with fly pupae (i.e. previously infested crops).
(b) destruction of infested foliage and plants: pinching the leaves will kill the maggots in the blisters.
(c) sprays of the following insecticides: DDT, dimethoate, malathion or trichlorphon.

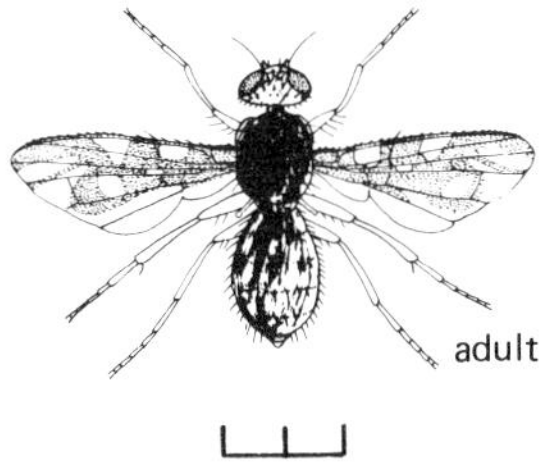

Rhagoletis spp.

Common name. Cherry Fruit Flies
Family. Tephritidae
Hosts (main). Cherries, both cultivated and wild.
(alternative). *Lonicera* (honeysuckle) fruits; other species of *Prunus*.
Damage. The boring maggots feed near the fruit stone; often 5–6 larvae in one fruit: infested fruits often malformed and unsaleable.
Pest status. These are serious pests of cultivated cherries, and in many orchards damage levels can be high.
Life history. Eggs are laid in the fruits, under the epidermis; each female fly lays about 300 eggs. Hatching takes 5–7 days.

The young maggots feed on the pulp of the fruit, usually near the stone; the three larval instars take some 2–3 weeks to complete, when the full-grown larvae measure 5–6 mm in length.

Pupation takes place within a yellow/brown puparium (3–4 mm long) in the soil under the tree, at a depth of 2–8 cm. The pupae overwinter in the soil.

Adults emerge in May to June, according to temperature, and females start egg-laying 7–10 days after emergence. The adults are dark-bodied flies, with yellowish head and legs, green eyes and mottled wings. In some species the black abdomen has lateral pale banding.

There is only one generation per year.

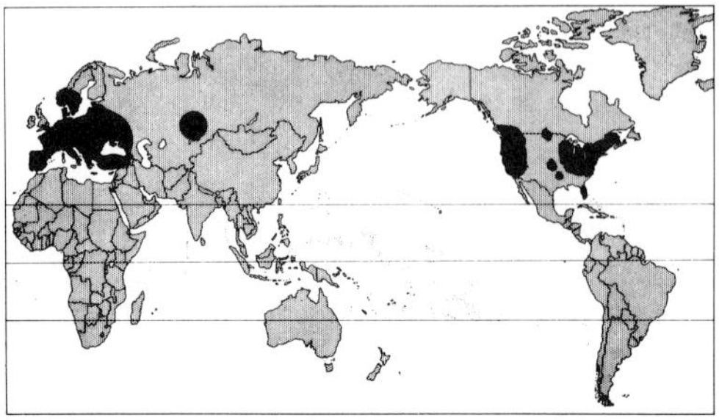

Distribution. The species concerned are:
Rhagoletis cerasi (L.) – (European Cherry Fruit Fly) Europe and W. Asia (not UK) (CIE map no. A.65).
Rhagoletis cingulata (Lw.) – (Cherry Fruit Fly) Canada, USA (CIE map no. A.159).
Rhagoletis fausta (O.S.) – (Black Cherry Fruit Fly) Canada, USA (CIE map no. A.160).
Rhagoletis indifferens Curran – (Western Cherry Fruit Fly) western USA.
Control. *Cultural control*: Picking and destruction of infested fruit is always recommended as a means of reducing the pest population. Cultivation of the ground under the trees will cause the death of many overwintering pupae.
Chemical control: Various insecticides have been used as sprays in cherry orchards in N. America, in order to kill the young adult flies before the females oviposit, including parathion, malathion, diazinon; 2–4 applications at 1–2-week intervals are usually made, starting when the first flies are trapped.

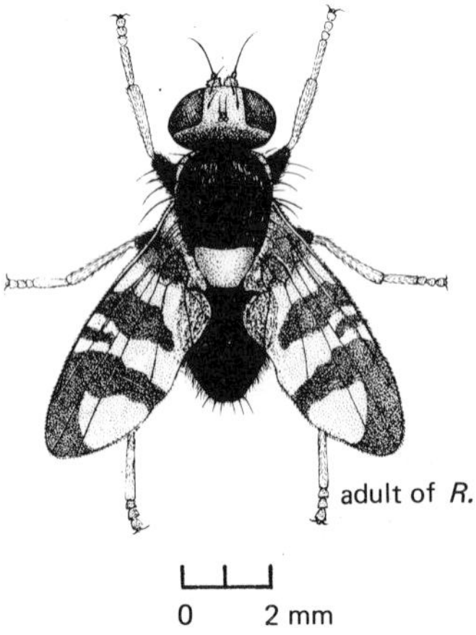

Rhagoletis pomonella (Walsh)

Common name. Apple Fruit Fly; Apple Maggot
Family. Tephritidae
Hosts (main). Apple
(alternative). Plum, pear, cherry, blueberry, huckleberry, hawthorn.
Damage. Maggots feed in the flesh of the fruit leaving irregular winding brown tunnels; heavily infested fruits often fall prematurely.
Pest status. A major pest of apple in N. America; in many regions regular pest-population monitoring is required, followed by routine spraying.
Life history. As with most other fruit flies the eggs are laid in the young fruits under the epidermis, each female fly laying several hundred eggs; development takes only a few days.

Larval development takes from 2–3 weeks in certain early fruit varieties, up to 2–3 months in some late-cropping varieties. Fully developed larvae leave the fruit to pupate in the soil where they overwinter in oval brown puparia.

Adults emerge in June and July, but a few do not emerge until the second summer. The small dark flies have a pale scutellum and pale bands across the abdomen, and conspicuously mottled wings. Females do not lay eggs for 7–10 days after emergence.

There is only one generation per year.

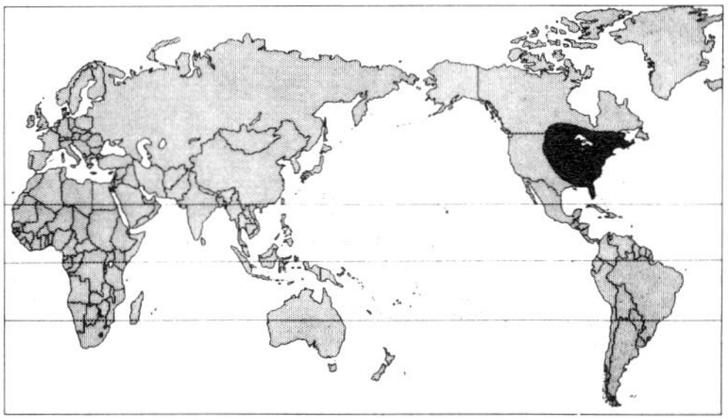

Distribution. This species is native to N. America and occurs throughout the eastern part of Canada and the USA (CIE map no. A.48).
Control.
Cultural control: This includes the collection and destruction of all infested fruits. Maggots in the fruits may be killed by exposure to a temperature of 0 °C for a period of 40 days. Removal of hawthorn bushes near the apple orchards may not be feasible, but would remove an important wild host.
Chemical control: The usual insecticide sprays employed against this pest have in the past included DDT, carbaryl, diazinon, and others, aimed at killing the young adults prior to oviposition.

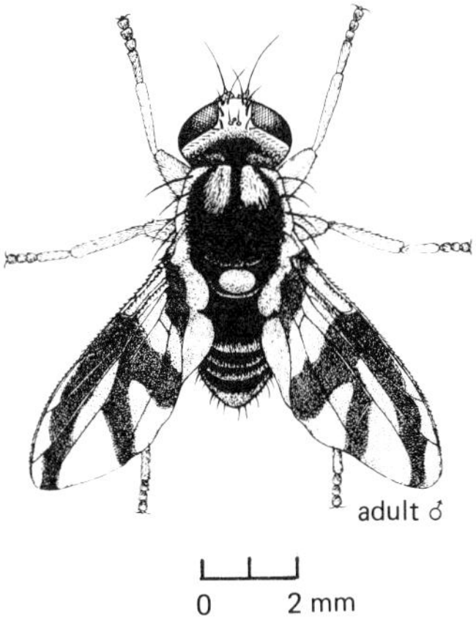

Phytomyza horticola Goureau
(= *P. atricornis* Mg.)

Common name. Pea Leaf Miner
Family. Agromyzidae
Hosts (main). Pea, brassicas, lettuce, onion.
(alternative). Flax, tomato, cucurbits; many other crops and wild plants; polyphagous.
Damage. The larvae make irregular linear mines in the leaves, in the mesophyll between the upper and lower epidermis. If many leaves are mined the yield (of peas) may be reduced; in *Brassica* the damage affects saleability: heavily infested leaves shrivel and wither.
Pest status. Widespread and abundant, recorded from many different crops; a pest of importance, but only needs controlling at very high infestation levels.
Life history. The adult flies feed by making small epidermal punctures in the leaves, and some of the feeding sites are used for oviposition, so the eggs are deposited into the leaf tissues. The number of eggs laid per female is recorded as 300–350, some 50 per day. As many as 150 feeding punctures have been counted in one leaf of *Pisum*, but eggs were only found in a few. Egg development takes 2–6 days.

The larva feeds within the leaf, making a long serpentine tunnel that may cross over itself in places; the feeding mine widens as the larva develops. Larval development takes 5–10 days, by which time the larva is 3–4 mm long, and greenish-white in colour. Pupation occurs at the end of the mine, and the brown puparium is clearly visible. The pupal period is from 7–15 days, but may be much longer at low temperatures.

The adult is a small blackish fly, 2–3 mm long, with a pale face, some yellow lateral markings and yellow 'knees', and is not a strong flier: first-generation adults are found in May in the UK.

There are 2–4 generations per year in Europe. In the Mediterranean and N. India the period of fly activity is during the winter months.
Distribution. Recorded throughout Europe, Asia, and parts of Africa (CIE map no. A.374; also A.205 as *P. atricornis*), as far north as Iceland.
Control. Many chalcids and ichneumonids are recorded parasitizing this species, and under normal conditions population control is effected by these natural enemies.

For further information on control see page 360.

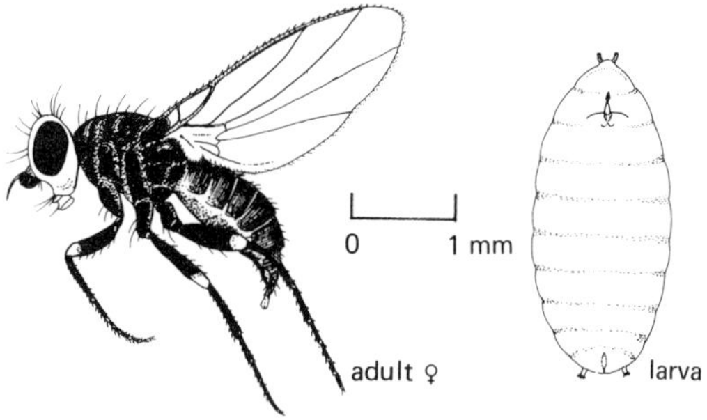

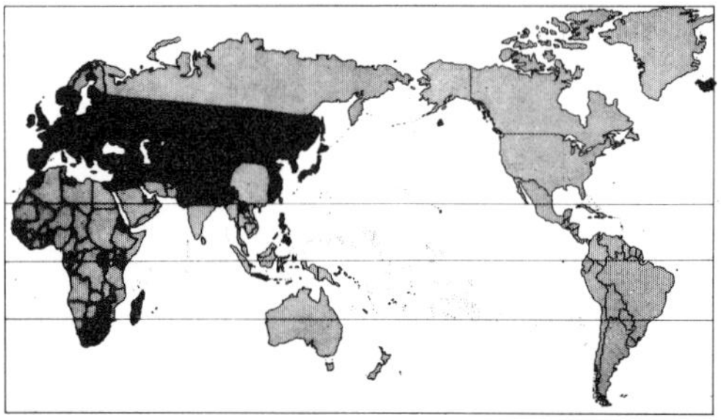

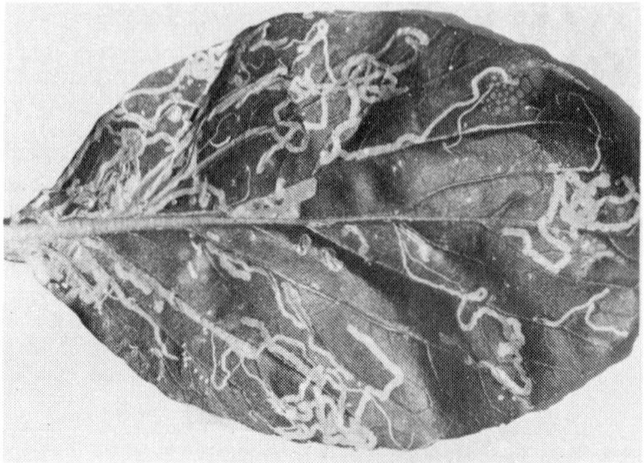

Infested leaf of *Brassica*

Leaf miners (Diptera; Agromyzidae)

A large and widespread group of small flies, most with phytophagous larvae attacking a wide range of plants, most as leaf miners, but some as stem borers, some as borers of leguminous pods, and a few as gall makers. Some species are cosmopolitan, others solely temperate and some restricted to the tropics. The range of host specificity is great, from complete polyphagy to restricted monophagy on a single genus of host plant (such as *Camellia*). About 150 species are regularly associated with cultivated plants, and these were the subject of a monograph by Spencer (1973); the total number of species recorded is about 1800.

The leaf-mining species are characterized by making long winding tunnels (mines) in the leaf lamina; the tunnel appears whitish because of light reflection from the air trapped in the mine; the larval faecal pellets are not very conspicuous as they are deposited at the side of the mine (most leaf-mining caterpillars leave a central black line of pellets). Some species make blotch mines, but this is generally more characteristic of other groups of leaf-mining flies. Pupation takes place usually in the mine at the end of the tunnel, with the two spiracular 'horns' projecting through the leaf epidermis, usually on the underside of the leaf. Some species, however, pupate in the soil, but this is not too common a practice.

Some crops are mined by different species of Agromyzidae in different parts of the world where they are allopatric in distribution, but some flies have overlapping distributions (i.e. sympatric), and some are cosmopolitan. The end-result is that in any one locality some crops are attacked by several very similar leaf miners simultaneously. The identification of Agromyzidae is extremely difficult and many species are really only distinguishable using the male genitalia, but at generic level there are some differences in wing venation and body coloration. The crops most likely to suffer multiple infestation are those belonging to the families Leguminosae, Gramineae, Solanaceae, and Compositae, and also the Cruciferae, Chenopodiaceae and Cucurbitaceae. On a worldwide basis *Beta vulgaris* is attacked by six species, five of which are *Liriomyza*; *Lactuca sativa* has seven species, *Pisum sativum* has 13 species in six different genera, *Hordeum vulgare* has 17 species, and *Triticum aestivum* is attacked by 18 species of Agromyzidae (Spencer, 1973).

The two species dealt with here in detail may be taken as typical of the leaf-mining group of Agromyzidae.

Some important leaf miner (Agromyzidae) pests

Agromyza ambigua Fall. – (Cereal Leaf Miner) Europe, N. America.

Agromyza oryzae (Mun.) – (Rice Leaf Miner) Japan, Java, E. Siberia.

Amauromyza maculosa (Mall.) – (Lettuce Leaf Miner) USA, S. America, Hawaii.

Cerodontha spp. (9) – (Cereal Leaf Miners) only Gramineae; worldwide.

Liriomyza brassicae (Riley) – (Cabbage Leaf Miner) Cruciferae mostly; cosmopolitan.

Liriomyza bryoniae (Kalt.) – (Tomato Leaf Miner) polyphagous; Europe, W. Asia.

Liriomyza cepae (Hering) – (Onion Leaf Miner) Europe (not UK).

Liriomyza chinensis (Kato) – (Onion Leaf Miner) Japan, China, Malaysia.

Liriomyza sativae Blanch. – polyphagous: Cucurbitaceae, Solanaceae, Leguminosae; USA, C. and S. America.

Liriomyza trifolii (Burgess) – ((American) Serpentine Leaf Miner) polyphagous; N. and S. America, introduced to UK.

Melanagromyza obtusa Mall. – (Bean Pod Fly) India, S.E. Asia.

Melanagromyza sojae (Zehn.) – (Bean Fly) Africa, S.E. Asia.

Ophiomyia phaseoli (Tryon) – (Bean Fly) Africa, Asia, Australasia (CIE map no. A.130).

Napomyza carotae Sp. – (Carrot Root Miner) Europe (not UK)

Phytobia spp. – (Cambium Borers) in twigs and trunks of many trees (apple, birch, willows, etc.); Holarctic.
Phytomyza horticola Goureau – (Pea Leaf Miner) polyphagous (see page 358).
Phytomyza rufipes Meig. – (Cabbage Leaf Miner) Europe, Canada, USA.
Phytomyza syngensiae (Hardy) – (Chrysanthemum Leaf Miner) polyphagous (see page 361).
Ptochomyza asparagi Hering – (Asparagus Leaf Miner) Europe, China.
Tropicomyia spp. (5) – (Tea Leaf Miners) some polyphagous; Africa, Asia, Japan, Australasia.

Control of leaf miners (Agromyzidae)

With many species the larval infestation is conspicuous, especially the leaf mines, but actual damage is often slight. The level of natural parasitism is often high and serves to keep the pest population in check; field observations of *Tropicomyia theae* on tea have shown that the level of parasitism is often as high as 70%. For many field crops where the damage consists of leaf mining, it is neither necessary nor really feasible to consider applying control measures.

Bean Fly (*Ophiomyia phaseoli*) is different from most in that it bores in the stems of bean seedlings, and here the usual method of control is to use seed dressings (formerly dieldrin, now phorate and disulfoton) or foliar sprays of dimethoate, permethrin, monocrotophos, omethoate, oxamyl and triazophos, applied twice, two and 12 days after crop emergence.

With glasshouse crops of high value, the most effective control measure is to apply aldicarb granules to the soil 2–4 weeks after planting. The systemic nature of this insecticide both deters adult flies from feeding on the leaves, and causes a reduction in the number of eggs laid, as well as killing the young larvae soon after hatching. An alternative method is to use foliar sprays of γHCH, diazinon, or nicotine, but these will not kill the pupae. Pirimicarb (used for aphid control) is also effective. It should be noted that chrysanthemums are sensitive to some insecticides so care should be taken when treating this crop in glasshouses.

Phytomyza syngenesiae (Hardy)

(= *P. atricornis* auctt.)

Common name. Chrysanthemum Leaf Miner
Family. Agromyzidae
Hosts (main). Chrysanthemum, lettuce, mint; various ornamentals (Compositae).

(alternative). Pea, carrot; and some other plants in the USA.

Damage. The feeding larvae make long winding tunnels in the leaf lamina; in heavy infestations the leaves wither and die; the female feeding scars may also be unsightly on the leaves and reduce market value of the crop.

Pest status. Primarily a pest on Compositae, but does feed on some other plants; probably most serious on glasshouse chrysanthemums in Europe, USA and the UK.

Life history. Eggs are laid on either surface of the leaf, but the upper is preferred; most are laid on shaded foliage. The holes in the leaves are made by the female ovipositor; she then sucks up sap exuding from the punctures, and in some of the sites then lays an egg. About 75 eggs are laid per female. Hatching takes 4–6 days.

The larvae feed and tunnel within the leaf, and along the mine the faecal frass is usually clearly evident as dark granules. Fully grown larvae are about 3 mm long, and take 7–10 days to develop.

Pupation takes place within a small blister; the puparium is pale brown, about 2.5 mm long; development takes a minimum of nine days but may be several months at low temperatures.

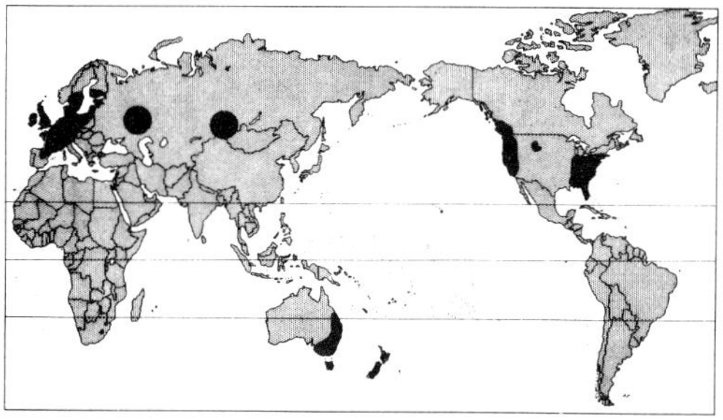

The adult is a small dark fly, about 2 mm long; identical in general appearance to *P. horticola*.

Under glass in Europe there is continuous breeding, but outdoor crops tend to have only 2–3 generations during the warmer part of the year.

Distribution. It is now thought that this is a European species (western Europe), recently introduced to Canada and the USA, and taken to New Zealand and eastern Australia by the early settlers (CIE map no. A.375).

Control. In Europe it is possible to buy commercially *Opius* wasps, which parasitize mature larvae, for release into glasshouses to exert pest population control.

For other control measures see page 360. Sprays of insecticides are recommended to be applied as soon as the first feeding/oviposition punctures are seen.

Infested chrysanthemum leaf

Psila rosae (F.)

Common name. Carrot Fly; Carrot Rust Fly (USA)
Family. Psilidae
Hosts (main). Carrot, parsnip, celery.
(alternative). Parsley, hemlock and various other wild Umbelliferae.

Damage. The larvae make surface tunnels in the root, and occasionally bore into the taproot cortex; the tunnels are usually infected with fungal rots. Young plants are usually killed after initial wilting (see figure).

Pest status. A serious pest of these crops wherever they are grown in Europe and N. America. Recently well controlled using dieldrin but, of late, resistance has become completely established and damage levels are high again.

Life history. Eggs are laid in small groups (3–4) in the soil near the young plants; they are white, elongate, with pronounced longitudinal ribbing (1 × 0.4 mm). Hatching takes 7–10 days. Each female lays about 30–90 eggs.

The young larvae burrow into the soil and feed on the surface of the root externally. The second- and third-instar larvae burrow into the root and make the characteristic brown tunnels. The fully grown maggots are elongate (8–10 mm) and creamy-white; larval development takes 4–6 weeks.

Pupation takes place in the soil, 2–5 cm from the damaged root, in an elongate brown puparium (5–6 × 1.2 mm); the pupal period is very variable.

The adult is a small fly, shiny black, with a reddish head capsule and yellow legs, about 4–6 mm long. The sexes are difficult to distinguish.

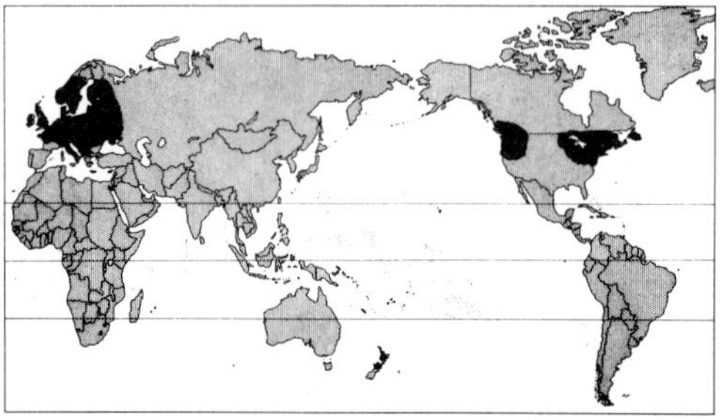

In the UK there are two distinct generations per year, and probably a partial third generation in the fens. Adults are found from May through to about November. Some larvae continue to feed in overwintering crops and finally pupate in the early spring.

Distribution. Widespread throughout Europe, including Fennoscandia, western USSR, New Zealand, Canada and the USA (CIE map no. A.84).

The closely related *Psila nigricornis* Meig. is the Chrysanthemum Stool Miner, which does also sometimes attack carrot and lettuce in Europe and in Canada.

Control. Control measures differ somewhat according to the crop concerned.

Cultural methods: These will not give control but will reduce the insect population somewhat.

(a) Crop rotation and spacing: Sometimes this is not successful as these crops require particular soil types for best growth. At high crop densities the insect population becomes very large, although individual root damage is less.
(b) Time of sowing: Late sowing may avoid most of the first-generation larvae.
(c) Reduction of shelter: The flies spend much time in the shelter around the field, so cutting the rank vegetation on the headlands disperses the flies.
(d) Varietal susceptibility: Some carrot varieties are considerably more severely damaged than others.
(e) Destruction of infested crops: Severely infested crops should be lifted and destroyed (i.e. fed to livestock); they should never be ploughed in!

Chemical control: The use of dieldrin seed dressings is generally no longer recommended owing to established resistance to this chemical.

Treatments at sowing time include the placement of granules 5–10 cm beneath the soil surface; band spraying may also be used so long as the soil is quickly turned to incorporate the insecticide. Chemicals left on the soil surface generally have little effect and rapidly degrade.

Present recommendations include granules of carbofuran, disulfoton, phorate, and chlorfenvinphos (and the latter also as a spray) applied preferably by the bow-wave method.

Treatments after sowing include mainly high-volume sprays of chlorfenvinphos, diazinon, pirimiphos-methyl or etrimfos, preferably applied to the crowns of the plants rather than just over the foliage. Generally this method gives less control than does the application of granules at sowing. With peat (muck) soils the dosage needs to be considerably increased (for chlorvenvinphos it needs to be doubled) because of chemical adsorption by the organic matter in the soil. With celery it may be feasible to use a root dip at the time of planting out, or a drench at a later date.

Generally the methods of growing carrots vary to such an extent that it is not possible to generalize about insecticide rates, but full details of recommendations are given in the MAFF Advisory Leaflet 68 (revised 1979).

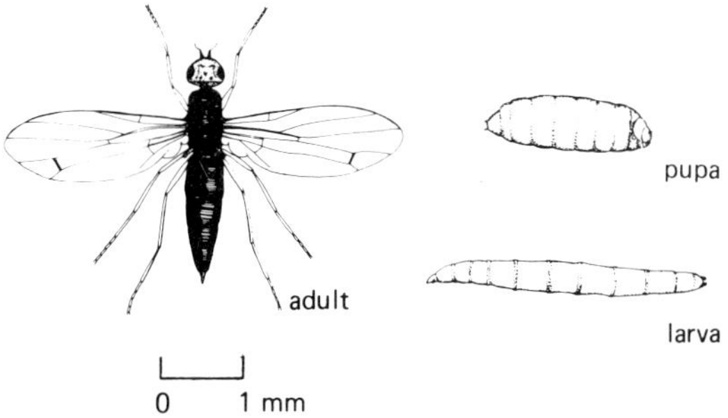

Infested carrots

Parsnip seedlings

Hydrellia griseola Fall.

Common name. Cereal Leaf Miner (Rice Whorl Maggot)
Family. Ephydridae
Hosts (main). Wheat, barley, oats.
(alternative). Rice, and many species of grasses and sedges, and some aquatic plants.
Damage. The maggots bore in the leaves, feeding on the mesophyll tissue; the mines are initially linear, later coalescing into a blotch. Damaged leaves shrivel and may die. Most damage is done under moist conditions.
Pest status. A widespread pest of cereals; most damaging on rice in California; occasionally serious on other cereals grown under damp conditions, i.e. where there are summer rains.
Life history. Eggs are laid singly on the leaves; each female laying 50–100 eggs. Hatching takes 3–5 days.

On hatching the maggots immedately bore into the leaf tissues; the feeding tunnel is initially a linear mine but soon coalesces into a blotch (whorl). Larval development under warm conditions takes 7–10 days, but is recorded as being as long as 40 days in northern regions.

Pupation takes place within the mine and the brown puparium is clearly visible; the pupal period is 5–40 days according to temperature.

The adult is a small grey fly with long legs (like a small house fly), with a shining grey frontal lunule; wingspan 2.5–3.2 mm; males are the smaller. Females start egg-laying three days after emergence, and can live for 3–4 months.

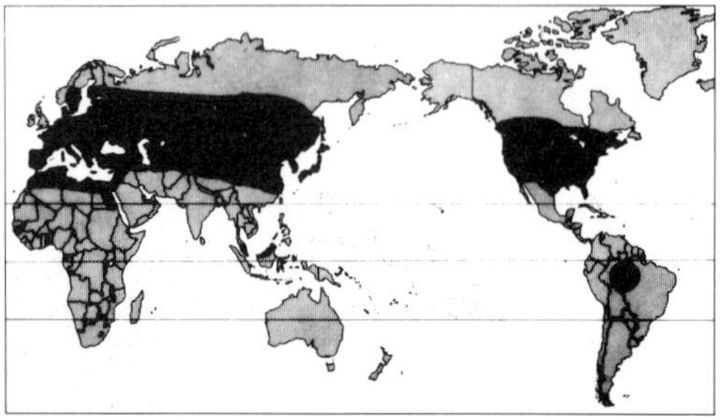

In the warmth of California there are 11 generations per year, but in northern Japan there are usually eight generations.
Distribution. Found throughout Europe (not UK) and temperate Asia, N. Africa, Egypt, Near East, Malaysia, China, Korea, Japan, USA and S. America.

Closely related species include:
Hydrellia philippina Ferino – (Rice Whorl Maggot) Philippines, Japan.
H. sasakii Y.et S. – (Paddy Stem Maggot) Japan.
H. tritici Coq. on wheat in temperate Australia.
Various species of *Notiphila* in China and Japan bore in the roots of rice plants. Species of *Ephydra* are rice pests in S. France, Spain, Egypt, Hungary, and C. Asia.

Other species of *Hydrellia* have larvae that mine the leaves of *Potamogeton*, watercress and other aquatic plants in freshwater habitats.
Control. The usual treatment when control is really necessary has been foliar sprays of dieldrin or heptachlor, which kills both adult flies and the mining maggots.

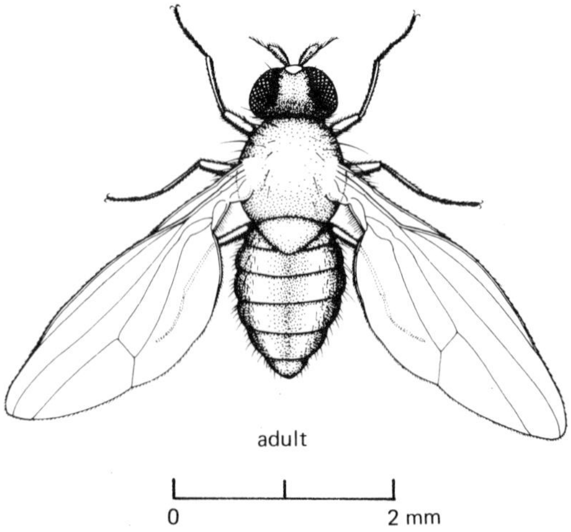

Chlorops pumilionis (Bjerk)

Common name. Gout Fly
Family. Chloropidae
Hosts (main). Barley
(alternative). Wheat, rye, and various grasses.

Damage. In autumn-sown cereals (especially barley) the feeding maggots in the tiller shoots cause them to remain short and swollen (gouty). Spring-sown barley, if attacked later, produces ears but they are deformed and have the grain on one side destroyed.

Pest status. A regular minor pest of some cereals and some grasses; sporadically serious.

Life history. Eggs are laid on the young shoots, generally one egg per shoot; and hatching takes 8–10 days. First-generation egg-laying occurs in May/June.

The emerging maggot burrows in to the shoot and then tunnels down one side of the stem, making a distinct feeding groove, down to the first node. After feeding for about one month, the creamy-white maggot (now 5–8 mm) moves up the groove and pupates.

The elongate, flattened brown puparium is about 5 mm long; pupation takes 4–5 weeks.

Adults emerge during August/September (second generation) and eggs are laid on young winter barley (or wheat) that was sown early, and on the wild grass hosts, usually in September and October. The second-generation larvae feed on the shoot over the winter period and usually pupate in March, giving rise to the first-generation adults in May/June. The adult is a small stout fly, distinctively patterned yellow and black, 4–5 mm long.

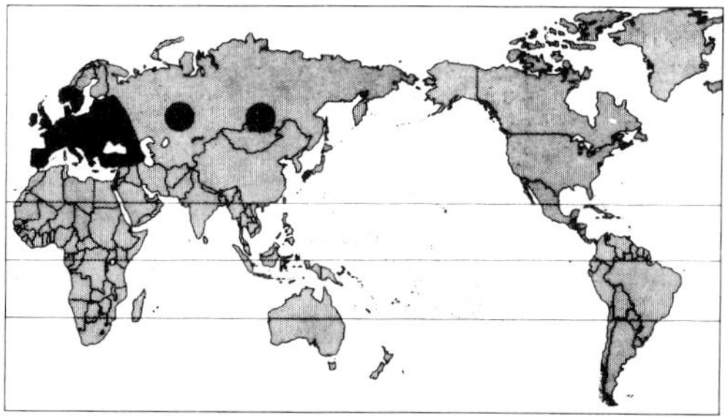

Distribution. This is a European species, found throughout the UK and Europe; in the UK there is a total of 18 species of *Chlorops* recorded. Important species in East Asia include: *Chlorops mugivorus* N.et K. – (Wheat Stem Maggot) Japan; *C. oryzae* Mats. – (Rice Stem Maggot) Japan; other species occur in China.

The closely related *Meromyza*, collectively known as grass flies, occurs as several different pest species.

Control. *Cultural control*: These methods include:

(a) early sowing of spring barley.
(b) delayed sowing of winter cereals.
(c) manuring of cereal crops to maximize growth rates.
(d) clean cultivation to remove alternative grass host plants.

Chemical control is not usually necessary.

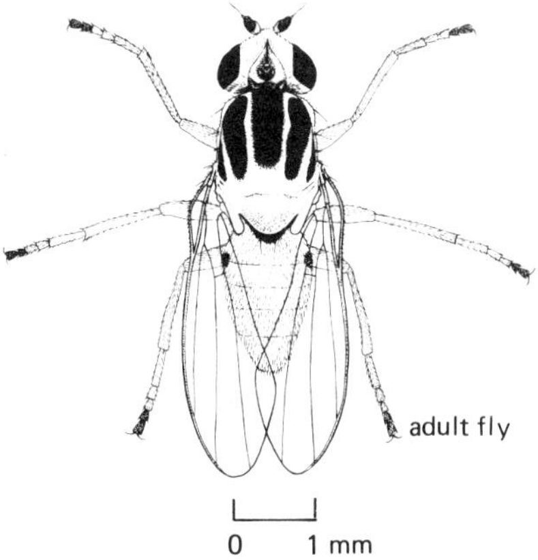

Infested couch grass shoots

Oscinella frit (L.)

Common name. Frit Fly
Family. Chloropidae
Hosts (main). Oats, wheat, barley.
(alternative). Rye, maize; various grasses, especially *Lolium*.

Damage. First-generation larvae bore the young shoot and cause a 'dead-heart': the shoot dies, turns brown and withers; profuse tillering results. Second-generation larvae feed in the ears and destroy some grains. In severe attacks 90% of the grains may be withered.

Pest status. In the UK a serious pest of oats; in Europe most serious on barley; and in N. America most serious on wheat.

Life history. In the spring eggs are laid on and at the base of the young shoots; most being laid in May and early June; each female fly lays about 100 eggs. Hatching requires 3–4 days, and the tiny maggots bore into the shoot and their feeding destroys the growing point. Very young plants are killed; older plants produce tillers, which may in turn become infested. Larval development takes about 14 days, after which a size of 3 mm is attained. Pupation takes place within the shoot, and requires about a further 14 days.

Second-generation flies emerge in June/July, at about the time oats are flowering, and eggs are laid on the oat husks (UK) and the larvae feed on the developing kernels, which shrivel. Wheat ears are also attacked, but apparently the panicles of grasses are not infested.

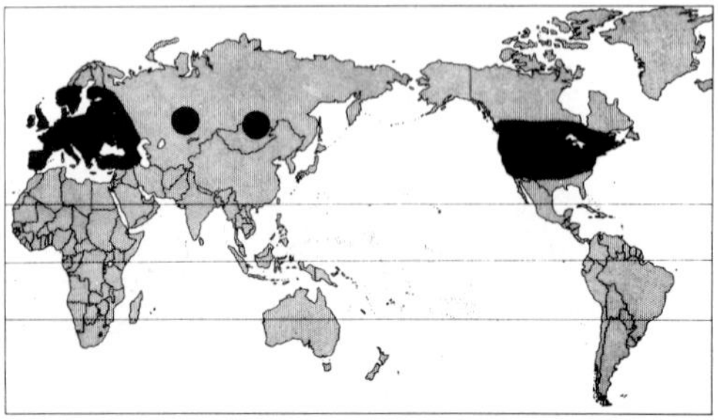

Second-generation adults emerge in the autumn and lay eggs of the autumn generation on grasses, especially perennial ryegrass and volunteer cereals. The larvae feed over winter, develop slowly, and pupate in the spring, giving rise to the adults that emerge in May.

The adult is a small black fly about 1.5 mm long, but the tarsi are yellow.

Distribution. It appears that what is called *Oscinella frit* may in fact be a group of closely related species, distributed throughout Europe, Canada and USA. In the UK 11 other species of *Oscinella* are recorded, one of which is also recorded from China (*O. pusilla* (Meig.)); in the USA the American Frit Fly (*O. soror* (Macq.)) is a cereal pest of minor importance, and several other species are of importance on cultivated grasses; in Canada there are at least five other species of *Oscinella* regularly recorded from cereals and cultivated grasses.

Control. *Cultural control*:

(a) early sowing of spring oats (UK).
(b) early and shallow sowing of maize.
(c) for winter corn after a ryegrass ley the land should be ploughed early: at least four weeks before sowing to allow time for larvae to die.
(d) for ryegrass leys the same treatment.
(e) use of resistant varieties wherever possible; some American varieties of wheat and some European varieties of ryegrass show resistance.

Chemical control: The most successful chemical previously used was DDT, but present recommendations include spraying of chlorpyrifos or triazophos, or else the systemic omethoate. Fenitrothion and pirimiphos-methyl are also used as sprays, and for forage maize in the UK granules of phorate, carbofuran or chlorfenvinphos may be applied at sowing. Control recommendations vary somewhat according to the crop concerned and local Ministry of Agriculture advice should be sought for details.

Delia antiqua (Meigen)

(= *Hylemya antiqua* (Mg.))

Common name. Onion Fly (Maggot)
Family. Anthomyiidae
Hosts (main). Onions (both ware and salad).
(alternative). Leek, shallot, garlic, chives (Amaryllidaceae); occasionally tulip bulbs.
Damage. The feeding larvae bore inside the bulb; seedlings and salad onions are destroyed, first wilting then dying. With bulbed onions there is leaf wilting, and the bulb usually rots; there may be up to 30 larvae per bulb.
Pest status. A serious pest of *Allium* species in many parts of the Holarctic region; infestations and crop losses are often very high, especially since resistance to dieldrin became widespread.
Life history. Eggs are laid in small groups (6–30) in the soil close to the young plants, or on the leaf sheath; they are white, about 1.2 mm long and longitudinally ridged. Hatching takes some three days. Egg-laying occurs in the latter half of May, and then later in July.

The young larvae burrow down and enter the plant from the underneath. They feed and grow rapidly and reach full size in about three weeks, when they measure 8–9 mm. With salad onions the large larvae move from plant to plant down the row, there usually being several maggots inside each stem; with large bulbs there may be up to 30 maggots per bulb.

Pupation takes place within an oval brown puparium a few centimetres away in the soil, and takes about 17 days. However, some pupae overwinter.

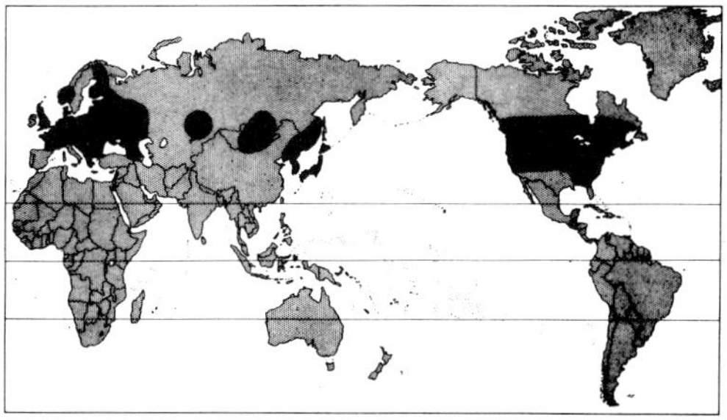

The adults are small grey flies, resembling the Common House Fly, 5–7 mm long. Second-generation adults emerge in early July; eggs are laid and the second-generation larvae feed until August when they pupate. In some locations there may be a partial third generation sometimes. Overwintering takes place inside the puparium. In the coldest regions there are usually only 1–2 generations annually.
Distribution. Onion Fly is a northern temperate species, found throughout Europe (as far north as Lapland), Near East, parts of Asia (Korea, Siberia), Canada and the USA (CIE map no. A.75).
Control. Destruction of infested crop residues is recommended, as is a good crop rotation, if possible.
(For further details, see page 369.)

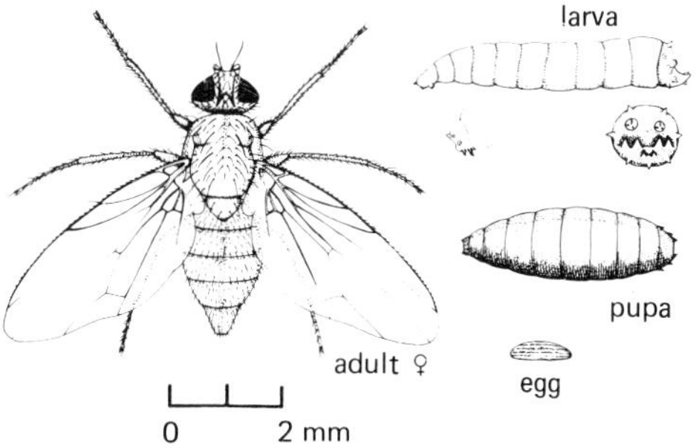

Infested salad onions

Root maggots etc. (Diptera; Anthomyiidae)

This is quite a large family of flies, closely related to the Muscidae (some species have been taxonomically shunted back and forth); the main taxonomic character seems to be that the anal vein extends right to the wing margin. Biologically and ecologically the two families can scarcely be separated, if at all; many larvae are saprophagous, feeding on dead organic matter both plant and animal, thus it is not surprising to find many synanthropic species. But some species are truly phytophagous and feed on living plant tissues.

The Anthomyiidae associated with cultivated plants can be grouped into four main assemblages, according to their larval habits, as follows.

(1) Root maggots:
   (a) phytophagous species: larvae eat and tunnel intact plant roots, and also intact sown seeds.
   (b) saprophagous species: larvae usually associated with the former species and feeding on damaged roots or on organic matter in the soil, and also damaging roots and eating sown seeds.

(2) Cereal shoot flies; larvae bore into the shoots of young cereals and grasses and destroy the growing point, causing a 'dead-heart'.

(3) Leaf miners: larvae make large blotch mines in the leaves of various plants.

A few species bore in plant shoots (shrubs), or stems, or leaf petioles, and their placement above is rather difficult. Some species, such as the 'Bean Seed Fly' complex show a mixture of larval feeding habits in being partly saprophagous and partly phytophagous.

As typical Cyclorrhapha the larvae (maggots) have only mouth-hooks with which to feed, but they clearly cope and are quite successful plant feeders.

Identification of larvae is not easy, but the shape of the mouth-hooks, and both anterior spiracle and posterior spiracular plate, is usually distinctive, and the pattern of sculpturing on the egg chorion is often recognizable. Field identification of crop infestations may be difficult because of the large number of saprophagous species of both Anthomyiidae and Muscidae to be found in soil rich in humus or organic debris (i.e. rotting crop residues), and these are only secondary pests (if that!) in that they are sometimes feeding on roots already damaged by the primary pests, and sometimes already infected by fungal rots. For example several species of *Muscina* (Muscidae) and *Pegohylemyia fugax* are regularly associated with root maggot infestations.

The genera *Delia* and *Hylemya* have been subject to recent taxonomic reappraisal and many of the crop pest species formerly regarded as being in *Hylemya*, or other genera, are now placed in *Delia*. The literature is rather confusing both in regard to generic placement and also species synonymy.

Important pest species of Anthomyiidae

Delia antiqua (Meig.) – (Onion Fly) Holarctic (see page 367).

Delia arambourgi (Seguy) – (Barley Fly) Africa.

Delia brunnescens (Zett.) – (Carnation Maggot) USA.

Delia coarctata (Fallen) – (Wheat Bulb Fly) Europe, W. Asia (see page 370).

Delia echinata (Seguy) – (Spinach Stem Fly (Carnation Tip Maggot)) Europe, Japan, USA.

Delia floralis (Fallen) – (Turnip Maggot) Europe, Canada, USA, China.

Delia florilega (Zett.) – ('Bean Seed Fly') Europe.

Delia pilipyga (Vill.) – (Turnip Maggot) China, Japan.

Delia planipalpis (Stein) – (Cruciferous Root Maggot) Canada.

Delia platura (Meig.) – (Bean Seed Fly) worldwide (see page 372).

Delia radicum (L.) – (Cabbage Root Fly) Holarctic (see page 373).

Hylemya cerealis (Gill.) – (Wheat Stem Maggot) Canada.

Hylemya depressa Stein } Found on roots of cruciferous
Hylemya nidicola (Ald.) } crops in Canada.

Hylemya flavibasis Stein – (Cereal Root Maggot) S. Europe, Asia Minor, N. Africa.

Hylemya planipalpis (Stein) – (Radish Root Maggot) Canada.
(Some of these species might now be placed in *Delia*, but this point is uncertain.)
Pegohylemyia fugax (Meig.) – saprophagous on cruciferous roots; Europe, Canada.
Pegohylemyia gnava (Meig.) – (Lettuce Seed Fly) Europe.
Pegomya dulcamarae Wood – (Potato Leaf Miner) Japan.
Pegomya hyoscyami (Panzer) – (Beet (Spinach) (Leaf Miner) Fly) Europe, China, Japan, USA, Canada (see page 375).
Pegomya mixta Vill. – (Beet Leaf Miner) Japan.
Pegomya rubivora (Coq.) – (Loganberry Cane Maggot) Europe, USA.
Pegomya ruficeps Stein – on cabbage; Canada.
Phorbia securis Tien. – (Late Wheat Shoot Fly) Europe.

To give an idea of the extent of the different species that may be encountered in some pest infestation situations the number of species in each genus now recorded from the UK are as follows: *Delia*, 34; *Hylemya*, 5; *Pegohylemyia*, 25; *Pegomya*, 41; *Phorbia*, 9. It should be remembered that most species within a genus have a similar biology, thus most *Pegomya* species are leaf miners, but in different groups of plants.

Control of Anthomyiidae

The species that are partially saprophagous (e.g. bean seed flies) are attracted by manure and rotting vegetable matter in the soil, or even just by recently turned soil; avoidance of such situations and the use of a 'stale' seedbed will tend to discourage infestation.

Many of the truly phytophagous species show striking population fluctuations, and even in areas 'at risk' in some years (often alternate years) populations are very low and damage negligible. Generally for these species it is preferable to 'scout and predict' by taking soil samples for eggs to estimate eventual local populations before deciding on control strategy.

It has been shown for several species of Anthomyiidae that natural levels of predation and parasitism are very high, and in many situations this serves to keep in check the pest populations; this is just as well for some pest species show tremendous powers of population increase. Predation of eggs and young larvae is mostly by Carabidae (Coleoptera) in the soil, and parasitism of the pupae by both Hymenoptera Parasitica and larvae of some Staphylinidae (Coleoptera).

The diversity of crops attacked makes generalization difficult and really means that each crop/pest situation needs to be evaluated separately, especially when considering *Brassica* crops, for sometimes the root is the crop (turnip, radish), other times the central shoot (cabbage), flowerhead (cauliflower) or lateral buds (Brussels sprouts) (also see page 74).

General control advice is given below, but it is recommended that local Ministry of Agriculture advice be sought for specific crop/pest situations.

(1) Root maggots
 (a) Seed dressings: formerly dieldrin gave excellent control, but resistance is now widespread so bromophos or pirimiphos-methyl are used.
 (b) Granules } (c) Sprays } carbofuran, chlorfenvinphos, chlorpyrifos or iodofenphos at sowing, transplanting or later, as dips or drenches.
(2) Shoot flies
 (a) Seed treatment: HCH, carbophenothion, or chlorfenvinphos.
 (b) Granules at sowing: chlorpyrifos or fonofos.
 (c) Sprays at sowing: chlorpyrifos.
 (d) At egg-hatch: sprays of chlorfenvinphos, chlorpyrifos or pirimiphos-methyl.
 (e) At first damage signs: sprays of systemic insecticides; dimethoate, formothion or omethoate.
(3) Leaf miners
 The chemicals recommended include acephate, dimethoate, formothion or trichlorphon.

Delia coarctata (Fallén)

(= *Leptohylemya coarctata* (Fall.))

Common name. Wheat Bulb Fly
Family. Anthomyiidae
Hosts (main). Wheat
(alternative). Rye, barley, and various grasses (couch, etc.).
Damage. The larvae bore into seedlings of winter wheat, etc., and destroy the growing point, causing a 'dead-heart'; one larva can destroy several tillers. In heavy infestations many plants are killed and the crop is distinctly patchy.
Pest status. A serious pest of winter wheat throughout the UK and Europe; attacks occasionally very heavy, but much variation in severity of attack.
Life history. An unusual pest in that eggs are laid in July and August and do not hatch until January and February. Oviposition occurs just under or on the soil surface on bare ground, or under a root crop. Hatching is delayed by frosty conditions. On hatching the young larvae die unless the field has been planted to cereals or grass; newly hatched larvae can survive for only 5–6 days without food. When the larvae find a young cereal plant they bore into the 'bulb' under the soil surface. After a few days feeding the central shoot starts to wither and then dies. Infestation becomes apparent in the UK about the end of February and in March. Large larvae may even move from plant to plant, not only from tiller to tiller, so damage is accumulative. By April the larvae are full-grown and about 10 mm long, creamy white in colour.

Pupation takes place in the soil a few centimetres from the plants, usually between mid-April and early May. Pupation takes some 5–6 weeks.

Adults emerge usually in the first half of June; there is only one generation per year. Adults look like smallish house flies, the males somewhat browner, and the females yellowish-grey; they may often be seen on the ears of wheat in calm weather.
Distribution. This species is native to western Europe and is found throughout Europe, western Asia and well into central Asia (CIE map no. A.115).
Control. In theory this pest may be easily controlled by crop rotation since the females lay their eggs on bare ground in late summer, and crops at risk are winter wheat taken after fallow or a root crop. However, there are practical limitations in that in these areas wheat is both traditional and often the most suitable crop to sow after early-harvested crops. Spring cereals are only at risk if sown very early, but if sowing is delayed much then yields will inevitably suffer. According to the type of soil, and the

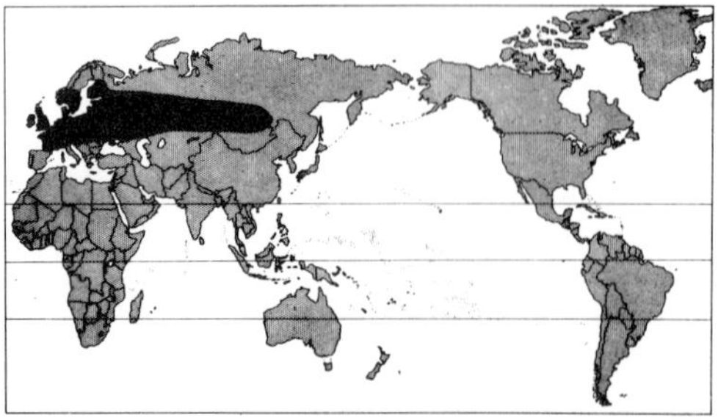

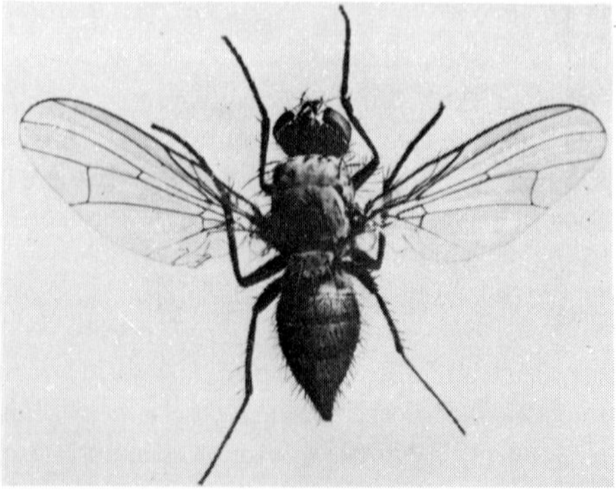

adult

0 2 r

crops to be grown, there are certain cultural practices that will reduce the level of attack; local Ministry of Agriculture staff should be consulted for details.

Chemical control: Despite much experimental work there is still no completely effective insecticidal treatment against this pest. At present a combination of methods is used: firstly, seed treatments using HCH, carbophenothion or chlorfenvinphos will deter larvae from entering the young plant. These chemicals are available both as dry and liquid formulations for seed treatment; they are only recommended for use on autumn-sown cereals. Granules of chlorpyrifos or fonofos applied at sowing has given good control in some situations. Sprays at egg-hatch in January/February, using chlorfenvinphos, chlorpyrifos or pirimiphos-methyl have given good results in Ministry trials. Sprays at the first sign of plant damage using dimethoate, formothion or omethoate have enabled some crops to recover; but if damage levels are too high then re-drilling the crop may be the only practical solution, leaving a post-ploughing period of at least 2–3 weeks before re-drilling.

Infested wheat seedling with 'dead heart'

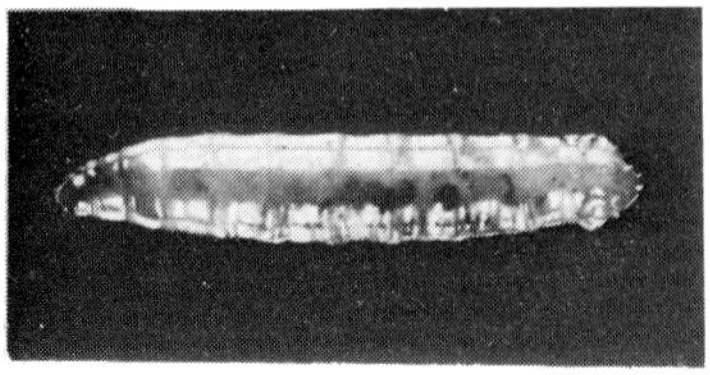

larva

Delia platura (Meigen)

(= *Hylemya platura* Mg.)
(= *Chortophila cilicrura* Rond.)

Common name. Bean Seed Fly; Corn Seed Maggot (USA)
Family. Anthomyiidae
Hosts (main). Sown seeds of beans and maize (and young seedlings).

(alternative). Onions, tobacco, cotton, marrow, cucumber, lettuce, peas, and crucifers.

Damage. The feeding maggots bore into the cotyledons of sown seeds, or into stems and roots of young seedlings; attacked seedlings usually die.

Pest status. Quite a serious pest (locally) in many areas of the world, on a wide range of crops. Basically a phytophagous pest, but saprophagous also.

Life history. Eggs are laid on disturbed soil, especially in the vicinity of rotting organic matter. The eggs are white (1 × 0.3 mm) with a distinctive reticulate pattern. Each female lays about 100 eggs, a few at a time, over a period of 3–4 weeks; hatching takes 2–4 days.

The larva is a typical white maggot, with three instars, taking about 3, 3 and 6–10 days for each stage of development (12–16 days in total). The anterior spiracle is fan-shaped with 5–8 processes.

Pupation takes place in the soil, a little way from the plant, at a depth of 2–4 cm. The puparia are oval, dark brown and 5 mm long. Under warm conditions pupation takes 2–3 weeks; in temperate regions the last-generation pupae overwinter.

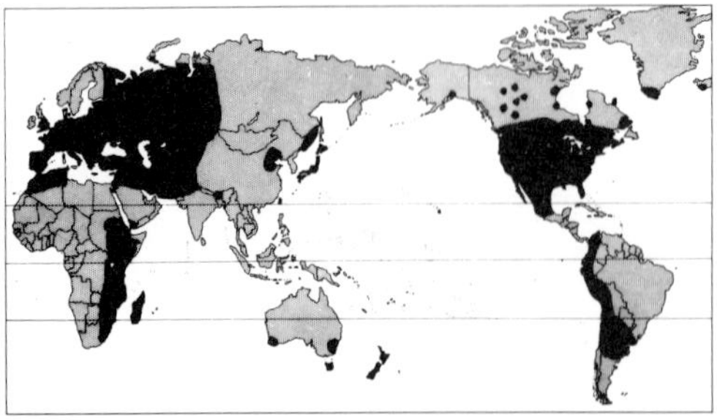

The adult is a small greyish-brown fly, about 5 mm long; the female has a pointed greyish abdomen, and the male a rounded blackish one. The female has a pre-oviposition period of 1–2 weeks, and she may lay eggs for as long as six weeks, and feeds on nectar from flowers.

In the UK there are usually 3–4 overlapping generations each year; elsewhere 2–5 generations annually according to temperature conditions. The life-cycle may be completed in 4–5 weeks under warm conditions.

Distribution. *Delia platura* is completely worldwide in distribution, extending right into the Arctic Circle in Asia, and recorded in Iceland and Greenland; not recorded from a few different localities (CIE map no. A.141). *Delia florilega* is regularly found associated with *D. platura* in Europe, and they are together known as 'Bean Seed Flies'.

Control. For control measures see page 369; resistance to dieldrin in the UK was first established in 1965; now the usual seed dressing is bromophos.

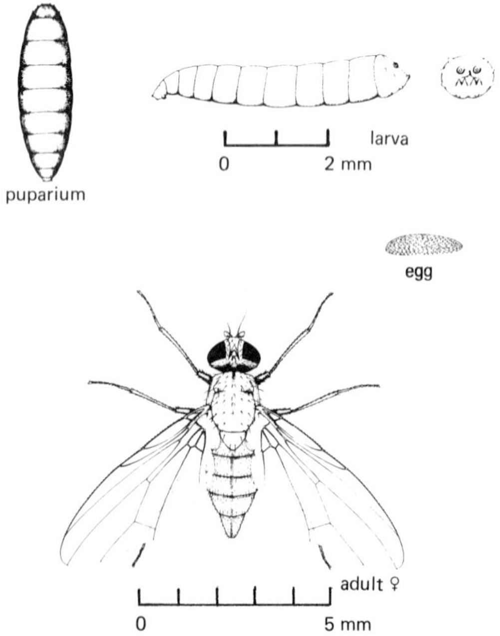

Delia radicum (L)

(= *Erioischia brassicae* (Bché.))

Common name. Cabbage Root Fly (Maggot)
Family. Anthomyiidae
Hosts (main). *Brassica* crops, and radish.
(alternative). Other Cruciferae.
Damage. Feeding larvae eat the lateral roots and then bore into the tap root, and sometimes the base of the stem. Sometimes found in Brussels sprout buttons. Attacked plants wilt, and the leaves turn bluish; the plant usually dies, or remains stunted.
Pest status. A serious pest of cultivated brassicas; losses of unprotected crops may be very high, both in Europe and in N. America.
Life history. Eggs are laid on the soil surface, in cracks close to the stem; some may be laid on the actual plant. In England the first eggs are laid in late April/early May, but later farther north. Each female fly lays about 100 eggs, over a period of time. Hatching takes 3–7 days.

The larvae burrow through the soil to the plant and feed close to the tap root and tunnel into the root. Larval development takes only about three weeks, then the larvae (about 8 mm long) move into the soil and pupate a few centimetres away. Pupation takes 15–35 days, then the second-generation flies emerge, at the end of June and in July. Eggs are laid and the second-generation larvae develop; they pupate in August. Many of these pupae overwinter in diapause, but some adults emerge in August/September in the UK, making a partial third generation that overwinters as larvae and pupates in the spring.

The adult is a dark grey fly, some 6 mm long, looking just like a house fly. They are long-lived (up to 50 days), and feed on nectar from hedgerow flowers; their emergence in the spring coincides with the flowering of cow parsley (*Anthriscus sylvestris*) a regular source of nectar for many flies and other insects.

Infested cabbage

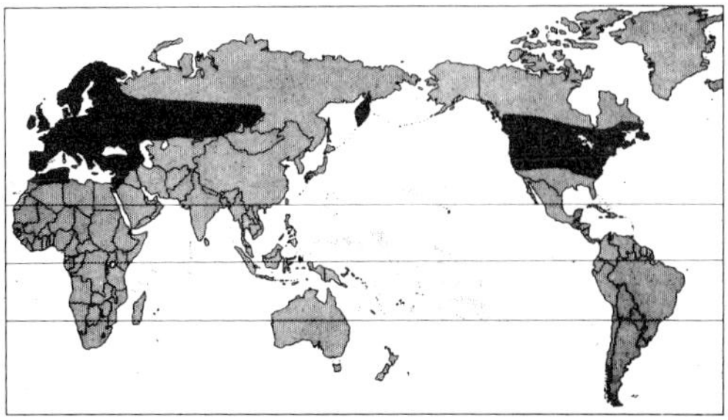

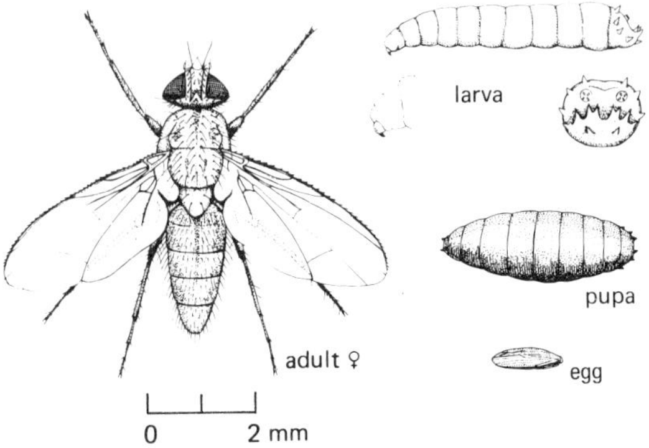

The three generations usually overlap considerably and most stages of this pest can be found most of the time during warmer weather.

Distribution. A widespread species found throughout Europe, N. Africa, W. Asia, Kola Peninsula (USSR), Canada and N. USA (CIE map no. A.83).

Control. At NVRS egg predation by carnivorous beetles (Carabidae) was about 90%, and pupal parasitism by staphylinid larvae (*Aleochara* spp.) is high (often 30%). For control see page 369; insecticide use has to be judicious so as to avoid destruction of natural enemies.

Left: Cabbage root fly damage on a radish

Brussels sprout cut open to show damage by maggots of cabbage root fly (M.A.F.F.)

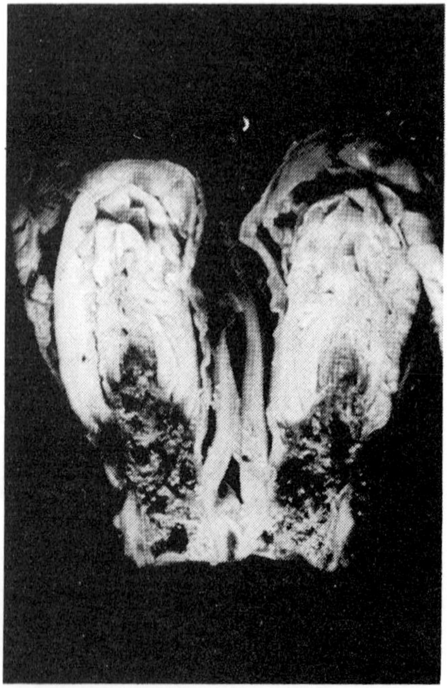

Pegomya hyoscyami (Panzer)

(= *P. betae* (Curtis))

Common name. Mangold Fly; Beet Leaf Miner
Family. Anthomyiidae
Hosts (main). Sugar beet, spinach, mangels (mangolds) (Chenopodiaceae).

(alternative). It is reported that a race of this species attacks Solanaceae.

Damage. The larvae mine in the leaves and cotyledons; seedlings may be destroyed, but established plants tolerate quite high levels of damage. Slight damage to spinach affects crop marketability considerably. The larvae make large blotch mines in the leaves, and there may be several larvae in one mine.

Pest status. A regular pest of chenopod crops in many temperate regions, but sporadic in occurrence; usually not very serious as most crops recover from the damage by compensatory growth. Sometimes serious on sugar beet in Europe.

Life history. Eggs are laid on the underside of the leaves, in batches of 2–3, but up to 20 are recorded; this takes place in the period from April to early June, with a peak at the end of May. The eggs are elongate (1 mm), white, and with a reticulate pattern; hatching takes 4–5 days. Each female lays up to 300 eggs.

The larvae bore immediately into the leaf tissues and start making a blotch mine. Development takes 10–15 days, and the large larvae (8 mm) drop from the leaf into the soil where they pupate in a brown oval puparium at a depth of 5–10 cm. Pupation takes 14–25 days.

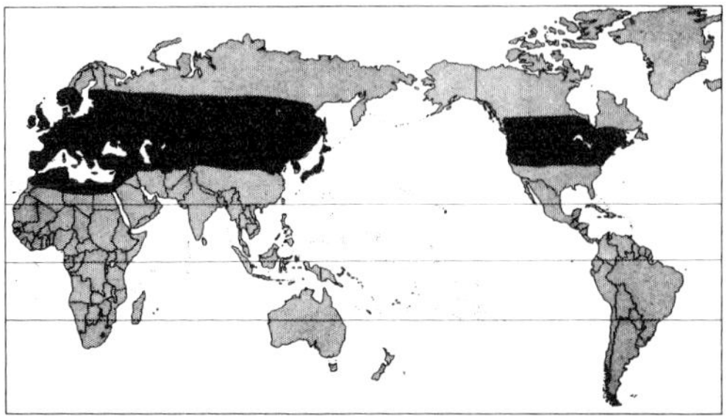

Second-generation adults emerge in late August/early September; but some of the pupae overwinter in diapause. Second-generation larvae are found in September. In the UK there are 2–3 overlapping generations per year; in Israel the two generations occur over the winter period.

The adult is a grey fly (5–6 mm) with paler legs, and usually lives for 2–4 weeks.

Distribution. Recorded throughout Europe, the Mediterranean Region, northern Asia, including China and Japan; introduced into N. America in the second half of the nineteenth century and now widespread throughout Canada and the USA.

Control. The usual recommendation is that chemical spraying is probably worthwhile if more than 4–5 eggs per leaf are found on plants before the eight-leaf stage. At a later stage it is probably seldom worth spraying, but is generally recommended if the total number of larvae per plant plus eggs, on average, exceeds the square of the number of rough leaves. However, this recommendation only applies to the first generation.

In central Europe, under dry conditions, crop growth is often retarded and then leaf miner damage may be very serious.

Sugar beet leaf with larval mines

Order

Lepidoptera

(moths and butterflies)

Families

Hepialidae (swift moths)
Microlepidoptera (several families)
Sesiidae (clearwing moths)
Yponomeutidae (ermine moths)
Oecophoridae
Gelechiidae
Cossidae (goat or leopard moths; carpenterworms)
Tortricidae (leaf rollers/budworms)
Pyralidae (grass moths)
Pieridae (white/yellow butterflies)
Lycaenidae (blue butterflies)
Geometridae (loopers)
Lasiocampidae (tent caterpillars)
Sphingidae (hawk moths; hornworms)
Noctuidae (owlet moths)
(Plusiinae (semiloopers))
Lymantriidae (tussock moths)

Hepialus spp.

(*H. humuli* (L.))
(*H. lupulinus* (L.))

Common name. Ghost Swift Moth; Common Swift Moth
Family. Hepialidae
Hosts. Totally polyphagous larvae in the soil eat the roots of many plants, both woody and fleshy, cultivated and wild; ecologically they are grassland insects.
Damage. The larvae in the soil eat the roots of many different plants, and bore into tubers of potato. Lettuce has the base of the crown and heart bored out, as do strawberry plants. Damaged plants wilt and often die.
Pest status. Locally important crop pests in many regions; often most serious in market gardens and smallholdings. Usually most serious in recently ploughed grassland or waste land. Most damaging to perennial crops in soil that is seldom disturbed.
Life history. The flying female moth drops eggs on to the ground, singly, up to 800 eggs in total; hatching takes 12–20 days. Eggs are laid throughout June, July and August.

The larvae (caterpillars) burrow into the ground and start feeding on fine rootlets. The caterpillars are elongate, white in colour, with a brown head capsule; quite characteristic in appearance. Fully grown the *H. humuli* and *H. lupulinus* larvae measure up to 50 mm and 35 mm respectively. Most individuals take nearly two years for larval development, but if food is abundant they can mature in one year. They feed throughout the winter, but more voraciously in the spring and summer.

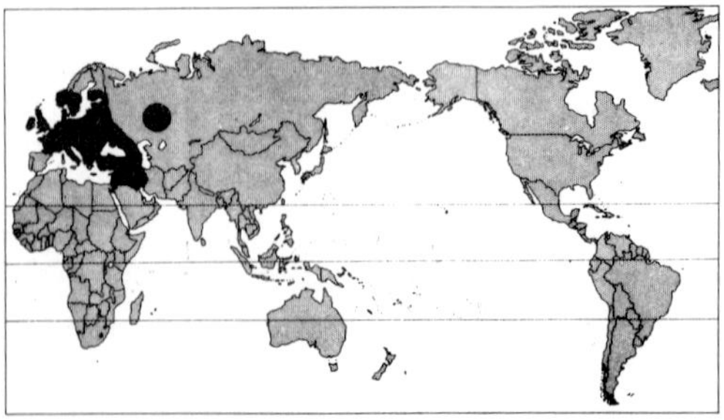

Pupation takes place in the spring in an earthen cell (usually April or May) under the damaged plants; pupation generally takes 2–3 weeks.

The smaller Common Swift more usually completes its development in a single year.

The adults are easily recognized by having two pairs of separate, equally sized wings; they fly slowly at dusk over grassland typically. Wingspan ranges from 50–70 mm and 25–45 mm respectively; males are the smaller.

Distribution. These two species occur throughout the UK, northern, central and southeastern Europe to the Near East, but the precise limits of their distribution is not known.

Some swift moths in Japan and China have larvae that bore grape vines and other woody stems; several species in Australia bore in *Acacia* and *Eucalyptus* trees.

Control. *Cultural control*: Recently ploughed grassland should not be planted up with perennial crops without first having a year of cleaning operations. Periodic cultivation exposes the larvae to birds and other predators.

Chemical control: Insecticides recommended are still DDT and HCH, preferably to be worked into the soil at the time of ploughing. On newly established grassland some control is achieved using an overall DDT spray.

Common Swift Moths

Ghost Swift Moths

Larva by lettuce seedling

Leaf-mining microlepidoptera

This is a large assemblage of tiny moths (3–8 mm total length), belonging to the heterogeneous group referred to as the microlepidoptera, whose larvae mine the leaves of both angiosperms and gymnosperms. The moths are seldom seen because of their tiny size, but the larval mines are often quite conspicuous. Some of the groups here included are not closely related, but are often considered en masse because of their tiny size and similarity of larval habits.

To the agriculturalist they are collectively of little importance (except for just a few species, *Leucoptera* spp.), but to the entomologists and ecologists they are of considerable interest (see Wilkinson, 1982).

The larvae of most species make simple mines, sometimes entirely linear winding tunnels, sometimes blotch mines, and sometimes the mine starts in a linear manner and terminates in a large blotch. The casebearers start as leaf miners, but the later larval instars leave the mine and build small cases in which the caterpillar lives while feeding externally on the leaf epidermis. The skeletonizers spend the first few instars (usually 1–3) internally as miners, but later their eating includes one epidermis so they become exposed as 'leaf skeletonizers'.

Pupation takes place either in the mine, or in leaf litter, or often the edge of the leaf is turned over dorsally with silken threads and pupation takes place under the fold; the casebearers pupate within the larval case.

Most species mine only in the leaves, but some prefer buds (which are, after all, just a collection of young leaves on a telescoped shoot); one or two species mine the bark of trees.

The host plants are mostly trees and woody shrubs, but some herbaceous plants are used; the vast majority of hosts are the local deciduous forest trees such as oaks, elms, beeches, willows, etc.

Sub-order Dacnonypha

Eriocraniidae

A small primitive family with larvae as leaf miners in angiosperms or in the seeds of gymnosperms; Holarctic and Australian in distribution, regular hosts include oaks, birch, hazel and hornbeam. The mine is a large blotch, sometimes starting as a linear mine; in the blotch are to be seen long threads of frass that are characteristic.

Sub-order Monotrysia

Nepticulidae

A largish, rather primitive family of tiny moths, about the smallest Lepidoptera known with wingspan 3–10 mm; worldwide in distribution; as an indication of their abundance there are 67 species recorded in the UK. Some larvae start as leaf tunnel (gallery) miners but end in making a blotch mine; others do not end in a blotch mine; some species mine other parts of the plant body, such as buds, stem or bark. Three British species mine leaf petioles or midribs. The shape of the mine and the nature of the frass track can be diagnostic with experience. Pupation usually takes place in the leaf litter inside a silken cocoon. The more important (pest) species include:

Stigmella aurella (F.) – (Rubus Leaf Miner) Europe, N. Africa, Near East.

Stigmella anomalella (Goeze) – (Rose Leaf Miner) throughout Europe.

Stigmella malella (Stainton) – (Apple Pygmy Moth) Europe, including Italy.

Stigmella pomella (Vaughan) – (Apple Leaf Miner) throughout Europe.

Stigmella gossypii (F. & L.) – (Cotton Leaf Miner) USA.

Stigmella juglandifoliella (Clemens) – (Pecan Serpentine Leaf Miner) USA.

Stigmella spp. – 70 other species in UK; mostly leaf miners in trees (fruit trees, hazelnut and ornamentals).

Nepticula is now regarded as a junior synonym of *Stigmella*.

Incurvariidae

A small, rather primitive family, with larvae that are leaf miners mostly in herbaceous plants (with a few bud borers); some emerge from the mines and construct a tiny case for further feeding and pupation. The few pest species include:

Lampronia capitella (Clerck) – (Currant Shoot Moth) Europe, Asia to E. Siberia.

Lampronia rubiella (Bjerk.) – (Raspberry (Bud) Moth) Europe, Asia, USA, Canada.

Tischeriidae

A small but widespread family, rather primitive; the larvae are leaf miners and many have legs reduced, they eject their frass from the tunnel. Formerly placed in the Lyonetiidae; most species occur in America, but some in Europe, India and S. Africa. The mine is a blotch lined with silk; pupation occurs in the mine, and the pupal exuvium protrudes after emergence. Few species are recorded as pests:

Tischeria malifoliella Clemens – (Apple Leaf Trumpet Miner) USA.

Tischeria marginea (How.) – (Rubus Leaf Miner) Europe, Near East, N. Africa.

Tischeria sp. – (Peach Leaf Miner) China.

Tischeria spp. – (Chestnut Leaf Miners) Europe.

Sub-order Ditrysia

Lyonetiidae

(In Imms, included in Gracillariidae)

Quite a large 'family' – regarded by some as a heterogeneous assemblage; worldwide in distribution; some species are leaf-miners throughout their larval life, but some are only leaf miners for instars 1–3, and the fourth instar becomes a skeletonizer by eating one epidermis. Pest species include:

Bedellia gossypii Tun. – (Cotton Leaf Miner) Australia.

B. orchilella Wals. – (Sweet Potato Leaf Miner) USA.

Bedellia somnulentella (Zeller) – (Sweet Potato Leaf Miner) pantropical.

Bucculatrix pyrivorella Kuroko – (Pear Leaf Miner) Japan.

Bucculatrix thurberiella Busck – (Cotton Leaf Perforator) USA, C. and S. America.

Crobylophora spp. – (Coffee Leaf Miners) C. Africa.

Leucoptera coffeella (Guer.) – (Coffee Leaf Miner) C. and S. America (CIE map no. A.315).

Leucoptera malifoliella (Costa) (Apple (Pear) Leaf Blister Moth) Europe, China.

Leucoptera spp. (4) – (Coffee Leaf Miners) Africa (CIE map no. A.316).

Lyonetia clerkella (L.) – (Apple (Peach) Leaf Miner) Europe, China, Japan.

L. prunifoliella Mats. – (Plum Leaf Miner) Japan, Europe, Asia Minor.

Opogona glycyphaga Meyr. – sugarcane, banana skins; Australia.

Opogona spp. – (Sugarcane Leaf Miners) Africa, S.E. Asia, Hawaii.

Gracillariidae (blotch miners)

A large and cosmopolitan family, with more than 1000 species, of tiny moths with narrow wings and long fringes whose larvae mostly make blotch mines in leaves. Some of the more important pest species include:

Acrocercops astourota Meyr. – (Pear Leaf Miner) China.

A. bifasciata Wlsm. – (Cotton Leaf Miner) Africa.

Callisto denticulella (Thnb.) – (Apple Leaf Miner) Europe.

Caloptilia spp. – on many ornamentals; USA.

Caloptilia soyella Dev. – (Soybean Leaf Roller) Japan.

Caloptilia theivora Wals. – (Tea Leaf Roller) Japan.

Cuphodes dispyrosella Issiki – (Persimmon Leaf Miner) Japan.

Gracillaria spp. – on many ornamentals; USA.

Marmara elotella (Busck) – (Apple Bark Miner) USA.

Marmara pomonella (Busck) – (Apple Fruit Miner) USA.

Phyllonorycter (= *Lithocolletis*) spp. – ((Oak) Leaf Miners) Europe, USA.

Phyllonorycter spp. (11) – on Malvaceae and Leguminosae; Australia.

P. blancardella (F.) – (Apple Leaf Miner) Europe.
P. crataegella (Clemens) – (Apple Blotch Leaf Miner) USA.
P. pomonella (zeller) – (Plum Leaf Miner) Europe.
P. ringoneella Mats. – (Apple Leaf Miner) Japan.
P. triflorella Reger – (Apple Leaf Miner) China.
Spulerina astaurcta Meyr. – (Pear Bark Miner) Japan.

Coleophoridae (= Eupistidae) (casebearer moths)
Quite a large group, more than 400 species recorded; Holarctic in distribution, with 95 British species. All the larvae are leaf miners in their first instars, some remain leaf miners and others become casebearers and feed externally; the leaf miners eat the entire mesophyll, leaving just upper and lower epidermis intact, so the mine looks like a window; most pupate inside a case on branches or the tree trunk. Most species are placed in the very large genus *Coleophora*; there are a few pest species to be noted:
Acrobasis caryae (Horn) – (Pecan Nut Casebearer) USA.
A. juglandis (Le Baron) – (Pecan Leaf Casebearer) USA.
C. anatipennella (Hübner) – (Cherry Pistol Casebearer) Europe.
C. coracipennella (Hübner) – (Apple and Plum Casebearer) Europe.
C. caryaefoliella Clemens – (Pecan Cigar Casebearer) USA.
C. ochroneura – larvae in flower head of white clover; Australia, New Zealand.
C. pruniella Clemens – (Cherry Casebearer) USA.
C. serratella L. – (Apple Casebearer) Japan.
C. ringoniella Oku – (Apple Pistol Casebearer) Japan.
Coleophora spp. – elms, etc.; Europe, Asia, USA.

Phyllocnistidae
This family is represented by a single large genus (50 species) of tiny moths, quite cosmopolitan in distribution; the larvae are all leaf miners and are characterized by being apodous. There is only one pest species of any consequence, although several species are quite abundant.
Phyllocnistis citrella (Stnt.) – (Citrus Leaf Miner) Africa and Asia (CIE map no. A.274).
Phyllocnistis toparca Meyr. – (Grapevine Leaf Miner) India, Japan.
Phyllocnistis spp. – various common trees and grapevine; Europe, Asia, India, Japan.

Control of leaf miners (Microlepidoptera)

Damage is seldom serious, although the leaf mines are usually quite conspicuous and often common, but sometimes economic injury is inflicted.

Natural parasitism levels are usually high; many species of Hymenoptera Parasitica are involved and any insecticide use should be done very carefully so as to avoid upsetting the high level of natural control.

Chemical control involves the use of insecticides with penetrant action in order to kill the larvae *in situ*, and some systemic chemicals are also effective. The other alternative is to use insecticides against the adult moths on the plant foliage prior to oviposition and this method requires very careful timing. The chemicals recommended include diazinon, chlorpyrifos, fenthion, fenitrothion, and phosphamidon.

Synanthedon salmachus (L.)

(= *Aegeria tipuliformis* L.)

Common name. Currant Clearwing (Moth)
Family. Sesiidae
Hosts (main). Black currant

(alternative). Red currant and gooseberry. (Other species on other crops.)

Damage. Larvae bore inside the stems, causing distal leaves to wilt, and fruit trusses fail to develop.

Pest status. Locally abundant, though seldom a serious pest; generally widespread. This species is being used as an example of the whole group (family Sesiidae) as it is quite representative.

Life history. Adults fly in June and July, and may be found sitting on the foliage of the currant bushes in sunny weather.

Eggs are laid singly on the stem, usually near a bud or a wound; they are yellow in colour and oval in shape. Hatching takes about one week.

On hatching the young larva bores straight away into the pith of the stem; the caterpillar is white with a brown head capsule; it usually burrows up towards the younger wood. The larva feeds and bores throughout the autumn and then the winter, and is usually fully grown by April, at about 15 mm long.

Pupation takes place in the tunnel, just under the bark, and after adult emergence the pupal case remains sticking out of the branch.

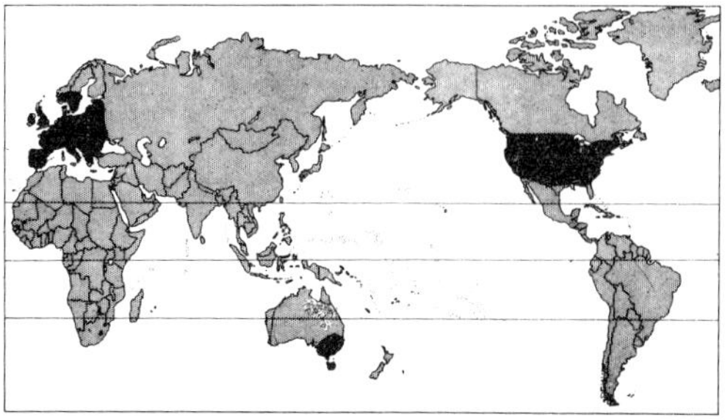

The adult is blackish in body colour, some 15–20 mm wingspan, with a dark border to the wings, and pale bands across the abdomen. In some species the body banding is bright yellow, and in others red.

This species is univoltine, but most larger species have a two-year life-cycle.

Distribution. Recorded throughout Europe to Siberia and now introduced into the USA, Australia (southeast, and Tasmania) and New Zealand.

Control. With these insect pests it is seldom that control measures are required; they are mostly species of academic interest. Should infestation levels be high the recommended action consists of:

(a) pruning and burning of infested branches.
(b) foliar sprays of DDT were formerly used, but now azinphos-methyl plus demeton-*S*-methyl sulphone, applied immediately after fruit picking, does give some control.

For the bark-boring species (e.g. Apple Clearwing) brushing a 10% tar oil on to the bark of infested trees is practised.

A very recent development was the virtual eradication of *S. salmachus* from 500000 black currant cuttings used to establish new plantations in Australia by use of the insect-parasitic nematode *Neoplectana bibionis* (Bedding, 1984).

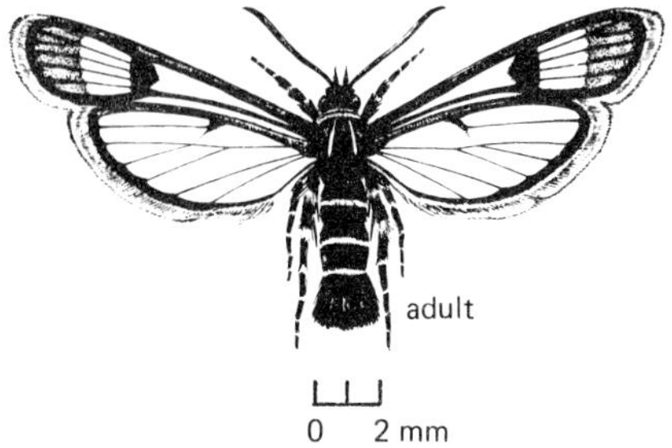

Clearwing moths (Lepidoptera: Sesiidae (=Aegeriidae))

This family is characterized by the adult moths having the forewings elongate and narrow, and both wings largely devoid of scales so they are clear; the overall effect is that they look remarkably like wasps, further enhanced in some species by the yellow banding on the black body. The banded abdomen terminates in a tuft of scales. Adults are diurnal and fly rapidly in warm sunshine; they are often to be seen sitting on exposed foliage on sunny days.

Some species are tropical, but the group is essentially a Holarctic (northern temperate) one. The tropical species have larvae that bore the vines of sweet potato and cucurbits, but most temperate species bore in the branches and trunks of trees and woody shrubs. However, a few non-pest species bore in the roots of herbaceous plants such as *Rumex*, *Armeria* and some legumes. The trees most favoured as hosts include willows, poplars, birches, oaks, but various fruit trees are also attacked.

Control is seldom really needed and most attacks are sporadic and local. Once the larvae are actually boring inside the branches they are extremely difficult to kill with insecticides as contact cannot be effected.

Some pest species of Sesiidae

Conopia hector Butler – (Cherry Tree Borer) Japan.
Melittia calabaza D. & E. – (Southwestern Squash Vine Borer) USA.
Melittia cucurbitae (Harris) – (Squash Vine Borer) USA.
Nokona regale Butler – (Grape Clearwing) Japan.
Paranthrenopsis constricta Butler – (Rose Clearwing) Japan.
Sanninoidea exitiosa (Say) – (Peach Tree Borer) USA.
Sannina uroceriformis Wlk. – (Persimmon Borer) USA.
Sesia apiformis (Clerk) – (Hornet Clearwing) poplars; Europe and USA.
Synanthedon bibionipennis (Bois.) – (Strawberry Crown Moth) USA.
Synanthedon myopaeformis (Bork.) – (Apple Clearwing) Europe, Asia Minor, USSR.
Synanthedon pictipes (G. & R.) – (Lesser Peach Tree Borer) USA.
Synanthedon pyri (Harris) – (Apple Bark Borer) USA.
Synanthedon salmachus (L.) – (Currant Clearwing) Europe, Asia, Australia, Canada, USA (see page 381).
Synanthedon vespiformis (L.) – (Yellow-legged Clearwing) chestnut, walnut, etc.; Europe, W. Asia.
Synanthedon spp. (3) – (Sweet Potato Clearwings) E. Africa.
Pennisetia marginata (Harris) – (Raspberry Crown Borer) USA.
Vitacea polistiformis (Harris) – (Grape Root Borer) USA.

Acrolepiopsis assectella (Zeller)

Common name. Leek Moth
Family. Yponomeutidae
Hosts (main). Leek
(alternative). Onions, garlic, chives.

Damage. Larvae bore through the folded leaves of the shoot; as the leek grows the leaves emerge with a series of small holes called 'windows', and rots often follow insect attack. In onions the larvae live inside the hollow leaves, but occasionally penetrate the bulb.

Pest status. Quite a serious pest in Europe, but only of minor status in the UK and Asia.

Life history. Eggs are laid singly at night at the base of the host plants; each female lays about 100 eggs. In the UK they are laid in early May; hatching takes 5–8 days.

The first-instar larvae are leaf miners, but after about five days they leave the mines and bore into the central shoot and eat the folded leaves; total larval development takes about 30 days, and the fully fed larvae are yellowish-green in colour and about 10 mm long.

Pupation takes place in a silken cocoon attached to the dead leaves of the host; development takes about two weeks.

In the UK there are 3–4 generations per year, and the pupae of the last generation hibernate to overwinter.

Adults emerge in April and eggs are laid in late April and early May. The adults are small, brownish-grey, nocturnal moths, with a wingspan of about 15 mm.

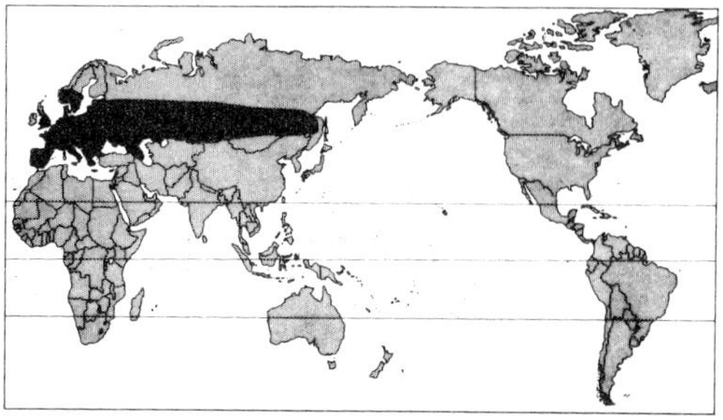

Distribution. Recorded throughout Europe (including England), up to southern Scandinavia, and right across Asia; also Hawaii (CIE map no. A.405).

Other pest species include:

Acrolepiopsis sapporensis Mats. – (Allium Leaf Miner) Japan.

Acrolepiopsis suzukiella Mats. – (Yam Leaf Miner) Japan.

Control. When infestations warrant control measures being taken, the most effective insecticide to date has been DDT, applied as a foliar spray, timed to kill the hatching larvae.

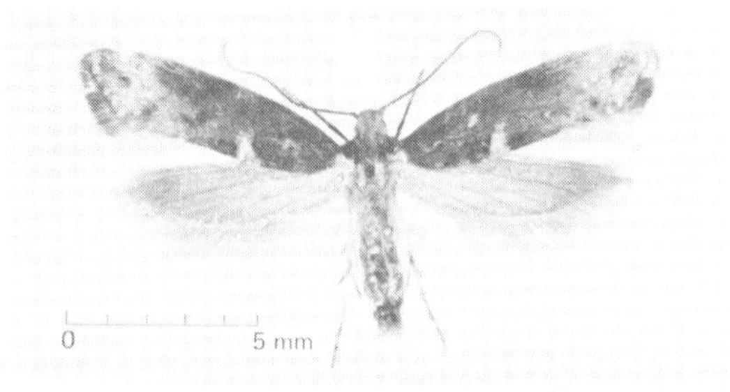

adult B.M. (N.H.)

Plutella xylostella (L.)

(= *Plutella maculipennis* (Curtis)

Common name. Diamond-back Moth
Family. Yponomeutidae
Hosts (main). *Brassica* species
(alternative). Cultivated and wild Cruciferae.
Damage. First-instar larvae mine the leaves, entering from the underside, but later instars also eat the lower epidermis making small 'windows' in the leaf; after a time the upper epidermis often ruptures so that a small hole results. With heavy infestations the entire plant is devastated.
Pest status. A very common and widespread pest of Cruciferae, often serious, especially in some of the warmer parts of the world. In hot dry weather attacks are particularly damaging.
Life history. The tiny yellow eggs are laid on the upper surface of the leaves, either singly or in small groups; hatching takes place after 3–8 days. Each female moth lays 50–150 eggs.

The caterpillar is pale green, with a tapering body widest in the middle; head is black when hatched but turns paler yellow when mature; length about 12 mm. If disturbed the larvae wriggle violently and may drop off the leaf suspended by a silken thread. Larval development takes 14–28 days.

Pupation takes place in a gauzy silken cocoon stuck to the plant foliage; the pupa is about 9 mm long; development takes 5–10 days under warm conditions.

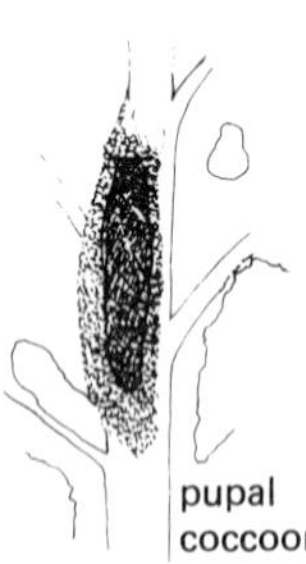

pupal coccoon

adults

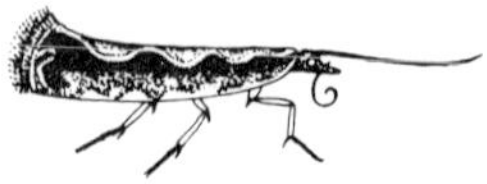

larva

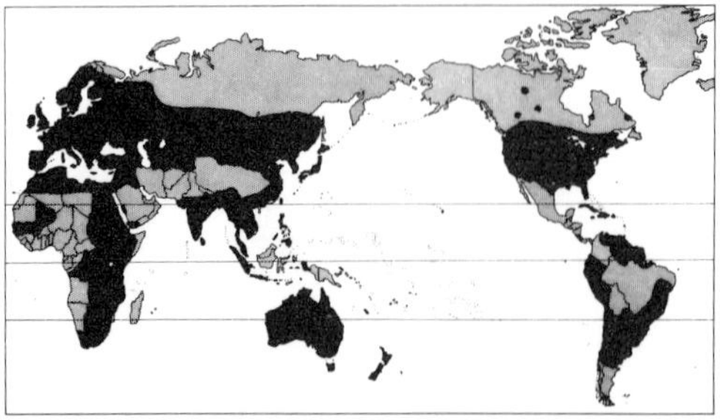

The adult is a small grey moth, about 6 mm long and wingspan 15 mm. Along the hind margin of each forewing are three pale triangular marks, and when the wings are closed the marks form a diamond pattern, hence the common name. The moths live for about two weeks, and regularly migrate in both Europe and N. America. In northern Canada they do not survive over winter; in the early summer migrants from the USA start new infestations. In British Columbia there are 2–3 generations per year; Ontario may have six; in lowland Malaysia 15 generations are recorded. Under warm conditions the life-cycle can be completed in 12–15 days.

Distribution. Totally cosmopolitan in distribution, in fact the most widespread pest species of Lepidoptera; almost completely worldwide, extending right into the Arctic Circle (CIE map no. A.32).

Control. In many countries this pest has developed resistance to the usual insecticides and so local Ministry of Agriculture advice should be sought. See page 412 for advice on the 'control of cabbage caterpillars'.

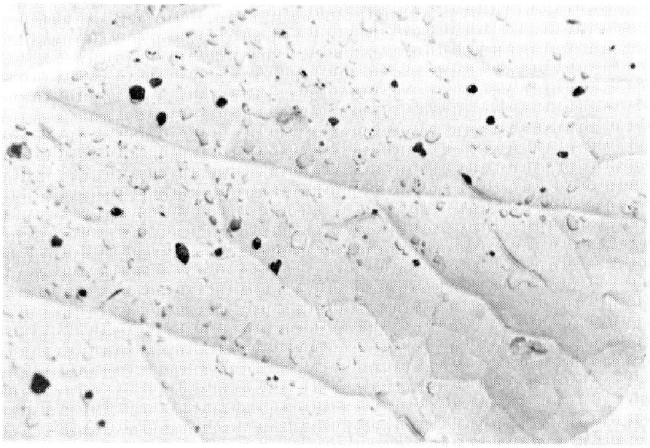

Damaged cabbage leaf

Yponomeuta spp.

Common name. Small Ermine Moths
Family. Yponomeutidae
Hosts (main). Apple
(alternative). Plum, other *Prunus* species, hawthorn (Rosaceae); also *Laurus* (laurel).
Damage. The gregarious caterpillars (grey with black spots) web the shoots and eat the leaves; the shoot may be destroyed, but defoliation is the usual result.
Pest status. These pests are widespread and very abundant; severe damage is mostly seen on hedges of hawthorn and blackthorn, orchard damage is only occasionally serious.
Life history. The small flat eggs are laid in batches on the branches of the food plant; the average batch size is about 50; egg-laying occurs in July and August. By September the larvae have hatched, but they remain under the egg-shield and overwinter in this stage. In the spring (usually May) the tiny larvae emerge and start to feed on the foliage. On apple (but not hawthorn) the first-instar larvae are leaf miners, making brown leaf blisters. Later the caterpillars live in colonies under an extensive silken web between the leaves. As the larvae grow the silken tent is enlarged. Sometimes the colony moves to another spur and makes a new web. With heavy infestations the webs coalesce and whole branches or a length of hedge may be covered by the silken webs. By the end of June, or early July, the mature larvae (12–15 mm) are grey with black spots, and they spin cocoons, often many together, and then pupate.

Adult moths emerge in July and August; they are small moths, about 8 mm long and 20–22 mm wingspan; forewings are silvery-white with black spots, and the hindwings uniformly grey.

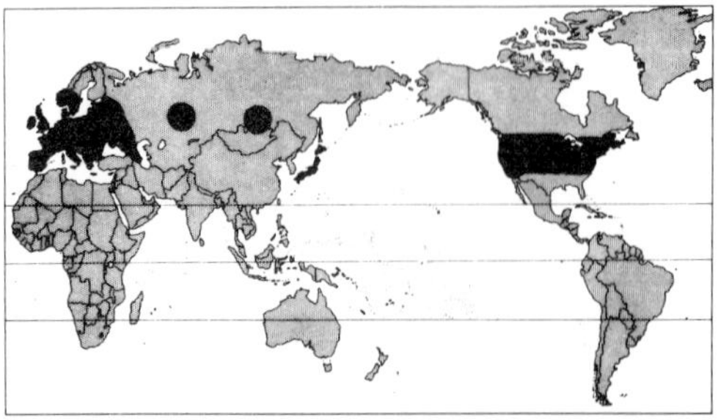

Distribution. There has long been a taxonomic problem with these pests, and it is now thought there are, in Europe, four species and five incipient species, all very similar in appearance but with quite definite host differences; the more important agricultural pests are:
Yponomeuta malinellus Zeller – (Common Small Ermine Moth) Europe and Japan.
Yponomeuta evonymella (L.) – (Bird-cherry Ermine Moth) Europe.
Yponomeuta padella (L.) – (Common Small Ermine Moth) Europe and USA.

Other pest species of Yponomeutidae of some importance include:
Acrolepiopsis spp. – Leek Moths, etc. (see page 383).
Argyresthia conjugella Zeller – (Apple Fruit Miner) Japan and Europe.
Argyresthia pruniella (Clerck) – (Cherry Fruit Moth) Europe.
Plutella xylostella (L.) – (Diamond-back Moth) cosmopolitan (see page 384).
Prays citri Mill. – (Citrus Flower Moth) Europe, Asia, Australasia.
Prays endocarpa Meyr. – (Citrus Rind Borer) India, Indonesia.
Prays oleae F. – (Olive Moth) Mediterranean, S. Africa (CIE map no. A.123).
Control. Natural control levels are often high; the larvae are eaten by birds, and are parasitized by several different wasps.

In orchards where winter washes are regularly applied this pest is seldom of economic importance. In gardens hand-picking of young colonies is recommended. If sprays are needed then carbaryl or trichlorphon as sprays of very fine droplets are successful.

Adult

0 5 mm

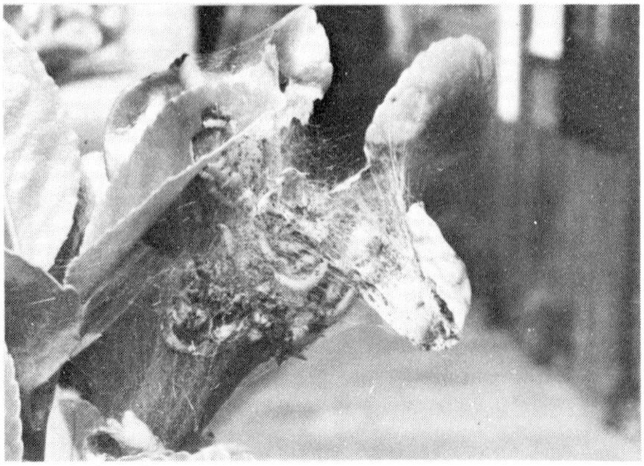
Larvae inside silken web

Depressaria pastinacella (Dup.)

(= *D. heracliana* auctt.)

Common name. Parsnip Moth
Family. Oecophoridae
Hosts (main). Parsnip

(alternative). Carrot, celery, other cultivated and wild Umbelliferae.

Damage. Feeding larvae web the seedheads and eat the leaves, flowers, bracts and especially the seeds. Some larvae bore in the main stem where they pupate.

Pest status. A regular pest of parsnip grown for seed; sometimes serious on other crops of Umbelliferae when seeding. Seed losses can be quite serious.

Life history. Eggs are laid in the developing flowerheads, and on foliage, in the late spring. The caterpillars are grey, yellowish laterally, with a brown head and many short setae and dark spots. They feed on the flowerheads mostly, under a web of silk; they grow to a length of 15 mm. Many of the large larvae bore into the plant stem where they pupate, but some pupate under the silken webs; pupation takes place mostly in July.

The adult moths emerge from the stems through the holes made by the entering larvae usually in September. Forewings are greyish in colour with pale markings; hindwings whitish; wingspan about 25 mm. This species is unusual in that the adults hibernate over winter, in leaf litter, under tree bark, in wood piles, sheds, etc. They resume activity in the spring and lay eggs.

There is only one generation per year.

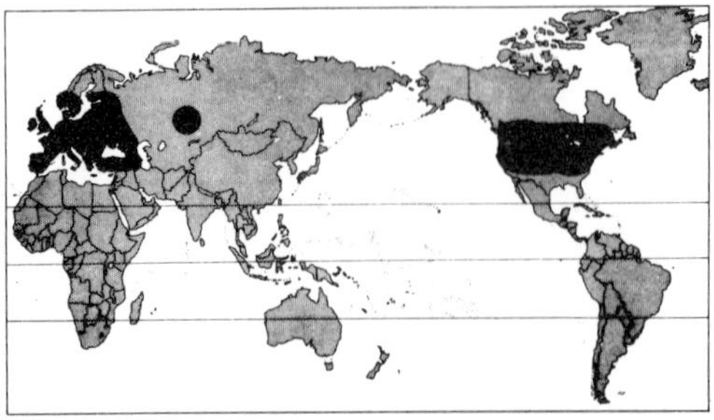

Distribution. Quite widely distributed throughout Europe, parts of Asia, Canada and the USA. It was first recorded in N. America in Ontario, in 1869, and is thought to have been accidentally introduced from Europe.

Other pest species of Oecophoridae of some importance include:

Carcina quercana (F.) – on oak, beech, apple, plum, pear, blackberry, etc.; Europe.
Depressaria daucella (D. & S.) – Europe.
Promalactis inonisema Butler – (Cotton Seedworm) Japan.
Psorostica melanocrepida Clarke – (Citrus Leaf Roller) Japan.
Psorostica zizyphi (St.) – on citrus; India.

There are also a few urban (domestic) pests of note:
Anchonoma xeraula Meyr. – (Grain Worm) Japan.
Endrosis sarcitrella (L.) – (Whiteshouldered House Moth) cosmopolitan.
Hofmannophila pseudospretella (Stain.) – (Brown House Moth) cosmopolitan.

Control. The most recent recommendations available are still DDT applied as a full foliar spray; but the destruction of infested flowerheads, and the removal of wild hosts from the vicinity of the crop will reduce the pest population locally.

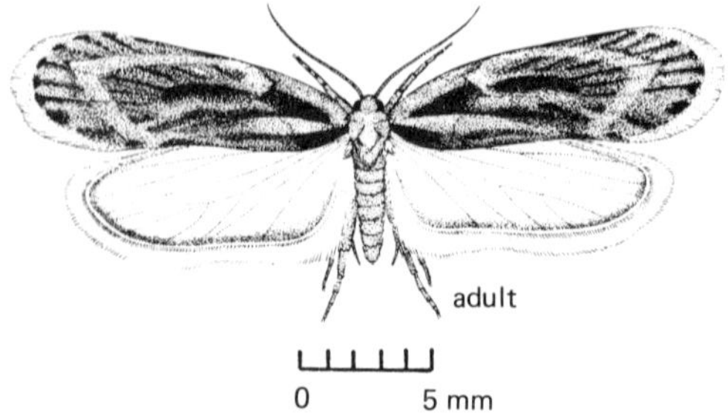

Anarsia lineatella (Zeller)

Common name. Peach Twig Borer
Family. Gelechiidae
Hosts (main). Peach
(alternative). Plum, apricot, almond, mango.

Damage. The larvae tunnel into young actively growing shoots; their feeding destroys the shoot and bushy lateral growth follows. Larvae also attack fruits, usually at the end of the season when the twigs have hardened; they feed on the flesh until fully grown when they leave the fruit through an exit hole from which there is a gummy exudate.

Pest status. Quite serious pests of peach and other stone fruits, with both shoot and fruit damage.

Life history. Small larvae overwinter in silk-lined crevices in the twig bark. In the spring when growth recommences these larvae emerge and bore into the new growth, either twigs or terminal buds. The feeding tunnel is usually short, stopping growth and often killing the shoot. The larvae move to other shoots to feed until fully grown. The red-coloured larvae then emerge and spin a cocoon on the bark where they pupate for about two weeks.

The adults emerge and soon eggs are laid on leaves and fruits; this generation of larvae feed on both twigs and fruits; the next generation feed almost entirely in the fruits. There are from 1–4 generations according to locality (temperature).

The adult is a tiny greyish moth with long wing fringes, body length 5–6 mm and wingspan 12–14 mm.

Distribution. Recorded throughout Europe (not UK), from southern Sweden to N. Africa, Near East, Middle East, China, Canada and the USA (CIE map no. A.103).

Other important pest species of Gelechiidae that bore shoots and fruits include:

Aristotelia fragariae Busck – (Strawberry Crownminer) USA.
Dichomeris ianthes (Meyr.) – (Alfalfa Leaftier) USA.
Keiferia lycopersicella (Wals.) – (Tomato Pinworm) USA, C. and S. America.
Pectinophora gossypiella (Saunders) – (Pink (Cotton) Bollworm) pantropical (CIE map no. A.13).
Platyedra subcinerea (Haw.) – (Cotton Stem Moth) USA.
Recurvaria nanella (D. & S.) – (Lesser (Fruit) Bud Moth) Europe, USA.
Recurvaria spp. – (Pine Needle Miners) USA and Canada.
Scrobipalpa heliopa (Lower) – (Tobacco Stem Borer) India, S. E. Asia.
Scrobipalpa ocellatella (Boyd) – (Beet Moth) Near East, Mediterranean Region.
Stegasta bosqueella (cham.) – (Rednecked Peanutworm) USA.
Tildenia inconspicuella (Murt.) – (Eggplant Leaf Miner) USA.

Control. In orchards regularly treated to winter washes of lime-sulphur treatments in the winter this pest is seldom serious. Otherwise the insecticides that have been used with success include carbaryl, diazinon, endosulfan, azinphos-methyl and parathion.

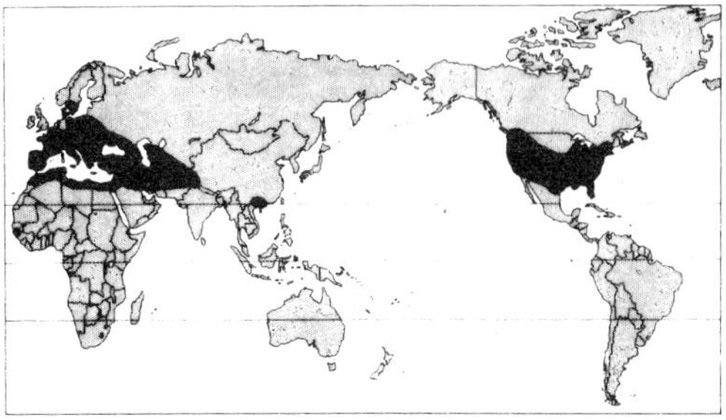

Adult moth B.M. (N.H.)

0 4 mm

Phthorimaea operculella (Zell.)

(= *Gnorimoschema operculella* (Zell.))

Common name. Potato Tuber Moth
Family. Gelechiidae
Hosts (main). Potato and tobacco.
(alternative). Tomato, eggplant, other Solanaceae; also *Beta vulgaris*.

Damage. Young larvae mine in the leaves, making silver blotches; later they tunnel in veins, petioles and stems, causing the stems to wilt. On potato the large larvae tunnel into the tubers underground and are eventually taken into the stores inside infested tubers. The tunnels usually become infected with fungi and bacteria, causing rots.

Pest status. A serious pest of potato in warmer countries, and of some importance on tobacco. Infestations start in the field and continue during storage of the tubers; infestations are spread through infested tubers.

Life history. The eggs are tiny, oval and yellow, 0.5 × 0.4 mm, laid singly on stored tubers near an 'eye', or on a sprout, or on the underside of a leaf. Each female lays some 150–200 eggs; hatching takes 3–15 days.

The young caterpillar bores into the leaf (or into the tuber in storage) and makes a blotch mine; it is pale green in colour. The larvae eventually eat their way into the leaf veins, down through the petiole and into the plant stem, and some go so far as into the young tubers. The larval period takes 9–33 days, and fully grown the larva measures 9–11 mm.

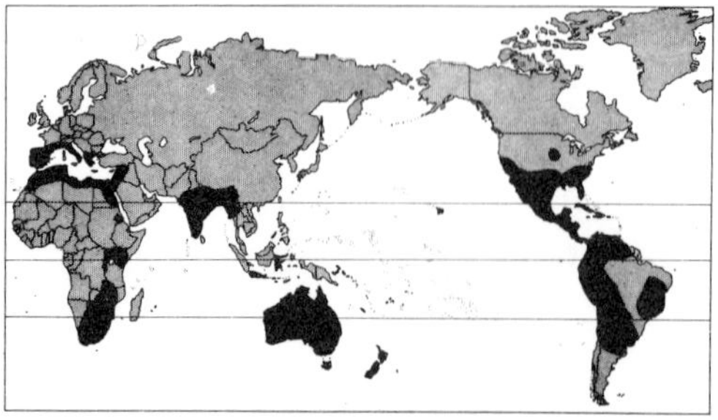

Pupation takes place either within the tuber or outside in the surface litter, and takes 6–26 days.

The adult is a small greyish moth with narrow fringed wings; wingspan 15 mm; short-lived.

One generation can be completed in 3–4 weeks, and in some localities 12 generations per year are recorded, but development rate is dependent upon temperature.

Distribution. This species is widely distributed throughout the warmer parts of the world; almost completely cosmopolitan, but as yet unrecorded from some major regions (CIE map no. A.10).

Control. The usual methods of control are insecticidal sprays applied to the crop at two-week intervals after the first leaf mines are observed; the chemicals used have been DDT, dicrotophos, dimethoate and parathion.

Crop hygiene is useful in reducing populations, and for tubers in store fumigation is usually effective.

eggs

larva

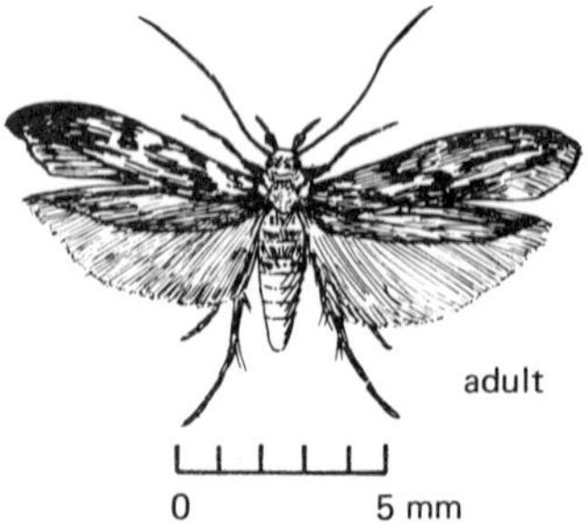

adult

pupa

larva in leaf mine

damaged potato

Sitotroga cerealella (Oliver)

Common name. Angoumois Grain Moth
Family. Gelechiidae
Hosts (main). Wheat and maize, both in stores and in the field.

(alternative). Sorghum, other stored grains, and dried fruits in store.

Damage. The larvae develop within individual grains, typically at the 'milky' stage in the field. Fully grown larvae and pupae lie under a small 'window' in the grain. The emerging adult moth pushes through the window leaving the small flap hinged to the grain characteristic of this pest.

Pest status. A serious pest of cereals in many warmer parts of the world; damage to growing crops is often extensive before the pest is taken into grain stores where it continues and worsens.

Life history. The pinkish, oval eggs are laid on the surface of the developing grains, usually when at the 'milky' stage. Each female lays about 100 eggs.

The larvae are dumpy little caterpillars, quite short, with reduced abdominal prolegs, each bearing only two crochets (which distinguishes them from all other caterpillars living on grain); dirty white in colour; covered with fine setae; fully grown at about 8 mm. The larval life is spent within a single grain, and it tunnels to the surface (leaving the epidermis intact) prior to pupation. In northern USA the larvae overwinters inside the grain.

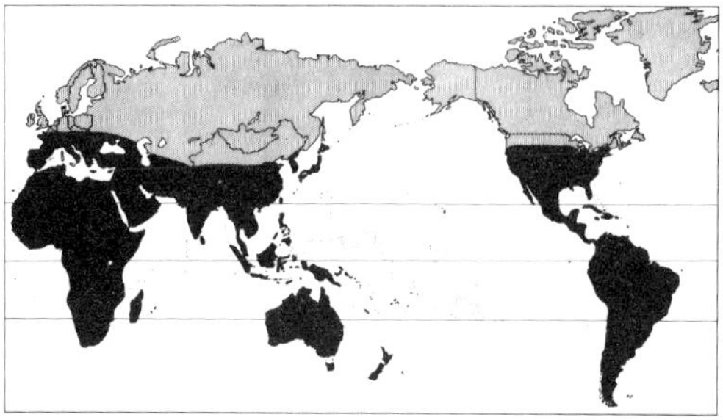

Pupation takes place within the grain under the circular 'window'; the brown pupa lies within a delicate spun cocoon.

The adult is a small, straw-coloured moth about 7 mm long (with wings folded) and wingspan of 15 mm. The wings are characteristically thin and long-fringed (Gelechiidae). The moths are short-lived but quite strong fliers, and females fly from the grain stores out to crops in the field.

The life-cycle takes about five weeks in warm locations where breeding may be more or less continuous.

Distribution. Now a completely cosmopolitan species found throughout the warmer parts of the world, and an occasional field pest in the summer in cooler temperate regions where it survives in heated grain stores. In the UK it seldom survives the winter out of doors or in unheated stores, but may persist in heated stores.

Control. Standard stored products preventative measures are effective against this pest (see pages 274 and 275).

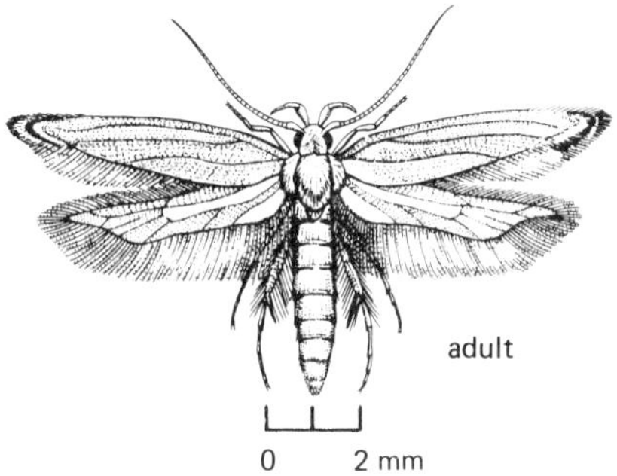

Zeuzera pyrina (L.)

Common name. Leopard Moth
Family. Cossidae
Hosts (main). Apple, pear, plum, cherry, walnut.
(alternative). Fig, olive, blackcurrant, willows, *Sorbus*, hawthorn, oak, maple, birch, ash, etc.
Damage. The larvae bore down the centre of branches and young stems of trees and woody shrubs. The young caterpillar bores into the current year's shoot near the tip; when this dies it enters older wood farther down the branch; usually a third entry into even older wood is made later. On old trees the final entry is made near the base of a branch; on young trees it enters the stem. Distal foliage of attacked trees dies and turns brown; branches may break.
Pest status. This polyphagous pest is very widespread, and occasionally serious on many different woody crops.
Life history. The orange eggs are laid singly in crevices in the bark; on young plants usually only one egg is laid. Each female lays 100–300 eggs. Hatching takes about two weeks.

The caterpillar is yellowish-white, head brown, and body with many black spots; fully grown it measures 4–5 cm. Larval development takes 2–3 years, in a series of three different tunnels in wood usually 4–6 cm in diameter (wood above 10 cm diameter almost never attacked).

Pupation takes place just under the bark, and the emerging young adult leaves the pupal exuvium protruding from the hole. Pupation takes several weeks in the late spring, and adults emerge in June and July (usually of the third year).

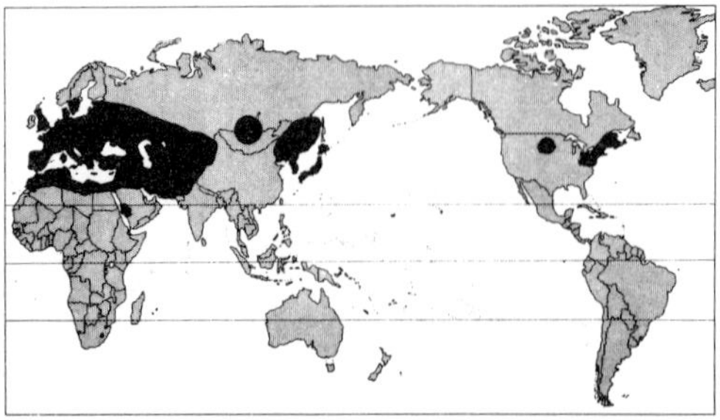

The adult is a handsome white moth with black markings; male wingspan is about 4 cm and in the female 5–6 cm. The male has bipectinate antennae. Adults live for several weeks.
Distribution. *Zeuzera pyrina* is recorded throughout Europe, the Mediterranean, much of Asia including Korea and Japan, and the USA (CIE map no. A.314).

Other pest species of Cossidae of some importance include (all essentially polyphagous):
Zeuzera coffeae Nietn. – (Red Coffee Borer) India, China, S.E. Asia (CIE map no. A.313).
Zeuzera leuconotum Butler – (Oriental Leopard Moth) Japan.
Cossus cossus (L.) – (Goat Moth) Europe, Asia, N. Africa.
Cossus japonica Gaede – (Oriental Goat Moth) Japan.
Cossula magnifica (Str.) – (Pecan Carpenterworm) USA.
Prionoxystus spp. – (Carpenterworms) USA.
Control. *Cultural control*: When frass or tunnel holes indicates an infestation, the infested shoot should be pruned and destroyed. Older caterpillars may be killed in the tunnel by pushing a springy wire up the hole.
Chemical control: A vapourizing insecticide can be injected into the tunnel, which is then sealed with clay or putty; HCH, paradichlorobenzene or carbon disulphide are all effective.

It appears that young larvae are most numerous only in very hot summers.

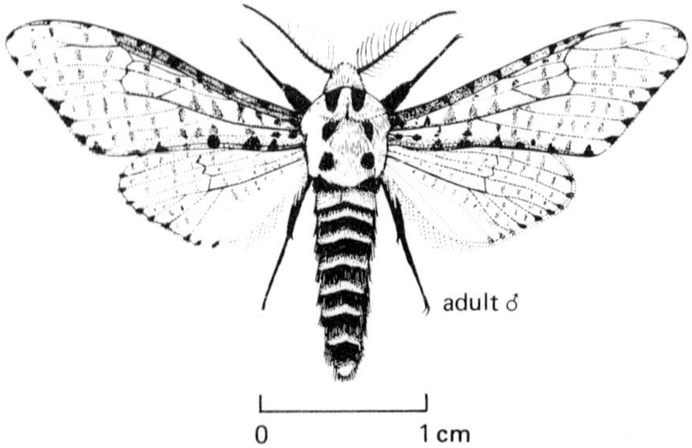

Acleris comariana (Zeller)

Common name. Strawberry Tortrix (Moth)
Family. Tortricidae
Hosts (main). Strawberry

(alternative). Wild strawberry, *Geum*, *Potentilla*; in Europe also on *Rhododendron* and *Azalea*.

Damage. The caterpillars feed inside folded leaves or spun shoots where they damage the flowers and prevent fruit-set.

Pest status. A regular pest on strawberry crops, sometimes causing serious damage; in practice several other tortricids are also found feeding on strawberry, and flower damage may be extensive; some of these overwinter as larvae.

Life history. Eggs are laid low on the plant, usually on stipules and petioles; eggs laid in the autumn overwinter on the plants and hatch in April/May.

The caterpillar is greenish with a darker back, very active and wriggles frantically when disturbed (as do most tortricid larvae); about 6 mm long.

Pupation takes place within the webbed leaves in June and July, inside a silken cocoon; pupation takes 2–3 weeks usually.

Adults emerge in late June and July (first generation); the second generation in September and October. The second-generation eggs overwinter on the strawberry plants. The adults are small moths of wingspan 13–18 mm; they are polymorphic, with eight different forms recognized (seven in UK), but no obvious sexual dimorphism; coloration is basically brown, but the extent of wing markings varies with the different forms. Adults fly at dusk.

This species is bivoltine.

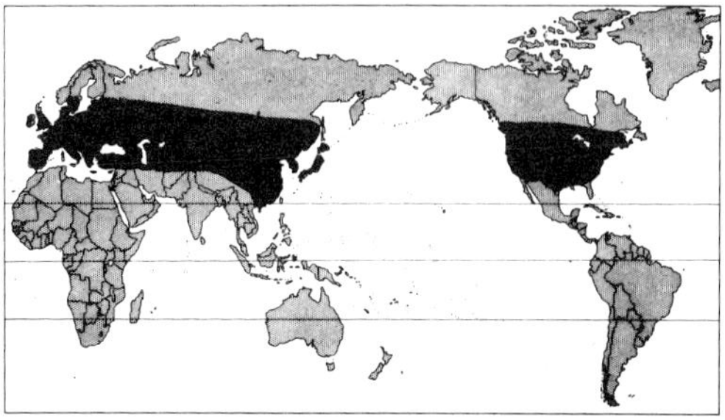

Distribution. Recorded throughout Europe, Russia, China, Japan, Canada and the USA.

Control. If control is required, the recommendation is a spray before flowering as soon as damage is observed; a second spray may be needed in the autumn, after picking, against second-generation caterpillars (for effective insecticides see page 397). If a spray is needed during flowering then a late evening application of chlorpyrifos, fenitrothion or cypermethrin is suggested, to minimize risk to pollinating insects.

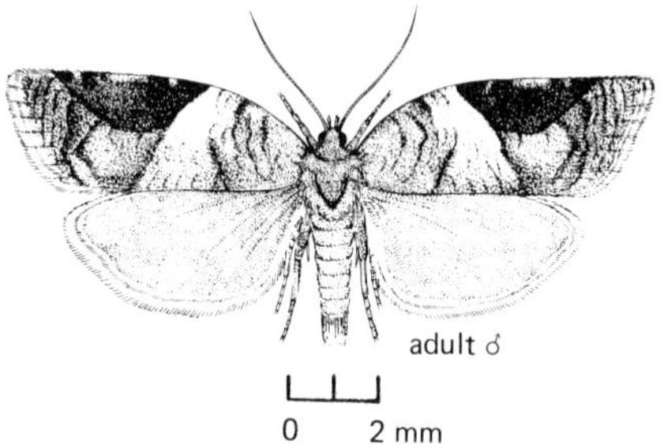

Archips podana (Scopoli)

Common name. Fruit Tree Tortrix
Family. Tortricidae
Hosts (main). Apple, and other fruit trees, walnut, blackberry, raspberry, hop.

(alternative). Polyphagous on many deciduous trees, shrubs, and occasionally conifers.

Damage. Large caterpillars feed on the leaves, and also eat deep cavities in apple fruitlets (other caterpillars cause similar fruit damage). Some caterpillars are carried into fruit stores where damage continues.

Pest status. A regular pest of many fruit and nut crops, in many regions, and damage to apples is often quite serious.

Life history. Young caterpillars overwinter, hibernating in a silken cocoon under a bud scale or a piece of dead leaf firmly stuck to the branch. This species develops slowly and adults do not usually emerge until mid-June (to mid-August). Eggs are laid in scale-like clusters, green in colour, on the leaves, in batches of 50–100, usually on the upper surface.

On hatching the larvae soon disperse and settle singly under a silk web at the side of the midrib on the underside of leaves where they feed. Sometimes they web adjacent leaves together (see figure page 397). After moulting once or twice some caterpillars stop feeding and hibernate; others continue feeding on foliage; others move to maturing fruit where they feed on the fruit surface, usually making a protective shelter by webbing a leaf to the fruit. Pupation takes place under the silken web in a cocoon. Second-generation adults emerge in September and the larvae that hatch from their eggs feed for a while, moulting once or twice, and then hibernate overwinter.

The adults are small reddish brown moths, sexually dimorphic; male wingspan 19–23 mm, female wingspan 20–28 mm; some variation in coloration.

One main generation annually, and a partial second.

Distribution. This species is recorded throughout Europe (north to Lapland), Asia Minor, most of northern Asia to Japan; it is thought to be introduced into N. America now.

In the USA and Canada the closely related *Archips argyrospila* (Wlk.) (Fruit Tree Leaf Roller) is a serious fruit orchard pest.

Control. In many orchards it is customary to combine sprays against Codling Moth with those against Fruit Tree Tortrix; but the latter usually emerge some 7–10 days later, and the long emergence period necessitates two sprays usually, the second one 1–2 weeks after the first (the first in late June). It is generally thought that the summer sprays give better control results than do pre-blossom applications in the spring.

Recommended chemicals are listed on page 397.

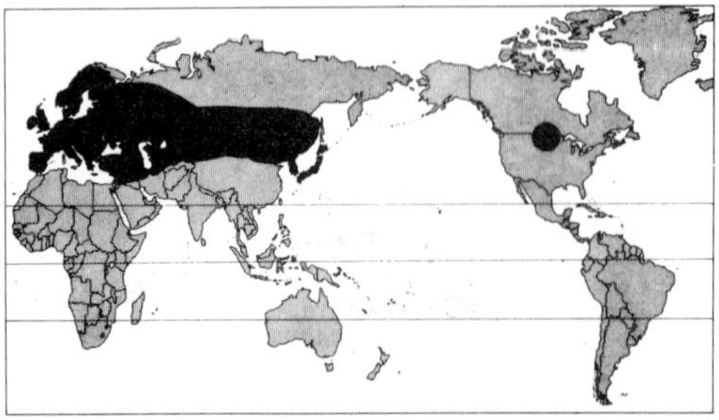

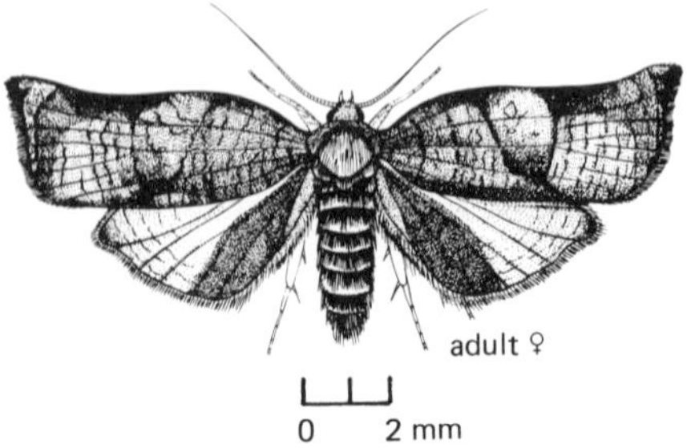

Fruit tree tortricids (and others) (Lepidoptera; Tortricoidea)

The present taxonomic arrangement accepted in the UK is that the superfamily Tortricoidea contains two families: the Cochylidae (= Phaloniidae), a small group, and the very large family Tortricidae. The Tortricidae now consists of two subfamilies, the Olethreutinae (containing what were formerly regarded as 11 different subfamilies) and the Tortricinae (containing three former subfamilies). In the literature there are sometimes as many as 6–8 different families/subfamilies listed for the Tortricoidea, so in some texts the taxa and the names used for this group are quite confusing, especially as the generic synonymy is extensive. The group is large and worldwide, with more than 4000 species, but more abundant in temperate regions. The adults are small dark-coloured moths with broad wings and usually crepuscular habits; most species respond to light traps and many to pheromones and sex attractants, which does enable their emergence to be monitored.

The eggs are typically small, oval and flattened, sometimes laid singly and sometimes in small groups, usually cryptically coloured, and sometimes covered with bristles or debris. The larvae are small and often pale-coloured rather slender little caterpillars (referred to in the USA as 'budworms') that usually live concealed inside folded or rolled leaves, or in shoots webbed with silk, though some bore into fruits and some bore into the stems and rootstocks of herbaceous plants. When exposed the caterpillars typically wriggle violently, often moving swiftly backwards, and fall off the plant hanging by a silken thread. This escape mechanism results in some caterpillars being accidentally transported from crop to crop (or glasshouse to glasshouse) on the clothing of workers. Some species are actually transported as first-instar larvae on their silken threads (as with young spiders) by air currents and winds. Larval coloration tends to vary considerably and in some species has been clearly demonstrated to vary according to the precise host plant inhabited and eaten.

Many species are pests of fruit trees (including *Rubus*, *Vaccinium* and *Fragaria*) and these are collectively referred to as the 'fruit tree tortricids'. Many others are damaging to forest trees (oak, larches, pines, etc.) and some are very serious pests in northern forests. Some are pests of ornamentals, especially various flowers and *Rosa*, and some are important pests of both tropical and temperate legume crops (pea moths, etc.). A few species damage some herbaceous annual crops. Ecologically they are basically woodland species, and it is not surprising to find that many species feed on hazel and *Lonicera* in Europe and N. America. Some species are quite polyphagous in diet but others are confined to a single genus of host plant. On some crops the total number of species recorded is large, and so field identification may be difficult sometimes; for example the number of tortricids recorded from some popular hosts are as follows (approximately): apple, 25; *Prunus*, 25; hazel, 15; *Vaccinium*, 20; strawberry, 10.

Important pest species of Tortricoidea

Family Cochylidae (= Phaloniidae)

Aethes dilucidana (Stephens) – parsnip flowerheads; Europe.

Aethes francillana (F.) – carrot flowerhead and stem; Europe.

Cochylis hospes Wal. – (Banded Sunflower Moth) USA.

Eupoecilia ambiguella – grapevine, *Prunus*, *Ribes*; Europe.

Lorita abnormana Busck – (Chrysanthemum Flower Borer) USA.

Family Tortricidae

Acleris comariana (L. & Z.) – (Strawberry Tortrix) Holarctic (see page 393).

Acleris cristana (D. & S.) – (a fruit tree tortrix) Europe, Asia, Japan.

Acleris rhombana (D. & S.) – (a fruit tree tortrix) Europe, Asia Minor, N. America.

Acleris variegana (D. & S.) – (a fruit tree tortrix) Europe, Asia, N. America.

Adoxophyes orana (F.von R.) – (Summer Fruit Tortrix) Europe, Asia, Japan.
Ancylis comptana (Froe.) – (Strawberry Leaf Roller) USA.
Archips argyrospila (Wlk.) – (Fruit Tree Leaf Roller) USA, Canada.
Archips crataegana (Hübner) – (a fruit tree tortrix) Europe, Asia, Japan.
Archips podana (Scopoli) – (Fruit Tree Tortrix) Europe, Asia (see page 394).
Archips rosana (L.) – (Rose Tortrix) polyphagous; Europe, Asia, N. America.
Archips spp. – (Apple Leaf Rollers) Japan.
Argyrotaenia citrana (Fern.) – (Orange Tortrix) USA.
Argyrotaenia juglandana (Fern.) – (Hickory Leaf Roller) USA.
Argyrotaenia pulchellana (Haw.) – (Polyphagous Leaf Roller) Europe, Asia Minor, Canada, USA.
Cacoecimorpha pronubana (Hübner) – (Carnation Leaf Roller) Europe, Asia, N. America (see page 398).
Choristoneura diversana (Hübner) – (Plum Tortrix) Europe, Asia, Japan.
Choristoneura hebenstreitella (Muller) – polyphagous; Europe, Asia Minor, Japan.
Clepsis spectrana (Treit.) – (Cyclamen Tortrix) N. and C. Europe, S.E. Russia.
Cnephasia interjectana (Haw.) – (Flax Tortrix) Europe, Asia, Canada.
Cnephasia longana (Haw.) – (Omnivorous Leaf Tier) Europe, USA (see page 399).
Croesia holmiana (L.) – (a fruit tree tortrix) Europe, Asia Minor.
Cryptophlebia leucotreta (Meyr.) – (False Codling Moth) Africa (CIE map no. A.352).
Cryptophlebia ombrodelta (Lower) – (Macadamia Nut Borer) Australia, USA, Africa, China, India (CIE map no. A.353).
Cydia funebrana (Treit.) – (Plum Fruit Maggot) Europe, Asia.
Cydia leucostoma (Meyr.) – (Tea Flushworm) India.
Cydia molesta (Busck) – (Oriental Fruit Moth) S. Europe, China, Japan, USA, S. America (see page 400).
Cydia nigricana (F.) – (Pea Moth) Europe, Japan, Canada, USA (see page 401).
Cydia pomonella (L.) – (Codling Moth) cosmopolitan (see page 402).
Cydia prunivora (Walsh) – (Lesser Appleworm) USA, Canada (CIE map no. A.421).
Cydia ptychora Meyr. – (African Pea Moth) most of Africa.
Cydia pyrivora (Dan.) – (Pear Tortrix) E. Europe, W. Asia (CIE map no. A.422).
Ditula angustiorana (Haw.) – (a fruit tree tortrix) Europe, Asia Minor, Canada, USA.
Epiphyas postvittana (Wlk.) – (Light Brown Apple Moth) Europe, Australasia, Hawaii, USA.
Gretchena bolliana (Sling). – (Pecan Bud Moth) USA.
Hedya nubiferana (Haw.) – (a fruit tree tortrix) Europe.
Homona coffearia (Nietn.) – (Tea (Coffee) Tortrix) India, China, Japan, S.E. Asia, Australasia.
Laspeyresia caryana (Fitch) – (Hickory Shuckworm) USA.
Laspeyresia glycinivorella Mats. – (Soybean Pod Borer) S.E. Asia, Japan.
Lobesia botrana (Schiff.) – (European Grape Berry Moth) Europe (not UK), Japan, E. Africa.
Lozotaenia forsterana (F.) – on strawberry, *Ribes*, etc.; Europe to S.E. Siberia.
Melissopus latiferreanus (Wals.) – (Filbertworm) USA.
Merophyas divulsana (Wlk.) – (Alfalfa Leaf Tier) Australia.
Olethreutes permundana (Clemens) – (Raspberry Leaf Roller) USA.
Olethreutes spp. – (fruit tree tortricids) Europe.
Pammene rhediella (Clerck) – (Fruitlet Mining Tortrix) Europe.
Pandemis cerasana (Hübner) – (Barred Fruit Tree Tortrix) Europe.
Pandemis corylana (F.) – polyphagous; Europe, Asia, Japan.

Pandemis heparana (D. & S.) – (a fruit tree tortrix) Europe, Asia, Japan.
Ptycholoma lecheana (L.) – (a fruit tree tortrix) Europe, Asia Minor.
Rhyacionia spp. – (Pine Shoot Moths) Europe, Asia, Canada, USA.
Spilonota ocellana (D. & S.) – (Eye-spotted Bud Moth) Holarctic (see page 403).
Suleima helianthana (Riley) – (Sunflower Bud Moth) USA.
Tetramoera schistaceana Snellen – (Grey Sugarcane Shoot Borer) Japan, China, S.E. Asia, Indonesia.
Tortrix dinota (Meyr.) – (Brown Tortrix) Africa.
Tortrix viridiana (L.) – (Green Tortrix) oak, blueberry, etc.; Europe, Asia Minor.

Control of fruit tree tortricids

For several of the more important species of pests there are now sex pheromones commercially available, and it has become standard practice to use ultra-violet light traps, sticky traps (and others) baited with sex attractants to monitor the emergence of the moths so as to permit precise timing of insecticide sprays. For a few species an economic threshold trap catch has been worked out, and for these species a lesser catch means that spraying is not recommended.

There are so many different species collectively referred to as 'fruit tree tortricids' with differing biologies and habits that it is somewhat difficult to generalize about details of control measures, so it is important to seek local advice when planning a pest control programme.

The insecticides currently regarded as 'effective' against a wide range of Tortricidae include the following: azinphos-methyl plus demeton-*S*-methyl sulphone, carbaryl, chlorpyrifos, demephion, dichlorvos, dimethoate, fenitrothion, formothion, heptenophos, malathion, mevinphos, oxydemeton-methyl, permethrin, phosalone, phosphamidon, thiometon, triazophos and vamidothion. Clearly some chemicals possess different qualities, and some are also effective against other pests; so the final choice of sprays depends upon several factors, one major factor being the nature of the pest complex locally. Some sprays are planned to kill the emerging adult moths, but most are aimed at the first-instar larvae hatching from the eggs. With the fruit-boring caterpillars spray timing is critical if the larvae are to be killed before they penetrate the fruits. Similarly with the legume pod borers the newly hatched larvae must be killed before they can bore into the pods.

In most fruit orchards it is not possible to predict the level of attack likely in any one year, so often the decision to apply control measures is based empirically on the previous history.

Apple leaves webbed and eaten

Pear leaf folded and cut

Rolled litchi leaves

Cacoecimorpha pronubana (Hb.)

Common name. Carnation Leaf Roller (Tortrix)
Family. Tortricidae
Hosts (main). Carnation, and many garden flowers.

(alternative). Polyphagous on tomato, bay, field beans, strawberry, privet, sea buckthorn, etc.

Damage. The caterpillars spin silk over a group of terminal leaves, binding them together, and feed on the webbed leaves; they also tunnel flower buds.

Pest status. A widespread and abundant species; occasionally serious as a pest, particularly on flower crops in glasshouses. Small larvae may be carried on silken threads from one glasshouse to another on clothing.

Life history. The flat green eggs are laid in small batches on the upper side of leaves; apparently up to 200 eggs have been recorded in a single batch; hatching requires 12–24 days. In UK glasshouses the two main egg-laying periods are May/June, and August/October.

The caterpillars vary in colour somewhat, according to the host plant (which may be true for most Tortricidae), but are generally a shade of yellowish-green with a blackish head capsule. Full-grown larvae are about 19 mm long; larval development takes about seven weeks in the summer, but larvae of the second generation overwinter.

Pupation takes place within a silken cocoon in the webbed foliage or in a hollowed flower bud, from April to June and again in August to October.

The adults are small brownish moths (wingspan in the male 14–18 mm and female 16–24 mm) with bright orange hindwings (characteristic); only slight sexual dimorphism; they fly during the day in sunshine.

In the UK this species is bivoltine, the generations overlapping.

Distribution. First recorded from the UK in 1905; now widespread throughout Europe, Asia Minor, N. and S. Africa, Canada and the USA.

Control. In glasshouses the hand-picking of infested shoots is recommended. When insecticidal sprays are used it is best to aim at killing the adult moths and newly hatched caterpillars; larger larvae inside webbed foliage are particularly difficult to kill. For recommended chemicals see page 397.

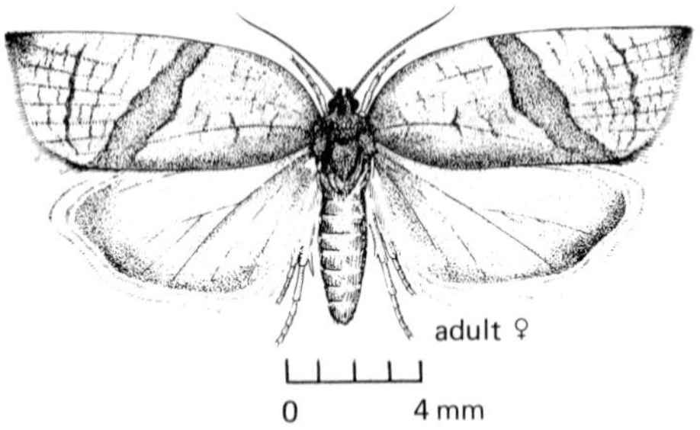

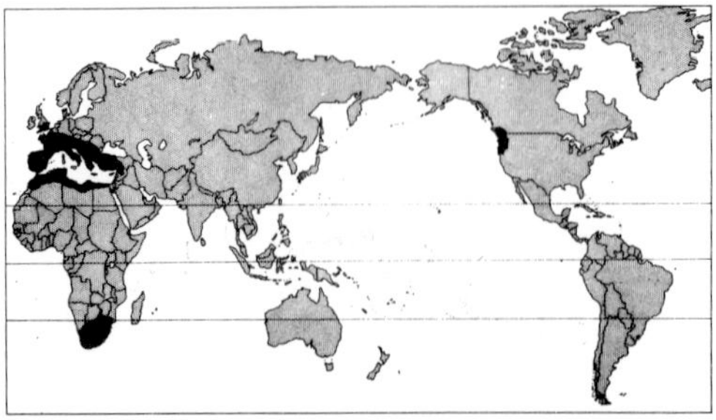

Cnephasia longana (Haw.)

Common name. Omnivorous Leaf Tier
Family. Tortricidae
Hosts (main). Chrysanthemum, flax, strawberry, apple, hop.

(alternative). Field beans, clovers, and many other plants.

Damage. Caterpillars web together the flower florets and feed within; they also bore into the fruit of strawberry; and infest terminal shoots of some plants.

Pest status. A polyphagous and widespread, but usually minor, pest on the crops it attacks; but together with the closely related polyphagous *C. interjectana* they are sometimes of economic importance. A more serious pest in N. America.

Life history. Eggs are laid in small batches on the food plant, sometimes singly; up to 200 eggs per female being laid, mostly in July; hatching requires 11–20 days.

The caterpillars are greenish-grey with faint longitudinal lines. In the UK it is recorded that immediately on hatching (in August) the young larvae immediately spin a shelter and enter hibernation near the egg site. They start feeding in the spring and are most frequently found in May and June on crop plants. First-instar larvae are usually leaf miners, but the later instars live in webbed foliage or flowers; the type of damage done does depend to some extent upon the host plant. In N. America oviposition often occurs in fence posts, tree trunks, and the like, and the first instars are wind-dispersed on silken threads from higher locations on to low-growing vegetation.

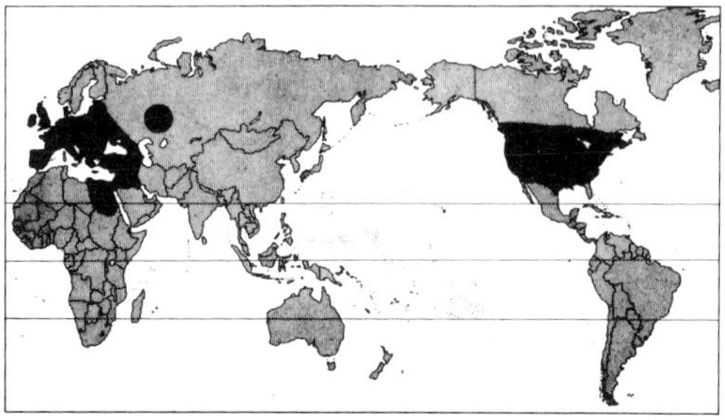

Pupation takes place inside a flimsy cocoon in surface debris; adults emerge in July.

The adults are small moths; wingspan 18–22 mm; sexually dimorphic, with male forewing unicolorous pale yellow-brown, and female forewing brownish with a distinct pattern of darker markings; in some specimens the ground colour may be much darker.

Distribution. This species is distributed throughout Europe to Asia Minor, northeast Africa, Canada and USA.

The closely related *Cnephasia interjectana* (Haw.) (Flax Tortrix) is a polyphagous pest throughout Europe, Asia and now established in Canada (more than 130 food plants have been recorded for this species).

Control. When control is needed, sprays of the insecticides listed on page 397 should prove effective.

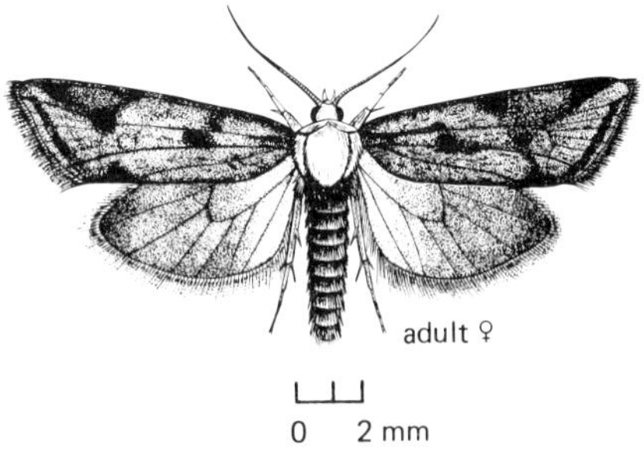

Cydia molesta (L.)

Common name. Oriental Fruit Moth
Family. Tortricidae
Hosts (main). Peach, and other stone fruits.
(alternative). Many other fruit trees, including apple.
Damage. The caterpillars damage both twigs and fruits. Early in the season the larvae bore into the tips of soft young shoots and cause them to wilt and die. Later, when the shoots have hardened and the fruits ripen, most of the caterpillars bore into the fruits. Shoot destruction causes bushy secondary growth.
Pest status. A serious pest of peaches and other stone fruits in various countries; of lesser importance on many other deciduous fruit trees.
Life history. Eggs are laid on the leaves and twigs (and later on fruits) just about flowering time.

The first caterpillars bore into soft green shoots and feed internally, but later generations tunnel into ripening fruits (like Codling Moth). Full-grown larvae are white with a pink tinge, 8–10 mm long, with a characteristic black, dorsal comb-like structure on the last abdominal segment bearing five teeth. Larval development can be completed in only two weeks; they leave the tunnels in order to pupate; those leaving fruits emerge from a sizeable hole that later exudes sap. They pupate inside silken cocoons stuck onto a solid substrate (the tree or ground). Pupation takes about ten days. The final-generation larvae overwinter in the pupal cocoons, and pupate in the spring.

The adults are small dark moths with distinctive costal strigulae; wingspan is about 12 mm. First-generation adults emerge at peach flowering time.

In southern USA the life-cycle is completed in about one month, and there are typically 4–5 generations annually.
Distribution. Found in southern Europe, China and Japan; now in S. Australia, S. America, Mauritius, the USA and Canada (CIE map no. A.8). It was introduced into the USA from Japan in 1913.
Control. In parts of N. America a high level of control is achieved using natural parasites, *Trichogramma minutum* on the eggs and several braconids on the larvae; natural populations are usually supplemented by reared releases.

When insecticidal sprays are needed the chemicals mentioned on page 397 are generally effective, but it is usual to apply 3–5 sprays through the season.

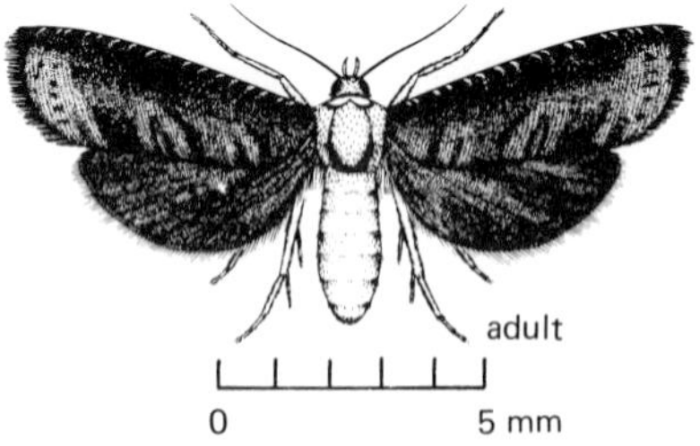

adult

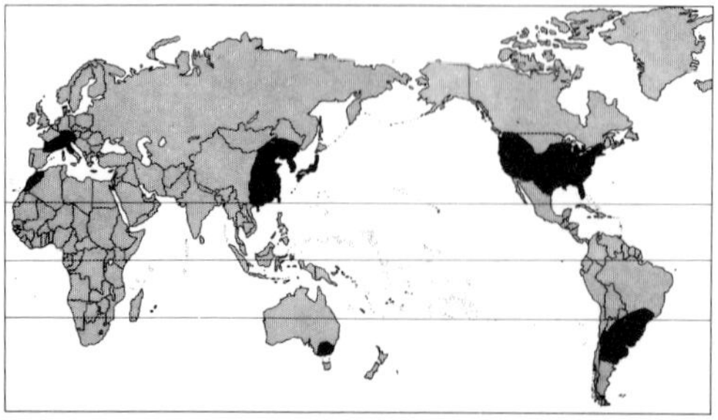

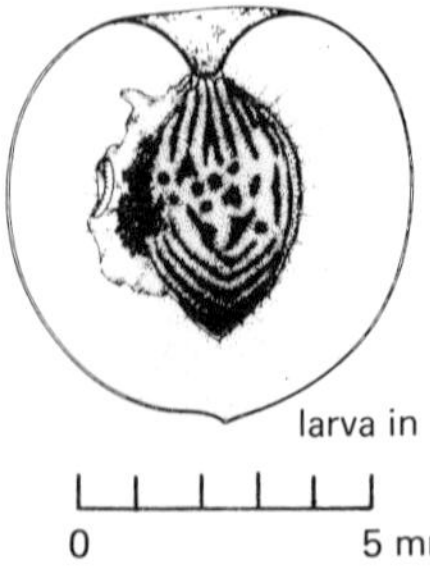

larva in fruit

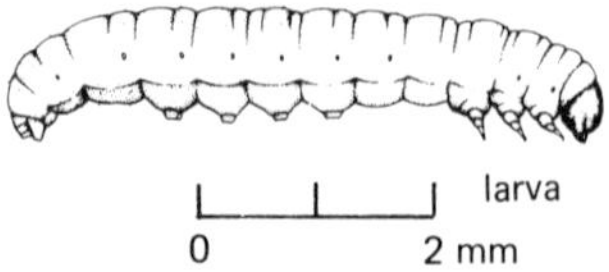

larva

Cydia nigricana (F.)

(= *Laspeyresia nigricana* F.)
(= *C. rusticella*)

Common name. Pea Moth
Family. Tortricidae
Hosts (main). Peas
(alternative). Some legumes, both cultivated and wild (not broad bean).
Damage. The larvae enter the pod and eat the developing seeds; sometimes several larvae develop in a single pod. Peas for freezing suffer greatest economic loss, but dried peas suffer greatest damage.
Pest status. A serious pest of peas, both in garden and field crops; damage may be very extensive.
Life history. Small flattened eggs are laid singly or in small groups, on the leaves and stipules, over the period mid-June to mid-August; hatching takes 7–8 days in Europe, but is recorded as 2–3 days in Canada.

The young caterpillars wander over the foliage for a while, then enter a young pod where they remain feeding upon the peas. The full-grown larva (fifth instar) is pale yellow, with a brown head and faint spotting, and 6–8 mm long; it leaves the pea pod to rest in the soil inside a tough silken cocoon. The larva overwinters in the cocoon and finally pupates in the spring or early summer, after having first left the overwintering cocoon and moved nearer to the soil surface to make a second cocoon in which actual pupation takes place.

Adults emerge in the UK about mid-June (earlier in a warm spring) and fly from then to about mid-August. There is only one generation per year. The adult is a small dark moth, with wingspan 12–15 mm and conspicuous costal strigulae along the distal wing margin.
Distribution. A widespread pest species throughout Europe, Japan, and since 1893 established in N. America (CIE map no. A.421).
Cydia ptychora Meyr. is the African Pea Moth, found throughout Africa.
Laspeyresia glycinivorella Mats. is the Soybean Pod Borer, found throughout S.E. Asia, China and Japan.
Control. Chemical control is often required, especially for crops grown for freezing or processing as only slight levels of damage can be tolerated in these crops. Spray timing is determined by the flight activity of the adults, nowadays usually monitored by pheromone sticky traps; two sprays are often required (see page 397). The economic threshold in the UK is ten or more moths per pheromone trap on two consecutive two-day periods.

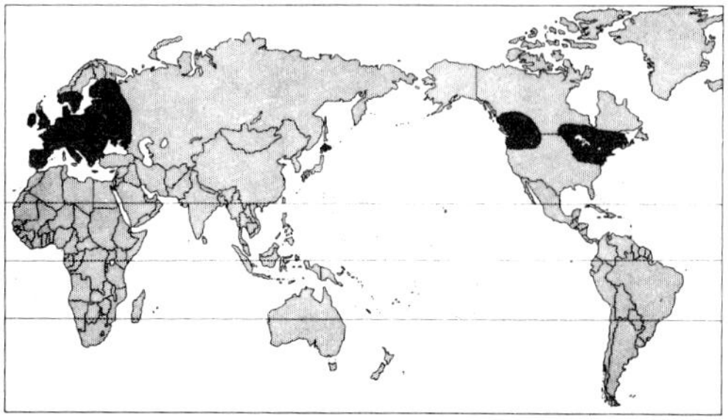

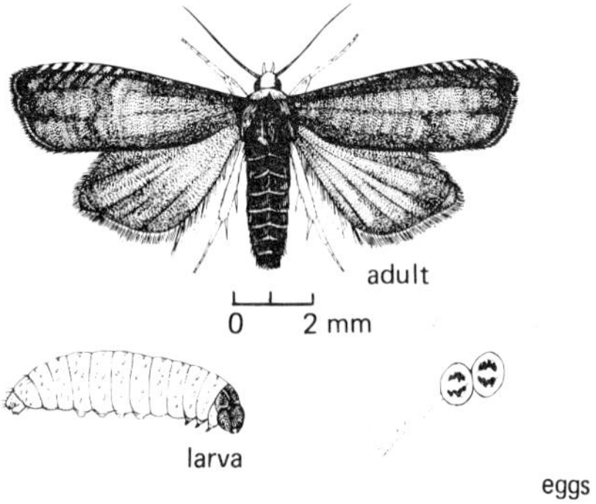

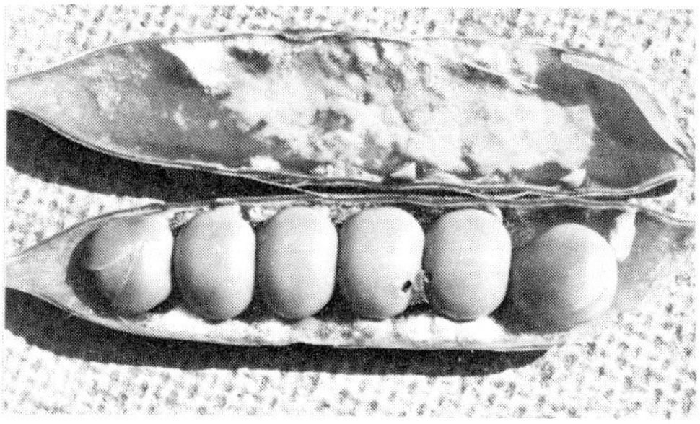
Infested pea pod

Cydia pomonella (L.)

Common name. Codling Moth
Family. Tortricidae
Hosts (main). Apple, pear
(alternative). Peach, quince, walnut, *Prunus* spp.

Damage. The feeding caterpillars tunnel inside ripening fruits, making a small entrance hole and a larger exit. The entrance hole has a surrounding dark red discoloration. Most attacked fruits fall prematurely.

Pest status. Possibly the single most important pest of apples worldwide; often most serious in gardens. Largest populations often unpredictable, but sometimes correlated with particularly warm and sunny conditions (optimum conditions recorded as 21–3 °C).

Life history. Eggs are laid singly, on foliage or fruits; they are flat, circular and translucent (1 mm diameter), and difficult to find; each female lays about 40–50 eggs, over a period from mid-June to early August in the UK. Hatching requires about 10–14 days usually.

Young larvae immediately search for a suitable entry point into the fruit, usually at the 'eye' or the side of the fruit. They bore into the flesh and feed, often finally tunnelling the core. The caterpillars are white when young but in the last instar turn pink (with a dark head) when they are about 12 mm long. Some larvae damage a second apple before ceasing to feed; larval development takes some 21–35 days. Full-grown larvae leave the fruit and spin cocoons under loose bark, or in crevices, and here they overwinter, pupating the following spring when pupation takes 14–28 days. In the UK there is usually just one generation annually, but in N. America there are two; in both areas in a warm season there may be a small additional generation in the autumn.

The adult is a small moth, 6–8 mm long, with wingspan 12–14 mm, with dark greyish wings having a copper-coloured spot distally; hindwings pale grey.

Distribution. Recorded from virtually all the apple-growing regions of the world; throughout Europe, Asia, Australia, New Zealand, S. Africa, Canada, USA and S. America (CIE map no. A.9).

Control. Where there is a recent history of damage it is usually necessary to use insecticides, but in many areas there is a tortrix complex to control, so choice of insecticide and precise timing may be governed by the nature of the local pest complex (see page 397). Codling Moth pheromone/ sticky traps are commercially available in most countries now.

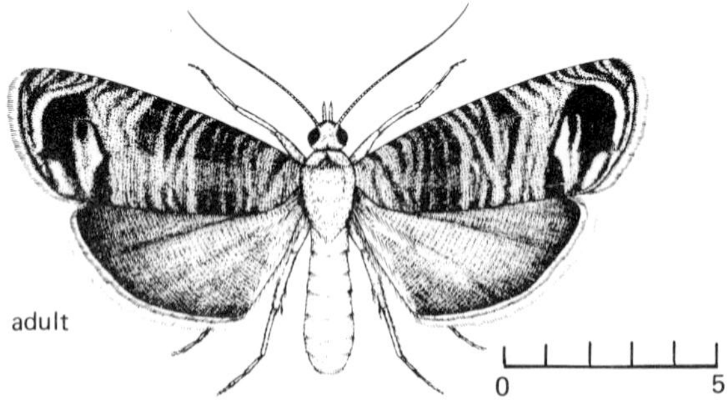

adult

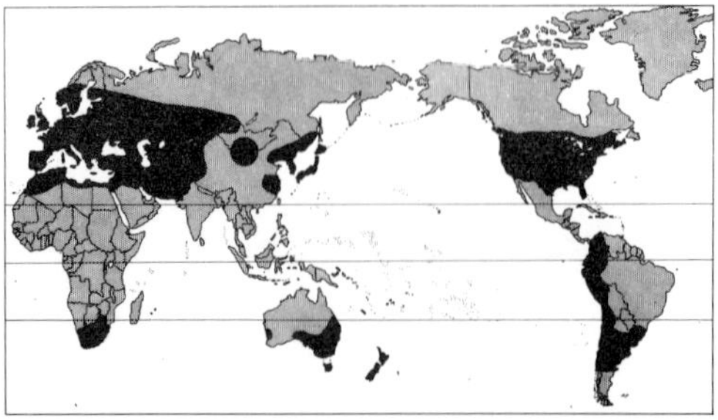

Infested fruit

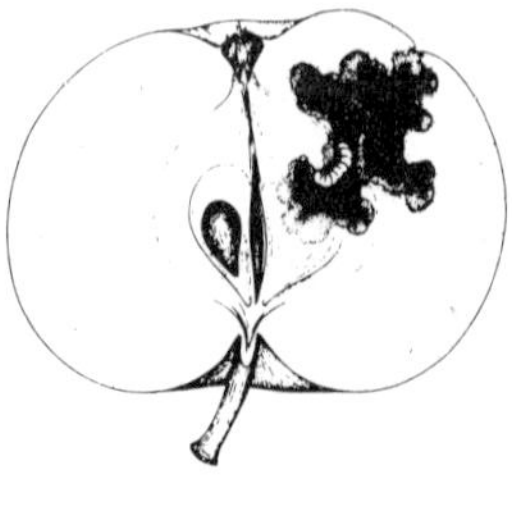

Larva inside fruit

Spilonota ocellana (D. & S.)

Common name. Eye-spotted Bud Moth
Family. Tortricidae
Hosts (main). Apple, blackberry.
(alternative). Pear, plum, cherry, quince, damson and raspberry.
Damage. Caterpillars bore into buds which wither and do not open; at flowering time they leave the buds and feed on foliage, webbing several leaves together. Some surface damage to fruit may be done in August and September.
Pest status. Not often a serious pest, but in parts of the USA and Europe damage of significance is done. Growth of young trees may be seriously affected by destruction of terminal buds.
Life history. Eggs are laid on the leaves, in small groups (1–4 usually). After hatching the young larvae feed on the underside of the leaves, and then construct a small open-ended silken tube in which they live individually.

A few larvae feed on fruits in August and September, but most prefer to remain on the leaves; they leave their shelter to feed and then return. In the autumn the larvae hibernate within a silken cocoon stuck to the bark of branches, although a few might already penetrate into dormant buds. In the spring, as growth commences, the larvae leave the cocoons and penetrate developing buds which are killed as feeding progresses. Large larvae are dark reddish-brown with a black head, and about 12 mm long. After destroying the buds the larvae leave them at flowering time and web blossom trusses and leaves; inside here they feed, and then in early summer they pupate. Adults start emerging in mid-June, and the univoltine life-cycle is completed. The adults are small moths, wingspan 10–12 mm, dark-coloured but with very conspicuous white patches on the forewings and small eye-spots.

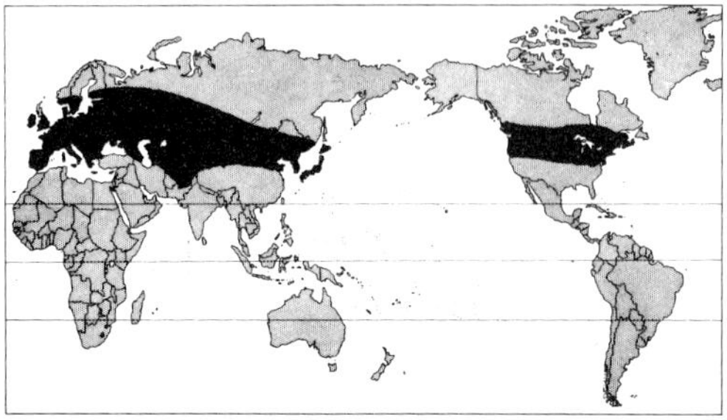

Distribution. A Holarctic species found throughout Europe, Asia and N. America (CIE map no. A.415).
Control. In the USA populations fluctuate considerably, and irregularly, in response to a high level of natural parasitism, together with susceptibility to low winter temperatures, especially frosts.

If control is needed then sprays of the chemicals listed on page 397 are usually recommended.

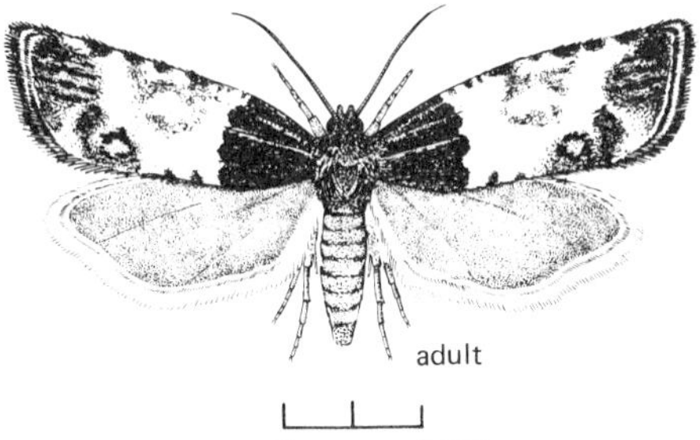

Chrysoteuchia culmella (L.) & spp. **Crambus** spp.

Common name. Grass Moths (Sod Webworms – USA, Canada)

Family. Crambidae (or Pyralidae)

Hosts (main). Grasslands, grass seed crops, quality turf.

(alternative). Maize, tobacco, etc., in recently ploughed grassland.

Damage. The larvae live in silken tubes on the soil surface and feed at the surface level or just below, destroying seedlings, and eating the leaves, stems and roots of many plants. In severe attacks whole patches of turf (or crops) are destroyed; light infestations are often unnoticed or thought to be drought effects.

Pest status. Several species are abundant in Europe but only occasionally is economic damage done; quite serious pests in Canada and the USA.

Life history. Eggs are either dropped at random on the soil, or laid by grass stems, usually in July and August; hatching requires about seven days.

The young larvae feed for a while and then hide in their silken tunnels to overwinter. In the spring they resume feeding and grow to a length of about 20 mm; they are stout-bodied, rather bristly, white in colour with dark spotting. In the USA it is reported that there may be 7–20 larval instars. Pupation takes place inside a silken cocoon within the webbing; pupation takes several weeks.

The adult is a smallish brown moth, of wingspan 20–5 mm, to be seen flying at dusk over grassland; several species are migratory and disperse at night (also attracted to lights).

Some species are univoltine, but others may have 2–3 generations per year.

Distribution. Many species are found throughout the Holarctic region; in Britain there are several species of both *Chrysoteuchia* and *Crambus*, but in the USA and Canada the 7–10 pest species are referred to as *Crambus*.

Control. Natural predation by birds is often heavy when populations are large, and parasitism levels are also usually high. But in N. America, where damage is more serious, it is often necessary to use insecticides; the chemicals previously used included aldrin, carbaryl, ethion, DDT, diazinon, heptachlor and toxaphene, and some of these are still recommended.

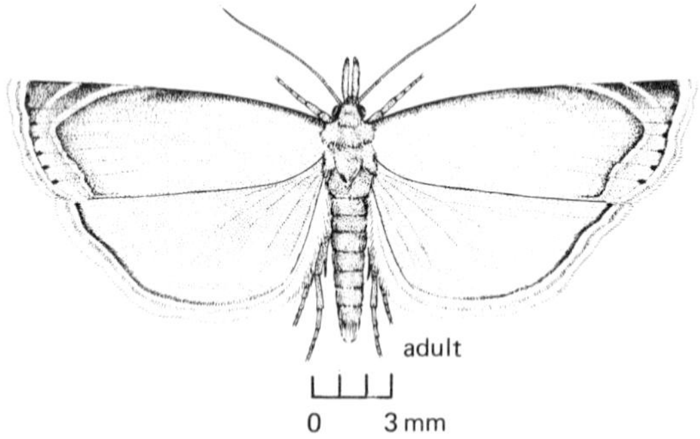

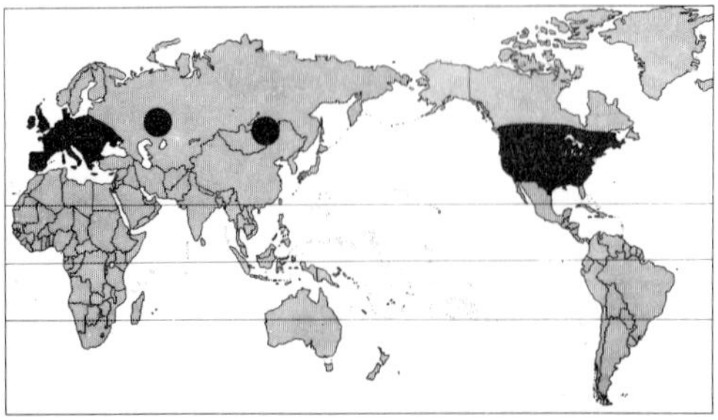

Ephestia cautella (Walker)

Common name. Almond Moth (Tropical Warehouse Moth; Dried Currant Moth; Date Moth)
Family. Pyralidae
Hosts. In the wild this and the other stored products Pyralidae are found on ripening fruits (including fallen and dried fruits) of date palm, and close relatives, and other fruits (cocoa pods etc.); but they are most serious as pests of stored products, both dried fruits and grains.
Damage. The larvae eat the fruit and the grains and produce a large amount of silk webbing over the feeding surfaces.
Pest status. A serious pest of stored products throughout the warmer parts of the world, and in heated stores in temperate regions; a wide range of stored produce is attacked, and damage may be very heavy.
Life history. Eggs are laid in the produce, or on the fruit bunches; each egg is globular, white turning orange; each female lays up to 250 eggs.

The larvae of *Ephestia* species are all very similar and can only be separated by detailed chaetotaxy; they are whitish-grey with many dark setae and small brown spots along the body; the head capsule is dark brown; full-grown size is about 12–15 mm. First-instar larvae generally feed on the seed germ, moving about freely in the produce. They pupate in crevices or where two surfaces touch (adjacent bags, fruits, etc.), in a silken cocoon.

The adults are small greyish moths, with rather indistinct markings on the forewings; body length about 7–8 mm and wingspan 14–18 mm; at rest the wings are folded along the abdomen. Adults live for about two weeks and fly quite strongly.

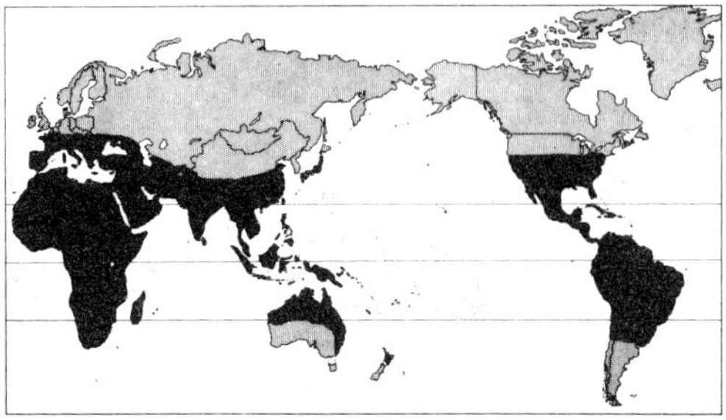

The life-cycle at 28 °C and 70% RH is 6–8 weeks; under warm conditions breeding may be continuous, and there may be many generations annually.
Distribution. A cosmopolitan pest species found throughout the warmer areas of the world and in heated stores in temperate regions.
Control. To kill the emerging moths before they can lay their eggs, sprays of pyrethroids are usually recommended in stores at serious risk, either as space sprays (to kill flying insects) or surface sprays. Dichlorvos slow-release strips are also successful for control of these moths.

General fumigation and the measures usually taken against stored products beetles are also usually effective against the stored products moths (see page 274).

Sex pheromones are now commercially available for the main species of *Ephestia* and can be used in stores both for monitoring moth populations and for trapping-out male moth populations if the stores are small.

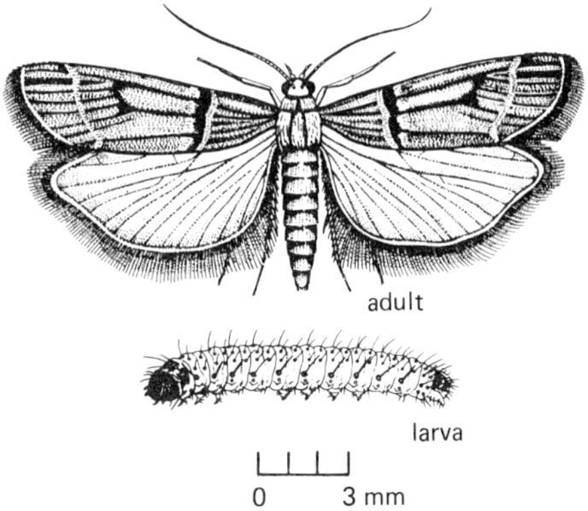

Ephestia elutella (Hübner)

Common name. Warehouse Moth (Cacao Moth; Stored Tobacco Moth)
Family. Pyralidae
Hosts (main). Stored cocoa beans, dried fruits, nuts and oil seeds.

(alternative). Stored grains, pulses, processed flours and foodstuffs, and tobacco.

Damage. Direct damage consists of the larvae eating the produce; this is another primary pest that also selectively eats the germinal part of the seeds; but indirect damage by contamination may often be of more economic consequence.

Pest status. A serious pest of stored produce throughout the subtropical and less cold temperate regions of the world; not abundant in the actual tropical regions where it is replaced by *E. cautella*.

Life history. Eggs are laid in crevices in the produce of sacks, and hatch after 10–14 days. The caterpillars burrow rapidly into the produce, and as they feed they spin silken webbing which binds the food together. They feed for several weeks, then when full grown (at about 12–14 mm) they usually have a small mass migration to seek favourable pupation sites, often leaving the produce to roam over the warehouse structure. In suitable crevices they spin silken cocoons in which they rest. In the UK many larvae overwinter, but some pupate immediately; in warmer regions they all pupate immediately, for a period of 1–3 weeks.

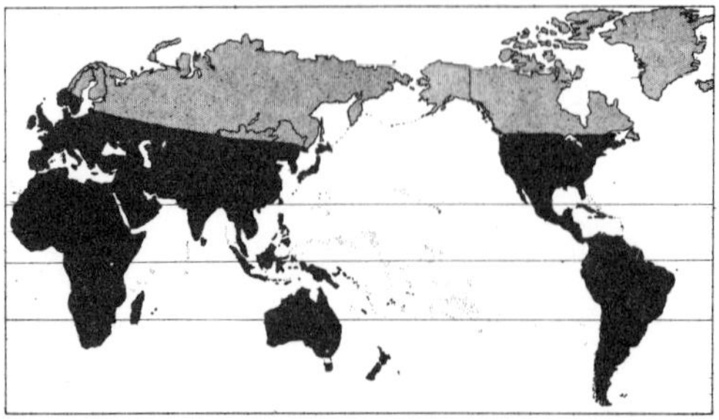

The adults are small brownish moths, about 5–9 mm long, wingspan 15–19 mm; the forewings have distinctive banding. They live for about two weeks, and the females lay most of their eggs (up to 500 recorded) in the first four days; they fly quite strongly and may start new infestations in buildings up to a kilometre away.

For these stored products pests the rate of development and final body size depends in part upon ambient conditions, and also on the type (quality) of food being eaten. Thus larval development may take from 20–120 days and final body size may vary from 8–15 mm long. Under optimum conditions the life-cycle can be completed in about 30 days (25° C and 75% RH); in the UK there is only one generation per year, but in more suitable climates breeding may be continuous.

Distribution. A cosmopolitan pest species, now dispersed completely throughout the world, but not abundant in the hotter tropical countries; it is basically a warm-temperate species.

Other species often found in association with this and the preceding species:

Ephestia calidella Guen. – (Date Moth) cosmopolitan in warmer regions.

Ephestia figulilella Greg. – (Raisin Moth) cosmopolitan in warmer regions.

Control. For control measures see pages 274 and 275.

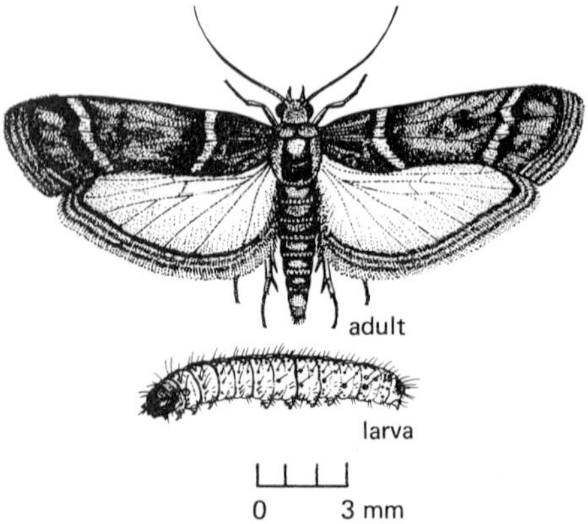

Etiella zinckenella (Treit.)

Common name. Pea Pod Borer (Lima Bean Pod Borer)
Family. Pyralidae
Hosts (main). Beans and pea.
(alternative). Pigeon pea, cowpea, and other pulses.
Damage. The larvae feed inside pods, eating the developing seeds. The partly grown caterpillar may leave the original pod and enter one or more fresh pods before ceasing to feed.
Pest status. A widespread and common pest of various legume crops, sometimes causing serious damage and economic losses.
Life history. The eggs are oval, shiny white, 0.6 × 0.3 mm; laid singly or in small groups (2–6) on young pods. Hatching requires 3–16 days, according to temperature. Each female lays 50–200 eggs.

The caterpillar is blue-green with a yellow head, and reaches a size of 12–17 mm long. On hatching the larva takes about 1.5 hours to actually penetrate the pod. Cannibalism often occurs if several caterpillars enter the same pod. If the pod is opened the caterpillar wriggles very violently. The larvae typically leave their frass inside the pod. Total larval development takes 3–5 weeks.

Pupation takes place in the soil, in a silken cocoon, at a depth of about 3 cm; development takes 2–4 weeks.

The adult is a small brown moth, about 24–27 mm wingspan; the dark forewing bears a characteristic small median mark and has a white leading edge. The adult female may lay her eggs throughout her 2–4-week life-span.

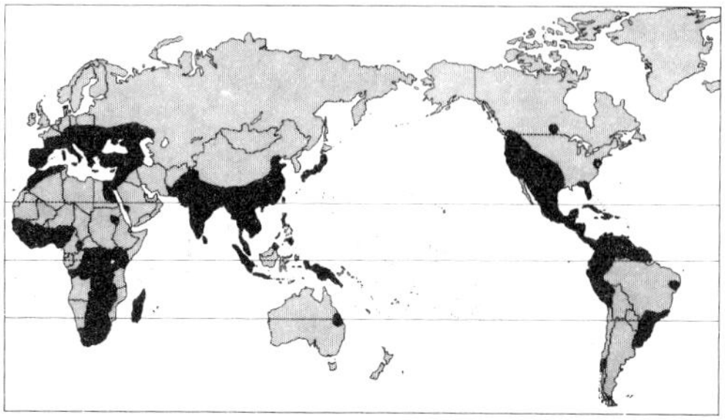

Distribution. A cosmopolitan species found throughout the world in warmer regions, but only one record from Australia (CIE map no. A.105).
Etiella behrii Zell. – (Lucerne Seed Web Moth) found in Australia feeding on soybean and alfalfa.
Control. A difficult pest to control in field infestations, for egg-laying occurs over a protracted period of time and repeated sprays of insecticides are required in order to achieve a reasonable level of control; this is usually not economic. If control was to be attempted, then weekly (or 10-day) sprays of the chemicals listed on page 412, after flowering starts, could be used.

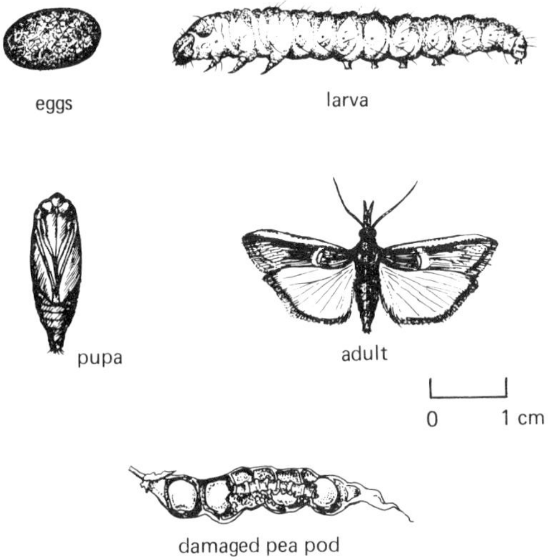

Evergestis spp.

forficalis (L.)
pallidata (Huf.)
rimosalis (Guen.) etc.

Common name. Cabbageworms
Family. Pyralidae
Hosts (main). *Brassica* species.
(alternative). Other Cruciferae, both cultivated and wild.
Damage. The caterpillars feed on the undersides of the leaves, or in the 'heart', under a webbing of silk. On cabbage they may hollow out the heart; on turnip the top of the root may be bored; the webbing is often a nuisance.
Pest status. A widespread group of closely related species; damage levels vary considerably, but economic damage is regularly done in some countries.
Life history. Eggs are laid in small batches (2–50 recorded) on the underside of leaves; they hatch in 4–8 days.

The caterpillars are yellowish-green, becoming glossy pale green, with a dark mid-dorsal stripe and yellowish lateral stripes; fully grown they measure 18–20 mm. Larval development can be as short as three weeks in duration, but is often longer.

Pupation takes place inside a silken cocoon in the soil, just under the surface. Overwintering larvae do not pupate until the spring.

In the UK this species is bivoltine, and adults emerge in May/June and again in August/September; in Canada the common species is univoltine and adults appear only in July/August. These moths are nocturnal in habits, yellowish with grey-brown markings; wingspan of about 25 mm.

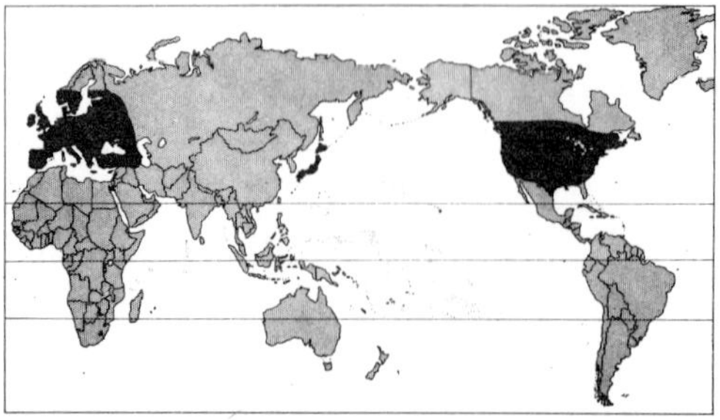

Distribution. This genus appears to be Euro-Asiatic in origin and has only recently (c. 1930) invaded N. America. The main pest species are:
Evergestis forficalis (L.) – (Garden Pebble Moth) Europe and Japan.
E. pallidata (Huf.) – (Purple-backed Cabbageworm) Europe, Canada, USA.
E. rimosalis (Guen.) – (Cross-striped Cabbageworm) Canada and USA.
E. extimalis Scopoli – (Rapeworm) Japan.
Control. When control is required it will often be as part of an overall programme against cabbage (*Brassica*) caterpillars; see page 412.

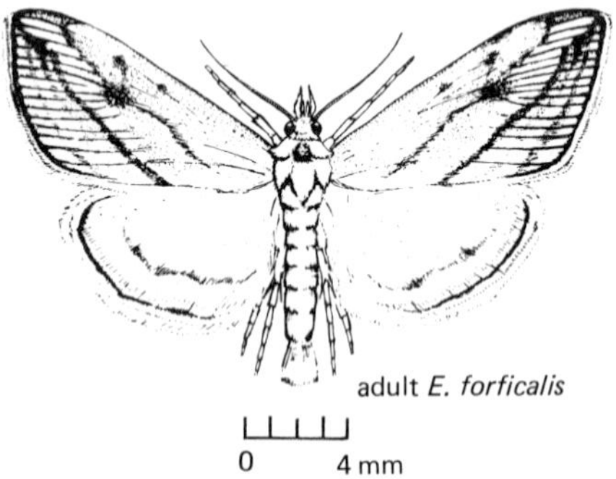

adult *E. forficalis*

Ostrinia nubilalis (Hübner)

Common name. European Corn Borer
Family. Pyralidae
Hosts (main). Maize
(alternative). Hop, Indian hemp, sorghum, some millets, capsicums, beet, tomato, soybean, beans, potato, oat, and many flowers and also many weeds.
Damage. The larvae tunnel up the plant stems, and on maize also inside the cob; bored stems often break. A symptom of early damage is leaf-windowing. Crop losses can be very heavy.
Pest status. A serious pest of maize in Europe and N. America (since 1909); totally polyphagous, but most damaging to maize crops in the USA and Canada.
Life history. Eggs are laid in clusters of 15–35 on the underside of the lower leaves; each female lays 500–1000 eggs; hatching takes 3–10 days.

The caterpillars initially feed on the leaf surface and in the midribs, causing leaf breakage; larger instars then bite through the folded leaves making a characteristic 'windowing' damage and finally enter the stem and hollow out the internode by their feeding. The larvae reach full size in 3–4 weeks of active feeding; they then cease feeding and overwinter inside the plant body in a state of diapause, during which time they can survive freezing conditions by supercooling. In tall stubble they overwinter very successfully. At this time they are pinkish with brown spots and 16–20 mm long. Pupation takes place either inside the stem or in the soil in late spring, and the adults emerge in June usually.

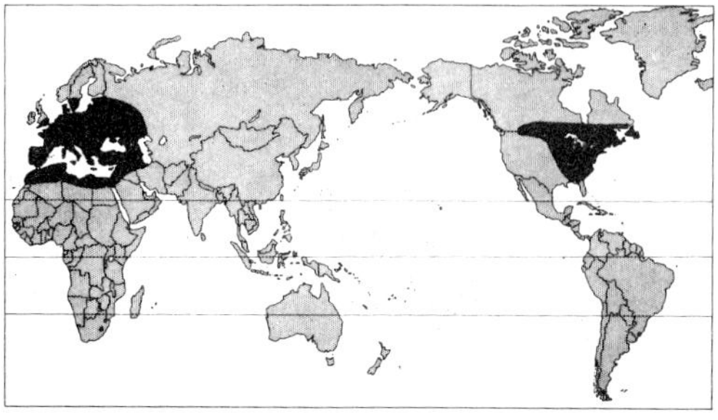

The adults are buff-coloured moths with brown markings on the wings; body length 12–14 mm and wingspan 25–30 mm; they are active nocturnal fliers and live for 10–24 days.

In the north this species is univoltine, but under warmer conditions there may be 2–3 generations.
Distribution. Endemic to Europe, from Sweden to N. Africa and Asia Minor, now in N. America where it is the most damaging (CIE map no. A.11 (revised)).
Control. Most control is achieved by cultural practices including: planting late, to avoid most of the egg-laying period; destruction of crop residues, to kill overwintering larvae; use of resistant varieties of maize, including tolerant varieties.

Biological control programmes using fungi, parasitic protozoa, and both tachinid and braconid parasites have met with some success.

Insecticidal control is aimed at killing the young larvae before they bore into the stem. Chemicals used include carbofuran, trichlorphon, diazinon and parathion, either as granules or sprays.

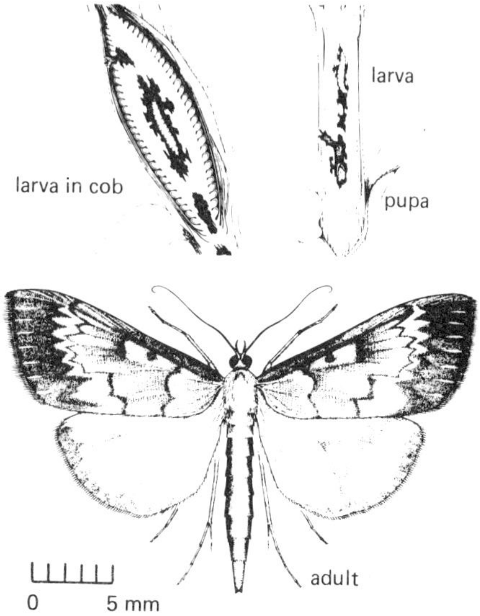

Plodia interpunctella (Hübner)

Common name. Indian Meal Moth
Family. Pyralidae
Hosts (main). Meals, flours and farinaceous products.
(alternative). Dried fruits, nuts, some pulses and some cereals in storage.
Damage. The caterpillars eat the stored produce, causing direct damage, especially when they selectively eat the germinal part of the seeds, and indirect damage in the contamination of the foodstuffs by frass, bodies and the silken webbing.
Pest status. A major pest of stored foodstuffs throughout the warmer parts of the world; regularly imported into Europe and N. America where it will survive while the temperature remains above about 18 °C (and RH above 40%).
Life history. Each female lays some 200–400 eggs stuck to the substrate (produce); hatching occurs after 4–6 days.

The caterpillars feed on the foodstuffs and grow; after some 12–20 days they are fully grown at 8–10 mm (under optimum conditions); at low temperatures or low humidities larval development may be very slow. The caterpillar is whitish, but does not have the dark spots characteristic of *Ephestia*. Pupation takes place in a silken cocoon stuck to the substrate.

The adult moth is small; body length 6–7 mm, wingspan 14–16 mm; but very distinctive in appearance with the outer half of the forewings a coppery-red separated from the creamy inner half by a dark grey band.

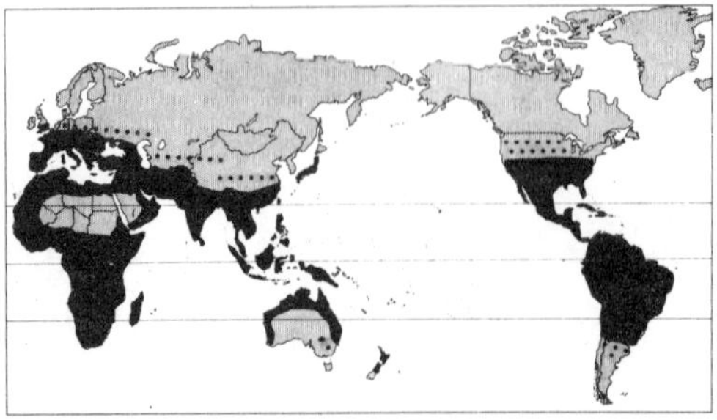

In Europe there may be only 1–2 generations per year, but in the tropics 6–8 is more usual; the total life-cycle can be as short as 24–30 days under warm conditions, but is much longer at lower temperatures; this is really a tropical insect that invades the warmer temperate regions from time to time.
Distribution. Worldwide in distribution throughout the tropics and subtropics, and in parts of the warm temperate regions where temperatures exceed some 18–20 °C.
Control. This pest is controlled by the usual practices of fumigation of food stores and the use of contact insecticides (see page 274).

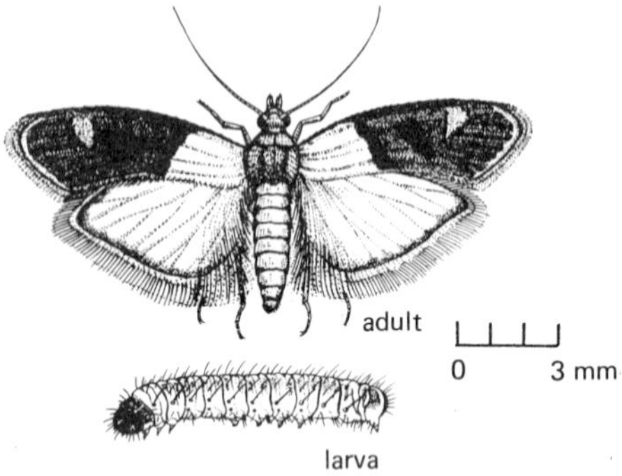

Pieris brassicae (L.)

Common name. Large White Butterfly
Family. Pieridae
Hosts (main). *Brassica* species.
(alternative). Cultivated and wild Cruciferae.

Damage. The caterpillars defoliate the plants and contaminate them with the large quantities of faecal matter produced; in heavy infestations leaves are reduced to the midrib and the plant killed. Garden and headland infestations are the most serious, as the adults prefer sheltered locations.

Pest status. A very serious pest of cultivated Cruciferae in the Old World; it is a strong migrant and may be scarce some years but very abundant in others when crop damage can be devastating.

Life history. Eggs are laid in groups of 20–100 on the leaves, mostly on the underneath; they are yellow, spindle-shaped with longitudinal ribs (1.5 mm long). Most eggs are laid in the UK in April and May; hatching requires 6–8 days.

The young larvae first eat their egg shells, and then start eating the leaf surface gregariously, keeping close together. As they grow and moult they gradually disperse over the whole plant. Usually there are several to many egg masses laid on each plant, so the total caterpillar population per plant may be very high, and they are easily seen feeding on both surfaces of the leaf; when populations are large it is usual to see plants completely defoliated with just leaf stalks and some midrib remaining. The larvae feed for 4–6 weeks, and reach their full size of about 25–40 mm; body coloration is a distinctive yellowish-green with large black spots and stout bristles, and a dorsal yellow line. They leave the plant to pupate on a solid substrate nearby (walls, fences, tree trunks, etc.) at about the end of June.

The chrysalis is positioned in an upright stance, resting on a silken pad with a band of silk around the thorax (as with other Pieridae). Pupation takes 10–15 days and the second-generation adults emerge in July (and August). Second-generation larvae pupate in late August/September, and the pupae overwinter. Damage by the second-generation larvae is often the more serious.

The adults are large white butterflies, with a wingspan 40–60 mm; the male has the forewings tipped black and a small black spot on the upper edge of the hind wing; the female has, in addition, two large black spots on the forewing and a marginal mark. A regular migrant species with strong flight powers; the adults may live for several weeks.

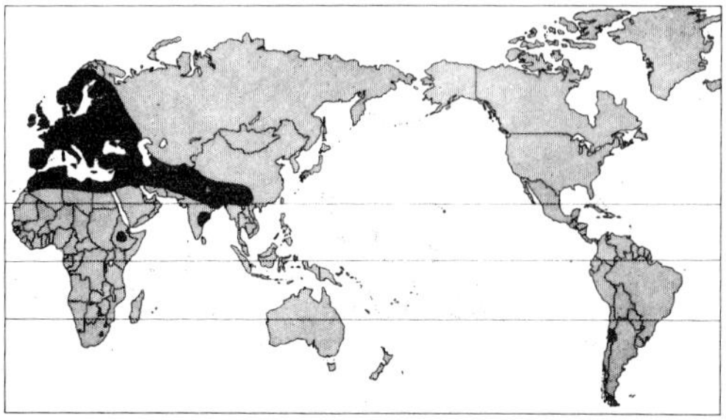

Adult ♂

Distribution. *P. brassicae* is an Old World species found throughout Europe, N. Africa, and southwest Asia, and now introduced into Chile (CIE map no. A.25).

In Europe this is one of the commonest caterpillars to be found on *Brassica* crops, but in point of fact, on a worldwide basis, there is a total of more than 20 species of caterpillars to be found eating the foliage of cabbage and other closely related crops, as well as sawfly larvae (*Athalia* spp.), so field identification of caterpillars is not always easy.

Control. These very conspicuous and strong-smelling caterpillars are eaten by some birds and other predators, and the natural level of parasitism by *Apanteles glomeratus* is often very high.

Hand-picking of egg masses and larvae in gardens is still recommended; females wanting to oviposit can be seen flying over the plants during the day and are very conspicuous.

Usually first-generation damage to a crop may not be important, for example defoliation of Brussels sprouts in late summer may not affect the crop yield at all.

The insecticides generally recommended for use against cabbage caterpillars include the following: derris, azinphos-methyl plus demeton-*S*-methyl, chlorpyrifos, etrimfos, iodofenphos, mevinphos, permethrin, triazophos and trichlorphon. Usually only one spray is needed, unless adult emergence is staggered when a second may be required. As the foliage of *Brassica* plants is very waxy, it may be necessary to add extra wetter to the spray formulation.

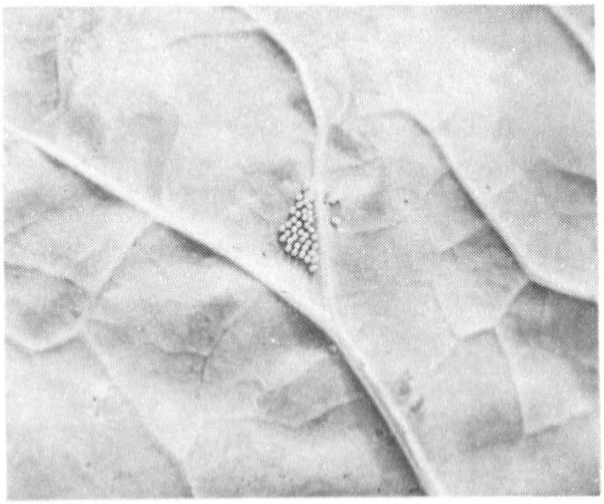
Egg mass

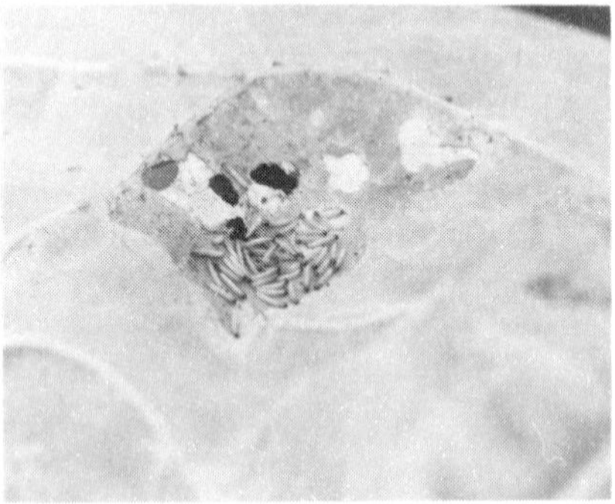
Young larvae

Damaged cabbage leaf

Mature larvae

Pieris rapae (L.)

Common name. Small White Butterfly (Imported Cabbageworm)
Family. Pieridae
Hosts (main). *Brassica* species.
(alternative). Cultivated and wild Cruciferae.
Damage. The caterpillars eat the leaves, and in heavy infestations they cause partial defoliation; but they are more solitary than *P. brassicae* so, although this pest species is generally more abundant, individual plant damage is usually less.
Pest status. A very serious, common and widespread pest of cruciferous crops worldwide; damage is often serious in that the 'heart' of the plant is preferred; faecal contamination is also an important factor.
Life history. Eggs are laid singly on the host plants, generally over a wide area; these butterflies are not confined to sheltered gardens and headlands, so entire commercial crops may be infested. Most eggs are laid on the underneath of leaves, as early as March and April; hatching requires about two weeks.

The caterpillar is green with many small setae and a yellow dorsal line, and it has a velvety appearance; fully grown it is about 25 mm long. There are often several caterpillars on the same leaf, although they are not at all gregarious: they apparently prefer to feed in the 'heart' of the cabbage.

Pupation takes place usually on the plant, and there is a typical pierid chrysalis coloured greenish-brown. Pupation takes place in June and July.

Second-generation adults emerge over a long period from July to September, and the pupae of this generation overwinter. Most damage is done by the second-generation larvae. Generally a bivoltine species in the UK, but has eight generations annually in Israel; under warm conditions breeding could be continuous.

The adults are typical 'white' butterflies, wingspan 40–5 mm; the female has spots on the forewing and the male does not.
Distribution. This species is widespread throughout Europe and much of Asia, and N. Africa, (down to the Tropic of Cancer); now introduced into Australia, New Zealand, Hawaii, Canada, USA and Mexico (CIE map no. A.19). In N. America a regular migrant, flying north during the warm summer period.
Pieris napi (L.) – (Green-veined White Butterfly) not scarce but more common on wild hosts.

A few other species (different genera) of Pieridae have larvae that also feed on Cruciferae, but these are mostly rather rare or uncommon, and seldom found.
Control. As for *Pieris brassicae*; also heavily parasitized by *Apanteles glomeratus* and *Pteromalus puparum* (as with *P. brassicae*), and the braconid *Meteorus versicolor*, so care has to be taken when applying insecticides.

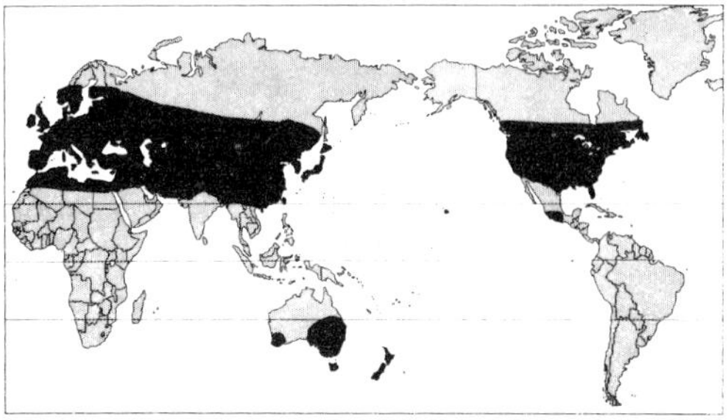

Adult ♀

0 1 cm

Larva

Lampides boeticus (L.)

Common name. Pea Blue Butterfly (Long-tailed Blue)
Family. Lycaenidae
Hosts (main). Various pulse crops (pea, cowpea, beans).
(alternative). Cultivated and wild species of Leguminosae.
Damage. The larvae bore the pods and eat the developing seeds; on small pods from the outside, but inside large pods.
Pest status. A widespread, but usually minor, pest of various legume crops in the Old World; there are in fact several Lycaenidae whose larvae feed on Leguminosae.
Life history. Eggs are laid singly on the shoots, on or near the young flowers; hatching takes 7–10 days. The short stubby caterpillar is green in colour and blends well with the plant foliage. Most hatch while the host plant pods are small and so feed on the developing seeds externally while sitting on the outside of the pod. Larval development takes 5–6 weeks; final length reaches 18 mm.

Pupation takes place within an oval cocoon, brown in colour and about 13 mm long, either on the plant or in leaf litter; development takes 7–10 days.

The adults are small blue butterflies with pronounced sexual dimorphism; the male has blue wings (brown underneath) with eyespots and single 'tails' on the hindwings; the female has wings more brownish (but blue near the body); wingspan 32–6 mm. (The female illustrated has lost her 'tails').

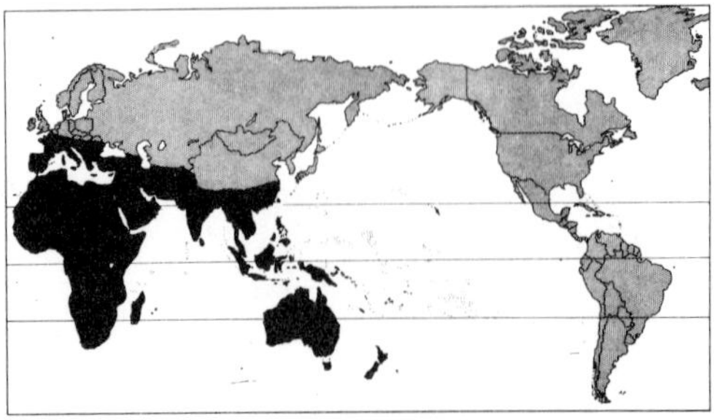

In Israel there are 3–4 generations per year, usually overlapping.
Distribution. Recorded throughout most of Africa, southern Europe, southern Asia, and Australasia.

Many of the Lycaenidae have larvae who feed on Leguminosae, but only a very few have any pest status; also on Leguminosae are several members of the Pieridae, notably the 'clouded yellows' and 'sulphurs' – (*Colias* spp.) both in the Old World and the New World.
Control. Generally control measures are not required as infestation levels are usually quite low.

Adult ♂ Adult ♀

Larva on pod (G. Johnston)

Abraxas grossulariata (L.)

Common name. Magpie Moth
Family. Geometridae
Hosts (main). Gooseberry and currants.
(alternative). Plum, sloe, apricot, hazel; also hawthorn, laurel, and other plants.
Damage. The single larvae are foliage eaters, and because of their size they occasionally defoliate part of a bush.
Pest status. Seldom a serious pest, but widespread and conspicuous, and in many respects a typical geometrid for teaching purposes.
Life history. Eggs are laid singly, or in small groups, on the underside of the leaves; smooth and round, yellow when laid, becoming black; hatching takes about two weeks.

The caterpillar is a typical looper (geometrid) and only has two pairs of functional prolegs on the abdomen (excluding terminal claspers), and it walks by 'looping' its elongate body; body coloration is distinctive black and white with a yellow lateral stripe; fully grown it measures 30–5 mm. Pupation takes place on a twig of the bush usually, and the pupa is held in place by silken strands making a loose cocoon (most Geometridae pupate in the soil).

The life-cycle is as follows. Eggs are laid in July and August; the young larvae feed on foliage in August/September, then while still quite small they hibernate in the leaf litter and crevices. In the spring, as new foliage emerges, the caterpillars emerge also and recommence feeding and growing; pupation takes place in June usually, and requires about four weeks; young adults emerge in July and August.

The adults are distinctively coloured, black and white moths with patches of orange; wingspan 35–40 mm; they are nocturnal in habits, but may be seen during the day resting on tree trunks and foliage.

There is generally only one generation per year, but in some areas it is thought to be bivoltine.

Distribution. Recorded throughout Europe and Asia to Siberia, China and Japan.

In Japan *Abraxas miranda* Butler also occurs.

Control. Control measures are not often required as it is most frequently a garden pest and hand-picking is successful. However, on occasions control is needed and then a spray of derris is recommended, to be applied after flowering.

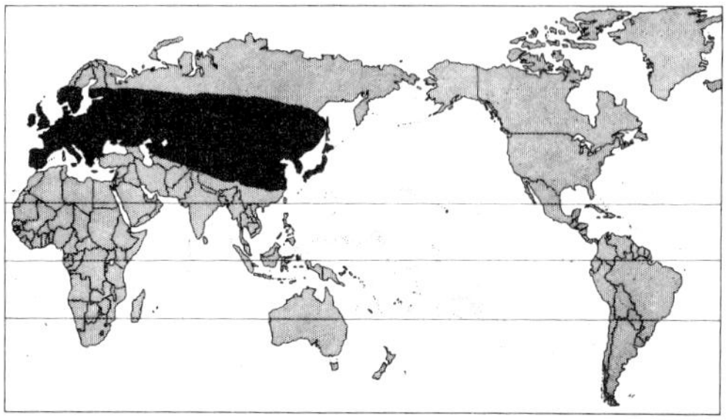

Adult

0 1 cm

Larva (on gooseberry bush)

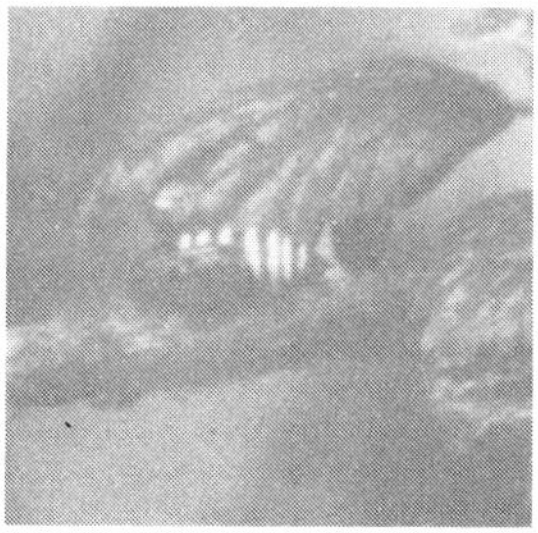

Pupa

Operophtera brumata (L.)

Common name. Winter Moth
Family. Geometridae
Hosts (main). Apple, pear, plum, walnut, hazel, currants.
(alternative). Other fruit and nut trees, and shrubs; many deciduous forest trees; totally polyphagous on deciduous woody plants.
Damage. The caterpillars feed on opening buds, young leaves and flower trusses; in the cold spring some caterpillars may tunnel into young buds and feed within. There is usually some light silk webbing over the feeding site. Sometimes they feed on the surface of young fruitlets.
Pest status. Quite a serious pest of most temperate fruit and nut trees, and being polyphagous it attacks a wide range of host plants. It is also here used as an example of a typical geometrid pest species, of which there are several.
Life history. The adult moths emerge from their cocoons over the winter, from October to the end of January; the males fly at night searching for the wingless females. Mating may take place either on the ground or on tree trunks. Later the females climb the trees and lay their eggs singly on the branches, in crevices in the bark, on spurs and on buds. The elongate eggs are green when laid, turning red, and hatch in the spring, usually the date of hatching being determined by ambient weather conditions. Each female lays 100–200 eggs.

The little caterpillars feed on the young leaves as they unfold, and they continue to feed and grow from March through April and May; most are fully grown by mid-June, at a length of 35–40 mm; body colour is green with pale longitudinal lines. The caterpillar then falls to the ground on a silken thread and pupates in an earthen cell some 7–10 cm deep in the soil.

In most years many of the adults emerge in October/November; they are greyish-brown in colour; male wingspan is some 30–5 mm; the female is quite plump-bodied with tiny vestigial wing buds.

There is only one generation per year.
Distribution. Recorded throughout Europe, up to Lapland, the USSR, Japan, N. Africa, and now introduced into Canada (CIE map no. A.69).

The closely related *Operophtera fagata* (Scharf.) (Northern Winter Moth) is in most respects virtually identical to the Winter Moth.
Control. Grease-banding the tree trunks used to be the standard practice against this and other moths with flightless females, and is still regularly practised with success. The economic threshold on apple is suggested to be 10% trusses infested with caterpillars in late spring.

Winter washes of tar oils (as used today) have little effect on the eggs of Winter Moth, but some of the DNOC/petroleum washes do kill the eggs.

Spring sprays of the following chemicals are however the more usual modern treatment: azinphos-methyl plus demeton-*S*-methyl sulphone, carbaryl, chlorpyrifos, diflubenzuron, dichlorvos, fenitrothion, permethrin and phosalone.

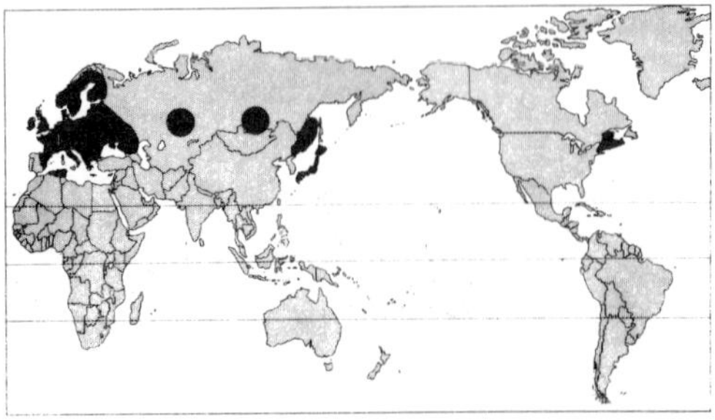

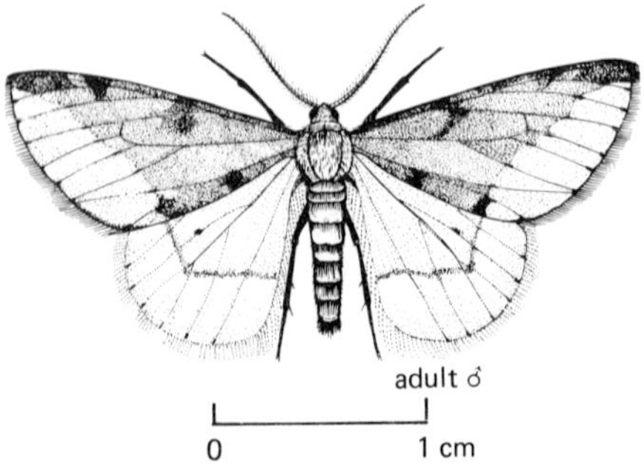

Looper/geometrid pests of economic importance (family Geometridae)

A very large group of moths with more than 12000 species, worldwide in distribution. Mostly they are woodland/forest species, and a number of species are orchard pests of some importance because of larval defoliation. The adults are slender-bodied, with large wings, nocturnal in habits, and may often be seen during the day sitting cryptically on tree trunks and walls with the wings extended laterally and pressed on to the substrate. One temperate group within the family is interesting in that adult activity is confined to the winter months, when most other insects are absent. These 'winter moths' have females either apterous or with vestigial wings; they depend upon female sex pheromones to bring the males to them for mating, and have to crawl up the tree trunks in order to reach suitable oviposition sites. The eggs are laid on the branches of the trees, to hatch in the spring when the buds open, and the defoliation done by the caterpillars in the spring to various fruit and nut trees can be very damaging.

The caterpillars are easily recognized by their elongate body form and reduced number of abdominal prolegs: most have only one pair of prolegs in addition to terminal claspers, though some have a reduced second pair. They locomote by bending the body in a vertical loop, hence the common name of 'loopers'. Some of the pest species have large brown larvae, 4–6 cm long and almost as thick as a pencil, some of which are quite polyphagous; they feed at night and spend the day resting cryptically alongside a twig or across two adjacent twigs, as shown in the photographs. Many loopers are quite long-lived, and the larger species have considerable impact as defoliators. In point of fact, the majority of species (to be found in woodlands) are quite small in size, and conform to H.C. Andersen's description of 'inchworm'. In the USA and Canada it is the usual practice to refer to the caterpillars as 'inchworms' or 'spanworms', for obvious reasons, and also as 'cankerworms', for less obvious reasons.

Within this winter-active group of geometrids, some are active at the start of the winter, some throughout the winter months, and some at the end of the winter, as is indicated by some of the common names listed below.

Another characteristic of many Geometridae is that the first-instar larvae dangle from the tree foliage on long strands of silk and may be carried by the wind to other trees, sometimes dispersing for quite long distances.

In the text on the preceding page the Winter Moth (*Operophtera brumata*) is reviewed in some detail, partly because of its importance as a polyphagous defoliator and also because it may be considered as quite typical of this group of low-temperature tolerant moths.

Illustrated here are adult males of two other important pest species, notably *Alsophila aescularia* (March Moth) and *Erannis defoliaria* (Mottled Umber Moth); there are also photographs of two caterpillars of the 'giant looper' type, thought to be *Ascotis selenaria*, on twigs of *Citrus*.

Important pest species of Geometridae

Abraxas grossulariata (L.) – (Magpie Moth) Europe, Asia (see page 415).

+ *Alsophila aescularia* (Schiff.) – (March Moth) Europe and Russia.

+ *Alsophila japonensis* Warren – (Pear Fall Cankerworm) Japan.

+ *Alsophila pometaria* (Harris) – (Fall Cankerworm) USA and Canada.

Ascotis selenaria (Wlk.) – (Giant Coffee (etc.) Looper) Africa, Asia, Japan.

Biston betularia L. – (Peppered Moth) Europe, Asia, Japan, Canada, USA.

Chloroclystis spp. (L.) – (Pug Moths) on apple, pear, acacia, blackberry, etc.; Europe, Asia, Egypt, Australia.

Epigynopteryx spp. (3) – (Coffee Loopers) East Africa.

+ *Epirrita dilutata* (D. & S.) – (November Moth) Europe.

+ *Erannis defoliaria* (Clerck) – (Mottled Umber Moth) Europe.

\+ *Erannis progemmaria* (Hübner) – (Dotted Border Moth) Europe and S.W. Asia.
\+ *Erannis* spp. – ('winter moths') Japan and USA.
Gelasma illiturata Wlk. – (Peach Geometrid) Japan.
Hyposidra spp. – (Coffee Loopers) S.E. Asia.
Inurois fletcheri Inoue – (Apple Fall Cankerworm) Japan.
Megabiston plumosaria Leech – (Tea Geometrid) Japan.
\+ *Operophtera brumata* (L.) – (Winter Moth) Europe, Asia, Japan, Canada, USA (see page 416).
\+ *Operophtera fagata* (Scharf) – (Northern Winter Moth) Europe to S.E. Russia.
\+ *Operophtera* spp. – ('winter moths') Asia and Japan.
\+ *Paleacrita vernata* (Peck) – (Spring Cankerworm) USA and Canada.
Phthonesema spp. – (Apple/Mulberry Loopers) Japan.
Pterotocera sinuosaria Leech – (Fruit Tree Looper) Japan.
\+ *Theria rupicapraria* (D. & S.) – (Early Moth) Europe and W. Asia.
Zamacra spp. – (Mulberry/Walnut Loopers) Japan.

\+ Species of the 'winter moths' group with flightless females.

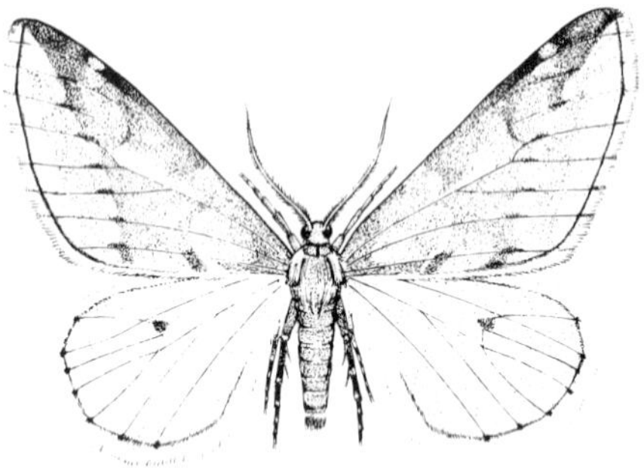

March Moth

adult ♂

0 1 cm

Giant Loopers (? *Ascotis*) on citrus

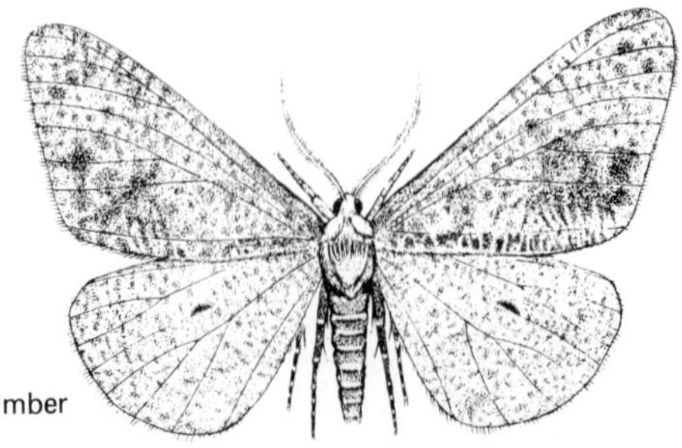

Mottled Umber Moth

adult ♂

Malacosoma spp.

americanum (F.)
neustria (L.), etc.

Common name. Lackey Moth (UK); Tent Caterpillars (USA)
Family. Lasiocampidae
Hosts (main). Apple, pear, plum, cherry.
(alternative). Hawthorn, rose; many deciduous trees and shrubs.
Damage. The gregarious caterpillars defoliate the trees; the colony inhabits a large silken nest or 'tent' when not feeding; severe attacks result in complete defoliation.
Pest status. A sporadic pest, sometimes serious, both in fruit orchards and in gardens.
Life history. Eggs are laid in a band around a twig, usually 100–200 per band; most are laid in August and September in the UK. The eggs overwinter and hatch in late April and May the following year.

The caterpillars are almost black when first hatched, but as they grow their colours develop to blue-grey with white and red stripes dorsally and laterally, and many reddish bristles; the head is slate-blue with two black spots dorsally. They are fully grown about the end of June/early July and measure 30–5 mm; at this time they disperse to pupate singly inside a silken cocoon in the foliage, herbage, or on the trunk. Pupation takes about three weeks.

The adult moths emerge from July to September; they are stout-bodied, brown in colour, about 3–4 cm in wingspan, and nocturnal in habits; there are usually two parallel cross-lines on the forewings that are distinctive.

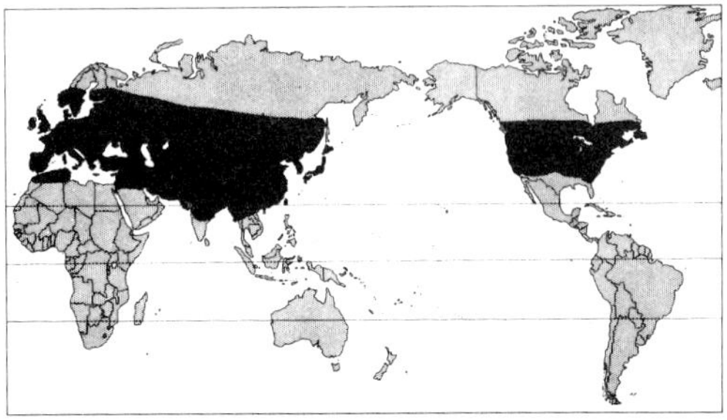

Distribution. *Malacosoma neustria* occurs throughout Europe and Asia as several different subspecies; a pest of fruit trees in China and Japan. Throughout this region (Palaearctic) there are several other closely related species of *Malacosoma*, some of which prefer saltmarsh habitats, although they are quite polyphagous.

In the USA the following species are pests:
Malacosoma americanum (F.) – Eastern Tent Caterpillar.
M. californicum (Packard) – California Tent Caterpillar.
M. disstria Hübner – Forest Tent Caterpillar.
M. fragile (Stretch) – Great Basin Tent Caterpillar.
M. lutescens (N. & D.) – Prairie Tent Caterpillar.
M. pluviale (Dyar) – Western Tent Caterpillar.
Malacosoma indica (Wlk.) is an equally polyphagous pest of fruit trees throughout southern Asia.
Control. Hand-collection of tents (nests) can be quite successful if only a few trees are affected. Sometimes it is feasible to collect the egg-masses before they hatch.

Egg-masses are resistant to the usual winter washes used on fruit trees, but the caterpillars may be killed using insecticides. In the past the most successful chemical was DDT; present recommendations use chlorpyrifos or diflubenzuron.

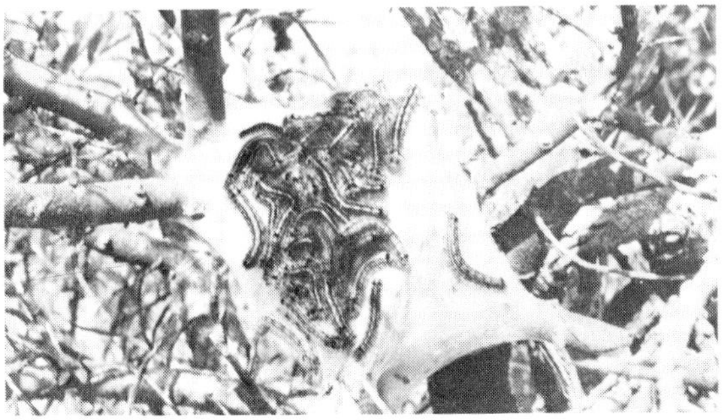

Larvae sunning themselves on the outside of their silken nest

0 3 cm

Acherontia atropos (L.)

Common name. Death's Head Hawk Moth
Family. Sphingidae
Hosts (main). Potato, tomato, tobacco, eggplant.
(alternative). Wild Solanaceae; and a few other plants (olive, sesame, etc.).
Damage. The solitary larvae eat leaves, but because of their size a few caterpillars can defoliate the host plant; often there is only one larva per plant though.
Pest status. Usually only a minor pest, but widespread and of regular occurence.
Life history. Eggs are laid singly on the plant foliage, usually only one egg per plant, often coloured red or blue.

The caterpillar is solitary, and coloured white initially, but as it grows through the five larval instars the colour changes to yellow, green or grey with diagonal blue body stripes. The 'horn' on the last abdominal segment (responsible for the collective name in the USA for all sphingid larvae of 'hornworms') is the same colour as the body and is characteristically hooked dorsally at the tip. Development of the larvae usually takes several months, by which time they eventually attain a body length of up to 10–12 cm.

As with all Sphingidae, pupation takes place in the soil within an earthen cocoon, 5–10 cm deep; in the UK this species is univoltine and the pupa overwinters, but further south there are usually two generations annually. The length of the pupa is about 8 cm.

The adult is a stout-bodied hawk moth of striking appearance, with long dark forewings, and short yellow hindwings with black barring. The dorsum bears the characteristic skull-like marking. Wingspan is 8–12 cm.
Distribution. This species, and several very close relatives, are the sphingid pests of Solanaceae in the Old World, and they are replaced by *Manduca* spp. in the New World.

Acherontia atropos is recorded throughout Europe, Africa, India, S.E. Asia, northwards up to China and Japan. In parts of Asia other species are sometimes recorded, including *Acherontia styx* (Westwood) and *A. lachesis* F. from India, Malaysia, China and Japan.
Control. Often control measures are not really required, but should they be needed then sprays of carbaryl should be effective against the younger larvae; the large larvae are very difficult to kill with any insecticides.

adult

0 2 cm

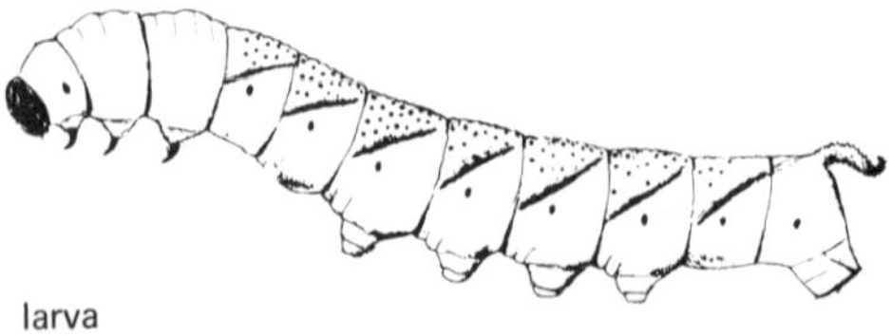

larva

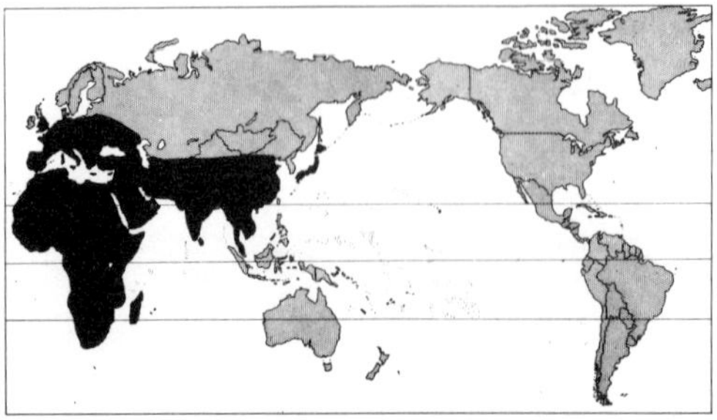

Manduca spp.
(*M. quinquemaculata* (Haw.))
(*M. sexta* (L.))

Common name. Tomato Hornworm; Tobacco Hornworm
Family. Sphingidae
Hosts (main). Tomato and tobacco; also potato.
(alternative). Other Solanaceae, both crops and wild plants.
Damage. The large larvae eat the leaves of the host plant and frequently cause defoliation; then the plant usually dies.
Pest status. Serious pests of solanaceous crops in the New World, due entirely to their voracious leaf-eating, which leads to defoliation, and their rapid development.
Life history. Eggs are laid singly or in small groups on the underside of leaves; they are spherical in shape, green in colour; hatching takes about five days.

The caterpillars differ somewhat in appearance: *M. sexta* is yellowish-green with seven lateral white, oblique lines, and the horn is curved and red; *M. quinquemaculata* has seven or eight lateral white stripes connected below the spiracles (giving a series of curved shapes), with the horn straight and black; brown larvae are sometimes seen. Fully grown larvae are 9–10 cm. Larval development through five instars takes only 3–4 weeks in southern USA. Pupation takes place in the soil and requires 2–4 weeks.

The adults are grey with fine black markings, and a series of yellow patches along the abdomen; wingspan is 10–12 cm. Like most of the Sphingidae they are nocturnal and take nectar from flowers at night.

Depending upon latitude, there are 1–4 generations annually.

Distribution. These two species are broadly sympatric, although they differ in abundance relatively, over an area from southern Canada (Ontario), through the USA into S. America.

Several other species of *Manduca* are known and a couple of them are also recorded as larvae feeding on Solanaceae.

In the Old World their niche is occupied by *Acherontia* (page 420).

Control. Various natural enemies are of importance in the USA, particularly the larval parasite *Apanteles congregatus* (Say). Deep ploughing will usually destroy many of the pupae in the soil.

In gardens hand-picking is recommended, but for commercial crops it is often necessary to use insecticides, although it must be noted that the large larvae are extremely difficult to kill with poisons. Spray timing is aimed at the young larvae, using carbaryl, endosulfan, parathion, toxaphene, also BTH and several other compounds.

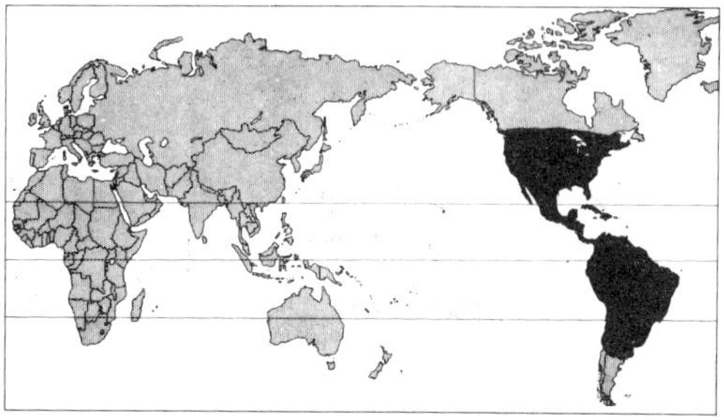

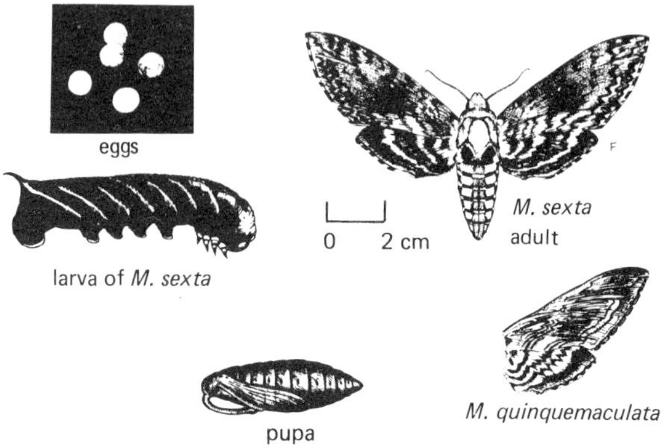

Hawk moths and hornworms
(Lepidoptera; Sphingidae)

A small group of spectacular moths, found worldwide, 1000 species recorded; probably to be regarded as essentially a tropical group but quite well represented in temperate regions. The adults are distinctive with their tapering body, elongate forewings, and fast flight. They feed by taking nectar from flowers with a long calyx, by hovering in front of the flower. Most are nocturnal and fly to lights at night, but a few smaller species are diurnal and look rather like large bumble bees at flowers. The proboscis is usually very long, and up to 20 cm is not at all unusual.

The larvae are very large, stout-bodied caterpillars, with a smooth skin, and a characteristic terminal dorsal 'hook'. They are leaf lamina feeders and some are pests purely because of the quantity of food they consume. They are essentially solitary, but in some species there may be several to many larvae per plant and defoliation may result. Most species are fairly specific as to their diet, and the majority are restricted to a single host family for food, but a few pest species are truly polyphagous. The Solanaceae and Convolvulaceae are well-represented agriculturally as host plants for Sphingidae, but by far the most popular host is grapevine with more than a dozen hawk moths regularly recorded worldwide.

Pupation always takes place in the soil inside an earthen cell or in leaf litter, and the pupa always has the long proboscis clearly defined externally, sometimes quite separated from the rest of the pupal body. At the time of pupation sometimes the entire population of large caterpillars literally disappears overnight as they all descend to the ground and burrow into the soil.

Pest species of Sphingidae

Acherontia atropos (L.) – (Death's Head Hawk Moth) Solanaceae; Old World (see page 420).
Acherontia styx (Westwood) – (Small Death's Head Hawk Moth) polyphagous?; Asia.
Agrius cingulatus (F.) – (Sweet Potato Hornworm) New World.
Agrius convolvuli (L.) – (Sweet Potato Hawk Moth) Old World, Hawaii (CIE map no. A.451).
Cephonodes hylas (L.) – (Coffee Hawk Moth) Africa, Asia, Japan.
Cressonia juglandis (J.E. Smith) – (Walnut Sphinx) USA.
Erinnyis spp. – guava and papaya; USA and S. America.
Eumorpha spp. – (Grapevine Sphinxes) USA.
Hippotion celerio (L.) – (Silver-striped Hawk Moth) polyphagous; S. Europe, Africa, S. Asia, Australia.
Hyles lineata (Esp.) – (Striped Hawk Moth) polyphagous; cosmopolitan (CIE map no. A.312).
Langia zeuzeroides Moore – deciduous fruit; N. India.
Manduca quinquemaculata (Haw.) – (Tomato Hornworm) Canada, USA, C. and S. America (see page 421).
Manduca sexta (L.) – (Tobacco Hornworm) Canada, USA, C. and S. America (see page 421).
Marumba gaschkewitschi (B. & G.) – (Peach Hornworm) China and Japan.
Paonias astylus (Drury) – (Huckleberry Sphinx) USA.
Smerinthus jamaicensis (Drury) – (Twin-spot Sphinx) USA.
Smerinthus ocellata (L.) – (Eyed Hawk Moth) Europe.
Smerinthus planus Wlk. – (Cherry Hornworm) Japan.
Sphinx spp. – (Fruit Tree Sphinxes) USA.
Theretra spp. – (Grapevine Moths) S. Europe, India, Asia, Japan.

Agrotis exclamationis (L.)

Common name. Heart and Dart Moth
Family. Noctuidae
Hosts. A common cutworm; this species shows little host specificity, attacking a wide range of cultivated crops and weed species, except cereals; sometimes particularly serious on sugar beet.
Damage. The larvae feeding at night cut off seedlings at ground level; on larger plants they feed on the roots and underground stem; on root crops the root may be hollowed out.
Pest status. One of the common cutworms throughout the UK and most of Europe; generally regarded as a serious polyphagous pest.
Life history. Eggs are laid in groups on the soil, from June to July, and sometimes also in October if there is a second generation; they are hemispherical and ribbed, whitish becoming pink. Hatching requires about 12 days (10–14).

The larvae are typical cutworms, body brownish above, greyish-white below; spiracles large and distinct, with a row of dark dorsal blotches along the body; fully grown about 38 mm long. The first hatched larvae feed and grow rapidly and are fully grown by July or August when they pupate and may give rise to a small second generation. Larvae from eggs laid later (July) develop more slowly and are univoltine; they continue to feed and grow throughout the autumn and early winter before burrowing deeper into the soil and making an earthen cell in which they hibernate. Pupation takes place within the cell between February and late April.

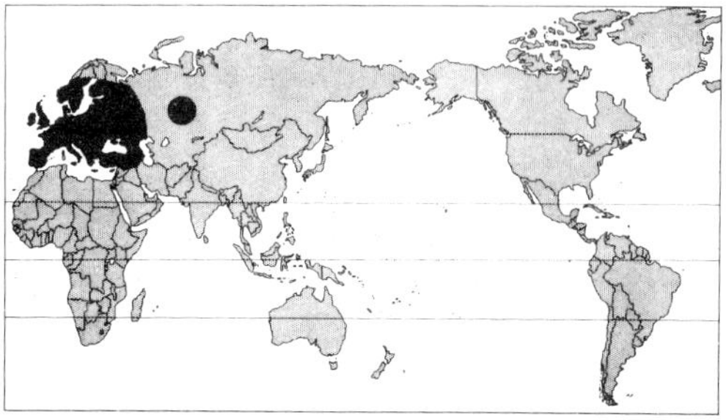

The adults are smallish brown moths with a variable wing coloration, but the pattern on the forewings is usually evident; wingspan 35–44 mm. Male hindwing usually white, females brownish. They are nocturnal and quite long-lived (up to several months), feeding on nectar, and are found flying from mid-May to late July; those seen in September/October are second generation.
Distribution. Recorded throughout Europe up to C. Scandinavia and Finland.
Control. See 'Control of cutworms' on page 425.

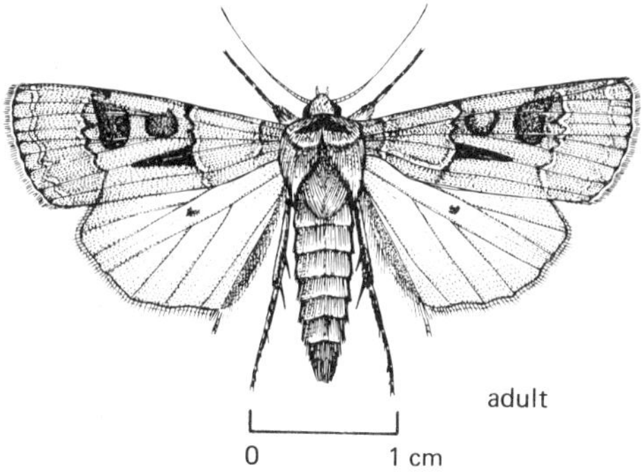

adult

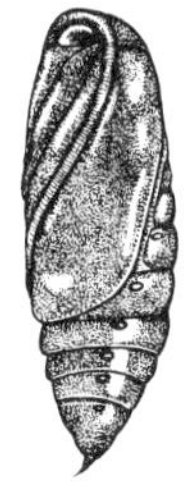
pupa

Cutworms (Lepidoptera; Noctuidae)

Cutworms are the larvae of various Noctuidae, that are so categorized purely on their behaviour and feeding biology and are all agricultural pests of some importance. They are large, quite stout-bodied caterpillars, 'typical' in appearance with four pairs of abdominal prolegs (excluding terminal claspers), usually drab in colour so as to blend with the soil. Most belong to the genera *Agrotis* and *Euxoa*, which are very closely related and separable only by minor esoteric taxonomic characters. The group is worldwide in distribution, but most cutworms are to be found in temperate regions rather than tropical. The larvae are nocturnal in habits and spend the day hiding in leaf litter or in the soil; some tend to remain in the soil (down to depths of 10 cm) but others come to the surface at night to feed on foliage; the subterranean ones tend to feed on roots and tubers. Typical damage to a root crop is a wide shallow hole, sometimes like slug damage, especially so in potatoes (see page 99). Vegetables with thick stems (lettuce and some brassicas) may have the underground part of the stem hollowed out. The most usual symptom of cutworm attack to an established crop is to see plants wilting. Their most serious damage is that done to seedlings, from which they derive their vernacular name, by surfacing at night and eating through the seedling stems (like field crickets), sometimes one caterpillar may destroy many plants along a single row in one night, sometimes only part of the stem is actually eaten. With field (and garden) crops, after thinning or after planting out, cutworm damage can be very serious in that the plant stand is greatly reduced; crops such as tobacco, field tomatoes, many brassicas, sugar beet and the like, are particularly affected. Tree and bush seedlings, or young freshly grafted stock, may also suffer from cutworm attacks. Close-planted crops such as carrot, lettuce, celery, red beet and some brassicas, probably suffer the greatest total damage. With the main root crops (potato, turnip, swede, parsnip, sweet potato) the level of damage is often not evident until harvest; root crops suffer most damage in light sandy soils where the caterpillars may burrow more easily, for these Noctuidae are quite soft-bodied and plump. Some aerial foliage damage may be done.

Another reason for the importance of these species is that the females usually lay a large number of eggs (1000–3000) in a series of batches, on leaves and stems of both weeds and crop plants, or else on the soil or surface debris. The first-instar larvae feed on the leaves of the plants (both weeds and crop), and as they grow they gradually develop the cutworm habits and descend to the soil. There are usually six larval instars, but sometimes five; under crowded conditions larval development is somewhat accelerated, but if very crowded then cannibalism occurs. Because of the progressive nature of insect growth, crop damage is often not noticed until the caterpillars are large; in point of fact, of the entire food intake of a cutworm, 80% is eaten during the last instar. Combined with this is the problem that the large caterpillars are notoriously difficult to kill with insecticides. It must be remembered when considering candidate insecticides for cutworm control that only the younger instars are really susceptible to poisons, the final instars are very difficult to kill. Thus lists of 'effective' insecticides for use in cutworm control have to be viewed carefully, if not with a little scepticism.

Some species are strongly migratory, and most are characterized by quite extreme population fluctuations, being very abundant in some years and scarce in others, although most are present most of the time. Population predictions are generally unsuccessful, although research into population monitoring is in progress, especially in Canada and the USA, now using pheromone traps in addition to ultra-violet light traps.

Although most species of cutworms are renowned for their polyphagous food preferences, many species do not attack cereals very often, and in turn there are a few species that show preference for Gramineae.

Control of cutworms

Suggested control methods of population control are as follows:

Cultural methods

(a) Weed destruction – these plants are often preferred sites for oviposition, and food for the first-instar larvae.
(b) Hand-collection of larvae – often more suitable for gardens and small-holdings.
(c) Flooding of the infested field may be feasible for some crops.
(d) Deep ploughing will bring larvae and pupae to soil surface for exposure to predators and sun.

Chemical methods

(a) High-volume sprays (at least 1000 l/ha) of insecticides.

DDT (25%)	4.2 l/ha
endrin (w.p.)	3.75 l/ha
chlorpyrifos (48% e.c.)	2.5 l/ha
triazophos (40% e.c.)	2 l/ha
cypermethrin	75 g a.i./ha

The spray should be directed along the plant rows, aiming at run-off to the soil below. Timing should ideally be aimed at the young caterpillars whilst still feeding on the plant leaves or on the soil surface.
(b) Soil application of bromophos (w.p.) or chlorpyrifos granules.
(c) Baits of moist bran mixed with DDT, γ-HCH, or endrin may be effective against older caterpillars.

Generally cutworms are extremely difficult to control, especially in 'boom' years, for by the time infestations are apparent the susceptible stages of the larvae are often past, and damage may be already quite serious. The sporadic nature of cutworm population outbreaks makes preventative treatments rather futile in most areas. Finally the soil-dwelling larvae, often under dense and continuous crop foliage, make targets difficult to 'hit' with insecticides, and in many areas such high-volume spraying is not feasible because of water shortages and equipment restrictions.

Agrotis ipsilon (Hyn.)

Common name. Black (Greasy) Cutworm (Dark Sword Grass)
Family. Noctuidae
Hosts. A polyphagous cutworm attacking many crops, both seedlings and established plants; in particular crucifers, lettuce, tobacco, potato, and sometimes cereals including rice; generally confined to herbaceous plants.
Damage. A typical cutworm, in that young larvae feed on the plant leaves, and the older caterpillars feed on stems, roots and tubers underground; surface damage only done at night.
Pest status. An abundant and completely cosmopolitan pest species, of sporadic importance on many different crops in different parts of the world. In some instances agriculturalists are faced with a cutworm complex and may be unable to identify the different species involved.
Life history. Eggs are laid near the base of the plants; they are small, conical and slightly ribbed; yellow becoming brownish-grey; each female may lay up to 1800 eggs; hatching requires 4–9 days usually (from July to September).

The larvae are grey-brown in colour with dark lateral stripes and a pale grey mid-dorsal line; the head capsule is very dark with two white spots. The caterpillars have a dark and greasy general appearance. Fully grown they measure 30–40 mm; development usually takes 28–34 days under warm conditions. In temperate countries larvae overwinter and pupate in the spring. The first two instars feed gregariously on the plant foliage, the third instar becomes solitary and becomes a cutworm. The first hatched larvae pupate in August in an earthen cell in the soil (in the UK), and development takes about 14 days; elsewhere the pupal period usually varies from 10–30 days. Thus some adults emerge in the UK in September, but it is thought they remain mostly immature and die in the first frosts. The majority of the larvae overwinter as such and pupate in the following spring.

The adults are large grey moths with a distinctive forewing pattern; wingspan 40–55 mm; hindwings are pale, almost white in the male.

In temperate regions this species is univoltine, but under warm conditions 4–5 generations or more are common: the life-cycle takes 32 days at 30 °C, 41 days at 26 °C, and 67 days at 20 °C.
Distribution. Completely worldwide in distribution, but to date absent from a few tropical regions (S. India, much of Brazil) (CIE map no. A.261).

In Canada and the USA there are recorded six other species of *Agrotis* and 12 species of *Euxoa* of importance as cutworms.
Control. See page 425 for advice on cutworm control.

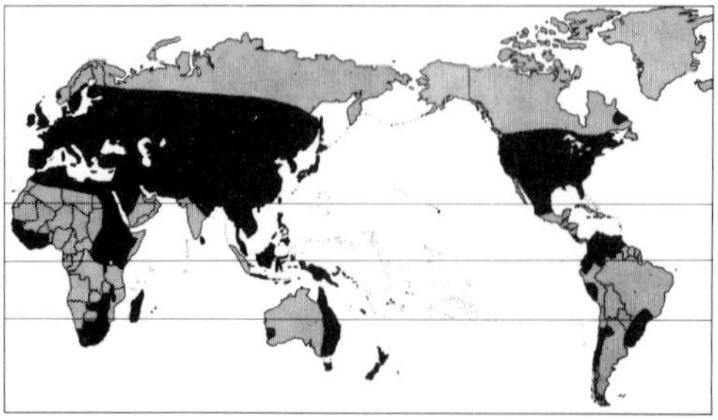

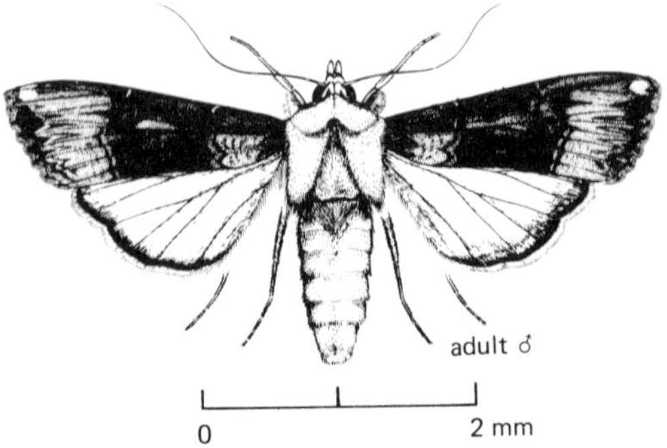

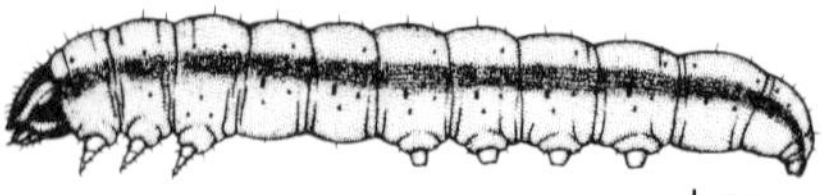

Agrotis segetum (Schiff.)

Common name. Turnip Moth (Common Cutworm)

Hosts. A polyphagous cutworm, attacking the seedlings of many different crops, and many vegetables and root crops.

Damage. Typical cutworm damage; seedlings cut off at ground level, and holes in the roots of turnips and other root crops, and potato; some young trees may suffer bark damage. Generally it is a more subterranean species than most.

Pest status. One of the most abundant cutworms in Europe, very common and widespread, and totally polyphagous; generally a serious pest in the Old World.

Life history. Eggs are laid in irregular clusters on the crop plants and weeds, or on the soil; they are globular, milky-white becoming cream-coloured, ribbed and reticulate; hatching requires 10–14 days. Each female lays 200–1000 eggs; oviposition is said to be inhibited at temperatures above 25°C.

The larvae are plump, glossy and smooth, grey-brown with two lateral lines; fully grown they measure 35–45 mm. The first instars feed on the plant foliage and later descend into the soil and become cutworms. Some of the larger caterpillars spend the day in the soil and emerge on to the surface at night to feed, but most of these caterpillars stay subterranean and feed underground. Most larvae hatch in the summer and feed up slowly over the autumn and winter, becoming fully fed in the spring; a few grow rapidly and produce a second generation of adults in the autumn, but it is not certain what happens to these moths.

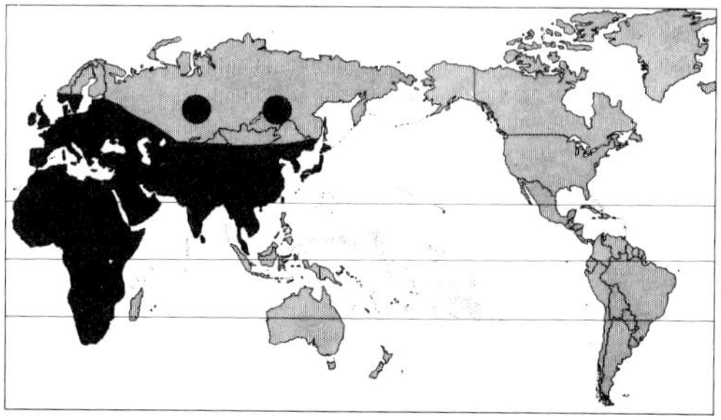

Pupation takes place in the soil in the spring, in an earthen cell, in the UK; the pupa is 12–20 mm long; pupal development is recorded as varying from 8–65 days according to temperature.

The adults are smaller than the Black Cutworm and measure 32–42 mm across the wings; forewing coloration is somewhat variable and may be quite dark, but the typical pattern is usually evident; the hindwing is almost white in the male but darker in the female. Adults fly in May and June; in the UK there may be a small second generation in September/October. In northern Europe this species is univoltine; in Africa there are 4–5 generations per year; under warm conditions one generation can be completed in 6–7 weeks.

Distribution. Widespread throughout Europe, Asia and Africa, but not recorded from Australasia.

Control. See page 425 for advice concerning control of cutworms.

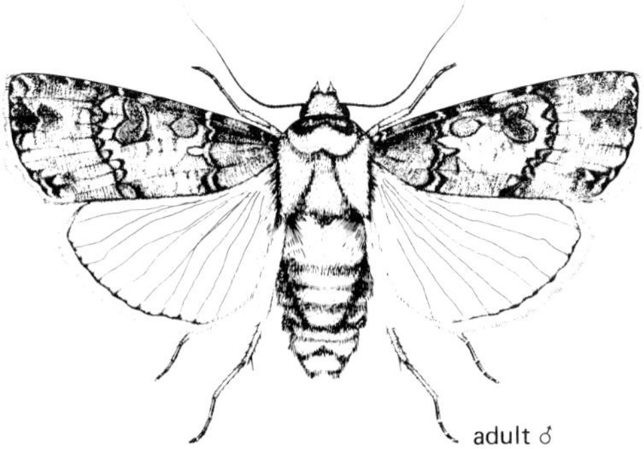

adult ♂

0 2 cm

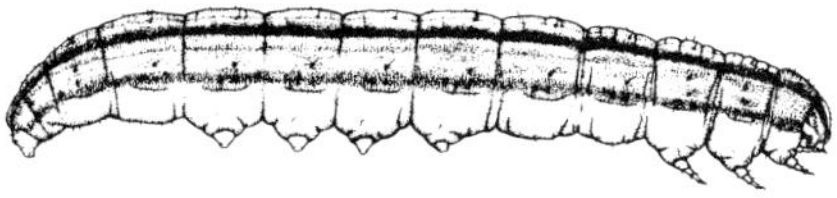

larva

Ceramica pisi (L.)

Common name. Broom Moth
Family. Noctuidae
Hosts (main). Sugarbeet, mangolds.

(alternative). Polyphagous on a wide range of deciduous trees, shrubs, herbaceous plants and ferns (bracken).

Damage. One of the less important cutworms in the UK and Europe, more damaging to foliage and less subterranean than most others; in habits partly a cutworm and partly a defoliator.

Pest status. Usually a minor cutworm pest, but quite widely distributed throughout Europe and the caterpillar is of distinctive appearance.

Life history. Eggs are laid on the food plant in untidy masses; flattened and spherical in shape, ribbed and reticulate, yellow slowly darkening; hatching requires about ten days.

The caterpillar is rather slender, somewhat variable in ground colour, but usually greenish or brown, with two distinctive pale lateral stripes, and pale ventrally. Fully grown larvae measure about 47 mm.

Pupation takes place in the soil within a fragile subterranean cocoon.

Adults fly in June and July and take nectar from flowers at night. They measure some 33–42 mm across the wings; the forewings are grey-brown; there is some colour variation recorded, but the subterminal line is usually quite distinct, as is the yellow pretornal blotch.

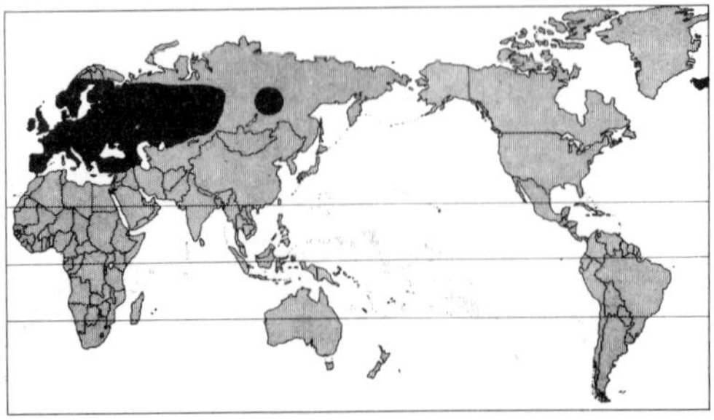

There is only one generation annually in the UK.

Distribution. Recorded throughout Europe to the Arctic Circle, and Asia through to Siberia, and Iceland.

Control. See ‘Control of cutworms’ on page 424.

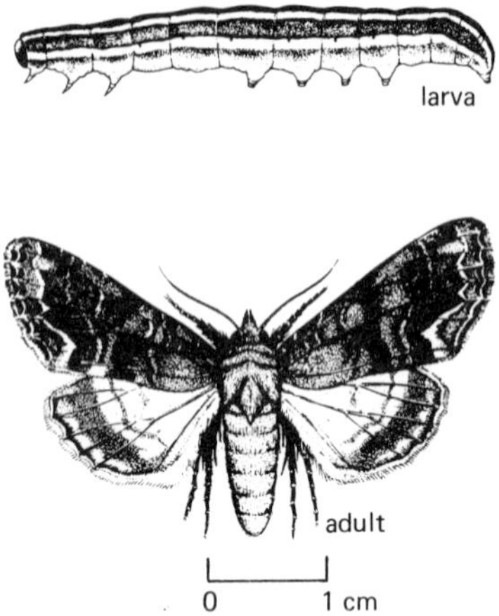

Cerapteryx graminis (L.)

Common name. Antler Moth (Cereal Armyworm)
Family. Noctuidae
Hosts (main). Cereals and various grasses (Gramineae).
(alternative). *Scirpus* and *Juncus* spp.

Damage. The feeding larvae destroy roots and shoots, especially in the early spring. At times population explosions occur, then the larvae behave as 'armyworms' and disperse en masse leaving areas of devastation behind.

Pest status. Not usually a serious pest, but in an outbreak year the damage can be quite spectacular. Although a European species, it is very similar to quite a large group of N. American species that act as both cutworms and armyworms, both on Gramineae and herbaceous crops.

Life history. Egg subspherical, finely sculptured, white turning pink then purple-grey; shell very hard. Eggs are dropped by the female in flight, mostly in August; each female lays about 200 eggs. The eggs overwinter.

Larvae hatch in the spring, usually March, although some may have hatched in the previous autumn and fed a while before hibernation; larval feeding continues until about June. Fully grown caterpillars measure about 35 mm (30–40); they are bronze-brown, and glossy, with dorsal, subdorsal and spiracular lines in pale yellow.

Pupation takes place in an earthen cell amongst the grass roots, and usually requires about 25 days.

Adult moths emerge over the period July to September, but are most abundant in August. They are a variable brown in coloration with quite a distinctive pattern on the forewings; the female is larger (35–9 mm wingspan) and paler than the male (27–32 mm wingspan).

There is only one generation per year.

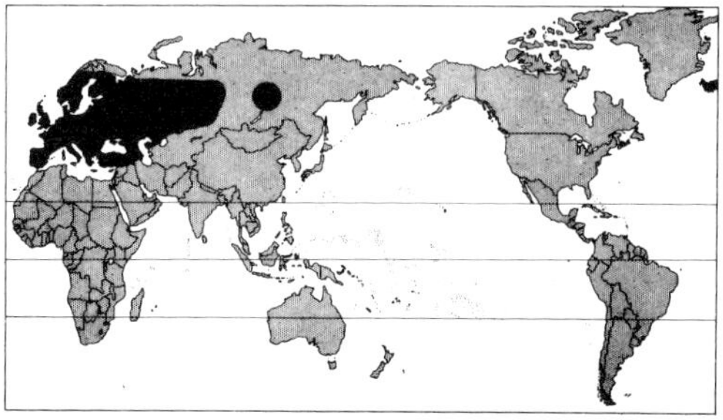

Distribution. Antler Moth is recorded throughout Europe, up to the Arctic Region, including Iceland, and western Asia to Siberia.

Similar species in N. America include:

Chorizagrotis auxiliaris Grote – (Army Cutworm) Canada.

Crymodes devastator (Brace) – (Glassy Cutworm) Canada and USA.

Dargida procincta (Grote) – (Olive-green Cutworm) Canada.

Faronta diffusa (Wlk.) – (Wheat Head Armyworm) Canada and USA.

Feltia spp. – (cutworms) Canada and USA.

Control. When control is required, in an outbreak year, then the chemicals suggested on page 425 should prove effective.

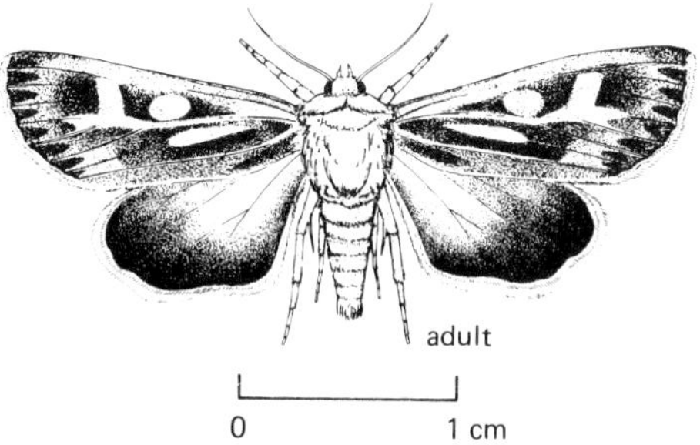

Euxoa nigricans (L.)

Common name. Garden Dart Moth
Family. Noctuidae
Hosts. A polyphagous cutworm recorded damaging a wide range of dicotyledenous herbaceous crops, particularly sugarbeet, carrot, potato, parsnip, onions and clovers; many wild hosts are also recorded, but not Gramineae.
Damage. Most serious damage is done to seedlings, which are cut off and left on the soil surface; sometimes a single cutworm destroys many seedlings in a single night.
Pest status. Not one of the most common cutworms in the UK, but still quite abundant and generally a serious crop pest, especially in lowland areas.
Life history. Eggs are globular, with a flat base, delicately reticulate, and glossy; most are laid in late summer but do not hatch until the following spring (about mid-April).

The larvae are yellowish-brown caterpillars, rather similar to Turnip Moth, smooth and shiny, with longitudinal greenish-grey lines, and a double whitish lateral stripe; fully grown the caterpillar is about 40 mm. It is thought that probably some eggs hatch in the autumn and some larvae overwinter, causing serious crop damage in the early spring. In June most of the larvae spin silken cocoons in the soil, close to the surface, where they pupate.

Adults emerge in July and August; they are mostly seen on the wing in August; forewings are dark to pale brown, generally very variable; wingspan is 32–40 mm. A species quite difficult to recognize.

There is only one generation per year.
Distribution. Recorded throughout Europe and Eurasia, from Portugal in the south to Finland in the north.

Other important species of *Euxoa* include:
E. tritici (L.) – (White-line Dart) a minor cutworm; Eurasia.
E. auxiliaris (Grote) – (Army Cutworm) USA.
E. detersa (Wlk.) – (Sandhill Cutworm) Canada.
E. messoria (Harris) – (Dark-sided Cutworm) Canada and USA.
E. ochrogaster (Guen.) – (Red-backed Cutworm) Canada and USA.
E. scandens (Riley) – (White Cutworm) Canada and USA.
E. tessellata (Harris) – (Striped (Spotted) Cutworm) Canada and USA.
E. tristicula (Morr.) – (Early Cutworm) Canada.
Control. For control of cutworms see page 425.

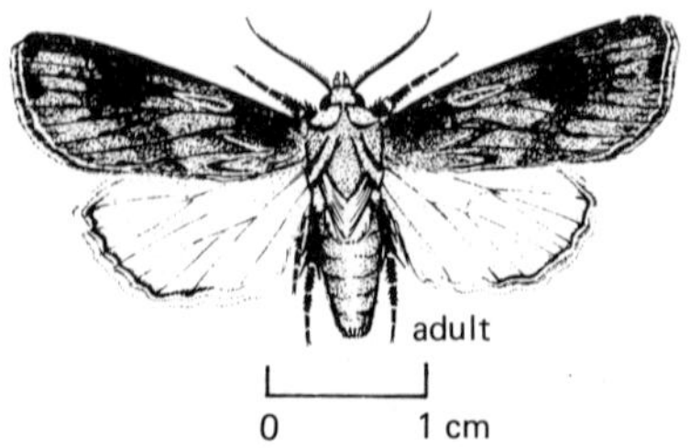

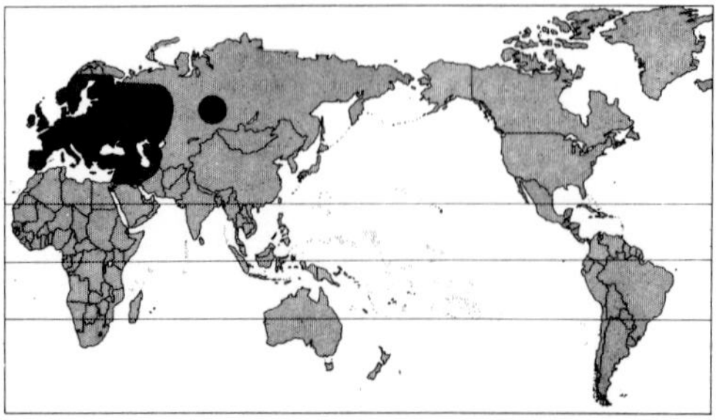

Hydraecia micacea (Esper)

(= *Gortyna micacea* Esp.)

Common name. Rosy Rustic Moth (Potato Stem Borer)
Family. Noctuidae
Hosts (main). Potato

(alternative). Tomato, sugarbeet, rhubarb, rape, hop, celery, beans, mangolds, wheat, barley, oats, strawberry and many wild plants; quite polyphagous.

Damage. The feeding larvae tunnel inside stems and thick leaf petioles, usually boring upwards but sometimes tunnelling down to the roots. Damaged plants wilt and usually die.

Pest status. One of the major temperate stem borers of herbaceous plants; possibly the most important pest species. Field infestations rarely exceed 10% of a crop.

Life history. Eggs are laid on the lower leaves of the plant, usually on wild grasses, in late summer and early autumn (fall); they overwinter and hatch the following spring (usually May).

The young larvae feed first on the grass leaves or weed leaves, then they disperse to larger plants and bore into their stems. The caterpillar is pinkish with a red dorsal stripe and large lateral brown spots; fully grown it measures 25–35 mm. In July pupation takes place in an earthen cell in the soil, and the time required is about five weeks.

Adults emerge in August and September, and egg-laying soon follows. The forewings are reddish-brown with a distinct colour pattern, and the hindwing is greyish; wingspan is about 25–35 mm.

There is only one generation per year.

Distribution. Widely distributed throughout the British Isles, throughout Europe and Asia, to Manchuria; also in the USA and Canada.

Hydraecia immanis (Guen.) is the Hop Vine Borer of Canada, boring hop vines and maize stems.

A similar temperate stem borer found in the UK and Europe, but mostly attacking cereals and grasses, is *Luperina testacea* (Schiff.), the Flounced Rustic Moth.

Control. Pests with this type of biology can be controlled somewhat by the destruction of weeds in the crop and around its edges, as these are the oviposition sites. Attacks in Europe are seldom sufficient to justify use of insecticides; usually infestations are few and scattered throughout the crop. Destruction of infested plants is always recommended.

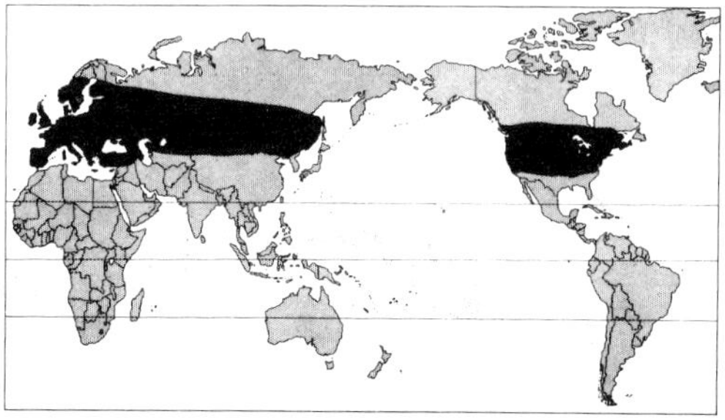

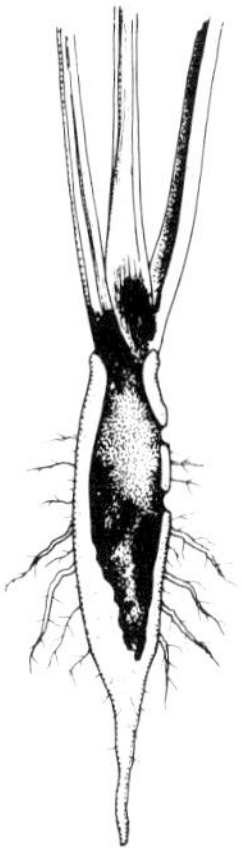

sugar beet root tunnelled by larva

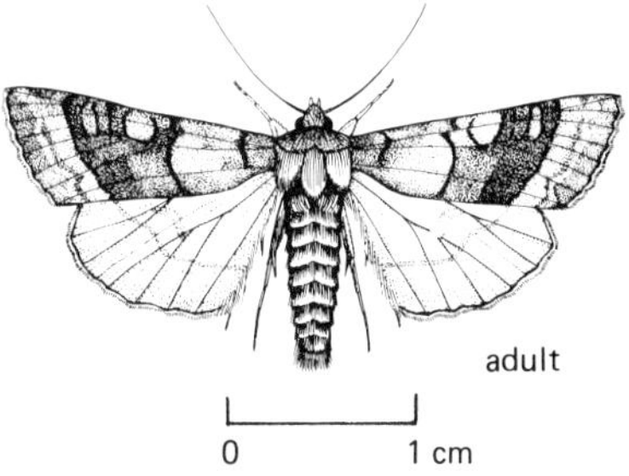

adult

Heliothis spp.

(*H. armigera* (Hubn.))
(*H. zea* (Boddie))
*(= *Helicoverpa* spp.)

Common name. American Bollworm; Corn Earworm, etc.
Family. Noctuidae
Hosts (main). Cotton, maize, tomato, beans.
(alternative). Tobacco, sorghum, cabbage, capsicums, okra, other legumes, other vegetables, other crops, and various wild hosts.
Damage. Essentially 'fruitworms' that bore into a wide range of fruits (in the botanical sense); most damaging when they bore cotton bolls, maize cobs, and tomatoes. Sometimes recorded as a leaf eater, more usually in buds.
Pest status. Very serious polyphagous pests, that together cover all the warmer parts of the world and regularly migrate into cooler temperate regions.
Life history. Eggs are laid singly, stuck on to the host plant; spherical in shape, 0.5 mm in diameter, yellow turning brown; hatching requires only 2–4 days. Each female lays 1000 or more eggs (2500 eggs recorded from one female).

The larvae are variable in colour, greenish or brown, with several longitudinal lines. After 5–6 instars they reach a body length of 40–5 mm, and are quite plump. Larval development takes 14–60 days, according to temperature.

Pupation takes place in the soil inside an earthen cell; the pupa measures 16–20 mm, and development is completed in 10–14 days in the tropics but in cooler climates the pupae overwinter.

The adult is a stout-bodied, brown moth, of wingspan 40–4 mm; the hindwing is pale but with a broad dark border. The two species are very similar in appearance, but genitalia studies have shown them to be allopatric, except for Hawaii where they both are recorded.

In the tropics the life-cycle can be completed in 25–30 days, and there are usually 5–6 generations annually, but the more northern populations have fewer generations, until in Canada and the UK the species only occurs as a summer migrant and does not usually overwinter successfully.

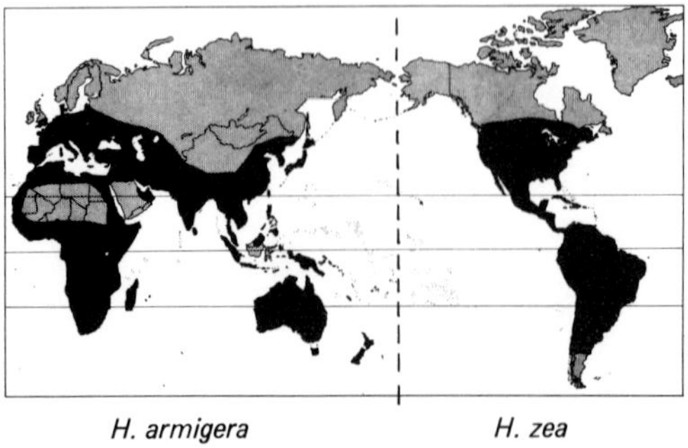

H. armigera *H. zea*

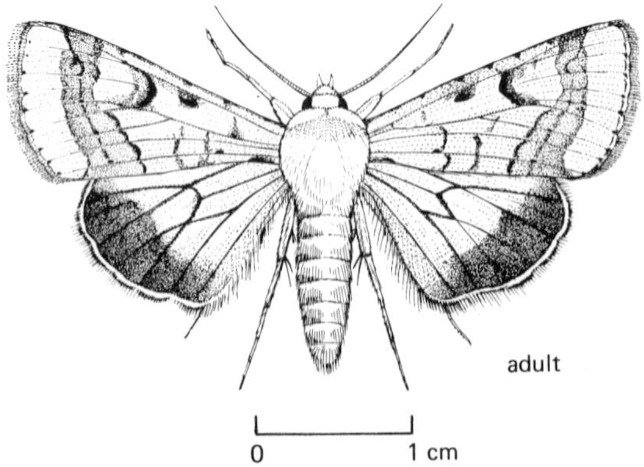

Distribution. The two species are allopatric, and together cover the warmer parts of the world; both species are recorded from Hawaii (CIE map nos. A.15 and A.239).

Other important species of *Heliothis* include the following:

Heliothis assulta Guen. – (Cape Gooseberry Budworm (Oriental Tobacco Budworm)) Old World Tropics and Japan (CIE map no. A.262).

Heliothis hawaiiensis (Q. & B.) – (Hawaiian Bud Moth) Hawaii.

Heliothis ononis (D. & S.) – (Flax Bollworm) Canada and USA.

Heliothis virescens (F.) – (Tobacco Budworm) N. and S. America (CIE map no. A.238).

Heliothis viriplaca Hufn. – (Flax Budworm) Japan.

Control. Insecticidal control is aimed at killing the young caterpillars whilst still exposed on the foliage; using DDT, HCH, carbaryl, endosulfan or toxaphene.

*Several of these well-known pest species are now referred to as *Helicoverpa* but are here retained in *Heliothis* for convenience.

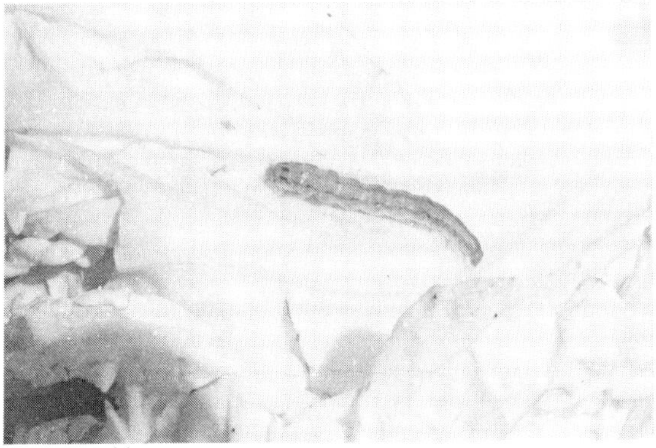

Larva on cabbage

Lacanobia oleracea (L.)

(= *Diataraxia oleracea* (L.))

Common name. Tomato Moth (Bright-Line Brown-Eye)
Family. Noctuidae
Hosts (main). Tomato (glasshouse and field crops)
(alternative). Cabbage, swede, sugarbeet, potato, spinach, many flowers, etc.; quite polyphagous but showing a definite preference for Chenopodiaceae and Polygonaceae.
Damage. The caterpillars feed on the leaves of many plants, but on tomato they eat leaves and bore into the ripening fruits; larvae are usually solitary.
Pest status. A widespread and polyphagous pest found in small numbers on many crop and garden plants.
Life history. Eggs are laid in batches on the food plant, in untidy clusters of 50 or more, usually on the underside of the leaves; coloured green, hemispherical and ribbed; hatching requires 7–10 days. Each female lays 300–800 eggs.

The caterpillars are variable in colour, ranging from green to brown, but with a pronounced lateral yellow stripe with a dark upper margin; fully grown they measure 40–5 mm. Larval development is rapid, and in the UK the time required is 4–6 weeks. Young larvae skeletonize the leaves, but larger larvae eat the entire lamina.

Pupation takes place in the soil inside a fragile silken cocoon; most larvae have pupated by the end of August, but some may overwinter and pupate in the spring; pupal development takes 3–6 weeks.

Adults emerge from late May to July and come to flowers at night for nectar; they are long-lived, and mostly inhabit gardens rather than open country; it is thought their preferred habitat is the margin of saltmarshes. The forewing is dark reddish brown with white markings; wingspan is 36–44 mm.

This species is univoltine in northern regions, but bivoltine further south.

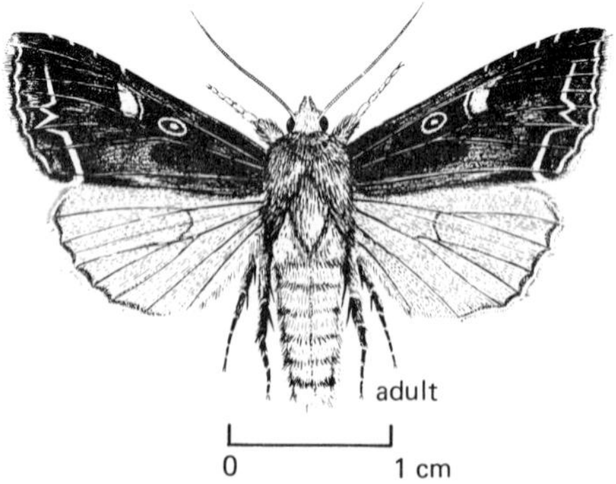

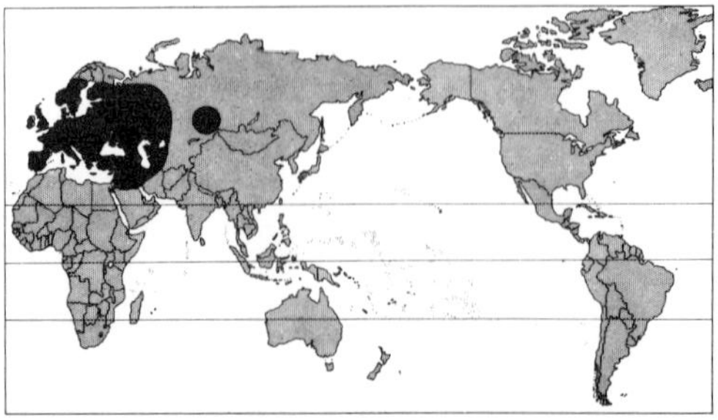

Distribution. Very abundant throughout the UK and Eurasia, from the south of Europe to central Fennoscandia.

Lacanobia legitima (Grote) is the Striped Garden Caterpillar of the USA.

Control. In the past the usual recommendation has been a spray of DDT against this pest, but the present recommendation is usually for carbaryl, BTH, dichlorvos, resmethrin or permethrin. In glasshouses smokes or fogs are often the most appropriate method of application; for field crops a high-volume spray over the foliage would be recommended.

Larva feeding on lettuce

Mamestra brassicae (L.)

Common name. Cabbage Moth
Family. Noctuidae
Hosts (main). Cabbage, and other Cruciferae.
(alternative). Lettuce, beet, onions, potato, pea, tomato, many flowers; forest trees (incl. beech and oaks); totally polyphagous.
Damage. Newly hatched larvae feed gregariously on the underside of outer leaves (causing skeletonization); larger larvae disperse over plant and eat the leaf lamina; some larvae show preference for tunnelling the cabbage heart; faecal fouling of vegetable crops is often more serious economically than actual damage.
Pest status. A serious pest of many crops, abundant and widespread; even small populations are quite damaging.
Life history. Eggs are hemispherical and ribbed, quite small, white turning pinkish, laid in neat batches up to 50 or more, usually on the underside of leaves; hatching requires about eight days. Most eggs are laid in May and June in the UK.

On hatching the young caterpillars feed gregariously on the leaf underside causing skeletonization; they remain thus for the first few instars, but the larger caterpillars disperse over the plant, and on cabbage show a preference for tunnelling into the heart. They produce copious quantities of faecal matter and contamination of crops is often very serious. The young larvae are greenish, but the fourth instar darkens to a brownish colour dorsally and yellowish green ventrally; some individuals are almost black. Full-grown larvae measure 40–50 mm, and have a characteristic small hump on abdominal segment 8; the border of the dark dorsal colour and pale ventral is quite sharp laterally. Larval development takes 4–5 weeks.

Pupation takes place in the soil in a flimsy cocoon in August. These pupae may overwinter, or some may give rise to a second generation whose larvae feed until October when they pupate.

Adults are recorded from every month of the year in the UK, but are most abundant from June to the end of September. They are large dark moths, of wingspan 34–50 mm, and characterized by having a curved dorsal spur on the tibia of the foreleg.

Univoltine or bivoltine, according to local conditions.
Distribution. This species is abundant throughout Europe (up to central Fennoscandia), and temperate Asia to India and Japan (CIE map no. A.467).

Other important species of *Mamestra* include:
Mamestra configurata Wlk. – (Bertha Armyworm) Canada and USA.
Mamestra illoba Butler – (Mulberry Caterpillar) Japan.
Mamestra persicariae L. – (Beet Caterpillar) Japan.
Control. Control measures are the same as for *Pieris brassicae*, see page 412.

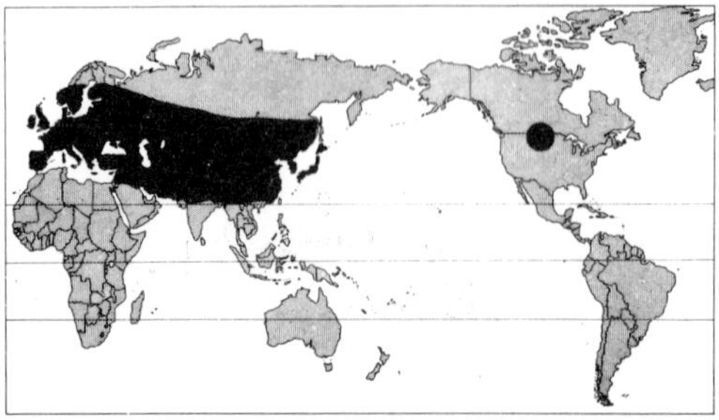

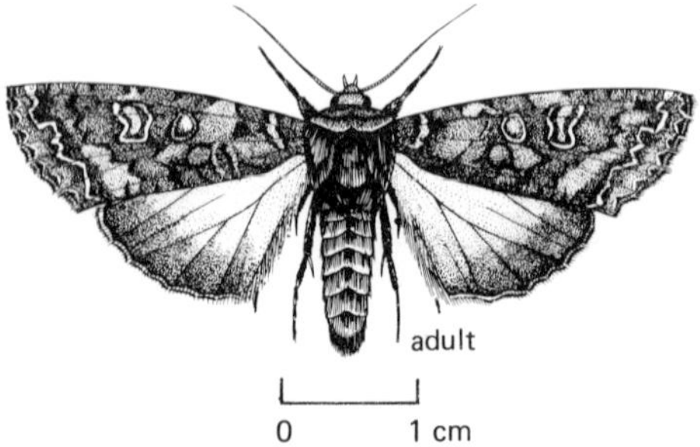

Melanchra persicariae (L.)

Common name. Dot Moth
Family. Noctuidae
Hosts. A polyphagous caterpillar recorded occasionally from many different low-growing crops, and many wild plants, also including elder, larch and willows; a common garden pest on flowers and fruit trees.
Damage. Larval feeding is both defoliation by eating the leaf lamina, and seedling destruction typical of cutworms; the larvae are nocturnal and hide in the foliage during the day.
Pest status. A minor crop pest but generally quite abundant in Europe, and being polyphagous it damages a wide range of crops.
Life history. The eggs are hemispherical, ribbed, whitish green becoming pinkish brown; laid singly or in small untidy masses on the foodplant, mostly in July; hatching requires about eight days.

The larvae have a variable body coloration, from pale grey-green to dark green or purple-brown; a pale dorsal line, and dark backward-pointing chevrons on the abdominal segments, together with a pronounced dorsal hump on segment 8; ventrally a yellow-brown colour. Full-grown larvae are 30–45 mm, and of a velvety texture. Larvae feed slowly and are not fully grown until September or October. Some larvae may overwinter, but most pupate in the soil in a cocoon in the autumn.

The adults are distinctive in being bluish-black with a white reniform stigma in the forewing; wingspan is 38–50 mm. Adults emerge over a lengthy period from late June to August; univoltine.
Distribution. This species is recorded throughout Europe (not Norway) and Asia to China and Japan.
Melanchra picta (Harris) (formerly *Ceramica picta*) is the polyphagous Zebra Caterpillar of Canada and the USA.
Control. Control for this species is seldom required, but should it be necessary then the recommendations for cutworms generally should suffice.

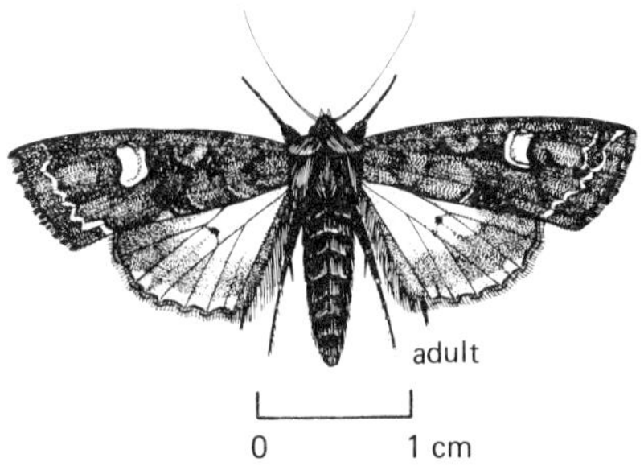

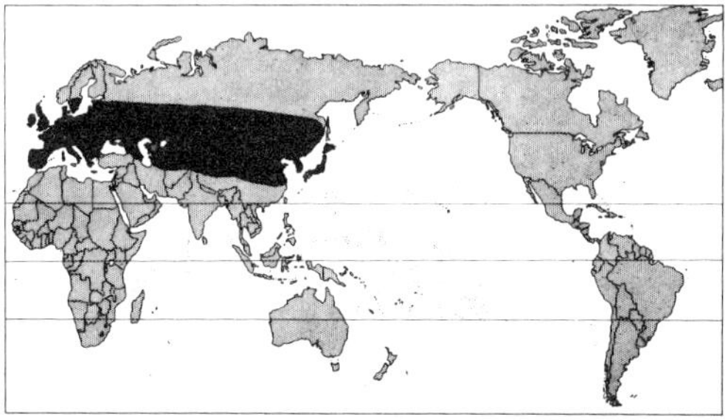

Larva on geranium leaf

Mythimna loreyi (Dup.)

(many synonyms)

Common name. Cereal Armyworm (The Cosmopolitan)
Family. Noctuidae
Hosts (main). Maize, rice, sorghum.
(alternative). Wheat and sugarcane; polyphagous within the Gramineae.

Damage. Young larvae skeletonize the leaves; older caterpillars are gregarious and eat entire leaves and sometimes the whole plant, feeding at night. Sometimes act as armyworms when damage may be devastating.

Pest status. A serious pest of cereal crops throughout the subtropics, tropics and warmer regions of the Old World; occasionally entire crops are destroyed.

Life history. Eggs are laid in the leaf sheath, in batches of up to 100; hatching takes about eight days.

The larvae are cylindrical; the head is dark brown (yellowish in *M. unipuncta*) and body coloration variable; first instars usually whitish or grey green, but later instars varying greatly from yellowish to dark brown above, always pale ventrally, dorsal line not very distinct (white and distinct in *M. unipuncta*) bounded by grey sub-dorsal lines; the typical *Mythimna* chevrons on each segment above the dorsal line are black and very distinct. There are usually six larval instars, and the larger larvae are gregarious and at times of high population density act as armyworms. Full-grown larvae measure 40–5 mm; larval development usually takes 14–22 days.

Pupation takes place in the soil, at a depth of about 4 cm, in an oval cocoon; the brown pupa is slim, 15–20 mm long; pupation takes about 18–20 days.

The adults are brown moths, with faint markings on the forewings, but the hindwings are white (dark in *M. unipuncta* and other *Mythimna* spp.); wingspan is 35–45 mm. The adults are nocturnal, and feed on nectar or honey-dew; they generally live for 1–2 weeks, and may be migratory.

Distribution. This species is cosmopolitan throughout the warmer parts of the Old World; it is thought to be of subtropical origin; recorded from Africa, Mediterranean Region, India, southern Asia, S.E. Asia, China and Japan (CIE map no. A.275). There are regular dispersal movements (migrations) farther north and this species has been recorded as an immigrant in the British Isles several dozen times.

Control. In an area at risk from this pest, crop rotation with non-graminaceous plants helps to reduce local populations, as also does grass weed elimination, destruction of crop residues (burning preferably) and deep ploughing.

As another sporadic pest, chemical control is difficult because the pest population is usually large and damaging by the time the seriousness of the attack is evident. Sprays or dusts of contact insecticides is generally recommended, involving DDT, HCH, endrin, parathion, dichlorvos, fenitrothion or trichlorphon.

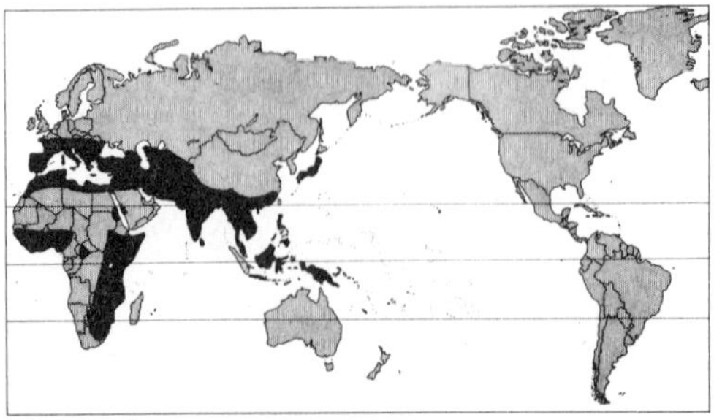

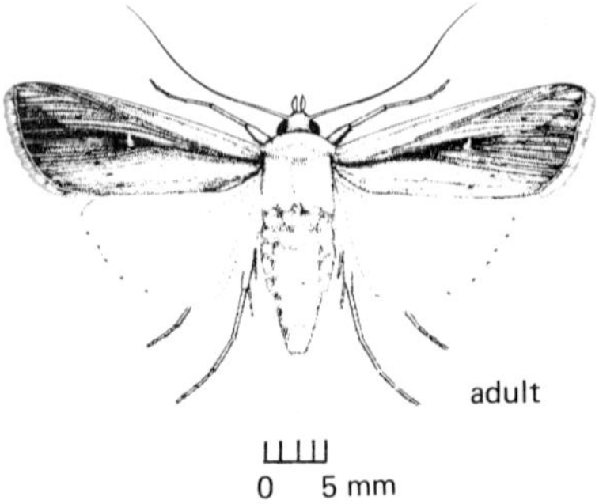

Mythimna unipunctata (Haw.)

(many synonyms)

Common name. Cereal Armyworm: Armyworm (USA) (White-speck; American Wainscot)
Family. Noctuidae
Hosts (main). Oats, wheat, barley, maize, millets, rye; various grasses (Gramineae).

(alternative). Alfalfa, clovers, sugarbeet, potato, cabbage, turnip, mangolds, flax, buckwheat and Jerusalem artichoke.

Damage. The feeding larvae eat leaves, and in peak years the gregarious swarms of caterpillars cause defoliation and even complete crop destruction. A field in Canada in which a crop of oats was completely destroyed contained about 200 pupae per square metre. Feeding takes place at night.

Pest status. In 'peak' years a very serious pest of cereal crops, and damaging to other crops; in Canada it is a summer migrant only, but can be very damaging.

Life history. Eggs are laid in small groups on the cereal or grass leaves, or under the sheath; the groups vary in size from 5–200 or more; they are hemispherical, white and glossy; hatching requires 8–10 days. Each female moth can lay 700–1400 or more eggs. Oviposition is sometimes communal, which is one reason for a population outbreak. Egg-laying occurs in May and June mostly.

The larvae are initially whitish or grey-green, but later vary from yellowish to dark purple-brown above, but pale ventrally. The white dorsal stripe and white or pink subdorsal lines are usually distinct, and the head is yellowish-brown. Full-grown larvae (sixth instar) are about 40 mm long. The large larvae are gregarious and swarm when numbers are high, preferring to feed on Gramineae, but damaging many other crops. Most of the damage is done in July and August. Larval development takes 3–4 weeks usually.

Pupation takes place inside a flimsy cocoon low down among the grass and cereal stems; during the summer, pupation takes 2–3 weeks; in cooler regions it appears that some larvae hibernate overwinter and others pupate and overwinter as pupae.

The adults have forewings reddish or yellowish brown, with a small white stigma; hindwings are dark grey; wingspan is 40–8 mm. The moths are strongly migratory, and in Canada it is thought they do not survive the winter but that there is an annual reinvasion from the USA in May; in Ontario there are 2–3 generations of this pest each summer; in warmer regions there may be 4–5 generations per year; one generation takes about 50 days.

Distribution. This species is recorded throughout the USA and part of Canada, in parts of S. America, southern Europe (immigrant in the UK), and parts of Africa and western Asia (CIE map no. A.231).

Control. For control see *Mythimna loreyi*, page 438.

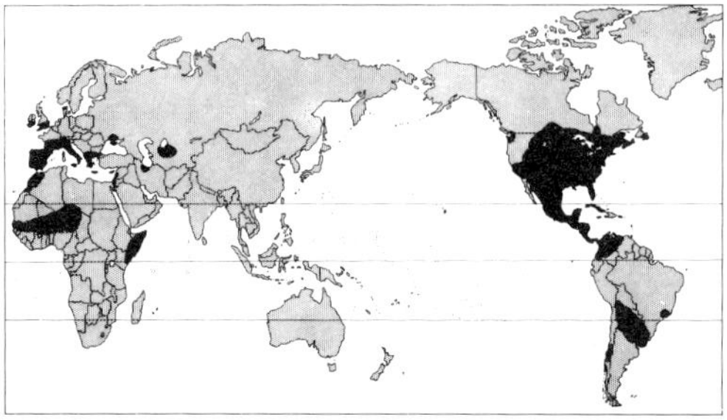

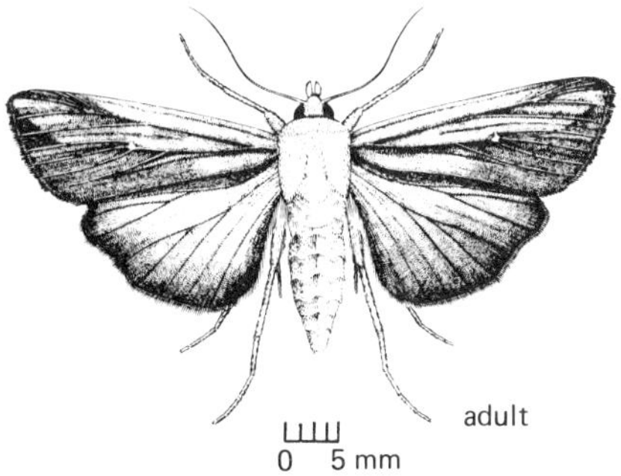

Noctua pronuba (L.)

Common name. Large Yellow Underwing
Family. Noctuidae
Hosts. A polyphagous cutworm feeding on a wide range of herbaceous plants, including grasses and cereals; most damaging to vegetables, especially brassicas and lettuce.
Damage. Typical cutworm damage; seedlings cut off at ground level at night; roots and tubers eaten underground and sometimes holed or even hollowed out. Lettuce stems may be hollowed out just below ground level.
Pest status. One of the commonest cutworms in Europe; very widespread and abundant and damaging to many garden and horticultural crops; not usually very serious on most agricultural crops.
Life history. Eggs are globular, ribbed and reticulate; creamy-white turning reddish-grey; laid in flat sheets under the leaves of the food plants; each female lays more than 1000 eggs over the period June to October; hatching requires about one month.

The larvae are typical cutworms, and more subterranean than most species. The caterpillar is stout and robust, variable in colour but mostly brown above and pale underneath; a narrow pale dorsal line, and subdorsal lines marked above by a series of black dashes on the abdominal segments, spiracles black. Full-grown larvae measure about 50 mm. Larvae feed throughout the winter and are quite active on mild days, but remain mostly in the soil. Larvae are found over the period July/August to the following May.

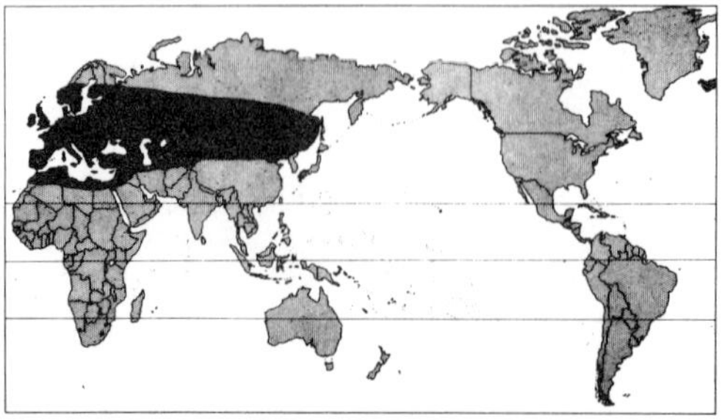

Pupation takes place in the soil in an earthen cell the pupa is large, being up to 25 mm long, shiny brown and with a very characteristic cremaster. Adults emerge over a prolonged period from June to September, but most in August. It appears that there is an obligatory pre-oviposition period of about one month in females. Adults are long-lived and feed from many flowers at night, usually resting on the ground by day. The moths are large and sexually dimorphic; wingspan is 50–60 mm, males are the smaller. Coloration is very variable and the forewing varies from pale brown to dark brown, and sometimes quite grey or else reddish; the hindwing is always yellow with a black border; generally the males are darker and the females paler.

An univoltine species in the UK, but 3–4 generations per year in Israel.
Distribution. *Noctua pronuba* is recorded as being Palaearctic, including Iceland and central Fennoscandia, and N. Africa.
Control. For control see page 424 on 'Control of cutworms'.

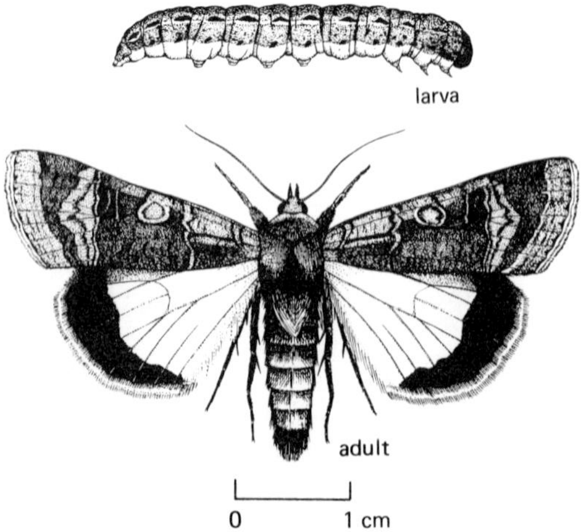

Spodoptera exigua (Hbst.)

Common name. Lesser (Beet) Armyworm
Family. Noctuidae
Hosts (main). Sugar beet, alfalfa, cotton, upland rice.

(alternative). Asparagus, tobacco, groundnut, tomato, cabbage, potato, strawberry; many other crops, many flowers, and wild plants; totally polyphagous.

Damage. The larvae are gregarious and may feed in large swarms, when defoliation may be serious; young plants are killed but older plants may recover, and cereals tiller.

Pest status. A sporadic pest, but widespread and very polyphagous; at times of population outbreak damage to crops may be devastating, and entire crops destroyed.

Life history. Eggs are laid in clusters, several layers thick, and covered with hairs from the female abdomen. Each cluster contains 50–300 eggs; one female lays 300–900 eggs (up to 1700); hatching requires only 2–4 days.

The larvae are very gregarious, at first green, later becoming variable green or brown usually with a lateral stripe, and grow to a length of about 30 mm. Young larvae skeletonize the underside of the leaves, but later as they grow they eat the entire lamina. Larvae develop through their six instars in only 10–13 days.

Pupation takes place in the soil in an earthen cell and requires about six days. The adults are small brownish moths of wingspan 24–30 mm. They are rather nondescript in appearance and not easy to recognize. In parts of Europe it does not move far from the place of emergence, but in Canada it is an annual migrant from the USA.

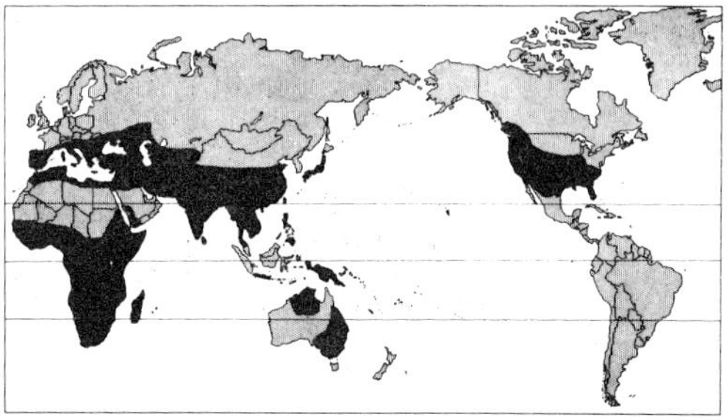

One generation can be produced in as little as 21–4 days; in Africa and Israel there are usually eight generations annually.

Distribution. Recorded throughout Africa, southern Europe, India and southern Asia, Japan, Australasia, USA and Canada (CIE map no. A.302).

Control. The sporadic nature of the very large populations makes this a difficult pest to control, as infestations are usually serious by the time they are clearly evident. The insecticides generally recommended include dichlorvos, endrin, trichlorphon and toxaphene; the other chemicals used against noctuid caterpillars should also be effective.

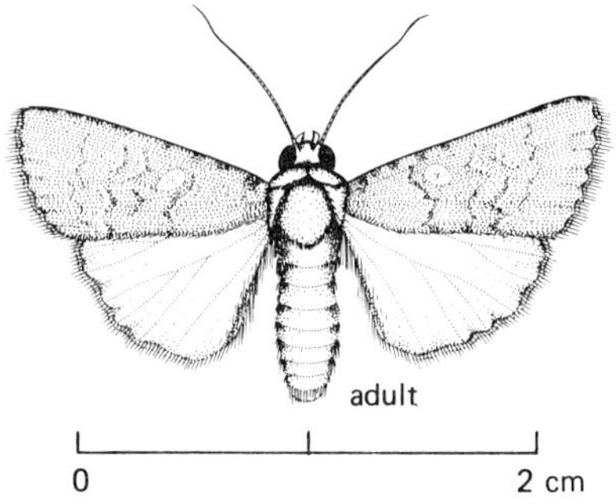

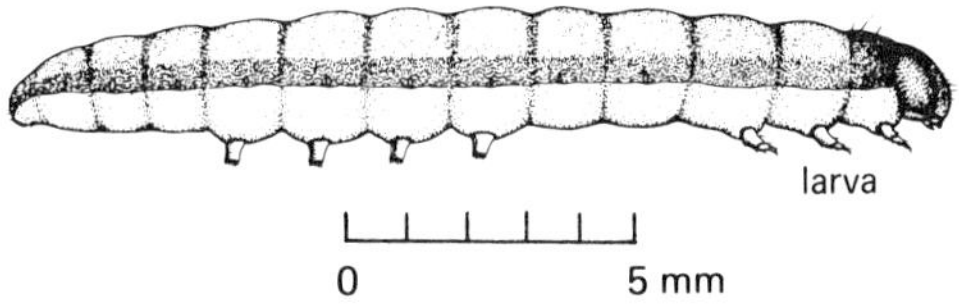

Spodoptera spp.
(*S. littoralis* (Boisd.))
(*S. litura* (F.))

Common name. Cotton Leafworms (Mediterranean Climbing Cutworm) etc.
Family. Noctuidae
Hosts. These two species are totally polyphagous, and feed on a very wide range of crops (herbaceous, tree and shrub) as well as Gramineae, ornamentals and wild plants.
Damage. The caterpillars are essentially leaf eaters, but they do occasionally adopt cutworm habits; heavy infestations result in crop defoliation. Young larvae cause leaf-skeletonization; later they disperse and become solitary.
Pest status. They are not too often serious pests on any one crop, but are frequently minor pests on a very wide range of local crops, and thus of considerable importance.
Life history. Eggs are spherical (0.3 mm diameter), and laid on the underside of leaves in batches of 100–300, and covered with hair-scales; each female lays some 1000–2000 eggs (3700 recorded); hatching requires 2–6 days.

Newly hatched larvae are gregarious and remain together and skeletonize the leaf on which they feed. Later they disperse and become more solitary and nocturnal, and are leaf lamina eaters. Development through the six instars takes 12–20 days under warm conditions, but is much longer under cooler conditions. The larvae are greenish at first, but later become more brown, with dark markings; there are distinctive lateral and dorsal lines in yellow; the lateral yellow line is interrupted on the first abdominal segment by a large black blotch, on the other segments it is bordered dorsally by a black semilunar mark. The larvae of the two species are not really separable by morphological characters, but they are allopatric. Full-grown larvae are mostly about 40–5 mm.

Pupation takes place in the soil in an earthen cell, and requires 5–30 days according to temperature.

The adults are pale brownish moths, with the yellow-brown forewing having a distinctive pale band medially: wingspan is 30–40 mm: they are short-lived.

One generation can be completed in as little as 24–30 days, and in the humid tropics there may be eight generations annually.
Distribution. The two species are allopatric, and together cover the warmer regions of the Old World (CIE maps nos. A.232 and A.61). In Northern Europe *S. littoralis* is occasionally introduced into glasshouses.
Control. The insecticides generally used against these pests include DDT, HCH, endosulfan, fenitrothion, malathion, tetrachlorvinphos and trichlorphon, usually as either high-volume sprays or ultra-low-volume applications.

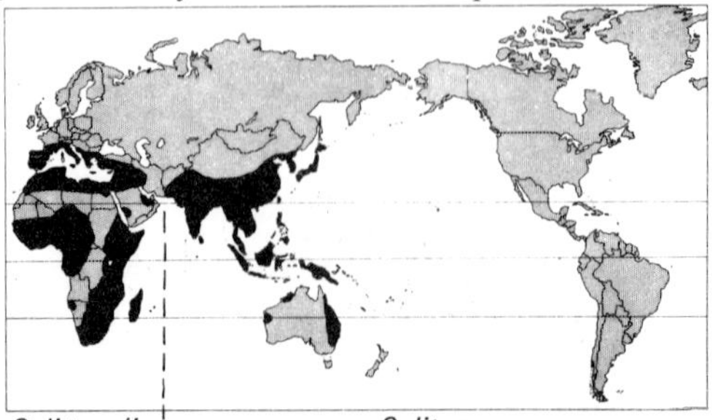

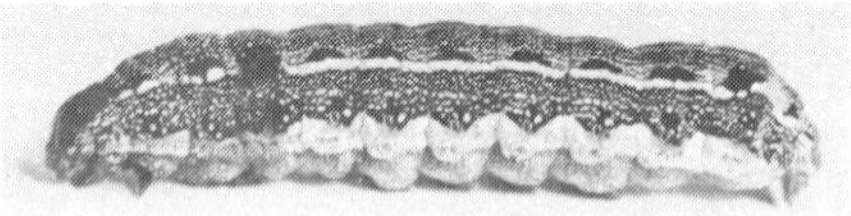
larva

adult

Xestia c-nigrum (L.)

(= *Amanthes c-nigrum* L.)

Common name. Spotted Cutworm (Setaceous Hebrew Character)

Family. Noctuidae

Hosts. A polyphagous cutworm showing no definite host specificity; recorded damaging celery, carrot, sugarbeet, cotton, Cruciferae, flax, grapevine, tobacco, tomato, various legumes, blueberry, chrysanthmum etc.; also many wild hosts including several common ruderals.

Damage. Feeding larvae eat the leaves of the host plants and also act as cutworms and destroy seedlings, as well as eating roots in the soil.

Pest status. One of the less important cutworms, but totally polyphagous and very widespread, and regularly encountered.

Life history. Eggs are laid singly on the food plant or on the surrounding soil; the average number laid per female is less than 100; hatching requires 8–9 days.

The larvae are bright green at first, becoming grey-brown or olive-brown, with a series of black abdominal marks subdorsally and a whitish transverse band on segment 8; spiracles are white, and a broad yellowish spiracular stripe. Fully grown the larvae measure about 37 mm. In habits the larvae are more similar to armyworms than to cutworms, but in general their biology is not well known. It is thought that most feed throughout the autumn and winter and again the following spring, pupating in April or May; adults are seen mostly in the autumn, but there is a small emergence usually in May and June.

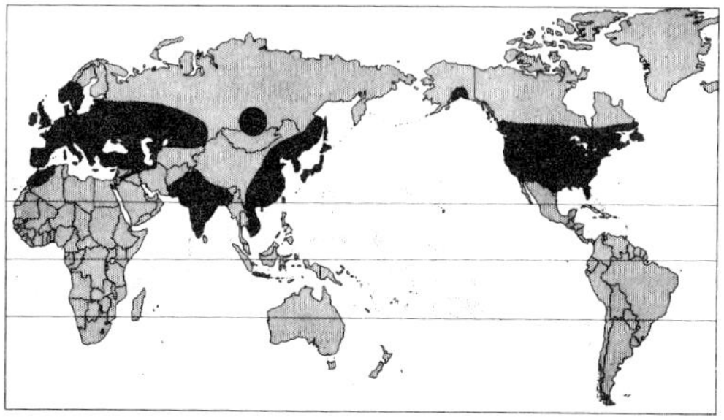

The adults are somewhat variable in colour, being usually greyish or brownish, with a distinctive pattern on the forewing; the hindwing is whitish but darker distally; wingspan is 35–45 mm.

In Canada and northern Europe they are probably bivoltine, but in warmer regions there may be many generations annually. In some regions this species is quite strongly migratory.

Distribution. Found throughout Europe (up to Finland), and Asia from India to Korea and Japan, also N. Africa, Java, and N. America (CIE map no. A.400).

Several other species of *Xestia* (= *Amanthes*, etc.) are recorded as crop pests throughout Europe and Asia.

Control. Natural levels of parasitism are reported to be very high in some locations. Control is not often required, but on these occasions sprays of the chemicals mentioned on the previous page should suffice.

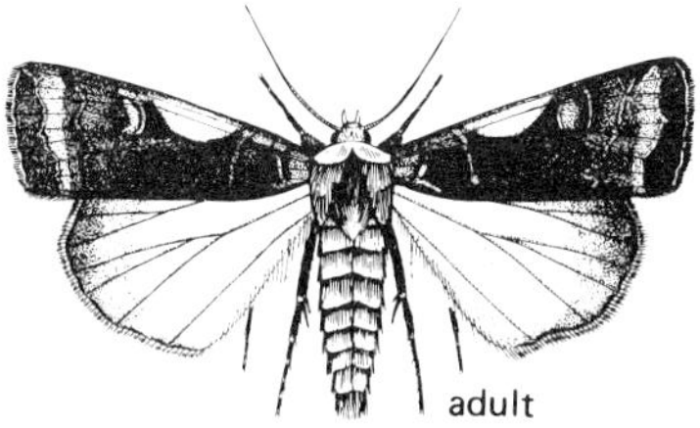

Autographa gamma (L.)

(= *Plusia gamma* L.)

Common name. Silver-Y Moth
Family. Noctuidae (Plusiinae)
Hosts (main). Sugarbeet, lettuce, cabbages, tomato, potato, beans, peas.

(alternative). Flax, clovers, alfalfa, strawberry, tobacco, cereals; many ornamentals, and many wild plants (ruderals); quite polyphagous.

Damage. The larvae are leaf eaters; when young they skeletonize the leaves, but older caterpillars eat the entire lamina. At times there are population explosions and large numbers of larvae destroy entire crops.

Pest status. An important and widespread pest, regularly encountered, but only occasionally is serious damage done.

Life history. The eggs are globular, ribbed and white, laid singly or in small groups, on the undersides of leaves of low-growing plants; each female lays 500–1000 eggs; hatching requires 10–12 days.

The larvae are 'semiloopers', with only two pairs of prolegs. Usually green in colour darker on the back, dorsal line white, subdorsal (lateral) line yellow. The head is small, brownish-green with black markings. Full-grown larvae are 20–30 mm and take 3–4 weeks to develop. When the larvae are disturbed they drop off the plant.

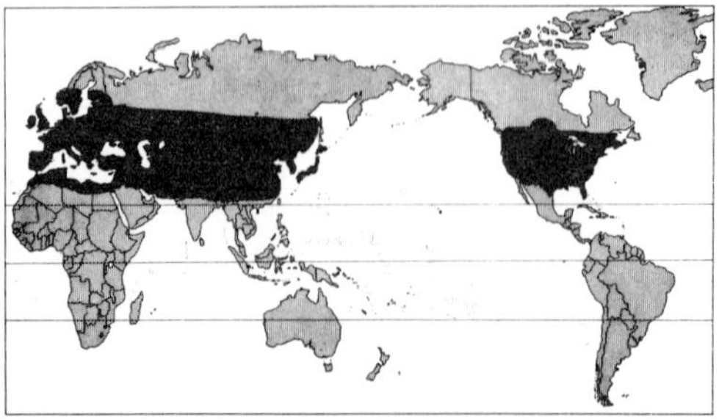

Pupation takes place on the plant at the feeding site; the blackish pupa (20 mm) is enclosed inside a whitish silken cocoon; pupation requires 12–14 days.

The adults are grey-coloured, with dark markings and the characteristic silvery comma in the centre of the forewings. Wingspan is 35–48 mm. They often do not live long, but are strongly migratory and take nectar from many flowers, often during the day. In the UK adults are seen in June (thought to be immigrants) and again in August and September.

In Europe this species is bivoltine usually, but in Israel there may be 4–5 generations per year; in the UK the life-cycle takes 45–50 days.

Distribution. Recorded throughout Europe, N. Africa, Asia and N. America (Holarctic); regular migrant up to the Arctic Circle, Iceland and Greenland.

Autographa californica (Speyer) is the Alfalfa Looper of the USA and Canada.

Autoplusia egena (Guen.) is the Bean Leaf Skeletonizer of the USA.

Chrysodeixis chalcites (Esp.) – (Golden Twin Spot Moth) Europe, Asia, USA.

Plusia festucae L. – (Rice Semilooper) Japan, Asia, Europe.

Plusia nigrisigna Wlk. – (Beet Semilooper) Japan.

Control. To control this pest the previous recommendations were to dust or spray with DDT or HCH against the young larvae. Additional chemicals now available include BTH, azinphos-methyl mixtures, chlorpyrifos, permethrin, pirimiphos-methyl, triazophos, trichlorphon and various others.

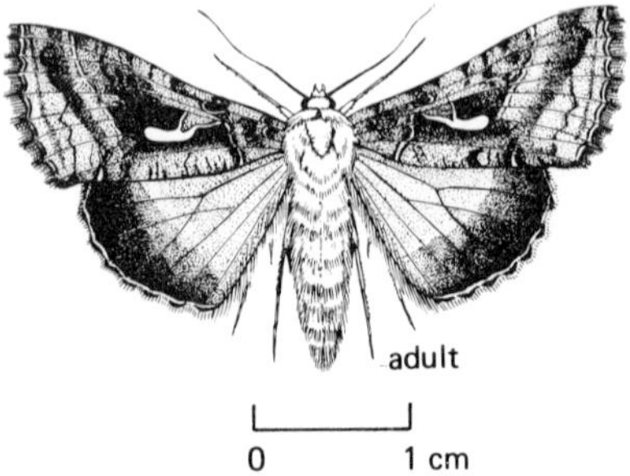

Trichoplusia ni (Hübner)

(= *Autographa brassicae* Riley)
(= *Plusia ni* Hb.)

Common name. Cabbage Semilooper (Ni Moth)
Family. Noctuidae (Plusiinae)
Hosts (main). Cabbage, and other Cruciferae, lettuce, tomato, celery.

(alternative). Potato, eggplant, beans, cucurbits, maize, cotton; totally polyphagous.

Damage. Young larvae skeletonize leaves, and the older larvae eat the entire leaf lamina, sometimes causing plant defoliation.

Pest status. Internationally regarded as the third most damaging pest of cabbage and other cultivated Cruciferae; although basically solitary the larvae sometimes occur in very large numbers; cabbage hearts may be bored.

Life history. Eggs are laid singly on the plant foliage, in May and June; each female lays up to 200 eggs; they are spherical and yellow, 0.5 mm in diameter, and hatch in 2–3 days.

The larvae are typical Plusiinae, yellowish-green in colour, with narrow dark dorsal lines and thick white lateral stripes; fully grown they measure 30–5 mm. Larval development takes some 10–30 days, according to temperature.

Pupation takes place inside a silken cocoon on the food plant; in cabbages it may be in the heart between adjacent leaves. The pupa is brown and about 20 mm long; development takes 12–15 days under optimum conditions.

The adult is a dark moth, with forewings a reddish grey-brown, with two small silvery marks in the centre; wingspan 30–40 mm. Adult life is usually about three weeks.

In warm regions there are usually five or more generations per year, but it is thought to be bivoltine in Canada and central Europe.

Distribution. A subtropical/warm temperate species, now established virtually worldwide; it does not thrive in the hot tropics. In parts of the hot tropics and the cooler temperate regions (Canada) there may be regular annual immigrations that maintain local populations. Not yet recorded from Australasia (CIE map no. A.328).

Control. A difficult pest to control, for sometimes there are population explosions and numbers are very high; it has also developed resistance to DDT, carbaryl, parathion, methomyl and others.

Natural levels of parasitism are usually high, as also is predation; both *Bacillus thuringiensis* and Nuclear Polyhedrosis Virus are effective controlling agents.

Both pheromone traps and u.v. light traps are used in the USA, but more as diagnostic tools than as control measures.

Before considering insecticidal use local advice should be sought as chemical resistance is quite widespread, and presumably increasing.

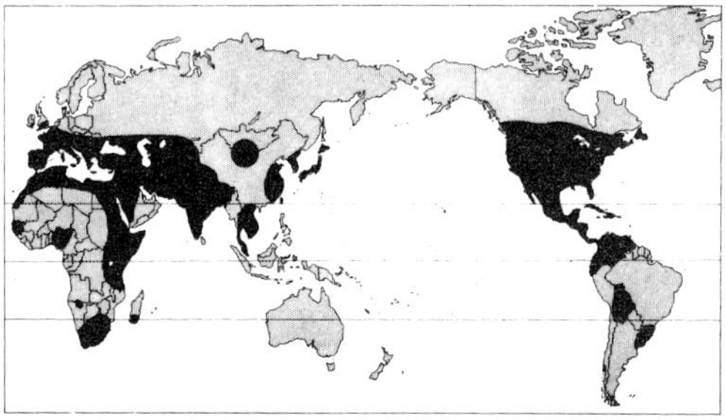

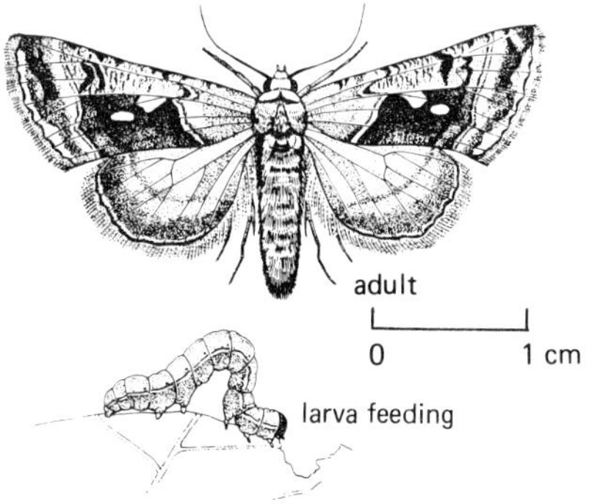

Orgyia antiqua (L.)

Common name. Vapourer Moth (Rusty Tussock Moth (USA))

Family. Lymantriidae

Hosts (main). Apple, plum, cherry, pear, raspberry, rose.

(alternative). Hazelnut, hawthorn, many forest trees and shrubs, and some herbaceous plants; quite polyphagous.

Damage. The feeding larvae are leaf eaters, and occasionally cause some defoliation, but they are usually more spectacular than seriously damaging. Fruit is seldom attacked.

Pest status. Widespread and polyphagous, and quite regularly encountered by entomologists, but usually a minor pest, except for occasional outbreaks.

Life history. Eggs are laid in a distinctive mass on the outside of the female pupal cocoon, on a twig or spur of the tree; usually 100–300 eggs in the one mass, they are round, with a rim and a central dorsal depression; brownish-grey in colour. Eggs are laid in the autumn in the UK, and remain on the tree over winter, hatching the following May and June; hatching is usually intermittent.

The larvae are spectacular caterpillars, bristly, coloured a mixture of red, yellow and brown, with a pair of long black hair pencils anteriorly, and dorsally four brushes of golden hairs (once seen, never forgotten). Full-grown larvae are about 25 mm long.

Pupation takes place inside a thin but tough cocoon of silk, attached to a twig, in a bark crevice, on a post, etc.; pupation takes about three weeks.

The adult male is a rich chestnut brown, with a white mark in the forewing, 25–30 mm across the wings, and bushy antennae. The female has vestigial wings, a fat brown body, and is totally flightless. She seldom leaves the cocoon. Because of the intermittent egg-hatching all stages of this species may be found over the greater part of the year; but there is usually only one generation per year.

Distribution. Recorded throughout Europe and Asia, including Asia Minor and Siberia; also N. America.

A total of 60 species of *Orgyia* are known, mostly in the Holarctic Region, and several are of regular occurrence on cultivated plants.

Control. In orchards or gardens at risk from this pest the hand-collection of egg masses is worthwhile; this can be done during winter-pruning operations.

Winter washes kill some of the eggs; caterpillars have been killed using sprays of DDT or derris, but control measures are seldom needed.

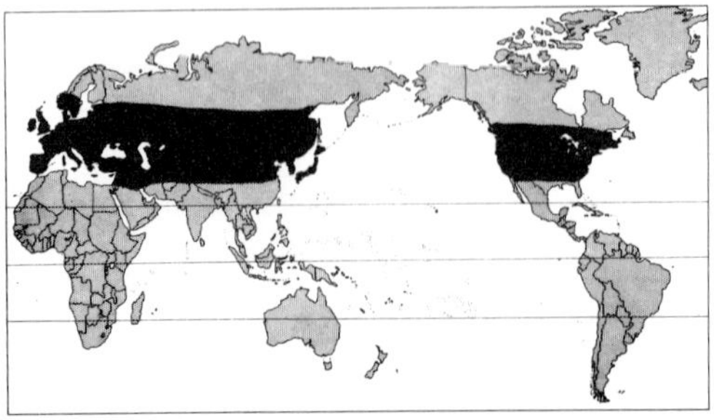

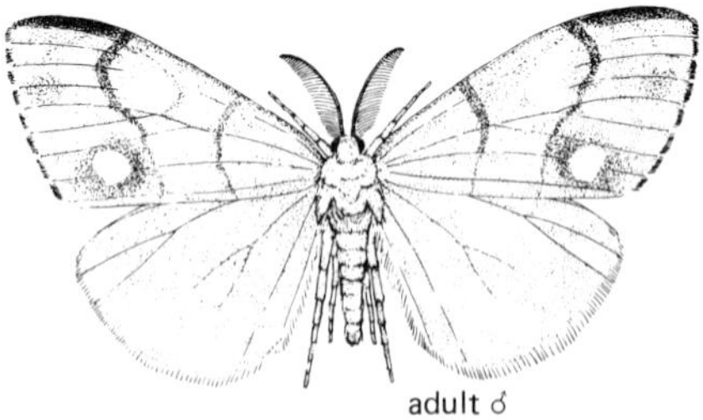

Tussock moths (Lepidoptera; Lymantriidae)

A family of moderate-sized moths, mostly of a drab or white coloration, hairy and heavy-bodied, the females typically bear a thick anal tuft of scale-hairs which are detached and used to cover the egg mass after oviposition. In a few species the female has wings vestigial or even absent (*Orgyia* spp.).

The larvae are stout-bodied and bristly caterpillars, sometimes strikingly coloured; they are general defoliators and some are very polyphagous. Most are damaging to forest trees (both conifers and deciduous trees), but some feed on fruit trees and various woody shrubs. In a few species the eggs are laid on the ground and the young larvae have to climb the trees in order to reach their food source (leaves). These species (of *Lymantria*) can clearly be controlled by either sticky bands on the tree trunk or by band-spraying the trunks with dieldrin.

Some of the more important agricultural pest species include:

Dasychira mendosa (Hb.) – polyphagous; India and southern Asia.
Dasychira spp. (400 species are known) – on many hosts; Old World.
Euproctis chrysorrhoea (L.) – (Brown-tail Moth) polyphagous; Europe, Asia, USA (CIE map no. A.362).
Euproctis similis (Fue.) – (Yellow-tail Moth) polyphagous; Europe and Asia (CIE map no. A.388).
Euproctis fraterna (Moore) – polyphagous; India.
Euproctis pseudoconspersa Strand – (Tea Tussock Moth) Japan.
Euproctis spp (600 species are known) – polyphagous; Old World.
Lymantria dispar (L.) – (Gypsy Moth) polyphagous; Asia, Europe, Canada and USA (CIE map no. A.26).
Lymantria lapidicola (H.-S.) – (Almond Tussock Moth) Asia Minor.
Lymantria monacha (L.) – (Nun Moth) polyphagous on trees; Europe and Asia (CIE map no. A.60).
Lymantria spp. (150 species recorded) several on cultivated plants; Old World.
Orgyia antiqua (L.) – (Vapourer Moth) polyphagous; Europe, Asia, USA (see page 446).
Orgyia spp. (60 species) several are polyphagous pests of cultivated plants; most known from the Holarctic Region.
Perina nuda (F.) – (Fig Tussock Moth) India, S.E. Asia, China.

In the figure is shown a larva of *Euproctis similis* on a leaf of *Rosa* sp.

Larva of *Euproctis similis* on leaf of *Rosa*

Order

Hymenoptera

(sawflies, ants, wasps, bees)

Families
Tenthredinidae (sawflies)
Vespidae (social wasps)

Athalia rosae (L.)

(= *A. colibri* F.)

Common name. Cabbage Sawfly; Turnip Sawfly
Family. Tenthredinidae
Hosts (main). Brassicas of all species.
(alternative). Other members of the Cruciferae.
Damage. The larvae eat the leaves, often only leaving the midrib intact. The numerous blackish larvae fall to the ground if the plant is shaken.
Pest status. A widespread, regular, and sporadically serious pest of cruciferous crops.
Life history. Eggs are laid singly in small cuts in the leaf lamina made by the female ovipositor; eggs are white and oval (1 mm long); hatching takes 5–12 days.

The larvae are caterpillars but with six pairs of prolegs (as distinct from the lepidopterous four or fewer); quite dark in colour (oily green or blackish). The head is black, and there is often a hump on the thorax; full-grown larvae are about 20–5 mm. First-instar larvae usually mine the leaves, and then skeletonize them before finally eating the entire lamina. Larval development takes about 20 days in the UK.

Pupation takes place in the soil inside a tough silken cocoon; the pupa is yellowish; development takes 15–20 days.

The adult has a black head and thorax and bright yellow-orange abdomen; body length about 6–8 mm, wingspan 15 mm. Adults may be seen flying slowly just above the crop.

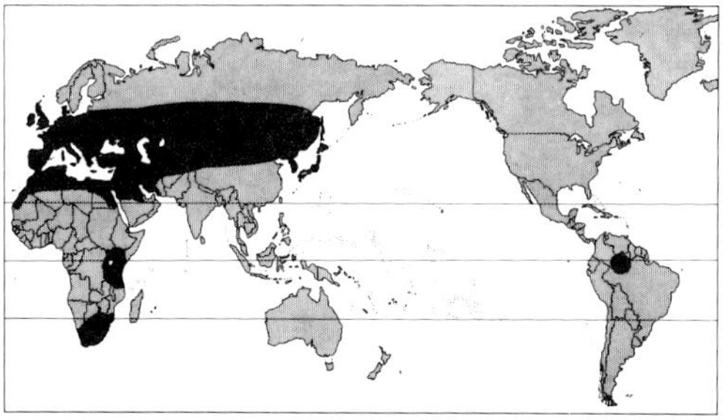

In the UK there are three generations annually, the last overwintering as pupae; these adults emerge in the following spring (May).
Distribution. This species is recorded throughout Europe and Asia, to China and Japan, N. Africa, E. and S. Africa, and also now in S. America.

In Japan there are a further two species recorded:
Athalia japonica (Klug) – (Cabbage Sawfly) Japan.
Athalia lugens (Klug) – (Cabbage Sawfly) Japan and Europe.

In the UK there is a total of ten species of *Athalia*.
Control. Control is not often required. Crop rotation and the destruction of cruciferous weeds on the headlands will help to keep the pest population down.

The pesticides found to be effective against this pest include DDT, carbaryl and the pyrethroids, as foliar sprays.

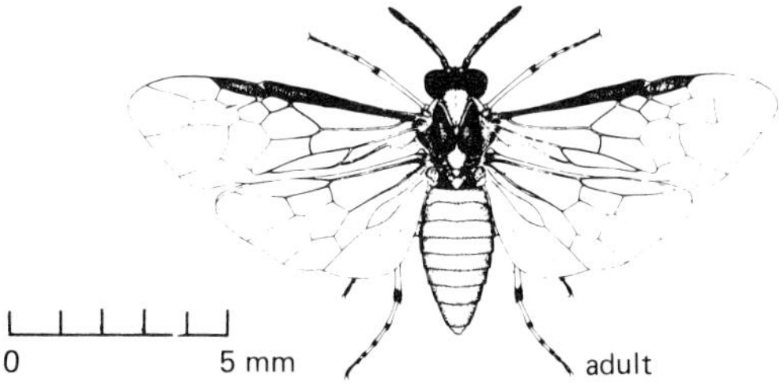

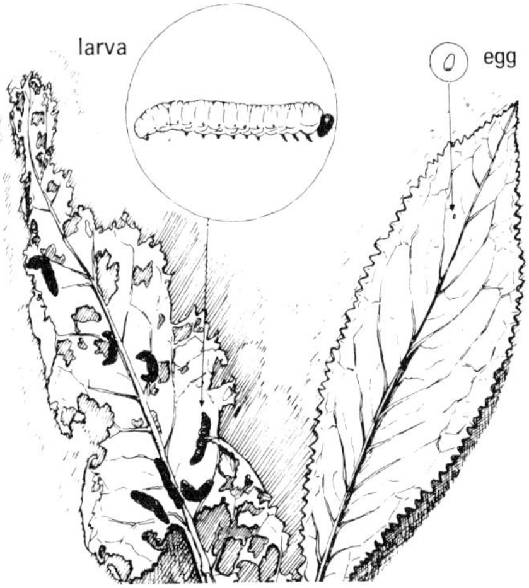

Caliroa cerasi (L.)

Common name. Pear Slug Sawfly (Cherry Slugworm)
Family. Tenthredinidae
Hosts (main). Pear, cherry, plum, peach, quince.
(alternative). Apple, almond, some ornamentals and wild hosts, all in the Rosaceae.
Damage. The slug-like larvae eat away the upper surface of the leaf, leaving only the veins and lower epidermis. On young trees the defoliation may be serious; but seldom important on older trees.
Pest status. A widespread and regularly encountered pest species on fruit trees in temperate regions, with easily recognized damage, but only occasionally an economic pest.
Life history. Eggs are laid singly into epidermal slits on the underside of leaves; hatching requires about two weeks.

The young larvae move round on to the upper leaf surface and start feeding. The larvae are at first whitish, but soon become coated with very dark green or black slime, when they look very slug-like. The body is broad anteriorly and tapers to the hind end; legs are reduced but there are ten pairs in all. Larval development takes 2–4 weeks, by which time they are 10 mm long; the slime covering is shed to reveal a wrinkled yellow skin, and they descend to the soil for pupation.

The pupa is formed in the soil, at a depth of 1–3 cm, inside a silken cocoon; development takes about two weeks.

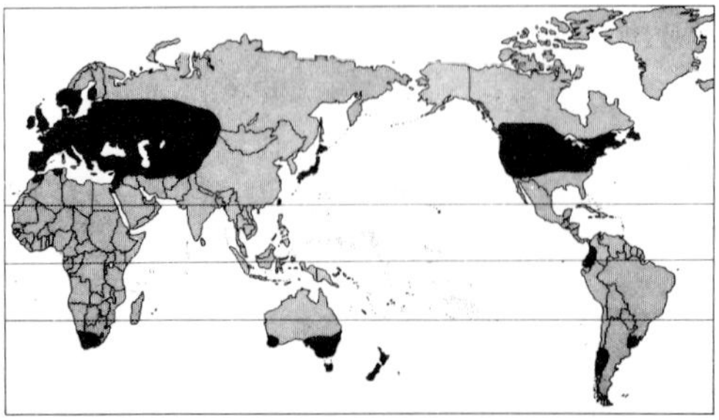

The adult is a small black sawfly, some 8 mm long. There are usually 2–3 generations per year; overwintering takes place by mature larvae inside the pupal cocoons.
Distribution. This species is recorded throughout Europe, Asia to Taiwan and Japan, Australia, New Zealand, S. Africa, N. and S. America (CIE map no. A.175).

Several other species of *Caliroa* are of some importance as pests, including:
Caliroa castaneae (Roh.) – (Chestnut Slug Sawfly) USA.
Caliroa matsumotonis Har. – (Peach Slug Sawfly) Japan.
Caliroa annulipes (Klug) – (Oak Slugworm) Europe.
Caliroa quercuscoccineae (Dyar) – (Scarlet Oak Sawfly) USA.

Several other species of 'slug sawflies' damage various ornamentals and fruit trees, but these larvae are not covered with excrement (slime); many are species of *Allantus*.
Control. If control is needed, foliar sprays of HCH, derris, dimethoate or oxydemeton-methyl should be effective.

Larva on pear leaf

Hoplocampa spp.

(*H. brevis* (Klug))
(*H. flava* (L.))
(*H. minuta* (Christ))
(*H. testudinea* (Klug))

Common name. Fruit Sawflies (see below)
Family. Tenthredinidae
Hosts. Pear; plum (and damson); plum; apple; (each species is confined to one particular host).
Damage. The larvae feed on and damage young fruits, firstly by tunnelling just under the skin, and later by boring straight into older fruits into the core.
Pest status. This group of species is collectively important on fruit trees over quite a large area; but regular orchard spraying has kept populations in check in most locations; often more serious as a garden pest.
Life history. Eggs are laid in open blossoms, in a slit in the receptacle (leaving a small spot of oozed sap). Each female lays about 30 eggs; hatching requires about 14 days.

The larva is a little white caterpillar with six pairs of prolegs and a black or brown head. After hatching it burrows into the fruit, starting within the calyx and moving outwards. On reaching the skin the larva turns and burrows just underneath the skin; later it leaves the surface and penetrates to the core where it eats the seeds. If the seeds are destroyed the fruitlet ceases to develop and falls, but if it fails to reach the core the scarred fruit continues to develop. Most larvae then penetrate a second fruit in which they tunnel directly into the core, leaving a large entry hole with frass. A few caterpillars will damage three separate fruits. Larval development takes about four weeks.

Pupation takes place deep in the soil (8–24 cm deep), inside a cocoon. The prepupa overwinters in the cocoon, and pupation takes place in the spring, and requires 3–4 weeks. Some prepupae do not develop the first spring and remain until the next spring.

Adults emerge just when the trees are in blossom; they are small, blackish, with red-yellow legs and sternum; wingspan is about 8–10 mm.
Distribution. The four most important pest species are:
Hoplocampa brevis – (Pear Sawfly) Europe and Asia (CIE map no. A.169).
Hoplocampa flava – (Plum Sawfly) Europe, Asia Minor (CIE map no. A.168).
Hoplocampa minuta – (Small Plum Sawfly) Europe, Asia Minor (CIE map no. A.167).
Hoplocampa testudinea – (Apple Sawfly) Europe, N. America (CIE map no. A.166).

For other species of some importance see page 453.
Control. See page 455.

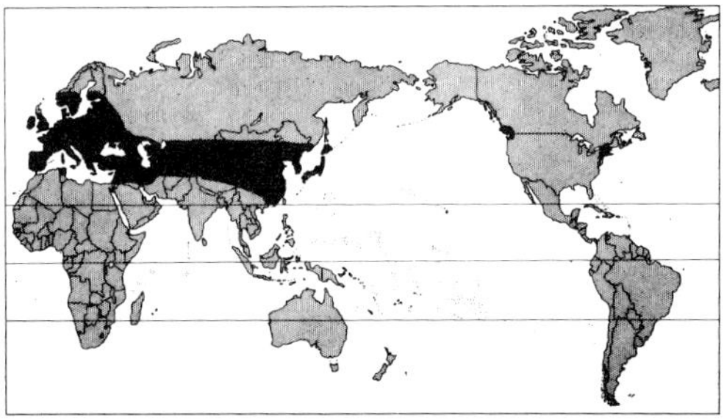

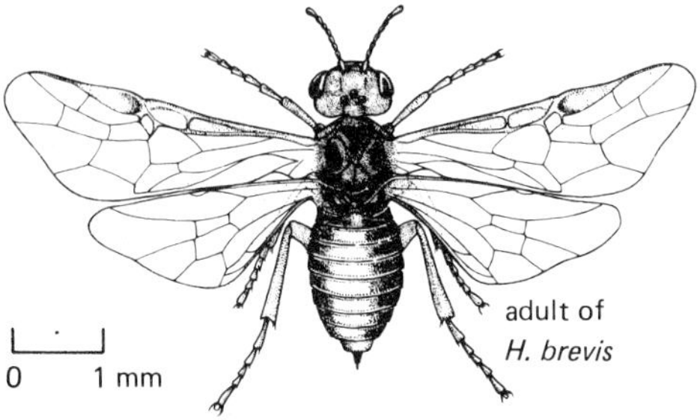

adult of *H. brevis*

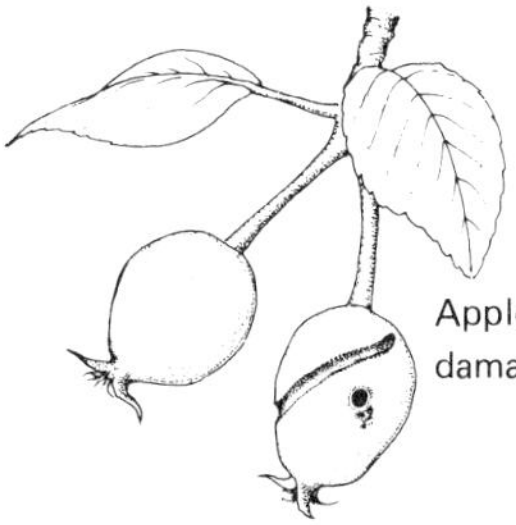
Apple Sawfly damage to fruitlet

Nematus spp.

(*N. olfasciens* Benson)
(*N. ribesii* (Scopoli))

Common name. Black Currant Sawfly; Gooseberry Sawfly
Family. Tenthredinidae
Hosts. Black, red and white currants; gooseberry, red and white currants (respectively).
Damage. The feeding larvae eat the leaves and cause defoliation, usually on one branch at a time, starting in the centre of the bush near to the ground; in heavy attacks the entire bush may be defoliated.
Pest status. A very large and widespread genus causing defoliation on a wide range of trees and bushes; these two species are quite damaging to currants and gooseberry bushes.
Life history. Eggs are laid on the underneath of leaves, in shallow slits along the main veins; 20–30 eggs may be found on a single leaf; the oval eggs are about 1 mm long and quite conspicuous; hatching requires about one week.

The young larvae emerge in May, and initially feed gregariously; but as they grow they gradually disperse over the bush. The body is green with small black spots, and the head is black (green in *olfasciens*). When fully grown (after 3–4 weeks) the larvae become pale green with an orange patch on the thorax; they leave the bush to pupate in a cocoon in the soil, at a depth of 5–8 cm; pupation takes 10–21 days. There are usually three generations per year in the UK, with considerable overlap so that all stages may be found on the same bush at the same time. The larvae of the last generation overwinter inside the pupal cocoons in the soil.

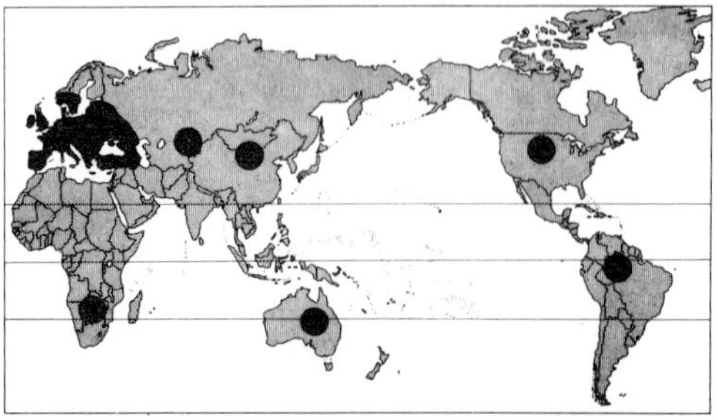

The adults are very similar, with a black head and thorax and yellow abdomen, and clear transparent wings 10–16 mm diameter.
Distribution. These two species are recorded from Europe, and *ribesii* is reputedly worldwide in distribution, but their precise ranges are not known.

Also included here are the other pest species of *Nematus*, namely:
Nematus leucotrochus Hart. – (Pale Spotted Gooseberry Sawfly) Europe.
Nematus salicis (L.) – (Willow Sawfly) Europe.
Nematus spiraeae Z. & B. – (Spiraea Sawfly) Europe.
Nematus spp. – on oak, poplars, willows, etc.; Europe and N. America.

There are at least 40 species of *Nematus* recorded from the UK alone, and a similar number in N. America.

On currants and gooseberry bushes, on both continents, there may be several other species of sawflies feeding on the foliage.
Control. When control is needed the usual recommendation is a spray of azinphos-methyl, derris, fenitrothion or malathion, etc., aimed at the centres of the bushes.

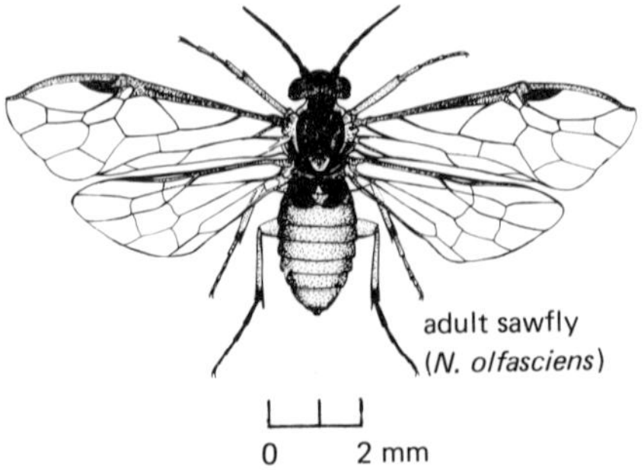

Sawflies (Hymenoptera; Symphyta)

This large group constitutes a sub-order within the Hymenoptera; the members possess various morphological and anatomical characters in common, and the distinctive larvae are phytophagous (except for a few parasitic on wood-boring beetle larvae), and quite a large number cause damage to cultivated plants. However, the majority are leaf eaters on trees in the north temperate deciduous forests and the taiga coniferous forests of the Holarctic Region. The group is worldwide in distribution, although mainly north temperate (Holarctic); representatives are to be found in every country; some species are actually sub-arctic, either in grass stems or eating conifer needles.

The adults are small to medium-sized, often black, black and yellow, or blue in body colour, with hyaline wings. The female has a saw-like ovipositor with which she deposits eggs singly into the host plant tissues (hence their common name of 'sawflies'); the ovipositor is usually concealed under the tip of the abdomen but in a few species of woodwasps it protrudes distinctively. In this group the abdomen is broadly sessile, and there is no constriction between segments 1 and 2.

The larvae are basically caterpillars (eruciform), but distinct from those of Lepidoptera in having 5–8 pairs of abdominal prolegs (in addition to terminal claspers); they are often spotted whereas Lepidoptera are more usually striped. Species living inside plant tissues usually have reduced legs (as with other insect borers), and a few of the leaf eaters are slug-like with tiny legs. Most of the larvae are solitary (after the first instars), but a few live and feed gregariously on leaves in a most characteristic manner (they are arranged around the leaf lamina edge and when disturbed they stop feeding and elevate the abdomen (see figure on page 455)). The species regarded behaviourally as 'solitary' may occur in dense populations which can defoliate entire trees, but there is no interaction between adjacent larvae.

The life-styles of the larvae vary considerably, and this directly relates to the type of damage done to the plants, as follows.

(a) Leaf eaters: solitary – most Tenthredinidae, Diprionidae, etc.
gregarious – *Croesus*, *Hemichroa*, etc.
(b) Leaf skeletonizers: *Caliroa*, *Endelomyia* (Caliroini).
(c) Leaf rollers: *Blennocampa*.
(d) Leaf miners: *Metallus*, *Fenusa*, etc.
(e) Leaf gall makers: *Pontania* on *Salix* leaves.
(f) Shoot and stem borers: in Gramineae – Cephidae.
in woody shrubs – Cephidae, etc.
(g) Fruit borers: *Hoplocampa* spp. (Tenthredinidae).
(h) Wood borers (tree trunks and branches):
evergreens – many Siricidae.
deciduous trees – *Tremex* (Siricidae)

Pupation takes place inside a cocoon, usually in the soil, but the wood wasps pupate inside the larval tunnel, as do the stem borers.

In the following list a few small families which do not contain pest species of importance have been omitted.

Sawfly pest species

Some of the more important pest species of sawflies are listed, in their usual taxonomic sequence, below.

Pamphilidae
Neurotoma saltuum L. – (Social Pear Sawfly) Europe.
Neurotoma spp. – (Plum (etc.) Sawflies) USA, Canada.
Pamphilius spp. – (Blackberry (etc.) Sawflies) USA, Canada.

Siricidae (wood wasps)
Sirex spp. (3+) – (Blue Woodwasps) conifers; Europe, Australia, USA, Canada.
Tremex spp. – deciduous trees; Europe, Asia, USA, Canada.
Urocerus gigas (L.) – (Giant Woodwasp) Europe, Asia.

Cephidae (stem sawflies)
Cephus cinctus Norton – (Wheat Stem Sawfly) Canada, USA (CIE map no. A.29).

Cephus pygmeus (L.) – (European Wheat Stem Sawfly) Europe, W. Asia, USA, Canada (CIE map no. A.30).
Hartigia trimaculata Say – (Blackberry Shoot Sawfly) USA, Canada.
Janus integer (Norton) – (Currant Stem Girdler) USA.
Syrista similis Mac. – (Rose Stem Sawfly) Japan.
Trachelus tabidus (F.) – (Black Grain Stem Sawfly) USA.

Argidae (rose sawflies)
Arge mali Tak. – (Apple Argid Sawfly) Japan.
Arge ochropus (Gm.) – (Large Rose Sawfly) Europe.
Arge pagana (Panzer) – (Rose Sawfly) Europe, Asia, Japan (see figure on page 455).
Arge spp. – (Rose (Birch, etc.) Sawflies) Europe, Asia, Japan, N. America.

Cimbicidae
Cimbex spp. – (Elm Sawflies) Europe, Asia, USA.
Palaecimbex carinulata Konow – (Pear Cimbicid Sawfly) Japan.

Diprionidae (conifer sawflies)
Diprion pini (L.) – (Pine Sawfly) Europe, Asia.
Diprion spp. – (Pine (etc.) Sawflies) Europe, Asia, Canada, USA.
Neodiprion spp. (15+) – (Pine (etc.) Sawflies) Europe, Asia, Japan, Canada, USA (CIE map no. A.98).

Tenthredinidae
Allantus albicinctus Mats. – (Strawberry Sawfly) Japan.
Allantus cinctus (L.) – (Banded Rose Sawfly) Europe, USA.
Allantus nakubusensis Tak. – (Cherry Sawfly) Japan.
Ametastegia glabrata (Fallen) – (Dock Sawfly) Europe, Canada.
Apethymus kuri Tak. – (Chestnut Sawfly) Japan.
Athalia japonica Klug – (Cabbage Sawfly) Japan.
Athalia lugens Marlatt – (Cabbage Sawfly) Japan.
Athalia rosae (L.) – (Cabbage Sawfly) Europe, Asia, Japan (see page 449).
Blennocampa caryae (Norton) – (Butternut Woollyworm) USA.
Blennocampa pusilla (Klug) – (Leaf-rolling Rose Sawfly) Europe (see figure on page 455).
Caliroa castaneae (Roh.) – (Chestnut Slug Sawfly) USA.
Caliroa cerasi (L.) – (Pear Slug Sawfly) Europe, Asia, Japan, Canada, USA (CIE map no. A.175; see page 450).
Caliroa matsumotonis (Har.) – (Peach Slug Sawfly) Japan.
Cladius spp. (4) – (Antler (Rose) Sawflies) Europe, Japan, USA.
Croesus castaneae Roh. – (Chestnut Sawfly) USA.
Croesus septentrionalis (L.) – (Hazel Sawfly) Europe, USA (see figure on page 455).
Croesus spp. – (Birch/Alder Sawflies) Europe, Asia, Japan, USA.
Dolerus ephippiatus Smith – (Wheat Sawfly) Japan.
Dolerus lewisi Cam. – (Wheat Sawfly) Japan.
Dolerus spp. – (Cereal and Grass Sawflies) Canada, USA.
Empria maculata (Norton) – (Strawberry Sawfly) USA, Canada.
Endelomyia aethiops (F.) – (Rose Slug Sawfly) Europe, USA.
Erythraspides vitis (Harris) – (Grape Sawfly) USA.
Fenusa spp. – (Birch/Elm Sawflies) USA.
Hemichroa crocea (Geo.) – (Striped Alder Sawfly) USA.
Hemichroa sp. – (Camphor Sawfly) China.
Hoplocampa brevis (Klug) – (Pear Sawfly) Europe (CIE map no. A.169; page 451).
Hoplocampa cookei (Clarke) – (Cherry Fruit Sawfly) USA.
Hoplocampa flava (L.) – (Plum Sawfly) Europe, Asia Minor (CIE map no. A.168; page 451).
Hoplocampa minuta (Christ) – (Small Plum Sawfly) Europe, Asia Minor (CIE map no. A.167; page 451).
Hoplocampa pyricola Roh. – (Pear Fruit Sawfly) Japan.

Fig 484 (*Opposite*) Different Types of Sawfly Infestations.

(*a*) *Arge pagana* – Rose Sawfly
(*b*) *Blennocampa pusilla* – Leaf-rolling Rose Sawfly
(*c*) *Croesus septentrionalis* – Hazel Sawfly – gregarious larvae

Arge pagana (Rose Sawfly)

Blennocampa pusila (Leaf-rolling Rose Sawfly)

Croesus septentrionalis (Hazel Sawfly)

Hoplocampa testudinea (Klug) – (Apple Sawfly) Europe, Asia, Canada, USA (CIE map no. A.166; page 451).
Metallus albipes (Cam.) – (Raspberry Leaf-mining Sawfly) Europe.
Metallus pumilus Klug – (Raspberry Leaf-mining Sawfly) Europe.
Monophadnoides genicultatus (Hartig) – (Raspberry (Geum) Sawfly) Europe, USA.
Nematus leucotrochus Hart. – (Pale Spotted Gooseberry Sawfly) Europe.
Nematus olfasciens Benson – (Black Currant Sawfly) Europe (page 452).
Nematus ribesii (Scop.) – (Gooseberry Sawfly) Europe (page 452).
Nematus spp. (many) – oaks, poplars, willows, etc.; Europe, Asia, USA, Canada.
Pachynematus spp. – (Wheat (and Grass) Sawflies) Holarctic (subarctic).
Pontania spp. – (Willow Redgall Sawflies) Europe, USA.
Priophorus pallipes Lep. – (Plum Leaf Sawfly) Europe.
Pristiphora abbreviata (Hartig) – (California Pear Sawfly) USA.
Pristiphora pallipes Lep. – (Small Gooseberry Sawfly) Europe, USA.
Pristiphora spp. – (Larch/Oak Sawflies) Europe, Asia, Japan, Canada, USA (CIE map no. A.97).
Pristiphora spp. – (Birch Sawflies) Europe, USA, Canada.
Takeuchiella pentagona Malaise – (Soybean Sawfly) Japan.
Tenthredo spp. – polyphagous; Europe, Canada, USA.
Trichiocampus pruni Tak. – (Cherry Sawfly) Japan.

Control of Sawflies

Of the sawfly species that are crop pests, it is almost invariably the larvae that do the damage, though some may be done to plant tissues by the action of the female ovipositor. So control measures are directed against the larvae on the crop plants. Most commercial orchards have a regular spraying programme with organophosphate materials, applied at or just after petal-fall, against the local pest complex, and this usually controls the leaf and fruit-eating sawfly larvae as well as other pests. Occasionally sawfly outbreaks do require specific treatment.

For many species it has been recently reported that γ-HCH still gives the best control, applied as a high-volume spray at petal-fall, unless resistance to this chemical has developed. Generally now resistance to HCH is quite widespread.

The chemicals generally available for control of sawfly larvae on foliage and fruits include the following: γ-HCH, derris, azinphos-methyl + demeton-*S*-methyl-sulphone, chlorpyrifos, dimethoate, malathion, oxydemeton-methyl, pirimiphos-methyl, and the pyrethroids cypermethrin, deltamethrin and permethrin.

Usually control recommendations require one of these chemicals to be applied within seven days, or thereabouts, of about 80% petal-fall, as a full-cover high-volume spray. Occasionally a second spray may be required 14 days after first application. On dense bushes or trees it is essential that the spray penetrates to the centre of the bush foliage. This treatment should kill most larvae, which at the time of the first spray should mostly be newly-hatched and small in size. If the fruitlets have already been penetrated by the young larvae then a systemic insecticide should be used. For the species that remain on the foliage, contact and stomach-acting insecticides are most effective.

Precise choice of insecticides to be used depends in part upon which species of sawfly is involved, and on which crop, and also on weather conditions at the time of spraying. If there is any doubt as to choice of candidate insecticide then the local Ministry of Agriculture staff should be consulted.

Vespa/Vespula spp.

Common name. Common Wasps; Yellowjackets (USA)
Family. Vespidae
Hosts. Soft fruits of all types (especially grapes, currants, plum, blackberry, raspberry, strawberry) and other fruits when ripe, including apple, pear, cherry, peach etc.
Damage. Direct damage consists of puncturing soft-skinned and ripe fruits to feed on the sugary flesh and sap. Indirect damage consists of nesting in the vicinity of orchards and harassing the workmen when they come near the nest.
Pest status. Very abundant and widespread pests, frequently encountered, although actual damage levels are usually low. It should be mentioned that these insects are basically carnivorous and prey on many pest species in the orchards and gardens and are thus also beneficial species.
Life history. These are social wasps and live in a large papery nest, built either underground, in a hollow tree, in a building, or sometimes hanging free in a tree or bush. The colony is essentially annual and dies in the early winter, survival being effected by the hibernating young queens.

The nest is spherical usually, consisting of an outer skin protecting a series of horizontal combs, each hexagonal cell of which is the home for one larva. Overall diameter is 20–40 cm, but sometimes the nest is longer than broad (see photograph). There are usually 6–8 horizontal combs inside each nest; the nest entrance is ventral. The whole nest is constructed of a papery material

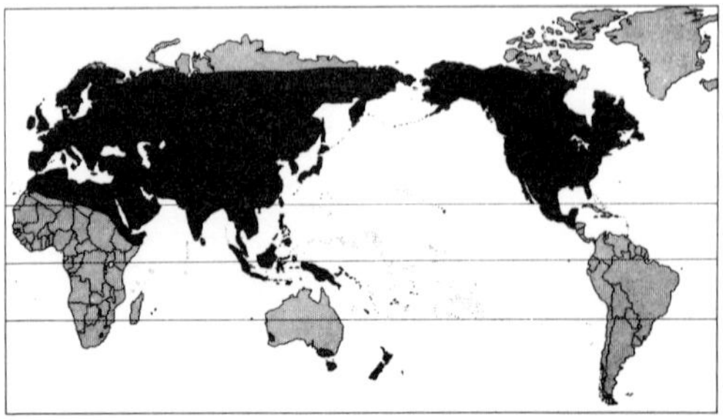

composed of chewed wood fragments mixed with saliva.

Eggs are laid singly by the queen into the comb cells, and the entire development takes place in the cell; the larva is fed on chewed insect remains, and sometimes honeydew or sugary solution obtained from fruits. The largest nests will contain several thousand individuals, but the colony is annual and the workers die in the early winter, as does the old queen; young queens, after mating, find hibernation sites where they overwinter.

Distribution. There are 10–20 species of *Vespa* and *Vespula*; they are completely cosmopolitan throughout the Northern Hemisphere, and now established in Australasia.

An additional genus that could be included in this category of pests is *Dolichovespula*, both throughout Europe, Asia and N. America.

Control. When control is needed there are various different approaches, as follows:

(a) direct nest destruction by physical means.
(b) nest poisoning, either by direct insecticide application to the nest entrance, or the use of poison baits of sugar or syrup with a slow-acting insecticide added, so that the bait is carried back to the nest.
(c) destruction of foraging workers using traps of different types; various synthetic attractants are now commercially available, including leptyl butyrate.

The chemicals regarded as effective against wasps include dieldrin, chlordane, iodofenphos, diazinon and fenitrothion, the latter two being also available in an encapsulated form for slow release.

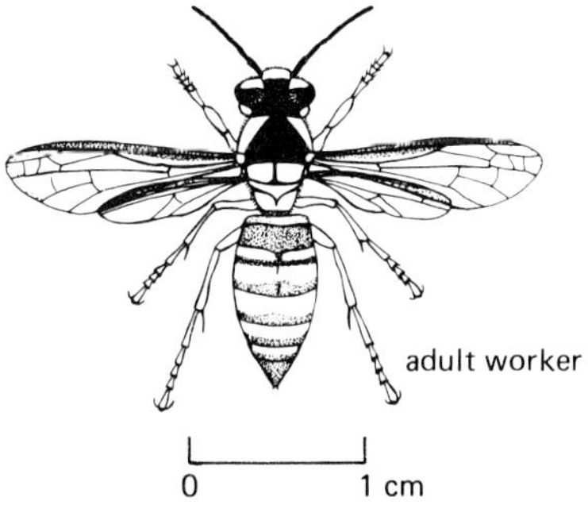

Large nest hanging free in a tree

Class

Arachnida

Order

Acarina

(mites)

Families
Tetranychidae (spider mites)
Tarsonemidae
Tenuipalpidae

Panonychus ulmi (Koch)

(= *Metatetranychus pilosus* (C. & F.))

Common name. Fruit Tree Red Spider Mite
Family. Tetranychidae
Hosts (main). Deciduous fruit trees.
(alternative). Elm and other deciduous forest trees.
Damage. Epidermal feeding causes the leaves to become scarified and speckled, later becoming bronzed and brittle in heavy infestations; many bronzed leaves fall prematurely; crop yields are reduced.
Pest status. A major pest species regularly causing serious damage if populations are unchecked.
Life history. The bright red winter eggs are laid on the bark, usually on the underside of spurs and twigs. They are laid from August to October, and hatch from late April/early May to June.

The newly hatched larvae migrate to the underside of the leaves and start feeding. The larval stage is followed by two nymphal stages, before the adult is reached, after about 14 days. The young adults then lay summer eggs, which are paler than the winter ones, underneath the leaves. These eggs hatch in about 2–3 weeks, and give rise to the second generation of mites. In the UK there are 4–5 generations per year; each generation takes about four weeks.

The adults are tiny (0.4 mm long), reddish in colour, with an oval body having several large bristles that are quite distinctive; immature stages and males are generally greenish but may also be reddish.

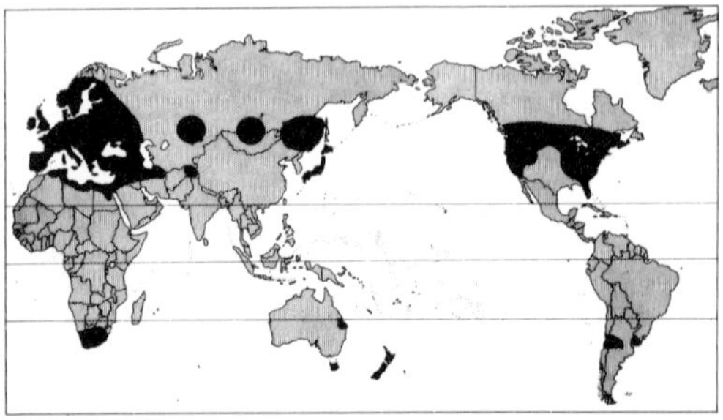

Distribution. This species is widely distributed throughout Europe, parts of Asia, Japan, N. Africa, S. Africa, Australia, New Zealand, Canada, the USA, and parts of S. America (CIE map no. A.31).

The closely related *Panonychus citri* (McG.) (Citrus Red Spider Mite) is also quite polyphagous, but rather more tropical in distribution (CIE map no. A.192).

Control. This and various other spider mites are usually preyed upon quite heavily by predacious mites and bugs, and it has proved to be very difficult to control the mite pest with chemicals without the destruction of the natural predators. In the days of widespread use of DDT in fruit orchards this mite became an ever-increasingly important pest as a result of the destruction of its natural enemies in relation to its own limited susceptibility to this pesticide. The present problem with this pest is its widespread development of resistance to the usual acaricides (see page 461 for details). The economic threshold is when more than seven leaves in a 50-leaf sample have more than four mites per leaf (in the UK).

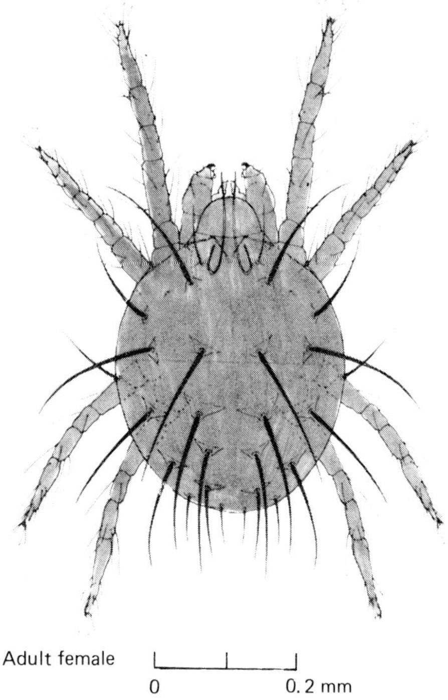

Adult female

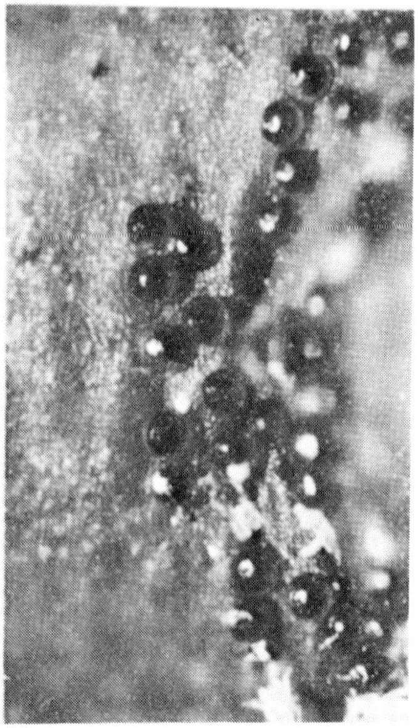
Winter eggs on a spur

Phytophagous mites (class Arachnida; order Acarina)

Mites are obviously not insects, but the phytophagous species are invariably regarded as the purview of the agricultural entomologist, together with a few other arthropods. Mites (order Acarina) are mostly small to tiny (1.0–0.1 mm) in size; typically they have a 'head' region (gnathosoma), with a large undivided posterior region (idiosoma). As typical arachnids, they have four pairs of walking legs; usually two pairs directed anteriorly and two backwards. However, the larva that hatches from the egg has only three pairs of legs; after the first moult into the nymphal stage it acquires the extra pair. In some families there has been body modification with the loss of legs; the Eriophyidae (gall mites) mostly have only the anterior two pairs of legs, and these are reduced in size, and in many species the body is elongated and worm-like.

In some respects it is difficult to generalize about the order as there are several quite different families (of phytophagous mites) with major differences in their biology, but in the present work there is insufficient space available to permit a more detailed treatment.

The phytophagous mites feed on sap, obtained by piercing the epidermal cells with the chelicerae that are modified into stylets. When epidermal cells are emptied of sap, the air that enters inside gives the leaf surface a silvery appearance (often referred to as scarified) which is typical of much mite damage (and also some thrips damage). When epidermal damage is extensive then the leaf suffers dehydration followed by desiccation and bronzing; severely bronzed leaves usually fall prematurely. The spider mites (Tetranychidae) typically produce leaf scarification (and bronzing) as their main damage symptom, sometimes accompanied by extensive foliage webbing with silk. Tarsonemidae and some other mites usually attack young leaves, buds and shoots, and their feeding causes growth deformation. The Eriophyidae, commonly called gall, rust, or blister mites, live in a far closer association with the host plant and they cause foliage distortion which provides a favourable microenvironment for their dwelling. Their infestation of young leaves results in leaf folding, leaf rolling, cupping, and the formation of erinea, and the tiny worm-like mites live within the proliferated epidermal tissues in a very sheltered microhabitat.

Most of the phytophagous species are quite host-specific, either to a genus or a family of plants; this is especially true of the Eriophyidae as their host plant relationship is more intimate than the others. However, a few species of spider mites are quite polyphagous and they are recorded from a very wide range of hosts, which accounts in part for their importance as crop pests. Some mites are fungivorous and others feed on decaying plant material, and the examination of a mite-damaged plant may be very confusing in that there may be several other species present that are not primary pests. There will also usually be predacious mites present, but these are usually distinguishable by their larger size and active nature (they often have longer legs). In the 'big-bud' condition induced on some plants by eriophyid mites there are usually some inquiline species present within the swollen bud. Many plant species are in fact attacked by a large number of different mites, in several different ways, in different parts of their distributional range, but some infestations are overlooked because of the tiny size of the mites and obscure symptoms; for example, the walnut group is recorded as host for about ten different phytophagous mite species worldwide. But for plant protection purposes we tend to concentrate on a relatively small number of mite species per crop plant, and in point of fact on some crops the mite pests are ignored completely in most literature.

One aspect of mite biology of importance in relation to their status as crop pests is their ability to survive periods of inclement weather in a state of diapause or aestivation. Diapause proper (i.e. overwinter) is practised by mature females and also as overwintering eggs. Aestivation is regularly encountered in warmer parts of the world, sometimes in the egg stage, but in the Eriophyidae by adult females only as these mites do not normally lay resistant eggs. Eggs of spider mites are typically laid on the leaves (except for overwintering temperate eggs) and most are large, red and globular and easily recognized through a hand lens; but the winter eggs of temperate species are laid on the spurs and woody twigs. Eggs of Eriophyidae are so minute that they are seldom seen in the field.

Field infestations are usually preyed upon heavily by predacious mites, various predacious bugs and a few Coccinellidae, and most pest populations are to some extent always controlled by these natural enemies. There have been many recorded cases where careless pesticide use has destroyed the natural enemies rather than the pests and has resulted in dramatic pest resurgences. The importance of mite predators can easily be seen through the present practice of commercially controlling red spider mites (*Tetranychus* spp.) in glasshouses using the predatory *Phytoseilus riegeli.* The family Phytoseiidae contains the most important species of predacious mites, and these mostly belong to the genera *Amblyseius*, *Phytoseilus* and *Typhlodromus*.

Resistance to acaricides is now very widespread for many species, especially the genera *Panonychus* and *Tetranychus* (Tetranychidae). The most striking aspect of this phenomenon has been the demonstration that some species of *Panonychus* actually thrived on a diet supplemented with DDT!

Some species of mites are agriculturally important in that they are known vectors of various virus diseases; for example *Eriophyes tulipae* carries three different wheat viruses, and *Cecidophyopsis ribis* carries the virus for black currant reversion disease. Some other species of mites regularly carry fungal spores on their bodies and thus are responsible for the spread of some fungal diseases.

The earlier use of the name *Tetranychus telarius* (L.) has been discontinued since it was established that it included two very closely related, but different species: *T. cinnabarinus* is the pantropical species (with red-coloured summer female), and *T. urticae* is the cosmopolitan temperate species (with green summer female); 'winter' females of both species are green and scarcely distinguishable.

Control of phytophagous mites

As already mentioned, the existing natural control through predacious mites and insects is usually important and care must be taken to ensure its continuation. The biological control of *T. urticae* in glasshouses in Europe and N. America using the predatory mite *Phytoseilus riegeli* is now a regular commercial practice and generally successful; the predator is available from commercial breeders. Release of predatory mites (and insects) is sometimes practised on field crops in the tropics, but usually only limited control is achieved.

Acaricides The earliest compound used for mite control was sulphur, and in fact this is still widely used. But susceptibility to sulphur varies somewhat; the Eriophyidae and Tenuipalidae are generally susceptible, but most spider mites are not. A complication is that some plants are sensitive to sulphur, and this phytotoxicity restricts the use of this chemical; these plants are sometimes referred to as 'sulphur-shy'.

The next effective compounds were the dormant sprays of tar oils, and they successfully killed the overwintering eggs of temperate spider mites. Now the recently developed refined 'summer oils' can be used on actual foliage for killing the active stages of the mites. The use of oils against mites has continued, despite the recent development of new acaricides, because there seems to be no development of resistance against these oils.

The range of pesticides effective against phytophagous mites is now very extensive and includes the following widely used chemicals.

sulphur	dicofol
tar oils	dimethoate
petroleum oils	disulfoton
DNOC	endosulfan
aldicarb	malathion
amitraz	oxydemeton-methyl

azinphos-methyl	pirimiphos-methyl
chlorpyrifos	quinomethionate
cyhexatin	tetradifon
derris	thiometon
demeton-*S*-methyl	triazophos
	vamidothion

They are usually employed in high-volume sprays, but other methods of application are also used, depending in part on the nature of the chemical being used. Some of these chemicals are more effective than others against a particular mite (some do not kill the eggs or hibernating females) and some acaricides are not recommended for use on certain crops, so local advice should always be sought for recommendations applicable to a particular pest/crop situation.

Important mite pests of agricultural crops

The total number of recorded mite pests attacking growing and stored crops is very large, as can be seen by reference to Jeppson, Keifer & Baker (1975); some of the more important species are listed here:

Tarsonemidae

Polyphagotarsonemus latus (Banks) – (Broad Mite) polyphagous; cosmopolitan (CIE map no. A.191; page 468).

Stenotarsonemus ananas (Tryon) – (Pineapple Tarsonemid) Australia, Hawaii.

Stenotarsonemus bancrofti (Mich.) – (Sugarcane Stalk Mite) Africa, S.E. Asia, N., C. and S. America.

Stenotarsonemus laticeps (Halb.) – (Bulb Scale Mite) Europe, USA.

Stenotarsonemus spirifex (Marchal) – (Oat Spiral Mite) Europe.

Tarsonemus pallidus Banks – (Strawberry (Cyclamen) Mite) Europe, Japan, USA (page 469) (now regarded as two distinct races – see below).

Tarsonemus pallidus fragariae Zimm. – (Strawberry Mite), *Tarsonemus p. pallidus* Banks – (Cyclamen Mite).

Tenuipalpidae (false spider mites)

Brevipalpus californicus (Banks) – (Scarlet Tea Mite) cosmopolitan (CIE map no. A.107).

Brevipalpus lewisi (McG.) – (Citrus Flat Mite) polyphagous; Near East, Australia, USA.

Brevipalpus obovatus Donn. – (Privet Mite) polyphagous; cosmopolitan (CIE map no. A.128) (see page 470).

Brevipalpus phoenicis (Geijskes) – (Red Crevice Tea Mite) polyphagous; cosmopolitan (CIE map no. A.106).

Cenopalpus spp. – on many fruit trees and other plants; Europe, W. and S. Asia.

Dolichotetranychus floridanus (Banks) – (Pineapple False Spider Mite) S.E. Asia, Japan, USA, C. America.

Raoiella indica Hirst. – (Date Palm Scarlet Mite) Egypt, India, S.E. Asia, C. America (CIE map no. A.210).

Tenuipalpus zhizhilashviliae Peck – (Persimmon False Spider Mite) Japan.

Penthaleidae

Halotydeus destructor (Tucker) – (Red-legged Earth Mite) Cyprus, S. Africa, New Zealand, Australia (CIE map no. A.119).

Penthaleus major Duges – (Winter Grain Mite) cosmopolitan.

Siteroptes cerealium (Kirch.) – (Grass and Cereal Mite) Europe.

Acaridae

Acarus siro L. – (Flour Mite) cosmopolitan.

Rhizoglyphus echinopus F.et R. – (Bulb Mite) Europe, Japan, USA.

Tryophagus spp. – grasses and cereals; cosmopolitan.

Tetranychidae (spider mites)

Bryobia cristata Duges – (Grass-Pear Bryobia Mite) Europe, N. Africa, Japan, Australasia.

Bryobia eharai P.et K. – (Chrysanthemum Spider Mite) Japan, Pakistan.

Bryobia graminum Schrank – (Grass Bryobia Mite) Europe.

Bryobia kissophila Eynd. – (Ivy (Clover) Bryobia Mite) Europe.

Bryobia praetiosa Koch – (Clover Mite) Europe, Asia, Africa, Australia, N. and S. America.
Bryobia ribis Thom. – (Gooseberry Bryobia Mite) Europe.
Bryobia rubrioculus (Sch.) – (Apple and Pear Bryobia (Brown) Mite) Europe, Asia, Japan, S. Africa, Australia, N. and S. America.
Eotetranychus boreus Ehara – (Apricot Spider Mite) Japan, USA.
Eotetranychus hicoriae (McG.) – (Pecan Leaf Scorch Mite) USA.
Eotetranychus pruni Oud. – (Chestnut Spider Mite) Europe, Asia, Japan, USA.
Eotetranychus sexmaculatus Riley – (Six-spotted Spider Mite) Japan, China, India, New Zealand, USA.
Eotetranychus uncatus Garman – (Walnut Spider Mite) Japan, USA.
Eutetranychus banksi (McG.) – (Texas Citrus Mite) USA.
Eutetranychus orientalis (Klein) – (Oriental Mite) Africa, Asia, USA.
Eutetranychus spp. – fruit trees; Africa, India.
Oligonychus afrasiaticus (McG.) – (Date Spider Mite) Africa, Near East.
Oligonychus coffeae (Niet.) – (Red Coffee (Tea) Mite) pantropical (CIE map no. A.165).
Oligonychus indicus (Hirst) – (Sugarcane Leaf Mite) India, USA.
Oligonychus mangiferus (R. & S.) – (Mango Spider Mite) Africa, India, USA, S. America (CIE map no. A.209).
Oligonychus orthius Rim. – (Sugarcane Spider Mite) Japan.
Oligonychus pratensis (Banks) – (Banks Grass Mite) Africa, USA, C. and S. America.
Oligonychus punicae (Hirst) – (Avocado Brown Mite) India, USA, C. and S. America.
Oligonychus shinkajii Ehara – (Rice Spider Mite) Japan.
Oligonychus yothersi (McG.) – (Avocado Red Mite) USA, C. and S. America.
Panonychus citri McG. – (Citrus Red Spider Mite) cosmopolitan (CIE map no. A.192).
Panonychus ulmi Koch – (Fruit Tree Red Spider Mite) cosmopolitan (CIE map no. A.31; see page 458).
Petrobia harti (Ew.) – (Oxalis Spider Mite) cosmopolitan.
Petrobia latens Muller – (Brown Wheat Mite) India, Japan, USA.
Schizotetranychus asparagi (Oud). – (Asparagus Spider Mite) Europe, USA, Hawaii.
Schizotetranychus celarius (Banks) – (Bamboo Spider Mite) China, Japan, USA.
Schizotetranychus spp. – (Citrus Spider Mites) India.
Tetranychus canadensis (McG.) – (Four-spotted Spider Mite) USA, Canada.
Tetranychus cinnabarinus Bois. – (Carmine Spider Mite) pantropical (see page 466).
Tetranychus ludeni Zacher – polyphagous; worldwide.
Tetranychus marianae McG. – (Solanum Spider Mite) S.E. Asia, Australia, USA, C. and S. America (CIE map no. A.403).
Tetranychus mcdanieli (McG.) – (McDaniel Spider Mite) Canada, USA.
Tetranychus neocaledonicus Andre – (Vegetable Spider Mite) pantropical.
Tetranychus turkestani V. & N. – (Strawberry Spider Mite) Europe, Asia, USA.
Tetranychus urticae Koch – (Two-spotted Spider Mite) cosmopolitan (temperate) (see page 467).
Tetranychus spp. (many) – (Spider Mites) many hosts; worldwide.

Eriophyidae (gall/blister/rust mites)
Abacarus hystrix (Nalepa) – (Grain Rust Mite) Europe, Asia, Canada, USA.
Acalitus essigi (Hassan) – (Redberry Mite) UK, New Zealand, USA.
Acalitus gossypii (Banks) – (Cotton Blister Mite) USA, C. and S. America.
Acalitus phloeocoptes (Nalepa) – (Almond and Plum Bud Gall Mite) Europe, Asia Minor, USA.
Acalitus vaccinii (Keiffer) – (Blueberry Bud Mite) USA.
Acaphylla theae Watt. – (Pink Tea Rust Mite) India, China, Japan.

Aceria sheldoni (Ewing) – (Citrus Bud Mite) cosmopolitan (CIE map no. A.127).
Aceria spp. (5) – fruit trees; India, S. America.
Aculops lycopersici (Massee) – (Tomato Russet Mite) worldwide (CIE map no. A.164).
Aculops pelekassi Keifer – (Pink Citrus Rust Mite) Japan.
Aculus cornutus (Banks) – (Peach Silver Mite) Europe, Asia, N., C. and S. America.
Aculus fockeui (N. & T.) – (Plum Rust Mite) Europe, Canada, USA.
Aculus schlechtendali (Nalepa) – (Apple Rust Mite) Europe, Asia, Canada, USA.
Calacarus carinatus (Green) – (Purple Tea Mite) Asia, Japan, Australasia, USA (CIE map no. A.324).
Calepitrimerus vitis (Nalepa) – (Grape Rust Mite) Japan.
Cecidophyopsis ribis (West.) – (Currant Bud Mite) Europe, Canada, USA (CIE map no. A.129).
Colomerus vitis Pag. – (Grape Gall Mite) worldwide.
Diptacus gigantorhynchus (Nalepa) – (Big-beaked Plum Mite) cosmopolitan.
Epitrimerus pyri Nalepa – (Pear Rust Mite) Japan, Canada, USA.
Eriophyes caryae Keiffer – (Pecan Leaf-roll Mite) USA.
Eriophyes erineus (Nalepa) – (Persian Walnut Blister Mite) Asia, USA.
Eriophyes ficus Cotte – (Fig Mite) USA
Eriophyes litchi Keiffer – (Litchi Mite) Hawaii, Pakistan.
Eriophyes lycopersici (Wolf.) – (Tomato Erineum Mite) pantropical.
Eriophyes mangiferae (Sayed) – (Mango Bud Mite) Near East, S. Asia, USA.
Eriophyes oleae Nalepa – (Olive Bud Mite) Mediterranean Region.
Eriophyes pyri (Pgst.) – (Pear Leaf Blister Mite) Europe, India, Canada.
Eriophyes rossettonis (Keifer) – (Cashew Bud Mite) S. America.
Eriophyes sheldoni Ewing – (Citrus Bud Mite) cosmopolitan (CIE map no. A.127).
Eriophyes tristriatus (Nalepa) – (Walnut Leaf Gall Mite) Europe, parts of Asia.
Eriophyes tulipae Keifer – (Wheat Curl Mite) Europe, Asia, N. Africa, Australasia, USA.
Eriophyes spp. – fruit trees and shrubs; worldwide.
Eriophyes spp. – grasses, cereals and sugarcane; worldwide.
Phyllocoptella avellaneae (Nalepa) – (Filbert Big Bud Mite) Europe, Asia, Australia, Canada, USA.
Phyllocoptes gracilis (Nalepa) – (Dryberry (Blackberry Leaf) Mite) Europe, USA, Canada.
Phyllocoptruta oleivora (Ash.) – (Citrus Rust Mite) pantropical (CIE map no. A.78).
Phytoptus pyri Pgst. – (Pear Leaf Blister Mite) Asia, S. Africa, USA.

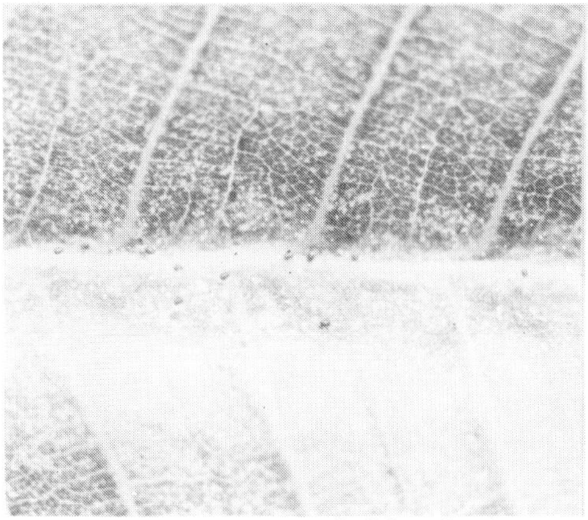
Eutetranychus orientalis on leaf of *Frangipani*

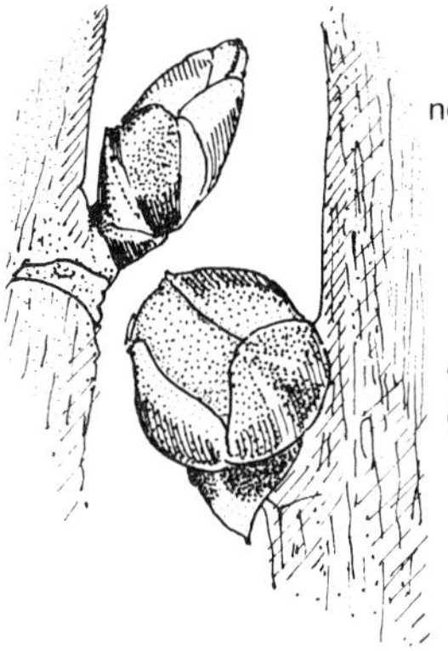

Cecidophyopsis ribis
(Black Currant Gall Mite)

Eriophyes tristriatus (Walnut Leaf Gall Mite)

Eriophys padi (Plum Pouch-gall Mite)

Tetranychus cinnabarinus (Boisd.)

(= *T. telarius* (L.) *part.*; etc.)

Common name. Carmine Spider Mite (Tropical Red Spider Mite)
Family. Tetranychidae
Hosts. Polyphagous on low-growing plants throughout the semitropical areas of the world; very frequently on cotton and citrus; in temperate regions mostly confined to glasshouses, on a wide range of crops; 100+ hosts recorded.
Damage. Infested leaves become spotted yellowish, then rust-coloured; finally they wither, die and are shed prematurely. Red or greenish mites are just visible to the unaided eye, underneath the leaves.
Pest status. A widespread and regular pest of many crops in warmer parts of the world and in greenhouses in temperate regions.
Life history. The eggs are spherical, pinkish, about 0.1 mm in diameter; laid singly on the underside of the leaves or stuck to the strands of silk spun by the adult mites. Hatching requires 4–7 days; each female may lay 100–200 eggs over about three weeks.

The tiny six-legged larva is pinkish; after 3–5 days it moults into the first nymphal stage (protonymph), and a few days later into the deuteronymph; total nymphal period is 6–10 days.

The female is red or green in colour, 0.4–0.5 mm long, with several dark body spots. The male is slightly smaller. Both adults spin silken webbing over the host plant foliage. All active stages tend to feed on the underside of the leaves between the main veins. These mites do not hibernate (diapause) but may pass into a state of extended quiescence if the temperature falls. Optimum temperature for development is about 32 °C. The life-cycle can be completed in as short a time as 8–12 days, and in warm climates there may be up to 20 generations per year.
Distribution. Recorded throughout much of Africa, southern Europe, southern Asia, Japan, Australia, New Zealand, USA, C. and S. America; farther north in temperate regions usually confined to greenhouses (CIE map no. A.390).
Control. In greenhouses in Europe and USA control is usually achieved using the predacious mite *Phytoseilus riegeli*. Many greenhouse infestations are often a mixture of this species together with *T. urticae*. For chemicals used to control this mite see page 461.

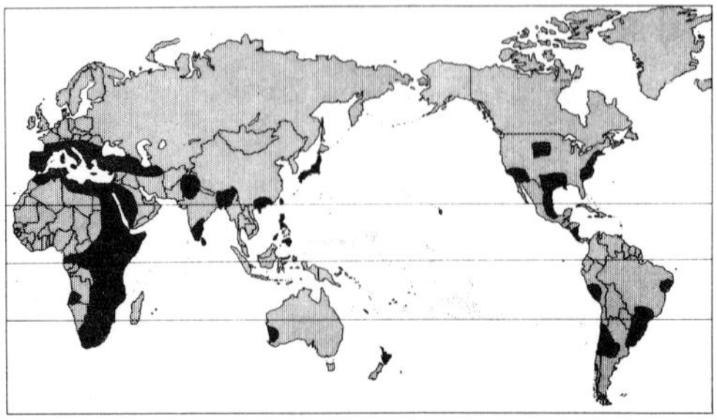

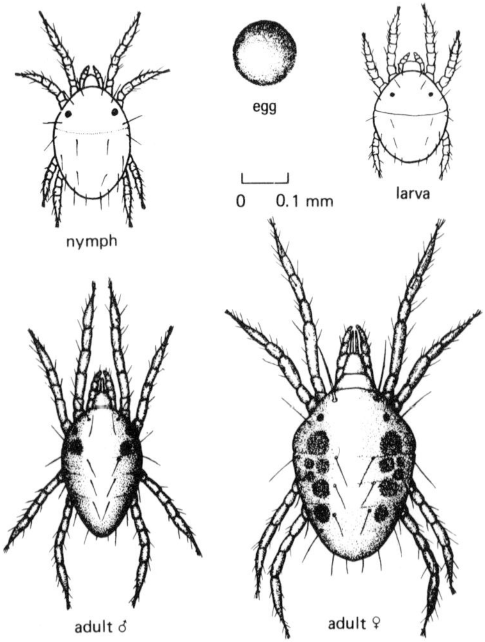

Tetranychus urticae (Koch)

(= *T. telarius* (L.) *part.*; etc.)

Common name. Two-spotted Red Spider Mite (Greenhouse (Temperate) Red Spider Mite)

Family. Tetranychidae

Hosts. A polyphagous pest on most greenhouse crops, and a wide range of field crops where temperatures permit, especially beans, strawberry; many ornamentals and pot plants; more than 150 plants of economic value are recorded as hosts.

Damage. Early damage is usually a fine and regular speckling on the upper leaf surface (caused by the mites feeding underneath the leaf); with heavy attacks the foliage becomes scarified, bronzed and it wilts and dies; heavy attacks are usually associated with thick silken webbing.

Pest status. A widespread, polyphagous, and common pest on a wide range of crop plants; regularly serious and very damaging.

Life history. Eggs are laid singly, usually on the underside of the leaf but on some plants on the upper surface (e.g. carnation); they are globular and whitish; each female lays 90–200 eggs. Hatching requires 3–5 days at 30–2°C (optimum temperature for development).

Development is closely controlled by temperature; at 35 °C the life-cycle is completed in only six days; at 21 °C the time is 12 days, and at 10°C it is 55 days. Under suitable conditions breeding is continuous, but on field crops there is an overwintering female that hibernates in diapause in leaf litter, crevices, hollow bamboo canes, bark crevices, etc.

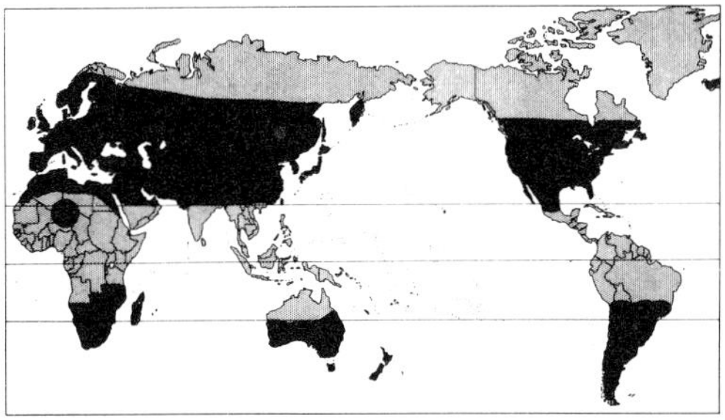

The adults are oval in shape; females about 0.8 mm long, males smaller; summer females are greenish with a pair of dark patches on the body (internal structures); males and immatures are also pale-greenish or yellowish; hibernating females are a bright red colour. Adult females usually live for about 30 days, during which time they lay their eggs.

Distribution. The complex of this and the preceding species (*T. cinnabarinus*), previously referred to as *T. telarius*, has a total of about 60 synonyms, each described from a different host or different geographical region.

T. urticae is recorded from all parts of the world where agriculture is practised, and it is probably completely worldwide in distribution: but many older records are of little use owing to the extensive synonymy and the previous confusion with *T. cinnabarinus*; it overlaps with the latter species in the subtropical parts of the world.

Control. For control see page 461.

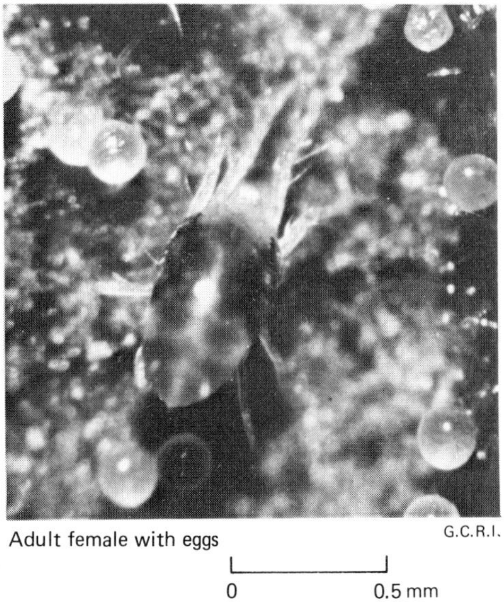

Adult female with eggs G.C.R.I.

Polyphagotarsonemus latus (Banks)

(= *Tarsonemus latus* Banks)

Common name. Broad Mite
Family. Tarsonemidae
Hosts. Totally polyphagous on a wide range of agricultural crops, ornamentals and wild plants (more than 50 recorded), including trees, shrubs and herbaceous plants.
Damage. Young leaves are cupped or distorted; with browned areas between the main veins, often bounded laterally by two distinct brown lines parallel to the main vein. With heavy infestations the flush growth is killed and herbaceous plants may die.
Pest status. A widespread pest of regular occurrence on many crops; usually a minor pest but occasionally serious.
Life history. Eggs are laid singly, stuck to the underside of flush leaves; they are oval in shape, flattened ventrally, and dorsally with rows of white tubercles; each egg is about 0.7 mm long. Hatching requires 2–3 days.

The larva is the only active immature stage; it is white and feeds near the old egg-shell; its development takes only 2–3 days. Nymphal development takes place within the larval skin, until the adult is formed, within 2–3 days. Female pupae are usually picked up by adult males and are carried on to newly opened young leaves attached to the abdomen tip. Adults are yellow and about 1.5 mm long. The female lives for about ten days and lays 2–4 eggs per day. All stages are found on the underside of flush leaves, usually in the grooves between the two halves of the leaf lamina before it has unfurled (causing the two parallel brown lines when the leaf is expanded).

Development can be completed in as short a time as 4–5 days in the summer, and 7–10 days in the winter.
Distribution. Distribution is very widespread, but records are sparse in some regions; more abundant in the tropics and subtropics, and in the colder temperate regions confined to greenhouses (CIE map no. A.191).
Control. This species is susceptible to sulphur, and dusting with sulphur is still recommended, as also is the use of dicofol, chlorobenzilate and carbophenothion in successive sprays.

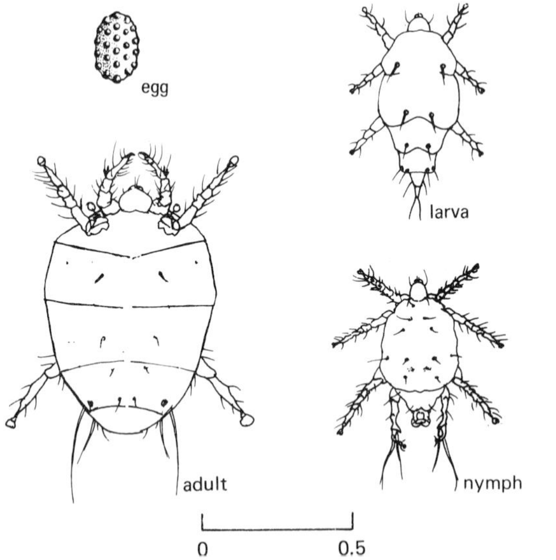

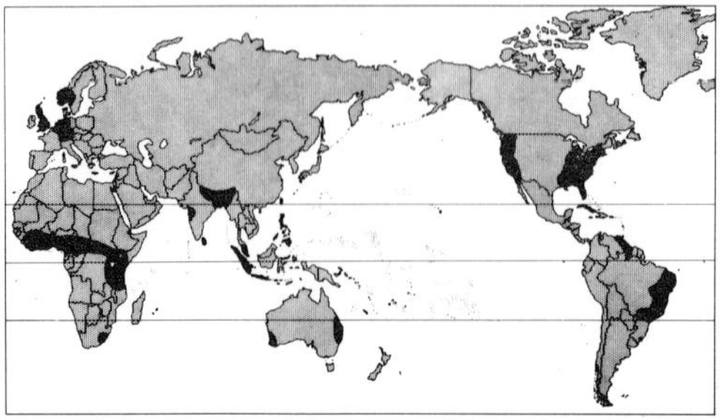

Tarsonemus pallidus Banks

(= *Stenotarsonemus pallidus* (Banks))

Common name. Strawberry (Cyclamen) Mite
Family. Tarsonemidae
Hosts (main). Strawberry, cyclamen.
(alternative). Watercress, many ornamental shrubs and flowers.
Damage. Feeding on young leaves causes leaf deformation; growth becomes restricted, and bud formation reduced; yield seriously reduced.
Pest status. A regular pest of strawberry in Europe, and apparently increasing in occurrence over recent years; equally serious on a wide range of ornamentals and pot plants.
Life history. Adult female mites overwinter in the crowns and lay eggs in the spring. The eggs are oval, white, and large (half as long as the adult female); laid in clusters; each female lays only 12–16 eggs; hatching takes 3–7 days. The larval and resting nymphal stage takes 3–11 days. In a heavy infestation there may be 300 mites per plant. The life-cycle may be completed in 1–3 weeks, and breeding is continuous throughout the warmer weather, mostly parthenogenetically, and males are scarce.

The worm-like adult female is only 0.25 mm long; the fourth pair of legs are reduced and very thin.
Distribution. A widespread pest species throughout Europe, Asia, Hawaii, Canada and USA.

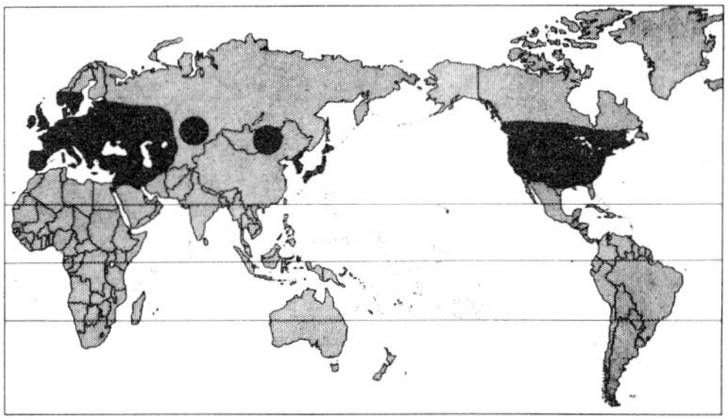

This is now regarded in the UK as being two distinct subspecies: *T. pallidus pallidus* Banks (Cyclamen Mite) and *T. pallidus fragariae* Zimm. (Strawberry Mite).
Control. Dispersal is probably by the mites crawling along the runners on strawberry, and by contamination of potting material in greenhouses. On field crops dicofol sprays are usually effective against light infestations, but for heavy attacks endosulfan is preferred, to be applied after the crop has been picked and after the removal of the old foliage. High-volume spraying is required, and the spray should be directed on to the crowns, for the mites are difficult to kill when hiding inside the folded leaflets.

Infested strawberry leaves D.V. Alford

Brevipalpus spp.

obovatus Donn.
phoenicis (Geijskes), etc.

Common name. Privet Mite; Red Crevice Mite
Family. Tenuipalpidae
Hosts. Polyphagous; but *B. obovatus* on citrus, privet, and many ornamentals (more than 50 genera); *B. phoenicis* on many different crops (citrus, tea, coffee, peach, papaya, rubber, coconut, apple, pear, guava, olive, fig, walnut, grape, date palm, cotton) and more than 50 genera of ornamentals.
Damage. These mites have toxic saliva and their feeding causes spotting on both foliage and fruits, but most serious damage is done to plants already stressed, which may be defoliated or even killed. Leaves of all ages are damaged.
Pest status. These are sporadic pests on a large number of crops in many parts of the world; occasionally serious damage is done.
Life history. Eggs are oval (0.1 mm long), bright red, and stuck to the underside of young leaves or in crevices in the bark of new wood; the incubation period is about ten days.

Larvae and nymphs are bright red, and the developmental period varies from ten days (30 °C) to 26 days (20 °C). These immature stages have periods of feeding followed by periods of inactivity.

Males are scarce, and generally reproduction is by parthenogenesis; females are about 0.3 mm long, the body is oval and divided clearly into propodosoma and hysterosoma; coloration varies from light orange to dark red. Females live for 20–60 days (according to temperature) and lay one egg per day usually. Adult females overwinter.

The life-cycle takes about six weeks in all.
Distribution. Together these species are quite worldwide, although *B.obovatus* appears to more temperate; in fact these two species do often occur in mixed infestations and are very difficult to distinguish (CIE maps nos. A.106 and A.128).

There are at least six other species of *Brevipalpus* found damaging a wide range of crop plants and ornamentals throughout the warmer parts of the world, and they all closely resemble these two species.
Control. *Brevipalpus* mites are susceptible to sulphur, dicofol and chlorobenzilate, but not to the organophosphorous or carbamate acaricides.

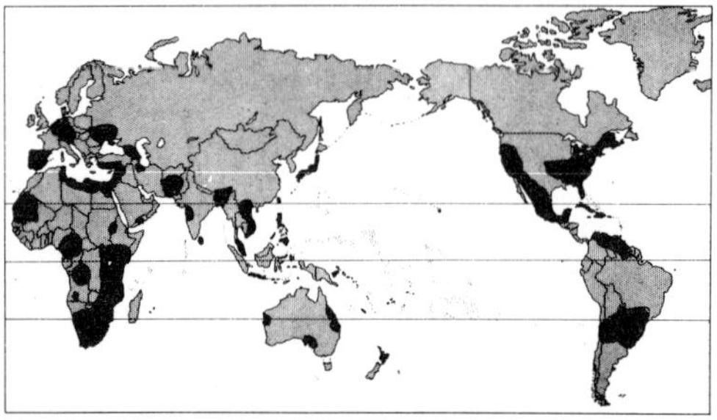

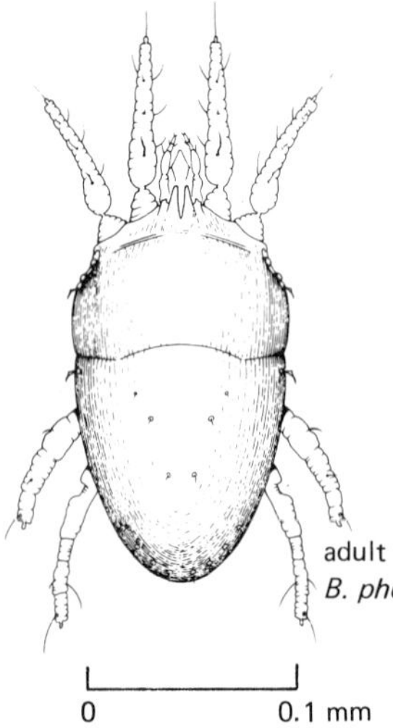

adult ♀
B. phoenicis

10 *Major temperate crops and their pest spectra*

In this chapter the 73 crops included are listed alphabetically for general ease of reference, with two terminal sections on the pests of seedlings and general pests, and the pests of stored products. These crops are temperate in distribution, as well as some from warmer regions that are technically not tropical. Thus there is a little overlap between the scope of this book and its predecessor *Agricultural insect pests of the tropics and their control*; the crops included in both volumes are those characteristic of the subtropics/warm temperate regions, or areas with a warm or hot summer.

The alternative to an alphabetical crop list would be one based on the product/commodity resulting from cultivation of that plant. Such a scheme has been proposed under the auspices of FAO and is called the *AGRIS classification scheme* (July, 1979), *Plants utilized by man*, pp. 51–9. The full title for the scheme is *International information system for the agricultural sciences and technology*. The category headings are listed below; the obvious main drawback with this scheme is the overlapping of categories, so that it is not immediately obvious to which category some crops belong. Thus potato, long regarded as a root vegetable crop is here placed as a starch crop, whereas cassava is classed as a root vegetable!

There is of course the third alternative of grouping the crops botanically, so that each family was grouped together, along with the other families that belonged to the same order. This approach would be of interest entomologically as many pests are oligophagous and to be found on most (or all) of the plants in the one family. But from an agricultural point of view such grouping would not be advantageous, so an alphabetical arrangement has been adopted.

AGRIS Classification Scheme

0010 *Plants utilized by man*
0100 Cereals
0200 Sugar and starch crops
0300 Oil crops
- Oil-producing trees
- Oil herbs and bushes

0400 Fibre plants
0600 Fruit trees and crops
- Temperate fruit trees and crops
- Cucurbit fruits
- Temperate berry plants and fruits
- Citrus trees and crops
- Tropical fruit trees and crops
- Edible nut trees and crops

1300 Vegetables
- Leguminous grains and vegetables
- Root vegetables
- Greens and leafy vegetables
- Bulbous vegetables
- Fruit vegetables

1900 Pastures and feed crops
2000 Cover crops
2100 Stimulant plants and crops
2200 Flavouring and perfume plants and crops
2400 Rubber and wax plants
2600 Tan and dye plants
2700 Drug plants
2800 Pesticide crops
3000 Forest and ornamental trees
- Conifers and other gymnosperms
- Angiosperm trees

3300 Ornamental plants
- Herbaceous ornamentals
- Bulbous or tuberous ornamentals
- Foliage plants and succulents
- Ornamental shrubs

3400 Various plants utilized by man

Pest spectra per crop

On the following pages are listed a representative selection of the more important insect and mite pests of the 73 crops included in this book. The selection of species to be included has been determined by the criteria mentioned on page 28, under the heading of 'Serious pests', with the over-riding editorial requirement of attempting to minimize parts of the pages left blank. In practice this arrangement actually works quite well. It must be remembered that throughout the entire cultivation range of a crop the total pest spectrum is very large (for example, cotton has 1360 species; cocoa 1400; sugarcane, 1300; etc.), but the vast majority of species recorded feeding/breeding on the crop are economically quite insignificant, and only a small proportion of pests warrant the term 'Major' (see page 28).

Crop and pest distributions

The pest distributions in respect to specific crops listed in this chapter are sometimes imprecisely known. In many instances it is known that a particular cosmopolitan pest feeds on a particular widespread crop, but it may not be known precisely just where the crop is attacked by this pest. Sometimes the pest concerned may be cosmopolitan whereas the crop may have a very restricted distribution, and of course *vice versa*. Frequently the major reference sources on distributional data are vague and not at all precise. Crop pest biogeographical studies should ideally be based firstly on detailed knowledge of the precise limits of cultivation of each crop worldwide, and also on phytogeography in relation to alternative and wild hosts; then the respective distributions of the more serious pests attacking each crop can be studied. Unfortunately such data are not readily available on a global basis, but some studies of this nature have been initiated both in the UK and in N. America, but only on a very restricted basis.

The major temperate crops included in this book are as follows:

Alfalfa
Almond
Apple
Apricot
Asparagus
Barley
Beans
Beetroot
Blackberry
Blueberry
Brassicas
Buckwheat
Capsicums
Carnation
Carrot
Celery
Cherry
Chestnut
Chrysanthemum
Clovers
Cranberry
Cucurbits
Currants
Dill
Eggplant
Field (broad) bean
Flax
Gladiolus
Globe artichoke
Gooseberry
Grapevine
Grass
Hazelnut
Hop
Horseradish
Jerusalem artichoke
Lettuce
Loganberry
Maize (and sweet corn)
Mint
Mulberry
Mushroom
Mustard
Narcissus
Oats
Onions
Parsley
Parsnip
Pea
Peach
Pear
Pecan
Persimmon
Plum
Pomegranate
Potato
Quince
Radish
Rape
Raspberry
Rhubarb
Rose
Rye
Spinach
Strawberry
Sugarbeet
Sunflower
Tobacco
Tomato
Tulip
Walnut
Watercress
Wheat

Pests of seedlings and general pests
Pests of stored products

ALFALFA (*Medicago sativa* – Leguminosae) (= Lucerne; Purple Medick)

An important forage legume, native to southwestern Asia, and probably the oldest cultivated forage plant. Now the most important forage plant in N. America, more than 7 million hectares being cultivated annually. A tall, semi-erect herb, up to 1 m in height, with typical clover-like appearance and large purple flowerheads. It is grown primarily for pasture, hay and silage, but also used widely as a green manure and for improving the soil. Many different varieties of this species have been bred over recent years. Other species of *Medicago* are generally referred to as medicks. This crop is dealt with separately from the other forage legumes (under the heading 'Clovers') because of its economic importance, and also its pest spectrum does show some important differences from the clovers.

MAJOR PESTS

Sminthurus viridis (L.)	Lucerne 'Flea'	Sminthuridae	Worldwide	Hole leaves; damage seedlings
Therioaphis maculata (Buckt.)	Spotted Alfalfa Aphid	Aphididae	Mediterranean, S. Asia, USA, Australia	Infest foliage; virus vector
Sitona spp.	Clover Weevils	Curculionidae	Cosmopolitan	Larvae eat root nodules; adults notch leaves

MINOR PESTS

Austroasca viridigrisea (Paoli)	Vegetable Jassid	Cicadellidae	Australia	Infest foliage; suck sap; some are virus vectors
Empoasca spp.	Green Leafhoppers	Cicadellidae	Cosmopolitan	
Stictocephala bubalis (F.)	Green Clover Treehopper	Membracidae	USA	
Acyrthosiphon kondoi Shinji	Blue–Green Alfalfa Aphid	Aphididae	Australia	Adults & nymphs infest foliage; suck sap; virus vectors
Acyrthosiphon pisum (Harris)	Pea Aphid	Aphididae	Cosmopolitan	
Aphis craccivora Koch	Groundnut Aphid	Aphididae	Cosmopolitan	
Aphis fabae Scopoli	Black Bean Aphid	Aphididae	Cosmopolitan	
Megoura viciae Buckt.	Vetch Aphid	Aphididae	Europe	
Adelphocoris spp.	Alfalfa Plant Bugs	Coreidae	Europe, Canada	Sap-suckers; toxic saliva; kill buds
Lygus spp.	Tarnished Plant Bugs	Miridae	Europe, USA, Canada	
Frankliniella spp.	Flower Thrips	Thripidae	Cosmopolitan	Scarify flowers & foliage
Merophyas divulsana Wkl.	Lucerne Leaf Roller	Tortricidae	Australia	Larvae roll & web foliage
Diacrisia spp.	Alfalfa Woollybears	Arctiidae	Canada	Larvae eat foliage
Etiella behrii Zell.	Lucerne Seed Web Moth	Pyralidae	Australia	Larvae eat flowers & seed pods
Loxostege sticticalis (L.)	Beet Webworm	Pyralidae	Cosmopolitan	Larvae eat foliage
Lampides boeticus (L.)	Pea Blue Butterfly	Lycaenidae	Europe, Africa, Asia, Australia	Larvae eat flowers, buds & pods
Zizina otis (F.)	Lesser Grass Blue	Lycaenidae	Asia, Australia	
Colias spp.	Clouded Yellow Butterflies	Pieridae	Canada, USA	Larvae eat foliage

Heliothis punctigera Wllgr.	Native Budworm	Noctuidae	Australia	Larvae defoliate
Spodoptera exigua (Hb.)	Lesser Armyworm	Noctuidae	Cosmopolitan in warm regions	Larvae eat foliage
Spodoptera frugiperda (Smith)	Fall Armyworm	Noctuidae	N., C. & S. America	
Autographa californica (Speyer)	Alfalfa Semilooper	Noctuidae	Canada	
Bruchophagus gibbus (Boh.)	Clover (Lucerne) Seed Wasp	Eurytomidae	Australia, Canada, USA	Larvae inside seeds
Asphondylia spp.	Alfalfa Gall Midges	Cecidomyiidae	Europe, USA	Larvae gall pods
Contarinia medicaginis Kieff.	Lucerne Flower Midge	Cecidomyiidae	Europe, Russia	Larvae in folded leaflets, flower heads & pods
Dasyneura spp.	'Clover' Midges	Cecidomyiidae	Europe	
Jaapiella medicaginis (Rüb.)	Lucerne Leaf Midge	Cecidomyiidae	Europe, Russia	
Tipula spp.	Leatherjackets	Tipulidae	Europe	Larvae eat roots; kill seedlings
Epicauta spp.	Black Blister Beetles	Meloidae	Asia, USA, Canada	Adults eat flowers & foliage
Bruchidius spp.	Clover Beetles	Bruchidae	Europe, Asia	Larvae bore seeds
Corrhenes stigmatica (Pascoe)	Lucerne Crown Borer	Cerambycidae	Australia	Larvae bore stems
Zygrita diva Thom.	Lucerne Crown Borer	Cerambycidae	Australia	
Apion spp.	Clover Seed Weevils	Apionidae	Europe	Larvae inside seeds
Graphognathus spp.	White-fringed Weevils	Curculionidae	Australia, S. Africa, USA, S. America	Larvae eat roots; adults eat leaves
Hypera postica (Gylh.)	Alfalfa Weevil	Curculionidae	Europe, Asia, Canada, USA	Larvae in shoots, buds, & flowerheads
Hypera spp.	Clover Leaf Weevils	Curculionidae	Europe, USA, Canada	
Otiorhynchus cribricollis Gylh.	Apple Weevil	Curculionidae	Mediterranean, Australia, USA	Larvae eat root nodules; adults eat leaves
Otiorhynchus ligustici (L.)	Alfalfa Weevil	Curculionidae	Europe, USA	
Philopedon plagiatum (Schall.)	Sand Weevil	Curculionidae	Europe	
Sitona spp.	Pea & Bean Weevils	Curculionidae	Cosmopolitan	
Halotydeus destructor (Tucker)	Red-legged Earth Mite	Penthaleidae	Europe, S. Africa, Australia, N.Z.	Adults & nymphs scarify foliage
Penthaleus major Duges	Winter Grain Mite	Penthaleidae	Cosmopolitan	

ALMOND (*Prunus amygdalus* – Rosaceae)

A native of eastern Mediterranean regions, it occurs as two distinct varieties, var. *dulcis* is the Sweet Almond, and var. *amara* is the Bitter Almond. The tree is small and closely resembles the near-relative peach. The trees are often cultivated as ornamentals for their convenient size and delicate blossoms. The nuts are very popular, and probably this is the nut sold in largest quantities in the world. The seed is eaten green, or more frequently roasted or salted, and also made into paste for cake-making. There are many different cultivars with varying shell thickness and seed flavour. The bitter almond contains a bitter glucoside (amygdalin) which readily breaks down into cyanic acid and so prevents its use as food. However, it is grown in southern Europe as a source of oil of bitter almond, which is used (after the acid is extracted) for flavouring. The bitter almond trees are also used as stock for grafting sweet almonds on to. Sweet almonds are grown throughout southern Europe, S. Africa, Australia and California.

MAJOR PESTS

Pterochloroides persicae Chol.	Peach Aphid	Aphididae	Mediterranean C. Asia, India	Infest foliage; suck sap
Cimbex quadrimaculatus Mull.	Sawfly	Cimbicidae	Mediterranean	Larvae defoliate
Eurytoma amygdali End.	Almond Stone Wasp	Eurytomidae	E. Mediterranean, S. Russia	Larvae bore in kernel of fruit
Cossus cossus L.	Goat Moth	Cossidae	Europe, Asia, Mediterranean	Larvae bore branches
Ephestia cautella (Hb.)	Almond Moth	Pyralidae	Cosmopolitan in warmer countries	Larvae feed on kernel
Paramyelois transitella (Wlk.)	Navel Orangeworm	Pyralidae	USA	Larvae bore fruits
Odinadiplosis amygdali (Anag.)	Almond Gall Midge	Cecidomyiidae	Lebanon, Greece	Larvae destroy flowers & fruits
Capnodis spp.	Flat-headed Borers	Buprestidae	W. Palaearctic	Larvae bore branches
Aceria phloeocoptes (Nal.)	Almond Bud Mite	Eriophyidae	Europe, C. Asia	Gall buds

MINOR PESTS

Brachycaudus helichrysi Kalt.	Peach Leaf-curl Aphid	Aphididae	India	Cause leaf-curl
Hyalopterus pruni Koch	Mealy Plum Aphid	Aphididae	Palaearctic	Infest foliage; sap-suckers; virus vectors
Myzus persicae (Sulz)	Peach–Potato Aphid	Aphididae	Cosmopolitan	
Didesmococcus onifasciatus	Soft Scale	Coccidae	Lebanon, S. Russia, Afghanistan	Adults & nymphs infest foliage; suck sap
Parthenolecanium corni (Bche.)	Plum Scale	Coccidae	Europe, W. Asia	
Icerya purchasi Mask.	Cottony Cushion Scale	Margarodidae	USA, Asia	
Hemiberlesia lataniae (Sign.)	Latania Scale	Diaspididae	Warmer countries	
Pseudaulacaspis pentagona T.	White Scale	Diaspididae	India	
Monosteira lobuliferia Reut.	Lace Bug	Tingidae	Lebanon, Libya, Syria	Sap-sucker; toxic saliva
Taeniothrips inconsequens (Uzel)	Pear Thrips	Thripidae	Europe, USA, Canada	Infest & scarify foliage
Synanthedon pictipes G. & R.	Lesser Peach Tree Borer	Sesiidae	USA	Larvae bore branches

Phyllonorycter cerasicolella H.S.	Leaf Blister Moth	Gracillariidae	S. Europe	Larvae make leaf mines
Anarsia lineatella Zell.	Peach Twig Borer	Gelechiidae	USA	Larvae bore shoots
Cydia funebrana (Treit.)	Red Plum Maggot	Tortricidae	Europe, Asia	Larvae bore fruits
Cydia molesta (Busck)	Oriental Fruit Moth	Tortricidae	USA	Larvae bore fruits & shoots
Malacosoma spp.	Tent Caterpillars	Lasiocampidae	Europe, Asia	Larvae defoliate
Trabala vishnou (Lef.)	Lappet Moth	Lasiocampidae	India	
Polyphylla fullo L.	Chafer Beetle	Scarabaeidae	Mediterranean, S. Russia	Adults eat leaves
Mimastra cyanura Hope	Leaf Beetle	Chrysomelidae	India	
Apate monachus F.	Black Borer	Bostrychidae	Mediterranean	Adults bore branches & trunk
Chrysobothris mali Horn	Flat-headed Apple Tree Borer	Buprestidae	USA	Larvae bore trunk and branches
Sphenoptera lafertei Thom.	Jewel Beetle	Buprestidae	India	
Cerambyx clux Fald.	Longhorn Beetle	Cerambycidae	Mediterranean	
Navomorpha sulcatus F.	Longhorn Beetle	Cerambycidae	Canada	
Myllocerus spp.	Grey Weevils	Curculionidae	India	Adults eat leaves; larvae in soil eat roots
Otiorhynchus cribricollis Gylh.	Apple Weevil	Curculionidae	Mediterranean, Australia, USA	
Tanymecus dilaticollis Gylh.	Southern Grey Weevil	Curculionidae	E. Europe	
Ruguloscolytus amygdali Guen.	Almond Bark Beetle	Scolytidae	Mediterranean	Adults bore wood under bark of trunk & branches to make breeding galleries
Ruguloscolytus spp.	Bark Beetles	Scolytidae	Mediterranean	
Bryobia rubrioculus Schent.	Bryobia (Brown) Mite	Tetranychidae	Mediterranean, Afghanistan	Scarify (bronze) foliage

APPLE (*Pyrus malus* – Rosaceae)

Apple is essentially a temperate fruit, native to E. Europe and W. Asia, and has been cultivated for more than 3000 years. Some 6500 horticultural forms are known, some varieties with distinct differences in their pest spectra. The crop is grown commercially in Britain, Europe, Canada, USA, Australia, New Zealand, and in some subtropical areas such as S. Africa, the Mediterranean Region and the cooler parts of India. There is a recent trend towards general crop diversification in most countries, and this has led to the introduction of apples as a crop in parts of Indonesia, Philippines and Kenya, in areas of high altitude where the climate is cooler.

MAJOR PESTS

Psylla mali Sch.	Apple Sucker	Psyllidae	Europe, India, N. America	Nymphs destroy flower trusses
Aphis pomi Deg.	Green Apple Aphid	Aphididae	Europe, W. Asia, N. America	Adults & nymphs infest foliage; cause leaf-curling (shoot death)
Dysaphis devecta (Wlk.)	Rosy Leaf-curling Aphid	Aphididae	Europe	
Dysaphis plantaginea (Pass.)	Rosy Apple Aphid	Aphididae	Europe, N. America, Asia	
Rhopalosiphum insertum (Wlk.)	Apple–Grass Aphid	Aphididae	Europe, N. America	
Eriosoma lanigerum (Ham.)	Woolly Apple Aphid	Pemphigidae	Cosmopolitan	Gregariously infest twigs & branches
Lepidosaphes ulmi (L.)	Mussel Scale	Diaspididae	Cosmopolitan	Infest twigs & trunk
Quadraspidiotus perniciosus (Comst.)	San José Scale	Diaspididae	Cosmopolitan	Infest foliage & fruit; very damaging
Lygocoris pabulinus	Common Green Capsid	Miridae	Europe	Sap-suckers; toxic saliva; scar fruits & may kill shoots
Plesiocoris rugicollis Fall.	Apple Capsid	Miridae	Europe	
Adoxophyes orana F.v.R.	Summer Fruit Tortrix	Tortricidae	Europe, Asia	Larvae eat leaves & surface of fruits
Archips podana (Scop.)	Fruit Tree Tortrix	Tortricidae	Europe	Larvae bore fruit buds & web shoots
Cydia pomonella (L.)	Codling Moth	Tortricidae	Cosmopolitan	Larvae bore fruits
Operophtera brumata (L.)	Winter Moth	Geometridae	Europe, Asia, N. America	Larvae eat leaves
Hoplocampa testudinea (Klug)	Apple Sawfly	Tenthredinidae	Europe, N. America	Larvae mine fruitlets
Chrysobothris spp.	Flat-headed Borers	Buprestidae	USA, Asia, Europe	Larvae bore branches under bark
Anthonomus pomorum (L.)	Apple Blossom Weevil	Curculionidae	Europe	Larvae 'cap' flower buds
Scolytus spp.	Fruit Bark Beetles	Scolytidae	Europe, Asia, N. & S. America	Adults bore under bark
Panonychus ulmi (Koch)	Fruit Tree Red Spider Mite	Tetranychidae	Europe, Asia, N. America	Scarify underside of leaves
Tetranychus spp.	Red Spider Mites	Tetranychidae	Cosmopolitan	Scarify & web foliage

MINOR PESTS

Forficula auricularia L.	Common Earwig	Forficulidae	Europe	Fruits holed
Empoasca spp.	Green Leafhoppers	Cicadellidae	Europe, Asia, N. & S. America	Adults & nymphs underneath leaves; sap-suckers
Euscelis obsoletus (Kir.)	Apple Leafhopper	Cicadellidae	Europe	
Typhlocyba spp.	Fruit Tree Leafhoppers	Cicadellidae	Europe, Canada, Australasia	
Cercopis vulnerata Ger.	Red & Black Froghopper	Cercopidae	Europe	Nymphs in spittle-mass on roots
Aphis spp.	Aphids	Aphididae	Europe, Asia	Infest foliage; curl leaves
Icerya purchasi Mask.	Cottony Cushion Scale	Margarodidae	India, USA	Adults & nymphs infest foliage; suck sap; may be with sooty moulds
Phenacoccus aceris (Sign.)	Apple Mealybug	Pseudococcidae	USA	
Pseudococcus comstocki (Kuw.)	Comstock Mealybug	Pseudococcidae	Asia, USA	
Eulecanium tiliae (L.)	Hazelnut Scale	Coccidae	Europe	
Parthenolecanium corni (Bche.)	Plum Scale	Coccidae	Europe, W. Asia	
Parthenolecanium persicae (F.)	Peach Scale	Coccidae	Cosmopolitan	
Pulvinaria spp.	Cottony Maple Scales	Coccidae	USA	
Aspidiotus hederae Bche.	Oleander Scale	Diaspididae	Cosmopolitan	Encrust foliage; sap-suckers
Parlatoria oleae (Colv.)	Olive Scale	Diaspididae	Europe, Asia, N. & S. America	
Quadraspidiotus ostraeformis (Curt.)	Oystershell Scale	Diaspididae	Europe	
Helopeltis antonii S.	Capsid Bug	Miridae	India	Sap-suckers; toxic saliva; feeding scars fruitlets & may kill shoots
Heterocordylus malinus Reut.	Dark Apple Redbug	Miridae	USA	
Lygidea mendax Reut.	Apple Redbug	Miridae	USA	
Lygus spp.	Capsid Bugs	Miridae	Europe, Asia, N. America	
Taeniothrips inconsequens Uzel	Pear Thrips	Thripidae	USA	Infest flowers & leaves; scarify foliage by feeding
Thrips spp.	Flower Thrips	Thripidae	Europe, Asia	
Stigmella malella (St.)	Apple Pygmy Moth	Nepticulidae	Europe	Larvae mine leaves
Tischeria malifoliella Clem.	Apple Leaf Trumpet Miner	Tischeriidae	USA	
Synanthedon myopiformis (Bork.)	Apple Clearwing Moth	Sesiidae	Europe	Larvae bore branches & trunk under bark
Thamnosphecia pyri (Harris)	Apple Bark Borer	Sesiidae	USA	
Anthophila pariana Clerck	Apple Leaf Skeletonizer	Glyphipterygidae	Europe, USA	Larvae skeletonize leaves
Coleophora nigricella Steph.	Casebearer Moth	Coleophoridae	Europe, Canada, USA	Larvae in cases on leaves; tiny in size
Eupista anatipennella Hbn.	Pistol Casebearer	Coleophoridae	Europe, USA	
Yponomeuta spp.	Small Ermine Moths	Yponomeutidae	Europe, Asia	Larvae spin webs & eat leaves
Blastodacna atra How.	Apple Pith Moth	Cosmopterygidae	Europe	Larvae bore shoots & flower truss stalks

Leucoptera scitella Zell.	Pear Leaf Blister Moth	Gracillariidae	Europe, China	Larvae mine leaves; blister mines
Lyonetia clerkella L.	Apple Leaf Miner	Gracillariidae	Europe	Larvae mine leaves
Marmara elotella (Busck)	Apple Bark Miner	Gracillariidae	USA	Larvae mine bark
Marmara pomonella Busck	Apple Fruit Miner	Gracillariidae	USA	Larvae mine fruitlets
Phyllonorycter spp.	Leaf Miners	Gracillariidae	Asia, Europe	Larvae mine leaves
Cossus cossus L.	Goat Moth	Cossidae	Europe, Asia	Larvae bore branches and trunks
Zeuzera spp.	Leopard Moths	Cossidae	Europe, Asia, USA	
Alsophila aescularia (Schiff.)	March Moth	Geometridae	Europe, Russia	Larvae eat leaves; may defoliate
Alsophila pometaria (Harris)	Fall Cankerworm	Geometridae	USA	
Biston betularia L.	Peppered Moth	Geometridae	Europe, Asia	
Erannis defoliaria (Clerck)	Mottled Umber Moth	Geometridae	Europe, W. Asia	
Psorosina hammondi (Riley)	Apple Leaf Skeletonizer	Phycitidae	USA	Larvae skeletonize leaves
Euzophera semifuneralis (Wlk.)	American Plum Borer	Phycitidae	USA	Larvae bore bark
Argyroploce pruniana Hbn.	Plum Tortrix	Tortricidae	Europe	Larvae bore shoots & eat leaves
Acroclita naevana Hbn.	Fruit Tree Tortrix	Tortricidae	Europe	Larvae eat leaves
Archips spp.	Fruit Tree Leaf Rollers	Tortricidae	Cosmopolitan	Larvae roll leaves
Cacaecia oporana L.	Fruit Tree Tortrix	Tortricidae	Europe	Larvae damage fruits
Cacoecimorpha pronubana (Hbn.)	Carnation Leaf Roller	Tortricidae	Europe, Mediterranean	Larvae roll leaves
Cnephasia longana Haw.	Omnivorous Leaf Tier	Tortricidae	Europe, Asia, N. America	
Cydia funebrana (Treit.)	Red Plum Maggot	Tortricidae	Europe, Asia	Larvae damage fruits
Cydia molesta (Busck)	Oriental Fruit Moth	Tortricidae	Asia, USA	
Cydia prunivora (Walsh)	Lesser Appleworm	Tortricidae	N. America	Larvae mine fruits under the skin
Epiphyas postvittana (Wlk.)	Light Brown Apple Moth	Tortricidae	Australasia (now Europe)	Larvae damage fruits & eat leaves
Pammene rhediella (Clerck)	Fruitlet Mining Tortrix	Tortricidae	Europe	Larvae eat fruitlets
Spilonota ocellana (D. & S.)	Eye-spotted Bud Moth	Tortricidae	Europe, Asia, N. America	Larvae bore buds
Arctias selene Hbn.	Moon Moth	Saturniidae	Asia	Larvae eat leaves
Smerinthus ocellata (L.)	Eyed Hawk Moth	Sphingidae	Europe	
Diloba caeruleocephala (L.)	Figure-of-Eight Moth	Notodontidae	Europe	Larvae eat leaves; occasionally may defoliate; some species also eat surface of fruitlets
Phalera bucephala L.	Bufftip Moth	Notodontidae	Europe, Asia	
Melanchra persicariae L.	Dot Moth	Noctuidae	Europe, Asia	
Orthosia inserta Hfn.	Clouded Drab	Noctuidae	Europe	
Orthosia spp.	Leaf-eating Caterpillars	Noctuidae	Europe, Asia	
Malacosoma spp.	Tent Caterpillars	Lasiocampidae	Europe, Asia	Gregarious larvae eat leaves
Euproctis spp.	'Brown-tail' Moths	Lymantriidae	Europe, Asia	
Orgyia spp.	Tussock Moths	Lymantriidae	Europe, Australia	Larvae eat leaves

Dasyneura mali (Kief.)	Apple Leaf Midge	Cecidomyiidae	Europe	Larvae roll leaves
Resseliella oculiperda (Rubs.)	Red Bud Borer	Cecidomyiidae	Europe	Larvae bore buds of young stock
Dacus spp.	Fruit Flies	Tephritidae	India	Larvae bore inside fruits
Rhagoletis pomonella (Walsh)	Apple Fruit Fly	Tephritidae	N. America	
Ametastegia glabrata (Fall.)	Dock Sawfly	Tenthredinidae	Europe	Larvae tunnel fruits & cut branches
Tremex columba L.	Horntail Wasp	Siricidae	USA	Larvae bore branches
Paravespula/*Vespa* spp.	Common Wasps	Vespidae	Cosmopolitan	Adults pierce ripe fruits
Syntomaspis druparum Boh.	Apple Seed Chalcid	Torymidae	N. America, Europe	Larvae eat seeds
Amphicerus bicaudatus (Say)	Apple Twig Borer	Bostrychidae	USA	Adults bore branches
Anomala spp.	Flower Beetles (Chafer Grubs)	Scarabaeidae	Asia	Adults eat foliage; may bite fruitlets; larvae in soil are polyphagous, but may damage young stock
Holotrichia spp.	Cockchafers	Scarabaeidae	Asia	
Melolontha spp.	Cockchafers	Scarabaeidae	Europe, Asia	
Phyllopertha horticola L.	Garden Chafer	Scarabaeidae	Europe	
Popillia spp.	Flower Beetles	Scarabaeidae	Asia, N. America	
Serica spp.	Cockchafers	Scarabaeidae	Europe, Asia	
Apthona euphorbiae Sk.	Large Flax Flea Beetle	Chrysomelidae	Europe	Adults eat leaves
Longitarsus parvulus Payk.	Flax Flea Beetle	Chrysomelidae	Europe	
Agrilus vittaticollis Rand.	Apple Root Borer	Buprestidae	USA	Larvae bore under bark; roots & trunk
Agrilus mali Mats.	Apple Buprestid	Buprestidae	China	
Anoplophora chinensis For.	Citrus Longhorn	Cerambycidae	China	Larvae bore in trunk and branches; sometimes in twigs
Apriona spp.	Longhorn Beetles	Cerambycidae	India, China	
Betula spp.	Longhorn Beetles	Cerambycidae	India	
Saperda candida F.	Longhorn Beetle	Cerambycidae	USA	
Tetrops praeusta L.	Little Longhorn Beetle	Cerambycidae	Europe	
Anthonomus cinctus Redt.	Apple Bud Weevil	Curculionidae	Europe	Larvae bore in buds
Caenorhinus aequatus (L.)	Apple Fruit Rhynchites	Curculionidae	Europe	Adults hole fruitlets
Conotrachelus nenuphar (Herbst)	Plum Curculio	Curculionidae	USA	Larvae inside fruits
Otiorhynchus singularis (L.)	Clay-coloured Weevil	Curculionidae	Europe	Adults eat leaves & bark
Otiorhynchus cribricollis Gylh.	Apple Weevil	Curculionidae	Mediterranean, Australia	Adults eat leaves & buds; larvae eat roots
Magdalis armigera Gf.	Boring Weevil	Curculionidae	Europe	Larvae bore twigs

Myllocerus spp.	Grey Weevils	Curculionidae	India	Adults eat leaves; larvae in soil eat roots
Phyllobius spp.	Leaf Weevils	Curculionidae	Europe	
Rhynchaenus pallicornis (Say)	Apple Flea Weevil	Curculionidae	USA	Larvae mine leaves
Rhynchites caeruleus (Deg.)	Apple Twig Cutter	Curculionidae	Europe	Adults cut off shoots
Tachypterellus quadrigibbus (Say)	Apple Curculio	Curculionidae	USA	Larvae inside fruits
Stephanoderes obscurus (F.)	Apple Twig Borer	Scolytidae	USA	Adults bore twigs
Xyleborus spp.	Shot-hole Borers	Scolytidae	Europe	Adults bore branches & trunk under bark
Aculus schlechtendali (Nal.)	Leaf & Bud Mite	Eriophyidae	Europe, N. America	Make blisters (erinea) on leaves
Eriophyes pyri (Pgst.)	Pear Leaf Blister Mite	Eriophyidae	Cosmopolitan	
Bryobia spp.	Bryobia (Brown) Mites	Tetranychidae	Europe, N. America, China	Scarify (bronze) leaves

APRICOT (*Prunus armeniaca* – Rosaceae)

A native of Asia, this small tree (6–10 m tall) has long been cultivated in India, China, Egypt and Iran, and is now grown in Europe, parts of Africa, and the warmer parts of the New World. The plant is susceptible to frost, so is grown in the subtropics and warmer parts of the temperate regions. The fruit is like a peach but the stone is smooth. Apricots are used as table fruits in the areas where they are grown, and are also dried, canned, frozen, candied, and made into a paste; an oil is extracted from the seeds.

MAJOR PESTS

Quadraspidiotus perniciosus (Comst.)	San José Scale	Diaspididae	Pantropical	Infest foliage
Anarsia lineatella Zell.	Peach Twig Borer	Gelechiidae	Cosmopolitan	Larvae bore buds & shoots
Tetranychus spp.	Red Spider Mites	Tetranychidae	Cosmopolitan	Scarify & web foliage

MINOR PESTS

Brachycaudus helichrysi Kalt.	Peach Leaf-curl Aphid	Aphididae	Europe, India	Infest foliage; suck sap; cause leaf-curl
Hyalopterus pruni (Geoff.)	Mealy Plum Aphid	Aphididae	Europe, India	
Myzus persicae (Sulz.)	Peach–Potato Aphid	Aphididae	Cosmopolitan	
Lecanium coryli L.	Hazelnut Scale	Coccidae	India	Adults & nymphs infest foliage; sap-suckers; some with sooty moulds
Parthenolecanium corni (Bché.)	Plum Scale	Coccidae	Europe, W. Asia	
Drosicha mangiferae (Green)	Giant Mealybug	Margarodidae	India	
Pseudococcus maritimus (Ehrh.)	Grape Mealybug	Pseudococcidae	Cosmopolitan	
Aspidiotus ostraeformis Curt.	Oystershell Scale	Diaspididae	Europe	
Pseudaulacaspis pentagona (Targ.)	White Peach Scale	Diaspididae	Cosmopolitan (not UK)	
Archips spp.	Tortrix Moths	Tortricidae	Asia	Larvae roll & eat leaves
Cacoecia spp.	Fruit Tree Tortrix Moths	Tortricidae	Europe, Asia	Larvae eat leaves & fruitlets
Eucosma ocellana Schiff.	Bud Moth	Tortricidae	India	Larvae bore buds
Cydia funebrana Treit.	Red Plum Maggot	Tortricidae	Europe, Asia	Larvae bore fruits
Abraxas grossulariata L.	Magpie Moth	Geometridae	Europe	Larvae eat leaves; may defoliate
Malacosoma spp.	Tent Caterpillars	Lasiocampidae	Europe, Asia	
Lymantria obfuscata Wlk.	Tussock Moth	Lymantriidae	India	
Dacus dorsalis Hend.	Oriental Fruit Fly	Tephritidae	India	Larvae inside fruits
Paravespula/Vespa spp.	Common Wasps	Vespidae	Cosmopolitan	Adults pierce fruits
Adoretus spp.	Flower Beetles	Scarabaeidae	India	Adults eat foliage; larvae in soil may damage roots
Anomala spp.	Flower Beetles		Asia	
Brahmina spp.	Cockchafers		India	
Melolontha spp.	Cockchafers		Europe, Asia	
Serica spp.	Cockchafers		Europe, Asia	
Sphenoptera lafertei Thom.	Jewel Beetle	Buprestidae	India	Larvae bore trunk
Aeolesthes holosericea F.	Apple Stem Borer	Cerambycidae	India	Larvae bore trunk & branches
Lophosternus hugelii Redt.	Longhorn Beetle	Cerambycidae	India	
Amblyrrhinus poricollis Boh.	Leaf-eating Weevil	Curculionidae	India	Adults eat leaves, larvae eat roots
Myllocerus spp.	Grey Weevils	Curculionidae	India	

Many of the pests listed under peach are also to be found attacking apricot.

ASPARAGUS (*Asparagus officinalis* – Liliaceae)

Asparagus is a native of temperate Europe and western Asia, and is still found in the wild in certain saline areas. It has been cultivated since Roman times as an epicurian item. The plant consists of a perennial root system and a tiny stem (crown) from which annually a tall (1–2 m) erect branching stem is produced. The asparagus of commerce consists of young succulent shoots (spears) as they issue from the crown and emerge above the soil; the later shoots are allowed to develop, for photosynthetic purposes, and the plant becomes a large woody bush; the flowers are axillary and small, and the fruit is a small round berry. Propagation can be either by seed, or by the use of one-year-old crowns. Once established an asparagus bed will continue to produce profitably for 15–20 years. The shoots are either sold fresh as a vegetable, or else canned, and sometimes frozen. Sometimes it is cultivated solely for the foliage as an ornamental; the usual species grown being *A. plumosus* and *A. sprengeri*. Some wild or ornamental species are scandent.

MAJOR PESTS

Crioceris asparagi (L.)	Asparagus Beetle	Chrysomelidae	Europe, N. America	Adults feed on shoots & fronds; larvae on fronds
Crioceris spp.	Asparagus Beetles	Chrysomelidae	Africa	

MINOR PESTS

Aphis gossypii Glov.	Cotton/Melon Aphid	Aphididae	Cosmopolitan	Infest foliage; suck sap
Myzus persicae (Sulz.)	Peach–Potato Aphid	Aphididae	Cosmopolitan	
Trialeurodes vaporariorum (Westw.)	Glasshouse Whitefly	Aleyrodidae	Europe, USA, Canada	Infest foliage (in greenhouses)
Pseudococcus adonidum (L.)	Longtailed Mealybug	Pseudococcidae	Cosmopolitan	Infest foliage
Thrips tabaci Lind.	Onion Thrips	Thripidae	Cosmopolitan	Infest & scarify foliage
Loxostege sticticalis (L.)	Beet Webworm	Pyralidae	Canada	Larvae eat foliage
Estigmene acrea (Drury)	Saltmarsh Caterpillar	Arctiidae	Canada	
Euxoa ochrogaster (Guen.)	Red-backed Cutworm	Noctuidae	Canada	Larvae are cutworms; cause 'goose-necking (bending) of spears by feeding
Euxoa spp.	Cutworms	Noctuidae	Canada	
Several species	Cutworms	Noctuidae	Canada	
Melanagromyza simplex (Loew)	Asparagus Miner	Agromyzidae	USA	Larvae mine stems
Ptochomyza asparagi Hering	Asparagus Leaf Miner	Agromyzidae	Europe, China	Larvae mine leaves
Platyparea poeciloptera (Schrank)	Asparagus Fly	Tephritidae	Continental Europe	Larvae mine stool
Crioceris duodecimpunctata (L.)	Spotted Asparagus Beetle	Chrysomelidae	USA, Europe	Larvae eat berries
Macrodactylus subspinosus (F.)	Rose Chafer	Scarabaeidae	Canada	Adults eat foliage; larvae eat roots
Tetranychus urticae (Koch)	Temperate Red Spider Mite	Tetranychidae	Cosmopolitan	Scarify & web foliage

BARLEY (*Hordeum vulgare* – Gramineae)

A crop of great antiquity, it has been regarded as the oldest of all cultivated plants. It is thought to have originated in the arid lands of S.W. Asia, N. Africa, and also S.E. Asia; it reached the Western Hemisphere in the sixteenth or seventeenth century. The plant is an annual grass, tending to become perennial, seldom taller than 1 m. The flower spike is conspicuously bearded in most varieties. The crop is hardy and has a short growing season, so that it can be grown at high latitudes and high altitudes. It is tolerant of a wide range of soil and climatic conditions, and is an important crop both in cold regions and some subtropical countries. Until the sixteenth century barley was the main source of bread flour, but has now been supplanted mostly by the more palatable wheat. The main use of barley now is as a feed for livestock (barley beef), and as a source of malt for brewing beer, whisky, and for malt extracts and breakfast foods. The main areas of production are USSR, China, Japan, Turkey, Rumania, W. Europe, USA, N. India, Canada and Spain.

MAJOR PESTS

Rhopalosiphum maidis (Fitch)	Corn Leaf Aphid	Aphididae	Cosmpolitan	Adults & nymphs infest foliage; suck sap; virus vectors
Rhopalosiphum padi (L.)	Bird-cherry Aphid	Aphididae	Europe, USA, Canada	
Schizaphis graminum Rond.	Wheat Aphid (Greenbug)	Aphididae	Africa, Asia, USA, S. America	
Sitobion avenae (F.)	Grain Aphid	Aphididae	Europe, N. America	
Oscinella frit (L.)	Frit Fly	Chloropidae	Europe	Larvae eat stems; make 'dead-hearts'

MINOR PESTS

Rhopalosiphum insertum (Wlk.)	Apple–Grass Aphid	Aphididae	Europe	Adults & nymphs infest foliage; suck sap; virus vectors
Macrosiphum fragariae Wlk.	Blackberry Aphid	Aphididae	Cosmopolitan	
Metapolophium dirhodum (Wlk.)	Rose–Grain Aphid	Aphididae	Europe	
Metapolophium festucae (Theob.)	Fescue Aphid	Aphididae	Europe	
Laodelphax striatella (Fall.)	Small Brown Planthopper	Delphacidae	Europe, Asia	Sap-suckers; virus vectors
Javasella pellucida (F.)	Cereal Planthopper	Delphacidae	Europe	
Eurygaster austriaca (Schr.)	Wheat Shield Bug	Pentatomidae	Europe, W. Asia	Sap-sucker; toxic saliva
Crambus spp.	Grass Moths	Pyralidae (Crambinae)	Europe, N. America	Larvae in silk webs eat leaves & stems
Hepialus spp.	Swift Moths	Hepialidae	Europe	Larvae in soil eat roots
Hydraecia micacea (Esp.)	Rosy Rustic Moth	Noctuidae	Europe	Larvae are stem borers
Mesapamea secalis (L.)	Common Rustic Moth	Noctuidae	Europe	
Sesamia spp.	Stalk Borers	Noctuidae	Asia	
Contarinia tritici (Kirby)	Lemon Wheat Blossom Midge	Cecidomyiidae	Europe, Asia	Larvae in flowerheads
Sitodiplosis mossellana (Gehin)	Orange Wheat Blossom Midge	Cecidomyiidae	Europe, Asia, N. America	

Haplodiplosis marginata (v.R.)	Saddle Gall Midge	Cecidomyiidae	Europe	Larvae gall stems
Mayetiola avenae (March.)	Oat Stem Midge	Cecidomyiidae	Europe	Larvae gall stems
Mayetiola destructor (Say)	Hessian Fly	Cecidomyiidae	Europe, Asia, N. America	
Bibio marci (L.)	St. Mark's Fly	Bibionidae	Europe	Larvae in soil eat roots
Tipula spp.	Leatherjackets	Tipulidae	Europe	
Nephrotoma spp.			Europe, N. America	
Geomyza spp.	Grass Flies	Opomyzidae	Europe, Asia, N. America	Larvae eat stems; cause 'dead-hearts'
Meromyza spp.				
Opomyza spp.				
Chlorops pumilionis (Bjerk.)	Gout Fly	Chloropidae	Europe	Larvae gall shoots
Agromyza ambigua Fall.	Cereal Leaf Miner	Agromyzidae	Europe, USA, Canada	Larvae mine leaves
Hydrellia griseola (Fall.)	Cereal Leaf Miner	Ephydridae	Europe, Asia, USA, S. America	Larvae mine leaves
Atherigona spp.	Cereal Shoot Flies	Muscidae	Africa, Asia	Larvae bore shoots
Delia platura (Meign.)	Bean Seed Fly	Anthomyiidae	Cosmopolitan	Larvae eat sown seeds
Delia coarctata (Fall.)	Wheat Bulb Fly	Anthomyiidae	Europe	Larvae bore shoots; make 'dead-hearts'
Phorbia securis Tiens.	Late Wheat Shoot Fly	Anthomyiidae	Europe	
Limothrips cerealium Hal.	Grain Thrips	Thripidae	Europe	
Stenothrips graminum Uzel	Oat Thrips	Thripidae	Europe	
Cephus cinctus Nort.	Western Wheat Stem Sawfly	Cephidae	W. USA	Larvae bore lower parts of stems
Cephus pygmaeus (L.)	Wheat Stem Sawfly	Cephidae	Europe, E. USA	
Dolerus spp.	Leaf Sawflies	Tenthredinidae	Europe	Larvae eat leaves
Agriotes spp.	Wireworms	Elateridae	Europe	Larvae eat hole in shoot underground
Helophorus nubilis F.	Wheat Shoot Beetle	Hydrophilidae	Europe	Larvae eat shoot base
Oulema melanopa (L.)	Cereal Leaf Beetle	Chrysomelidae	Europe, Asia	Adults & larvae eat leaf strips
Crepidodera ferruginea (Scop.)	Wheat Flea Beetle	Chrysomelidae	Europe	Larvae feed in shoots
Phyllotreta vittula Redt.	Barley Flea Beetle	Chrysomelidae	Europe	Adults eat holes in leaves; larvae bore stem
Melolontha spp.	Cockchafers	Scarabaeidae	Europe	Larvae in soil eat roots; may destroy seedlings
Phyllophaga spp.	June Beetles	Scarabaeidae	Canada, USA	
Schizonycha spp.	Chafer Grubs	Scarabaeidae	Europe, Asia	
Tanymecus dilaticollis Gylh.	Southern Grey Weevil	Curculionidae	E. Europe	Adults eat leaves
Siteroptes graminum (Reut.)	Grass & Cereal Mite	Penthaleidae	Europe	Cause 'silver-top'

BEANS (*Phaseolus* spp. – Leguminosae)

P. angularis – Adzuki Bean
P. aureus – Mung Bean, Green Gram
P. calcaratus – Rice Bean
P. coccineus – Scarlet Runner Bean
P. limensis – Lima Bean
P. lunatus – Sieva Bean
P. mungo – Black Gram
P. vulgaris – Common, Haricot, French, Kidney Bean

The genus contains several quite distinct species of beans with different crop qualities, and different habits, which are now grown in different parts of the world; but in general the pest spectra are very similar, and from this point of view scarcely worthwhile considering separately. The centre of origin for most species is either C. or S. America, but Scarlet Runner and Common Bean are now grown very widely throughout temperate regions of the world. The others tend to be grown in rather warmer regions. The plants are annuals, either scandent or climbing or with a shrubby habit. The fruit is always a pod containing a number of seeds. They form the single largest group of pulse legumes and are an important source of vegetable protein. In some species the pods are harvested immature as a vegetable, or else the ripe seeds collected as a pulse. The plant remains are often used as forage or for silage.
The other species tend to be very local, and more tropical.

MAJOR PESTS

Aphis fabae Scop.	Black Bean Aphid	Aphididae	Cosmopolitan (not Australia)	Infest foliage; suck sap
Clavigralla spp.	Spiny Brown Bugs	Coreidae	Africa	Sap-suckers; toxic saliva
Lamprosema spp.	Bean Leaf Rollers	Pyralidae	India, S.E. Asia	Larvae roll leaves
Maruca testulalis (Geyer)	Mung Moth	Pyralidae	Pantropical	Larvae bore pods to eat seeds
Heliothis armigera (Hb.)	American Bollworm	Noctuidae	Cosmopolitan in Old World	Larvae bore pods to eat seeds
Taeniothrips sjostedti (Tryb.)	Bean Flower Thrips	Thripidae	Africa	Infest flowers
Ophiomyia phaseoli (Tryon)	Bean Fly	Agromyzidae	Europe, Africa, India, S.E. Asia, Australasia	Larvae bore inside swollen stem or leaf petioles
Delia platura (Meign.)	Bean Seed Fly	Anthomyiidae	Cosmopolitan	Larvae bore sown seed
Coryna spp.	Pollen Beetles	Meloidae	Africa	Adults eat pollen
Mylabris spp.	Banded Blister Beetles	Meloidae	Africa, India, S.E. Asia	Adults eat flowers & foliage
Epicauta spp.	Black Blister Beetles	Meloidae	Africa, Asia, N. & S. America	Adults eat flowers & foliage
Epilachna varievestis Muls.	Mexican Bean Beetle	Coccinellidae	USA, Mexico, C. America	Adults & larvae defoliate
Acanthoscelides obtectus (Say)	Bean Bruchid	Bruchidae	Cosmopolitan	Larvae bore pods & eat seeds
Callosobruchus spp.	Cowpea Bruchids	Bruchidae	Cosmopolitan in warmer regions	Larvae eat ripe seeds in pods
Zabrotes subfasciatus (Boh.)	Mexican Bean Beetle	Bruchidae	Africa, USA, C. & S. America	
Colaspis brunnea (F.)	Grape Colaspis	Chrysomelidae	Southern USA	Larvae feed on roots
Oötheca mutabilis (Sahlb.)	Brown Leaf Beetle	Chrysomelidae	Africa	Adults eat leaves
Apion spp.	Apion Weevils	Apionidae	Cosmopolitan	Infest flowers
Tetranychus spp.	Red Spider Mites	Tetranychidae	Cosmopolitan	Scarify foliage

MINOR PESTS

Empoasca spp.	Green Leafhoppers	Cicadellidae	Cosmopolitan	Adults & nymphs infest foliage; suck sap; virus vectors
Aphis craccivora Koch	Groundnut Aphid	Aphididae	Cosmopolitan	
Macrosiphum spp.	Aphids	Aphididae	Africa, Asia	
Coccus spp.	Soft Scales	Coccidae	Cosmopolitan	Infest foliage; suck sap
Ferrisia virgata (Ckll.)	Striped Mealybug	Pseudococcidae	Pantropical	
Amblypelta spp.	Coreid Bugs	Coreidae	S.E. Asia	Adults & nymphs infest foliage; sap-suckers with toxic saliva; feeding causes necrotic spots (speckling and distorsion of foliage)
Anaplocnemis horrida Germ.	Coreid Bug	Coreidae	Africa	
Leptoglossus australis (F.)	Leaf-footed Plant Bug	Coreidae	Africa, India, S.E. Asia, Australia	
Riptortus pedestris F.	Coreid Bug	Coreidae	India	
Lygus spp.	Capsid Bugs	Miridae	Europe, Africa, N. America	
Calocoris norvegicus (Gmel.)	Potato Capsid	Miridae	Europe	
Halticus tibialis	Capsid Bug	Miridae	S.E. Asia	
Lygocoris pabulinus (L.)	Common Green Capsid	Miridae	Europe	
Nezara viridula (L.)	Green Stink Bug	Pentatomidae	Cosmopolitan	
Taeniothrips cinctipennis (Bagn.)	Thrips	Thripidae	S.E. Asia	Adults & nymphs infest foliage; scarify leaves, destroy flowers
Thrips tabaci Lind.	Onion Thrips	Thripidae	Cosmopolitan	
Kakothrips robustus (Uzel)	Pea Thrips	Thripidae	Europe	
Homona coffearia (Niet.)	Tea Tortrix	Tortricidae	S.E. Asia	Larvae roll & eat leaves
Eucosma melanaula (Meyr.)	Leaf Roller	Tortricidae	India	
Cnephasia spp.	Leaf Tiers	Tortricidae	Europe, Asia	
Matsumuraeses phaseoli (Mats.)	Adzuki Pod Worm	Tortricidae	Japan, China	Larvae bore pods to eat seeds
Lampides boeticus L.	Pea Blue Butterfly	Lycaenidae	Asia	
Agrius convolvuli (L.)	Convolvulus Hawk Moth	Sphingidae	Asia, Europe	Larvae defoliate
Amsacta spp.	Red Hairy Caterpillars	Arctiidae	Africa, Asia	
Diacrisia spp.	Tiger Moths	Arctiidae	India, Canada	
Heliothis spp.	Podborers	Noctuidae	Asia, N. America	Larvae bore pods
Anticarsia gemmatalis (Hb.)	Velvetbean Caterpillar	Noctuidae	USA, S. America	
Anticarsia irrorata B.	Green Caterpillar	Noctuidae	India	Larvae eat leaves, may defoliate
Spodoptera litura (F.)	Fall Armyworm	Noctuidae	Asia, India, Australasia	
Spodoptera littoralis (Boisd.)	Cotton Leafworm	Noctuidae	Africa	
Achaea spp.	Semiloopers	Noctuidae	Africa, Asia	
Plusia spp.	Semiloopers	Noctuidae	Europe, Asia	
Ophiomyia spp.	Bean Flies	Agromyzidae	S.E. Asia	Larvae bore stem
Phytomyza horticola Gour.	Pea Leaf Miner	Agromyzidae	Cosmopolitan in Old World	Larvae mine leaves
Delia spp.	Bean Seed Flies	Anthomyiidae	Europe, Asia, N. America	Larvae bore sown seed
Tipula spp.	Leatherjackets	Tipulidae	Europe, Asia	Larvae damage seedlings

Epilachna spp.	Epilachna Beetles	Coccinellidae	Africa, Asia, C. America	Adults & larvae defoliate
Bruchus spp.	Bean Beetles	Bruchidae	Europe, Asia, Canada, USA	Larvae eat seeds in young pods
Monolepta elegantula (Boh.)	Leaf Beetle	Chrysomelidae	S.E. Asia	Adults defoliate
Alcidodes spp.	Striped Weevils	Curculionidae	Africa, India	Adults girdle stems
Graphognathus spp.	White-fringed Weevils	Curculionidae	Australia, USA, S. America	Larvae eat roots; adults eat leaves
Nematocerus spp.	Nematocerus Weevils	Curculionidae	Africa	Adults eat leaves; larvae in soil may eat roots
Oribius spp.	Leaf Weevils	Curculionidae	S.E. Asia	
Sitona spp.	Pea & Bean Weevils	Curculionidae	Europe, Asia	

BEETROOT (*Beta vulgaris* – Chenopodiaceae) (= Red Beet; Spinach: Chard; Mangolds)

The various beets now in cultivation, including Sugarbeet, are all referred to the single botanical species, which was undoubtedly derived from the wild beet (*Beta maritima*) of the sea coasts of southern Europe. The oldest type of beet was Chard (*Beta vulgaris* var. *cicla*) which has been known since 300 BC. Initially it was grown for its root; later the tender leaves were eaten; now it is grown on a limited scale only for use as a pot herb, with large leaves and an insignificant root. Beetroot is obviously grown for the root, as is mangold, whereas the close relative spinach (and spinach beet) is cultivated solely for its leaves. Beetroots are boiled and then pickled, and eaten with salads; there is a limited amount of canning and preservation in glass jars. Tender young leaves are often used as a pot herb. Mangolds are grown solely as cattle food; either used dry or made into silage; the roots contain 3–8% sugar.

MAJOR PESTS

Circulifer spp.	Beet Leafhoppers	Cicadellidae	Europe, Mediterranean, N. America	Adults & nymphs infest foliage; virus vectors
Aphis fabae Scop.	Black Bean Aphid	Aphididae	Cosmopolitan (not Australia)	
Pegomya hyoscyami (Panz.)	Mangold (Beet) Fly	Anthomyiidae	Europe, USA, Canada	Larvae mine leaves

MINOR PESTS

Myzus ascalonicus Don.	Shallot Aphid	Aphididae	Europe, N. America	Adults & nymphs infest foliage; virus vectors
Myzus persicae (Sulz.)	Peach–Potato Aphid	Aphididae	Cosmopolitan	
Piesma quadrata (Fieb.)	Beet Lace Bug	Tingidae	Europe, Asia	Sap-suckers; toxic saliva
Nysius ericae (Schill.)	False Chinch Bug	Lygaeidae	USA	
Thrips tabaci Lind.	Onion Thrips	Thripidae	Cosmopolitan	Scarify foliage
Scrobipalpa ocellatella (Boyd.)	Beet Moth	Gelechiidae	Europe, Near East	Larvae eat leaves; some webbing of young leaves; may defoliate
Hymenia perspectalis (Hbn.)	Spotted Beet Webworm	Pyralidae	Cosmopolitan	
Loxostege sticticalis (L.)	Beet Webworm	Pyralidae	N. America	
Ceramica pisi L.	Broom Moth	Noctuidae	Europe	Larvae eat leaves; sometimes defoliate
Lacanobia oleracea (L.)	Tomato Moth	Noctuidae	Europe	
Spodoptera exigua (Hbst.)	Lesser (Beet) Armyworm	Noctuidae	Cosmopolitan	
Anagrapha falcifera (Kirby)	Celery Semilooper	Noctuidae	USA	
Autographa gamma (L.)	Silver-Y Moth	Noctuidae	Europe	
Delia echinata (Seguy)	Spinach Stem Fly	Anthomyiidae	Europe, USA	Larvae bore stems
Disonycha xanthomelas (Dalm.)	Spinach Flea Beetle	Chrysomelidae	USA	Adults & larvae eat foliage
Erynephala puncticollis (Say)	Beet Leaf Beetle	Chrysomelidae	USA	
Bothynoderes punctiventris (Germ.)	Beet Weevil	Curculionidae	Mediterranean, W. Asia	Larvae bore in root & leaf petioles; adults eat leaves
Lixus junci Boh.	Beet Weevil	Curculionidae	Mediterranean	
Otiorhynchus ligustici L.	Alfalfa Weevil	Curculionidae	Europe, USA	Larvae eat roots; adults eat leaves

BLACKBERRY (*Rubus* spp. – Rosaceae)

The cultivated species are mostly American in origin, but many species are found wild in different parts of the world, including the common Dewberry throughout Europe. It is not an important crop generally, but can be of local importance. The fruit is sold fresh, or used for canning, jams and cordials. The 'berry' is in fact a compound fruit, each fruitlet containing its own seed. Seedless varieties are now being cultivated. In habit it is an erect, decumbent or creeping shrub, usually bearing thorns, with a perennial rootstock from which new stems are produced annually; flowers are white and borne in clusters, followed by velvety-black fruits.

MAJOR PESTS

Sitobion fragariae (Wlk.)	Blackberry (Cereal) Aphid	Aphididae	Europe	Cause leaf-curl
Lygocoris pabulinus (L.)	Common Green Capsid	Miridae	Europe	Sap-sucker; toxic saliva
Spilonota ocellana Schiff.	Eye-spotted Bud Moth	Tortricidae	Europe, Asia, N. America	Larvae bore buds
Notocelia uddmanniana (L.)	Bramble Shoot Moth	Tortricidae	Europe	Larvae eat flowers & web shoots
Byturus tomentosus (Deg.)	Raspberry Beetle	Byturidae	Europe	Larvae bore fruits

MINOR PESTS

Typhlocyba spp.	Rubus Leafhoppers	Cicadellidae	Europe, Asia	Sap-suckers under leaves
Cercopis vulnerata Ger.	Red & Black Froghopper	Cercopidae	Europe	Larvae in spittlemass on roots
Trioza tripunctata (Fitch)	Blackberry Psyllid	Psyllidae	USA	Nymphs distort leaves
Amphorophora rubi (Kalt.)	Rubus Aphid	Aphididae	Europe	Adults & nymphs infest foliage; cause leaf-curl
Macrosiphum funestrum Macch.	Scarce Blackberry Aphid	Aphididae	Europe	
Aulacaspis rosae (Bche.)	Scurfy Scale	Diaspididae	Europe	Encrust stems
Thrips flavus Schr.	Honeysuckle Thrips	Thripidae	Europe	Adults & nymphs infest flowers & fruitlets
Thrips tabaci Lind.	Onion Thrips	Thripidae	Cosmopolitan	
Stigmella aurella (F.)	Rubus Leafminer	Nepticulidae	Europe	Larvae mine leaves
Cnephasia longana Haw.	Omnivorous Leaf Tier	Tortricidae	USA	Larvae eat leaves
Phalera bucephala L.	Buff-tip Moth	Notodontidae	Europe	Larvae defoliate
Contarinia rubicola Rübs.	Blackberry Flower Midge	Cecidomyiidae	Europe	Larvae inside flower buds
Dasyneura plicatrix H.Lw.	Blackberry Leaf Midge	Cecidomyiidae	Europe	Larvae curl leaves
Pegomyia rubivora Coq.	Loganberry Cane Fly	Anthomyiidae	Europe, Canada	Larvae mine & girdle shoots
Agromyza spiraeae Kalt.	Leaf Miner	Agromyzidae	Europe, Canada	Larvae mine leaves

Metallus spp.	Raspberry Leaf-mining Sawflies	Tenthredinidae	Europe, USA, Canada	Larvae mine leaves
Priophorus varipes (Lep.)	Plum Leaf Sawfly	Tenthredinidae	Europe	Larvae defoliate
Vespa/Vespula spp.	Common Wasps	Vespidae	Cosmopolitan	Adults eat ripe fruits
Diastrophus turgidus Bassett	Blackberry Gall Wasp	Cynipidae	USA	Make stem galls
Batophila spp.	Raspberry Flea Beetles	Chrysomelidae	Europe	Adults hole leaves
Agrilus ruficolis (F.)	Red-necked Cane Borer	Buprestidae	USA, Canada	Larvae bore canes
Otiorhynchus spp.	Clay-coloured Weevils	Curculionidae	Europe, USA	Adults gnaw stems & foliage; larvae eat roots
Rhynchites germanicus	Strawberry Rhynchites	Curculionidae	Europe	Adults girdle shoots & flower truss stems
Metatetranychus ulmi Koch	Fruit Tree Red Spider Mite	Tetranychidae	Europe, Asia, N. America	Adults & nymphs scarify foliage
Tetranychus urticae (Koch)	Temperate Red Spider Mite	Tetranychidae	Cosmopolitan	
Acalitis essigi (Hassan)	Blackberry Mite	Eriophyidae	Europe, USA	Cause 'redberry'
Phytoptus gracilis Nal.	Raspberry Leaf & Bud Mite	Eriophyidae	Europe	Found in buds & on leaves

BLUEBERRY (*Vaccinium* spp. – Ericaceae)

Native to N. America, wild plants are still the source of most of the supply of blueberries, but of late some cultivation on sandy acid soils is taking place, with a remarkable increase in size and yield. The plant is a small ericaceous shrub; the globular, blue fruit is a true berry. Blueberries are eaten fresh or cooked, mostly in pies; they are also canned and frozen. About seven different species are involved, some classed as low-bush forms the others as high-bush.

MAJOR PESTS

Rhagoletis mendax Curran	Blueberry Maggot	Tephritidae	USA	Larvae inside fruits

MINOR PESTS

Frankliniella vaccinii Morgan	Blueberry Thrips	Thripidae	USA	Infest flowers
Choristoneura hebenstreitella (Muller)	Tortricid	Tortricidae	Europe, Asia Minor, Japan	Larvae eat leaves
Grapholitha packardi Zell.	Cherry Fruitworm	Tortricidae	USA	Larvae bore fruits
Several species (8+)	Tortricids	Tortricidae	Europe, Asia	Larvae roll & eat leaves
Actebia fennica Tausch.	Black Army Cutworm	Noctuidae	Canada, USA	Larvae eat flowers & buds
Itame ribearia (Fitch)	Currant Spanworm	Geometridae	USA	Larvae eat leaves, may defoliate
Datana major G. & R.	Azalea Caterpillar	Notodontidae	USA	Larvae eat leaves, may defoliate
Paonias astylus (Drury)	Huckleberry Sphinx	Sphingidae	USA	Larvae eat leaves, may defoliate
Contarinia vaccinii Felt	Blueberry Tip Midge	Cecidomyiidae	Europe	Larvae in terminal buds
Rhagoletis pomonella (Walsh)	Apple Maggot	Tephritidae	N. America	Larvae inside fruits

BRASSICAS (*Brassica* spp. – Cruciferae) (Cabbage, Cauliflower, Broccoli, Brussels Sprout, Kale, Swede, Turnip, etc.)

An agriculturally diverse group of vegetables of European origin, and of great antiquity. They are cultivated from the Arctic to the subtropics, and at higher altitudes in the tropics. Certain species and varieties are more adapted to the tropics than others. Some of the heat-tolerant forms are quick-growing, for example the different types of Chinese Flowering Cabbage. As a group they do tend to have a similar spectrum of pests in most regions, but here Radish is considered separately as it is a very quick-growing crop, and Rape and Mustard are also separated because they are seed crops and, as such, differ from vegetables.

MAJOR PESTS

Brevicoryne brassicae (L.)	Cabbage Aphid	Aphididae	Cosmopolitan in cooler regions	Adults & nymphs infest foliage; suck sap; virus vectors
Lipaphis erysimi (Kalt.)	Turnip Aphid	Aphididae	Cosmopolitan in warmer regions	Adults & nymphs infest foliage; suck sap; virus vectors
Aleyrodes proletella (L.)	Brassica Whitefly	Aleyrodidae	Europe, Asia, N.Z., Brazil	Infest foliage
Bagrada spp.	Harlequin Bugs	Pentatomidae	Africa, Asia	Sap-suckers; toxic saliva
Plutella xylostella (L.)	Diamond-back Moth	Yponomeutidae	Cosmopolitan	Larvae hole & window leaves
Pieris brassicae (L.)	Large White Butterfly	Pieridae	Europe, W. Asia, N. America	Larvae eat leaves; may defoliate
Pieris canidia (L.)	Small White Butterfly	Pieridae	E. Asia	Larvae eat leaves; may defoliate
Pieris rapae (L.)	Small White Butterfly	Pieridae	Europe, Asia	Larvae eat leaves; may defoliate
Agrotis ipsilon (Hfn.)	Black Cutworm	Noctuidae	Cosmopolitan	Larvae are cutworms; may destroy seedlings
Agrotis segetum (D. & S.)	Common Cutworm	Noctuidae	Cosmopolitan in Old World	Larvae are cutworms; may destroy seedlings
Delia radicum (L.)	Cabbage Root Fly	Anthomyiidae	Europe, N. America	Larvae eat roots
Athalia spp.	Cabbage Sawflies	Tenthredinidae	Cosmopolitan	Larvae defoliate
Phyllotreta spp.	Cabbage Flea Beetles	Chrysomelidae	Cosmopolitan	Adults hole leaves; may kill seedlings

MINOR PESTS

Acheta spp.	Field Crickets	Gryllidae	Cosmopolitan	Roots & seedlings eaten
Gryllotalpa spp.	Mole Crickets	Gryllotalpidae	Cosmopolitan	Roots & seedlings eaten
Bemisia spp.	Whiteflies	Aleyrodidae	Asia	Infest foliage
Aphis gossypii Glov.	Cotton Aphid	Aphididae	Cosmopolitan	Infest foliage; suck sap; virus vectors
Myzus persicae (Sulz.)	Peach–Potato Aphid	Aphididae	Cosmopolitan	Infest foliage; suck sap; virus vectors
Eurydema pulchrum (Westw.)	Harlequin Bug	Pentatomidae	S.E. Asia	Sap-suckers; toxic saliva; feeding causes necrotic spots
Murgantia histrionica (Hahn)	Harlequin Bug	Pentatomidae	USA	Sap-suckers; toxic saliva; feeding causes necrotic spots
Nezara viridula (L.)	Green Stink Bug	Pentatomidae	Cosmopolitan in warmer regions	Sap-suckers; toxic saliva; feeding causes necrotic spots
Dysdercus spp.	Cotton Stainers	Pyrrhocoridae	Pantropical	Sap-suckers; toxic saliva; feeding causes necrotic spots

Nysius ericae (Schill.)	False Chinch Bug	Lygaeidae	USA	Sap-sucker; toxic saliva
Thrips angusticeps Uzel	Cabbage Thrips	Thripidae	Europe	Infest foliage; feeding causes speckling
Thrips tabaci Lind.	Onion Thrips	Thripidae	Cosmopolitan	
Hepialus spp.	Swift Moths	Hepialidae	Europe	Larvae in soil eat roots
Pieris napi (L.)	Green-veined White Butterfly	Pieridae	Europe	Larvae eat leaves; may defoliate
Pieris spp.	White Butterflies	Pieridae	Cosmopolitan	
Crocidolomia binotalis Zell.	Cabbage Moth	Pyralidae	Africa, Asia, Australasia	Larvae eat leaves; sometimes defoliate; some species make silk webbing
Evergestis spp.	Cabbageworms	Pyralidae	Europe, Asia, N. America	
Hellula phidilealis Wlk.	Cabbage Webworm	Pyralidae	Africa, C. & S. America	
Hellula undalis (F.)	Oriental Cabbage Webworm	Pyralidae	Africa, Asia, S. Europe, Australasia	
Agrotis spp.	Cutworms	Noctuidae	Cosmopolitan	Larvae destroy seedlings
Heliothis armigera (Hub.)	American Bollworm	Noctuidae	Old World	Larvae eat leaves; may defoliate; may destroy seedlings; some species behave as cutworms, some as armyworms; faecal contamination may be serious (collectively these caterpillars are serious pests)
Mamestra brassicae (L.)	Cabbage Moth	Noctuidae	Europe	
Mamestra configurata Wlk.	Bertha Armyworm	Noctuidae	Canada, USA	
Mythimna spp.	Cereal Armyworms	Noctuidae	Cosmopolitan	
Melanchra persicariae (L.)	Dot Moth	Noctuidae	Europe	
Melanchra picta (Harris)	Zebra Caterpillar	Noctuidae	Canada, USA	
Spodoptera exigua (Hub.)	Lesser Armyworm	Noctuidae	Cosmopolitan	
Spodoptera littoralis (Boisd.)	Cotton Leafworm	Noctuidae	Mediterranean, Africa	
Spodoptera litura (F.)	Fall Armyworm	Noctuidae	Asia	
Spodoptera spp.	Armyworms, etc.	Noctuidae	Cosmopolitan	
Xestia c-nigrum (L.)	Spotted Cutworm	Noctuidae	Europe, Asia, N. America	
Chrysodeixis chalcites (Esp.)	Green Semilooper	Noctuidae (Plusiinae)	Mediterranean, Africa, Asia	Larvae are semiloopers & eat leaves; may defoliate
Plusia orichalcea (Hub.)	Semilooper	Noctuidae	Tropical Asia	
Trichoplusia ni (Hub.)	Cabbage Semilooper	Noctuidae	Asia, USA, Canada	
Autographa gamma (L.)	Silver-Y Moth	Noctuidae	Europe, Asia	
Tipula spp.	Leatherjackets	Tipulidae	Europe	Larvae in soil eat roots
Contarinia nasturtii (Kieff.)	Swede Midge	Cecidomyiidae	Europe	Larvae distort flowers & shoots
Dasyneura brassicae (Winn.)	Brassica Pod Midge	Cecidomyiidae	Europe	Larvae inside damaged pods
Gephyraulus raphanistri (Kieff.)	Brassica Flower Midge	Cecidomyiidae	Europe	Larvae inside closed swollen flowers
Scaptomyza apicalis Hardy	Leaf Miner	Drosophilidae	Europe	Larvae mine leaves
Liriomyza brassicae Riley	Cabbage Leaf Miner	Agromyzidae	Cosmopolitan (not UK)	
Phytomyza horticola Gour.	Pea Leaf Miner	Agromyzidae	Cosmopolitan in Old World	

Phytomyza rufipes Meig.	Cabbage Leaf Miner	Agromyzidae	Europe, Asia, N. America	Larvae mine leaf petioles & stems
Phytomyza spp.	Leaf Miners	Agromyzidae	Europe, Asia	Larvae mine leaves
Delia floralis (Fall.)	Turnip Root Fly	Anthomyiidae	Europe	Larvae eat fine roots, & bore large roots & also stems
Delia platura (Meign.)	Bean Seed Fly	Anthomyiidae	Cosmopolitan	
Delia spp.	Root Maggots	Anthomyiidae	N. America	
Agriotes spp.	Wireworms	Elateridae	Europe	Larvae eat roots
Megempleurus spp.	Turnip Mud Beetles	Hydrophilidae	Europe	Larvae bore into stems & roots
Meligethes aeneus F.	Blossom Beetle	Nitidulidae	Europe	Adults infest flowers
Entomoscelis americana Brown	Red Turnip Beetle	Chrysomelidae	Canada, USA	Adults & larvae eat foliage
Phaedon spp.	Mustard Beetles	Chrysomelidae	Europe	
Psylliodes chrysocephala (L.)	Cabbage Stem Flea Beetle	Chrysomelidae	Europe	Larvae gall stems
Adoretus spp. *Anomala* spp. *Leucopholis* spp.	White Grubs	Scarabaeidae	S.E. Asia	Larvae in soil eat roots; may destroy seedlings
Ceutorhynchus assimilis (Payk.)	Cabbage Seed Weevil	Curculionidae	Europe, USA, Canada	Larvae gall seeds in pods
Ceutorhynchus pleurostigmata (March.)	Turnip Gall Weevil	Curculionidae	Europe	Larvae gall roots
Ceutorhynchus quadridens (Panz.)	Cabbage Stem Weevil	Curculionidae	Europe	Larvae bore petioles & stems
Ceutorhynchus spp.	Turnip Weevils	Curculionidae	N. America	Adults eat leaves; larvae eat roots
Lixus anguinus L.	Cabbage Weevil	Curculionidae	Mediterranean	Larvae mine stem & petioles
Listroderes costirostris (Klug)	Vegetable Weevil	Curculionidae	S. America, USA	Adults eat foliage; larvae eat roots

BUCKWHEAT (*Fagopyrum esculentum* – Polygonaceae) (= *F. sagittatum*)

One of the so-called 'pseudocereals', buckwheat is native to Central Asia, and still grows wild in Siberia and Manchuria. It is only of recent use, the earliest records being from China in the tenth century; introduced into Europe in the Middle Ages and first cultivated in 1436. Now widely grown on the Continent, especially USSR, where some 2 million hectares are cultivated as one of the peasant staples. Also grown in quantity in France, Germany and the USA. The plant is a small branching annual, preferring a cool, moist temperate climate, with well-drained sandy soil. It will also grow in arid regions with poor soil and drainage. The fruit is a three-cornered achene, resembling a beechnut, hence the common name. The seeds (groats) are ground and the flour used for soups, pancakes, porridge; the seeds are also used for feeding livestock and poultry, and the straw for feed and bedding. The flowers are an important source of honey, and the crop is also grown for cover and green manure.

MAJOR PESTS

Euxoa tritici (L.)	White-line Dart	Noctuidae	Eurasia	Larvae are cutworms
Mythimna unipuncta (Haw.)	American Armyworm	Noctuidae	Europe, Africa, N., C. & S. America	Larvae defoliate

MINOR PESTS

Teleogryllus emma O.et M.	Field Cricket	Gryllidae	Japan	Destroy seedlings
Aulacorthum solani (Kalt.)	Glasshouse–Potato Aphid	Aphididae	Cosmopolitan	Adults & nymphs infest foliage; suck sap; virus vectors
Macrosiphum euphorbiae (Thos.)	Potato Aphid	Aphididae	Cosmopolitan	
Nezara antennata Scott	Green Stink Bug	Pentatomidae	Japan	Sap-sucker; toxic saliva
Timandra griseata Pet.	Looper Caterpillar	Geometridae	Japan	Larvae eat foliage
Mamestra brassicae (L.)	Cabbage Moth	Noctuidae	Europe, Asia	Larvae eat foliage
Pyrrhia umbra (Hfn.)	Tobacco Striped Caterpillar	Noctuidae	Japan	
Hyles lineata (F.)	Striped Hawk Moth	Sphingidae	Cosmopolitan	Larvae eat foliage
Galerucella vittaticollis Baly	Strawberry Leaf Beetle	Chrysomelidae	Japan	Adults & larvae eat foliage
Phyllophaga spp.	June Beetles	Scarabaeidae	Canada	Larvae in soil eat roots
Lixus acutipennis Roel.	Weevil	Curculionidae	Japan	Adults eat leaves; larvae inside stems

CAPSICUMS (*Capsicum* spp. – Solanaceae) (= Sweet Peppers; Chillies)

The centre of origin is uncertain, but probably Peru; they were spread throughout the New World very early and are now grown widely throughout the warmer parts of the world, and under glass (or polythene tunnels) in temperate regions. They can be grown from sea level to 2000 m or more in the tropics, preferably with a rainfall of 60–120 cm per annum. They are sensitive to frost, waterlogging and too much rain. In habit the plant is a very variable herb, erect, many-branched, and is grown as an annual. Sweet Peppers are large (up to 10–15 cm long) and green, turning red as they ripen, and used fresh in salads or cooked as vegetables. Chillies are small (2–6 cm), pungent, and bright red, used in curries or dried to make cayenne pepper and paprika. The main areas of production are India, Thailand, Indonesia, Japan, Mexico, Uganda, Kenya, Nigeria and Sudan.

MAJOR PESTS

Aphis gossypii Glov.	Melon/Cotton Aphid	Aphididae	Cosmopolitan	Adults & nymphs infest foliage; sap-suckers; virus vectors
Aulacorthum solani (Kltb.)	Glasshouse–Potato Aphid	Aphididae	Europe, Japan, N. America	
Myzus persicae (Sulz.)	Peach–Potato Aphid	Aphididae	Cosmopolitan	
Macrosiphum euphorbiae (Thos.)	Potato Aphid	Aphididae	Cosmopolitan	
Heliothis armigera (Hb.)	American Bollworm	Noctuidae	Africa, Asia	Larvae bore fruits
Epilachna spp.	Epilachna Beetles	Coccinellidae	Africa, Asia, Mexico, C. America	Adults & larvae eat leaves
Epicauta spp.	Black Blister Beetles	Meloidae	Africa, Asia, N. America	Adults eat leaves & flowers
Polyphagotarsonemus latus (Banks)	Yellow Tea Mite	Tarsonemidae	Cosmopolitan	Foliage scarified
Tetranychus cinnabarinus (Boisd.)	Tropical Red Spider Mite	Tetranychidae	Pantropical	Scarify & web foliage
Tetranychus urticae (Koch)	Temperate Red Spider Mite	Tetranychidae	Cosmopolitan	

MINOR PESTS

Brachytrupes spp.	Large Brown Crickets	Gryllidae	Africa, Asia	Destroy seedlings; eat roots & stems
Gryllotalpa spp.	Mole Crickets	Gryllotalpidae	Africa, Asia, S. Europe, N. America	
Empoasca spp.	Green Leafhoppers	Cicadellidae	Cosmopolitan	Infest foliage; sap-suckers; virus vectors
Bemisia tabaci (Genn.)	Cotton Whitefly	Aleyrodidae	Africa, Asia	
Trialeurodes vaporarorium (Westw.)	Glasshouse Whitefly	Aleyrodidae	Europe	
Saissetia coffeae (Wlk.)	Helmet Scale	Coccidae	India	Infest foliage; sap-suckers; usually with sooty moulds
Ferrisia virgata (Ckll.)	Striped Mealybug	Pseudococcidae	India, S.E. Asia	
Pseudococcus adonidum (L.)	Long-tailed Mealybug	Pseudococcidae	Cosmopolitan	
Aspidiotus destructor Sign.	Coconut Scale	Diaspididae	Pantropical	
Cyrtopeltis tenuis Reut.	Tomato Mirid	Miridae	S.E. Asia	Infest foliage; sap-suckers; toxic saliva
Helopeltis spp.	Capsid Bugs	Miridae	Africa, Asia	
Acanthocoris spp.	Coreid Bugs	Coreidae	S.E. Asia	
Scirtothrips dorsalis Hood	Chilli Thrips	Thripidae	India, S.E. Asia	Adults & nymphs infest foliage; scarify leaves
Thrips tabaci Lind.	Onion Thrips	Thripidae	Cosmopolitan	

Agrius convolvuli (L.)	Convolvulus Hawk Moth	Sphingidae	Africa, Asia	Larvae eat leaves; may defoliate
Acherontia spp.	Death's Head Hawk Moths	Sphingidae	Old World	
Manduca spp.	Tomato (etc.) Hornworms	Sphingidae	New World	
Agrotis ipsilon (Hfn.)	Black Cutworm	Noctuidae	India, S.E. Asia	Larvae are cutworms
Mythimna spp.	Cereal Armyworms	Noctuidae	Asia	Larvae eat leaves and sometimes bore into fruits
Spodoptera exigua (Hub.)	Beet Armyworm	Noctuidae	Asia	
Spodoptera littoralis (Boisd.)	Cotton Leafworm	Noctuidae	Mediterranean, Africa	
Spodoptera litura (F.)	Fall Armyworm	Noctuidae	Asia	
Tiracola plagiata (Wlk.)	Banana Fruit Caterpillar	Noctuidae	S.E. Asia	
Plusia spp.	Semiloopers	Noctuidae	S.E. Asia	
Asphondylia capsici Barnes	Capsicum Gall Midge	Cecidomyiidae	Mediterranean	Larvae gall fruits
Dacus spp.	Fruit Flies	Tephritidae	S.E. Asia	Larvae inside fruits
Zonosemata electa (Say)	Pepper Maggot	Tephritidae	Canada, USA	
Liriomyza bryoniae (Kalt)	Tomato Leaf Miner	Agromyzidae	Europe, USSR	Larvae mine seedlings (glasshouses)
Liriomyza sativae Bl.	Leaf Miner	Agromyzidae	N. & S. America	Larvae mine leaves
Anomala spp.	White Grubs	Scarabaeidae	India, S.E. Asia	Larvae in soil eat roots; may destroy seedlings
Leucopholis spp.	White Grubs	Scarabaeidae	S.E. Asia	
Epitrex spp.	Flea Beetles	Chrysomelidae	USA	Adults hole leaves; larvae eat roots
Psylliodes spp.	Flea Beetles	Chrysomelidae	S.E. Asia	
Monolepta signata Oliv.	Leaf Beetle	Chrysomelidae	India	Adults eat leaves
Orthaulaca similis Oliv.	Leaf Beetle	Chrysomelidae	Philippines	
Anthonomus eugenii Cano	Pepper Weevil	Curculionidae	USA, C. America	Larvae bore fruits
Tarsonemus translucens Green	Leaf Mite	Tarsonemidae	India	Scarify foliage
Calacarus carinatus (Green)	Purple Mite	Eriophyidae	Asia, USA, Australasia	Distort foliage

CARNATION (*Dianthus caryophyllum* – Caryophyllaceae)

An erect perennial herb, cultivated mostly as a cut flower, grown as an annual, with over 2000 horticultural varieties. It is native to southern Europe and northern Africa. The plant is also cultivated as a source of oil for the perfumery industry, the less highly cultivated strains having the strongest scent and being used as the source of oil. In temperate regions most of the flower cultivation involves protection using greenhouses or polythene shelters.

MAJOR PESTS

Myzus persicae (Sulz.)	Peach–Potato Aphid	Aphididae	Cosmopolitan	Infest foliage & flowers
Tetranychus urticae (Kock)	Temperate Red Spider Mite	Tetranychidae	Cosmopolitan in cooler regions	Scarify & web foliage

MINOR PESTS

Forficula auricularia L.	Common Earwig	Forficulidae	Cosmopolitan	Bite holes in flowers and foliage
Aphis sambuci L.	Elder Aphid	Aphididae	Europe	Adults & nymphs infest foliage; suck sap
Aulacorthum solani (Kltb.)	Glasshouse–Potato Aphid	Aphididae	Europe	
Saissetia coffeae (Wlk.)	Helmet Scale	Coccidae	Cosmopolitan	Infest foliage
Taeniothrips atratus (Hal.)	Carnation Thrips	Thripidae	Europe	Leaves & petals silvered & distorted
Thrips tabaci Lind.	Onion Thrips	Thripidae	Cosmopolitan	
Cacoecimorpha pronubana (Hbn.)	Carnation Tortrix Moth	Tortricidae	Europe	Larvae eat shoots & bore buds
Epichoristodes acerbella (Wlk.)	South African Carnation Worm	Tortricidae	S. Africa, Italy	
Lacanobia oleracea (L.)	Tomato Moth	Noctuidae	Europe	Larvae eat foliage
Bradysia paupera Tuom.	Sciarid Fly	Sciaridae	Europe	Larvae in soil attack pot plants
Delia cardui (Meig.)	Carnation Fly	Anthomyiidae	Europe, USA	Larvae mine leaves & bore stems
Agriotes spp.	Wireworms	Elateridae	Europe	Larvae in soil eat roots
Tetranychus cinnabarinus (Boisd.)	Tropical Red Spider Mite	Tetranychidae	Cosmopolitan (in temperate greenhouses	Scarify & web foliage
Tetranychus spp.	Red Spider Mites	Tetranychidae	Cosmopolitan	
Siteroptes cerealium (Kirch.)	Grass & Cereal Mite	Penthaleidae	Europe	Spread bud rot fungus

CARROT (*Daucus carota* – Umbelliferae)

The carrot has been cultivated for more than 2000 years, first by the Greeks and Romans, and early reached Europe. In the seventeenth century it was taken to N. America, and now is found in all parts of the world. The plant is biennial, but the swollen fleshy taproot matures in one season. Many different cultivars are known, and grown for different purposes, but in general the varietal characteristics are loose. Some varieties appear to be more susceptible to insect pests than others. The preferred soil is deep sandy loam, or else a peaty loam with good drainage. The root is eaten raw or cooked, and is often used for flavouring soups and stews. In Europe and N. America many crops are now canned or frozen, and in some areas continuous cultivation is practised (using different varieties) to provide a continuous supply for local markets.

MAJOR PESTS

Cavariella aegopodii (Scop.)	Willow–Carrot Aphid	Aphididae	Europe, Australia, Canada USA	Sap-sucker; virus vector
Psila rosae (F.)	Carrot (Rust) Fly	Psilidae	Europe, USA, Canada	Larvae mine root

MINOR PESTS

Macrosteles fascifrons (Stal)	Six-spotted Leafhopper	Cicadellidae	Canada, USA	
Trioza spp.	Carrot Psyllids	Psyllidae	Europe, Asia	Adults & nymphs infest foliage; sap-suckers; virus vectors
Cavariella konoi Tak.	Celery Aphid	Aphididae	Canada	
Cavariella pastinaceae (L.)	Parsnip Aphid	Aphididae	Canada	
Dysaphis crataegi (Klth.)	Hawthorn–Carrot Aphid	Aphididae	Europe	
Orthops campestris (L.)	Stack Bug	Miridae	Europe, USA, Canada	Sap-suckers; toxic saliva
Nysius ericae (Schill.)	False Chinch Bug	Lygaeidae	Canada, USA	
Lasioptera carophila Loew	Gall Midge	Cecidomyiidae	Europe, Russia	Larvae gall seed-head
Napomyza carotae Spencer	Carrot Root Miner	Agromyzidae	Europe (not UK)	
Psila nigricornis Meig.	Chrysanthemum Stool Miner	Psilidae	Europe, Canada	Larvae mine root
Hepialus spp.	Swift Moths	Hepialidae	Europe	Larvae eat roots
Depressaria heracliana (L.)	Parsnip Moth	Oecophoridae	Europe, Canada	Larvae eat foliage
Aethes francillana (F.)	Carrot Flower Tortricid	Cochylidae	Europe	Larvae web flowerhead
Papilio polyxene asterias Stoll	Black Swallowtail Butterfly	Papilionidae	Canada, USA	
Papilio machaon Hbn.	Swallowtail Butterfly	Papilionidae	Israel	Larvae eat foliage
Loxostege sticticalis (L.)	Beet Webworm	Pyralidae	Cosmopolitan	
Agrotis segetum (D. & S.)	Common Cutworm	Noctuidae	Europe, Asia, Africa	
Agrotis ipsilon (Hfn.)	Black Cutworm	Noctuidae	Cosmopolitan	Larvae are cutworms; eat holes in root; may kill seedlings
Noctua pronuba (L.)	Large Yellow Underwing	Noctuidae	Europe	
Xestia c-nigrum (L.)	Spotted Cutworm	Noctuidae	Europe, Asia, N. America	

Spodoptera littoralis (Boisd.)	Cotton Leafworm	Noctuidae	Israel	Larvae eat foliage; may defoliate
Autographa gamma (L.)	Silver-Y Moth	Noctuidae	Europe	
Anagrapha falcifera (Kirby)	Celery Semilooper	Noctuidae	USA	
Bothynus gibbosus (De Geer)	Carrot Beetle	Scarabaeidae	USA	Adults eat foliage, larvae eat roots
Phytoecia geniculata Muls.	Carrot Borer	Cerambycidae	Asia Minor	Larvae bore root
Apion aestivum Germ.	Clover Weevil	Apionidae	Israel	Adults eat leaves
Listroderes costirostris (Klug)	Vegetable Weevil	Curculionidae	USA, S. America, Australia	Adults eat foliage, larvae eat roots
Listronotus oregonensis (Le C.)	Carrot Weevil	Curculionidae	Canada, USA	Larvae bore root
Eriophyes pseucedani (Can.)	Carrot Bud Mite	Eriophyidae	Europe, USA, Canada	Deform foliage

CELERY (*Apium graveolens* – Umbelliferae)

A native of temperate Europe, from England to Asia Minor. Under cultivation it is a biennial which matures for harvest in one year, or less. The large succulent leaf-stalks are the celery of commerce. The crop requires a rich sandy, or peaty, loam with plenty of water. Some varieties are self-blanching, but with other varieties earthing-up or other methods of shielding the leaf-stalks from light are used to produce blanching. The inner leaf-stalks are used as a salad vegetable; the tougher outer stalks are used as the basis of celery soup; celery seeds are grown for use as a savoury and for flavouring. Celeriac is *A. graveolens* var. *rapaceum*, but is less commonly grown.

MAJOR PESTS

Cavariella aegopodii (Scop.)	Willow–Carrot Aphid	Aphididae	Europe, Australia, USA, Canada	Infest foliage; virus vector
Psila rosae (F.)	Carrot Fly	Psilidae	Europe, USA, Canada	Larvae mine stem and roots
Philophylla heraclei (L.)	Celery Fly	Tephritidae	Europe	Larvae mine leaves
Liriomyza sativae Blanch.	Vegetable Leaf Miner	Agromyzidae	USA	

MINOR PESTS

Forficula auricularia L.	Common Earwig	Forficulidae	Cosmopolitan	Make holes in leaves
Macrosteles fascifrons (Stal)	Six-spotted Leafhopper	Cicadellidae	Canada, USA	Infest foliage; sap-suckers; virus vectors
Cavariella konoi Tak.	Celery Aphid	Aphididae	Canada	
Dysaphis petroselini (C.B.)	Hawthorn–Parsley Aphid	Aphididae	Europe	
Lygus rugulipennis Popp.	Tarnished Plant Bug	Miridae	Europe	Adults & nymphs infest foliage; suck sap; toxic saliva causes foliage speckling
Orthops campestris (L.)	Stack Bug	Miridae	Europe, USA, Canada	
Nysius ericae (Schill.)	False Chinch Bug	Lygaeidae	Canada, USA	
Depressaria heracliana (L.)	Parsnip Moth	Oecophoridae	Europe, Canada	Larvae eat foliage & seedheads
Udea rubigalis (Gn.)	Celery Leaf Tier	Pyralidae	Canada, USA	Larvae web & eat leaves
Papilio polyxena asterias Stoll.	Black Swallowtail	Papilionidae	Canada, USA	Larvae eat foliage
Agrotis spp.	Cutworms	Noctuidae	Cosmopolitan	Larvae are cutworms; eat leaf bases & destroy seedlings
Xestia c-nigrum (L.)	Spotted Cutworm	Noctuidae	Europe, Asia, N. America	
Hydraecia micacea (Esp.)	Rosy Rustic Moth	Noctuidae	Europe	Larvae bore stalks
Autographa gamma (L.)	Silver-Y Moth	Noctuidae	Europe	Larvae eat foliage
Anagrapha falcifera (Kirby)	Celery Semilooper	Noctuidae	USA	
Bothynus gibbosus (De Geer)	Carrot Beetle	Scarabaeidae	USA	Adults eat foliage; larvae eat roots
Listronotus oregonensis (Le C.)	Carrot Weevil	Curculionidae	USA, Canada	Larvae bore stem & stalks

CHERRY (*Prunus avium* – Rosaceae)

Often called the Sweet Cherry to distinguish it from the Sour Cherry (*Prunus cerasus*) which is usually only used for canning. This is a temperate fruit, and the main areas of production are USA, Germany, Italy, and other southern European countries. The centre of origin was Eurasia, but it has been cultivated in the USA since colonial times. The tree is of moderate size, and the fruits are round, small, fleshy, with a smooth stone and long peduncle. Cherries are used as table fruits, and cooked in pies, also canned; the juice is used for syrup and jelly.

MAJOR PESTS

Myzus cerasi (F.)	Cherry Blackfly	Aphididae	Europe	Leaves severely curled
Quadraspidiotus perniciosus (Comst.)	San José Scale	Diaspididae	USA, Canada	Encrust foliage
Argyresthia pruniella (Clerck)	Cherry Fruit Moth	Yponomeutidae	Europe	Larvae bore buds in spring, feed on flowers & fruitlets
Operophtera brumata (L.)	Winter Moth	Geometridae	Europe, Asia, Canada, USA	Larvae defoliate
Rhagoletis cerasi (L.)	Cherry Fruit Fly	Tephritidae	Europe (not UK)	Maggots bore inside fruits
Rhagoletis cingulata (Loew)	Cherry Fruit Fly	Tephritidae	USA	
Caliroa cerasi (L.)	Plum Slug Sawfly	Tenthredinidae	Cosmopolitan	Larvae skeletonize leaves
Phyllobius spp.	Leaf Weevils	Curculionidae	Europe	Adults eat leaves

MINOR PESTS

Eupteryx stellulata Brum.	Cherry Leafhopper	Cicadellidae	Europe	Infest foliage
Cercopis vulnerata Ger.	Red & Black Froghopper	Cercopidae	Europe	Nymphs in spittle-mass on roots
Parthenolecanium corni (Bche.)	Plum Scale	Coccidae	Europe, W. Asia	Encrust twigs; suck sap
Aspidiotus spp.	Armoured Scales	Diaspididae	USA, Canada	
Lepidosaphes ulmi L.	Mussel Scale	Diaspididae	Europe, USA	
Lygocoris pabulinus (L.)	Common Green Capsid	Miridae	Europe	Sap-suckers; toxic saliva
Corythucha pruni (O. & D.)	Cherry Lacebug	Tingidae	USA	
Pentatoma rufipes L.	Forest Bug	Pentatomidae	Europe	
Coleophora nigricella Steph.	Apple & Plum Casebearer	Coleophoridae	Europe	Larvae in tiny cases
Anthophila pariana Clerck	Apple Leaf Skeletonizer	Glyphipterygidae	Europe	Larvae skeletonize leaves
Cossus cossus L.	Goat Moth	Cossidae	Europe	Larvae bore branches & trunk
Zeuzera pyrina L.	Leopard Moth	Cossidae	Europe, Asia	
Argyroploce pruniana Hbn.	Plum Tortrix	Tortricidae	Europe	Larvae eat leaves & bore shoots
Cacoecia oporana L.	Fruit Tree Tortrix	Tortricidae	Europe	Larvae eat leaves & damage fruits
Cydia prunivora (Walsh)	Lesser Appleworm	Tortricidae	N. America	Larvae bore fruits
Enarmonia formosana (Scop.)	Cherry Bark Tortrix Moth	Tortricidae	Europe	Larvae bore under bark

Spilonota ocellana (D. & S.)	Eye-spotted Bud Moth	Tortricidae	Europe, Asia, N. America	Larvae bore buds
Alsophila spp.	Cankerworms	Geometridae	Europe, Asia, Canada, USA	
Biston betularia L.	Peppered Moth	Geometridae	Europe	Looper larvae eat leaves; may defoliate
Erannis defoliaria Clerck	Mottled Umber Moth	Geometridae	Europe	
Operophtera spp.	Winter Moths	Geometridae	Europe, Asia, Canada, USA	
Diloba caeruleocephala (L.)	Figure-of-eight Moth	Notodontidae	Europe	Larvae eat leaves
Phalera bucephala L.	Buff-tip Moth	Notodontidae	Europe	
Malacosoma neustria L.	Lackey Moth	Lasiocampidae	Europe	Gregarious larvae defoliate
Euproctis chrysorrhoea L.	Yellow-tail Moth	Lymantriidae	Europe	
Orgyia antiqua L.	Vapourer Moth	Lymantriidae	Europe	Larvae eat leaves
Paravespula/Vespa spp.	Common Wasps	Vespidae	Cosmopolitan	Adults pierce ripe fruits
Rhagoletis fausta (O-S.)	Black Cherry Fruit Fly	Tephritidae	N. America	Larvae inside fruits
Chrysobothris spp.	Flat-headed Borers	Buprestidae	USA, Asia	Larvae bore branches
Aeolesthes holosericea F.	Cherry Stem Borer	Cerambycidae	India, Pakistan	Larvae bore trunk
Scolytus spp.	Fruit Bark Beetles	Scolytidae	Europe, N. America	Adults bore twigs & under bark making breeding galleries
Xyleborus dispar (F.)	Shot-hole Borer	Scolytidae	Europe	
Panonychus ulmi (Koch)	Fruit Tree Red Spider Mite	Tetranychidae	Europe, Asia	Scarify foliage

CHESTNUT (*Castanea* spp. – Fagaceae) (=Sweet Chestnuts)

C. crenata – Japanese Chestnut
C. dentata – American Chestnut
C. sativa – European Chestnut

These are tall handsome trees, up to 20 m in height, with elongate leaves having serrated margins. The three cultivated species of Sweet Chestnut are listed above; many varieties are known. The European Chestnut has large fruits, and has long been cultivated, often grown on hillsides too dry for other cultivation in southern Europe. The American species has smaller fruits, and was nearly exterminated in the USA by the ravages of Chestnut Blight disease during the last 50 years. The Japanese species is immune to Blight and is being introduced into the USA. The familiar nuts are eaten raw or roasted, and may sometimes be otherwise cooked. The tree is also a source of valuable timber, and is often grown as an ornamental. Wild species of *Castanea* occur in Asia, where there are many species of *Castanopsis*; the pest spectrum is also shared with *Fagus* (beeches) and *Quercus* (oaks) to some extent.

MAJOR PESTS

None of the pests recorded appear to be particularly serious individually

MINOR PESTS

Glossonotus acuminatus (F.)	Chestnut Treehopper	Membracidae	USA	Adults and nymphs infest foliage; suck sap; sooty moulds often present
Cinalathura folial (Theo.)	Big Aphid	Aphididae	China	
Cinara pinea Mord.	Black Pine Aphid	Aphididae	China	
Longistigma caryae (Harris)	Giant Bark Aphid	Aphididae	USA	
Pterochlorus tropicalis V.d.G.	Chestnut Aphid	Aphididae	China	
Eulecanium tiliae (L.)	Hazel Scale	Coccidae	S. Europe	Adults & nymphs infest twigs; suck sap
Parthenolecanium corni (Bche.)	Plum Scale	Coccidae	Europe, Asia	
Melanaspis obscura (Cmst.)	Obscure Scale	Diaspididae	USA	
Zeuzera pyrina L.	Leopard Moth	Cossidae	Europe, USA	Larvae bore branches
Tischeria spp.	Leaf Miners	Tischeriidae	Europe	Larvae mine leaves
Coleophora spp.	Casebearers	Coleophoridae	S. Europe	Defoliating larvae in tiny cases
Synanthedon vespiformis (L.)	Yellow-legged Clearwing	Sesiidae	Europe	Larvae bore wood under bark
Phyllonorycter messaniella (Zeller)	Blister Moth	Gracillariidae	Europe	Larvae make leaf blister mines
Cydia splendana (Hbn.)	Tortricid	Tortricidae	Europe	Larvae bore fruits
Datana ministra (Drury)	Yellow-necked Caterpillar	Notodontidae	USA	Larvae eat leaves, may defoliate
Lymantria dispar (L.)	Gypsy Moth	Lymantriidae	Europe, Asia, N. America	
Apethymus kuri Tak.	Chestnut Sawfly	Tenthredinidae	Japan	
Caliroa castaneae (Roh.)	Chestnut Slug Sawfly	Tenthredinidae	USA	Larvae skeletonize leaves
Croesus castaneae Roh.	Chestnut Sawfly	Tenthredinidae	USA	Larvae defoliate
Dryocosmus kuriphilus Yas.	Chestnut Gall Wasp	Cynipidae	China	Larvae gall flowers

Rhopalomyia castaneae Felt	Chestnut Bud Gall Midge	Cecidomyiidae	USA	Larvae gall buds
Melittomma sericeum (Harris)	Chestnut Timberworm	Lymexylidae	USA	Larvae bore trunk
Sp. indet.	Longhorn Beetle	Cerambycidae	Yugoslavia	Larvae bore in trunk and branches, under the bark
Agrilus bilineatus Weber	Two-lined Chestnut Borer	Buprestidae	USA	
Agrilus spp.	Jewel Beetles	Buprestidae	Europe, Asia, N. America	
Chrysobothris spp.	Flat-headed Borers	Buprestidae	Europe, Asia, N. America	
Curculio elephas Gyll.	Elephant Weevil	Curculionidae	Europe, Asia	Larvae bore nuts
Curculio spp.	Chestnut Weevils	Curculionidae	USA	
Cyrtepistomus castaneus (Roe.)	Chestnut (Asiatic Oak) Weevil	Curculionidae	Asia, USA	Adults eat leaves; larvae bore nuts
Phyllobius pyri L.	Common Leaf Weevil	Curculionidae	Europe	Adults eat leaves; larvae in soil eat roots
Phyllobius spp.	Leaf Weevils	Curculionidae	Europe	
Xyleborus dispar (F.)	Twig Borer	Scolytidae	Europe	Adults bore twigs
Oligonychus bicolor (Banks)	Oak Red Mite	Tetranychidae	USA	Adults & nymphs scarify (scorch) foliage by feeding
Eotetranychus hicoriae (McG.)	Hickory Scorch Mite	Tetranychidae	USA	
Panonychus ulmi (Koch)	Fruit Tree Red Spider Mite	Tetranychidae	Europe, Asia, N. America	

CHRYSANTHEMUM (*Chrysanthemum* spp. – Compositae)

An erect perennial herb, growing to a height of 1 m, cultivated as an annual for its cut flower. Essentially a temperate plant, it may be cultivated under glass (or polythene) or in open fields. It is a short-day flower and blooms in the winter; propagation is usually by cuttings. There are several different species under cultivation, and many different varieties; some varieties show marked differences in their susceptibilities to pests and dieseases, and in their tolerance to pesticides.

MAJOR PESTS

Myzus persicae (Sulz.)	Peach–Potato Aphid	Aphididae	Cosmopolitan	Infest foliage; virus vector
Thrips nigropilosus Uzel	Chrysanthemum Thrips	Thripidae	Europe, Japan, E. Africa, N. America	Infest flowers & foliage
Phytomyza horticola Gour.	Pea Leaf Miner	Agromyzidae	Cosmopolitan in Old World	Larvae mine leaves
Phytomyza syngenesiae (Hardy)	Chrysanthemum Leaf Miner	Agromyzidae	Europe, N. America	
Cnephasia longana (Haw.)	Omnivorous Leaf Tier	Tortricidae	Europe, USA	Larvae feed in flowerhead
Tetranychus urticae (Koch)	Temperate Red Spider Mite	Tetranychidae	Cosmopolitan	Scarify & web foliage

MINOR PESTS

Forficula auricularia L.	Common Earwig	Forficulidae	Cosmopolitan	Bite holes in foliage
Trialeurodes vaporariorum (Westw.)	Glasshouse Whitefly	Aleyrodidae	Europe	Infest foliage (in greenhouses)
Aulacorthum circumflexum (Bckt.)	Mottled Arum Aphid	Aphididae	Europe	Adults & nymphs infest foliage; suck sap; virus vectors
Brachycaudus helichrysi (Kltb.)	Leaf-curling Plum Aphid	Aphididae	Europe	
Macrosiphoniella sanborni (Gill.)	Chrysanthemum Stem Aphid	Aphididae	Europe, Asia	
Myzus ascalonicus Don.	Shallot Aphid	Aphididae	Europe, N. America	
Lygocoris pabulinus (L.)	Common Green Capsid	Miridae	Europe	Sap-suckers; toxic saliva; cause necrotic spots (speckling)
Lygus rugulipennis Popp.	Tarnished Plant Bug	Miridae	Europe	
Thrips tabaci Lind.	Onion Thrips	Thripidae	Cosmopolitan	Infest flowerheads
Lorita abornana Busck	Chrysanthemum Flower Borer	Cochylidae	USA	Larvae bore flowerhead
Epichoristodes acerbella (Wlk.)	South African Carnation Worm	Tortricidae	S. Africa, Italy	Larvae eat shoots & buds
Cnephasia interjectana (Haw)	Flax Tortrix	Tortricidae	Europe, Asia, Canada	
Phlogophora meticulosa (L.)	Angleshades Moth	Noctuidae	Europe	Larvae eat leaves & flowerheads

Spodoptera littoralis (Boisd.)	Cotton Leafworm	Noctuidae	S. Europe	Larvae eat foliage
Diarthronomyia chrysanthemi Ahlberg	Chrysanthemum Midge	Cecidomyiidae	Europe, N. America, N.Z.	Larvae in flowerhead
Bradysia paupera Tuom.	Sciarid Fly	Sciaridae	Europe	Larvae in soil damage seedlings
Liriomyza trifolii (Burg.)	Serpentine Leaf Miner	Agromyzidae	N. & S. America, now UK	Larvae mine leaves
Psila nigricornis Meig.	Chrysanthemum Stool Miner	Psilidae	Europe, Canada	Larvae mine stool
Euribia zoe (Meig.)	Chrysanthemum Blotch Miner	Tephritidae	Europe	Larvae mine leaves
Paroxyna misella (Lw.)	Chrysanthemum Stem Fly	Tephritidae	Europe	Larvae gall shoot apex
Trypeta trifasciata Shiraki	Chrysanthemum Fruit Fly	Tephritidae	Japan	Larvae in flowerhead
Gonipterus gibberus (Boisd.)	Eucalyptus Weevil	Curculionidae	Australia, S. America	Adults & larvae feed on foliage
Paraphytoptus chrysanthemi K.	Chrysanthmum Rust Mite	Eriophyidae	Europe, USA, Canada	Scarify & curl leaves; cause stunting

CLOVERS (Forage Legumes – Leguminosae)

Lathyrus spp. – Vetchlings
Lespedeza spp. – Bush Clovers
Lotus spp. – Trefoils
Trifolium pratense – Red Clover
Trifolium repens – White Clover
Vicia spp. – Vetches, Tares, etc.

Also several other genera and species

The forage legumes are a large group, mainly temperate in distribution, grown for forage purposes, either mixed with grass or on their own. They are cultivated either for hay, or short-term leys for grazing, or for silage. Apart from their nutritive value as fodder, the symbiotic bacteria in the root nodules convert (fix) atmospheric nitrogen into soluble nitrogenous compounds in the soil, so these crops are used in rotations for renewing worn-out soils. Many new species and varieties are constantly being introduced and developed for agricultural purposes, particularly as green manure and as cover crops in orchards and plantations, and for grazing and soil improvement in semi-arid areas and regions with impoverished soils.

MAJOR PESTS

Smintherus spp.	Lucerne 'Fleas'	Sminthuridae	Cosmopolitan	Hole leaves
Acyrthosiphon pisum (Harris)	Pea Aphid	Aphididae	Holarctic	Infest foliage; suck sap; virus vectors
Megoura viciae Bckt.	Vetch Aphid	Aphididae	Europe	
Bruchidius spp.	Clover Beetles	Bruchidae	Europe, Asia	Larvae eat seeds inside pods
Apion spp.	Clover Seed Weevils	Apionidae	Europe, Asia, Africa	
Sitona spp.	Clover Weevils	Curculionidae	Cosmopolitan	Adults eat leaves; larvae in root nodules

MINOR PESTS

Austroasca viridigrisea (Paoli)	Vegetable Jassid	Cicadellidae	Australia	Sap-sucker
Acyrthosiphon kondoi Shinji	Blue-green Alfalfa Aphid	Aphididae	Australia	Adults & nymphs infest foliage; suck sap; virus vectors
Aphis craccivora Koch	Groundnut Aphid	Aphididae	Cosmopolitan	
Aphis fabae Scopoli	Black Bean Aphid	Aphididae	Cosmopolitan (not Australia)	
Therioaphis maculata (Bckt.)	Spotted Alfalfa Aphid	Aphididae	Cosmopolitan	
Lygus spp.	Capsid Bugs	Miridae	Europe, USA, Canada	Speckle foliage
Coleophora frischella (L.)	Clover Seed Moth	Coleophoridae	Australia	Larvae in flowerheads
Cydia nigricana (F.)	Pea Moth	Tortricidae	Europe	Larvae in seed pods
Colias spp.	Clouded Yellows	Pieridae	Canada, USA, Europe, Asia	Larvae eat foliage & bore pods to eat seeds inside
Lampides boeticus (L.)	Pea Blue Butterfly	Lycaenidae	Europe, Asia, Africa	
Zizina otis (F.)	Lesser Grass Blue	Lycaenidae	Asia, Australia	
Mythimna spp.	Cereal Armyworms	Noctuidae	Cosmopolitan	Larvae eat leaves, may defoliate
Xestia c-nigrum (L.)	Spotted Cutworm	Noctuidae	Europe, USA, Canada	
Contarinia medicaginis Kiell.	Lucerne Flower Midge	Cecidomyiidae	Europe, Russia	Larvae (yellow) gall flowers

Dasineura leguminicola (Lint.)	Clover Seed Midge	Cecidomyiidae	Europe, USA, Canada	Larvae (pink) in flowerheads
Dasineura trifolii (Loew)	Clover Leaf Midge	Cecidomyiidae	Europe, Russia, USA, Canada	Larvae (yellow) inside leaflet galls
Dasineura viciae (Kieff.)	Vetch Leaf Midge	Cecidomyiidae	Europe	Larvae (white) in leaflets & shoots
Tipula spp.	Leatherjackets	Tipulidae	Europe	Larvae in soil eat roots & seedlings
Bruchophagus gibbus (Boh.)	Clover Seed Wasp	Eurytomidae	Australia, USA, Canada	Larvae inside seeds
Phyllophaga spp.	June Beetles (White Grubs)	Scarabaeidae	Canada, USA	Larvae eat roots
Epicauta spp.	Black Blister Beetles	Meloidae	Canada, USA	Adults eat foliage
Colaspis brunnea (F.)	Grape Colaspis	Chrysomelidae	USA	Larvae eat roots
Systena spp.	Flea Beetles	Chrysomelidae	Canada, USA	Adults hole leaves
Amnemus quadrituberculatus (Boh.)	Clover Root Weevil	Curculionidae	Australia	Adults eat leaves; larvae in soil eat roots
Graphognathus spp.	White-fringed Weevils	Curculionidae	Australia, USA, S. America	
Hypera spp. (5+)	Clover Weevils	Curculionidae	Europe, USA, Canada	Larvae feed in shoots & on young leaves
Hyalstinus obscurus (Marsh)	Clover Root Borer	Curculionidae	Canada	Larvae bore roots
Miccotrogus picirostris (F.)	Clover Weevil	Curculionidae	Europe, USA, Canada	Larvae inside seeds
Sitona spp.	Pea & Bean Weevils	Curculionidae	Cosmopolitan	Larvae eat root nodules; adults notch leaves
Halotydeus destructor (Tucker)	Red-legged Earth Mite	Penthaleidae	Australia, N.Z., S. Africa, Europe	Adults & nymphs scarify foliage; cause wilting
Bryobia praetiosa Koch	Clover Mite	Tetranychidae	Worldwide	

A total of about 40 species of Cecidomyiidae are listed by Barnes (1946) to be found in the flowers, pods, and folded leaflets of the forage legumes as a group. Some of the pests listed above show distinct preferences for certain genera, and even species, of legumes as host (there are demonstrable differences in pest spectrum between White and Red Clover).

CRANBERRY (*Vaccinium macrocarpon* – Ericaceae)

A low trailing woody plant, characteristic of bogs and wet acid soils throughout northeastern America and northern Europe, and has been cultivated in these regions for more than 100 years. Much of the crop (a small purple berry) is canned as sauce or jelly, or made into a beverage. The smaller Mountain Cranberry (*V. vitis-idaea*) is a boreal and arctic–alpine species found in northern Europe (especially Scandinavia) with a firmer and more spicy fruit; the American form (var. *minus*) is equally desirable.

MAJOR PESTS

MINOR PESTS

Acrobasis vacinii Riley	Cranberry Fruitworm	Pyralidae	USA, Canada	Larvae eat fruits
Chrysoteuchia topiaria (Zeller)	Cranberry Girdler	Pyralidae	USA	Larvae girdle stem
Rhabdopterus picipes (Ol.)	Cranberry Rootworm	Chrysomelidae	USA	Larvae damage roots
Anthonomus musculus Say	Cranberry Weevil	Curculionidae	USA	Larvae inside fruits

CUCURBITS (Several species – Cucurbitaceae)

Cucumis spp. – Cucumber; Gherkin; Melons

Cucurbita spp. – Squash; Gourds; Marrow; etc.

Important cultivated species belong to nine separate genera within this family, and agriculturally the crops are very diverse, but are biologically similar and have very similar pest spectra. Different species are native to different parts of the world, for example watermelon to Africa, marrow to the New World, loofah to Asia. They are tendril-climbing or prostrate annuals with soft stems, simple broad, deeply-cut leaves, and with large fleshy fruits. Some of the fruits are eaten as table fruits, others as vegetables, some form gourds, and others are loofahs. In temperate countries most cucurbits are cultivated under glass or in polythene tunnels.

MAJOR PESTS

Trialeurodes vaporariorum (Westw.)	Glasshouse Whitefly	Aleyrodidae	Europe, N. America	Infest foliage (in greenhouses)
Leptoglossus spp.	Leaf-footed Plant Bugs	Coreidae	Africa, Asia, Australasia	Sap-suckers; toxic saliva
Dacus cucurbitae Coq.	Melon Fly	Tephritidae	Africa, Asia, Australasia	Larvae inside fruits
Dacus spp.	Fruit Flies	Tephritidae		
Epilachna spp.	Epilachna Beetles	Coccinellidae	Africa, Asia, USA	Adults & larvae eat leaves
Diabrotica undecimpunctata Mann.	Spotted Cucumber Beetle	Chrysomelidae	N. America	Adults defoliate; larvae eat roots
Aulacophora spp.	Cucumber Beetles	Chrysomelidae	Old World	
Raphidopalpa foveicollis (Lucus)	Red Pumpkin Beetle	Chrysomelidae	Mediterranean, Asia, Australia	Defoliate
Tetranychus cinnabarinus (Boisd.)	Carmine Spider Mite	Tetranychidae	Cosmopolitan in warmer regions	Scarify & web foliage
Tetranychus urticae (Koch)	Temperate Red Spider Mite	Tetranychidae	Cosmopolitan in cooler regions	Scarify & web foliage (in greenhouses)

MINOR PESTS

Onychiurus spp.	Springtails	Onychiuridae	Europe	Damage seedlings (in greenhouses)
Circulifer tenellus (Baker)	Beet Leafhopper	Cicadellidae	Mediterranean, USA	Adults & nymphs infest foliage; sap-suckers; virus vectors
Empoasca spp.	Green Leafhoppers	Cicadellidae	Africa, Asia	
Zygina pallidifrons (J. Ed.)	Tomato (Glasshouse) Leafhopper	Cicadellidae	Europe	
Bothrogonia ferruginea	Large Brown Leafhopper	Cicadellidae	S.E. Asia	
Aphis gossypii Glov.	Melon/Cotton Aphid	Aphididae	Cosmopolitan	Infest foliage; sap suckers; virus vectors
Myzus persicae (Sulz.)	Peach–Potato Aphid	Aphididae	Cosmopolitan	

Ferrisia virgata (Ckll.)	Striped Mealybug	Pseudococcidae	Asia, Africa	Infest foliage
Anasa tristis (De Geer)	Squash Bug	Coreidae	Canada, USA	Adults & nymphs infest foliage; suck sap; toxic saliva; leaves may wilt & die
Anoplocnemis phasiana F.	Coreid Bug	Coreidae	Philippines	
Nezara viridula L.	Green Stink Bug	Pentatomidae	Cosmopolitan	
Aspongopus spp.	Cucurbit Stink Bugs	Pentatomidae	India	
Piezosternum calidum	Shield Bug	Pentatomidae	Africa	
Thrips fuscipennis Hal.	Rose Thrips	Thripidae	Europe	Infest flowers & leaves
Thrips tabaci Lind.	Onion Thrips	Thripidae	Cosmopolitan	
Melittia cucurbitae (Harris)	Squash Vine Borer	Sesiidae	N. & S. America	Larvae bore vines
Sphenarches caffer Zell.	Plume Moth	Pterophoridae	Africa, Asia, W. Indies	Larvae eat leaves
Adoxophyes privatana (Wlk.)	Leaf Roller	Tortricidae	S.E. Asia	Larvae roll & eat leaves
Palpita indica (Saund.)	Cucumber Moth	Pyralidae	Asia, Australia	
Diaphania hyalinata (L.)	Melonworm	Pyralidae	USA	
Diaphania nitidalis (Stoll)	Pickleworm	Pyralidae	USA	Larvae bore fruits & eat foliage
Hellula undalis F.	Oriental Cabbage Webworm	Pyralidae	China	Larvae eat foliage
Agrotis spp.	Cutworms	Noctuidae	Cosmopolitan	Larvae destroy young plants
Euxoa spp.	Cutworms	Noctuidae	Canada, USA, Asia	
Plusia spp.	Semiloopers	Noctuidae	Asia	Larvae eat leaves
Lasioptera falcata F.	Gall Midge	Cecidomyiidae	India	Larvae gall stems
Sciara spp.	Sciarid Flies	Sciaridae	Europe	Larvae damage roots (in greenhouses)
Phytomyza horticola Gour.	Pea Leaf Miner	Agromyzidae	Cosmopolitan in Old World	Larvae mine leaves
Myiopardalis pardalina (Big.)	Baluchistan Melon Fly	Tephritidae	Near & Middle East	Larvae inside fruits
Anomala spp.	White Grubs	Scarabaeidae	Asia	Larvae in soil eat roots; may destroy seedlings
Leucopholis irrorata (Chevr.)	White Grub	Scarabaeidae	Philippines	
Epicauta spp.	Black Blister Beetles	Meloidae	Asia, USA, Canada	Adults eat flowers; damage foliage
Mylabris spp.	Banded Blister Beetles	Meloidae	Africa, Asia	
Acalymma vittatum (F.)	Striped Cucumber Beetle	Chrysomelidae	Canada, USA	Adults eat foliage; also virus vector
Copa kunowi Weise	Brown Flower Beetle	Chrysomelidae	Africa	Adults eat flowers
Ceratia frontalis	Leaf Beetle	Chrysomelidae	S.E. Asia, Philippines	Adults eat foliage
Diabrotica spp.	Spotted Cucumber Beetles	Chrysomelidae	Canada, USA	Adults eat foliage; larvae eat roots
Monolepta bifasciata (Hornst.)	Leaf Beetle	Chrysomelidae	S.E. Asia	Adults eat foliage
Phyllotreta crucifera (Goeze)	Cabbage Flea Beetle	Chrysomelidae	Africa, Asia	Adult hole leaves
Apomecyna spp.	Vine Borers	Cerambycidae	Africa, India, Philippines	Larvae bore vines
Bryobia spp.	Bryobia (Brown) Mites	Tetranychidae	Europe	(Greenhouse pests) Scarify & web foliage
Eutetranychus orientalis (Klein)	Oriental Mite	Tetranychidae	Africa, Asia	
Tyrophagus dimidiatus (Herm.)	Mushroom Mite	Acaridae	Europe	Damage seedlings (in greenhouses)

CURRANTS (*Ribes* spp. – Saxifragaceae)

R. nigrum – Blackcurrant

R. sativum – Redcurrant; Whitecurrant

Low bushy shrubs, thought to be native to Eurasia, but now naturalized in temperate regions throughout both Old and New Worlds; they are well-adapted to cold climates. The fruits are borne in racemes, and are mostly used for making jams, jellies, and juice. The white form of currant tends to occur as a garden oddity and is seldom cultivated on any scale. The two species do show differences in pest spectra, as indicated by the pest common names.

MAJOR PESTS

Cryptomyzus ribis (L.)	Redcurrant Blister Aphid	Aphididae	Europe, USA	Cause red leaf-blisters
Hyperomyzus lactucae (L.)	Currant–Sowthistle Aphid	Aphididae	Europe	Cause leaf-curl
Quadraspidiotus perniciosus (Comst.)	San José Scale	Diaspididae	USA	Encrust stems
Lygocoris pabulinus (L.)	Common Green Capsid	Miridae	Europe	Sap-sucker; toxic saliva
Dasyneura tetensi (Rubs.)	Blackcurrant Leaf Midge	Cecidomyiidae	Europe	Larvae gall leaves
Nematus olfasciens Benson	Blackcurrant Sawfly	Tenthredinidae	Europe	Larvae defoliate
Tetranychus urticae (Koch)	Temperate Red Spider Mite	Tetranychidae	Europe	Scarify & web foliage
Cecidophyopsis ribis (Westw.)	Blackcurrant Gall Mite	Eriophyidae	Europe, W. Asia, Canada	Cause 'big-bud'

MINOR PESTS

Cercopis vulnerata Ger.	Red & Black Froghopper	Cercopidae	Europe	Nymphs in spittle-mass on roots
Typhlocyba spp.	Currant Leafhoppers	Cicadellidae	Europe	Infest foliage
Aphis grossulariae Kltb.	Gooseberry Aphid	Aphididae	Europe	}
Aphis schneideri (Borner)	Permanent Currant Aphid	Aphididae	Europe	Adults & nymphs infest foliage; suck sap; cause leaf-curl
Cryptomyzus galeopsidis (Kltb.)	Blackcurrant Aphid	Aphididae	Europe	}
Nasonovia ribis-nigri (Mosley)	Lettuce Aphid	Aphididae	Europe	}
Schizoneura ulmi (L.)	Currant Root Aphid	Aphididae	Europe	Infest roots; suck sap
Parthenolecanium corni (Bche.)	Plum Scale	Coccidae	Europe	}
Pulvinaria vitis L.	Woolly Currant Scale	Coccidae	Europe	Infest stems; suck sap
Aspidiotus spp.	Armoured Scales	Diaspididae	USA	Encrust stems; suck sap
Lepidosaphes ulmi L.	Mussel Scale	Diaspididae	Europe	}
Plesiocoris rugicollis Fall.	Apple Capsid	Miridae	Europe	Infest foliage; sap suckers; toxic saliva
Poecilocapsus lineatus (F.)	Four-lined Plant Bug	Miridae	USA	}
Lampronia capitella Cl.	Currant Shoot Borer	Incurvariidae	Europe	Larvae bore shoots & buds

Synanthedon salmachus (L.)	Currant Clearwing Moth	Sesiidae	Europe, USA, Canada, Australasia	Larvae bore stems
Zophodia convolutella (Hb.)	Gooseberry Fruitworm	Pyralidae	Canada, USA	Larvae eat fruits
Abraxas grossulariata (L.)	Magpie Moth	Geometridae	Europe	Larvae eat leaves; may defoliate
Itame ribearia (Fitch)	Currant Spanworm	Geometridae	Canada, USA	Larvae eat leaves; may defoliate
Erannis defoliaria Clerck	Mottled Umber Moth	Geometridae	Europe	Larvae eat leaves; may defoliate
Operophtera brumata (L.)	Winter Moth	Geometridae	Europe	Larvae eat leaves; may defoliate
Thamnonoma wauaria (L.)	V-Moth	Geometridae	Europe	Larvae eat leaves; may defoliate
Melanchra persicariae L.	Dot Moth	Noctuidae	Europe	Larvae eat leaves
Janus integer (Norton)	Currant Stem Girdler	Cephidae	Canada, USA	Oviposition punctures girdle stem; larvae bore stem
Nematus ribesii (Scop.)	Gooseberry Sawfly	Tenthredinidae	Europe, USA, Canada	Larvae defoliate (not blackcurrant)
Asphondylia ribesii (Meig.)	Redcurrant Leaf Midge	Cecidomyiidae	Europe	Larvae crinkle leaves
Dasyneura ribis Barnes	Blackcurrant Flower Midge	Cecidomyiidae	Europe	Larvae gall flowers
Epochra canadensis (Loew)	Currant Fruit Fly	Tephritidae	Canada, USA	Larvae inside fruits
Otiorhynchus singularis (L.)	Clay-coloured Weevil	Curculionidae	Europe	Adults eat bark & leaves; larvae in soil eat roots
Bryobia ribis Thom.	Gooseberry Bryobia Mite	Tetranychidae	Europe	Scarify and web foliage
Panonychus ulmi (Koch)	Fruit Tree Red Spider Mite	Tetranychidae	Europe	Scarify and web foliage

DILL (*Anthenum graveolens* – Umbelliferae)

A small annual or biennial herb with pale green leaves and yellow flowers, native to Eurasia, cultivated for its seeds. The 'seeds' are oval in shape, pale brown in colour, and compressed, and they are used for flavouring pickles (USA) and for flavouring soups, stews and sauces, and for other culinary purposes. Both the seeds and the oil extracted from them are used in medicine. In some countries (India, Europe, USSR) the leaves are also utilized. The main areas of cultivation are Europe, USA and India.

MAJOR PESTS

MINOR PESTS

Cavariella aegopodii (Scop.)	Willow–Carrot Aphid	Aphididae	Europe, N. America	Adults & nymphs infest foliage; suck sap; virus vectors
Cavariella konoi Tak.	Celery Aphid	Aphididae	Canada	
Orthops campestris (L.)	Stack Bug	Miridae	Europe, USA, Canada	Sap-suckers; toxic saliva
Nysius ericae (Schill.)	False Chinch Bug	Lygaeidae	Canada, USA	
Papilio polyxenes asterias Stoll	Black Swallowtail Butterfly	Papilionidae	Canada, USA	Larvae eat foliage
Papilio machaon sphyrus Hbn.	Common Swallowtail	Papilionidae	Palaearctic	
Lacanobia oleracea (L.)	Tomato Moth	Noctuidae	Europe	
Listronotus oregonensis (LeC.)	Carrot Weevil	Curculionidae	USA	Larvae bore stem & eat roots

EGGPLANT (*Solanum melongena* – Solanaceae) (=Brinjal; fruit called Aubergine)

Found wild and first cultivated in India; now widely grown throughout the tropics and warmer parts of the world, especially southern USA and the Mediterranean Region. More recently cultivated in southern Europe and in glasshouses in temperate Europe. It grows well up to 1000 m in altitude on light soils, but requires high temperatures. A perennial, weakly erect herb, 0.5–1.5 m in height, with the fruit a large pendant berry, ovoid or oblong, 5–15 cm long, smooth in texture, usually black or purple in colour. Several distinct varieties are recognized. The fruit is eaten as a vegetable, boiled, fried or stuffed. Propagation is by seed. It is grown throughout the warmer regions of the world for local consumption, but some countries have developed an export trade with more temperate countries.

MAJOR PESTS

Bemisia tabaci (Genn.)	Tobacco Whitefly	Aleyrodidae	Cosmopolitan	Infest foliage
Phthorimaea operculella (Zell.)	Potato Tuber Moth	Gelechiidae	Cosmopolitan	Larvae bore stems & mine leaves
Leucinodes orbonalis Gn.	Eggplant-boring Caterpillar	Pyralidae	Tropical Asia	Larvae bore stems
Epicauta spp.	Black Blister Beetles	Meloidae	Asia, USA, Africa	Adults eat flowers
Epilachna spp.	Epilachna Beetles	Coccinellidae	Tropical Asia, China	Adults & larvae eat foliage

MINOR PESTS

Aphis gossypii Glov.	Cotton Aphid	Aphididae	Cosmopolitan	Infest foliage; suck sap; virus vectors
Macrosiphum euphorbiae (Thos.)	Potato Aphid	Aphididae	Cosmopolitan	
Myzus persicae (Sulz.)	Peach–Potato Aphid	Aphididae	Cosmopolitan	
Empoasca spp.	Green Leafhoppers	Cicadellidae	Africa, Asia, USA	Adult & nymphs infest foliage; suck sap
Ferrisia virgata (Ckll.)	Striped Mealybug	Pseudococcidae	Africa, S. Asia, USA, C. & S. America	
Orthezia insignis Browne	Jacaranda Bug	Orthezidae	Africa, Asia, N., C. & S. America	
Nezara viridula (L.)	Green Stink Bug	Pentatomidae	Africa, Asia	Sap-suckers; toxic saliva; feeding causes necrosis
Dysdercus spp.	Cotton Stainers	Pyrrhocoridae	Africa, Asia	
Gargaphia solani Heid.	Eggplant Lace Bug	Tingidae	USA	
Urentius spp.	Brinjal Lacebugs	Tingidae	India	
Corythaica passiflorae Berg.	Lacebug	Tingidae	W. Indies	
Thrips tabaci Lind.	Onion Thrips	Thripidae	Cosmopolitan	Adults & nymphs infest flowers & foliage; scarify
Thrips palmi Karny	Palm Thrips	Thripidae	China	
Keiferia glochinella (Zell.)	Eggplant Leaf Miner	Gelechiidae	USA	Larvae mine leaves & bore stems
Scrobipalpa heliopa (Lower)	Tobacco Stem Borer	Gelechiidae	Asia	Larvae bore stems
Agrotis spp.	Cutworms	Noctuidae	Cosmopolitan	Larvae destroy young plants
Heliothis spp.	Budworms	Noctuidae	Cosmopolitan	Larvae bore buds, fruits & eat leaves
Spodoptera spp.	Leafworms	Noctuidae	Cosmopolitan	Larvae eat foliage
Mythimna spp.	Cereal Armyworms	Noctuidae	Cosmopolitan	
Plusia spp.	Semiloopers	Noctuidae	Cosmopolitan	
Acherontia spp.	Death's Head Hawk Moths	Sphingidae	Europe, Africa, Asia	
Manduca spp.	Tomato Hornworm, etc.	Sphingidae	N., C. & S. America	
Zonosemata electa (Say)	Pepper Maggot	Tephritidae	USA	Larvae inside fruits
Leptinotarsa decemlineata (Say)	Colorado Beetle	Chrysomelidae	Europe, N. & C. America	Adults & larvae defoliate
Oulema spp.	Leaf Beetles	Chrysomelidae	Asia, USA	
Epitrix fuscula Crotch	Eggplant Flea Beetle	Chrysomelidae	USA	Adults damage leaves; larvae in soil eat roots; seedlings killed
Epitrix spp.	Potato Flea Beetles	Chrysomelidae	USA, C. America	
Psylliodes spp.	Flea Beetles	Chrysomelidae	Europe, Asia, N. America	
Anomala spp.	White Grubs	Scarabaeidae	Asia	Larvae in soil eat roots & may kill seedlings; adults damage leaves
Holotrichia spp.	Cockchafers	Scarabaeidae	India	
Cotinis spp.	June Beetles	Scarabaeidae	USA	
Myllocerus spp.	Grey Weevils	Curculionidae	India	Adults eat leaves
Trichobaris trinotata (Say)	Potato Stalk Borer	Curculionidae	USA	Larvae bore stalks
Polyphagotarsonemus latus (Banks)	Yellow Tea Mite	Tarsonemidae	Asia	Scarify foliage
Tetranychus spp.	Red Spider Mites	Tetranychidae	Cosmopolitan	Scarify & web foliage

The more restricted tropical pest species have been omitted from this list.

FIELD BEAN (*Vicia faba* – Leguminosae) (= Broad, Pigeon and Horse Bean)

A strong erect annual, from 0.5–1.5 m in height; cultivated since prehistoric times, and probably originated in southwest Asia. Essentially a temperate crop, but widely grown in the Near and Middle East; it does not tolerate heat or dryness well. In Europe many different varieties are grown; those with the largest seeds are called Horse Beans, and the smallest are Pigeon Beans, and are used as animal feed. Broad Beans are grown as vegetables, and the beans (seeds) are used fresh and green, frozen or dried; the plant body may be used as fodder or for silage.

MAJOR PESTS

Aphis fabae (Scop.)	Black Bean Aphid	Aphididae	Cosmopolitan	Infest foliage; virus vector
Apion vorax Hbst.	Seed Weevil	Apionidae	Europe, N. Africa	Larvae in ovaries; adults carry virus
Sitona lineatus (L.)	Pea & Bean Weevil	Curculionidae	Europe	Larvae eat root nodules; adults notch leaves

MINOR PESTS

Sminthurus spp. etc.	Springtails	Collembola	Europe, Asia	Damage seedlings
Forficula auricularia L.	Common Earwig	Forficulidae	Cosmopolitan	Infest flowers
Bemisia tabaci (Genn.)	Tobacco Whitefly	Aleyrodidae	Cosmopolitan	Adults & nymphs infest leaves (underneath); suck sap; may stunt growth
Balclutha hebe (Kirk.)	Leafhopper	Cicadellidae	Middle East	
Empoasca spp.	Green Leafhoppers	Cicadellidae	Cosmopolitan	
Zygina lubiae (China)	Leafhopper	Cicadellidae	Middle East	
Acyrthosiphon pisum (Harris)	Pea Aphid	Aphididae	Cosmopolitan	Adults & nymphs infest foliage; suck sap; virus vectors; often with sooty moulds
Megoura viciae Bukt.	Vetch Aphid	Aphididae	Europe	
Several species	(Legume Aphids)	Aphididae	Worldwide	
Lygus spp.	Capsid Bugs	Miridae	Europe, Asia	Suck sap; toxic saliva causes necrotic spots
Macrotylus nigricornis Fischer	Capsid Bug	Miridae	Middle East	
Coptosoma punctatissimum Mont.	Shield Bug	Plataspidae	Japan	
Kakothrips robustus (Uzel)	Pea Thrips	Thripidae	Europe	Adults & nymphs infest flowers; cause flower-fall & pod deformation
Taeniothrips spp.	Flower Thrips	Thripidae	Europe, Asia	
Thrips spp.	Thrips	Thripidae	Europe, Asia	
Cacoecimorpha pronubana (Hb.)	Carnation Leaf Roller	Tortricidae	Europe, Asia, N. America	Larvae roll & eat leaves
Cnephasia spp.	Leaf Tiers	Tortricidae	Palaearctic	
Cydia fabivora (Meyr.)	Pod Moth	Tortricidae	S. America	Larvae bore into pods & eat seeds
Cydia nigricana F.	Pea Moth	Tortricidae	Europe, USA	

Etiella zinckenella (Treit.)	Pea Pod Borer	Pyralidae	Pantropical & subtropics	Larvae bore into pods & eat seeds inside
Maruca testulalis (Geyer)	Mung Moth	Pyralidae	Pantropical	
Lampides boeticus (L.)	Pea Blue Butterfly	Lycaenidae	Africa, Asia, S. Europe	
Virachola livia (Klug.)	Pomegranate Butterfly	Lycaenidae	Near East	
Agrotis spp.	Cutworms	Noctuidae	Cosmopolitan	Larvae destroy seedlings
Mamestra brassicae L.	Cabbage Moth	Noctuidae	Europe, Asia	Caterpillars eat leaves; sometimes defoliate
Spodoptera spp.	Leafworms	Noctuidae	Europe, Asia	
Autographa gamma (L.)	Silver-Y Moth	Noctuidae	Europe, Asia	
Resseliella sp.	Gall Midge	Cecidomyiidae	UK	Larvae gall stem
Contarinia pisi (Winn.)	Pea Midge	Cecidomyiidae	Europe	Larvae inside pods
Liriomyza spp.	Leaf Miners	Agromyzidae	Europe, Asia, S. America, USA, Hawaii	Larvae mine in leaves; sometimes in stems
Melanagromyza fabae Spencer	Bean Stem Fly	Agromyzidae	England	
Melanagromyza spp.	Bean Leaf Miners	Agromyzidae	Europe, Asia	
Phytomyza horticola Gour.	Pea Leaf Miner	Agromyzidae	Africa, Asia, Europe	
Bruchus rufimanus (Boh.)	Broadbean Bruchid	Bruchidae	USA	Larvae inside pods eat seeds
Bruchus spp.	Bruchids	Bruchidae	Middle East	
Callosobruchus spp.	Cowpea Bruchids	Bruchidae	S. Asia	
Hypera spp.	Clover Leaf Weevils	Curculionidae	Middle East	Adults eat leaves; larvae in soil eat roots
Sitona spp.	Pea & Bean Weevils	Curculionidae	Europe, Asia, N. America	
Otiorhynchus ligustici L.	Broad-nosed Weevil	Curculionidae	Europe	
Lixus algirus L.	Stem-boring Weevil	Curculionidae	Mediterranean, Middle East	Larvae bore stems; adults eat leaves
Tetranychus spp.	Red Spider Mites	Tetranychidae	Cosmopolitan	Scarify & web foliage

FLAX (*Linum usitatissimum* – Linaceae) (= Linseed)

At one time flax was the most useful and valuable of the fibre crops, but now is second to cotton, and possibly jute; the fabric is finer than cotton and superior in quality, and is called linen. The region of origin is not known, for the plant has been widely cultivated since prehistoric times, more than 5000 years. The plant is an annual herb from 30–100 cm in height, with small leaves and blue or white flowers. The fibres are formed in the pericycle by aggregates of long pointed cells with thick cellulose walls, making tough stringy strands from 30–90 cm in length. The best growing conditions are a soil rich in moisture and organic matter, in a cool temperate climate. The cost of preparation of flax is high and so the linen is more costly than cotton. Flax fibres are renowned for their great strength, length, fineness and durability, and are used to make linen cloth, thread, canvas, carpets, fine quality fishing lines and nets, and the finest writing paper.

The seeds are the source of linseed oil, an important drying oil, used in making varnishes, paints, linoleum, soft soap and printer's ink; for these purposes the oil is first boiled. Cold-pressed linseed oil is produced in eastern Europe for edible purposes.

MAJOR PESTS

No individual species appears to be particularly serious as a pest

MINOR PESTS

Bemisia tabaci (Genn.)	Cotton Whitefly	Aleyrodidae	India	Sap-suckers
Peritrechus saskatchewanensis Barb.	Lygaeid Bug	Lygaeidae	Canada	Adults & nymphs suck sap; toxic saliva; may kill plants
Nysius ericae (Schill.)	False Chinch Bug	Lygaeidae	Canada, USA	
Thrips angusticeps (Uzel)	Cabbage Thrips	Thripidae	Europe	Suck sap; infest foliage; retard growth, shoots swollen
Thrips linarius Uzel	Flax Thrips	Thripidae	Europe (not UK), Russia	
Thrips nigropilosus Uzel	Chrysanthemum Thrips	Thripidae	Europe, N. America	
Cnephasia longana Haw.	Omnivorous Leaf Tier	Tortricidae	Europe, N. America	Larvae eat leaves & bore shoots
Cnephasia interjectana (Haw.)	Flax Tortrix Moth	Tortricidae	Europe, Asia, Canada	
Diacrisia obliqua Wlk.	Tiger Moth	Arctiidae	India	Larvae defoliate
Agrotis orthogonia (Morr.)	Pale Western Cutworm	Noctuidae	Canada	Larvae are cutworms; destroy seedlings & eat foliage
Agrotis spp.	Cutworms	Noctuidae	Cosmopolitan	
Euxoa spp.	Cutworms	Noctuidae	Canada, USA	
Xestia c-nigrum (L.)	Spotted Cutworm	Noctuidae	Europe, Asia, N. America	
Heliothis ononis (Wlk.)	Flax Bollworm	Noctuidae	Canada	Larvae bore buds & eat seeds; may also eat leaves
Heliothis punctigera Wllgr.	Native Budworm	Noctuidae	Australia	
Heliothis spp.	Budworms/ Bollworms	Noctuidae	Cosmopolitan	

Mamestra configurata (Wlk.)	Bertha Armyworm	Noctuidae	Canada	Larvae eat foliage
Autographa gamma (L.)	Silver-Y Moth	Noctuidae	Europe	
Plusia orichalcea F.	Slender Burnished Brass Moth	Noctuidae	E. Africa	
Dasychira mendosa Hb.	Tussock Caterpillar	Lymantriidae	India	
Dasineura lini Barnes	Linseed Flower Midge	Cecidomyiidae	Europe, India	Larvae feed in flowers
Phytomyza atricornis (Mgn.)	Pea Leaf Miner	Agromyzidae	India	Larvae mine leaves
Epicauta spp.	Black Blister Beetles	Meloidae	Canada, USA	Adults eat foliage
Aphthona euphorbiae (Schrank.)	Large Flax Beetle	Chrysomelidae	Europe	Adults eat holes in leaves & damage seedhead; larvae in soil eat roots
Longitarsus parvulus (Fayk.)	Flax Flea Beetle	Chrysomelidae	Europe	
Halotydeus destructor (Tucker)	Red-legged Earth Mite	Penthaleidae	Australia	Scarify (silver) & wilt leaves

GLADIOLUS (*Gladiolus gandavensis* – Iridaceae)

A tall erect lily, up to 1 m in height, with a series of large beautiful flowers arranged on both sides of the long flower spike; many different varieties with different coloured flowers. The plant body grows from a corm underground, and it is grown as an annual for cut flowers and for the corms for horticultural ornamentation. It is basically a temperate plant which flowers in the summer in temperate regions and in the winter in subtropical areas.

MAJOR PESTS

Dysaphis tulipae (Fonsc.)	Tulip Bulb Aphid	Aphididae	Europe, USA	Infest young shoots
Taeniothrips simplex (Mor.)	Gladiolus Thrips	Thripidae	Europe, USA	Infest leaf bases
Phlogophora meticulosa (L.)	Angleshades Moth	Noctuidae	Eurasiatic	Larvae eat buds & flowers

MINOR PESTS

Aphis gossypii Glov.	Melon/Cotton Aphid	Aphididae	Cosmopolitan	Infest foliage; suck sap; virus vectors
Macrosiphum euphorbiae (Thos.)	Potato Aphid	Aphididae	Cosmopolitan	
Pseudococcus maritimus (Ehrh.)	Grape Mealybug	Pseudococcidae	Widespread	Infest foliage
Lygus lineolaris (Beauv.)	Tarnished Plant Bug	Miridae	Canada, USA, Mexico	Sap sucker; toxic saliva
Frankliniella spp.	Flower Thrips	Thripidae	Cosmopolitan	Infest flowers & foliage; feeding causes scarification & distortion
Heliothrips haemorrhoidalis (Bche.)	Greenhouse Thrips	Thripidae	Cosmopolitan	
Hercinothrips femoralis (Reuter)	Banded Greenhouse Thrips	Thripidae	Cosmopolitan	
Hepialus spp.	Swift Moths	Hepialidae	Europe	Larvae damage corms in soil
Ostrinia nubilalis (Hbn.)	European Corn Borer	Pyralidae	Europe, USA	Larvae eat foliage above stem
Phyllophaga spp. etc.	White Grubs	Scarabaeidae	USA, Canada	Larvae in soil damage corms
Several species	Wireworms	Elateridae	Europe, USA	
Rhizoglyphus spp.	Bulb Mites	Acaridae	Cosmopolitan	Infest corm scales & spread fungi
Tetranychus spp.	Red Spider Mites	Tetranychidae	Cosmopolitan	Scarify & web foliage

GLOBE ARTICHOKE (*Cynara scolymus* – Compositae)

A native of the Mediterranean Region and the Canary Islands, this plant resembles a large semi-prostrate thistle; the flower stalks terminate in a large globular inflorescence covered with many involucral bracts. The immature flowerhead is eaten as a vegetable, either raw or cooked, and sometimes canned; the head together with the thickened receptacle and the fleshy bases of the involucral leaves are eaten. The plant grows best in low ground near the sea coast; the main areas of cultivation are central and southern Europe, and California in the USA.

MAJOR PESTS

Aphis fabae Scopoli	Black Bean Aphid	Aphididae	Cosmopolitan	Infest flower bracts
Platyptilia carduidactyla (Riley)	Artichoke Plume Moth	Pterophoridae	USA	Larvae bore plant stem

MINOR PESTS

Agromyza apfelbecki Str.	Leaf Miner	Agromyzidae	S. Europe, Chile	Larvae mine veins
Spodoptera littoralis (Boid.)	Cotton Leafworm	Noctuidae	Near East	Larvae eat foliage
Sphaeroderma rubidum Graëlls	Artichoke Beetle	Chrysomelidae	Europe & Mediterranean	Adults eat foliage; larvae mine leaves
Cassida palaestina Reiche	Palestine Tortoise Beetle	Chrysomelidae	Near East	Larvae window leaves
Tetranychus spp.	Red Spider Mites	Tetranychidae	Asia	Scarify foliage

GOOSEBERRY (*Ribes grossularia* – Saxifragaceae) (=*Grossularia* spp.)

Sometimes referred to as the European Gooseberry, this is native to Eurasia, and is widely grown in the cooler regions of Europe and America, including Canada. The plant is a small woody shrub with many long thorns on the stems and twigs, well adapted to cold climates, with flowers and fruits that are solitary. The fruits are globular, green when unripe and either yellow or red when fully ripe, and hairy or smooth according to the variety. Gooseberries are used mostly for cooking as sweets, tarts, or puddings, sometimes mixed with other fruits as gooseberries picked unripe are rather tart in taste. Not an important fruit but widely cultivated in temperate regions.

MAJOR PESTS

Aphis grossulariae Kltb.	Gooseberry Aphid	Aphididae	Europe	Adults & nymphs infest foliage; suck sap & cause leaf curl (all three)
Nasonovia ribis-nigri (Mosley)	Currant–Lettuce Aphid	Aphididae	Europe	
Hyperomyzus pallidus HRL	Gooseberry–Sowthistle Aphid	Aphididae	Europe	

MINOR PESTS

Empoasca spp.	Green Leafhoppers	Cicadellidae	Europe, Asia	Sap-suckers; infest foliage (both)
Typhlocyba spp.	Leafhoppers	Cicadellidae	Europe	
Lepidosaphes ulmi (L.)	Mussel Scale	Diaspididae	Europe	Encrust twigs
Pulvinaria ribesiae Sign.	Woolly Currant Scale	Coccidae	Europe	Infest twigs
Lygocoris pabulinus (L.)	Common Green Capsid	Miridae	Europe	Sap-sucker; toxic saliva
Synanthedon salmachus (L.)	Currant Clearwing Moth	Sesiidae	Europe, USA, Australia	Larvae bore stems
Zophodia convolutella (Hbn.)	Gooseberry Fruitworm	Pyralidae	Canada, USA	Larvae eat fruits
Abraxas grossulariata (L.)	Magpie Moth	Geometridae	Europe	Larvae eat leaves; may defoliate (both)
Itame ribearia (Fitch)	Currant Spanworm	Geometridae	Canada, USA	
Nematus ribesii (Scop.)	Gooseberry Sawfly	Tenthredinidae	Europe, USA	Larvae eat leaves; sometimes defoliate bush (both)
Nematus leucotrochus Hart.	Pale-spotted Gooseberry Sawfly	Tenthredinidae	Europe	
Cecidomyia grossulariae Fitch	Gooseberry Midge	Cecidomyiidae	USA, Canada	Larvae infest fruits
Contarinia ribis (Kieffer)	Gooseberry Flower Midge	Cecidomyiidae	Europe (not UK)	Larvae gall flowers
Dasyneura ribicola (Kieffer)	Gooseberry Leaf Midge	Cecidomyiidae	Europe	Larvae crinkle leaves
Rhopalomyia grossulariae Felt	Gooseberry Bud Midge	Cecidomyiidae	England, USA	Larvae gall buds
Epochra canadensis (Loew)	Currant Fruit Fly	Tephritidae	Canada, USA	Larvae inside fruits
Agrilus spp.	Flat-headed Borers	Buprestidae	Canada, USA	Larvae bore stems
Bryobia ribis Thos.	Gooseberry Bryobia Mite	Tetranychidae	Europe	Scarify foliage

GRAPEVINE (*Vitis vinifer* – Vitaceae) (=Wine Grape; European Grape)

One of the oldest cultivated plants, probably originating near the Caspian Sea of W. Asia. It was spread all over Europe by the Romans, and is now found in all temperate regions and the higher parts of the many tropical countries. There are several native American species with larger and more hardy fruit, and high resistance to various diseases. Most wine grapes are grafted on to stock of American species to reduce disease incidence. The plants are woody, climbing, tendril-bearing vines, with large palmate leaves, and small insignificant flowers leading to large clusters of fruits. The fruits are technically berries, and are either eaten fresh, dried as raisins or currants, or the juice is pressed out to make wine. Many different varieties are grown commercially. Raisins are dried wine grapes of high quality, and may be seedless like the Sultana variety. Currants are small dried grapes from a variety that grows in Greece. The chief grape-growing regions are Europe, USA, Argentina, Chile, Australia, and S. Africa. Propagation is by stem cuttings.

MAJOR PESTS

Erythroneura spp.	Grape Leafhoppers	Cicadellidae	USA, Canada, China	Infest foliage, under leaves
Viteus vitifolii (Fitch)	Grape Phylloxera	Phylloxeridae	Europe, USA	Leaves galled, roots callused
Planococcus spp.	Root Mealybugs	Pseudococcidae	Cosmopolitan	Infest rootstocks & foliage
Pseudococcus maritimus (Ehrh.)	Grape Mealybug	Pseudococcidae	Europe, Asia, USA	Infest foliage

MINOR PESTS

Odontotermes obesus (Ramb.)	Scavenging Termite	Termitidae	India	Damage roots
Aleurocanthus spiniferus Quaint	Orange Spiny Whitefly	Aleyrodidae	India, S.E. Asia, China, Japan	Adults & nymphs infest foliage; suck sap; often with sooty moulds on honey-dew
Aleurocanthus woglumi (Ashby)	Citrus Blackfly	Aleyrodidae	India	
Trialeurodes vittatus (Quaint)	Grape Whitefly	Aleyrodidae	USA	
Aphis illinoisensis Shimer	Grapevine Aphid	Aphididae	USA	
Aphis gossypii (Glov.)	Cotton Aphid	Aphididae	India	
Lecanium spp.	Soft Scales	Coccidae	India	Infest foliage; suck sap; often with sooty moulds
Ceroplastes spp.	Waxy Scales	Coccidae	Asia, USA	
Saissetia spp.	Brown Scales	Coccidae	Asia	
Parasaissetia nigra (Niet.)	Nigra Scale	Coccidae	India	
Parthenolecanium corni (Bche.)	Plum Scale	Coccidae	Europe	
Parthenolecanium persicae (F.)	Peach Scale	Coccidae	Cosmopolitan	
Ferrisia virgata (Ckll.)	Striped Mealybug	Pseudococcidae	India	Infest foliage
Maconellicoccus hirsutus (Green)	Hibiscus Mealybug	Pseudococcidae	India, China	Infest foliage; stunt & kill shoots
Kerria lacca (Kerr.)	Lac Insect	Lacciferidae	India	Encrust stems
Aspidiotus spp.	Grape Scales	Diaspididae	Asia, USA	Encrust foliage
Anasa tristis (Deg.)	Squash Bug	Coreidae	Canada, USA	Sap-sucker; toxic saliva
Retithrips syriacus (Mayet)	Black Vine Thrips	Thripidae	Mediterranean, India	Infest & scarify foliage
Scirtothrips dorsalis Hood	Chilli Thrips	Thripidae	India	

Sciapteron regale Butler	Grape Clearwing Moth	Sesiidae	China	Larvae bore roots
Vitacea polistiformis (Harris)	Grape Root Borer	Sesiidae	USA	
Cnephasia longana (Clem.)	Omnivorous Leaf Tier	Tortricidae	Switzerland, USA	Larvae roll leaves
Cydia molesta (Busck.)	Oriental Fruit Moth	Tortricidae	China	Larvae in fruits
Epiphyas postvittana (Wlk.)	Light-brown Apple Moth	Tortricidae	Australasia	Larvae eat leaves
Lobesia botrana (Schiff.)	Grape Berry Moth	Tortricidae	Europe, E. Africa, Japan	Larvae destroy fruits
Paralobesia viteana (Clem.)	Grape Berry Moth	Tortricidae	Canada, USA	
Clysia ambiguella (Hb.)	Vine Moth	Cochylidae	Europe, Asia, S. America	Larvae bore shoots & stems
Pterophorus periscelidactylus Fitch	Grape Plume Moth	Pterophoridae	USA	Larvae web & eat shoots
Hyles lineata (F.)	Silver-striped Hawk Moth	Sphingidae	Pantropical	Larvae eat leaves; may defoliate
Eumorpha spp.	Grapevine Sphinx Moths	Sphingidae	USA	
Theretra spp.	Grapevine Hawk Moths	Sphingidae	Europe, Asia	
Desmia funeralis (Hub.)	Grape Leaf Folder	Pyralidae	USA	Larvae fold & eat leaves
Paramyelois transitella (Wlk.)	Navel Orangeworm	Pyralidae	USA	Larvae damage fruits
Sylepta lunalis (Guen.)	Grape Leaf Roller	Pyralidae	India	Larvae roll & eat leaves
Harrisina americana (G.-M.)	Grape Leaf Skeletonizer	Zygaenidae	USA	Larvae skeletonize leaves
Lygris diversilineata (Hub.)	Grapevine Looper	Geometridae	USA	Larvae eat leaves
Achaea spp.	Fruit-piercing Moths	Noctuidae	Asia	Adult moths pierce ripe fruits
Othreis fullonia (Cl.)	Fruit-piercing Moth	Noctuidae	Africa, Asia	
Spodoptera litura (F.)	Fall Armyworm	Noctuidae	Asia	Larvae eat foliage
Xestes c-nigrum (L.)	Spotted Cutworm	Noctuidae	Europe, Asia, N. America	
Euproctis spp.	Tussock Moths	Lymantriidae	Asia	
Contarinia spp.	Grape Flower Midges	Cecidomyiidae	Europe, USA	Larvae gall flowers
Lasioptera vitis (O.S.)	Grape Tomato Gall Midge	Cecidomyiidae	USA	Small red leaf-galls
Dacus spp.	Fruit Flies	Tephritidae	Pantropical	Larvae inside fruits
Polistes spp.	Paper Wasps	Vespidae	Pantropical	Adults pierce ripe fruits
Vespa/Vespula spp.	Common Wasps	Vespidae	Cosmopolitan	
Erythraspides vitis Hams.	Grape Sawfly	Tenthredinidae	USA	Larvae defoliate
Evoxysoma vitis (Saunders)	Grape Seed Chalcid	Eurytomidae	USA	Larvae gall seeds inside fruits
Adoretus spp.	Cockchafers	Scarabaeidae	Asia	Adults eat leaves; larvae in soil eat roots
Anomala spp.	Flower Beetles	Scarabaeidae	Asia	
Holotrichia spp.	Cockchafers	Scarabaeidae	Asia	

Apate monachus F.	Black Borer	Bostrychidae	Israel	Adults bore stock
Schistocerus bimaculatus Oliv.	Grape Cane Borer	Bostrychidae	Mediterranean	
Altica chalybaea (Illig.)	Grape Flea Beetle	Chrysomelidae	Canada, USA	Adults eat buds; larvae skeletonize leaves
Colaspis brunnea (F.)	Grape Colaspis	Chrysomelidae	USA	Adults eat leaves
Fidia viticida Walsh	Grape Rootworm	Chrysomelidae	USA	Larvae eat roots; adults skeletonize
Monolepta spp.	Spotted Leaf Beetles	Chrysomelidae	Australia, India	Adults eat foliage
Nodostoma spp.	Leaf Beetles	Chrysomelidae	India	Adults defoliate
Oides decempunctata Bill.	Leaf Beetle	Chrysomelidae	China, India	Adults & larvae defoliate
Scelodonta strigicollis (Mot.)	Flea Beetle	Chrysomelidae	India	Adults eat buds & shoots; larvae in soil eat roots
Cerasphorus albofasciatus (L. & G.)	Grape Trunk Borer	Cerambycidae	USA	Larvae bore stock (stem)
Chlorophorus varius	Longhorn Beetle	Cerambycidae	Asia	
Sthenias grisator F.	Girdler Beetle	Cerambycidae	India	
Ampeloglypter ater (Le. C.)	Grape Cane Girdler	Curculionidae	USA	Adults girdle stems (ovipositing); larvae in stem pith
Ampeloglypter sesostris (Le. C.)	Grape Cane Gall Maker	Curculionidae	USA	Larvae gall stems
Craponius inaequalis (Say)	Grape Curculio	Curculionidae	USA	Larvae inside fruits
Myllocerus spp.	Grey Weevils	Curculionidae	India	Adults eat leaves
Orthorhinus klugi Boh.	Vine Weevil	Curculionidae	Australia	Larvae bore vines; adults eat foliage
Orthorhinus cylindrirostris (F.)	Elephant Weevil	Curculionidae	Australia	Larvae bore stem & rootstock
Otiorhynchus cribricollis Gylh.	Apple Weevil	Curculionidae	Mediterranean, Australia, USA	Larvae in soil eat roots; adults eat foliage
Otiorhynchus sulcatus (F.)	Black Vine Weevil	Curculionidae	Cosmopolitan	
Xyleborus semiopacus Eich.	Black Twig Borer	Scolytidae	India	Adults bore stems
Eutetranychus orientalis (Klein)	Oriental Mite	Tetranychidae	Asia	Adults & nymphs infest foliage & scarify (silver) leaves; sometimes make extensive silk webbing
Tetranychus spp.	Red Spider Mites	Tetranychidae	Cosmopolitan	
Oligonychus spp.	Red Spider Mites	Tetranychidae	Asia	
Brevipalpus californicus (Banks)	Scarlet Tea Mite	Tenuipalpidae	Cosmopolitan	
Calepitrimerus vitis (Can.)	Grape Rust Mite	Eriophyidae	Australia, USA, S. Africa	Scarify foliage
Eriophyes vitis (Pgst.)	Grape Erineum Mite	Eriophyidae	Australia, China, USA	Make erinea on leaves

GRASS (Many species – Gramineae)

It is not feasible to consider the different genera and species of grasses separately because of the number involved. Some of the grass pests are polyphagous soil-dwelling insects, but many are species confined to the Gramineae as host plants. Most of the pests listed under the different graminaceous crops will also be found on different grasses as these are the natural alternative hosts for the pests. However, certain pests, for example some gall midges, are quite host-specific and different species will be found attacking different species of grasses.

THE MORE IMPORTANT PEST SPECIES

Gryllus spp.	Field Crickets	Gryllidae	Cosmopolitan	Defoliate & destroy roots & seedlings
Gryllotalpa spp.	Mole Crickets	Gryllotalpidae	Cosmopolitan	Destroy roots
Melanoplus spp.	Grasshoppers	Acrididae	Canada, USA	Adults & nymphs eat the leaf lamina; sometimes present in swarms
Locusta migratoria sspp.	Migratory Locusts	Acrididae	Africa, Asia	
Camnula pellucida (Scudd.)	Clearwinged Grasshopper	Acrididae	Canada	
Many species	Short-horned Grasshoppers	Acrididae	Cosmopolitan	
Nephotettix spp.	Green Leafhoppers	Cicadellidae	Africa, Asia	Adults & nymphs infest foliage; suck sap; very important as virus vectors
Empoasca spp.	Leafhoppers	Cicadellidae	Cosmopolitan	
Many species	Planthoppers	Delphacidae	Cosmopolitan	
Some species	Spittlebugs	Cercopidae	Cosmopolitan	
Hysteroneura setariae (Th.)	Rusty Plum Aphid	Aphididae	Pantropical & USA	Adults & nymphs infest foliage; suck sap; virus vectors
Macrosiphum spp.	Grass Aphids	Aphididae	Cosmopolitan	
Metapolophium spp.	Grass Aphids	Aphididae	Cosmopolitan	
Rhopalosiphum spp.	Grass Aphids	Aphididae	Cosmopolitan	
Schizaphis graminum (Rond.)	Wheat Aphid	Aphididae	Cosmopolitan	
Brevennia rehi (Ldgr.)	Rice Mealybug	Pseudococcidae	S. Asia, S. USA	Infest roots & stems
Saissetia oleae (Ol.)	Black Scale	Coccidae	Subtropical	Infest foliage
Antonia graminis (Mask.)	Rhodes Grass Scale	Diaspididae	Pantropical	Infest foliage
Aptinothrips spp.	Grass Thrips	Thripidae	Europe	Infest foliage; suck sap & scarify (fleck) foliage
Limothrips cerealium Hal.	Grain Thrips	Thripidae	Europe	
Stenothrips graminum Uzel	Oat Thrips	Thripidae	Europe	
Hepialus spp. *Oncopera* spp.	Swift Moths	Hepialidae	Palaearctic Australia	Larvae in soil eat roots
Crambus spp.	Grass Moths (Sod Webworms)	Crambidae	Cosmopolitan	Larvae cut stems & eat leaves
Diatraea spp.	Stalk Borers	Pyralidae	New World	Larvae bore stems
Ostrinia spp.	Corn Stalk Borers	Pyralidae	Cosmopolitan	
Chilo spp.	Stem Borers	Pyralidae	Tropics & subtropics	
Spodoptera spp.	Armyworms	Noctuidae	Cosmopolitan in warmer parts	Larvae defoliate
Cerapteryx graminis (L.)	Antler Moth	Noctuidae	Europe	Larvae defoliate
Several species	Skippers	Hesperiidae	Cosmopolitan	Larvae fold or roll leaves

Amaurosoma spp.	Grass Midges	Cecidomyiidae	Europe	Larvae gall flowerheads & panicles
Contarinia spp.	Grass Flower Midge	Cecidomyiidae	Europe	
Dasyneura spp.	Grass Flower Midges	Cecidomyiidae	Europe	
Mayetiola spp.	'Flax' Midges	Cecidomyiidae	Europe, Asia, N. America, N.Z.	Larvae gall stems; cause lodging
Tipula spp.	Common Craneflies (Leatherjackets)	Tipulidae	Palaearctic	Larvae in soil eat roots; destroy seedlings
Nephrotoma spp.	Spotted Craneflies	Tipulidae	Holarctic	
Geomyza spp.	Grass Flies	Opomyzidae	Holarctic	Larvae bore shoots; cause 'dead-hearts'
Meromyza spp.				
Opomyza spp.				
Chlorops spp.	'Gout' Flies	Chloropidae	Palaearctic	Larvae gall stems
Oscinella spp.	'Frit' Flies	Chloropidae	Palaearctic	Larvae kill shoots
Atherigona spp.	Shoot Flies	Muscidae	Cosmopolitan	Larvae kill shoots
Several species	Leaf-cutting Ants	Formicidae	S. USA, C. & S. America	Foraging workers cut pieces of leaf
Agriotes spp.	Wireworms	Elateridae	Palaearctic	Larvae in soil eat roots
Athous spp.	Garden Wireworms	Elateridae	Europe	
Corymbites spp.	Upland Wireworms	Elateridae	Europe	
Lacon spp.	Tropical Wireworms	Elateridae	Pantropical	
Anomala spp.	Flower Beetles	Scarabaeidae	Cosmopolitan	Larvae in soil called 'white grubs' or 'chafers'; eat roots; polyphagous mostly but most abundant in grassland
Adoretus spp.	Flower Beetles	Scarabaeidae	Pantropical	
Cetonia spp.	Rose Beetles	Scarabaeidae	Cosmopolitan	
Heteronychus spp.	Cereal Beetles	Scarabaeidae	Pantropical	
Melolontha spp.	Cockchafers	Scarabaeidae	Europe, Asia	
Phyllophaga spp.	June Beetles	Scarabaeidae	Canada, USA	
Popillia japonica	Japanese Beetle	Scarabaeidae	China, Japan, USA	
Protaetia spp.	Rose Chafers	Scarabaeidae	Pantropical	
Schizonycha spp.	White Grubs	Scarabaeidae	Africa, Europe	
Serica spp.	Brown Chafers	Scarabaeidae	Europe, Asia	
Oulema spp.	Cereal Leaf Beetles	Chrysomelidae	Cosmopolitan	Adults & larvae eat leaves
Phyllobius spp.	Leaf Weevils	Curculionidae	Palaearctic	Larvae in soil eat roots; adults eat leaves
Tanymecus spp.	Grey Weevils	Curculionidae	Palaearctic	
Sphenophorus spp.	Billbugs	Curculionidae	N. America	
Siteroptes cerealium (Kirch.)	Grass & Cereal Mite	Penthaleidae	Europe	Cause 'silvertop'
Halotydeus destructor (Tucker)	Red-legged Earth Mite	Penthaleidae	Europe, N. America, Australia, S. Africa	Scarify (silver) & wilt leaves
Tyrophagus spp.	Grass Mites	Acaridae	Cosmopolitan	Infest leaf-sheaths

HAZELNUT (*Corylus avellana* – Betulaceae) (=Cob Nut; Filbert; Barcelona Nut)

This is a temperate nut found in both hemispheres. The native American species are *C. americana* and *C. cornuta*, and they produce small palatable nuts of little commercial importance. The larger European species, and also *C. maxima*, are the source of the commercial cob nuts, filberts and Barcelona nuts. The fruit is one of the few so-called 'nuts' which is in fact botanically a nut, being a one-celled, one-seeded dry fruit with a hard pericarp (shell). The main areas of commercial hazelnut production are southern Europe, and parts of the USA.

MAJOR PESTS

Corylobium avellanae (Schr.)	Large Hazelnut Aphid	Aphididae	Europe	Adults & nymphs infest foliage; suck sap
Myzocallis coryli (Goeze)	Hazel Aphid	Aphididae	Europe, USA	
Lepidosaphes ulmi (L.)	Mussel Scale	Diaspididae	Cosmopolitan	Encrust twigs & bark
Nezara viridula (L.)	Green Stink Bug	Pentatomidae	Cosmopolitan	Sap-sucker; toxic saliva
Operophtera brumata (L.)	Winter Moth	Geometridae	Europe	Larvae defoliate
Curculio nucum (L.)	Nut Weevil	Curculionidae	Europe	Larvae eat kernel; adults kill young nuts
Phyllobius spp.	Leaf Weevils	Curculionidae	S. Europe	Adults eat leaves
Phytoptus avellanae Nal.	Hazel 'Big-bud' Mite	Eriophyidae	Europe, USA	Cause 'big-bud' & infest catkins

MINOR PESTS

Cicadella viridis L.	Green Leafhopper	Cicadellidae	S. Europe	Sap-suckers; eggs laid in slits in twigs
Ledra aurita (L.)	Leafhopper	Cicadellidae	Europe	
Typhlocyba spp.	Leafhoppers	Cicadellidae	Europe	
Eulecanium tiliae (L.)	Hazelnut Scale	Coccidae	S. Europe, UK	Infest twigs
Gonocerus acuteangulatus Goeze	Lygaeid Bug	Lygaeidae	S. Europe	Sap-suckers; toxic saliva; feeding causes necrosis
Deraeocoris spp.	Capsid Bugs	Miridae	S. Europe	
Zeuzera pyrina L.	Leopard Moth	Cossidae	Europe	Larvae bore trunk & branches
Coleophora spp. (10)	Case-bearers	Coleophoridae	Europe	Larvae in tiny cases mine leaves
Phyllonorycter coryli (Nic.)	Nut Leaf Blister Moths	Gracillariidae	Europe	Larvae blotch-mine leaves
Phyllonorycter spp.		Gracillariidae	Europe	
Melissopus latiferreanus (Wals.)	Filbertworm	Olethreutidae	USA	Larvae bore in nuts
Archips podana (Scop.)	Fruit Tree Tortrix	Tortricidae	Europe	Larvae roll & eat leaves
Archips spp.	Tortrix Moths	Tortricidae	Europe	
Epinota penkleriana (F.R.)	Nut Bud Moth	Tortricidae	Europe	Larvae eat buds, catkins & leaves
Alsophila aescularia (Schiff.)	March Moth	Geometridae	Europe	Looper larvae eat leaves & may defoliate
Erannis defoliaria (Clerck)	Mottled Umber Moth	Geometridae	Europe	

Phalera bucephala L.	Buff-tip Moth	Notodontidae	Europe	Larvae eat leaves; may defoliate
Malacosoma spp.	Tent Caterpillars	Lasiocampidae	Europe	
Lymantria dispar L.	Gypsy Moth	Lymantriidae	S. Europe	
Contarinia corylina Loew	Hazel Catkin Midge	Cecidomyiidae	Europe	Larvae gall catkins
Croesus septentrionalis (L.)	Hazel Sawfly	Tenthredinidae	Europe	Larvae defoliate
Melolontha spp.	Cockchafers	Scarabaeidae	Europe	Larvae in soil eat roots, damage young plants
Haplidia etrusca Kr.	Cockchafer	Scarabaeidae	S. Europe	
Altica spp.	Flea Beetles	Chrysomelidae	Europe	Adults & larvae eat leaves
Agrilus viridis L.	Jewel Beetle	Buprestidae	S. Europe	Larvae bore under bark of branches & trunk
Agrilus laticornis Ill.	Jewel Beetle	Buprestidae	Europe	
Byctiscus betulae (L.)	Hazel Leaf-roller Weevil	Curculionidae	Europe	Larvae roll leaf edge
Curculio neocorylus Gibson	Hazelnut Weevil	Curculionidae	USA	Larvae eat nut kernels
Curculio occidentis (Casey)	Filbert Weevil	Curculionidae	USA	
Strophosomus melanogrammus Fo.	Hazelnut Leaf Weevil	Curculionidae	Europe	Adults eat leaves
Xyleborus dispar L.	Twig Borer	Scolytidae	Europe	Adults bore twigs & branches
Panonychus ulmi Koch.	Fruit Tree Red Spider Mite	Tetranychidae	Europe, Asia, N. America	Scarify foliage

HOP (*Humulus lupulus* – Moraceae)

Hop is native to the north temperate regions of both hemispheres; it was known to the Romans and has been grown extensively in parts of Europe since the ninth century. It is a climbing herb with perennial roots, which annually send up several rough, weak, angular stems (bines), with deeply lobed leaves and dioecious flowers. The female flowers are produced in scaly, cone-like catkins (cones) covered with glandular hairs, and they contain resin and various bitter aromatic principles, chief of which is lupulin. The dried hops are used mainly for beer brewing, in that they impart the characteristic bitter taste to the brew, and also improve the general flavour and prevent bacterial action. In cultivation the hop plants are trained on poles or trellises; the hops are picked in the autumn and dried carefully in a special kiln. The main production areas are the USA, Europe, Australia and parts of S. America.

MAJOR PESTS

Phorodon humuli (Schrank)	Damson–Hop Aphid	Aphididae	Europe, USA, Canada	Infest foliage; stunt plant growth

MINOR PESTS

Evacanthus interruptus (L.)	Hop Leafhopper	Cicadellidae	Europe	Infest foliage; suck sap
Typhlocyba spp.	Leafhoppers	Cicadellidae	Europe	
Calocoris fulvomaculatus (Deg.)	Hop Capsid	Miridae	Europe	Sap-suckers; toxic saliva
Taedia hawleyi (Knight)	Hop Plant Bug	Miridae	USA	
Archips rosaceana (Harr.)	Oblique-banded Leaf Roller	Tortricidae	Canada	Larvae roll & eat leaves
Ostrinia nubilalis (Hb.)	European Corn Borer	Pyralidae	Europe, USA, Canada	Larvae bore bines (vines; stems)
Hydraecia micacea (Esp.)	Rosy Rustic Moth	Noctuidae	Europe	
Hydraecia immanis (Guen.)	Hop Vine Borer	Noctuidae	Canada	
Noctua pronuba (L.)	Large Yellow Underwing	Noctuidae	Europe	Larvae are cutworms
Hypena humuli (Harris)	Hop Semilooper	Noctuidae	Canada, USA	Larvae eat foliage
Contarinia humuli (Theo.)	Hop Strig Midge	Cecidomyiidae	Europe	Larvae in cones
Melolontha spp.	Cockchafers	Scarabaeidae	Europe	Larvae in soil eat roots
Psylliodes attenuata (Koch)	Hop Flea Beetle	Chrysomelidae	Europe	Adults damage leaves & cones
Psylliodes punctulata Melsh.	Hop Flea Beetle	Chrysomelidae	Canada, USA	
Agriotes spp.	Wireworms	Elateridae	Europe	Larvae in soil eat roots
Plinthus caliginosus (F.)	Hop Root Weevil	Curculionidae	Europe	Larvae tunnel root-stock
Otiorhynchus singularis (L.)	Clay-coloured Weevil	Curculionidae	Europe	Larvae in soil eat roots
Tetranychus urticae (Koch)	Temperate Red Spider Mite	Tetranychidae	Europe	Scarify & web foliage

HORSERADISH (*Armoracia lapathifolia* – Cruciferae)

A native of southeastern Europe, this plant is a weed in many places but is extensively grown in many parts of Europe and America. The plant is tall and hardy with long glossy green leaves, with toothed margins, and masses of small white flowers. The roots are large and fleshy, white and cylindrical, and are harvested in late summer or autumn. They are grated or scraped, usually mixed with vinegar (but sometimes used raw) and used as a condiment. It is a valuable condiment, which apparently aids digestion and prevents scurvy, and has been used for many years in Europe. The pungent taste is due to a glucoside (sinigrin) which breaks down in water by enzyme action.

MAJOR PESTS

Mamestra brassicae (L.)	Cabbage Moth	Noctuidae	Europe	Larvae defoliate

MINOR PESTS

Murgantia histrionica (Hahn)	Harlequin Bug	Pentatomidae	USA	Sap-sucker; toxic saliva
Disonycha xanthomelana (Dalm.)	Spinach Flea Beetle	Chrysomelidae	Canada	Adults hole leaves
Phyllotreta armoraciae (Koch)	Horseradish Flea Beetle	Chrysomelidae	USA, Canada	Larvae mine stems & leaf veins

Some, but not all, of the pests listed under 'Brassicas' will also attack this member of the Cruciferae. In many parts of Europe foliage damage by slugs and snails is very extensive.

JERUSALEM ARTICHOKE (*Helianthus tuberosus* – Compositae)

This plant is native to N. America; it is a hardy perennial sunflower growing to a height of 2–4 m, and was introduced to Europe in 1616 where it is now widely cultivated. Many varieties have been established, adapted to a wide range of climate. The tubers develop on the roots of the plant and resemble potatoes; they are cooked, pickled or eaten raw. The plants are also grown as a forage crop and a weed eradicator.

MAJOR PESTS

MINOR PESTS

Aphis fabae (Scop.)	Black Bean Aphid	Aphididae	Cosmopolitan	Adults & nymphs infest foliage, especially flowers
Aphis gossypii Glov.	Melon/Cotton Aphid	Aphididae	Cosmopolitan	
Cassida spp.	Tortoise Beetles	Chrysomelidae	Europe	Adults & larvae eat foliage

It is to be expected that many of the pests of 'Sunflower' will also attack this crop, as both crops are species of *Helianthus*.

LETTUCE (*Lactuca sativa* – Compositae)

This is native to southern Europe and western Asia, and is descended from the wild lettuce (*Lactuca scariola*), a common wasteland and roadside weed (ruderal) in both Old and New Worlds. It is another herbage vegetable of great antiquity; at the present there are many varieties showing different horticultural characters. It has a milky sap (latex), and grows as a basal rosette of leaves, producing later in the season a stalk bearing the flowers. It has little food value in itself but does contain vitamins and iron salts. It grows best in a light sandy or loamy soil with a rather cool climate, and not too much sunshine. Among the principal types grown are head, cos, romains, and cut-leaf forms.

MAJOR PESTS

Macrosteles fascifrons (Stal)	Six-spotted Leafhopper	Cicadellidae	Canada, USA	Adults & nymphs infest foliage; sap-suckers; virus vectors; stunt growth
Aphis gossypii Glov.	Melon/Cotton Aphid	Aphididae	Cosmopolitan	
Macrosiphum euphorbiae (Thos.)	Potato Aphid	Aphididae	Europe, USA, Canada	
Myzus ascalonicus Don.	Shallot Aphid	Aphididae	Europe, USA, Canada	
Myzus persicae (Sulz.)	Peach–Potato Aphid	Aphididae	Cosmopolitan	
Nasonovia ribis-nigri (Mosley)	Lettuce Aphid	Aphididae	Europe	
Pemphigus bursarius (L.)	Lettuce Root Aphid	Pemphigidae	Europe	Infest roots underground

MINOR PESTS

Onychiurus spp.	Springtails	Onychiuridae	Europe	Damage seedlings
Nysius ericae (Sch.)	False Chinch Bug	Lygaeidae	Canada	Sap-sucker; toxic saliva
Hepialus spp.	Swift Moths	Hepialidae	Europe	Larvae eat roots
Cnephasia spp	Tortrix Moths	Tortricidae	Europe	Larvae eat leaves; may spin silk
Udea rubigalis (Guen.)	Greenhouse Leaf Tier	Tortricidae	Canada	
Agrotis spp.	Cutworms	Noctuidae	Cosmopolitan	Larvae are cutworms; eat roots & sometimes eat through stem underground
Euxoa spp.	Cutworms	Noctuidae	Cosmopolitan	
Noctua pronuba (L.)	Large Yellow Underwing	Noctuidae	Europe, Asia	
Mamestra brassicae (L.)	Cabbage Moth	Noctuidae	Europe	Larvae eat leaves; may defoliate
Anagrapha falcifera (Kirby)	Celery Semilooper	Noctuidae	USA	
Autographa gamma (L.)	Silver-Y Moth	Noctuidae	Europe	
Autographa californica (Speyer)	Alfalfa Semilooper	Noctuidae	Canada	
Plusia orichalcea (F.)	Lesser Burnished Brass	Noctuidae	Asia	
Nephrotoma spp.	Leatherjackets	Tipulidae	Cosmopolitan	Larvae in soil eat roots; may kill small plants
Tipula spp.	Leatherjackets	Tipulidae	Cosmopolitan	

Amauromyza maculosa (Mall.)	Lettuce Leaf Miner	Agromyzidae	N. & S. America	Larvae mine leaves
Liriomyza sativa Bl.	Vegetable Leaf Miner	Agromyzidae	USA	
Phytomyza syngenesiae Hardy	Chrysanthemum Leaf Miner	Agromyzidae	Cosmopolitan	
Phytomyza horticola Gour.	Pea Leaf Miner	Agromyzidae	Cosmopolitan	
Psila nigricornis Meig.	Chrysanthemum Stool Miner	Psilidae	Europe	Larvae mine roots
Trupanea amoena Von Frau.	Lettuce Fruit Fly	Tephritidae	Japan	Larvae in flowerhead
Pegohylemyia gnava (Meig.)	Lettuce Seed Fly	Anthomyiidae	Europe	
Megempleurus spp.	Turnip Mud Beetles	Hydrophilidae	Europe	Larvae in soil eat roots & may tunnel in stem
Agriotes spp.	Wireworms	Elateridae	Europe	
Melolontha spp.	Cockchafers	Scarabaeidae	Europe, Asia	

LOGANBERRY (*Rubus loganobaccus* – Rosaceae)

An erect, creeping or decumbent shrub, armed with thorns and prickles on the foliage. The long semi-erect 'canes' die back to the ground after a few years, but are replaced annually by new stems from the rootstock. It has been suggested that loganberry is the result of a natural cross between blackberry and raspberry. Agriculturally these plants are usually grown sprawled over a fence or wire framework. The large globular compound fruit is rather poor in flavour and is usually grown for canning, both the fruit and juice being used.

MAJOR PESTS

Notocelia uddmanniana (L.)	Bramble Shoot Moth	Tortricidae	Europe	Larvae eat flowers & web shoots
Byturus tomentosus (Deg.)	Raspberry Beetle	Byturidae	Europe	Larvae bore fruits

MINOR PESTS

Typhlocyba spp.	Rubus Leafhoppers	Cicadellidae	Europe	Sap-suckers under leaves
Amphorophora idaei v.d.G.	Raspberry Aphid	Aphididae	Europe	Infest foliage; curl leaves; virus vectors
Macrosiphum euphorbiae Thos.	Potato Aphid	Aphididae	Europe	
Macrosiphum fragariae Wlk.	Blackberry Aphid	Aphididae	Europe	
Aulacaspis rosae Bche.	Scurfy Scale	Diaspididae	Europe	Encrust stems
Lygocoris pabulinus (L.)	Common Green Capsid	Miridae	Europe	Sap-sucker; toxic saliva
Thrips flavus Schr.	Honeysuckle Thrips	Thripidae	Europe	Adults & nymphs infest flowers
Thrips tabaci Lind.	Onion Thrips	Thripidae	Cosmopolitan	
Spilonota ocellana (D. & S.)	Eye-spotted Bud Moth	Tortricidae	Europe, Asia, N. America	Larvae bore buds
Lampronia rubiella (Bjerk.)	Raspberry Moth	Tinaeidae	Europe	Larvae bore plug; later shoots (April)
Phalera bucephala (L.)	Buff-tip Moth	Notodontidae	Europe	Larvae defoliate
Pegomyia rubivora Coq.	Loganberry Cane Fly	Anthomyiidae	Europe	Larvae mine & girdle shoots
Metallus spp.	Raspberry Leaf-mining Sawflies	Tenthredinidae	Europe	Larvae mine leaves
Byturus spp.	Raspberry Fruitworms	Byturidae	USA	Larvae bore fruits
Caenorhinus germanicus (Herbst)	Strawberry Rhynchites	Curculionidae	Europe	Adults girdle shoots and flower stalks
Otiorhynchus singularis (L.)	Clay-coloured Weevil	Curculionidae	Europe	Adults damage stems & leaves
Metatetranychus ulmi Koch	Fruit Tree Red Spider Mite	Tetranychidae	Europe, N. America	Adults & nymphs scarify foliage
Tetranychus urticae (Koch)	Temperate Red Spider Mite	Tetranychidae	Cosmopolitan	
Acalitis essigi (Hassan)	Blackberry Mite	Eriophyidae	Europe	Cause 'redberry'
Phyllocoptes gracilis (Nal.)	Raspberry Bud Mite	Eriophyidae	Europe	Found in buds & on leaves

MAIZE (*Zea mays* – Gramineae) (=Sweet Corn, when unripe; Corn (in USA))

Maize originated in America and is now the principal cereal in the tropics and subtropics, and is also being grown for fodder and as a vegetable (Sweet Corn) in Europe and northern N. America. It needs a good summer temperature for the grain to ripen (Britain is generally too cool), and grows best in lowlands with a good soil; it can withstand some drought once established. It is a tall, broad-leaved cereal, a single stem usually, 1–2 m in height (UK), with the male flower terminal and 1–2 cobs per stalk. Some varieties tiller more than others. The main production areas are S. America, parts of the USA, E. and S. Africa. In temperate regions the grain is far less attacked by pests than in the tropics.

MAJOR PESTS

Rhopalosiphum maidis (Fitch)	Corn Leaf Aphid	Aphididae	Cosmopolitan	Infest foliage
Chilo spp.	Stalk Borers	Pyralidae	Warmer parts of the Old World	Larvae bore stems
Ostrinia spp.	Corn Borers	Pyralidae	Cosmopolitan	Larvae bore stems & cobs
Diatraea spp.	Stalk Borers	Pyralidae	N., C. & S. America	Larvae bore stems
Marasmia trapezalis (Gn.)	Maize Webworm	Pyralidae	Africa, Asia, Australasia, C. & S. America	Larvae web foliage
Heliothis spp.	Corn Earworms	Noctuidae	Cosmopolitan	Larvae eat cobs
Agrotis spp. *Euxoa* spp.	Cutworms	Noctuidae	Cosmopolitan	Larvae eat roots; may destroy seedlings
Spodoptera spp.	Armyworms	Noctuidae	Pantropical	Larvae defoliate
Mythimna spp.	Cereal Armyworms	Noctuidae	Cosmopolitan in warmer areas	Larvae defoliate
Oscinella frit (L.)	Frit Fly	Chloropidae	Europe	Larvae eat stem; cause 'dead-heart'
Atherigona spp.	Shoot Flies	Muscidae	Africa, Asia	Larvae bore stem; cause 'dead-heart'
Delia platura (Meign.)	Bean Seed Fly	Anthomyiidae	Europe, Asia, N. America	Larvae bore sown seeds & seedlings
Melolontha spp. *Schizonycha* spp.	Chafer Grubs	Scarabaeidae	Cosmopolitan	Larvae in soil eat roots & seedlings
Sitophilus spp.	Grain Weevils	Curculionidae	Cosmopolitan	Attack ripe grains in field & stores

MINOR PESTS

Acheta spp.	Field Crickets	Gryllidae	Pantropical	Pest of seedlings
Melanoplus spp.	Grasshoppers	Acrididae	Canada, USA	Defoliate
Dalbulus maidis D. & W.	Corn Leafhopper	Cicadellidae	USA, C. & S. America	Sap-suckers; virus vectors
Peregrinus maidis (Ashm.)	Maize Planthopper	Delphacidae	Pantropical	Sap-suckers; virus vectors
Laodelphax striatella (Fall.)	Small Brown Planthopper	Delphacidae	Europe, Asia	Sap-suckers; virus vectors

Schizaphis graminum (Rond.)	Wheat Aphid	Aphididae	Old World	Infest foliage; suck sap
Blissus leucopterus (Say)	Chinch Bug	Lygaeidae	Canada, USA	Adults & nymphs infest foliage; suck sap; toxic saliva; suck milky grains
Eurygaster austriaca (Schr.)	Wheat Shield Bug	Pentatomidae	S. Europe, W. Asia	
Nezara viridula (L.)	Green Stink Bug	Pentatomidae	Cosmopolitan	
Crambus spp.	Grass Moths (Corn Webworms)	Pyralidae (Crambidae)	Canada, USA	Larvae eat leaves & stems
Elasmopalpus lignosellus (Zell.)	Lesser Corn-stalk Borer	Pyralidae	N. & S. America	Larvae cause 'dead-hearts' in seedlings; bore stems of older plants
Hydraecia micacea (Esp.)	Rosy Rustic Moth	Noctuidae	Europe	
Mesapamaea secalis (L.)	Common Rustic Moth	Noctuidae	Europe	
Sesamia spp.	Stalk Borers	Noctuidae	Mediterranean, Africa, India, Asia, Australasia	
Borbo spp.	Skipper Butterflies	Hesperiidae	Asia	Larvae fold leaves
Limothrips cerealium (Hal.)	Cereal Thrips	Thripidae	USA	Infest panicle
Mylabris spp.	Banded Blister Beetles	Meloidae	Europe, Asia, Africa	Adults eat flower tassels
Carpophilus spp.	Sap Beetles	Nitidulidae	N. America	Damage ears & grains
Anomala spp.	White Grubs	Scarabaeidae	Cosmopolitan Europe, Asia	Larvae in soil eat roots; may destroy seedlings; adults may damage flowers
Melolontha spp.	(Chafer (Grubs))	Scarabaeidae		
Phyllophaga spp.	White Grubs	Scarabaeidae	Canada, USA	
Diabrotica spp.	Corn Rootworms	Chrysomelidae	Canada, USA	Larvae bore stems underground
Chaetocnema spp.	Corn Flea Beetles	Chrysomelidae	Canada	Adults eat leaves; vectors of bacterial wilt
Graphognathus spp.	White-fringed Weevils	Curculionidae	Australia, USA, S. America	Larvae eat roots; adults eat leaves
Sphenophorus maidis Chitt.	Maize Billbug	Curculionidae	USA	
Tanymecus spp.	Grey Weevils	Curculionidae	E. Europe, Asia	Adults eat leaves
Tetranychus urticae (Koch)	Temperate Red Spider Mite	Tetranychidae	Cosmopolitan	Scarify foliage

The more tropical pest species recorded attacking this crop are omitted from this list.

Some of the pests listed under the other different cereal crops (Gramineae) and grasses will also feed on and damage maize (sweet corn) plants in different parts of the world.

MINT (*Mentha piperita* – Labiatae) (=Peppermint)

One of the more important of the aromatic herbs; a small perennial herb found wild in moist ground in the temperate parts of Europe, Asia and N. America. It is widely cultivated in both Europe and the USA. The leaves are used to some extent for flavouring in cooking, but the crop is largely grown for its oil which is steam-distilled from the dried plant body. The oil is used to flavour chewing gum, sweets, toothpaste and various pharmaceutical products. More than one million pounds of oil are produced annually in the USA. Menthol is a derivative of peppermint oil, and is a valuable antiseptic and used for treatment of colds. Most menthol production is actually from Japanese Peppermint (*M. arvensis* var. *piperascens*) in Japan, Brazil and the USA, but both the oil and the menthol (Peppermint Camphor) are bitter and so less valuable.

MAJOR PESTS

Spodoptera litura (F.)	Fall Armyworm	Noctuidae	S. China	Larvae defoliate

MINOR PESTS

Philaenus spumarius (L.)	Common Spittlebug	Cercopidae	Europe, N. America	Spittlemass in foliage
Ovatus crataegarius (Wlk.)	Mint Aphid	Aphididae	USA	Infest foliage & suck sap; most under leaves
Aleyrodes lonicerae Wlk.	Strawberry Whitefly	Aleyrodidae	Europe	
Crambus spp.	Sod Webworms	Crambidae	Canada	Larvae eat stems, roots and leaves
Lacanobia oleracea (L.)	Tomato Moth	Noctuidae	Europe	Larvae defoliate
Cnephasia spp.	Leaf-tiers	Tortricidae	Europe	Larvae fold & eat leaves
Asphondylia menthae Pierre	Mint Flower Midge	Cecidomyiidae	Europe (not UK)	Larvae gall flowers
Dasyneura aromaticae Felt	Spearmint Gall Midge	Cecidomyiidae	USA	Larvae gall shoots
Dasyneura piperita (Felt)	Peppermint Gall Midge	Cecidomyiidae	USA	
Phytomyza horticola Goureau	Pea Leaf Miner	Agromyzidae	Europe, India, Japan	Larvae mine leaves
Phytomyza petoei Hering	Mint Leaf Miner	Agromyzidae	Europe, Asia	
Cassida viridis L.	Green Tortoise Beetle	Chrysomelidae	Europe	Larvae & adults hole leaves
Ceutorhynchus melanostictus (Mm.)	Mint Weevil	Curculionidae	Europe	Larvae tunnel stems; adults hole leaves
Baris menthae Kono	Mint Weevil	Curculionidae	Japan	

MULBERRY (*Morus* spp. – Moraceae)

Morus alba – White Mulberry

White Mulberry is native to China and grown in parts of the tropics for its edible fruits, for its leaves as food for silkworms, and for its wood used for making certain sports goods such as hockey sticks and tennis racquets. A small tree, up to 5 m in height, and the fruit is a syncarp. Black Mulberry is native to Iran and is more temperate, growing well in the tropics only at higher altitudes.

Morus nigra – Black Mulberry

MAJOR PESTS

Pealius mori (Tak.)	Mulberry Whitefly	Aleyrodidae	S.E. Asia	Infest leaves
Apriona spp.	Jackfruit Longhorns	Cerambycidae	Asia	Larvae bore trunks & branches
Batocera spp.	Spotted Longhorn Beetles	Cerambycidae	Asia, Africa	

MINOR PESTS

Aleurolobus marlatti (Quaint.)	Marlatt Blackfly	Aleyrodidae	India	Infest leaves & suck sap
Tetraleurodes mori (Quaint.)	Mulberry Whitefly	Aleyrodidae	USA	
Erythroneura mori Mats.	Blood Spot Leafhopper	Cicadellidae	China	
Lecanium spp.	Soft Scales	Coccidae	USA	Infest foliage & suck sap
Pulvinaria innumerabilis (Rath.)	Cottony Maple Scale	Coccidae	USA	
Drosicha spp.	Giant Mealybugs	Margarodidae	India, China	Infest foliage & suck sap
Icerya aegyptica (Dgl.)	Egyptian Fluted Scale	Margarodidae	Africa, Asia	
Maconellicoccus hirsutus (Green)	Hibiscus Mealybug	Pseudococcidae	S.E. Asia	Infest & stunt (may kill) shoots
Nipaecoccus nipae (Mask.)	Nipa Mealybug	Pseudococcidae	Asia, USA	Infest foliage; suck sap; sooty moulds often present
Perissopneumon tamarinda (Green)	Tamarind Mealybug	Pseudococcidae	India	
Pseudococcus comstocki (Kuw.)	Comstock Mealybug	Pseudococcidae	Asia, USA	
Aonidiella aurantii (Mask.)	Red Scale	Diaspididae	India, USA	Encrust twigs & suck sap; some on leaves
Chrysomphalus spp.	Armoured Scales	Diaspididae	India, USA	
Lepidosaphes spp.	Mussel Scales	Diaspididae	Asia, USA	
Pseudaulacaspis pentagona (Targ.)	White Peach Scale	Diaspididae	Cosmopolitan	
Quadraspidiotus perniciosus (Comst.)	San José Scale	Diaspididae	India, USA	
Halys dentatus F.	Mulberry Bug	Pentatomidae	India	Sap-sucker; toxic saliva
Indarbela spp.	Wood-borer Moths	Metarbelidae	India, China	Larvae eat bark & bore branches
Archips micaceana (Wlk.)	Leaf Roller	Tortricidae	S.E. Asia	Larvae roll & eat leaves
Latoia lepida (Cramer)	Blue-striped Nettlegrub	Limacodidae	India	Larvae eat leaves
Hemerophila atrilineata Butl.	Mulberry Looper	Geometridae	China	
Euzophera semifuneralis (Wlk.)	American Plum Borer	Phycitidae	USA	Larvae bore inner bark

Dichocrocis punctiferalis (Guen.)	Shoot Borer	Pyralidae	India	Larvae bore shoots & fruits
Spodoptera litura F.	Fall Armyworm	Noctuidae	India	Larvae eat leaves, may defoliate
Bombyx mori (L.)	Silkworm	Bombycidae	S.E. Asia, China, Japan	
Euproctis chrysorrhoea (L.)	Brown-tail Moth	Lymantriidae	China	
Porthesia similis Fuess.	Mulberry Lymantriid	Lymantriidae	China	
Dacus tau (Wlk.)	Fruit Fly	Tephritidae	India, S.E. Asia	Larvae inside fruits
Vespa/Vespula spp.	Common Wasps	Vespidae	Asia	Adults pierce ripe fruits
Mimastra cyanura Hope	Almond (Leaf) Beetle	Chrysomelidae	India, China	Adults eat leaves
Sthenias grisator F.	Stem Girdler	Cerambycidae	India	Larvae girdle stem
Baris deplanata Roc.	Stem Weevil	Curculionidae	China	Larvae bore stem
Hypomeces squamosus (F.)	Gold-dust Weevil	Curculionidae	S.E. Asia	Adults eat leaves
Phloeotribus liminaris Harris	Peach Bark Beetle	Scolytidae	USA	Adults bore under bark

Hanson (1963) reports that 170 species of insect pests are recorded attacking mulberry in China.

MUSHROOM (*Agaricus campestris* – Agaricaceae (Fungi)) (=Field or Meadow Mushroom)

Mushrooms occur naturally in fields, pastures and woods, and they represent the reproductive stage of certain higher fungi. The vegetative stage consists of masses of fine thread-like hyphae which form the mycelium. Cultivation started at the beginning of the seventeenth century, and today the common mushroom remains the most popular species. Various wild mushrooms are edible, but others are toxic and may be fatal if eaten. Most commercial cultivation takes place in dark sheds or buildings from which light is excluded, where the mycelium is grown in a rich organic compost (stable manure probably still being the best medium). In France much cultivation takes place in abandoned mines, caves and quarries, which abound in certain southern regions. Most of the mushrooms are harvested at the button stage but some are allowed to open and darken before harvest. The main areas of commercial production are France, England, eastern USA and China.

The oriental Straw Mushroom is *Volvariella volvulacea*, and the expensive dried Black or Shiitake Mushroom is *Lentinus elodes* (both also in the Agaricaceae).

MAJOR PESTS

Sciara spp. *Lycoriella* spp.	Sciarid Flies	Sciaridae	Cosmopolitan	Larvae tunnel in stalk and cap
Tyrophagus spp.	French 'Fly' (Mushroom Mites)	Acaridae	Cosmopolitan	Destroy mycelium & young mushrooms

MINOR PESTS

Xenylla spp.	Mushroom Springtails	Onychiuridae	Cosmopolitan	Feed on mycelium & small mushrooms
Henria spp. *Heteropeza* spp. *Lastrimia* spp. *Mycophila* spp.	Mushroom Midges	Cecidomyiidae	Asia Cosmopolitan Asia Cosmopolitan	Larvae in compost eat mycelium and may climb mushroom stalk and cap
Megascelia spp.	Scuttle Flies	Phoridae	Cosmopolitan	Larvae bore stalk & cap of mushroom
Tarsonemus myceliophagus Hussey	Mushroom Mite	Tarsonemidae	Europe	Discolour stalks

MUSTARD (*Brassica hirta* (White Mustard); *Brassica nigra* (Black Mustard) – Cruciferae)

A crop of great antiquity, native to Eurasia, grown as a field crop now in most temperate countries. Cultivated mainly for the seeds, it is sometimes grown as green manure, and the tops can be used as pot herbs and salad greens. Both species are freely branching annual herbs, about 1 m in height, with yellow flowers and later long thin pods with a pronounced beak. The small rounded seeds are either yellow (white mustard) or dark-brown (black mustard) in colour, and contain a glucoside, sinalbin (or sinigrin). The seeds are ground, and when mixed with water the glucoside breaks down yielding a sulphur compound with the characteristic pungent, aromatic odour and flavour. Black mustard is generally more pungent and is used to make 'English Mustard'. In addition to being used as a condiment, mustard oils have medicinal uses and some industrial uses.

Indian (Leaf) Mustard is *Brassica juncea*, and used in India and parts of Europe as a spice and for cooking, and has properties similar to black mustard.

MAJOR PESTS

Myzus persicae (Sulz.)	Peach–Potato Aphid	Aphididae	Cosmopolitan	Infest foliage; sap sucker
Pieris spp.	White Butterflies	Pieridae	Cosmopolitan	Larvae defoliate
Phaedon cochleariae (F.)	Mustard Beetle	Chrysomelidae	Europe	Adults & larvae eat foliage

MINOR PESTS

Nysius ericae (Schill.)	False Chinch Bug	Lygaeidae	Canada, USA	Sap-sucker; toxic saliva
Euxoa ochrogaster (Guen.)	Red-backed Cutworm	Noctuidae	Canada	Larvae are cutworms
Dasyneura brassicae Winn.	Brassica Pod Midge	Cecidomyiidae	Europe	Larvae gall pods (siliquas) of Black Mustard only
Phytomyza spp.	Leaf Miners	Agromyzidae	Europe, Asia	Larvae mine leaves
Meligethes spp.	Blossom Beetles	Nitidulidae	Europe	Adults eat flowers
Entomoscelis americana Brown	Red Turnip Beetle	Chrysomelidae	Canada, USA	Adults & larvae eat foliage
Phaedon spp.	Mustard Beetles	Chrysomelidae	Europe	Adults & larvae eat foliage
Ceutorhynchus quadridens (Panz.)	Cabbage Stem Weevil	Curculionidae	Europe	Larvae in stems & petioles; adults hole leaves
Agriotes spp.	Wireworms	Elateridae	Europe	Larvae in soil eat roots

Many of the pests listed under 'Brassicas' (page 493) may also be found on mustards, but are usually of less importance on these crops than on the more leafy vegetables.

NARCISSUS (*Narcissus tazetta* – Amaryllidaceae)

A small lily-like plant with annual foliage growing from a bulb, and a conspicuous yellow or white flower with a distinct trumpet-shaped corolla. A temperate species which flowers in early spring in northern Europe and N. America. Cultivation is extensive both as a cut-flower and for the bulbs which are planted in beds as ornamentals. There is some cultivation in Europe for its perfume. The closely related Jonquil is *Narcissus jonquilla*, with its much smaller, more delicate and often multiple flowers.

MAJOR PESTS

Merodon equestris (F.)	Large Narcissus Fly	Syrphidae	Europe, USA	Larvae infest bulb
Eumerus spp.	Small Narcissus Flies	Syrphidae	Europe, USA, Asia	Larvae infest bulb

MINOR PESTS

Dysaphis tulipae (Fonsc.)	Tulip Bulb Aphid	Aphididae	Europe, USA	Infest bulb in store & young shoots
Myzus ascalonicus Doncaster	Shallot Aphid	Aphididae	Europe	Infest bulb scale
Hepialus spp.	Swift Moths	Hepialidae	Europe	Larvae eat into bulbs
Merodon geniculata Strobl.	Mediterranean Narcissus Bulb Fly	Syrphidae	Mediterranean	Larvae infest bulbs
Bryobia spp.	Bryobia Mites	Tetranychidae	Europe	Silver foliage
Stenotarsonemus laticeps (Halb.)	Bulb Scale Mite	Tarsonemidae	Europe, USA	Infest bulb & distort foliage
Rhizoglyphus echinopus (F. & R.)	Bulb Mite	Tryoglyphidae	Europe, USA	Infest bulb

OATS (*Avena sativa*, etc. – Gramineae)

An ancient crop of uncertain origin in that the plant has never been found in a truly wild state; they may be of multiple origin, some from the Mediterranean, some from Abyssinia, and another from China. The plant is an annual grass, about 1 m in height, with bluish-green leaves and characteristic spreading panicle, usually erect. Other species cultivated include the Side Oat (*A. orientalis*), Red Oat (*A. byzantina*), Naked Oat (*A. nuda*) and the Short Oat (*A. brevis*). Oats can be grown farther north than any other cereal except rye, and they like cool moist climates. They reach a latitude of 69 °N in Alaska and 65 °N in Norway. It is the most nutritious of all cereals for human use as the seeds have a high fat, protein and mineral content, and oatmeal has long been a popular food. Most of the oats grown are used on the farms as grain food for horses and other livestock (except pigs); the main areas of cultivation are the USA, USSR, UK, Canada, Germany, France and Poland.

MAJOR PESTS

Schizaphis graminum Rond.	Wheat Aphid	Aphididae	Cosmopolitan	Infest foliage; suck sap; virus vectors
Sitobium avenae (F.)	Grain Aphid	Aphididae	Europe, N. America	
Rhopalosiphum padi (L.)	Bird-cherry Aphid	Aphididae	Europe, USA, Canada	
Oscinella frit (L.)	Frit Fly	Chloropidae	Europe	Larvae eat stem; cause 'dead-heart'
Oulema melanopa (L.)	Cereal Leaf Beetle	Chrysomelidae	Europe, USA	Adults & larvae eat leaf strips

MINOR PESTS

Rhopalosiphum maidis (Fitch)	Corn Leaf Aphid	Aphididae	Cosmopolitan	Adults & nymphs infest foliage; suck sap; virus vectors
Rhopalosiphum insertum (Wlk.)	Apple–Grass Aphid	Aphididae	Europe	
Macrosiphum fragariae Wlk.	Blackberry Aphid	Aphididae	Cosmopolitan	
Metapolophium dirhodum (Wlk.)	Rose–Grain Aphid	Aphididae	Europe	
Metapolophium festucae (Theob.)	Fescue Aphid	Aphididae	Europe	
Laodelphax striatella (Fall.)	Small Brown Planthopper	Delphacidae	Europe, Asia	Sap suckers; virus vectors
Javasella pellucida (F.)	Cereal Planthopper	Delphacidae	Europe	
Crambus spp.	Grass Moths	Pyralidae	Europe, N. America	Larvae eat roots, stems, leaves
Hepialus spp.	Swift Moths	Hepialidae	Europe	Larvae in soil eat roots
Luperina testacea (Schiff.)	Flounced Rustic Moth	Noctuidae	Europe	Larvae eat grains
Mesapamea secalis (L.)	Common Rustic Moth	Noctuidae	Europe	Larvae bore stems
Haplodiplosis marginata (v.R.)	Saddle Gall Midge	Cecidomyiidae	Europe	Larvae gall stems
Mayetiola avenae (March.)	Oat Stem Midge	Cecidomyiidae	Europe	
Sitodiplosis mossellana (Gehin)	Orange Wheat Blossom Midge	Cecidomyiidae	Europe, Asia, N. America	Larvae in flowerhead

Tipula spp.	Leatherjackets	Tipulidae	Europe	Larvae in soil eat roots; may destroy seedlings
Nephrotoma spp.	Leatherjackets	Tipulidae	Europe, N. America	
Bibio marci (L.)	St. Mark's Fly	Bibionidae	Europe	Larvae in soil eat roots
Hydrellia griseola (Fall.)	Cereal Leaf Miner	Ephydridae	Cosmopolitan	Larvae mine leaves
Agromyza ambigua Fall.	Cereal Leaf Miner	Agromyzidae	Europe, N. America	
Geomyza spp. *Meromyza* spp. *Opomyza* spp.	Grass Flies	Opomyzidae	Europe	Larvae bore stems; make 'dead-hearts'
Delia platura (Meign.)	Bean Seed Fly	Anthomyiidae	Cosmopolitan	Larvae eat sown seeds
Limothrips cerealium Hal.	Grain Thrips	Thripidae	Europe	Adults & nymphs infest foliage; fleck leaves
Stenothrips graminum Uzel	Oat Thrips	Thripidae	Europe	
Thrips nigropilosus Uzel	Chrysanthemum Thrips	Thripidae	Europe, N. America	
Agriotes spp.	Wireworms	Elateridae	Europe	Larvae in soil eat roots; may kill seedlings
Helophorus nubilus F.	Wheat Shoot Beetle	Hydrophilidae	Europe	Larvae damage stems of seedlings
Crepidodera ferruginea (Scop.)	Wheat Flea Beetle	Chrysomelidae	Europe	Larvae feed in stem; make 'dead-heart'
Phyllotreta vittula Redt.	Barley Flea Beetle	Chrysomelidae	Europe	Adults eat leaves
Melolontha spp.	Cockchafers	Scarabaeidae	Europe, Asia	Larvae in soil eat roots; may destroy seedlings
Phyllophaga spp.	June Beetles	Scarabaeidae	Canada, USA	
Schizonycha spp.	Chafer Grubs	Scarabaeidae	Europe, Asia	
Tanymecus dilaticollis Gylh.	Southern Grey Weevil	Curculionidae	E. Europe	Adults eat leaves
Stenotarsonemus spirifex (March.)	Oat Spiral Mite	Tarsonemidae	Europe	Damage flowerhead

ONIONS (*Allium* spp. – Amaryllidaceae) (Onions, Chives, Garlic, Shallot, Leek)

Onions are a crop of great antiquity, unknown in the wild state, but probably originating in southern Asia or the Mediterranean Region. They are mainly temperate crops but are quite widely grown in subtropical areas. They prefer light sandy soils in cool moist regions. Onions are grown either as bulbs for drying (ware onions), smaller for pickling, or as salad onions. Leeks are more temperate and grown as a vegetable for cooking. Chives are grown for the leaves, used in cooking for flavouring and garnishing. Garlic, used in cooking for its strong flavour, is grown mostly in southern Europe and the highlands of N. Thailand, and widely exported. Several different species of onions and chives are grown in eastern Asia.

MAJOR PESTS

Thrips tabaci Lind.	Onion Thrips	Thripidae	Cosmopolitan	Infest & silver leaves
Delia antiqua (Meign.)	Onion Fly	Anthomyiidae	Cosmopolitan	Larvae invade bulb
Delia platura (Meign.)	Bean Seed Fly	Anthomyiidae	Cosmopolitan	Larvae destroy seed or invade bulb
Spodoptera exigua (Hbn.)	Lesser (Beet) Armyworm	Noctuidae	Mediterranean, Asia	Larvae eat leaves

MINOR PESTS

Euborellia annulipes Lucas	Groundnut Earwig	Forficulidae	India	Damage seedlings
Lipaphis erysimi (Kalt.)	Turnip Aphid	Aphididae	Cosmopolitan	Adults & nymphs infest foliage; suck sap
Myzus ascalonicus Don.	Shallot Aphid	Aphididae	Europe	
Myzus persicae (Sulz.)	Peach–Potato Aphid	Aphididae	Cosmopolitan	
Aeolothrips spp.	Thrips	Thripidae	India	Scarify (silver) foliage
Caliothrips indicus (Bag.)	Groundnut Thrips	Thripidae	India	
Thrips angusticeps (Uzel)	Cabbage Thrips	Thripidae	Europe	
Acrolepiopsis assectella (Zell.)	Leek Moth	Yponomeutidae	Europe, Asia	Larvae mine leaves
Cnephasia spp.	Leaf Tiers	Tortricidae	Europe	Larvae eat leaves
Agrotis spp.	Cutworms	Noctuidae	Cosmopolitan	Larvae are cutworms; destroy seedlings; eat leaves & stems
Euxoa spp.	Cutworms	Noctuidae	Cosmopolitan	
Noctua pronuba (L.)	Large Yellow Underwing	Noctuidae	Europe	
Heliothis spp.	Leafworms	Noctuidae	Cosmopolitan	Larvae eat leaves; may feed inside onion leaves
Mamestra brassicae (L.)	Cabbage Moth	Noctuidae	Europe, Asia	
Spodoptera litura F.	Fall Armyworm	Noctuidae	India, S.E. Asia, China	
Spodoptera littoralis (Boisd.)	Cotton Leafworm	Noctuidae	Africa, S. Europe	
Phytobia cepae (Her.)	Onion Leaf Miner	Agromyzidae	C. Europe, Asia	Larvae mine leaves
Eumerus spp.	Narcissus Flies	Syrphidae	Canada	Larvae invade bulb
Agriotes spp.	Wireworms	Elateridae	Europe, Asia	Larvae in soil eat roots
Aceria tulipae (K.)	Onion (Bulb) Eriophyid	Eriophyidae	Cosmopolitan	Infest bulbs & spread fungi causing rots
Rhyzoglyphus spp.	Bulb Mites	Acaridae	Cosmopolitan	

PARSLEY (*Petroselinum crispum* – Umbelliferae)

One of the most familiar garden herbs; very widely cultivated in Europe and N. America and most places with moist cool climates. It is a biennial herb, or short-lived perennial, producing a dense tuft of dark green finely divided leaves. The leaves are used for garnishing, and for flavouring soups, omelettes, and in stuffing; they have a high concentration of vitamin C. The plant is native to the rocky shores of the Mediterranean.

MAJOR PESTS

None of the usual pests of cultivated Umbelliferae appear to be serious on this crop.

MINOR PESTS

Cavariella aegopodii (Scopoli)	Willow–Carrot Aphid	Aphididae	Europe, Canada, USA	Adults & nymphs infest foliage; suck sap; virus vectors
Dysaphis petroselini (C.B.)	Hawthorn–Parsley Aphid	Aphididae	Europe	(as above)
Papilio polyxenes asterias Stoll.	Black Swallowtail Butterfly	Papilionidae	Canada, USA	Larvae eat foliage
Lasioptera carophila Loew	Gall Midge	Cecidomyiidae	Mediterranean	Larvae gall seedheads
Philophylla heraclei (L.)	Celery Fly	Tephritidae	Europe	Larvae make leaf blister mines

It is to be expected that the pests of other cultivated Umbelliferae (carrot, celery, etc.) will also attack parsley, but records to date are sparse.

PARSNIP (*Pastinaca sativa* – Umbelliferae)

Parsnip is a temperate vegetable cultivated for its root; first used by the Greeks and Romans, and cultivated throughout Europe since then. Wild forms are found in Europe, and in general the cultivated varieties do easily revert to the wild type with a woody dry root. It was taken to the West Indies in 1564, and to Virginia in 1609. The enlarged fleshy root has quite a high sugar content, and even some fat, and is eaten by man as a vegetable, and used for livestock food. It is a biennial herb, 30–150 cm high; it does not succeed in tropical lowlands, but is sometimes grown at higher altitudes in the tropics.

MAJOR PESTS

Psila rosae (F.)	Carrot Fly	Psilidae	Europe, USA, Canada	Larvae mine root

MINOR PESTS

Cavariella pastinacea (L.)	Willow–Parsnip Aphid	Aphididae	Europe, Canada	Infest foliage; virus vector
Philophylla heraclei (L.)	Celery Fly	Tephritidae	Europe	Larvae mine leaves with blister mines
Napomyza carotae Spencer	Carrot Root Miner	Agromyzidae	Europe (not UK)	Larvae mine root
Depressaria heracliana (L.)	Parsnip Moth	Oecophoridae	Europe, Canada	Larvae eat foliage & seedheads
Aethes dilucidana (Stephens)	Parsnip Flower Moth	Cochylidae	Europe	Larvae web flowerhead
Papilio polyxenes asterias Stoll.	Black Swallowtail Butterfly	Papilionidae	Canada	Larvae eat foliage
Agrotis spp.	Cutworms	Noctuidae	Cosmopolitan	Larvae in soil eat holes in root
Euxoa spp.	Cutworms	Noctuidae	Cosmopolitan	
Anagrapha falcifera (Kirby)	Celery Semilooper	Noctuidae	USA	Larvae eat foliage
Autographa gamma (L.)	Silver-Y Moth	Noctuidae	Europe	
Bothynus gibbosus (De Geer)	Carrot Beetle	Scarabaeidae	USA	Larvae in soil eat roots; adults eat foliage
Listronotus oregonensis (Le C.)	Carrot Weevil	Curculionidae	Canada, USA	Larvae bore root

PEA (*Pisum sativum* – Leguminosae) (Garden Pea; Field Pea)

The common pea is native to southern Europe and has been cultivated since prehistoric times, and was taken to America by the earliest colonists. Although indigenous to warm climates, it grows well under cool moist summer conditions, and thrives in N. Europe, N. America and Canada. The two main groups of varieties are field peas, grown for their dried seeds and the plants used for silage, forage and green manuring, and garden peas grown as a vegetable (the seeds are eaten green, either fresh or frozen, or used in canning, and in some varieties (sugar peas) the young pod is cooked whole). The plant body is used as livestock feed. Most peas are now grown commercially as field crops, with a bushy habit, and harvested mechanically. Peas are important field and garden crops in all temperate countries, and in some tropical countries in the cool season or at high altitudes.

MAJOR PESTS

Acyrthosiphon pisum (Harris)	Pea Aphid	Aphididae	Cosmopolitan	Infest foliage
Kakothrips robustus (Uzel)	Pea Thrips	Thripidae	Europe	Scarify young pods
Cydia nigricana (F.)	Pea Moth	Tortricidae	Europe, USA, Canada	Larvae inside pods eat seeds
Maruca testulalis (Geyer)	Mung Moth	Pyralidae	Pantropical	
Etiella zinckenella (Triet.)	Pea Pod Borer	Pyralidae	Cosmopolitan	
Contarinia pisi (Winn.)	Pea Midge	Cecidomyiidae	Europe	Larvae infest shoots (also pods)
Delia platura (Meig.)	Bean Seed Fly	Anthomyiidae	Cosmopolitan	Larvae destroy seeds in soil
Sitona spp.	Pea & Bean Weevils	Curculionidae	Europe, USA, Canada	Adults eat leaves; larvae eat roots

MINOR PESTS

Aphis craccivora Koch.	Groundnut Aphid	Aphididae	Cosmopolitan	Adults & nymphs infest foliage; suck sap
Bemisia tabaci (Genn.)	Tobacco Whitefly	Aleyrodidae	Asia	
Empoasca spp.	Green Leafhoppers	Cicadellidae	Cosmopolitan	
Creontiades spp.	Capsid Bugs	Miridae	India	Sap-suckers; toxic saliva
Thrips angusticeps Uzel	Cabbage Thrips	Thripidae	Europe	Infest flowers & shoots
Cnephasia spp.	Leaf Rollers	Tortricidae	Europe, USA, Canada	Larvae roll leaves
Colias spp.	Clouded Yellow Butterflies	Pieridae	Canada, USA	Larvae eat leaves
Lampides boeticus (L.)	Pea Blue Butterfly	Lycaenidae	Asia	Larvae bore pods to eat seeds
Euchrysops cnejus (F.)	Blue Butterfly	Lycaenidae	India	
Loxostege sticticalis (L.)	Beet Webworm	Pyralidae	Cosmopolitan	Larvae eat foliage
Leucinodes orbonalis Gn.	Eggplant Boring Caterpillar	Pyralidae	India	Larvae bore shoots
Heliothis spp.	Cotton Bollworms	Noctuidae	Cosmopolitan	Larvae eat foliage & bore pods

Mythimna spp.	Cereal Armyworms	Noctuidae	Cosmopolitan	Larvae eat leaves
Spodoptera spp.	Leafworms	Noctuidae	Cosmopolitan	
Plusia spp.	Semiloopers	Noctuidae	Cosmopolitan	
Diacrisia obliqua Wlk.	Tiger Moth	Arctiidae	India	Larvae defoliate
Phytomyza horticola Gour.	Pea Leaf Miner	Agromyzidae	Old World	Larvae mine leaves
Liriomyza huidobrensis (Blan.)	Pea Leaf Miner	Agromyzidae	USA, S. America	Larvae mine leaves & scarify pod
Ophiomyia phaseoli (Tryon)	Bean Fly	Agromyzidae	India	Larvae mine leaf petioles & stems
Tipula spp.	Leatherjackets	Tipulidae	Europe	Larvae in soil eat roots
Anomala spp.	White Grubs	Scarabaeidae	Asia	Larvae in soil eat roots
Epicauta spp.	Black Blister Beetles	Meloidae	Asia, USA, Canada	Adults eat flowers
Bruchus pisorum (L.)	Pea Bruchid	Bruchidae	Europe, Asia, Canada	Larvae bore seeds in young pods
Callosobruchus chinensis L.	Oriental Cowpea Bruchid	Bruchidae	S. Asia	Attack ripe pods
Chaetocnema concinnipennis Baly	Flea Beetle	Chrysomelidae	S. Asia	Adults hole leaves
Cerotoma trifurcata (For.)	Bean Leaf Beetle	Chrysomelidae	USA	Adults eat leaves; larvae eat roots
Alcidodes spp.	Striped Weevils	Curculionidae	S. Asia	Larvae bore stems; adults damage stems
Tanymecus indicus Fst.	Surface Weevil	Curculionidae	S. Asia	Adults cut seedling stems
Apion spp.	Seed Weevils	Apionidae	Cosmopolitan	Adults infest flowers & shoots
Hylastinus obscurus (Mar.)	Clover Root Borer	Scolytidae	USA	Larvae bore roots
Tetranychus spp.	Red Spider Mites	Tetranychidae	Cosmopolitan	Scarify & web foliage

PEACH (*Prunus persicae* – Rosaceae)

This is a tree native to China, now grown commercially in temperate and subtropical areas. The more important centres of cultivation are S. USA, southern Europe, S. Africa, Australia and Japan. The tree is small, rather short-lived, and susceptible to frost injury and low temperatures. The fruit has a soft velvety skin, and a pitted, compressed stone, and is used mainly as a table fruit. Because of their delicate perishable nature the fruits are difficult to transport and store, but they are the most popular fruit for canning, and large quantities are also dried. Nectarine is var. *nectarina* and clearly closely allied to peach, but with a smaller, smooth fruit.

MAJOR PESTS

Myzus persicae (Sulz.)	Peach–Potato Aphid	Aphididae	Cosmopolitan	Infest foliage; virus vector
Quadraspidiotus perniciosus (Comst.)	San José Scale	Diaspididae	Pantropical	Encrust foliage
Cydia molesta (Busck.)	Oriental Fruit Moth	Tortricidae	Cosmopolitan	Larvae bore shoots & fruits
Cydia pomonella (L.)	Codling Moth	Tortricidae	Cosmopolitan	Larvae bore fruits
Ceratitis capitata (Wied.)	Medfly	Tephritidae	Cosmopolitan	Larvae inside fruits
Ceratitis rosa Karsch	Natal Fruit Fly	Tephritidae	S. & E. Africa	
Panonychus citri (McG.)	Citrus Red Spider Mite	Tetranychidae	Cosmopolitan	Scarify & web foliage
Tetrannychus spp.	Red Spider Mites	Tetranychidae	Cosmopolitan	

MINOR PESTS

Empoasca spp.	Green Leafhoppers	Cicadellidae	Cosmopolitan	Infest foliage; suck sap
Typhlocyba spp.	Fruit Tree Leafhoppers	Cicadellidae	Cosmopolitan	
Aphis spiraecola Patch	Spiraea Aphid	Aphididae	Cosmopolitan	Adults & nymphs infest foliage; suck sap; cause leaf-curling
Appelia schwartzi Borner	Peach Aphid	Aphididae	Europe	
Brachycaudus persicae (Pass.)	Black Peach Aphid	Aphididae	Europe	
Hyalopterus arundinis F.	Large-tailed Peach Aphid	Aphididae	China	
Hyalopterus pruni Geoff.	Mealy Plum Aphid	Aphididae	Europe, Asia	
Hysteroneura setariae (Thomas)	Rusty Plum Aphid	Aphididae	Pantropical	
Myzus momonis Mats.	Peach Gall Aphid	Aphididae	China, Japan	
Pterochloroides persicae (Chol.)	Peach Aphid	Aphididae	Mediterranean, C. Asia	
Parthenolecanium corni (Bche.)	Plum Scale	Coccidae	Europe, W. Asia	Infest foliage; mostly on twigs
Parthenolecanium persicae (F.)	Peach Scale	Coccidae	Cosmopolitan	
Pseudococcus maritimus (Ehrh.)	Grape Mealybug	Pseudococcidae	Cosmopolitan	
Icerya purchasi Mask.	Cottony Cushion Scale	Margarodidae	Cosmopolitan	
Drosicha spp.	Giant Mealybugs	Margarodidae	Asia	

Aspidiotus ostraeiformis Curt.	Oystershell Scale	Diaspididae	Europe	Encrust twigs
Pseudaulacaspis pentagona (Targ.)	White Peach Scale	Diaspididae	Europe	
Lygocoris pabulinus (L.)	Common Green Capsid	Miridae	Europe	Sap-suckers; toxic saliva; feeding causes necrotic spots
Lygus oblineatus (Say)	Tarnished Plant Bug	Miridae	Canada, USA	
Lygus spp.	Plant Bugs	Miridae	Cosmopolitan	
Anarsia lineatella Zell.	Peach Twig Borer	Gelechiidae	Europe, Asia, N. America	Larvae bore twigs
Sanninoidea exitiosa (Say)	Peach Tree Borer	Sesiidae	Canada, USA	Larvae bore under bark & may girdle stem
Synanthedon pictipes (G. & R.)	Lesser Peach Tree Borer	Sesiidae	Canada, USA	Larvae bore branches
Cydia prunivora (Walsh)	Lesser Appleworm	Tortricidae	Canada, USA	Larvae bore fruits
Ernarmonia formosana (Scop.)	Cherry Bark Tortrix	Tortricidae	Europe	Larvae bore bark
Dichocrosis punctiferalis (Guen.)	Peach Moth	Pyralidae	Asia	Larvae bore fruits
Paramyelois transitella (Wlk.)	Navel Orangeworm	Pyralidae	USA	
Erannis defoliaria Clerck	Mottled Umber Moth	Geometridae	Europe	Larvae eat leaves; may defoliate
Operophtera brumata L.	Winter Moth	Geometridae	Europe	
Phalera bucephala L.	Buff-tip Moth	Notodontidae	Europe	Larvae eat leaves, may defoliate
Malacosoma spp.	Tent Caterpillars	Lasiocampidae	Europe, Asia	
Dacus spp.	Fruit Flies	Tephritidae	Asia	Larvae inside fruits
Ceratitis cosyra (Wlk.)	Mango Fruit Fly	Tephritidae	Africa	
Anatrepha fraterculus (Wied.)	–	Tephritidae	C. & S. America	
Pardalaspis quinaria Bez.	Rhodesian Fruit Fly	Tephritidae	Africa	
Rhagoletis completa Cress.	Walnut Husk Fly	Tephritidae	USA	
Hoplocampa spp.	Sawflies	Tenthredinidae	Europe, Asia	Larvae defoliate
Conotrachelus nenuphar (Hbst.)	Plum Curculio	Curculionidae	USA	Larvae bore fruits
Phloeotribus liminaris (Harris)	Peach Bark Beetle	Scolytidae	USA	Adults bore under bark
Panonychus ulmi (Koch)	Fruit Tree Red Spider Mite	Tetranychidae	Europe, Asia, N. America	Scarify foliage
Aculus cornutus (Banks)	Peach Silver Mite	Eriophyidae	USA	Scarify (silver) foliage

PEAR (*Pyrus communis* – Rosaceae)

The common pear is a native of Eurasia, and is a tall upright tree, with white flowers that appear with the leaves. The fruit is of a characteristic pyriform shape, sweet and juicy, but the flesh contains a number of stone cells. They prefer heavy soils with plenty of humus and good drainage and an equable climate. Over 5000 kinds are recognised although most varieties fall into a few well-defined categories (William, Conference, etc.). The main production areas are France, USA, Argentina, S. Africa, Australia and New Zealand. The fruit is typically picked before fully ripe. The European or Common Pear is the source of most dessert pears. The Chinese or Sand Pear is *Pyrus pyrifolia* var. *culta*, and is a native of China; this is the source of many varieties of cooking pear.

MAJOR PESTS

Dysaphis pyri (Fonsc.)	Pear–Bedstraw Aphid	Aphididae	Europe	Infest foliage; suck sap
Psylla pyricola (Forst.)	Pear Sucker	Psyllidae	Europe, China, USA	
Lygocoris pabulinus (L.)	Common Green Capsid	Miridae	Europe	Sap-sucker; toxic saliva
Operophtera brumata (L.)	Winter Moth	Geometridae	Europe	Larvae defoliate
Caliroa cerasi (L.)	Pear Slug Sawfly	Tenthredinidae	Cosmopolitan	Larvae skeletonize leaves
Bryobia spp.	Brown (Bryobia) Mites	Tetranychidae	Cosmopolitan	Scarify (speckle) & web foliage
Panonychus ulmi (Koch)	Fruit Tree Red Spider Mite	Tetranychidae	Cosmopolitan	
Tetranychus spp.	Red Spider Mites	Tetranychidae	Cosmopolitan	

MINOR PESTS

Aleurocanthus spiniferus Quaint.	Orange Spiny Whitefly	Aleyrodidae	Africa, Asia	Infest foliage; suck sap
Psylla pyri (L.)	Pear Psylla	Psyllidae	Europe	
Anuraphis farfarae (Koch)	Pear–Coltsfoot Aphid	Aphididae	Europe	Infest foliage; sap-suckers; cause leaf-curl
Aphis pomi Deg.	Green Apple Aphid	Aphididae	Cosmopolitan	
Rhopalosiphum insertum (Wlk.)	Apple–Grass Aphid	Aphididae	Europe	
Longiunguis pyrarium (Pass.)	Pear Aphid	Aphididae	Europe	
Toxoptera pyricola Mats.	Pear Aphid	Aphididae	China, Japan	
Aphanostigma piri Chol.	Pear Phylloxera	Phylloxeridae	Israel	Infest leaves
Eriosoma lanigerum (Hsm.)	Woolly Apple Aphid	Pemphigidae	Cosmopolitan	Infest twigs & branches
Ceroplastes sinensis Del G.	Chinese Waxy Scale	Coccidae	Cosmopolitan	Infest foliage
Lepidosaphes ulmi (L.)	Mussel Scale	Diaspididae	Cosmopolitan	Encrust twigs & fruits; suck sap
Parlatoria oleae (Colv.)	Olive Scale	Diaspididae	Cosmopolitan	
Quadraspidiotus perniciosus (Cmst.)	San José Scale	Diaspididae	Cosmopolitan	
Quadraspidiotus ostraeiformis (Curt.)	Oystershell Scale	Diaspididae	Europe	

Neolygus communis Knight	Pear Plant Bug	Miridae	USA	Sap-suckers; toxic saliva
Stephanitis ambigua Hor.	Pear Lace Bug	Tingidae	China	
Cryptotympana atrata (F.)	Black Cicada	Cicadidae	China	Sucks sap from twigs
Acleris spp.	Leaf Rollers	Tortricidae	Europe	Larvae roll leaves
Archips podana (Scop.)	Fruit Tree Tortrix	Tortricidae	Europe	Larvae bore fruit buds & web shoots
Archips spp.	Fruit Tree Leaf Rollers	Tortricidae	Cosmopolitan	
Cacaecimorpha pronubana (Hb.)	Carnation Leaf Roller	Tortricidae	Mediterranean, Europe	Larvae roll & eat leaves
Cydia molesta Busck	Oriental Fruit Moth	Tortricidae	China, Japan	Larvae bore shoots & fruits
Cydia pomonella (L.)	Codling Moth	Tortricidae	Cosmopolitan	Larvae bore fruits
Cydia pyrivora (Dan.)	Pear Tortricid	Tortricidae	E. Europe, W. Asia	
Spilonota ocellana (D. & S.)	Eye-spotted Bud Moth	Tortricidae	Europe, Asia, N. America	Larvae feed in buds
Nephosteryx rubizonella Rag.	Pear Fruit Borer	Phycitidae	USA	Larvae bore fruits
Paramyelois transitella (Wlk.)	Navel Orangeworm	Pyralidae	USA	
Arctias selene Hbn.	Moon Moth	Saturniidae	China	Larvae eat leaves
Hoplocampa brevis (Klug)	Pear Sawfly	Tenthredinidae	Europe, Asia	
Vespa/Paravespula spp.	Common Wasps	Vespidae	Cosmopolitan	Adults pierce ripe fruits
Contarinia pyrivora (Riley)	Pear Midge	Cecidomyiidae	Europe, USA	Larvae invade & destroy fruitlets
Dasyneura pyri (Bche.)	Pear Leaf Midge	Cecidomyiidae	Europe	Larvae make leaf galls
Thomasiniana oculiperda (Rubs.)	Red Bud Borer	Cecidomyiidae	Europe	Larvae bore buds
Rhagoletis pomonella (Walsh)	Apple Fruit Fly	Tephritidae	Canada, USA	Larvae bore fruits
Agrilus sinuatus (Oliv.)	Sinuate Pear Tree Borer	Buprestidae	USA	Larvae bore trunk under bark
Lampra spp.	Golden Jewel Beetles	Buprestidae	China	
Anthonomus pomorum (L.)	Apple Blossom Weevil	Curculionidae	Europe, Asia	Larvae destroy flowers
Otiorhynchus singularis (L.)	Clay-coloured Weevil	Curculionidae	Europe	Larvae eat roots; adults eat leaves
Conotrachelus nenuphar (Hbst.)	Plum Weevil	Curculionidae	Canada, USA	Larvae bore in fruit
Rhynchites caeruleus (Deg.)	Apple Twig Cutter	Curculionidae	Europe	Adults cut through young shoots
Tachypterellus quadrigibbus (Say)	Apple Curculio	Curculionidae	USA	Larvae inside fruit
Aculus schlechtendali (Nal.)	Leaf & Bud Mite	Eriophyidae	Europe, USA	Erinea on leaves
Epitrimerus pyri (Nal.)	Pear Rust Mite	Eriophyidae	USA	Scarify foliage
Phytoptus pyri (Pgst.)	Pear Leaf Blister Mite	Eriophyidae	Cosmopolitan	Erinea on leaves; fruit damaged

These are the more common pests found attacking pear; many of the pests listed under 'Apple' will also be found on pear.

PECAN (*Carya illinoensis* – Juglandaceae) (=Hickory)

A native of southwestern USA and Mexico. Originally the nuts were harvested from wild trees, but demand is now so great that the trees are extensively cultivated in the southern States (especially Texas and Oklahoma), and with development of new varieties the area of cultivation is spreading farther northwards. The trees start to bear nuts within four years of setting out, and paper-shelled varieties are now established. About half the crop in the USA is now marketed in the shell. Pecans have a higher fat content than any other plant product: over 70%. The nuts are used for dessert, and in ice cream, candy, cakes etc.

MAJOR PESTS

Acrobasis nuxvorella Neun.	Pecan Nut Casebearer	Coleophoridae	USA, Mexico	Larvae in tiny cases mine leaves
Cydia caryana Fitch	Hickory Shuckworm	Tortricidae	USA, Mexico	Larvae eat kernel inside nut
Curculio caryae (Horn)	Pecan Weevil	Curculionidae	USA, Mexico	

MINOR PESTS

Clastoptera achatina Germ.	Pecan Spittlebug	Cercopidae	USA, Mexico	Nymphs in spittlemass on shoots
Empoasca decedens (Paoli)	Green Leafhopper	Cicadellidae	Israel	Sap sucker
Melanocallis caryaefoliae (Davis)	Black Pecan Aphid	Aphididae	USA	Nymphs & adults on foliage; suck sap
Monellia caryella (Fitch)	Yellow Aphid	Aphididae	USA	
Phylloxera notabilis Perg.	Pecan Leaf Phylloxera	Phylloxeridae	USA, Mexico	Make leaf-galls on veins & petioles
Icerya purchasi Mask.	Cottony Cushion Scale	Margarodidae	USA	Infest foliage
Lecanium caryae Fitch	Hickory Lecanium	Coccidae	USA	Infest foliage; suck sap; with sooty moulds usually
Parthenolecanium corni Bche.	Plum Scale	Coccidae	USA	
Chrysomphalus aonidium (L.)	Florida Red Scale	Diaspididae	Israel	Infest leaves & twigs
Lygocoris caryae (Knight)	Hickory Plant Bug	Miridae	USA	Sap-suckers; toxic saliva; feeding causes necrosis & death of tissues
Leptoglossus spp.	Leaf-footed Plant Bugs	Coreidae	USA	
Several species	Shield Bugs	Pentatomidae	USA	
Synanthedon scitula Harris	Pecan Borer	Sesiidae	USA	Larvae bore branches & trunk
Cossula magnifica (Streck.)	Pecan Carpenterworm	Cossidae	USA	
Stigmella juglandifoliella (Clemens)	Pecan Serpentine Leaf Miner	Nepticulidae	USA	Larvae mine leaves
Acrobasis caryae Grote	Pecan Nut Casebearer	Coleophoridae	USA	
Acrobasis juglandis (Le Baron)	Pecan Leaf Casebearer	Coleophoridae	USA	Larvae in tiny cases mine leaves
Coleophora caryaefoliella Clemens	Pecan Cigar Casebearer	Coleophoridae	USA	

Argyrotaenia juglandana (Fern.)	Hickory Leaf Roller	Tortricidae	USA	Larvae roll leaves
Gretchena bolliana (Sling.)	Pecan Bud Borer	Tortricidae	USA	Larvae bore buds
Hyphantria cunea (Drury)	Fall Webworm	Arctiidae	USA, Mexico	
Lophocampa caryae (Harris)	Hickory Tussock Moth	Arctiidae	USA, Mexico	Larvae eat leaves; may defoliate
Datana integerrima (G. & R.)	Walnut Caterpillar	Notodontidae	USA, Mexico	
Alsophila pometaria (Harris)	Fall Cankerworm	Geometridae	USA	
Paramyelois transitella (Wlk.)	Navel Orangeworm	Pyralidae	USA	Larvae bore young fruits
Euzophora semifuneralis (Wlk.)	American Plum Borer	Phycitidae	USA	Larvae bore inner bark
Tremex columba L.	Horntail	Siricidae	USA	Larvae bore branches & trunk
Eriocampa juglandis (Fitch)	Butternut Woolly Sawfly	Tenthredinidae	USA	Larvae eat leaves; may defoliate
Periclista marginicollis	Pecan Sawfly	Tenthredinidae	USA	
Apate monachus F.	Black Borer	Bostrychidae	Israel	Adults bore branches
Elaphidionoides villosus (F.)	Oak Twig Pruner	Cerambycidae	USA	Larvae bore twigs
Neoclytus acuminatus (F.)	Red-headed Ash Borer	Cerambycidae	USA	Larvae bore branches
Oberea ulmicola Chitt.	Hickory Borer	Cerambycidae	USA	
Oncideres cingulata (Say)	Twig Girdler	Cerambycidae	USA	Larvae bore & girdle twigs
Conotrachelus spp.	Nut Curculios	Curculionidae	USA	Larvae bore & eat nut kernel
Cyrtepistomus castaneus (Roe.)	Asiatic Oak Weevil	Curculionidae	USA	Adults eat leaves
Scolytus quadrispinosus Say	Hickory Bark Beetle	Scolytidae	USA	Adults bore under bark
Eotetranychus caryae Reeves	Pecan Spider Mite	Tetranychidae	USA	Adults & nymphs scarify foliage
Eotetranychus hickoriae (McG.)	Hickory Scorch Mite	Tetranychidae	USA	
Aceria caryae Perg.	Pecan Leafroll Mite	Eriophyidae	USA	Distort young leaves

PERSIMMON (*Diospyros virginiana, Diospyros kaki* – Ebenaceae)

This plant is native to eastern USA; a temperate species grown for the fruit. The fruits when fully ripe are of delicious flavour and high quality, but are difficult to transport; unripe they are acid and not very palatable. The tree is large, up to 12 m in height, and the fruits up to 5 cm in diameter, orange-red in colour, and technically they are berries with an enlarged calyx at the base; they are eaten fresh or dried.

The Japanese Persimmon or Kaki (*Diospyros kaki*) is native to China, widely cultivated in Japan, and now introduced into France and other Mediterranean countries, as well as Florida, California and Texas in the USA, where it is popular because of the larger fruit, and its firmer texture makes it easier to transport.

MAJOR PESTS

At present, it is not apparent which, if any, of the pests recorded are particularly serious

MINOR PESTS

Geisha distinctissima Wlk.	Green Flattid Planthopper	Flattidae	Japan	Infest foliage
Trioza diospyri (Ash.)	Persimmon Psylla	Psyllidae	USA	Deform leaves
Erythroneura mori Mats.	Blood-spot Leafhopper	Cicadellidae	China	Infest foliage
Acanthococcus kaki Kuw.	Woolly Persimmon Scale	Coccidae	China	
Coccus mangiferae Green	Mango Soft Scale	Coccidae	Israel	Adults & nymphs infest foliage; suck sap; often with sooty moulds
Ceroplastes floridiensis Com.	Florida Wax Scale	Coccidae	Israel	
Phenacoccus pergandei Cock.	Persimmon Mealybug	Pseudococcidae	China	
Planococcus citri (Risso)	Citrus Mealybug	Pseudococcidae	China, Israel	
Aleurocanthus spiniferus Quaint.	Orange Spiny Whitefly	Aleyrodidae	India, S.E. Asia, China, Japan	
Drosicha corpulenta (Kuw.)	Giant Mealybug	Margarodidae	China	
Hemiberlesia lataniae (Sign.)	Latania Scale	Diaspididae	Israel	Infest twigs
Leptoglossus spp.	Leaf-footed Plant Bugs	Coreidae	USA	
Oxycarenus hyalinipennis (Costa)	Cotton Seed Bug	Lygaeidae	Israel	Sap-suckers; toxic saliva
Nezara antennata Scott	Green Stink Bug	Pentatomidae	Japan	
Heliothrips haemorrhoidalis (Bche)	Black Tea Thrips	Thripidae	Israel	Scarify foliage
Kakivoria flavofasciatus Nag.	Persimmon Budworm	Tortricidae	China	Larvae bore buds
Culcula panterinaria B.et G.	Looper Caterpillar	Geometridae	China	
Percnia giraffata Gn.	Spotted Persimmon Looper	Geometridae	China, Japan	
Latoia consocia Wlk.	Green Stinging Caterpillar	Limacodidae	Japan	Larvae eat leaves; may defoliate
Lymantria dispar L.	Gypsy Moth	Lymantriidae	China, Japan	
Hyphantria cunea Drury	Fall Webworm	Arctiidae	Japan	
Ceratitis capitata (Wied.)	Medfly	Tephritidae	Asia	Larvae inside fruits

Anomala spp.	Chafer Beetles	Scarabaeidae	Japan	Adults eat leaves
Oncideres cingulata (Say)	Twig Girdler	Cerambycidae	USA	Larvae bore & girdle twigs
Apate monachus F.	Black Borer	Bostrychidae	Near East	Adults bore branches
Xyleborus rubricollis Eich.	Black Twig Borer	Scolytidae	Japan	Adults bore twigs
Tetranychus spp.	Red Spider Mites	Tetranychidae	Asia	Scarify foliage

In Japan some 120 species of insect and mite pests of Persimmon are recorded.

PLUM (*Prunus domestica* – Rosaceae) (= European Plum)

The plums of commerce come from three main sources, the European Plum, native American species (*Prunus americana*, etc.), and the Japanese species (*P. salicina*). However, the bulk of fruit comes from the European Plum which is now very widely cultivated throughout the temperate parts of the world. It has been cultivated for over 2000 years, and was taken to America by the colonists. The tree is large, 10–16 m in height, with variously coloured fruits; over 900 varieties are now cultivated, although certain varieties are only used for cooking or for drying as prunes. The fruit is soft and fleshy, with a smooth, flattened stone. The variety *P. domestica damascena* is the Damson which has an almost identical pest spectrum to the plum.

MAJOR PESTS

Brachycaudus helichrysi (Kalt.)	Leaf-curling Plum Aphid	Aphididae	Europe, Australia	Infest foliage; suck sap; cause leaf-curling; some shoots killed
Hyalopterus pruni (Geoff.)	Mealy Plum Aphid	Aphididae	Europe	
Phorodon humuli (Schr.)	Damson–Hop Aphid	Aphididae	Europe, USA, Canada	
Cydia funebrana (Treits.)	Red Plum Maggot	Tortricidae	Europe, Asia	Larvae bore fruits
Operophtera brumata (L.)	Winter Moth	Geometridae	Europe	Larvae defoliate
Caliroa cerasi (L.)	Pear Slug Sawfly	Tenthredinidae	Cosmopolitan	Larvae skeletonize leaves
Hoplocampa flava (L.)	Plum Sawfly	Tenthredinidae	Europe	Larvae defoliate
Panonychus ulmi (Koch)	Fruit Tree Red Spider Mite	Tetranychidae	Cosmopolitan	Scarify & web foliage

MINOR PESTS

Eupteryx stellulata Burm.	Cherry Leafhopper	Cicadellidae	Europe	Infest foliage; suck sap
Macropsis trimaculata (Fitch)	Plum Leafhopper	Cicadellidae	USA	
Typhlocyba spp.	Fruit Tree Leafhoppers	Cicadellidae	Cosmopolitan	
Hysteroneura setariae (Th.)	Rusty Plum Aphid	Aphididae	Pantropical & USA	Infest foliage
Parthenolecanium corni (Bche.)	Plum Scale	Coccidae	Europe, W. Asia	Infest foliage; mostly twigs; suck sap
Lepidosaphes ulmi (L.)	Mussel Scale	Diaspididae	Cosmopolitan	
Quadraspidiotus perniciosus (Comst.)	San José Scale	Diaspididae	Cosmopolitan	
Quadraspidiotus ostraeiformis (Curt.)	Oystershell Scale	Diaspididae	Europe	
Lygocoris pabulinus (L.)	Common Green Capsid	Miridae	Europe	Sap-sucker; toxic saliva
Taeniothrips inconsequens (Uzel)	Pear Thrips	Thripidae	Europe	Infest foliage
Acleris spp.	Leaf Rollers	Tortricidae	Europe	Larvae roll leaves
Cydia prunivora (Walsh)	Lesser Appleworm	Tortricidae	Canada, USA	Larvae bore fruits
Hedya pruniana (Hbn.)	Plum Tortrix Moth	Tortricidae	Europe	Larvae eat leaves & bore shoots
Spilonota ocellana (D. & S.)	Eye-spotted Bud Moth	Tortricidae	Europe, Asia, N. America	Larvae bore buds
Euzophera semifuneralis (Wlk.)	American Plum Borer	Phycitidae	USA	Larvae tunnel bark

Paramyelois transitella (Wlk.)	Navel Orangeworm	Pyralidae	USA	Larvae bore fruits
Anthera polyphemus (Cramer)	Emperor Moth	Saturniidae	USA	Larvae eat leaves, may defoliate
Euproctis spp.	Brown-tail Moths	Lymantriidae	Cosmopolitan	
Hoplocampa minuta (Christ)	Plum Sawfly	Tenthredinidae	Europe, W. Asia	Larvae eat leaves; may defoliate
Neurotoma inconspicua (Nort.)	Plum Web-spinning Sawfly	Tenthredinidae	USA	
Paravespula/Vespa spp.	Common Wasps	Vespidae	Cosmopolitan	Adults pierce ripe fruits
Anthonomus scutellaris Le Conte	Plum Gouger	Curculionidae	USA	Larvae develop inside fruits
Conotrachelus nenuphar Hbst.	Plum Weevil	Curculionidae	USA	
Otiorhynchus cribricollis Gylh.	Apple Weevil	Curculionidae	Mediterranean, Australia, USA	Adults eat leaves; larvae in soil eat roots
Scolytus mali Bht.	Large Fruit Bark Beetle	Scolytidae	Europe	Adults bore under bark to make breeding galleries
Scolytus rugulosus (Muller)	Fruit Bark Beetle	Scolytidae	Eurasia, N. & S. America	
Phytoptus spp.	Plum Leaf Bead Gall Mites	Eriophyidae	Europe	Make erinia on leaf edges
Vasates fockeui (N. & T.)	Plum Nursery Mite	Eriophyidae	Canada, USA	Distort leaves

POMEGRANATE (*Punica granatum* – Punicaceae)

A native of Iran, this plant has been cultivated in the Mediterranean Region for centuries, and was early taken to India, China and S.E. Asia. It is now grown in most parts of the subtropics where the climate is not too humid; the best quality fruits are produced in areas with a hot dry summer and cool winter. The plant is a bush or small tree, 2–4 m in height, deciduous in cooler regions; the fruit is a round brown berry, 5–12 cm diameter, containing many seeds embedded in a pink juicy pulp. The acid pulp is the edible part of the fruit, and may be eaten as a dessert fruit, as a salad, or in beverages. The roots, seeds and fruit rinds are used medicinally. Propagation is usually by cuttings. The main areas of commercial production are California, Arizona, New Mexico and parts of the Mediterranean.

MAJOR PESTS

Indarbela spp.	Bark Caterpillars	Metarbelidae	India	Larvae eat bark
Virachola isocrates (F.)	Pomegranate Butterfly	Lycaenidae	India, Israel	Larvae bore fruits

MINOR PESTS

Aleurocanthus woglumi Ashby	Citrus Blackfly	Aleyrodidae	Pantropical	Adults & nymphs infest foliage; suck sap; often with sooty moulds
Siphoninus phillyreae (Hal.)	Pomegranate Whitefly	Aleyrodidae	India, Israel	
Aphis punicae Pass.	Pomegranate Aphid	Aphididae	India, Israel	
Icerya purchasi Mask.	Cottony Cushion Scale	Margarodidae	USA	Adults & nymphs infest foliage; suck sap; often with sooty moulds
Drosicha mangiferae (Green)	Mango Giant Mealybug	Margarodidae	India	
Ferrisia virgata (Ckll.)	Striped Mealybug	Pseudococcidae	India	
Planococcus lilacinus (Ckll.)	Cocoa Mealybug	Pseudococcidae	India, Israel	
Saissetia oleae (Bern.)	Black Scale	Coccidae	USA, Israel	
Aspidiotus hederae Vall.	Oleander Scale	Diaspididae	USA	Adults & nymphs infest twigs mostly; suck sap
Aspidiotus rossi (Mask.)	Pomegranate Scale	Diaspididae	India	
Hemiberlesia rapax (Com.)	Greedy Scale	Diaspididae	USA	
Parlatoria oleae (Bern.)	Olive Scale	Diaspididae	India	
Retithrips syriacus (Mayet)	Thrips	Thripidae	India	Infest foliage
Zeuzera coffeae Nietn.	Red Coffee Borer	Cossidae	India	Larvae bore stems
Clania crameri Westw.	Bagworm	Psychidae	India	Larvae eat leaves
Deudorix epijarbas (Moore)	Pomegranate Borer	Pyralidae	India	Larvae bore fruits
Dichocrosis punctiferalis (Guen.)	Castor Capsule Borer	Pyralidae	India	
Trabala vishnou (Lef.)	Lappet Moth	Lasiocampidae	India	Larvae eat leaves; sometimes defoliate
Latoia lepida (Cram.)	Blue-striped Nettlegrub	Limacodidae	India	
Euproctis spp.	Browntail Moths	Lymantriidae	India	
Othreis fullonica (L.)	Fruit-piercing Moth	Noctuidae	India	Adults pierce fruit
Dacus spp.	Fruit Flies	Tephritidae	India	Larvae inside fruit
Anomala spp.	Flower (Chafer) Beetles	Scarabaeidae	India	Adults eat leaves
Olenecamptus bilobus F.	Stem Borer	Cerambycidae	India	Larvae bore stems
Myllocerus spp.	Grey (Leaf) Weevils	Curculionidae	India	Adults eat leaves

POTATO (*Solanum tuberosum* – Solanaceae) (=Irish Potato)

Wild species are found from S. USA to S. Chile, with the centre of diversity in the Andes between 10 °N and 20 °S at altitudes above 2000 m. They spread slowly, to the Philippines and India in the seventeenth century; later to Europe, Japan, Java and E. Africa. The crop is essentially temperate, but can be grown in the tropics at higher altitudes, using short-day cultivars. The usual cultivars are long-day forms. Optimum temperatures for tuber development are about 15 °C (not above 25 °C), and the crop can be grown successfully under irrigation. The plant is an annual, herbaceous, and with a branched habit, 0.3–1.0 m in height. The swollen stem tubers (potatoes) contain about 2% protein and 17% starch; the crop is grown more universally than any other crop.

MAJOR PESTS

Empoasca fabae (Harris)	Potato Leafhopper	Cicadellidae	USA	Infest foliage; suck sap; virus vectors
Aulacorthum solani (Kalt.)	Potato (Glasshouse) Aphid	Aphididae	Cosmopolitan	
Myzus persicae (Sulz.)	Peach–Potato Aphid	Aphididae	Cosmopolitan	
Phthorimaea operculella (Zell.)	Potato Tuber Moth	Gelechiidae	Tropics & subtropics	Larvae bore stems & tubers
Epilachna spp.	Epilachna Beetles	Coccinellidae	Africa, India, China	Adults & larvae eat foliage
Epicauta spp.	Black Blister Beetles	Meloidae	Canada, USA, Asia, Africa	Adults eat flowers & leaves
Leptinotarsa decemlineata (Say)	Colorado Beetle	Chrysomelidae	N. & C. America, Europe (not UK)	Adults & larvae defoliate
Agriotes spp.	Wireworms	Elateridae	Europe, Asia, USA	Larvae in soil bore tubers

MINOR PESTS

Amrasca devastans (Dist.)	Cotton Jassid	Cicadellidae	S. Asia	Infest foliage; suck sap; virus vectors
Eupteryx aurata (L.)	Potato Leafhopper	Cicadellidae	Europe	
Empoasca spp.	Green Leafhoppers	Cicadellidae	Cosmopolitan	
Typhlocyba jucunda H.-S.	Potato Leafhopper	Cicadellidae	Europe	
Bemisia spp.	Whiteflies	Aleyrodidae	Pantropical	
Paratrioza cockerelli (Sulc.)	Potato Psyllid	Psyllidae	Canada, USA	Cause leaf-curl (psyllid yellows)
Aphis nasturtii (Borner)	Buckthorn–Potato Aphid	Aphididae	Europe, India	Infest foliage; suck sap; virus vectors
Macrosiphum euphorbiae (Thos.)	Potato Aphid	Aphididae	Cosmopolitan	
Rhopalosiphoninus latysiphon (Davids.)	Bulb & Potato Aphid	Aphididae	Europe, India	
Ferrisia virgata (Ckll.)	Striped Mealybug	Pseudococcidae	Pantropical	Infest foliage
Planococcus citri (Risso)	Root Mealybug	Pseudococcidae	Pantropical	Infest roots or foliage
Pseudococcus maritimus (Ehrh.)	Grape Mealybug	Pseudococcidae	Pantropical	

Calocoris norvegicus (Gmel.)	Potato Capsid	Miridae	Europe, Australia	Sap-suckers; toxic saliva; feeding causes leaf-speckling & tattering
Dicyphus errans (Wolff)	Capsid Bug	Miridae	Europe	
Lygocoris pabulinus (L.)	Common Green Capsid	Miridae	Europe	
Lygus rugulipennis Popp.	Tarnished Plant Bug	Miridae	Europe	
Nysius ericae (Schill.)	False Chinch Bug	Lygaeidae	Canada, USA	
Bagrada spp.	Harlequin Bugs	Pentatomidae	Africa, Asia	
Nezara viridula (L.)	Green Stink Bug	Pentatomidae	Cosmopolitan	
Hepialus spp.	Swift Moths	Hepialidae	Europe	Larvae in soil eat roots & hole tubers
Leucinodes orbonalis (Guen.)	Eggplant Fruit-Borer	Pyralidae	Africa, S. Asia	Larvae eat foliage
Agrotis spp.	Cutworms	Noctuidae	Cosmopolitan	Larvae are cutworms; live in soil, eat roots and hole tubers
Euxoa spp.	Cutworms	Noctuidae	Cosmopolitan	
Noctua pronuba (L.)	Large Yellow Underwing	Noctuidae	Europe	
Heliothis spp.	Leafworms	Noctuidae	Cosmopolitan	Larvae eat leaves
Hydraecia micacea (Esp.)	Rosy Rustic Moth (Potato Stem Borer)	Noctuidae	Europe, Asia, USA	Larvae bore stems
Phlogophora meticulosa (L.)	Angleshade Moth	Noctuidae	Europe	Larvae eat leaves
Acherontia atropos (L.)	Death's Head Hawk Moth	Sphingidae	Europe, Asia, Africa	
Tipula spp.	Leatherjackets	Tipulidae	Europe	Larvae in soil eat roots and damage tubers
Nephrotoma spp.	Leatherjackets	Tipulidae	Europe, Asia, Canada	
Pnyxia scabei (Hopkins)	Potato Scab Gnat	Sciaridae	USA	
Melolontha spp.	Cockchafers	Scarabaeidae	Europe	Larvae in soil eat roots and make holes in tubers
Phyllophaga spp.	June Beetles	Scarabaeidae	Canada, USA	
Serica spp.	White Grubs	Scarabaeidae	Europe, Asia	
Cetonia spp.	Rose Chafers	Scarabaeidae	Europe, Asia	
Phyllopertha horticola (L.)	Garden Chafer	Scarabaeidae	Europe	
Amphimallon solstitialis (L.)	Summer Chafer	Scarabaeidae	Europe	
Aspidomorpha spp.	Tortoise Beetles	Chrysomelidae	Africa	Adults & larvae eat leaves
Oulema trilineata (Ol.)	Three-lined Potato Beetle	Chrysomelidae	Canada, USA	
Epitrix cucumeris (Harris)	Potato Flea Beetle	Chrysomelidae	Canada, USA	Adults hole leaves; larvae in soil eat roots
Epitrix spp.	Flea Beetles	Chrysomelidae	Canada, USA	
Psylliodes affinis (Payk.)	Potato Flea Beetle	Chrysomelidae	Europe	
Conoderus spp.			USA	Larvae in soil eat roots and bore into tubers
Ctenicera spp.			Canada	
Drasterius spp.	Wireworms	Elateridae	S. Asia	
Limonius spp.			Canada, USA	
Hypolithus spp.			Canada, USA	
Listroderes costirostris Sch.	Vegetable Weevil	Curculionidae	Australia, USA, S. America	Adults eat foliage; larvae eat roots
Trichobaris trinotata (Say)	Potato Stalk Borer	Curculionidae	Canada, USA	Larvae bore stalks
Polyphagotarsonemus latus (Banks)	Yellow Tea Mite	Tarsonemidae	Cosmopolitan	Scarify foliage
Tetranychus spp.	Red Spider Mites	Tetranychidae	Cosmopolitan	

QUINCE (*Cydonia oblonga* – Rosaceae)

A native of western Asia, this small deciduous tree, 5–7 m in height, often bushy in habit, has been widely cultivated since very ancient times. After thousands of years of cultivation it differs little from the wild plants still to be found. It is grown for the fruit, which is quite large and oblong in shape, with hard and rather unpalatable flesh. The fruit is mostly used to make jelly or marmalade, and the seeds in their mucilaginous covering are used in medicine. The main areas of commercial cultivation are in the USA. It is sometimes cultivated as an ornamental for the bright red flowers, and can be trained to grow densely over a wall.

MAJOR PESTS

Quadraspidiotus perniciosus (Comst.)	San José Scale	Diaspididae	USA	Encrust foliage
Cydia molesta (Busck)	Oriental Fruit Moth	Tortricidae	China, Japan, N. America	Larvae bore in fruits
Cydia pomonella (L.)	Codling Moth	Tortricidae	Europe, N. America	Larvae bore in fruits

MINOR PESTS

Aphis pomi Deg.	Green Apple Aphid	Aphididae	Europe, W. Asia, N. America	Adults & nymphs infest foliage; suck sap; often with sooty moulds
Myzus persicae (Sulz.)	Peach–Potato Aphid	Aphididae	Cosmopolitan	Adults & nymphs infest foliage; suck sap; often with sooty moulds
Aleurocanthus woglumi Ashby	Citrus Blackfly	Aleyrodidae	India	Adults & nymphs infest foliage; suck sap; often with sooty moulds
Eriosoma lanigerum (Hsm.)	Woolly Apple Aphid	Pemphigidae	Cosmopolitan	Adults & nymphs infest foliage; suck sap; often with sooty moulds
Eulecanium corylil L.	Soft Scale	Coccidae	India	Adults & nymphs infest foliage; suck sap; often with sooty moulds
Parthenolecanium persicae (F.)	Peach Scale	Coccidae	Cosmopolitan	Adults & nymphs infest foliage; suck sap; often with sooty moulds
Pseudococcus maritimus (Ehrh.)	Grape Mealybug	Pseudococcidae	Widespread	Adults & nymphs infest foliage; suck sap; often with sooty moulds
Aspidiotus spp.	Armoured Scales	Diaspididae	USA	Infest foliage
Chionaspis furfura (Fitch)	Scurfy Scale	Diaspididae	USA	Infest foliage
Cacaecia sarcostega Meyr.	Fruit Tree Tortrix	Tortricidae	India	Larvae damage fruits
Spilonota ocellana (D. & S.)	Eye-spotted Bud Moth	Tortricidae	Europe, Asia, N. America	Larvae bore buds
Zeuzera spp.	Leopard Moths	Cossidae	Europe, India, Asia	Larvae bore branches
Euproctis fraterna (Moore)	Tussock Moth	Lymantriidae	India	Larvae defoliate
Ceratitis capitata (Wied.)	Medfly	Tephritidae	Mediterranean	Larvae inside fruits
Dacus dorsalis Hend.	Oriental Fruit Fly	Tephritidae	India	Larvae inside fruits
Apate monachus F.	Black Borer	Bostrychidae	Israel	Adults bore branches
Conotrachelus crataegi Walsh	Quince Curculio	Curculionidae	USA	Larvae bore in fruits
Conotrachelus nenuphar (Herbst)	Plum Weevil	Curculionidae	USA	Larvae bore in fruits
Aeolesthes sarta Solsky	Longhorn Beetle	Cerambycidae	India	Larvae bore trunk & branches
Cerambyx dux Fald.	Duke Beetle	Cerambycidae	Mediterranean	Larvae bore trunk & branches

RADISH (*Rhaphanus sativus* – Cruciferae)

A small annual or biennial herb with a rosette of small leaves and a fleshy taproot, cultivated for more than 2000 years but still very similar to the ancestral type, with a dry woody root, to which it easily reverts. They are grown worldwide, and eaten raw in salads usually, but may be cooked like any other vegetable; esteemed for their pungent flavour. Many different varieties are now grown, differing greatly in size, colour and shape of the root.

The very large white-rooted Chinese Radish is *R. sativus* var. *longipinnatus*, extensively grown throughout China and Japan.

MAJOR PESTS

Myzus persicae (Sulz.)	Peach–Potato Aphid	Aphididae	Cosmopolitan	Infest foliage; suck sap; virus vector
Delia radicum (L.)	Cabbage Root Fly	Anthomyiidae	Europe, N. America	Larvae bore root
Phyllotreta spp.	Cabbage Flea Beetles	Chrysomelidae	Cosmopolitan	Adults hole leaves; kill seedlings

MINOR PESTS

Bourletiella hortensis (Fitch)	Garden Springtail	Sminthuridae	Europe, USA, Canada	Adults bite holes in leaves
Empoasca spp.	Green Leafhoppers	Cicadellidae	Asia	Infest foliage
Nysius ericae (Schill.)	False Chinch Bug	Lygaeidae	Canada, USA	Sap-sucker; toxic saliva
Hellula undalis (F.)	Oriental Cabbage Webworm	Pyralidae	Old World	Larvae eat leaves
Liriomyza brassicae Riley	Cabbage Leaf Miner	Agromyzidae	Cosmopolitan (not UK)	Larvae mine leaves
Hylemya planipalpis (Stein)	Radish Root Maggot	Anthomyiidae	Canada	Larvae bore root
Entomoscelis americana Brown	Red Turnip Beetle	Chrysomelidae	N. America	Adults eat leaves
Listroderes costirostris Sch.	Vegetable Weevil	Curculionidae	Australia, USA, S. America	Adults eat leaves; larvae eat roots
Tetranychus spp.	Red Spider Mites	Tetranychidae	Mediterranean	Scarify foliage

Many of the pests listed under 'Brassicas' may be found attacking radish plants.

RAPE (*Brassica napus* – Cruciferae) (=Oilseed Rape; Cole)

An annual or biennial herb, with erect branching stems up to a metre tall, bearing small yellow cruciferous flowers and later elongate brown pods containing the seeds, for which the crop is grown. As a crop it is often undersown with wheat or other cereals, as it has a lengthy growing period. The tiny spherical seeds are pressed for their oil content, which is used for various culinary purposes in Europe and India mostly. The seed-cake is used as cattle food. The intact seeds are used for feeding cage birds. Rape has been cultivated since ancient times, and its origin is not known. The plant is often cultivated as a source of spring greens or winter salads, and is widely grown for feeding to cattle. It is also grown in small patches on some farms as a source of food and shelter for game birds. The oil content of the seed is about 30–45%, and the main areas of commercial cultivation (for oil) are China, Japan, India, and parts of Europe. Now being grown very extensively in the UK.

As the crop is grown for seed production, the cultivation period is lengthy and the flowers and pods susceptible to a range of pests that usually do not attack crops of the other cultivated species of *Brassica*.

MAJOR PESTS

Brevicoryne brassicae (L.)	Cabbage Aphid	Aphididae	Cosmopolitan	Infest foliage
Dasyneura brassicae (Winn.)	Brassica Pod Midge	Cecidomyiidae	Europe	Larvae inside damaged pods
Meligethes spp.	Blossom Beetles	Nitidulidae	Europe	Adults & larvae in flowers
Entomoscelis americana Brown	Red Turnip Beetle	Chrysomelidae	Canada, USA	Adults & larvae eat leaves
Phyllotreta spp.	Flea Beetles	Chrysomelidae	Cosmopolitan	Adults hole leaves
Psylliodes chrysocephala (L.)	Cabbage Stem Flea Beetle	Chrysomelidae	Europe	Larvae tunnel leaf stalks & stems
Ceutorhynchus assimilis (Payk.)	Cabbage Seed Weevil	Curculionidae	Europe	Larvae eat seeds inside pods
Ceutorhynchus quadridens (Panz.)	Cabbage Stem Weevil	Curculionidae	Europe	Larvae tunnel stem & petioles; adults eat seedling leaves

MINOR PESTS

Hydraecia micacea (Esper)	Rosy Rustic Moth	Noctuidae	Europe	Larvae tunnel stem
Euxoa ochrogaster (Guen.)	Red-backed Cutworm	Noctuidae	Canada	Larvae are cutworms; destroy young plants
Mamestra configurata (Wlk.)	Bertha Armyworm	Noctuidae	Canada	Larvae eat leaves & pods
Ceutorhynchus picitarsis Gyll.	Rape Winter Stem Weevil	Curculionidae	Europe	Larvae tunnel stems

The species listed above are those particularly damaging to rape, but almost all of the pests listed under 'Brassicas' will be found feeding on rape plants, usually as minor pests.

RASPBERRY (*Rubus idaeus*, etc. – Rosaceae)

Although botanically very closely related to Blackberry, some of the pests which attack the plant are different. A small shrub, usually of an erect vigorous habit, with small spines. The erect 'canes' die annually and are replaced from the rootstalks; the aggregate fruit separates from the receptacle (plug) when ripe. The common Red Raspberry is derived from the European *Rubus idaeus*, and the unusual Black Raspberry is from the American *Rubus occidentalis*. The European Raspberry was cultivated by both the Greeks and Romans, is of great antiquity, and can be found in an apparently wild state in many woodland areas in Britain. It is one of the hardiest fruits and can be grown as far north as Alaska, southern Scandinavia and Canada. The stems (canes) are typically annual and the usual cultivation practice is to cut them down close to the ground after fruiting; propagation is by underground shoots. The fruit is used fresh, canned, frozen, or cooked into jams, vinegar, etc.

MAJOR PESTS

Amphorophora idaei (Borner)	Large Raspberry Aphid	Aphididae	Europe	Sap-suckers; cause leaf-curl
Amphorophora rubi (Kalt.)	Blackberry Aphid	Aphididae	Europe	
Thomasiniana theobaldi Barnes	Raspberry Cane Midge	Cecidomyiidae	Europe	Larvae under stem rind
Byturus tomentosus (Deg.)	Raspberry Beetle	Byturidae	Europe	Larvae bore fruits
Otiorhynchus spp.	'Raspberry' Weevils	Curculionidae	Europe	Adults damage flowers & fruit buds; larvae eat roots

MINOR PESTS

Oecanthus spp.	Tree Crickets	Gryllidae	N. America	Egg-punctures split canes
Amphorophora spp.	Raspberry Aphids	Aphididae	N. America	Sap-suckers; cause leaf-curl
Macrosiphum euphorbiae Thos.	Potato Aphid	Aphididae	Europe	
Macrosiphum fragariae Wlk.	Blackberry Aphid	Aphididae	Europe	
Eulecanium corni Bche.	Brown Scale	Coccidae	Europe	Encrust stems
Lygocoris pabulinus (L.)	Common Green Capsid	Miridae	Europe	Sap-suckers; toxic saliva; tatter foliage
Pentatoma rufipes L.	Forest Bug	Pentatomidae	Europe	
Hepialus humili L.	Ghost Swift Moth	Hepialidae	Europe	Larvae eat roots
Cacoecia oporana L.	Fruit Tree Tortrix	Tortricidae	Europe	Larvae web leaves & shoots
Spilonota ocellana Schiff.	Eye-spotted Bud Moth	Tortricidae	Europe, Asia, N. America	Larvae bore buds
Notocelia uddmanniana (L.)	Bramble Shoot Moth	Eucosmidae	Europe	Larvae eat flowers & web shoots
Lampronia rubiella (Bjerk.)	Raspberry Moth	Incurvariidae	Europe, Canada	Larvae bore plug; later shoots (April)

Thamnonoma wauaris (L.)	V-Moth	Geometridae	Europe	
Acronycta psi (L.)	Grey Dagger Moth	Noctuidae	Europe	
Melanchra persicariae (L.)	Dot Moth	Noctuidae	Europe	Larvae eat leaves; may defoliate
Phalera bucephala (L.)	Buff-tip Moth	Notodontidae	Europe	
Euproctis chrysorrhoea Don.	Yellow-tail Moth	Lymantriidae	Europe	
Orgyia antiqua (L.)	Vapourer Moth	Lymantriidae	Europe	
Dasyneura plicatrix H.Lw.	Blackberry Leaf Midge	Cecidomyiidae	Europe	Larvae distort leaves
Lasioptera rubi Heeg.	Raspberry Stem Gall Midge	Cecidomyiidae	Europe	Larvae gall stems
Tipula spp.	Common Crane Flies (Leatherjackets)	Tipulidae	Europe	Larvae eat roots
Pegomyia rubivora (Coq.)	Loganberry Cane Fly	Anthomyiidae	Europe, USA	Larvae mine & girdle shoots
Allantus cinctus (L.)	Banded Rose Sawfly	Tenthredinidae	Europe	Larvae defoliate
Metallus spp.	Raspberry Leaf-mining Sawflies	Tenthredinidae	Europe, USA	Larvae mine leaves
Monophadnoides geniculatus (Htg.)	Raspberry Sawfly	Tenthredinidae	USA, Canada	Larvae defoliate
Melolontha spp.	Cockchafers	Scarabaeidae	Europe, Asia, Canada, USA	Larvae in soil eat roots
Phyllophaga spp.	June Beetles	Scarabaeidae	Canada, USA	
Phyllopertha horticola (L.)	Garden Chafer	Scarabaeidae	Europe	
Cetonia spp.	Rose Chafers	Scarabaeidae	Cosmopolitan	
Byturus spp.	Raspberry Fruitworms	Byturidae	USA	Larvae in fruits
Batophila spp.	Raspberry Flea Beetles	Chrysomelidae	Europe	Adults hole leaves
Agrilus ruficollis (F.)	Red-necked Cane Borer	Buprestidae	USA, Canada	Larvae bore canes
Rhynchites germanicus Herbst	Strawberry Rhynchites	Curculionidae	Europe	Adults girdle shoots & truss stems
Otiorhynchus clavipes (Boisd.)	Red-legged Weevil	Curculionidae	Europe	Adults damage stems & leaves
Otiorhynchus singularis (L.)	Clay-coloured Weevil	Curculionidae	Europe	
Metatetranychus ulmi Koch	Fruit Tree Red Spider Mite	Tetranychidae	Europe, Asia, N. America	Scarify & web leaves
Tetranychus urticae (Koch)	Temperate Red Spider Mite	Tetranychidae	Europe, N. America	
Phytopterus gracilis Nal.	Raspberry Leaf & Bud Mite	Eriophyidae	Europe	Found in buds & on leaves

RHUBARB (*Rheum rhaponticum* – Polygonaceae)

This is a native of Asia where it still grows wild, and is now grown widely in the temperate regions of Europe and America. The portion used as food is the succulent acid leaf stalk; and it is used for pies, tarts, and sauce. The plant is perennial, growing annually from an underground rhizome; leaves are large, with long thick petioles; later in the season a flowering stalk bears masses of tiny whitish flowers. It is sometimes grown as an ornamental. The petioles are nearly 95% water, with a little sugar, and salts of oxalic and malic acids.

MAJOR PESTS

Aphis fabae Scop.	Black Bean Aphid	Aphididae	Cosmopolitan	Suck sap; virus vector

MINOR PESTS

Empoasca fabae (Harris)	Potato Leafhopper	Cicadellidae	Canada, USA, C. & S. America	Suck sap; cause 'hopperburn'
Hepialus spp.	Swift Moths	Hepialidae	Europe	Larvae in soil eat roots
Clepsis spectrana (Treit.)	Tortricid	Tortricidae	Europe	Larvae eat foliage
Euxoa spp.	Cutworms	Noctuidae	Canada	Larvae are cutworms
Hydraecia micacea (Esp.)	Rosy Rustic Moth	Noctuidae	Europe, Canada	Larvae bore petioles
Lacanobia oleracea (L.)	Tomato Moth	Noctuidae	Europe	Larvae eat leaves
Spodoptera littoralis (Boisd.)	Cotton Leafworm	Noctuidae	Israel	Larvae eat leaves
Papaipema spp.	Stalk Borers	Noctuidae	Canada	Larvae bore petioles
Xestia c-nigrum (L.)	Spotted Cutworm	Noctuidae	Canada	Larvae are cutworms
Lixus concavus Say	Rhubarb Curculio	Curculionidae	USA	Adults bore stalks (to oviposit) & eat leaves
Tetranychus spp.	Red Spider Mites	Tetranychidae	Asia	Scarify foliage

ROSE (*Rosa* spp. – Rosaceae)

Roses are native to northern temperate regions; 250 species are recognized, as well as many varieties and commercial hybrids. Cultivation has been practised since early historical times, partly for decoration, and partly for medicinal and religious purposes. At the present time cultivation for ornamental purposes is widespread and a large-scale industry. The main types are hybrid tea, bush roses, floribunda and ramblers, each with a different life-form and habit. Wild species are abundant in temperate regions. The other commercial product is an essential oil extracted from the petals of the flowers called attar of roses, or rose oil. Rose oil is a favourite perfume, but is often adulterated commercially with geranium oil which is cheaper to produce. The main areas of cultivation for oil are Bulgaria, India, France, Italy, N. Africa and Asia Minor.

MAJOR PESTS

Macrosiphum rosae (L.)	Rose Aphid	Aphididae	Europe, USA	Infest foliage; suck sap; virus vectors; often with ants
Myzus persicae (Sulz.)	Peach–Potato Aphid	Aphididae	Cosmopolitan	
Parthenolecanium corni (Bche.)	Brown Scale	Coccidae	Europe	Infest stems; suck sap
Lepidosaphes ulmi (L.)	Mussel Scale	Diaspididae	Cosmopolitan	
Arge spp.	Large Rose Sawflies	Argidae	Europe, Asia	Larvae defoliate

MINOR PESTS

Edwardsiana rosae (L.)	Rose Leafhopper	Cicadellidae	USA, Europe	Infest foliage; suck sap
Zygina pallidifrons	Tomato Leafhopper	Cicadellidae	Europe	
Macrosiphum euphorbiae (Thos.)	Potato Aphid	Aphididae	Cosmopolitan	Infest foliage; cause leaf-curling
Saissetia oleae (Bern.)	Black Scale	Coccidae	Cosmopolitan	Infest foliage; suck sap
Orthezia insignis Browne	Jacaranda Bug	Orthezidae	Cosmopolitan	
Icerya purchasi Mask.	Cottony Cushion Scale	Margarodidae	Cosmopolitan	
Aspidiotus spp.	Armoured Scales	Diaspididae	Cosmopolitan	Infest foliage; suck sap
Aulacaspis rosae (Mask.)	Scurfy Scale	Diaspididae	Europe, USA	
Aonidiella aurantii (Mask.)	Red Scale	Diaspididae	Cosmopolitan	
Chrysomphalus dictyospermi (Morg.)	Dictyospermum Scale	Diaspididae	Cosmopolitan (not UK)	
Quadraspidiotus perniciosus (Comst.)	San José Scale	Diaspididae	Cosmopolitan (not UK)	
Hemiberlesia lantaniae (Sign.)	Latania Scale	Diaspididae	Cosmopolitan (not UK)	
Lygocoris pabulinus (L.)	Common Green Capsid	Miridae	Europe	Sap-suckers; toxic saliva
Lygus rugulipennis Popp.	Tarnished Plant Bug	Miridae	Europe	
Frankliniella spp.	Flower Thrips	Thripidae	Cosmopolitan	Infest flowers
Thrips fuscipennis Hal.	Rose Thrips	Thripidae	Europe	
Paraleucoptera heinrichi Jones	Cambium Borer	Gracillariidae	USA	Larvae mine cambium

Archips spp.	Rose Tortrix Moths	Tortricidae	Europe, Asia, USA	Larvae roll & eat leaves
Hedya nubiferana (Hubner)	Fruit Tree Tortrix	Tortricidae	Europe	
Cydia prunivora (Walsh)	Lesser Appleworm	Tortricidae	Canada, USA	Larvae eat fruits
Operophtera brumata (L.)	Winter Moth	Geometridae	Europe	
Automeris io (F.)	Io Moth	Saturniidae	USA	Larvae eat leaves; sometimes defoliate
Malacosoma spp.	Tent Caterpillars	Lasiocampidae	Europe, USA	
Orgyia antiqua (L.)	Vapourer Moth	Lymantriidae	Europe	
Allantus cinctus (L.)	Banded Rose Sawfly	Tenthredinidae	Europe, USA	
Blennocampa pusilla (Klug)	Rose Leaf-rolling Sawfly	Tenthredinidae	Europe	Larvae roll leaves
Cladius isomeres Norton	Bristly Rose Slug Sawfly	Tenthredinidae	USA	Larvae eat leaves
Endelomyia aethiops (F.)	Rose Slug Sawfly	Tenthredinidae	Europe, USA	Larvae skeletonize leaves
Hartigia spp.	Sawfly Borers	Tenthredinidae	USA	Larvae bore canes & girdle tips
Diplolepis radicum (O.S.)	Rose Root Gall Wasp	Cynipidae	USA	Larvae gall roots
Diplolepis rosae (L.)	Mossy Rose Gall	Cynipidae	Cosmopolitan	Larvae gall twigs
Diplolepis dichlorcerus Harris	Long Rose Gall	Cynipidae	USA	Larvae make long twig gall
Megachile spp.	Leaf-cutting Bees	Megachilidae	Cosmopolitan	Adults cut pieces of leaf lamina
Dasyneura rhodophaga (Coq.)	Rose Bud Midge	Cecidomyiidae	Canada, USA	Larvae infest flower & leaf buds
Thomasiniana oculiperda (Rubs.)	Red Bud Borer	Cecidomyiidae	Europe	Larvae under grafted buds
Wachtliella rosarum (Hardy)	Rose Leaf Midge	Cecidomyiidae	Europe, USA	Larvae in leaflets
Cetonia spp.	Rose Chafers	Scarabaeidae	Cosmopolitan	Adults eat flowers; larvae in soil eat roots
Pachnoda spp.	Rose Beetles	Scarabaeidae	Pantropical	
Macrodactylus subspinosus (F.)	Rose Chafer	Scarabaeidae	USA	
Mylabris spp.	Banded Blister Beetles	Meloidae	Pantropical	Adults eat flowers
Nodonta puncticollis (Say)	Rose Leaf Beetle	Chrysomelidae	USA	Adults eat foliage
Popillia japonica New.	Japanese Beetle	Chrysomelidae	Canada, USA	
Agrilus rubicola Ab.	Rose Stem Girdler	Buprestidae	USA	Larvae bore stem under bark
Chrysobothris femorata Oliv.	Flat-headed Apple Tree Borer	Buprestidae	USA	Larvae bore stems
Chrysobothris mali (Horn)	Pacific Flat-headed Borer	Buprestidae	USA	
Phyllobius spp.	Leaf Weevils	Curculionidae	Europe, Asia	Adults eat leaves
Rhynchites bicolor (F.)	Rose Curculio	Curculionidae	USA	Adults eat flower buds
Tetranychus urticae (Koch)	Temperate Red Spider Mite	Tetranychidae	Cosmopolitan	Scarify & web foliage

RYE (*Secale cereale* – Gramineae)

A cereal of recent origin, thought to have come from the Black Sea region of Central Eurasia. It is related to both wheat and barley, being a tall grass, with bluish foliage, and a head of many spikelets. A very adaptable cereal which will produce crops in regions with severe winters and at high altitudes and poor soils, as well as in arid regions. It is basically a European crop used for the making of black (rye) bread. The grain is also used for stock feed, and as a source of whisky. The main areas of cultivation are the USSR, Germany, Poland, Czechoslovakia, and the USA.

MAJOR PESTS

Rhopalosiphum padi (L.)	Bird-cherry Aphid	Aphididae	Europe, Near East	Adults & nymphs infest foliage; suck sap; virus vectors
Schizaphis graminum Rond.	Wheat Aphid	Aphididae	Africa, Asia, USA, S. America	
Sitobion avenae (F.)	Grain Aphid	Aphididae	Europe	
Oscinella frit (L.)	Frit Fly	Chloropidae	Europe, USA	Larvae bore shoot; cause 'dead-heart'

MINOR PESTS

Rhopalosiphum maidis (Fitch)	Corn Leaf Aphid	Aphididae	Cosmopolitan	Adults & nymphs infest foliage; suck sap; virus vectors
Rhopalosiphum insertum (Wlk.)	Apple–Grass Aphid	Aphididae	Europe	
Macrosiphum fragariae Wlk.	Blackberry Aphid	Aphididae	Cosmopolitan	
Metapolophium dirhodum (Wlk.)	Rose-Grain Aphid	Aphididae	Europe	
Metapolophium festucae (Theob.)	Fescue Aphid	Aphididae	Europe	
Laodelphax striatella (Fall.)	Small Brown Planthopper	Delphacidae	Europe, Asia	Infest foliage; suck sap; virus vectors
Javasella pellucida (F.)	Cereal Planthopper	Delphacidae	Europe	
Eurygaster austriaca (Schr.)	Wheat Shield Bug	Pentatomidae	S. Europe, W. Asia	Sap-sucker; toxic saliva
Crambus spp.	Grass Moths (Sod Webworms)	Pyralidae	Europe, Asia, N. America	Larvae in silk tubes eat foliage & stems
Mesapamea secalis (L.)	Common Rustic Moth	Noctuidae	Europe	Larvae bore shoots; cause 'dead-hearts'
Contarinia tritici (Kirby)	Lemon Wheat Blossom Midge	Cecidomyiidae	Europe, Asia	Larvae infest flowerheads
Sitodiplosis mossellana (Gehin)	Orange Wheat Blossom Midge	Cecidomyiidae	Europe, N. America	
Haplodiplosis marginata (v.R.)	Saddle Gall Midge	Cecidomyiidae	Europe	Larvae infest stems & make galls
Mayetiola spp.	'Hessian' Flies	Cecidomyiidae	Cosmopolitan	
Tipula spp.; *Nephrotoma* spp.	Leatherjackets	Tipulidae	Europe; Europe, N. America	Larvae in soil eat roots; kill seedlings
Geomyza spp.; *Meromyza* spp.; *Opomyza* spp.	Grass Flies	Opomyzidae	Holarctic	Larvae bore shoots; cause 'dead-hearts'
Chlorops pumilionis (Bjerk.)	Gout Fly	Chloropidae	Europe	Larvae gall stems
Hydrellia griseola (Fall.)	Cereal Leaf Miner	Ephydridae	Europe, Asia, USA	Larvae mine leaves
Delia coarctata (Fall.)	Wheat Bulb Fly	Anthomyiidae	Europe	(Larvae fail to develop completely)

Agromyza ambigua Fall.	Cereal Leaf Miner	Agromyzidae	Europe, USA, Canada	Larvae mine leaves
Limothrips cerealium Hal.	Grain Thrips	Thripidae	Europe	Infest foliage, cause speckling of leaves
Stenothrips graminum Uzel	Oat Thrips	Thripidae	Europe	Infest foliage, cause speckling of leaves
Thrips angusticeps (Uzel.)	Cabbage Thrips	Thripidae	Europe	Infest foliage, cause speckling of leaves
Cephus cinctus Nort.	Western Wheat-stem Sawfly	Cephidae	W. USA	Larvae bore lower parts of stems
Cephus pygmaeus (L.)	Wheat-stem Sawfly	Cephidae	Europe, Canada, E. USA	Larvae bore lower parts of stems
Dolerus spp.	Leaf Sawflies	Tenthredinidae	Europe	Larvae eat leaves
Tetramesa secale (Fitch)	Rye Jointworm	Eurytomidae	USA, Canada	Larvae gall stems
Agriotes spp.	Wireworms	Elateridae	Europe	Larvae eat roots
Melolontha spp.	Cockchafers	Scarabaeidae	Europe	Larvae in soil eat roots; may destroy seedlings
Phyllophaga spp.	June Beetles	Scarabaeidae	Canada, USA	Larvae in soil eat roots; may destroy seedlings
Schizonycha spp.	Chafer Grubs	Scarabaeidae	Europe, Asia	Larvae in soil eat roots; may destroy seedlings
Helophorus nubilis F.	Wheat Shoot Beetle	Hydrophilidae	Europe	Larvae damage seedling stems
Oulema melanopus (L.)	Cereal Leaf Beetle	Chrysomelidae	Europe, USA	Adults & larvae eat leaf strips
Crepidodera ferruginea (Scop.)	Wheat Flea Beetle	Chrysomelidae	Europe	Larvae feed in shoots
Phyllotreta vittula Redt.	Barley Flea Beetle	Chrysomelidae	Europe	Adults eat leaf strips

SPINACH (*Spinacia oleracea* – Chenopodiaceae)

A native of southwestern Asia, this is possibly the commonest herbage vegetable used for greens. It is widely cultivated in cool regions where there is plenty of water. An annual herb, it first produces a large number of basal leaves, and later in the season the flower stem. Used mainly as a pot herb, mostly fresh, but now canned in large quantities, and also frozen fresh.

Other types of vegetable referred to as 'spinach' include the biennial Spinach Beet (*Beta vulgaris*), and New Zealand Spinach which is *Tetragonia expansa* and is only half-hardy, being unable to survive frost.

MAJOR PESTS

Circulifer tenellus (Baker)	Beet Leafhopper	Cicadellidae	Mediterranean, USA, S. Africa	Infest foliage; suck sap; virus vectors
Aphis fabae Scop.	Black Bean Aphid	Aphididae	Cosmopolitan (not Australia)	
Myzus persicae (Sulz.)	Peach–Potato Aphid	Aphididae	Cosmopolitan	

MINOR PESTS

Lygocoris pabulinus (L.)	Common Green Capsid	Miridae	Europe	Suck sap; toxic saliva; foliage speckled & distorted
Lygus rugulipennis Popp.	Tarnished Plant Bug	Miridae	Europe	
Piesma quadrata (Fieb.)	Beet Lace Bug	Tingidae	Europe, Asia	
Thrips tabaci Lind.	Onion Thrips	Thripidae	Cosmopolitan	Scarify foliage
Loxostege sticticalis (L.)	Beet Webworm	Pyralidae	Cosmopolitan	Larvae eat foliage
Udea rubigalis (Gn.)	Celery Leaf Tier	Pyralidae	USA	
Delia echinata (Séguy)	Spinach Stem Fly	Anthomyiidae	Europe	Larvae mine stem & leaves
Pegomya betae (Curt.)	Mangold Fly	Anthomyiidae	Europe, USA, Canada	Larvae mine leaves
Epicauta spp.	Black Blister Beetles	Meloidae	Canada, USA	Adults eat foliage
Atomaria linearis Steph.	Pygmy Mangold Beetle	Cryptophagidae	Europe	Adults damage stems of seedlings
Agriotes spp.	Wireworms	Elateridae	Europe	Larvae in soil eat roots
Silpha bituberosa Lec.	Spinach Carrion Beetle	Silphidae	Canada	Larvae eat foliage
Erynephala puncticollis (Say)	Beet Leaf Beetle	Chrysomelidae	USA	Adults & larvae eat leaves
Disonycha xanthomelana (Dalm.)	Spinach Flea Beetle	Chrysomelidae	Canada, USA	
Lixus junci Boh.	Beet Weevil	Curculionidae	Mediterranean	Larvae bore plant

A number of the pests listed under 'Sugarbeet' can be also found attacking spinach.

STRAWBERRY (*Fragaria* spp. – Rosaceae)

This is an important temperate fruit of delicate flavour, but of a perishable nature. The fruit is not a berry, but an aggregate accessory fruit, consisting of a swollen receptacle with a number of small dry achenes embedded on the surface. It is a low perennial herb, with a very short thick stem (crown) and trifoliate leaves; propagation is by the many runners which root at the tip. It has been cultivated in Europe since the fourteenth century, and in America since Colonial days. It requires a good soil and plenty of sunshine, and is now widely grown throughout cooler regions. There are three main sources of cultivated strawberry; the main one is the American *F. virginiana*, although to a lesser extent there is the dioecious *F. chiloensis*, and the European *F. vesca* which is the source of the ever-bearing varieties. Hundreds of varieties are now cultivated. Strawberries are primarily a fresh dessert fruit, but are also frozen, canned, and made into jams and preserves, and used for flavouring.

MAJOR PESTS

Chaetosiphon fragaefolii (Ckll.)	Strawberry Aphid	Aphididae	Europe, USA	Sap-sucker; virus vector
Acleris comariana (Zell.)	Strawberry Tortrix	Tortricidae	Europe	Larvae eat leaves & flowers
Anthonomus rubi (Herbst.)	Strawberry Blossom Weevil	Curculionidae	Europe }	Adults cut flower stalks; larvae in flower buds
Anthonomus signatus Say	Strawberry Weevil	Curculionidae	USA }	
Tarsonemus pallidus Banks	Strawberry Mite	Tarsonemidae	Europe	Stunt & wrinkle leaves
Tetranychus urticae (Koch)	Temperate Red Spider Mite	Tetranychidae	Cosmopolitan	Scarify & web foliage

MINOR PESTS

Aphis forbesi Weed	Strawberry Root Aphid	Aphididae	USA	Adults & nymphs infest roots
Myzus ascalonicus Don.	Shallot Aphid	Aphididae	Europe }	Infest foliage; suck sap; cause leaf-curling
Myzus ornatus (Laing)	Violet Aphid	Aphididae	Cosmopolitan }	
Macrosiphum euphorbiae (Thos.)	Potato Aphid	Aphididae	Europe }	
Macrosiphum fragariae Wlk.	Blackberry Aphid	Aphididae	Europe }	
Cercopis vulnerata Ger.	Red & Black Froghopper	Cercopidae	Europe	Larvae in spittlemass on roots
Philaenus spumarius (L.)	Common Spittlebug	Cercopidae	Europe, Asia, USA	Larvae in spittlemass on foliage
Aphrodes spp.	Leafhoppers	Cicadellidae	Europe }	Adults & nymphs infest foliage; suck sap; virus vectors
Euscelis spp.	Leafhoppers	Cicadellidae	Europe }	
Typhlocyba spp.	Strawberry Leafhoppers	Cicadellidae	Europe }	
Aleyrodes lonicerae Wlk.	Strawberry Whitefly	Aleyrodidae	Europe, Asia }	Infest foliage; suck sap
Trialeurodes packardi (Morr.)	Strawberry Whitefly	Aleyrodidae	USA }	
Lygocoris pabulinus (L.)	Common Green Capsid	Miridae	Europe	Sap-sucker; toxic saliva
Synanthedon bibionipennis (Bois.)	Strawberry Crown Moth	Sesiidae	USA	Larvae bore crown

Hepialus spp.	Swift Moths	Hepialidae	Europe	Larvae in soil eat roots & crown
Aristotelia fragariae Busck	Strawberry Crown Miner	Gelechiidae	USA	Larvae bore in crown
Ancylis fragariae (W. & R.)	Strawberry Leaf Roller	Tortricidae	USA	Larvae roll & eat leaves
Olethreutes lacunana (Schiff.)	Fruit Tree Tortrix	Tortricidae	Europe	
Cacoecimorpha pronubana (Hbn.)	Carnation Tortrix Moth	Tortricidae	Europe	
Clepsis spectrana (Treit.)	Cyclamen Tortrix	Tortricidae	Europe, S.E. Russia	Larvae web shoots & leaves; eat leaves & flowers
Cnephasia interjectana (Haw.)	Flax Tortrix Moth	Tortricidae	Europe	
Cnephasia longana (Haw.)	Omnivorous Leaf Tier	Tortricidae	Europe, USA, Canada	
Lozotaenia forsterana (F.)	–	Tortricidae	Europe, S.E. Siberia	
Agrotis spp.	Cutworms	Noctuidae	Cosmopolitan	Cutworms in soil eat roots & may damage crown
Noctua pronuba (L.)	Large Yellow Underwing	Noctuidae	Europe	
Xestia c-nigrum (L.)	Spotted Cutworm	Noctuidae	Europe	
Tipula spp.	Leatherjackets	Tipulidae	Europe	Larvae in soil eat roots
Agromyza spiraeae Kalt.	Leaf Miner	Agromyzidae	Europe, Canada	Larvae mine leaves
Feronia spp.	Strawberry Ground Beetles	Carabidae	Europe	Adults remove seeds from fruits
Harpalus rufipes (Deg.)	Strawberry Seed Beetle	Carabidae	Europe	
Pterostichus spp.	Strawberry Ground Beetles	Carabidae	Europe	Adults eat fruits
Agriotes spp.	Wireworms	Elateridae	Europe	Larvae eat roots & damage crowns
Cetonia spp.	White Grubs	Scarabaeidae	Europe, Asia	Larvae in soil eat roots & may damage crowns
Melolontha spp.	Cockchafers	Scarabaeidae	Europe	
Phyllophaga spp.	White Grubs	Scarabaeidae	Canada, USA	
Paria fragariae Wil.	Strawberry Rootworm	Chrysomelidae	USA	Larvae in soil eat roots & crown
Otiorhynchus singularis (L.)	Clay-coloured Weevil	Curculionidae	Europe, USA	Larvae in soil eat roots & damage crown; adults eat leaves
Otiorhynchus sulcatus (F.)	Black Vine Weevil	Curculionidae	Europe, USA	
Otiorhynchus ovatus (L.)	Strawberry Root Weevils	Curculionidae	Europe, USA, Canada	
Otiorhynchus rugostriatus (Goeze)			Europe, USA	
Rhynchites germanicus (Hbst.)	Strawberry Rhynchites	Curculionidae	Europe	Adults cut fruit truss stalks & petioles
Tyloderma fragariae (Riley)	Strawberry Crown Borer	Curculionidae	USA	Larvae in soil bore crown
Tetranychus atlanticus McG.	Strawberry Spider Mite	Tetranychidae	USA	Scarify & web foliage

SUGARBEET (*Beta vulgaris* – Chenopodiaceae)

Sugarbeet is the temperate source of sugar, as opposed to the tropical cane. This beet was not used for sugar extraction until modern times, and production did not really commence in Europe until the middle of the nineteenth century, and in the USA at the end of that century. The plant is a white-rooted biennial which grows best in areas where the summer temperature is about 21 °C, in a wide variety of soils, and also will grow in semi-arid areas with irrigation. For sugar it is grown as an annual. As a crop it lends itself to complete agricultural mechanization, particularly now so with pelleted seeds and precision drilling. The main areas of production are the USSR, USA, France, Germany, UK, Poland and Czechoslovakia.

MAJOR PESTS

Circulifer tenellus (Baker)	Beet Leafhopper	Cicadellidae	Mediterranean, USA	Adults & nymphs infest foliage; suck sap; virus vectors
Aphis fabae Scopoli	Black Bean Aphid	Aphididae	Cosmopolitan	
Myzus persicae Sulz.	Peach–Potato Aphid	Aphididae	Cosmopolitan	
Pegomyia betae (Curt.)	Mangold Fly	Anthomyiidae	Europe, USA, Asia	Larvae mine leaves
Autographa gamma (L.)	Silver-Y Moth	Noctuidae	Europe	Larvae defoliate

MINOR PESTS

Onychiurus spp.	Springtails	Onychiuridae	Europe	Damage seedlings
Forficula auricularia L.	Common Earwig	Forficulidae	Cosmopolitan	Damage seedling leaves
Circulifer opacipennis (Leth.)	Beet Leafhopper	Cicadellidae	Mediterranean	Damage seedlings
Philaenus spumarius (L.)	Common Froghopper	Cercopidae	Europe	Nymphs in spittlemass
Macrosiphum euphorbiae (Thos.)	Potato Aphid	Aphididae	Europe	Adults & nymphs infest foliage; suck sap; virus vectors
Myzus ascalonicus Don.	Shallot Aphid	Aphididae	Europe	
Rhopalosiphoninus staphyleae (Koch)	Mangel Aphid	Aphididae	Europe	
Pemphigus populivenae Fitch	Sugarbeet Root Aphid	Pemphigidae	USA, Canada	Infest roots
Calocoris norvegicus (Gmel.)	Potato Capsid	Miridae	Europe	Sap-suckers; infest foliage; toxic saliva causes tattered leaves
Lygus rugulipennis Popp.	Tarnished Plant Bug	Miridae	Europe	
Lygocoris pabulinus (L.)	Common Green Capsid	Miridae	Europe	
Orthotylus flavosparsus Sahl.	Sugar Beet Leaf Bug	Miridae	Japan	
Piesma quadrata (Fieb.)	Beet Lace Bug	Tingidae	Europe	
Hercinothrips femoralis (Reut.)	Banded Greenhouse Thrips	Thripidae	Cosmopolitan	Infest foliage & scarify leaves
Thrips angusticeps Uzel	Cabbage Thrips	Thripidae	Europe	
Hepialus spp.	Swift Moths	Hepialidae	Europe	Larvae in soil eat roots
Scrobipalpa ocellatella (Boyd)	Beet Moth	Gelechiidae	Europe, Near East	Larvae bore root
Plutella xylostella (L.)	Diamond-back Moth	Yponomeutidae	Europe, Russia	Larvae eat leaves
Cnephasia spp.	Leaf Tiers	Tortricidae	Europe	Larvae roll leaves

Hymenia perspectalis (Hbn.)	Spotted Beet Webworm	Pyralidae	Cosmopolitan	Larvae eat & web leaves; may defoliate
Loxostege sticticalis (L.)	Beet Webworm	Pyralidae	Cosmopolitan	
Agrotis segetum (D. & S.)	Common Cutworm	Noctuidae	Cosmopolitan in Old World	Larvae are cutworms; eat roots & destroy seedlings
Ceramica pisi L.	Broom Moth	Noctuidae	Europe, Asia	
Euxoa nigricans L.	Garden Dart Moth	Noctuidae	Europe	
Hydraecia micacea (Esp.)	Rosy Rustic Moth	Noctuidae	Europe	Larvae burrow stem & root
Spodoptera exigua (Hb.)	Beet Armyworm	Noctuidae	Cosmopolitan (not UK or N. Europe)	Larvae eat leaves; may defoliate
Spodoptera litura F.	Fall Armyworm	Noctuidae	China	
Tipula paludosa Meig.	Marsh Crane Fly	Tipulidae	Europe, W. Asia	Larvae in soil eat roots & seedlings
Tetanops myopaeformis (Röder)	Sugarbeet Root Maggot	Otitidae	Canada, USA	Larvae mine root
Aclypea opaca (L.)	Beet Carrion Beetle	Silphidae	Europe	Adults & larvae eat foliage
Amphimallon solstitialis (L.)	Summer Chafer	Scarabaeidae	Europe	Larvae in soil eat roots
Melolontha spp.	Cockchafers	Scarabaeidae	Europe, Asia	
Phyllophaga spp.	June Beetles	Scarabaeidae	Canada, USA	
Chaetocnema concinna (Marsh.)	Beet Flea Beetle	Chrysomelidae	Europe, Asia	Adults hole leaves
Systena spp.	Flea Beetles	Chrysomelidae	Canada	Adults eat leaves
Erynephala puncticollis (Say)	Beet Leaf Beetle	Chrysomelidae	USA	Adults & larvae eat foliage
Cassida spp.	Tortoise Beetles	Chrysomelidae	Europe, China	
Disonycha xanthomelas (Dalm.)	Spinach Flea Beetle	Chrysomelidae	USA	
Atomaria linearis Steph.	Pygmy Mangold Beetle	Cryptophagidae	Europe	Adults gnaw hypocotyl & leaves
Epicauta spp.	Black Blister Beetles	Meloidae	Canada, USA	Adults eat foliage
Agriotes spp.	Wireworms	Elateridae	Europe, USA	Larvae in soil eat roots; destroy seedlings
Limonius californicus (Mann.)	Sugarbeet Wireworm	Elateridae	Canada, USA	
Bothynoderes punctiventris (Germ.)	Beet Weevil	Curculionidae	Mediterranean, Asia	Adults eat leaves
Lixus junci Boh.	Beet Weevil	Curculionidae	Mediterranean	Larvae bore plant; adults eat leaves
Philopedon plagiatum (Schall.)	Sand Weevil	Curculionidae	Europe	Adults eat leaves
Tanymecus dilaticollis Gylh.	Southern Grey Weevil	Curculionidae	E. Europe	Adults eat leaves; may destroy seedlings
Tanymecus palliatus (F.)	Beet Leaf Weevil	Curculionidae	Europe, China	
Tetranychus urticae (Koch)	Two-spotted Red Spider Mite	Tetranychidae	Cosmopolitan	Scarify & web foliage

SUNFLOWER (*Helianthus annuus* – Compositae)

Sunflower is not known in the truly wild state, but possibly originated in southern USA; taken to Europe in 1510 and to Russia in the eighteenth century; now grown throughout the world, from the Equator to about 55°N; it can withstand a slight frost. It can be grown in very dry areas in a range of soils. A variable annual herb, 0.7–3.5 m tall. Giant, semi-dwarf, and dwarf varieties are now grown. The large flowerhead, with yellow petals, produces a disc of ovoid achenes, white, black or striped in colour. The seeds are rich in oil (25–35%) and linoleic acid, and 13–20% protein. The decorticated cake is also rich in protein. Seeds can be eaten raw, roasted or fed to stock; the oil is used for cooking and in margarine. The main production areas are the USSR, Argentina, Rumania, Bulgaria, Hungary, Yugoslavia, Turkey, S. Africa, Uruguay, Tanzania, Kenya, Zimbabwe and Australia.

MAJOR PESTS

Calidea spp.	Blue Bugs	Pentatomidae	Africa	Suck sap from seeds; toxic saliva
Nezara viridula (L.)	Green Stink Bug	Pentatomidae	Cosmopolitan	
Heliothis armigera (Hub.)	American Bollworm	Noctuidae	Old World	Larvae eat seeds
Schizonycha spp.	Chafer Grubs	Scarabaeidae	Africa	Larvae eat roots

MINOR PESTS

Aphis gossypii Glov.	Melon/Cotton Aphid	Aphididae	Cosmopolitan	Infest foliage; suck sap; virus vectors
Macrosiphum euphorbiae (Thos.)	Potato Aphid	Aphididae	Cosmopolitan	
Clastoptera xanthocephala Germ.	Sunflower Spittlebug	Cercopidae	USA	Suck sap
Amrasca terraereginae (Paoli)	Leafhopper	Cicadellidae	Australia	Under leaves; suck sap
Empoasca spp.	Green Leafhoppers	Cicadellidae	Cosmopolitan	
Nysius spp.	Grey Cluster Bugs	Lygaeidae	Australia	Sap-suckers; toxic saliva
Agonoscelis pubescens Thnb.	Cluster Bug	Pentatomidae	Africa	
Cochylis spp.	Sunflower Tortricids	Cochylidae	Canada, USA	Larvae bore in flowerhead
Homoeosoma electellum (Hulst)	Sunflower Moth	Phycitidae	Canada, USA	
Vanessa cardui (L.)	Painted Lady Butterfly	Nymphalidae	Holarctic	Larvae eat foliage; sometimes defoliate
Agrius convolvuli (L.)	Sweet Potato Moth	Sphingidae	Europe, Asia, Africa	
Hydraecia micacea (Esp.)	Rosy Rustic Moth	Noctuidae	Europe, Asia, USA	Larvae bore stem
Heliothis punctigera Wllgr.	Native Budworm	Noctuidae	Australia	Larvae bore buds
Spodoptera littoralis (Boisd.)	Cotton Leafworm	Noctuidae	Africa	Larvae eat foliage
Spodoptera litura (F.)	Rice Cutworm	Noctuidae	Asia, Australasia	
Lasioptera murtfeldtiana Felt	Sunflower Seed Midge	Cecidomyiidae	USA	Larvae gall seeds
Dacus cucurbitae Coq.	Melon Fly	Tephritidae	Africa, Asia, Australasia	Larvae in flowerhead
Strauzia longipennis (Wied.)	Sunflower Maggot	Tephritidae	USA	Larvae bore stem pith & flowerhead
Epilachna spp.	Epilachna Beetles	Coccinellidae	Africa	Adults & larvae eat foliage
Zygospila exclamationis (F.)	Sunflower Beetle	Chrysomelidae	Canada, USA	Larvae eat foliage
Gonocephalum macleayi Blkb.	False Wireworm	Tenebrionidae	Australia	Larvae eat roots
Tanymecus dilaticollis Gylh.	Southern Grey Weevil	Curculionidae	E. Europe	Adults eat leaves

TOBACCO (*Nicotiana tabacum* – Solanaceae)

Probably tobacco originated in northwestern Argentina, but it was cultivated in pre-Columbian times in the W. Indies, Mexico, C. and S. America. By the seventeenth century it had been spread to India, Africa, Japan, Philippines and the Middle East. It is now very widely cultivated throughout the warmer parts of the world, from central Sweden in the north down to southern Australia. The crop needs 90–120 frost-free days from transplanting to harvest. The optimum mean temperature for growth is 20–26 °C; strong illumination is needed; it can grow in as little as 25 cm of rain, but prefers 50 cm. Dry weather is essential for ripening and harvest. The plant is a perennial herb, 1–3 m tall, usually grown as an annual for its leaves which are cured to make tobacco and snuff. The main production areas are USA, Brazil, Japan, Canada, Pakistan, India, Greece, Turkey and Zimbabwe.

MAJOR PESTS

Brachytrupes spp.	Tobacco Crickets	Gryllidae	Africa, Asia	Eat roots & destroy seedlings
Gryllotalpa spp.	Mole Crickets	Gryllotalpidae	Cosmopolitan	
Bemisia tabaci (Genn.)	Tobacco Whitefly	Aleyrodidae	Cosmopolitan	Infest leaves
Thrips tabaci Lind.	Onion Thrips	Thripidae	Cosmopolitan	Scarify leaves
Heliothis armigera (Hbn.)	American Bollworm	Noctuidae	Cosmopolitan in Old World	Larvae bore buds & eat foliage
Heliothis zea (Boddie)	Cotton Bollworm	Noctuidae	N., C. & S. America	
Agrotis ipsilon (Hfn.)	Black Cutworm	Noctuidae	Cosmopolitan	Larvae cutworms
Manduca sexta (L.)	Tobacco Hornworm	Sphingidae	USA, C. & S. America	Larvae defoliate
Lasioderma serricorne (F.)	Tobacco Beetle (Cigarette Beetle)	Anobiidae	Pantropical	Damage stored leaf only

MINOR PESTS

Myzus ascalonicus Don.	Shallot Aphid	Aphididae	Europe, USA, Canada	Adults & nymphs infest foliage; suck sap; virus vectors
Myzus persicae (Sulz.)	Peach–Potato Aphid	Aphididae	Cosmopolitan	
Aphis gossypii Glov.	Cotton Aphid	Aphididae	Cosmopolitan	
Aulacorthum solani (Malt.)	Potato Aphid	Aphididae	Cosmopolitan	
Rhopalosiphum maidis (Fitch)	Corn Leaf Aphid	Aphididae	Cosmopolitan	
Empoasca spp.	Green Leafhoppers	Cicadellidae	Asia	Infest foliage & suck sap
Ferrisia virgata (Ckll.)	Striped Mealybug	Pseudococcidae	Asia	
Planococcus citri Risso	Citrus (Root) Mealybug	Pseudococcidae	Cosmopolitan	
Saissetia coffeae (Wlk.)	Helmet Scale	Coccidae	Asia	
Cyrtopeltis spp.	Tomato Mirids	Miridae	Asia, USA	Sap-suckers; toxic saliva; feeding causes necrosis
Acanthocoris spp.	Coreid Bugs	Coreidae	S.E. Asia	
Nezara viridula (L.)	Green Stink Bug	Pentatomidae	Cosmopolitan	
Frankliniella spp.	Flower Thrips	Thripidae	Cosmopolitan	Infest flowers & foliage
Lamprosema diamenalis Guen.	–	Pyralidae	Malaysia	Larvae eat leaves
Maruca testulalis (Geyer)	Mung Moth	Pyralidae	Pantropical	

Sylepta derogata (F.)	Cotton Leaf Roller	Pyralidae	S.E. Asia	Larvae roll leaves
Agrotis segetum (D. & S.)	Common Cutworm	Noctuidae	India, S.E. Asia	Larvae are cutworms
Xestia c-nigrum (L.)	Spotted Cutworm	Noctuidae	Cosmopolitan	
Tiracola plagiata (Wlk.)	Banana Fruit Caterpillar	Noctuidae	S.E. Asia	Larvae defoliate
Plusia spp.	Semiloopers	Noctuidae	Cosmopolitan	
Chrysodeixis chalcites (Esp.)	Cabbage Semilooper	Noctuidae	Old World	
Heliothis assulta Gn.	Cape Gooseberry Budworm	Noctuidae	India, S.E. Asia	Larvae bore & eat buds & leaves
Heliothis virescens (F.)	Tobacco Budworm	Noctuidae	N., C. & S. America	
Heliothis punctigera Wllgr.	Native Budworm	Noctuidae	Australia	
Mythimna spp.	Cereal Armyworms	Noctuidae	Pantropical	Larvae are leaf eaters; may defoliate
Spodoptera ornithogalli (Gn.)	Yellow-striped Armyworm	Noctuidae	N., C. & S. America	
Spodoptera litura (F.)	Rice Cutworm	Noctuidae	S.E. Asia	
Spodoptera littoralis (Boisd.)	Cotton Leafworm	Noctuidae	Africa, S. Europe	
Spodoptera exigua (Hb.)	Lesser Armyworm	Noctuidae	Cosmopolitan	
Manduca quinquemaculata (Haw.)	Tomato Hornworm	Sphingidae	N., C. & S. America	
Agrius convolvuli (L.)	Convolvulus Hawk Moth	Sphingidae	Asia, Europe, Africa, Australasia	
Phthorimaea operculella (Zell.)	Potato Tuber Moth	Gelechiidae	Cosmopolitan	Larvae mine leaves and bore stem
Scrobipalpa heliopa (Lower)	Tobacco Stem Borer	Gelechiidae	India, S.E. Asia, Africa	
Scrobipalpula absoluta Meyr.	Tomato Leaf Miner	Gelechiidae	S. America	
Delia platura (Meign.)	Bean Seed Fly	Anthomyiidae	Cosmopolitan	Larvae bore sown seeds & seedlings
Solenopsis geminata (F.)	Fire Ant	Formicidae	Asia	Attack workers
Gonocephalum spp.	Dusty Brown Beetles	Tenebrionidae	Africa, Asia	Adults damage stem
Epilachna spp.	Elipachna Beetles	Coccinellidae	Asia, USA	Adults & larvae defoliate
Psylliodes spp.	Tobacco Flea Beetles	Chrysomelidae	Asia	Adults hole leaves
Leptinotarsa decemlineata Say	Colorado Beetle	Chrysomelidae	Europe, USA, C. America	Adults & larvae defoliate
Epitrix hirtipennis (Melsh.)	Tobacco Flea Beetle	Chrysomelidae	USA (Florida)	Adults eat leaves
Oulema bilineata (Germ.)	Tobacco Leaf Beetle	Chrysomelidae	Africa, S. America	
Orthaulaca similis Ol.	Tobacco Leaf Beetle	Chrysomelidae	Philippines	
Agriotes spp.	Wireworms	Elateridae	Cosmopolitan	Larvae in soil eat roots & stem base
Conoderus spp.	Tobacco Wireworms	Elateridae	USA	
Anomala spp.	White Grubs	Scarabaeidae	Asia	
Cotinis spp.	June Beetles	Scarabaeidae	USA, Canada	
Leucopholis irrorata (Chevr.)	White Grub	Scarabaeidae	Philippines	
Polyphagotarsonemus latus (Banks)	Yellow Tea Mite	Tarsonemidae	S.E. Asia	Scarify leaves
Tetranychus cinnabarinus (Boisd.)	Tropical Red Spider Mite	Tetranychidae	Pantropical	

TOMATO (*Lycopersicum esculentum* – Solanaceae)

Tomato originated in S. America in the Peru/Ecuador region, and was taken to the Philippines and Malaya by 1650. For some time it has been cultivated in the temperate regions of America and Europe, but it was not cultivated in the tropics until the twentieth century. It is now grown very widely throughout the world. It can be grown in the open wherever there are more than three months of frost-free weather, but it needs even rainfall and long sunny periods for best results. It can be grown at sea level but usually does better at higher altitudes. It is a variable annual herb, 0.7–2 m high, and the fruit for which it is grown is a fleshy berry, red or yellow when ripe, containing vitamins A and C. The fruit is used raw as a vegetable, or cooked, made into soup, sauce, juice, ketchup, paste, puree, powder, or may be canned; also used unripe in chutneys. The main production areas are in the USA, Mexico and Italy, but most countries have a large local production and consumption; in cooler temperate countries most production is under glass or polythene shelters.

MAJOR PESTS

Brachytrupes spp.	Large Brown Crickets	Gryllidae	Africa, Asia	Destroy seedlings
Bemisia tabaci (Genn.)	Tobacco Whitefly	Aleyrodidae	Cosmopolitan	Infest foliage; suck sap; virus vectors
Trialeurodes vaporariorum (Westw.)	Glasshouse Whitefly	Aleyrodidae	Europe, USA	
Cyrtopeltis tenuis Reut.	Tomato Mirid	Miridae	S. & E. Asia	Sap-suckers; toxic saliva; foliage & fruit spotted
Nezara viridula (L.)	Green Stink Bug	Pentatomidae	Cosmopolitan	
Scrobipalpula absoluta Meyr.	Tomato Leafminer	Gelechiidae	S. America	Larvae mine leaves, etc.
Heliothis spp.	'Tomato Fruitworms'	Noctuidae	Cosmopolitan	Larvae bore fruits
Thrips tabaci Lind.	Onion Thrips	Thripidae	Cosmopolitan	Infest foliage
Tetranychus urticae (Koch)	Temperate Red Spider Mite	Tetranychidae	Europe, Asia, N. America	Scarify & web foliage (glasshouse pest)

MINOR PESTS

Gryllotalpa spp.	Mole Crickets	Gryllotalpidae	Cosmopolitan	Eat roots; may kill seedlings
Paratrioza cockerelli (Sulc.)	Potato Psyllid	Psyllidae	Canada, USA	Cause leaf-curling
Empoasca spp.	Green Leafhoppers	Cicadellidae	Cosmopolitan	Infest foliage; sap suckers; virus vectors
Zygina pallidifrons (J.Ed.)	Glasshouse Leafhopper	Cicadellidae	Europe	
Aphis gossypii Glov.	Cotton/Melon Aphid	Aphididae	Cosmopolitan	
Aulacorthum solani (Kalt.)	Glasshouse–Potato Aphid	Aphididae	Europe, USA	Infest foliage; sap suckers; virus vectors
Macrosiphum euphorbiae (Thos.)	Potato Aphid	Aphididae	Cosmopolitan	
Myzus persicae (Sulz.)	Peach–Potato Aphid	Aphididae	Cosmopolitan	
Ferrisia virgata (Ckll.)	Striped Mealybug	Pseudococcidae	Pantropical	Infest foliage; suck sap
Pinnaspis minor Mask.	Armoured Scale	Diaspididae	S.E. Asia	
Anthocoris spp.	Coreid Bugs	Coreidae	S.E. Asia	Sap-suckers; toxic saliva
Frankliniella spp.	Flower Thrips	Thripidae	Cosmopolitan	Infest flowers & foliage

Keiferia lycopersicella (Wals.)	Tomato Pinworm	Gelechiidae	USA, C. & S. America	Larvae bore in leaves & fruits
Phthorimaea operculella (Zell.)	Potato Tuber Moth	Gelechiidae	Cosmopolitan	Larvae bore stem
Cnephasia spp.	Leaf Tiers	Tortricidae	Europe	Larvae eat leaves
Leucinodes orbonalis Gn.	Eggplant Boring Caterpillar	Pyralidae	Africa, Asia	Larvae bore fruits
Acherontia atropos (L.)	Death's Head Hawk Moth	Sphingidae	Europe, Asia, Africa	Larvae eat leaves; sometimes defoliate
Manduca quinquemaculata (Haw.)	Tomato Hornworm	Sphingidae	Canada, USA, C. & S. America	
Manduca sexta (L.)	Tobacco Hornworm			
Agrotis spp.	Cutworms	Noctuidae	Cosmopolitan	Cutworm larvae kill seedlings
Hydraecia micacea (Esp.)	Rosy Rustic Moth	Noctuidae	Europe, Asia, USA	Larvae bore stem
Lacanobia oleracea (L.)	Tomato Moth	Noctuidae	Europe	Larvae bore fruits
Mythimna spp.	Cereal Armyworms	Noctuidae	Asia	Larvae defoliate
Othreis fullonia (Cl.)	Fruit-piercing Moth	Noctuidae	Old World	Adults pierce fruits
Spodoptera spp.	'Armyworms'	Noctuidae	Cosmopolitan	Larvae eat foliage
Xestia c-nigrum (L.)	Spotted Cutworm	Noctuidae	Europe, Asia, N. America	Larvae damage young plants
Anomis flava (F.)	Cotton Semilooper	Noctuidae	Old World tropics	Larvae are semiloopers; eat foliage
Chrysodeixis chalcites (Esp.)	Cabbage Semilooper	Noctuidae	Old World	
Plusia spp.	Semiloopers	Noctuidae	Cosmopolitan	
Contarinia lycopersici Felt	Tomato Flower Midge	Cecidomyiidae	C. & S. America	Larvae destroy flowers
Dacus spp.	Fruit Flies	Tephritidae	Asia	Larvae inside fruits
Liriomyza bryoniae (Kalt.)	Tomato Leaf Miner	Agromyzidae	Europe, N. Africa, USSR	Larvae mine leaves
Liriomyza solani (Her.)	Tomato Leaf Miner	Agromyzidae	Europe	
Liriomyza trifolii (Burg.)	Serpentine Leaf Miner	Agromyzidae	N. & S. America, now UK	Larvae mine leaves
Anomala spp.	White Grubs	Scarabaeidae	Asia	Larvae eat roots
Epilachna spp.	Epilachna Beetles	Coccinellidae	Asia	Larvae & adults eat foliage
Epicauta spp.	Black Blister Beetles	Meloidae	Asia, USA, Canada, Africa	Adults eat flowers & foliage
Agriotes spp.	Wireworms	Elateridae	Europe, Asia	Larvae in soil eat roots
Epitrix spp.	Flea Beetles	Chrysomelidae	USA	Adults eat leaves; larvae eat roots
Psylliodes spp.	Tobacco Flea Beetles	Chrysomelidae	Europe, Asia	
Leptinotarsa decemlineata (Say)	Colorado Beetle	Chrysomelidae	Europe (not UK), N. & C. America	Adults & larvae defoliate
Listroderes costirostris (Sch.)	Vegetable Weevil	Curculionidae	Australia, USA, S. America	Adults eat foliage; larvae eat roots
Aculus lycopersici (Massee)	Tomato Russet Mite	Eriophyidae	Cosmopolitan (not UK)	Scarify foliage
Eriophyes lycopersici (Wolf.)	Tomato Gall Mite	Eriophyidae	Pantropical	Erinea on leaves
Polyphagotarsonemus latus (Banks)	Yellow Tea Mite	Tarsonemidae	Cosmopolitan	Scarify foliage
Tetranychus cinnabarinus (Boisd.)	Carmine Spider Mite	Tetranychidae	Pantropical	Scarify & web foliage
Tetranychus spp.	Red Spider Mites	Tetranychidae	Cosmopolitan	

TULIP (*Tulipa* spp. – Liliaceae)

About 100 species are cultivated for their flowers as ornamentals; field-scale cultivation for the bulbs is widespread and common especially in the Netherlands and eastern England. The genus is indigenous to Turkey. The brightly coloured goblet-shaped flowers grow from rounded or ovoid bulbs, which are susceptible to many diseases, nematodes and other soil pests; generally not attacked by many insect pests.

MAJOR PESTS

The more important pests of Tulip appear to be nematodes rather than insects

MINOR PESTS

Aulacorthum circumflexum (Buckt.)	Mottled Arum Aphid	Aphididae	Europe, USA, Japan	Adults & nymphs infest both bulbs in store & foliage; suck sap & virus vectors
Dysaphis tulipae (Fonsc.)	Tulip Aphid	Aphididae	Europe, USA, Japan	
Myzus persicae (Sulz.)	Peach–Potato Aphid	Aphididae	Cosmopolitan	
Rhopalosiphoninus staphyleae ssp. *tulipaellus* (Theo.)	Tulip Leaf Aphid	Aphididae	Europe, USA	
Merodon equestris (F.)	Narcissus Bulb Fly	Syrphidae	Europe, USA, Japan	Maggots inside bulb
Delia antiqua (Meig.)	Onion Fly	Anthomyiidae	Europe, USA	Larvae inside bulbs
Rhizoglyphus echinopus (F. & R.)	Bulb Mite	Acaridae	Europe, USA	Infest bulbs; may cause rots
Eriophyes tulipae Keif.	Tulip Bulb Mite	Eriophyidae	Europe, USA	

WALNUT (*Juglans regia* – Juglandaceae) (= English Walnut)

Despite the name this is native to Iran, now extensively cultivated in Europe (especially France), China, parts of Asia and also the USA. The trees are also used in an additional capacity as ornamentals as they are large (up to 20 m) and attractive. The kernels are furrowed and easily freed from the pericarp (husk); they represent the cotyledons of the seed, no endosperm being present. The kernels are eaten raw, in cakes, in confectionery, and as a source of oil which is excellent for table use. *J. nigra* is the Black Walnut of the USA, but the shell is so hard that its use is more or less confined to the confectionery industry. *J. cinerea* is the Butternut, also of eastern USA and Canada, with a higher fat content, used mainly for confectionery.

MAJOR PESTS

Paramyelois transitella Wlk.	Navel Orangeworm	Pyralidae	USA	Larvae bore fruits
Batocera horsfieldi Hope	Longhorn Beetle	Cerambycidae	India	Larvae bore trunk

MINOR PESTS

Arytania fasciata Laing	Walnut Psylla	Psyllidae	India	Nymphs gall leaves
Aphis pomi de Geer	Apple Aphid	Aphididae	India	Adults & nymphs infest foliage; suck sap & curl leaves
Callipterus juglandis Goeze	Walnut Aphid	Aphididae	Europe, India	
Chromaphis juglandicola (Kltb.)	Walnut Aphid	Aphididae	Europe, Asia, USA	
Lecanium spp.	Soft Scales	Coccidae	USA	Adults & nymphs infest foliage; suck sap; often with sooty moulds
Parthenolecanium corni (Bche.)	Plum Scale	Coccidae	Europe, Asia	
Icerya purchasi Mask.	Cottony Cushion Scale	Margarodidae	Asia, USA	
Pseudococcus maritimus (Ehrh.)	Grape Mealybug	Pseudococcidae	Cosmopolitan	
Aonidiella aurantii (Mask.)	California Red Scale	Diaspididae	USA	Infest bark of twigs, branches & trunk
Aspidiotus juglans-regiae Comst.	Walnut Scale	Diaspididae	USA	
Quadraspidiotus perniciosus (Comst.)	San José Scale	Diaspididae	Asia, USA	
Corythucha juglandis (Fitch)	Walnut Lacebug	Tingidae	Canada, USA	Sap-sucker; toxic saliva
Coleophora laticornella Clem.	Pecan Cigar Casebearer	Coleophoridae	USA	Larvae in cases mine leaves
Zeuzera spp.	Leopard Moths	Cossidae	Europe, Asia, USA	Larvae bore branches
Cacoecia oporana L.	Fruit Tree Tortrix	Tortricidae	Europe	Larvae eat foliage
Cydia funebrana (Treit.)	Red Plum Maggot	Tortricidae	Europe, Asia	Larvae bore fruits
Cydia pomonella (L.)	Codling Moth	Tortricidae	Cosmopolitan	
Datana integerrima G. & R.	Walnut Caterpillar	Notodontidae	Canada, USA	Larvae eat leaves, may defoliate
Datana spp.	–	Notodontidae	USA	
Phalera bucephala L.	Buff-tip Moth	Notodontidae	Europe	
Schizura concinna (J.E. Sm.)	Red-humped Caterpillar	Notodontidae	USA	
Actias selene (Hb.)	Moon Moth	Saturniidae	India	Larvae eat leaves, may defoliate
Cressonia juglandis (J.E. Sm.)	Walnut Sphinx	Sphingidae	USA	
Malacosoma spp.	Tent Caterpillars	Lasiocampidae	Europe, Asia	Larvae defoliate gregariously
Rhagoletis completa Cresson	Walnut Husk Fly	Tephritidae	USA	Larvae bore fruit husk

Eriocampa juglandis (Fitch)	Butternut Woolly Sawfly	Tenthredinidae	USA	Larvae defoliate
Holotrichia longipennis (Blanch.)	Cockchafer	Scarabaeidae	India	Adults eat leaves; larvae in soil eat roots & damage young trees
Anomala spp.	Flower Beetles	Scarabaeidae	India	
Mimela pusilla Hope	Flower Beetle	Scarabaeidae	India	
Dorysthenus hugelii Redt.	Root Borer	Scarabaeidae	India	
Apate monachus F.	Black Borer	Bostrychidae	Israel	Adults bore branches
Hispa dama Chap.	Hispid Beetle	Chrysomelidae	India	Larvae mine leaves; adults eat leaves
Altica cerulescens (Baly)	Flea Beetle	Chrysomelidae	India	Adults & larvae eat leaves
Monolepta erythrocephala Baly	Leaf Beetle	Chrysomelidae	India	Defoliate
Agrilus spp.	Walnut Twig Borers	Buprestidae	China	Larvae bore twigs
Aeolesthes holoserica F.	Cherry Stem Borer	Cerambycidae	Pakistan, India	Larvae bore trunk & branches
Aeolesthes sarta Solsky	Quetta Borer	Cerambycidae	Pakistan, India	
Batocera rufomaculata (de Geer)	Red-spotted Longhorn	Cerambycidae	India, China	
Alcidodes porrectirostris Mshll.	Walnut Weevil	Curculionidae	India	Larvae bore fruits
Conotrachelus juglandis Lec.	Butternut Curculio	Curculionidae	N. America	Larvae bore shoots & fruits
Myllocerus viridianus F.	Grey Weevil	Curculionidae	India	Adults eat leaves
Conotrachelus retentus Say	Black Walnut Curculio	Curculionidae	USA	Larvae bore shoots
Phyllobius spp.	Leaf Weevils	Curculionidae	Europe	Adults eat leaves
Diapus pusillimus Chapuis	Walnut Pinhole Borer	Scolytidae	Australia	Adults bore under bark of branches & trunk
Scolytus juglandis	Walnut Pinhole Borer	Scolytidae	India	
Eotetranychus hicoriae (McG.)	Hickory Scorch Mite	Tetranychidae	USA	Scarify foliage
Eriophyes spp. (6)	Walnut Gall Mites	Eriophyidae	Europe, Asia, USA	Make erinea, leaf galls & roll leaves
Eriophyes brachytarsus (Keif.)	California Black Walnut Pouch Gall Mite	Eriophyidae	USA	Make erinea under leaves
Aceria erinea (Nal.)	Walnut Blister Mite	Eriophyidae	Europe, USA, Australia	

WATERCRESS (*Nasturtium officinale* – Cruciferеae)

This is regarded as one of the minor herbage vegetables, grown widely but only for immediate local consumption. It occurs in the wild state in the UK and S. Europe, and is now introduced into many parts of the world. The plant is an aquatic perennial herb, and the tips of the leafy stems are used as salad, or it may be cooked as a vegetable. It grows best in clear-running shallow water; the distal part of the plant usually protrudes from the water, and flowers are aerial.

MAJOR PESTS

Plutella xylostella L.	Diamond-back Moth	Yponomeutidae	S. E. Asia	Larvae eat leaves

MINOR PESTS

Draeculacephala mollipes (Say)	Watercress Sharpshooter	Cicadellidae	USA	Adults & nymphs infest aerial foliage; suck sap
Aphis nasturtii Kalt.	Buckthorn–Potato Aphid	Aphididae	Europe, Asia, N. America	
Myzus ascalonicus Don.	Shallot Aphid	Aphididae	Europe, Asia, N. & S. America	
Myzus persicae (Sulz.)	Peach–Potato Aphid	Aphididae	S. China	
Lipaphis erysimi (Kalt.)	Turnip Aphid	Aphididae	S. China	
Rhopalosiphum rufiabdominalis (Sasaki)	Rice Root Aphid	Aphididae	S. China	
Phaedon aeruginosus Suff.	Watercress Leaf Beetle	Chrysomelidae	USA	Adults & larvae eat aerial foliage
Phaedon spp.	Mustard Beetles	Chrysomelidae	Europe	
Tarsonemus pallidus Banks	Strawberry Mite	Tarsonemidae	Europe, Asia, N. America	Infest aerial foliage

WHEAT (*Triticum sativum* – Gramineae)

The chief cereal of temperate regions, and the most widely grown cereal; of great antiquity as a crop, of uncertain origin but thought to be from C. or S.W. Asia. Introduced into the New World in 1529 by the Spaniards who took it to Mexico. Wheat is an annual grass, 0.5–1 m in height, with the inflorescence as a terminal spike. Many wild species are known, and many cultivars have been bred including a number of polyploid species which are important agriculturally. It is adapted to moderately dry temperate climates, where it requires a growing season of at least 90 days and 20 cm rainfall; more than 70 cm of rain is detrimental. The main production areas are the vast temperate grasslands, that is the plains of S. Russia, N.W. Europe, N.W. India, N.W. China, S.E. Australia, and the central plains of Canada and the USA, and the Argentine.

MAJOR PESTS

Rhopalosiphum padi (L.)	Bird-cherry Aphid	Aphididae	Europe, Australia	Adults & nymphs infest foliage; suck sap; virus vectors
Schizaphis graminum Rond.	Wheat Aphid	Aphididae	Cosmopolitan	
Sitobion avenae (F.)	Grain Aphid	Aphididae	Europe, N. America	
Eurygaster integriceps Put.	Senn Pest	Pentatomidae	W. Europe, Near & Middle East	Sap-sucker; toxic saliva
Mayetiola destructor (Say)	Hessian Fly	Cecidomyiidae	Europe, USA	Larvae gall stems & cause breaks
Delia coarctata (Fall.)	Wheat Bulb Fly	Anthomyiidae	Europe	Larvae bore seedling stems; dead-heart

MINOR PESTS

Rhopalosiphum maidis (Fitch)	Corn Leaf Aphid	Aphididae	Cosmopolitan	Adults & nymphs infest foliage; suck sap; virus vectors; may be honey-dew & sooty moulds
Rhopalosiphum insertum (Wlk.)	Apple–Grass Aphid	Aphididae	Europe	
Sipha spp.	Aphids	Aphididae	Europe, Asia, N. America	
Macrosiphum fragariae Wlk.	Blackberry Aphid	Aphididae	Cosmopolitan	
Metapolophium dirhodum (Wlk.)	Rose–Grain Aphid	Aphididae	Europe	
Metapolophium festucae (Theob.)	Fescue Aphid	Aphididae	Europe	
Forda spp.	Root Aphids	Pemphigidae	Europe, W. Asia	Infest roots
Javasella pellucida (F.)	Cereal Planthopper	Delphacidae	Europe	Infest foliage; suck sap; virus vectors
Laodelphax striatella (Fall.)	Small Brown Planthopper	Delphacidae	Europe, Asia	
Stenodema calcaratum Fall.	Wheat Leaf Bug	Miridae	Japan	Sap-suckers; toxic saliva
Blissus leucopterus (Say)	Chinch Bug	Lygaeidae	N., C. & S. America	
Blissus pallipes Dist.	Chinch Bug	Lygaeidae	China	
Eurygaster austriaca (Schr.)	Wheat Shield Bug	Pentatomidae	S. Europe, W. Asia	
Aelia spp.	Wheat Shield Bugs	Pentatomidae		
Agriphila straminella (D. & S.)	Grass Moth	Pyralidae (Crambinac)	Europe	Larvae bite stems and eat leaves; often in silken tubes on ground
Crambus spp.	Grass Moths (Sod Webworms)	Pyralidae (Crambinae)	Europe, USA, Canada	
Sitotroga cerealella (Ol.)	Angoumois Grain Moth	Gelechiidae	Cosmopolitan	Larvae inside grains, in field & stores

Hepialus spp.	Swift Moths	Hepialidae	Europe	Larvae in soil eat roots
Cnephasia longana Haw.	Omnivorous Leaf Tier	Tortricidae	Europe, N. America	Larvae tie & eat leaves
Marasmia trapezalis (Gn.)	Maize Webworm	Pyralidae	Africa, Asia, Australasia, C. & S. America	Larvae feed on young seedhead
Agrotis spp.	Cutworms	Noctuidae	Cosmopolitan	Larvae are cutworms
Hydraecia micacea (Esp.)	Rosy Rustic Moth	Noctuidae	Europe, N. Asia, USA	
Mesapamea secalis (L.)	Common Rustic Moth	Noctuidae	Europe, Asia	Larvae bore stems; cause 'dead-hearts'
Oria musculosa (Hbn.)	Brighton Wainscot	Noctuidae	S. Europe, Asia Minor	
Luperina testacea (Schiff.)	Flounced Rustic Moth	Noctuidae	Europe	
Mythimna loreyi (Dup.)	Cereal Armyworm	Noctuidae	Old World	
Mythimna unipuncta (Haw.)	American Army-worm	Noctuidae	USA, Canada	Larvae (sometimes gregarious) eat leaves & defoliate
Spodoptera frugiperda (J.E.Sm.)	Fall Armyworm	Noctuidae	N. & S. America	
Sesamia spp.	Stalk Borers	Noctuidae	Mediterranean, Africa, Asia	Larvae bore stems
Contarinia tritici (Kirby)	Lemon Wheat Blossom Midge	Cecidomyiidae	Europe, Asia	
Sitodiplosis mossellana (Gehin)	Orange Wheat Blossom Midge	Cecidomyiidae	Europe, Asia, N. America	Larvae in flowerheads
Haplodiplosis marginata (v.R.)	Saddle Gall Midge	Cecidomyiidae	Europe	
Mayetiola avenae (March.)	Oat Stem Midge	Cecidomyiidae	Europe	
Mayetiola spp.	Gall Midges	Cecidomyiidae	Europe, Asia, N. America	Larvae gall stems
Nephrotoma spp. *Tipula* spp.	Leatherjackets	Tipulidae	Europe, Asia, Canada, USA	Larvae in soil eat roots; may destroy seedlings
Bibio marci (L.)	St. Mark's Fly	Bibionidae	Europe	
Geomyza spp. *Meromyza* spp. *Opomyza* spp.	Grass Flies	Opomyzidae	Europe, Asia, N. America	Larvae eat stems & make 'dead-hearts'
Chlorops pumilionis (Bjerk.)	Gout Fly	Chloropidae	Europe, Asia	
Oscinella frit (L.)	Frit Fly	Chloropidae	Europe	Larvae gall stems
Oscinella spp.	'Frit' Flies	Chloropidae	Cosmopolitan	
Hydrellia griseola (Fall.)	Cereal Leaf Miner	Ephydridae	Europe, Asia, USA, S. America	
Agromyza ambigua Fall.	Cereal Leaf Miner	Agromyzidae	Europe, USA, Canada	Larvae mine leaves
Atherigona spp.	Cereal Shoot Flies	Muscidae	Cosmopolitan	
Phorbia genitalis Tiens.	Late Wheat Shoot Fly	Anthomyiidae	Europe	Larvae eat shoots; cause 'dead-hearts'

Delia platura (Meign.)	Bean Seed Fly	Anthomyiidae	Cosmopolitan	Larvae eat sown seed; may kill seedlings
Anaphothrips obscurus (Mull.)	Grain Thrips	Thripidae	Canada, USA	Infest flowers; reduce seed yield
Aptinothrips spp.	Grass Thrips	Thripidae	Europe, USA, Canada	Adults & nymphs infest foliage & scarify (fleck) leaves
Limothrips cerealium Hal.	Grain Thrips	Thripidae	Europe, USA, Canada, Australia	
Stenothrips graminum Uzel	Oat Thrips	Thripidae	Europe	
Thrips nigropilosus Uzel	Chrysanthemum Thrips	Thripidae	Europe, N. America	
Haplothrips tritici (Kurd.)	Wheat Thrips	Phlaeothripidae	Europe (not UK)	Fold leaves
Cephus cinctus Nort.	Western Wheat-stem Sawfly	Cephidae	N. America	Larvae bore lower parts of stems
Cephus pygmaeus (L.)	Wheat-stem Sawfly (European)	Cephidae	Europe, Canada, E. USA	
Dolerus spp.	Leaf Sawflies	Tenthredinidae	Europe, China	Larvae eat leaves
Tetramesa spp.	Wheat Jointworms	Eurytomidae	Canada, USA	Larvae gall stems
Agriotes spp.	Wireworms	Elateridae	Europe, USA, Asia	Larvae in soil eat roots; may destroy seedlings
Athous spp.	Garden Wireworms	Elateridae	Europe	
Ctenicera spp.	American Wireworms	Elateridae	Europe, N. America	
Helophorus nubilus F.	Wheat Shoot Beetle	Hydrophilidae	Europe	Larvae damage seedling stems
Anomala spp.	White Grubs	Scarabaeidae	Asia	Larvae in soil eat roots & may destroy seedlings
Melolontha spp.	Cockchafers	Scarabaeidae	Europe	
Phyllophaga spp.	June Beetles	Scarabaeidae	Canada, USA	
Schizonycha spp.	Chafer Grubs	Scarabaeidae	Europe, Africa	
Amphimallon solstitialis (L.)	Summer Chafer	Scarabaeidae	Europe	
Serica spp.	Brown Chafers	Scarabaeidae	Europe, Asia	
Oulema melanopa (L.)	Cereal Leaf Beetle	Chrysomelidae	Europe, USA	Adults & larvae eat strips along leaves
Chaetocnema hortensis (Geoff.)	Cereal Flea Beetle	Chrysomelidae	Europe	Adults hole leaves
Crepidodera ferruginea (Scop.)	Wheat Flea Beetle	Chrysomelidae	Europe	Larvae bore central shoot
Phyllotreta vittula Redt.	Barley Flea Beetle	Chrysomelidae	Europe	Larvae mine stems; adults hole leaves
Tanymecus dilaticollis Gylh.	Southern Grey Weevil	Curculionidae	E. Europe	Adults notch leaves
Siteroptes cerealium (Kirch.)	Grass & Cereal Mite	Penthaleidae	Europe	Cause 'silvertop'
Stenotarsonemus spirifex (March)	Oat Spiral Mite	Tarsonemidae	Europe	Damage flowerhead
Tyrophagus spp.	Grass Mites	Acaridae	Cosmopolitan	Infest leaf sheaths

PESTS OF SEEDLINGS & GENERAL PESTS

Many of the pests already referred to are polyphagous and recorded damaging many different crops, and most are particularly damaging to young plants and seedlings which may be destroyed by their feeding. In general soil-dwelling pests tend not to be very host-specific and usually attack (eat) the roots and underground stems of anything not too woody growing in that soil. Similarly, some leaf-eaters (Acrididae; Noctuidae; Curculionidae) and some sap-suckers (Hemiptera; Tetranychidae) do not appear to be very selective as to host plant. Some small insects with weak mouthparts (Collembola) only cause damage to seedlings, because of their size. Some of the general pests, such as locusts, are more restricted geographically than by particular host preferences.

THE MORE IMPORTANT PESTS

Sminthurus spp.	Garden Springtails	Sminthuridae	Cosmopolitan	Damage seedlings, especially ones with a soft stem
Bourletiella spp.	Garden Springtails	Sminthuridae	Cosmopolitan	
Onychiurus spp.	Springtails	Onychiuridae	Cosmopolitan	
Chortoicetes terminifera (Wlk.)	Australian Plague Locust	Acrididae	Australia	General defoliators, both adults & nymphs; often show preference for Gramineae as food
Dociostaurus maroccanus (Thnb.)	Mediterranean Locust	Acrididae	Mediterranean	
Locusta m. migratoria	Asiatic Migratory Locust	Acrididae	C. Asia	
Locusta m. migratorioides (R. & F.)	African Migratory Locust	Acrididae	Africa	
Locusta m. maniliensis (Meyr.)	Oriental Migratory Locust	Acrididae	E. & S.E. Asia	
Melanoplus spp.	Grasshoppers	Acrididae	Canada, USA	
Nomadacris septemfasciata (Serv.)	Red Locust	Acrididae	Africa	
Patanga succincta (L.)	Bombay Locust	Acrididae	S. & E. Asia	
Schistocerca gregaria (Forsk.)	Desert Locust	Acrididae	Africa to India	
Zonocerus spp.	Variegated Grasshoppers	Acrididae	Africa	
Gryllotalpa spp.	Mole Crickets	Gryllotalpidae	Cosmopolitan	Live underground; nocturnal; eat roots and destroy seedlings at night
Brachytrupes spp.	Large Brown Crickets	Gryllidae	Asia, Africa	
Gryllus spp., *Acheta* spp.	Field Crickets	Gryllidae	Cosmopolitan	
Forficula auricularia L.	Common Earwig	Forficulidae	Cosmopolitan	Bite holes; may kill soft seedlings
Aphis spp., *Myzus* spp.	Aphids	Aphididae	Cosmopolitan	Infest foliage; sap-suckers; virus vectors
Empoasca spp.	Green Leafhoppers	Cicadellidae	Cosmopolitan	
Coccus spp.	Soft Green Scales	Coccidae	Cosmopolitan	Infest foliage; suck sap
Saissetia spp.	Brown Scales	Coccidae	Cosmopolitan	
Pseudococcus spp.	Mealybugs	Pseudococcidae	Cosmopolitan	
Aspidiotus spp.	Armoured Scales	Diaspididae	Cosmopolitan	Infest twigs mostly
Lygus spp.	Tarnished Plant Bugs	Miridae	Cosmopolitan	Sap-suckers; toxic saliva; feeding causes necrosis
Lygocoris pabulinus (L.)	Common Green Capsid	Miridae	Europe	
Nezara viridula (L.)	Green Stink Bug	Pentatomidae	Cosmopolitan	

Frankliniella spp.	Flower Thrips	Thripidae	Cosmopolitan	Infest flowers & plant foliage
Thrips spp.	Thrips	Thripidae	Cosmopolitan	
Hepialus spp.	Swift Moths	Hepialidae	Europe, Asia	Larvae in soil eat roots
Cnephasia spp.	Polyphagous Leaf Tiers	Tortricidae	Cosmopolitan	Larvae roll & eat leaves
Agrotis spp.				Larvae are cutworms; live in soil; eat stems of seedlings; damage stems & roots of older plants
Euxoa spp.	Cutworms	Noctuidae	Cosmopolitan	
Spodoptera spp.				
Noctua pronubà (L.)	Large Yellow Underwing	Noctuidae	Europe	
Heliothis spp.	Leafworms, etc.	Noctuidae	Cosmopolitan	Larvae are polyphagous leaf eaters
Spodoptera spp.				
Mythimna spp.	Armyworms	Noctuidae	Cosmopolitan	Larvae gregarious polyphagous leaf eaters
Spodoptera spp.				
Plusia spp.	Semiloopers	Noctuidae	Cosmopolitan	Larvae defoliate
Liriomyza spp.	Leaf Miners	Agromyzidae	Cosmopolitan	Larvae mine leaves
Phytomyza spp.				
Delia spp.	Bean Seed Flies	Anthomyiidae	Cosmopolitan	Larvae eat sown seeds & seedlings
Tipula spp.	Leatherjackets	Tipulidae	Cosmopolitan	Larvae in soil eat roots & seedlings
Nephrotoma spp.				
Atta spp.	Leaf-cutting Ants	Formicidae	S. USA, C. & S. America	Adults polyphagous defoliators
Acromyrmex spp.				
Vespa/Vespula spp.	Common Wasps	Vespidae	Cosmopolitan	Adults pierce ripe fruits
Polistes spp.	Paper Wasps	Vespidae	Pantropical	
Agriotes spp.	Wireworms	Elateridae	Europe, USA, Canada	Larvae polyphagous soil pests
Limonius spp.				
Anomala spp.	Flower Beetles	Scarabaeidae	Cosmopolitan	Larvae in soil eat roots, kill young plants; adults eat leaves & fruits
Cetonia spp.	Rose Chafers		Cosmopolitan	
Holotrichia spp.	White Grubs		Europe, Asia	
Melolontha spp.	Cockchafers		Europe, Asia	
Phyllophaga spp.	June Beetles		Canada, USA	
Popillia japonica Newm.	Japanese Beetle		Asia, N. America	Adults defoliate
Otiorhynchus spp.	Weevils	Curculionidae	Cosmopolitan	Adults eat leaves; larvae in soil eat roots
Phyllobius spp.	Leaf Weevils	Curculionidae	Europe	
Sphenophorous spp.	Billbugs	Curculionidae	Canada, USA	
Hypomeces squamosus (F.)	Gold-dust Weevil	Curculionidae	Asia	Adults notch (eat) edges of leaves
Myllocerus spp.	Grey Weevils	Curculionidae	S. Asia	
Systates spp.	Systates Weevils	Curculionidae	Africa	
Bryobia spp.	Brown Mites	Tetranychidae	Cosmopolitan	Scarify (silver) & web foliage
Tetranychus spp.	Red Spider Mites	Tetranychidae	Cosmopolitan	
Panonychus ulmi (Koch)	Fruit Tree Red Spider Mite	Tetranychidae	Cosmopolitan in warmer countries	

INSECT PESTS OF STORED PRODUCTS

The types of produce stored is very great, but in this book the major concern is for on-farm storage, rather than warehouse storage in towns and regional depots, as the latter is seldom the purview of the agricultural entomologist. The most serious pests and the most extensive damage is to be found in the tropics where conditions are suitable for continuous breeding of insects and for rapid infection by fungi and bacteria. In temperate regions there is generally a cold winter and most insect pests have evolved diapausal mechanisms to permit their survival, and rates of development are low. However, in the colder temperate regions many food stores are heated and here many of the more tropical pest species may survive, but it is not usual for the on-farm stores to be heated. The overall range of stored products pests (insects and mites) encountered in the different parts of the world is much the same, irrespective of the precise locality, but the dominant species will probably differ. Some species are broadly polyphagous (omnivorous), but some only feed on grains, or on pulses, or flours, or dried animal material. Some of the most important species are regularly found in field infestations and then carried into the produce stores where the breeding cycle continues (see pages 335 and 405). Some species are primary pests in that they can penetrate intact seeds and grains; the secondary pests are unable to do this and can only feed on grains already damaged. Some pests are carried into produce stores after harvest, and development may continue there but there will be no breeding (such pests include Pea Moth larvae and Codling Moth larvae in infested apples); this type of insect is not included in this section.

MAJOR PESTS

Sitotroga cerealella (Ol.)	Angoumois Grain Moth	Gelechiidae	Cosmopolitan	On grains & foodstuffs
Nemapogon granella (L.)	Corn Moth	Tinaeidae	Cosmopolitan	Stored grains
Ephestia cautella (Hb.)	Dried Currant Moth	Pyralidae	Cosmopolitan	Dried fruits, etc.
Ephestia elutella (Hb.)	Warehouse Moth	Pyralidae	Cosmopolitan	Dried fruits, cocoa beans, tobacco, etc.
Lasioderma serricorne (F.)	Tobacco Beetle (Cigarette Beetle)	Anobiidae	Pantropical	Dried tobacco, foodstuffs, etc.
Cryptolestes ferrugineus (Steph.)	Rust-red Grain Beetle	Cucujidae	Cosmopolitan	Stored grains
Tribolium castaneum (Hbst.)	Red Flour Beetle	Tenebrionidae	Cosmopolitan	Flours & grain products
Oryzaephilus mercator (Fauvel)	Merchant Grain Beetle	Silvanidae	Pantropical	Secondary pests on most stored products
Oryzaephilus surinamensis (L.)	Saw-toothed Grain Beetle	Silvanidae	Pantropical	
Rhizopertha dominica (F.)	Lesser Grain Borer	Bostrychidae	Pantropical	Stored grains
Bruchus spp.	Pea & Bean Bruchids	Bruchidae	Europe, Asia, Canada, USA	Pulses in field & in stores
Dermestes lardarius L.	Larder Beetle	Dermestidae	Cosmopolitan	Feed on dried animal material mostly
Dermestes maculatus Deg.	Hide Beetle	Dermestidae	Cosmopolitan	
Trogoderma granarium Everts.	Khapra Beetle	Dermestidae	Pantropical	Stored grains, etc.
Sitophilus granarius (L.)	Grain Weevil	Curculionidae	Cosmopolitan (Temperate)	Wheat, other grains & foodstuffs

MINOR PESTS

Lepisma saccharina L.	'Silverfish'	Lepismatidae	Cosmopolitan	Scavengers; omnivorous
Acheta domesticus (L.)	House Cricket	Gryllidae	Cosmopolitan	
Blatta orientalis L.	Oriental Cockroach	Blattidae	Cosmopolitan (temperate)	Scavengers; polyphagous
Blatella germanica (L.)	German Cockroach	Blattidae	Cosmopolitan	
Periplaneta americana (L.)	American Cockroach	Blattidae	Cosmopolitan (tropical)	
Ephestia kuehniella (Zell.)	Mediterranean Flour Moth	Pyralidae	Subtropical	Infest fruits & berries, in field & in stores
Ephestia spp.	Date & Raisin Moths	Pyralidae	Subtropical	
Plodia interpunctella (Hbn.)	Indian Meal Moth	Pyralidae	Pantropical	Dried fruits, flours, meals, etc.
Stegobium paniceum L.	Biscuit Beetle	Anobiidae	Cosmopolitan	Seeds, produce, etc.
Carpophilus hemipterus (L.)	Dried Fruit Beetle	Nitidulidae	Cosmopolitan	Dried fruits
Attagenus piceus Ol.	Black Carpet Beetle	Dermestidae	Cosmopolitan	Dried animal matter
Ahasverus advena (Waltl.)	Foreign Grain Beetle	Silvanidae	Cosmopolitan	Feeds on fungus
Prostephanus truncatus (Horn)	Greater Grain Borer	Bostrychidae	S. America, E. Africa	Stored maize, cassava, rice, etc.
Necrobia rufipes (Deg.)	Copra Beetle	Cleridae	Pantropical	Copra, oil seeds, dried meats
Acanthoscelides obtectus (Say)	Bean Bruchid	Bruchidae	Pantropical	Attack pulses in the field & in stores
Callosobruchus chinensis (L.)	Oriental Cowpea Bruchid	Bruchidae	Pantropical	
Callosobruchus maculatus (F.)	Spotted Cowpea Bruchid	Bruchidae	Pantropical	
Caryedon serratus (Ol.)	Groundnut Borer	Bruchidae	Pantropical	Groundnuts, pulses
Typhaea stercorea (L.)	Hairy Fungus Beetle	Mycetophagidae	Cosmopolitan	Feeds on fungus
Ptinus spp.	Spider Beetles	Ptinidae	Pantropical	Foodstuffs of many types
Tenebrio molitor L.	Yellow Mealworm Beetle	Tenebrionidae	Cosmopolitan	Flours, meals, etc.
Tribolium confusum J.du V.	Confused Flour Beetle	Tenebrionidae	Cosmopolitan	Flours mostly
Tenebriodes mauritanicus (L.)	Cadelle	Tenebrionidae	Cosmopolitan	Many foodstuffs
Araecerus fasciculatus (Deg.)	Coffee Bean Weevil	Anthribidae	Cosmopolitan	Seeds of many types
Sitophilus oryzae (L.)	Rice Weevil	Curculionidae	Pantropical	Rice, maize, etc.
Sitophilus zeamais (Mot.)	Maize Weevil	Curculionidae	Pantropical	Maize, rice, etc.
Acarus siro L.	Flour Mite	Acaridae	Cosmopolitan	Flours, meals, etc.

11 *General bibliography*

A comprehensive bibliography (in the proper sense of the word) of crop pests (insects and mites), pesticides, and all the different aspects of pest ecology, biology, and control, relevant to the temperate regions of the world would be immense and quite inappropriate for such a text. As presented here the 'Bibliography' contains all the more important sources of information used in this compilation with a selection of other publications thought to be of possible use to users of this book.

AAB (1983 & 1984). Tests of agrochemicals and cultivars. *Ann. appl. Biol.* **102**, Suppl. No. 4, 144 pp.
104, Suppl. No. 5, 135 pp.

Akehurst, B. C. (1968). *Tobacco*, 551 pp. Longmans: London.

Alford, D. V. (1984). *A Colour Atlas of Fruit Pests – their recognition, biology and control*, 320 pp. Wolfe Publishing: London.

Allee, W. C., A. E. Emerson, O. Park, T. Park & K. P. Schmidt (1955). *Principles of Animal Ecology*, 835 pp. W. B. Saunders: Philadelphia.

Amsden, R. C. & C. P. Lewins (1966). Assessment of wettability of leaves by dipping in crystal violet. *World Rev. Pest Control* **5**, 187–94.

Ananthakrishnan, T. N. (1979). Biosystematics of Thysanoptera. *Ann. Rev. Entomol.* **24**, 159–83.

Andrewartha, H. G. & L. C. Birch (1954 & 1961). *The Distribution and Abundance of Animals*, 782 pp. University of Chicago Press: Chicago & London.

Annecke, D. P. & V. C. Moran (1982). *Insects and Mites of Cultivated Plants in South Africa*, 382 pp. Butterworths: London.

Anon. (1952). *Agriculture (Poisonous Substances) Act, 1952*, 9 pp. HMSO: London.

Anon. (1961). Farm sprayers and their use. *MAFF Bulletin no. 182*, 99 pp. HMSO: London.

Anon. (1964). *Bibliography on Insect Pest Resistance in Plants*, 39 pp. Imp. Bur. Pl. Breed. Genetics: Cambridge.

Anon. (1965). Conversion tables for research workers in forestry and agriculture. *Forestry Commission Booklet no. 5*, 64 pp. HMSO: London.

Anti-Locust Research Centre (1966). *The Locust Handbook*, 276 pp. Anti-Locust Research Centre: London.

Apple, J. L. & R. F. Smith (1976). *Integrated Pest Management*, 200 pp. Plenum Publishing Corporation: New York.

Ashworth, R. de B. & G. A. Lloyd (1961). Laboratory and field tests for evaluating the efficiency of wetting agents used in agriculture. *J. Sci. Food Agric.* **12**, 234–40.

Avidoz, Z. & I. Harpaz (1969). *Plant Pests of Israel*, 549 pp. Israel University Press: Jerusalem.

Bailey, S. F. (1938). Thrips of economic importance in California. *Agric. Expt. Sta., Berkeley, California*, circ. 346, 77 pp.

Bailey, S. F. (1964). A revision of the genus *Scirtothrips* Shull (Thysanoptera: Thripidae). *Hilgardia* **35**, 329–62.

Baker, E. W. & A. E. Pritchard (1960). The Tetranychoid mites of Africa. *Hilgardia* **29**, 455–574.

Baker, E. W. & G. W. Wharton (1964). *An Introduction to Acarology*, 465 pp. Macmillan: New York.

Balachowsky, A. S. (ed.) (1962). *Entomologie Appliquée à l'Agriculture*, 8 vols. Masson: Paris.

Balachowsky, A. S. & L. Mesnil (1935). *Les Insectes Nuisibles aux Plantes Cultivées*, 2 vols., 1921 pp. Min. Agric.: Paris.

Bale, J. S. (1984). Bud burst and success of the beech weevil, *Rhynchaenus fagi*: feeding and oviposition. *Ecol. Ent.* **9**, 139–48.

Bals, E. J. (1970). Ultra low volume and ultra low dosage spraying. *Cott. Gr. Rev.* **47**, 217–21.

Banerjee, B. (1981). An analysis of the effect of latitude, age and area on the number of arthropod pest species of tea. *J. Appl. Ecol.* **18**, 339–42.

Bardner, R. & K. E. Fletcher (1979). Larvae of the pea and bean weevil, *Sitona lineatus*, and the yield of field beans. *J. agric. Sci., Camb.* **92**, 109–12.

Bardner, R. & K. E. Fletcher (1974). Insect infestations and their effects on the growth and yield of field crops: a review. *Bull. ent. Res.* **64**, 141–60.

Bardner, R., R. A. French & M. J. Dupuch (1981). Agricultural benefits of the Rothamsted aphid bulletin. *Rothamsted Report (1980)*, pp. 21–39.

Barnes, H. F. (1946–59). *Gall Midges of Economic Importance*, Crosby Lockwood & Son Ltd.: London.
(1946) Vol. I, *Root and Vegetable Crops*, 104 pp.
(1946) Vol. II, *Fodder Crops*, 160 pp.
(1948) Vol. III, *Fruit*, 184 pp.
(1948) Vol. IV, *Ornamental Plants and Shrubs*, 165 pp.
(1951) Vol. V, *Trees*, 270 pp.
(1949) Vol. VI, *Miscellaneous Crops*, 229 pp.
(1956) Vol. VII, *Cereal Crops*, 261 pp.
(W. Nijveldt) (1969) Vol. VIII, *Miscellaneous*, 221 pp.

Bateman, M. A. (1972). The ecology of fruit flies. *Ann. Rev. Entomol.* **17**, 493–518.

'Bayer' (1968). *Bayer Crop Protection Compendium*, 2 vols., 511 pp. Bayer Co.: Leverkusen.

'Bayer' (1981). *Catalogue of principal noxious and beneficial animals, with synonyms, common names and abbreviations.* (3rd edition). Part 1, pp. 1177; Part 2, pp. 1193; Bayer A. G.: Leverkusen.

Beardsley, J. W. & R. H. Gonzalez (1975). The biology and ecology of armoured scales. *Ann. Rev. Entomol.* **20**, 47–73.

Beaver, R. A. (1977). Bark and Ambrosia Beetles in tropical forests. *Biotrop. Spec. Pub.* No. 2, 133–47.

Becker, P. (1974). *Pests of Ornamental Plants*, 175 pp. MAFF Bulletin 97, HMSO: London.

Bedding, R. A. (1984). Large scale production, storage and transport of the insect-parasitic nematodes *Neoaplectana* spp. and *Heterorhabditis* spp. *Ann. appl. Biol.* **104**, 117–20.

Beeman, R. W. (1982). Recent advances in mode of action of insecticides. *Ann. Rev. Entomol.* **27**, 253–81.

Beirne, B. P. (1967). *Pest Management*, 123 pp. Leonard Hill: London.

Beirne, B. P. (1971). Pest insects of annual crops in Canada. I. Lepidoptera; II. Diptera; III. Coleoptera. *Mem. Ent. Soc. Canada* No. **78**, 124 pp.

Beirne, B. P. (1972). Pest insects of annual crops in Canada. IV. Hemiptera – Homoptera; V. Orthoptera; VI. Other Groups. *Mem. Ent. Soc. Canada* No. **85**. 73 pp.

Bell, T. R. D. & F. B. Scott (1937). *The Fauna of British India*, Vol. 5. *Moths*, 537 pp. Taylor & Francis: London.

Bell, W. J. & R. T. Cardé (1984). *Chemical Ecology of Insects*, 536 pp. Assoc. Book Pub.: London.

Berger, R. S. (1968). Sex pheromone of the Cotton Leafworm. *J. econ. Ent.* **61**, 326–7.

Berger, R. S., J. M. McGough & D. F. Martin (1965). Sex attractants of *Heliothis zea* and *H. virescens*. *J. econ. Ent.* **58**, 1023–4.

Beroza, M. (1964). Insect sex attractants and their use. *Proc. 2nd Int. Congr. Endocrinol.* 203–8.

Beroza, M. (ed.) (1970). *Chemicals Controlling Insect Behaviour*, 182 pp. Academic Press: New York.

Birch, M. C. & K. F. Haynes (1982). *Insect Pheromones* (Studies in Biology No. 147) 60 pp. Edward Arnold: London.

Blackman, R. L. & V. F. Eastop (eds.) (1984). *Aphids on the World's Crops*, 480 pp. John Wiley: Chichester.

Bleszynski S. (1970). A revision of the world species of *Chilo* Zincken (Lep.: Pyralidae). *Bull. Br. Mus. Nat. His.* (B) **25**, 97 pp.

Boller, E. F. & R. J. Prokopy (1976). Bionomics and management of *Rhagoletis*. *Ann. Rev. Entomol.* **21**, 223–46.

Borror, D. J. & D. M. DeLong (1971). *An Introduction to the Study of Insects*, 3rd ed., 812 pp. Holt, Rinehart & Winston: New York.

Bottrell, D. R. (1979). *Integrated Pest Management*, (Council on Environmental Quality), 120 pp. US Govt. Printing Office: Washington.

Bournier, A. (1977). Grape insects. *Ann. Rev. Entomol.* **22**, 355–76.

Bowden, J., J. Cochrane, B. J. Emmett, T. E. Minall & P. L. Sherlock (1983). A survey of cutworm attacks in England and Wales, and a descriptive population model for *Agrotis segetum* (Lepidoptera: Noctuidae). *Ann. appl. Biol.* **102**, 29–47.

Brader, L. (1979). Integrated pest control in the developing world. *Ann. Rev. Entomol.* **24**, 225–54.

Bradley, J. D. (1967). Some Lepidoptera of economic importance in Commonwealth countries. *Acta Universitatis Agriculturae* **15**, 501–19.

Bradley, J. D., W. G. Tremewan & A. Smith (1973). *British Tortricoid Moths Cochylidae and Tortricidae: Tortricinae*, (Ray Society No. 147), 251 pp. The Ray Society; London.

Braun-Blanquet, J. (1927). *Pflanzensoziologie*. Springer: Wien.

Brooks (1980). *See* Int. Congress of Entomology (1980).

Brooks, A. R. (1951). Identification of the root maggots (Diptera: Anthomyiidae) attacking cruciferous garden crops in Canada, with notes on biology and control. *Canad. Ent.* **83**, 109–20.

Brown, A. W. A. & R. Pal (1971). *Insecticide Resistance in Arthropods*. 2nd ed. WHO: Geneva.

Brown, E. S. (1962). *The African Armyworm* Spodoptera exempta (*Walker*) (*Lepidoptera, Noctuidae*): *a review of the literature*, 69 pp. CIE: London.

Brown, E. S. (1972). Armyworm control. *PANS* **18**, 197–204.

Brown, E. S. & C. F. Dewhurst (1975). The genus *Spodoptera* (Lepidoptera, Noctuidae) in Africa and the Near East. *Bull. ent. Res.* **65**, 221–62.

Brown F. G. (1968). *Pests and Diseases of Forest Plantation Trees: an annotated list of the principle species occurring in the British Commonwealth*, 1330 pp. Oxford: Clarendon Press.

BSI (1969). *Recommended Common Names for Pesticides*, 4th revision, 108 pp. Brit. Stand. Inst.: London.

Bucher, G. E. & H. H. Cheng (1970). Use of trap crops for attracting cutworm larvae. *Canad. Ent.* **102**, 797–8.

Buczacki, S. T. & K. M. Harris (1981). *Collins Guide to the Pests, Diseases and Disorders of Garden Plants*, 512 pp. Collins: London.

Bullock J. A. (1965). The control of *Hylemya arambourgi* Seguy (Dipt., Anthomyiidae) on barley. *Bull. ent. Res.* **55**, 645–61.

Burges, H. D. (ed.) (1981). *Microbial Control of Pests and Plant Diseases 1970–1980*, 914 pp. Academic Press: London.

Burges, H. D. & N. W. Hussey (1971). *Microbial Control of Insects and Mites*, 861 pp. Academic Press: London.

Burke, H. R. (1976). Bionomics of the Anthomine weevils. *Ann. Rev. Entomol.* **21**, 283–303.

Busvine, J. R. (1966). *Insects and Hygiene*, 2nd ed. 467 pp. Methuen: London.

Busvine, J. R. (1971*a*). *A Critical Review of the Techniques for Testing Insecticides*, 2nd ed., 345 pp. CIE: London.

Busvine, J. R. (1971*b*). The biochemical and genetic bases of insecticidal resistance. *PANS* **17**, 135–46.

Busvine, J. R. (1980). *Recommended Methods for Measurement of Pest Resistance to Pesticides*, (FAO Pl. Prod. Prot. Paper –21), 132 pp. FAO: Rome.

Butani, D. K. (1979). *Insects and Fruits*, 415 pp. Periodical Expert Book Agency: New Delhi.

Buyckx, E. J. E. (1962). *Précis des Maladies et des Insectes nuisibles recontrées sur les Plants Cultivées au Congo, au Rwanda et au Burundi*, 708 pp. INEAC.

Byass, J. B. & J. Holroyd (eds.) (1970). Proceedings of a Symposium for Research Workers on Pesticide Application. *Br. Crop. Prot. Council, Mon. no. 2*, 139 pp. Boots Pure Drug Co., Nottingham.

CAB (1951–85). *Distribution Maps of Insect Pests, Series A* (*Agricultural*), nos. 1–472, with index (1–430). CIE: London.

CAB (1961–7). *CIBC Technical Bulletins, 1–8*. CAB: London.

CAB (1980). *Perspectives in World Agriculture*, 532 pp. CAB: Slough.

CAB (1981). *List of Research Workers in the Agricultural Sciences in the Commonwealth and in the Republic of Ireland*, 4th ed., 658 pp. CAB: London.

Caltagirone, L. E. (1981). Landmark examples in classical biological control. *Ann. Rev. Entomol.* **26**, 213–32.

Campion, D. G. (1972) Insect chemosterilants: a review. *Bull. ent. Res.* **61**, 577–635.

Caresche, L., G. S. Cotterell, J. E. Peachey, R. W. Rayner & H. Jacques-Felix (1969). *Handbook for Phytosanitary Inspectors in Africa*, 444 pp. OAU/STRC: Lagos.

Caswell, G. H. (1962). *Agricultural Entomology in the Tropics*, 152 pp. Edward Arnold: London.

Chapman, P. J. (1973). Bionomics of apple-feeding Tortricidae. *Ann. Rev. Entomol.* **18**, 73–96.

Chapman, R. F. (1970). *The Insects – structure and function*, 819 pp. English Universities Press: London.

Chapman, R. F. (1974). *Feeding in Leaf-eating Insects*, (Oxford Biology Reader 69), 16 pp. Oxford University Press: Oxford.

Cherrett, J. M., J. B. Ford, I. V. Herbert & A. J. Probert (1971). *The Control of Injurious Animals*, 210 pp. English Universities Press: London.

Cherrett, J. M. & T. Lewis (1974). *Control of insects by exploiting their behaviour*, pp. 130–46. In: *Biology in Pest and Disease Control*, eds. Price Jones, D. & M. E. Solomon, Blackwell: Oxford.

Cherrett, J. M. & G. R. Sagar (1977). *Origins of Pest, Parasite, Disease and Weed Problems*, (18th Symp. Brit. Ecol. Soc.) 413 pp. Blackwell: Oxford.

Chiang, H. C. (1973). Bionomics of the Northern and Western Corn Rootworms. *Ann. Rev. Entomol.* **18**, 47–72.

Chiang, H. C. (1978). Pest management in corn. *Ann. Rev. Entomol.* **23**, 101–23.

Chu (1980). *See* Int. Congress of Entomology (1980).

Clark, L. R., P. W. Geier, R. D. Hughes & R. F. Morris (1967). *The Ecology of Insect Populations in Theory and Practice*, 232 pp. Methuen: London.

Clausen, C. P. (1940). *Entomophagous Insects*, 1st ed., 688 pp. McGraw-Hill: New York.
Clearwater, J. R. (1981). Practical identification of the females of five species of *Atherigona* Rondani (Diptera, Muscidae) in Kenya. *Tropical Pest Management* (formerly PANS) **27**, 303–12.
Coaker, T. H. (1977). Crop pest problems resulting from chemical control. pp. 313–28. In: Cherrett & Sagar (eds.) (1977).
Coaker, T. H. & S. Finch (1971). The cabbage root fly. *Erioischia brassicae* (Bouché). *Rep. natn. Veg. Res. Stn.* 1970, 23–42.
Coffee, R. (1973). Electrostatic crop spraying. *New Scientist* **84**, 194–6.
Cole, R. A. (1980*a*). Volatile components produced during ontogeny of some cultivated crucifers. *J. Sci. Food Agric.* **31**, 549–57.
Cole, R. A. (1980*b*). The use of porous polymers for the collection of plant volatiles. *J. Sci. Food Agric.* **31**, 1242–9.
Conway, G. R. (1972). *Ecological Aspects of Pest Control in Malaysia*. In: J. Milton (ed.), *The Careless Technology; Ecological Aspects of International Development*, Nat. Hist. Press.
Cope, O. B. (1971). Interactions between pesticides and wild life. *Ann Rev. Entomol.* **16**, 325–64.
Coppock, L. J. (1974). Notes on the biology of Carrot Fly in Eastern England. *Plant Pathology* **23**, 93–100.
COPR (1982). *The Locust and Grasshopper Manual*, 690 pp. COPR: London.
COPR (1983). *Pest Control in Tropical Tomatoes*, 130 pp. COPR: London.
Coulson, R. N. (1979). Population dynamics of bark beetles. *Ann. Rev. Entomol.* **24**, 417–47.
Corpuz, L. R. (1969). The biology, host range, and natural enemies of *Nezara viridula* L. (Pentatomidae, Hemiptera). *Philipp. Ent.* **1**, 227–39.
Cramer, H. H. (1967). *Plant Protection and World Crop Production*, 254 pp. Bayer Pflanzenschutz: Leverkusen.
Crane, E. & P. Walker (1983). *The Impact of Pest Management on Bees and Pollination*, (Int. Bee Res. Assoc.), 73 pp. TDRI: London.
Crowson, R. A. (1981). *The Biology of the Coleoptera*, 802 pp. Academic Press: London.
CSCPRC (1977). *Insect Control in the People's Republic of China*, 218 pp. National Academy of Sciences: Washington.
CSIRO (1973). Scientific and common names of insects and allied forms occurring in Australia. *CSIRO Bull.* 287, 47 pp.
CSIRO (*see also* Mackerras).
Danilevskii, A. S. (1965). *Photoperiodism and Seasonal Development of Insects*, 283 pp. (Translation of 1961 edition in Russian.) Oliver & Boyd: London.
Danthanarayana, W. (1983). Population ecology of the light brown apple moth, *Epiphyas postvittana* (Lepidoptera: Tortricidae). *J. Anim. Ecol.* **52**, 1–33.
Darlington, A. (1968). *The Pocket Encyclopedia of Plant Galls in Colour*, 191 pp. Blandford Press: London.
Davidson, R. H. & L. M. Peairs (1966). *Insect Pests of Farm, Garden and Orchard*, 6th ed., 675 pp. John Wiley: New York.
DeBach, P. (1964). *Biological Control of Insect Pests and Weeds*, 844 pp. Chapman & Hall: London.
DeBach, P. (1971). *Fortuitous biological control from ecesis of natural enemies*. In: *Entomological Essays to Commemorate the Retirement of Professor K. Yasumatsu*. pp. 293–307, Hokuryukan Pub. Co. Ltd.: Tokyo.
DeBach, P. (1974). *Biological Control by Natural Enemies*, 323 pp. Cambridge University Press: Cambridge.
DeBach, P. & C. B. Huffaker (1971). Experimental techniques for evaluation of the effectiveness of natural enemies. In: C. B. Huffaker (ed.), *Biological Control*, pp. 113–40. Plenum: New York.
DeBach, P., D. Rosen & C. E. Kennett (1971). Biological control of coccids by introduced natural enemies. In: C. B. Huffaker (ed.), *Biological Control*, pp. 165–94. Plenum: New York.
De Long, D. (1971). The bionomics of leafhoppers. *Ann. Rev. Entomol.* **16**, 179–210.
Delucchi, V. L. (ed.) (1976). *Studies in Biological Control*, (IBP–9), 304 pp. Cambridge University Press: Cambridge.
Den Otter (1980). *See* Int. Congress of Entomology (1980).
Drew, R. A. I., G. H. S. Hooper & M. A. Bateman (1978). *Economic Fruit Flies of the South Pacific Region*, 137 pp. Dept. Primary Industries: Queensland.
Duval, C. T. (1970). Some introductory aspects of the chemical relationships and nomenclature of synthetic organic insecticides. *PANS* **16**, 11–35.

Eastop, V. F. (1966). A taxonomic study of Australian Aphidoidea (Homoptera). *Aust. J. Zool.* **14**; 399–592.

Eastop, V. F. (1971). Keys for identification of *Acryrthosiphon* (Hemiptera: Aphididae). *Bull. Br. Mus. Nat. Hist., Ent.* **26**, 1–115.

Ebbels, D. L. & J. E. King (1979). *Plant Health*, 322 pp. Blackwell: Oxford.

Ebeling, W. (1959). *Subtropical Fruit Pests*, 2nd ed., 436 pp. University of California Press: California.

Ebeling, W. (1971). Sorptive dusts for pest control. *Ann. Rev. Entomol.* **16**, 123–58.

Ebeling, W. (1975). *Urban Entomology*, 695 pp. University of California, Division of Agricultural Sciences: California.

Edwards, C. A. (1970*a*). Problem of insecticidal residues in agricultural soils. *PANS* **16**, 271–6.

Edwards, C. A. (1970*b*). *Persistent Pesticides in the Environment*, 77pp. Butterworths: London.

Edwards, C. A. & G. W. Heath (1964). *Principles of Agricultural Entomology*, 418 pp. Chapman & Hall: London.

Elliott, M., N. F. James & C. Potter (1978). The future of pyrethroids in insect control. *Ann. Rev. Entomol.* **23**, 443–69.

Ellis, P. R., J. A. Hardman, J. C. Jackson & B. D. Dowker (1980). Screening of carrots for their susceptibility to carrot fly attack. *J. natn. Inst. agric. Bot.* **15**, 294–302.

Ellis, P. R., R. A. Cole, P. Crisp & J. A. Hardman (1980). The relationship between Cabbage Root Fly egg laying and volatile hydrolysis products of radish. *Ann. appl. Biol.* **95**, 283–9.

Elton, C. S. (1958). *The Ecology of Invasions by Animals and Plants*, 181 pp. Methuen: London.

Emden, H. F. van (ed.) (1972). *Insect/Plant Relationships*, (Symp. Roy. Ent. Soc. Lond., No. 6) 213 pp. Blackwell: Oxford.

Emden, H. F. van, V. F. Eastop, R. D. Hughes & M. J. Way (1969). The ecology of *Myzus persicae*. *Ann. Rev. Entomol.* **14**, 197–270.

Emmett, B. J. (1980). Key for the identification of lepidopterous larvae infesting brassica crops. *Plant Pathology* **29**, 122–3.

EPA (1976). *List of Insects and other Organisms*, (3rd edition) Parts I, II, III, and IV. EPA: Washington.

EPPO (1970). Report of the International Conference on Methods for Forecasting, Warning, Pest Assessment and Detection of Infestation. *EPPO Pub. Ser.* no. 57, 206 pp. EPPO: Paris.

Evans, J. W. (1952). *Injurious Insects of the British Commonwealth*, 242 pp. CIE: London.

FAO (1966). *Proceedings of the FAO Symposium on Integrated Pest Control* (11–15 October 1965). 1, 91 pp. 2, 186 pp. 3, 129 pp. FAO: Rome.

FAO (1974). *Proceedings of the FAO Conference on Ecology in relation to Plant Pest Control (Rome, Italy, 11–15 December, 1972)*. 326 pp. FAO: Rome.

FAO (1979). *Guidelines for Integrated Control of Rice Insect Pests*, (FAO Pl. Prod. Prot. Paper – 14), 115 pp. FAO: Rome.

FAO (1979). *Guidelines for Integrated Control of Maize Pests*, (FAO Pl. Prod. Prot. Paper – 18), 91 pp. FAO: Rome.

FAO (1979). *Elements of Integrated Control of Sorghum Pests*, (FAO Pl. Prod. Prot. Paper – 19), 159 pp. FAO: Rome.

FAO/CAB (1971). *Crop Loss Assessments Methods*. FAO Manual on the evaluation of losses by pests, diseases and weeds, *c*. 130 pp. FAO: Rome. Supplement 1 (1973), 2 (1977), 3 (1981).

Feltwell, J. (1981). *The Large White Butterfly. The biology, biochemistry and physiology of* Pieris brassicae, 550 pp.

Ferro, D. N. (ed.) (1976). *New Zealand Insect Pests*, 311 pp. Lincoln Univ. Coll. Agric.: Canterbury.

Ferron, P. (1978). Biological control of insect pests by entomogenous fungi. *Ann. Rev. Entomol.* **23**, 409–42.

Fichter, G. S. (1968). *Insect Pests*, 160 pp. Paul Hamlyn: London.

Fletcher, W. W. (1974). *The Pest War*, 218 pp. Blackwell: Oxford.

Florkin, M. & B. T. Scheer (1970). *Chemical Zoology*, Vol V. *Arthropoda*, A, 460 pp. Academic Press: New York.

Florkin, M. & B. T. Scheer (1971). *Chemical Zoology*, Vol. VI. *Arthropoda*, B, 484 pp. Academic Press: New York.

Fowler, S. V. & J. H. Lawton (1984). Trees don't talk: do they even murmur? *Antenna*, **8**, 69–71.

Fox Wilson, G. (1960). *Horticultural Pests – Detection and Control*, 2nd ed., 240 pp. Crosby Lockwood: London.

Free, J. B. (1970). *Insect Pollination of Crops*, 544 pp. Academic Press: London.

Free, J. B. & I. H. Williams (1977). *The Pollination of Crops by Bees*, 14 pp. Apimondia: Bucharest & Int. Bee Res. Assoc., UK.

Freeman, G. H. (1967). Problems in plant pathology and entomology field trials. *Exp. Agric.* **3**, 351–8.

Freeman, P. (1940). A contribution to the study of the genus *Nezara* A. & S. (Hemiptera, Pentatomidae). *Trans. R. ent. Soc. Lond.* **90**, 351–74.

Frohlich, G. & W. Rodewald (1970). *General Pests and Diseases of Tropical Crops and their Control*, 366 pp. Pergamon Press: London.

Gair, R., (1968). The conduct of field variety trials involving fertilisers and pesticides or both. *PANS* (A) **14**, 216–30.

Gair, R., J. E. E. Jenkins & E. Lester (1983). *Cereal Pests and Diseases*, 3rd ed., 259 pp. Farming Press Ltd.: Ipswich.

Gay F. J. (ed.) (1966). *Scientific and Common Names of Insects and Allied Forms Occurring in Australia*, Bull. No. 285, 52 pp. CSIRO: Melbourne.

Geier, P. W. (1966). Management of insect pests. *Ann. Rev. Entomol.* **11**, 471–90.

Geier, P. W., L. R. Clark, D. J. Anderson & H. A. Nix (eds.) (1973). *Insects: studies in population management*, 294 pp. Ecol. Soc. Australia Memoir 1: Canberra.

Getz, W. M. & A. P. Gutierrez (1982). A perspective of systems analysis in crop production and insect pest management. *Ann. Rev. Entomol.* **27**, 447–66.

Ghauri, M. S. K. (1971). Revision of the genus *Nephotettix* Matsumura (Homoptera: Cicadelloidea: Euscelidae) based upon the type material. *Bull. ent. Res.* **60**, 481–512.

Glass, E. H. (co-ordinator) (1975). *Integrated Pest Management: rationale, potential, needs and implementation*, 141 pp. Ent. Soc. Amer., Special Pub. 75–2.

Glen, D. M. & C. C. Payne (1984). Production and field evaluation of codling moth granulosis virus for control of *Cydia pomonella* in the United Kingdom. *Ann. appl. Biol.* **104**, 87–98.

Gray, B. (1972). Economic tropical forest entomology. *Ann. Rev. Entomol.* **17**, 313–54.

Greathead, D. J. (1971). A review of biological control in the Ethiopian Region. *Tech. Commun. CIBC*, no. 5, 162 pp. CAB: London.

Gram, E., P. Bovien & C. Stapel (1969). *Recognition of Diseases and Pests of Farm Crops*, 2nd ed., 128 pp. Blandford Press: London.

Gruys, P. & A. K. Minks (eds.) (1979). *Integrated Control of Pests in the Netherlands*, 300 pp. Pudoc: Netherlands.

Hagen, K. S. & R. van den Bosch (1968). Impact of pathogens, parasites and predators on aphids. *Ann. Rev. Entomol.* **13**, 325–84.

Halstead, D. G. H. (1964). The separation of *Sitophilus oryzae* (L.) and *S. zeamais* Motschulsky (Col., Curculionidae), with a summary of their distribution. *Ent. mon. Mag.* **99**, 72–4.

Halstead, D. G. H. (1980). A revision of the genus *Oryzaephilus* Ganglbauer, including descriptions of related genera (Coleoptera: Silvanidae). *Zool. J. Linn. Soc.* **69**, 271–374.

Hanover, J. W. (1975). Physiology of tree resistance to insects. *Ann. Rev. Entomol.* **20**, 75–95.

Hanson, H. C. (1963). *Diseases and Pests of Economic Plants of Central and South China, Hong Kong and Taiwan (Formosa)*, 184 pp. Amer. Inst. Crop Ecology: Washington DC.

Harcourt, D. G. (1963). Major mortality factors in the population dynamics of the Diamondback moth, *Plutella maculipennis* (Curt.) (Lepidoptera: Plutellidae). *Mem. ent. Soc. Canada* **32**, 55–66.

Harcourt, D. G. (1969). The development and use of life tables in the study of natural insect populations. *Ann. Rev. Entomol.* **14**, 175–96.

Hardman, J. A. & P. R. Ellis (1982). An investigation of the host range of the carrot fly. *Ann. appl. Biol.* **100**, 1–9.

Hardwick, D. F. (1965). The Corn Earworm complex. *Mem. ent. Soc. Canada* **40**, 1–247.

Harris, C. R. (1972). Factors influencing the effectiveness of soil insecticides. *Ann. Rev. Entomol.* **17**, 177–98.

Harris, K. F. & K. Maramorosch (eds.) (1977). *Aphids as Virus Vectors*, 570 pp. Academic Press: New York.

Harris, K. F. & K. Maramorosch (eds.) (1980). *Vectors of Plant Pathogens*, 480 pp. Academic Press: New York.

Harris, K. M. (1966). Gall midge genera of economic importance (Diptera: Cecidomyiinae). Part I: Introduction and subfamily Cecidomyiinae: supertribe Cecidomyiidi. *Trans. R. ent. Soc. Lond.* **118**, 313–58.

Harris, K. M. (1968). A systematic revision and biological review of the cecidomyiid predators (Diptera: Cecidomyiidae) on world Coccoidea (Hemiptera: Homoptera). *Trans. R. ent. Soc. Lond.* **119**, 401–94.

Harris, M. K. (1983). Integrated pest management of pecans. *Ann. Rev. Entomol.* **28**, 291–318.

Hartley, G. S. & R. T. Brunskill (1958). Reflection of water drops from surfaces. In: *Surface Phenomena in Chemistry & Biology*, Danielli, J. F. *et al.*, pp. 214–23. Pergamon Press: London.

Hartley. G. S. & T. F. West (1969). *Chemicals for Pest Control*, 316 pp. Pergamon Press: London.

Hassan, E. (1977). *Major Insect and Mite Pests of Australian Crops*, 238 pp. Ento Press: Queensland.

Hassell, M. P. & T. R. E. Southwood (1978). Foraging strategies of insects. *Ann. Rev. Ecol. Syst.* **9**, 75–98.

Haynes, D. L. & S. H. Gage (1981). The Cereal Leaf Beetle in North America. *Ann. Rev. Entomol.* **26**, 259–87.

Headley, J. C. (1972). Economics of agricultural pest control. *Ann. Rev. Entomol.* **17**, 273–86.

Heath, J. & A. Maitland Emmet (eds.) (1976, 1979 & 1983). *The Moths and Butterflies of Great Britain and Ireland*, Vol. 1. *Micropterigidae – Heliozelidae*, 343 pp. (1976). Vol. 9. *Sphingidae – Noctuidae (Part I)*, 288 pp. (1979). Vol. 10. *Noctuidae (Part II) and Agaristidae*, 459 pp. (1983). Curwen Press: London.

Henry, J. E. (1981). Natural and applied control of insects by protozoa. *Ann. Rev. Entomol.* **26**, 49–73.

Hensley (1980). *See* Int. Congress of Entomology (1980).

Hill, A. F. (1952). *Economic Botany*, 560 pp. McGraw-Hill: New York.

Hill, D. S. (1974*a*). *Synoptic Catalogue of Insect and Mite Pests of Agricultural and Horticultural Crops*, 150 pp. Department of Zoology, Hong Kong University; Occasional Papers No. 1.

Hill, D. S. (1974*b*). Susceptibilities of carrot cultivars to carrot fly (*Psila rosae* (F.)). *Plant Pathology* **23**, 36–9.

Hill, D. S. (1982). *Hong Kong Insects, Volume II*, 144 pp. Urban Council: Hong Kong.

Hill, D. S. (1983). *Agricultural Insect Pests of the Tropics and their Control*, (2nd ed.), 746 pp. Cambridge University Press: Cambridge.

Hill, D. S., P. Hore & I. W. B. Thornton (1982). *Insects of Hong Kong*, 503 pp. Hong Kong University Press: Hong Kong.

Hill, D. S. & J. M. Waller (1982). *Pests and Diseases of Tropical Crops. Volume I: Principles and Methods of Control*, 170 pp. Longmans: London.

Hill, D. S. & J. M. Waller (1986). *Pests and Diseases of Tropical Crops. Volume II: Field Handbook*, Longmans: London.

Hinton, H. E. & A. S. Corbet (1955). *Common Insect Pests of Stored Food Products*, 3rd ed., 61 pp. Econ. Ser., no. 15, British Museum (NH): London.

HMSO (1964). *Review of the Persistent Organochlorine Pesticides*. HMSO: London.

HMSO (1984). *Approved Products for Farmers and Growers*, 253 pp. Agric. Chem. Approval Scheme, HMSO: London

Hodkinson, I. D. & M. K. Hughes (1982). *Insect Herbivory* (Outline Studies in Ecology), 77 pp. Chapman & Hall: London.

Hodkinson, I. D. & I. M. White (1979). *Handbooks for the Identification of British Insects: Homoptera Psylloidea*. Vol. II, part 5(a), 98 pp. RESL: London.

Homeyer, B. (1970). Present state of soil insect pest control. *Pflanz.-Nachr. Bayer* **23**, 224–30.

Howe, R. W. (1957). A laboratory study of the Cigarette Beetle, *Lasioderma serricorne* (F.) (Col., Anobiidae) with a critical review of the literature on its biology. *Bull. ent. Res.* **48**, 9–56.

Howe, R. W. (1965). A summary of estimates of optimal and minimal conditions for population increase of some stored products insects. *J. Stored Prod. Res.* **1**, 177–84.

Huffaker, C. B. (ed.) (1971). *Biological Control*, 511 pp. Plenum: New York.

Huffaker, C. B., J. A. McMurtry & M. Van de Vrie (1969). The ecology of tetranychid mites and their natural control. *Ann. Rev. Entomol.* **14**, 125–74.

Hughes, K. M. (1957). An annotated list and bibliography of insects reported to have virus diseases. *Hilgardia* **26**, 597–629.

Hussey, N. W., W. H. Read & J. J. Hesling (1969). *The Pests of Protected Cultivation*, 416 pp. Edward Arnold: London.

Imms, A. D. (1960). *A General Textbook of Entomology*, 9th ed., 886 pp. Methuen: London. Revised by O. W. Richards & R. G. Davis.

Imms, A. D. (1967). *Outlines of Entomology*, 5th ed., 224 pp. Methuen: London. Revised by O. W. Richards & R. G. Davis. (see Richards & Davies, 1977).

Int. Congress of Entomology (XVI) (1980). *Abstracts*, 480 pp. Kyoto: Japan.

Int. Pest Control (1981*a*). Neem – pesticide potential. *Int. Pest Control*, Vol. 1981 (3), 68–70.

Int. Pest Control (1981*b*). *International Pesticide Directory*, suppl. to *Int. Pest Control*, Sept./Oct. 1981, 70 pp.

Int. Pest Control (1984). *International Pesticide Directory*, (4th edition) suppl. to *Int. Pest Control*, Sept./Oct. 1984, 91 pp.

Jacobson, M. (1965). *Insect Sex Attractants*, John Wiley: New York.

Jacobson, M. (1966). Chemical insect attractants and repellants. *Ann. Rev. Entomol.* **11**, 403–22.

Jacobson, M. & M. Beroza (1963). Chemical sex attractants. *Science* **140**, 1367–73.

Jacobson, M. & D. G. Crosby (1971). *Naturally Occurring Insecticides*, 585 pp. Dekker: New York.

Janzen, D. H. & P. G. Waterman (1984). A seasonal census of phenolics, fibre and alkaloids in foliage of forest trees in Costa Rica; some factors influencing their distribution and relation to host selection by Sphingidae and Saturniidae. *Biol. J. Linn. Soc.* **21**, 439–54.

Japan Plant Protection Assoc. (1980). *Major Insect and other Pests of Economic Plants in Japan*, 307 pp. Jap. Pl. Prot. Assoc.: Tokyo.

Jeppson, L. R., H. H. Keifer & E. W. Baker (1975). *Mites Injurious to Economic Plants*, 614 pp. Univ. California Press; Berkeley.

Johnson, W. T. & H. H. Lyon (1976). *Insects that Feed on Trees and Shrubs* (*An illustrated practical guide*), 464 pp. Cornell University Press: Ithaca.

Jones, F. G. W. & M. Jones (1974). *Pests of Field Crops*, 2nd Edition, 448 pp. Edward Arnold: London.

Kenaga, E. E. & R. W. Morgan (1978). Commercial and experimental organic insecticides (1978 revision). *Ent. Soc. Amer., Sp. Pub.* **78** – 1, 79 pp.

Kennedy, C. E. J. & T. R. E. Southwood (1984). The number of species of insects associated with British trees: a re-analysis. *J. Anim. Ecol.* **53**, 455–78.

Kennedy, J. S. (1966). Mechanisms of host plant selection. *Proc. Assoc. Appl. Biol.* 317–22.

Kennedy, J. S., M. F. Day & V. F. Eastop (1962). *A Conspectus of Aphids as Vectors of Plant Viruses*, 144 pp. CIE: London.

Kevan, P. G. & H. G. Baker (1983). Insects as flower visitors and pollinators. *Ann. Rev. Entomol.* **28**, 407–53.

Kilgore, W. W. & R. L. Doutt (1967). *Pest Control – biological, physical, and selected chemical methods*, 477 pp. Academic Press: New York & London.

Kiritani, K. (1979). Pest management in rice. *Ann. Rev. Entomol.* **24**, 279–312.

Kloet, G. S. & W. D. Hincks (1964–78). *A Check List of British Insects* (2nd ed. – revised). Part 1: Small Orders and Hemiptera (1964). Part 2: Lepidoptera (1972). Part 3: Coleoptera and Strepsiptera (1977). Part 4: Hymenoptera (1978). Part 5: Diptera and Siphonaptera (1976). Royal Entomological Society of London.

Knipling, E. F. (1963). *Alternative Methods in Pest Control.* Symposium on New Developments and Problems in the Use of Pesticides, pp. 23–38. Fd. Nut. Bd., Nat. Acad. Sci.: Washington.

Knipling, E. F. & D. A. Spencer (1963). *Protection from Insect and Vertebrate Pests in Relation to Crop Production.* UN Conference on Application of Science and Technology for the Benefit of the Less Developed Areas, pp. 160–74. UN: Geneva.

Krantz, G. W. & E. E. Lindquist (1979). Evolution of phytophagous mites (Acari). *Ann. Rev. Entomol.* **24**, 121–58.

Kranz, J., H. Schmütterer & W. Koch (1979). *Diseases, Pests, and Weeds in Tropical Crops*, 666 pp. John Wiley: Chichester.

Kring, J. B. (1972). Flight behaviour of aphids. *Ann. Rev. Entomol.* **17**, 461–92.

Kulman, H. M. (1971). Effects of insect defoliation on growth and mortality of trees. *Ann. Rev. Entomol.* **16**, 289–324.

Labeyrie, V. (1981). The ecology of Bruchids attacking legumes (pulses). *Proc. Symposium (1980)*, 249 pp.

La Brecque, G. C. & C. N. Smith (1968). *Principles of Insect Chemosterilisation*, 354 pp. Appleton-Century-Crofts: New York.

Laffoon, J. L. (1960). Common names of insects – approved by the Entomological Society of America. *Bull. Ent. Soc. Amer.* **6**, 175–211.

Lange, W. H. & L. Bronson (1981). Insect pests of tomatoes. *Ann. Rev. Entomol.* **26**, 345–71.

Large, E. C. (1966). Measuring plant disease. *Ann. Rev. Phytopathology*, **4**, 9–28.

Lashomb, J. H. & R. A. Casagrande (eds.) (1982). *Advances in Potato Pest Management*, 304 pp. Academic Press: New York.

Laveglia, J. & P. A. Dahm (1977). Degradation of organophosphorus and carbamate insecticides in the soil and by soil microorganisms. *Ann. Rev. Entomol.* **22**, 483–513.

Lawton, J. H. & D. Strong (1981). Community patterns and competition in folivorous insects. *American Naturalist* **118**, 317–38.

Le Clerq, E. L., W. H. Leonard & A. G. Clark (1966). *Field Plot Technique*, 2nd ed., 373 pp. Burgess: Minneapolis.

Levins, R. & M. Wilson (1980). Ecological theory and pest management. *Ann. Rev. Entomol.* **25**, 287–308.

Lewis T. (1973). *Thrips: their biology, ecology and economic importance*, 350 pp. Academic Press: London.

Li, Li-ying (1980). *See* Int. Congress of Entomology (1980).

Lincoln, R. J., G. A. Boxshall and P. F. Clark (1982). *A Dictionary of Ecology, Evolution and Systematics*, 298 pp. Cambridge University Press: Cambridge.

Maas, W. (1971). *ULV Application and Formulation Techniques*, 164 pp. Philips-Duphar: Amsterdam.

McCallan, E. (1959). Some aspects of the geographical distribution of insect pests. *J. ent. Soc. S. Afr.* **22**, 3–12.

Mackerras, I. M. (ed.) (CSIRO) (1969). *The Insects of Australia*, 1029 pp. Melbourne University Press: Victoria.

McKinley, D. J. (1971). An introduction to the use and preparation of artificial diets with special emphasis on diets for phytophagous Lepidoptera. *PANS* **17**, 421–4.

Madsen, H. F. (1971). Integrated control of the Codling Moth. *PANS* **17**, 417–20.

Madsen, H. F. & J. C. Arrand (1977). *The Recognition and Life History of the Major Orchard Insects and Mites in British Columbia*, 32 pp. Min. of Agric.: Victoria, B.C.

Madsen, H. F. & C. V. G. Morgan (1970). Pome fruit pests and their control. *Ann. Rev. Entomol.* **15**, 295–320.

Maeta & Kitamura (1980). *See* Int Congress of Entomology (1980).

MAFF (1971). *Pesticides safety precautions scheme agreed between Government departments and industry*, 149 pp. MAFF: London.

MAFF (1965–84). *Advisory Leaflets* (on crop pests, etc.), Nos. 1–650 approx. Now termed '*Leaflets*', Nos. 1–760 approx. Also various other short-term leaflets concerned with crop pests, etc.

MAFF (1965–84). *Bulletins*, and *Technical Bulletins*, (on specific crops and groups of crop pests, etc.), Nos. 1–210 approx.

MAFF (1981–84). *Identification Cards* (for important UK pests and diseases), IC/1–66. HMSO: UK.

MAFF (1984). *List of Approved Products and their uses for farmers and growers*, 253 pp. HMSO: London.

Maramorosch, K. & K. F. Harris (eds.) (1979). *Leafhopper Vectors and Plant Disease Agents*, 650 pp. Academic Press: New York.

Maramorosch, K. & K. F. Harris (eds.) (1981). *Plant Diseases and Vectors: Ecology and Epidemiology*, 360 pp. Academic Press: New York.

Marsh, R. W. (1969). Glossary of terms used in the application of crop protection measures. *Scient. Hort.* **21**, 147–55.

Martin, E. C. & S. E. McGregor (1973). Changing trends in insect pollination of commercial crops. *Ann. Rev. Entomol.* **18**, 207–26.

Martin, H. (1961). *Guide to the Chemicals Used in Crop Protection*, 4th ed., 387 pp. Canada Dept. of Agric.: London, Ont.

Martin, H. (1964). *The Scientific Principles of Crop Protection*, 384 pp. Edward Arnold: London.

Martin, H. (1969). *Insecticide and Fungicide Handbook for Crop Protection*, 3rd ed., 387 pp. Blackwell: Oxford.

Martin, H. (1970). *Pesticide Manual*, 2nd ed., 464 pp. Brit. Crop Prot. Council.

Martin, H. (1972*a*). *Pesticide Manual*, 3rd ed., 535 pp. Brit. Crop Prot. Council.

Martin, H. (1972*b*). *Insecticide and Fungicide Handbook for Crop Protection*, 4th ed., 415 pp. Blackwell: Oxford.

Martin, H. (1973). *The Scientific Principles of Crop Protection*, 6th ed. Edward Arnold: London.

Martin, H. & C. R. Worthing (1976). *Insecticide and Fungicide Handbook*, 5th ed. 427 pp. Blackwell: Oxford.

Massee, A. M. (1954). *The Pests of Fruit and Hops*, 3rd ed., 325 pp. Crosby Lockwood: London.

Matthews, G. A. (1977). C.d.a. – Controlled Droplet Application. *PANS* **23**, 387–94.

Matthews, G. A. (1979). *Pesticide Application Methods*, 334 pp. Longmans: London.

Matthews, G. A. (1981). Improved systems of pesticide application. *Phil. Trans. R. Soc. Lond. B.* **295**, 163–73.

Matthews, G. A. (1984). *Pest Management*. 231 pp. Longmans: London.

Matteson, P. C., M. A. Altieri & W. C. Gagne (1984). Modification of small farmer practices for better pest management. *Ann. Rev. Entomol.* **29**, 383–402.
May, R. M. (ed.) (1976). *Theoretical Ecology – Principles and Applications*, 317 pp. Blackwell: Oxford.
Menken, S. B. J. (1980). Allozyme polymorphism and the speciation process in small ermine moths (Lepidoptera, Yponomeutidae), (*Studies in Yponomeuta* **2**), 119 pp. J. H. Pasmans B. V.: Gravenhage.
Menzie, C. M. (1969). *Metabolism of Pesticides*. Bur. Sport, Fish, Wildlife; Sp. Sci. Rep. – Wildlife no. 127, 487 pp.
Menzie, C. M. (1972). Fate of pesticides in the environment. *Ann. Rev. Entomol.* **17**, 199–222.
Mercer, S. L. (1981). Tropical Pest Management Pesticide Index – 1981 Edition. *Trop. Pest Man.* (1981) 1–66.
Metcalf, C. L., W. P. Flint & R. L. Metcalf (1962). *Destructive and Useful Insects*, 1087 pp. McGraw-Hill: New York.
Metcalf, R. L. (ed.) (1957–68). *Advances in Pest Control Research*, vols 1–8. Interscience: London & New York.
Metcalf, R. L. (1980). Changing role of insecticides in crop protection. *Ann. Rev. Entomol.* **25**, 219–56.
Metcalf, R. L. & W. H. Luckmann (eds.) (1975). *Introduction to Insect Pest Management*, 587 pp. John Wiley: New York.
Miles M. (1950). Studies of British Anthomyiid flies. I. Biology and habits of the Bean Seed Flies, *Chortophila cilicrura* (Rond.) and *C. tridactyla* (Rond.) *Bull. ent. Res.* **41**, 343–54.
Miller, D. R. & M. Kosztarab (1979). Recent advances in the study of scale insects. *Ann. Rev. Entomol.* **24**, 1–27.
Monro, H. A. V. (1980). *Manual of Fumigation for Insect Control*, 2nd ed., 381 pp. FAO: Rome.
Moore, D. (1984). The role of silica in protecting Italian ryegrass (*Lolium multiflorum*) from attack by dipterous stem-boring larvae (*Oscinella frit* and other related species). *Ann. appl. Biol.* **104**, 161–6.
Moran, V. C. (1983). The phytophagous insects and mites of cultivated plants in South Africa: patterns and pest status. *J. appl. Ecol.* **20**, 439–50.
Mound, L. A. (1965). An introduction to the Aleyrodidae of Western Africa (Homoptera). *Bull. Brit. Mus. Nat. Hist., Ent.* **17**, 113–60.
Mound, L. A. & S. H. Halsey (1977). *Whiteflies of the World: a systematic catalogue of the Aleyrodidae (Homoptera) with host plant and natural enemy data*, 336 pp. John Wiley: London.
Mumford, J. D. & G. A. Norton (1984). Economics of decision making in pest management. *Ann. Rev. Entomol.* **29**, 157–74.
Munro, J. W. (1966). *Pests of Stored Products*, 234 pp. Hutchinson: London.
Nat. Acad. Sci. US (1969). *Principles of Plant and Animal Pest Control*: Vol 3, *Insect-pest management and control*, 508 pp.; Vol. 4, *Control of plant parasitic nematodes*, 172 pp. *Pub. Nat. Acad. Sci. US*, no. 1695 (Washington DC).
Nayar, K. K., T. N. Ananthakrishnan & B. V. David (1976). *General and Applied Entomology*, 589 pp. Tata McGraw-Hill: New Delhi.
Needham, J. G. (1959). *Culture Methods for Invertebrate Animals*, 509 pp. Dover: New York.
Nixon, G. E. J. (1951). *The Association of Ants with Aphids and Coccids*, 36 pp. CIE: London.
Noble-Nesbitt, J. (1970). Structural aspects of penetration through insects cuticles. *Pesticide Sci.* **1**, 204–8.
Norton, G. A. & G. R. Conway (1977). The economic and social context of pest, disease and weed problems. pp. 205–26 In: Cherrett & Sagar (eds.) (1977).
Nye, I. W. B. (1958). The external morphology of some of the dipterous larvae living in the Gramineae of Britain. *Trans. R. ent. Soc. London* **110**, 411–87.
O'Brien, R. D. (1967). *Insecticides, Action and Metabolism*, 332 pp. Academic Press: New York.
O'Brien, R. D. (ed.) (1970). *Biochemical Toxicology of Insecticides*, 218 pp. Academic Press: New York.
Odiyo, P. O. (1975). Seasonal distribution and migrations of *Agrotis ipsilon* (Hufnagel) (Lepidoptera, Noctuidae), *Trop. Pest Bull.* **4**, 26 pp. COPR: London.
Odum, E. P. (1959). *Fundamentals of Ecology*, 2nd ed., 546 pp. Saunders: London.
Oldfield, G. N. (1970). Mite transmission of plant viruses. *Ann. Rev. Entomol.* **15**, 343–80.
Oldroyd, H. (1958). *Preserving and Studying Insects*, 327 pp. Hutchinson: London.
Oldroyd, H. (1968). *Elements of Entomology*, 312 pp. Weidenfeld & Nicolson: London.
Ordish, G. (1967). *Biological Methods in Crop Pest Control*, London. 242 pp.

Ordish, G. (1976). *The Constant Pest*, 240 pp. Peter Davies: London.

Ordish, G. *et al.* (1966). Current papers on integrated pest control. *PANS* A **12**, 35–72.

Padwick, G. W. (1956). Losses caused by plant diseases in the colonies. *Phytopath. Papers*, no. 1, 60 pp. CMI Survey.

Painter, R. H. (1951). *Insect Resistance in Crop Plants*, MacMillan: New York.

Painter, R. H. (1958). Resistance of plants to insects. *Ann. Rev. Entomol.* **3**, 267–90.

Pan (1980). *See* Int. Congress of Entomology (1980).

Papavizas, G. C. (ed.) (1981). *Biological Control in Crop Production* (Beltsville Symp. Agric. Res. (5), 461 pp. Allanheld, Osmund: Toronto.

Parkin, E. A. (1956). Stored Product Entomology (the assessment and reduction of losses caused by insects to stored foodstuffs). *Ann. Rev. Entomol.* **1**, 233–40.

Perring, F. H. & K. Mellanby (eds.) (1977). *Ecological Effects of Pesticides*, 193 pp. Linn. Soc. Symp. Series No. 5, Academic Press: London.

Perry, J. N., E. D. M. Macaulay & B. J. Emmett (1981). Phenological and geographical relationship between catches of pea moth in sex-attractant traps. *Ann. appl. Biol.* **97**, 17–26.

PESTDOC (1974). *Organism Thesaurus*, Vol. 1 *Animal Organisms*, 1317 pp. Ciba-Geigy: Basle, & Derwent Pub.: London.

Peterson, A. (1953). *Entomological Techniques*. Edwards Bros: Ann Arbor, Michigan.

Pfadt, R. E. (1962). *Fundamentals of Applied Entomology*, 668 pp. Macmillan: New York.

Pimentel, D. (1977). The ecological basis of insect pest, pathogen and weed problems. In: Cherrett & Sagar (eds.) (1977). 3–31.

Pimentel, D. (ed.) (1982). *CRC Handbook of Pest Management in Agriculture* Vol. I. 597 pp. Vol. II. 501 pp. Vol. III. 656 pp. CRC Press, Boca Raton: Florida.

Pirone, P. P. (1978). *Diseases and Pests of Ornamental Plants* (5th ed.), 566 pp. John Wiley: New York.

Poe, S. L. (1973). Tomato Pinworm, *Keiferia lycopersicella* (Walshingham) (Lepidoptera: Gelechiidae) in Florida. *Florida Dept. Agric., Ent. Circ. No. 131*, 2 pp.

Popham, W. L. & D. G. Hall (1958). Insect eradication programs. *Ann. Rev. Entomol.* **3**, 335–54.

Price Jones, D. & M. E. Solomon (1974). *Biology in Pest and Disease Control*, 398 pp. 13th Symp. Brit. Ecol. Soc., Blackwell: Oxford.

Pritchard, G. (1983). Biology of Tipulidae. *Ann. Rev. Entomol.* **28**, 1–22.

Prokopy, R. J. & E. D. Owens (1983). Visual detection of plants by herbivorous insects. *Ann. Rev. Entomol.* **28**, 337–64.

Proverbs, M. D. (1969). Induced sterilisation and control of insects. *Ann. Rev. Entomol.* **14**, 81–102.

Pschorn-Walcher, H. (1977). Biological control of forest insects. *Ann. Rev. Entomol.* **22**, 1–22.

Purseglove, J. W. (1968). *Tropical Crops. Dicotyledons*, vols. I & II, 719 pp. Longmans: London.

Purseglove, J. W. (1972). *Tropical Crops. Monocotyledons*, vols. I & II, 607 pp. Longmans: London.

Rabb, R. L. & F. E. Guthrie (1970). *Concepts of Pest Management*. Proceedings of a Conference held at N.C. State University at Raleigh, N.C., 25–27 March 1970, 242 pp. North Carolina State University: Raleigh, NC.

Rabb, R. L., F. A. Todd & H. C. Ellis (1976). *Tobacco Pest Management*, pp. 71–106. In: Apple & Smith (1976). *Integrated Pest Management*, Plenum Press: New York.

Radcliffe, E. B. (1982). Insect pests of potato. *Ann. Rev. Entomol.* **27**, 173–204.

Rajamohan, N. (1976). Pest complex of sunflower – a bibliography. *PANS* **22**, 546–63.

Raunkiaer, C. (1934). *The Life Forms of Plants and Statistical Plant Geography*. Oxford University Press: Oxford. (Translated from Danish.)

Raw, F. (1967). Some aspects of the wheat bulb fly problem. *Ann. appl. Biol.* **59**, 155–73.

Reay, R. C. (1969). *Insects and Insecticides*, 152 pp. Oliver & Boyd: Edinburgh.

Reddy, D. B. (1968). *Plant Protection in India*, 454 pp. Allied Pub.: Bombay.

Richards, O. W. & R. G. Davis (1977). *Imm's General Textbook of Entomology*, 10th ed. Vol. 1. *Structure Physiology and Development*, 418 pp. Vol. 2. *Classification and Biology*, 1354 pp. Chapman & Hall: London.

Riechert, S. E. & T. Lockley (1984). Spiders as biological control agents. *Ann. Rev. Entomol.* **29**, 299–320.

Roelofs, W. L. (ed.) (1979). *Establishing Efficacy of Sex Attractants and Disruptants for Insect Control*, 97 pp. Ent. Soc. Amer.: Washington.

Roelofs, W. L. & R. T. Cardé (1977). Responses of Lepidoptera to synthetic sex pheromone chemicals and their analogues. *Ann. Rev. Entomol.* **22**, 377–405.
Rose, D. J. W. (1972). Times and sizes of dispersal flights by *Cicadulina* species (Homoptera: Cicadellidae) vectors of Maize Streak Disease. *J. Anim. Ecol.* **41**, 495–506.
Rose, G. (1963). *Crop Protection*, 2nd ed., 490 pp. Leonard Hill: London.
Russell, G. E. (1978). *Plant Breeding for Pest and Disease Resistance*, 485 pp. Butterworths: London.
Schneider, D. (1961). The olfactory sense of insects. *Drapco Report* **8**, 135–46.
Schoohoven, L. M. (1968). Chemosensory basis of host plant selection. *Ann. Rev. Entomol.* **13**, 115–36.
Schreck, C. E. (1977). Techniques for the evaluation of insect repellants: a critical review. *Ann. Rev. Entomol.* **22**, 101–119.
SCI (1969). The effect of rain on plants, pests and pesticides. *Chemistry and Industry*, 1495–504.
Scopes, N. & M. Ledieu (eds.) (1979). *Pest and Disease Control Handbook*, BCPC Pub.: London.
Seabrook, W. D. (1978). Neurobiological contributions to understanding insect pheromone systems. *Ann. Rev. Entomol.* **23**, 471–85.
Seymour, P. R. (1979). *Invertebrates of Economic Importance in Britain* (formerly MAFF *Tech. Bull.* No. 6), 132 pp. HMSO: London.
Shorey, H. H. (1976). *Animal Communication by Pheromones*, Academic Press: New York.
Short, L. R. T. (1963). *Introduction to Applied Entomology*, 235 pp. Longmans: London.
Simmonds, F. J. & D. J. Greathead (1977). Introductions and pest and weed problems. In: Cherrett & Sagar (eds.) (1977). pp. 109–24.
Singh, J.P. (1970). *Elements of Vegetable Pests*, 275 pp. Vora & Co.: Bombay.
Singh, S. R., H. F. van Emden & T. A. Taylor (eds.) (1978). *Pests of Grain Legumes: ecology and control*, Academic Press: London.
Singh, S. R. & H. F. van Emden (1979). Insect pests of grain legumes. *Ann. Rev. Entomol.* **24**, 255–78.
Smith, C. N. (1966). *Insect Colonization and Mass Production*. Academic Press: New York.
Smith, K. M. (1951). *Agricultural Entomology*, 2nd ed., 289 pp. Cambridge University Press: London.
Smith, R. F. & K. S. Hagen (1959). Impact of commercial insecticide treatments. *Hilgardia* **29**, 131–54.
Southgate, B. J. (1979). Biology of the Bruchidae. *Ann. Rev. Entomol.* **24**, 449–73.
Southwood, T. R. E. (ed.) (1968). *Insect Abundance, a symposium*, 160 pp. R. ent. Soc.: London.
Southwood, T. R. E. (1972). The insect/plant relationship – an evolutionary perspective. In: H. F. van Emden (ed.), *Insect/Plant Relationships*. pp. 3–30.
Southwood, T. R. E. (1977). *The relevance of population dynamic theory to pest status*. pp. 35–54. In: Cherrett & Sagar (1977). *Origins of Pest, Parasite, Disease and Weed Problems*, Blackwell: Oxford.
Southwood, T. R. E. (1978). *Ecological Methods*, 2nd ed. 524 pp. Chapman & Hall: London.
Southwood, T. R. E. & H. N. Comins (1976). A synoptic population model. *J. Anim. Ecol.* **45**, 949–65.
Southwood, T. R. E., V. C. Moran & C. E. J. Kennedy (1982). The richness, abundance and biomass of the arthropod communities on trees. *J. Anim. Ecol.* **51**, 635–49.
Spencer K. A. (1973). *Agromyzidae (Diptera) of Economic Importance*, 418 pp. W. Junk: The Hague. (Vol. 9 of Series Entomologica.)
Spradbery, J. P. (1973). *Wasps*, Sidgwick & Jackson: London.
Stapley, J. H. & F. C. H. Gayner (1969). *World Crop Protection*. Vol. I, *Pests and Diseases*, 270 pp.; Vol. II, *Pesticides*, 249 pp. (K. A. Hassell).
Stern, V. M. (1973). Economic thresholds. *Ann. Rev. Entomol.* **18**, 259–80.
Stern, V. M., R. F. Smith, R. van den Bosch & K. S. Hagen (1959). The integrated control concept. *Hilgardia* **29**, 81–101.
Stern, V. M. & R. van den Bosch (1959). Field experiments on the effects of insecticides. *Hilgardia* **29**, 103–30.
Stinner, R. E., C. S. Barfield, J. L. Stimac & L. Dohse (1983). Dispersal and movement of insect pests. *Ann. Rev. Entomol.* **28**, 319–36.
Storey, H. H. (1961). Vector relationships of plant viruses. *E. Afr. Med. J.*. **38**, 215–20.
Strickland, A. H. (1967). Some problems in the economic integration of crop loss control. *Proc. 4th Br. Ins. Fung. Conf.* **2**, 478–91.

Strickland, A. H. (1971). The actual status of crop loss assessment. *EPPO Bull.* **1**, 39–51.
Strong, D. R., J. H. Lawton & T. R. E. Southwood (1984). *Insects on Plants: community patterns and mechanisms.* 320 pp. Blackwell Sci. Pub.: Oxford.
Sweetman, H. L. (1958). *Principles of Biological Control*, 560 pp. W. C. Brown: Iowa.
Tahori, A. S. (1971). *Pesticide Terminal Residues*, 374 pp. Butterworth: London.
Tait, E. J. (ed.) (1981). Perception and management of pests and pesticides: guidelines for research, 42 pp. (Working Paper EPR – 8) Inst. Environ. Studies: Univ. of Toronto.
Tait, J. (1981). The flow of pesticides: industrial and farming perspectives. Chapter 8, pp. 219–50. In: O'Riordan, T. & R. K. Turner (eds.) (1981). *Progress in Resource Management and Environmental Planning, Volume 3.*
Talhouk, A. M. S. (1969). *Insects and Mites Injurious to Crops in Middle Eastern Countries*, Monographien zur angew. Entomologie, Nr. 21, 230 pp. Verlag Paul Pary: Hamburg & Berlin.
TDRI Information Service Annotated Bibliographies (1981). No. 1. Insect Pests of Pre-harvest Wheat and their Control in the Developing World: 1975–80 (compiled by E. Southam & P. Schofield.)
(1983). No. 2 *Heliothis* Dispersal and Migration. (compiled by N. W. Widmer & P. Schofield).
Thompson, W. T. (1969). *The Ornamental Pesticide Application Guide*, 471 pp. Thompson: Fresno, California.
Thompson, *A Catalogue of the Parasites and Predators on Insect Pests.* Sec. I, Parasite Host Catalogue – Parts 1–11. Sec. II, Host Parasite Catalogue – Parts 1–5. Sec. III, Predator Host Catalogue. Sec. IV. Host Predator Catalogue. CAB (CIBC): London.
Tinsley, T. W. (1979). The potential of insect pathogenic viruses as pesticidal agents. *Ann. Rev. Entomol.* **24**, 63–87.
Trought, T. E. T. (1965). *Farm Pests*, 32 pp. Blackwell: Oxford.
Turnipseed, S. G. & M. Kogan (1976). Soybean entomology. *Ann. Rev. Entomol.* **21**, 247–82.
Tuttle, D. M. & E. W. Baker (1968). *Spider Mites of South-Western United States and a revision of the family Tetranychidae.* University of Arizona Press.
Tzanakakis, M. E. (1959). An ecological study of the Indian Meal Moth *Plodia interpunctella* (Hübner) with emphasis on diapause. *Hilgardia* **29**, 205–46.
USDA (1968). *Suggested Guide for the use of insecticides to control insects affecting crops, livestock, households, stored products, forests and forest products*, 273 pp. US Dept. of Agric.
USDA (1952). *The Yearbook of Agriculture (1952) Insects*, 780 pp. US Dept. of Agric.
Uvarov, B. (1966). *Grasshoppers and Locusts*, Cambridge University Press: London.
Van Emden, H. F. *et al.* (1969). The ecology of *Myzus persicae*. *Ann. Rev. Entomol.* **14**, 197–270.
Van Emden, H. F. (ed.) (1972). *Aphid Technology*, 344 pp. Academic Press: London.
Van Emden, H. F. (1974). *Pest Control and its Ecology*, 60 pp. Inst. Biol. Studies in Biology, No. 50 Edward Arnold: London.
Wallbank, B. E. & G. A. Wheatley (1979). Some responses of Cabbage Root Fly (*Delia brassicae*) to allyl isothiocyanate and other volatile constituents of crucifers. *Ann. appl. Biol.* **91**, 1–12.
Walthers, H. J. (1969). Beetle transmission of plant viruses. *Adv. Virus Res.* **15**, 339–63.
Waters, W. E. & R. W. Stark (1980). Forest pest management: concept and reality. *Ann. Rev. Entomol.* **25**, 479–509.
Watson. M. A. & R. T. Plumb (1972). Transmission of plant-pathogenic viruses by aphids. *Ann. Rev. Entomol.* **17**, 425–52.
Watson, T. F., L. Moore & G. W. Ware (1975). *Practical Insect Pest Management*, 196 pp. W. H. Freeman & Co: San Francisco.
Way, M. J. (1977). Pest and disease status in mixed stands vs. monocultures; the relevance of ecosystem stability. pp. 127–38. In: Cherrett & Sagar (eds.) (1977).
Way, M. J., M. E. Cammell, L. R. Taylor & I. P. Woiwod (1981). The use of egg counts and suction trap samples to forecast the infestation of spring-sown field beans, *Vicia faba*, by the black bean aphid, *Aphis fabae*. *Ann. appl. Biol.* **98**, 21–34.
Welling, W. (1977). Dynamic aspects of insect–insecticide interactions. *Ann. Rev. Entomol.* **22**, 53–78.
Werner, F. G. (1982). *Common Names of Insects and Related Organisms*, 132 pp. Ent. Soc. Amer.: Washington, DC.
Westcott, R. L. (1983). Revision of the *aerea* group of *Chrysobothris* (Coleoptera: Buprestidae). *Syst. Ent.* **8**, 339–59.

Whalon, M. E. & B. A. Croft (1984). Apple IPM implication in North America. *Ann. Rev. Entomol.* **29**, 435–70.

Wheatley, G. A. (1971). Pest control in vegetables: some further limitations in insecticides for Cabbage Root Fly and Carrot Fly control. *Proc. 6th Br. Ins. Fung. Conf.* **2**, 386–95.

Wheatley, G. A. & T. H. Coaker (1969). Pest control objectives in relation to changing practices in agricultural crop production. *Tech. Econ. Crop Prot. Pest Control*, Mon. **36**, 42–55.

White, R. E. (1964). Injurious beetles of the genus *Diabrotica* (Coleoptera: Chrysomelidae). *Florida Dept. Agric., Ent. Circ. No. 27*, 2 pp.

Wilkinson, C. (1982). Systematic entomology in a Dutch university. *Antenna*, **6**, 158–61.

Williams, D. J. (1969). The family-group of the scale insects (Hemiptera: Coccoidea). *Bull. Br. Mus. Nat. Hist., Ent.* **32**, 315–41.

Williams, D. J. (1971). Synoptic discussion of *Lepidosaphes* Shimer and its allies with a key to genera (Homoptera, Coccoidea, Diaspididae). *Bull. ent. Res.* **61**, 7–11.

Williams, G. C. (1964). The life history of the Indian Meal Moth, *Plodia interpunctella* Hbn. (Lepidoptera: Phycitidae) in a warehouse in Britain and on different foods. *Ann. appl. Biol.* **53**, 459–75.

Wilson, A. (1969). *Further Review of Certain Persistent Organochlorine Pesticides in Great Britain*, 148 pp. HMSO: London.

Wilson, F. (1960). *A review of the Biological Control of Insects and Weeds in Australia and Australian New Guinea.* Tech. Comm. no. 1, 102 pp. CIBC: Ottawa.

Wilson, F. (1971). *Biotic Agents of Pest Control as an Important Natural Resource.* Gooding Memorial Lecture, 12 pp. Cent. Assoc. Bee-keepers: London.

Wilson, J. W. (1931). The two-spotted mite (*Tetranychus telarius* L.) on *Asparagus plumosus. University Florida Agric. Expt. Sta. Bull. no. 234*, 20 pp.

Winteringham, F. P. W. (1969). Mechanisms of selective insecticidal action. *Ann. Rev. Entomol.* **14**, 409–41.

Wood, B. J. (1971). The importance of ecological studies to pest control in Malaysian plantations. *PANS* **17**, 411–16.

Wood, D. L., R. M. Silverstein & M. Nakajima (1970). *Control of Insect Behaviour by Natural Products*, 346 pp. Academic Press: New York.

Wood, R. S. K. (ed.) (1971). Altering the resistance of plants to pests and diseases. *PANS* **17**, 240–57.

Worthing, C. R. (1983). *The Pesticide Manual: a world compendium* 7th ed. 695 pp. BCPC Pub.: London.

Wright, R. H. (1970). Some alternatives to insecticides. *Pesticide Sci.* **1**, 24–7.

Wyniger, R. (1962, 1968). *Pests of Crops in Warm Climates and their Control.* Supplement, *Control Measures*, 2nd ed., 555 & 162 pp. Basel, Switzerland.

Yamamoto, I. (1970). Mode of action of pyrethroids, nicotinoids and rotenoids. *Ann. Rev. Entomol.* **15**, 257–72.

Young, W. R. & G. L. Teetes (1977). Sorghum entomology. *Ann. Rev. Entomol.* **22**, 193–218.

Additional References

Curtis, C.F. (1985). Genetic control of insect pests: growth industry or lead balloon. *Biol. J. Linn. Soc.* **26**, 359–74.

Dunn, P.E. (1986). Biochemical aspects of insect immunology. *Ann. Rev. Entomol.* **31**, 321–9.

Pedigo, L.P., S.H. Hutchins & L.G. Higley (1986). Economic injury levels in theory and practice. *Ann. Rev. Entomol.* **31**, 341–68.

Pimental, D. (1985). Insect pest management. *Antenna*, **9** (4), 168–71.

Staal, G.B. (1986). Anti-juvenile hormone agents. *Ann. Rev. Entomol.* **31**, 391–429.

Taylor, C.E. (1986). Genetics and evolution of resistance to insecticides. *Biol. J. Linn. Soc.* **27**, 103–12.

Visser, J.H. (1986). Host odour perception in phytophagous insects. *Ann. Rev. Entomol.* **31**, 121–44.

Appendices

A List of pesticides cited and some of their trade names

Approved names are on the left; trade names follow in italic; only a selection of the more widely used trade names are included, for further names refer to the Tropical Pest Management (formerly PANS) Pesticide Index, 1984 edition, pp. 1–91. The categorization of a few of the pesticides listed below may be open to dispute.

Chlorinated hydrocarbons

1 Aldrin *Aldrex, Aldrite, Drinox, Toxadrin*, etc.
2 Chlordane *Sydane* (many trade names)
3 Chlorobenzilate *Akar, Folbex*
4 DDT *DDT* (many trade names)
5 Dicofol *Acarin, Childion, Kelthane, Mitigan*
6 Dieldrin *Alvit, Dieldrex, Dilstan, Endosil*, etc.
7 Endosulfan *Cyclodan, Thiodan, Thiofor, Thionex*
8 Endrin *Endrex, Hexadrin, Mendrin*
9 HCH, gamma (γ) (= BHC) *Lindane*, etc.
10 Heptachlor *Drinox, Heptamul, Velsicol*
11 Mirex *Dechlorane*
12 Tetradifon *Duphar, Tedion*
13 Tetrasul *Animert V-101*
14 TDE *DDD, Rhothane*

Substituted phenols

15 Binapacryl *Acaricide, Ambox, Dapacryl, Morocide*
16 Dinocap *Karathane*
17 DNOC *Cresofin, Detal, Dinitrol, Sandolin, Sinox*
18 Pentachlorophenol *Dowicide, Pentacide, Santophen*

Organophosphorous compounds

19 Acephate *Orthene, Ortran, Tornado*
20 Azinphos-methyl *Benthion, Carfene, Gusathion*
21 Azinphos-methyl + demeton-*S*-methyl sulphone *Gusathion MS*
22 Bromophos *Brofene, Bromovur, Nexion, Pluridox*
23 Bromopropylate *Acarol, Neoron*
24 Carbophenothion *Dagadip, Garrathion, Trithion*
25 Chlorfenvinphos *Birlane, Sapecron, Supona*
26 Chlorpyrifos *Dursban, Lorsban, Spannit*
27 Chlorpyrifos-methyl *Reldan*
28 Cyanofenphos *Surecide, Watathion*
29 Demephion *Cymetox, Pyracide*
30 Demeton *Solvirex, Systemox, Systox*
31 Demeton-*S*-methyl *Demetox, Metasystox 55*
32 Diazinon *Basudin, Diazitol, DBD, Neocid*, etc.
33 Dichlorvos *Dedevap, Nogos, Nuvan, Vapona*
34 Dimefox *Hanane, Terra-sytam*
35 Dimethoate *Cygon, Dantox, Rogor, Roxion*
36 Disulfoton *Disyston, Murvin 50, Parsolin, Solvirex*
37 Ethion *Embathion, Hylemox, Nialate, Rhodocide*
38 Ethoate-methyl *Fitios*
39 Etrimfos *Ekamet*
40 Fenitrothion *Accothion, Folithion, Sumithion, Dicofen*
41 Fenthion *Baytex, Lebaycid, Queleton, Tiguvon*
42 Fonofos *Dyfonate*
43 Formothion *Aflix, Anthio*
44 Heptenophos *Hostaquick, Ragadan*
45 Iodofenphos *Alfacron, Elocril, Nuvanol*
46 Malathion *Malathion, Malastan, Malathexo*
47 Mecarbam *Afos, Murfotox, Pestan*
48 Menazon *Aphex, Saphicol, Saphizon, Sayfos*
49 Mephosfolan *Cytro-lane*
50 Methidathion *Supracide, Ultracide*
51 Mevinphos *Fastac, Menite, Phosdrin, Phosfene*
52 Monocrotophos *Azodrin, Monocron, Nuvacron*
53 Naled *Bromex, Dibrom, Ortho*
54 Omethoate *Folimat*
55 Oxydemeton-methyl *Metasystox-R*
56 Oxydisulfoton *Disyston-S*
57 Parathion *Bladan, Folidol, Fosferno, Fosfex, Thiophos*
58 Parathion-methyl *Dalf, Folidol-M, Metacide, Nitrox-80*
59 Phenisobromolate *Acarol, Neoron*
60 Phenthoate *Cidial, Elsan, Papthion, Tanone*
61 Phorate *Granutox, Rampart, Thimet, Timet*
62 Phosalone *Embacide, Rubitox, Zolone*

63 Phosfolan *Cyalane, Cyolon*
64 Phosmet *Appa, Germisan, Imidan, Prolate*
65 Phosphamidon *Dicron, Dimecron, Famfos*
66 Phoxim *Baythion, Valexon, Volaton*
67 Pirimiphos-ethyl *Fernex, Primicid, Primotec*
68 Pirimiphos-methyl *Actellic, Actellifog, Blex*
69 Profenofos *Curacron*
70 Prothoate *Fac, Fostin, Oleofac, Telefos*
71 Quinalphos *Savall*
72 Quinomethionate *Erade, Forstan, Morestan*
73 Schradan *Pestox 3, Sytam*
74 TEPP *Bladen, Fosvex, Nifos T, Tetron, Vapotone*
75 Terbufos *Counter*
76 Tetrachlorvinphos *Gardona, Ostabil, Rabon, Ravap*
77 Thiometon *Ekatin, Intrathion*
78 Thionazin *Nemafos, Nemasol, Zinophos*
79 Thioquinox *Eradex, Eraditon*
80 Triazophos *Hostathion*
81 Trichloronate *Agritox, Agrisil, Phytosol*
82 Trichlorphon *Anthon, Chlorofos, Dipterex, Tugon*
83 Vamidothion *Kilval, Vamidoate, Vation*

Carbamates and related compounds

84 Aldicarb *Temik*
85 Bendiocarb *Garvox*
86 Bufencarb *Bux*
87 Carbaryl *Carbaryl 85, Murvin, Septon, Sevin*
88 Carbofuran *Curaterr, Furadan, Yaltox*
89 Ethiofencarb *Croneton*
90 Methiocarb *Baysol, Draza, Mesurol*
91 Methomyl *Halvard, Lannate, Nudrin*
92 Oxamyl *Tumex, Vydate*
93 Pirimicarb *Aphox, Fernos, Pirimor*
94 Promecarb *Carbamult, Minacide*
95 Propoxur *Baygon, Blattanex, Suncide, Unden*
96 Thiofanox *Decamox*

Miscellaneous compounds and fumigants

97 Aluminium phosphide *Phostoxin*
98 Amitraz *Azaform, Baam, Mitac, Taktic, Triatox*
99 Carbon disulphide *Weevil-Tox*
100 Copper acetoarsenite *Paris Green*
101 Cryolite (Sodium aluminofluoride) —
102 Cyhexatin *Plictran*
103 Ethylene dibromide *Bromofume, Dowfume-W*, etc.
104 Lead arsenate *Gypsine, Soprabel*
105 Mercurous chloride *Calomel, Cyclostan*
106 Methyl bromide *Bromogas, Embafume*, etc.
107 Sulphur (Lime-sulphur) *Cosan, Hexasul, Thiovit*, etc.

Natural organic compounds and pyrethroids

108 Bioallethrin *D-Trans, Esbiol*
109 Cypermethrin *Cymbush, Cyperkill, Ripcord*
110 Deltamethrin *Decis*
111 Fenvalerate *Belmark, Pydrin, Sumicidin*
112 Nicotine *Nicofume*
113 Permethrin *Ambush, Kalfil, Talcord*
114 Pyrethrum (Pyrethrins) —
115 Resmethrin *Chryson, For-Syn, Synthrin*
116 Rotenone *Cube, Derris, Rotacide*

Organic oils

117 Petroleum Oils (White oils) *Volck* etc.
118 Tar oils —

Biological compounds

119 *Bacillus thuringiensis* *Biotrol, BTB-183, BTH, Thuricide*
120 Codling Moth GV *Cp GV*
121 *Heliothis* PHV *Elcar, Viron H*
122 *Trichoplusia* PHV *Viron T*
123 *Verticillium lecanii* *Mycotal, Vertalec*

Insect growth regulators

124 Diflubenzuron *Dimilin*
125 Methoprene *Altosid, Kabat, Manta*
126 Kinoprene *Enstar*

B Glossary of terms used in applied entomology and crop protection

Acaricide Material toxic to mites (Acarina).

Activator Chemical added to a pesticide to increase its toxicity.

Active ingredient (a.i.) Toxic component of a formulated pesticide.

Adherence The ability of a material to stick to a particular surface.

Adhesive (= Sticker) Material added to increase pesticide retention; different commercial preparations of methyl cellulose are available for this purpose.

Adjuvant A spray additive to improve either physical or chemical properties (*see also* Supplement, Sticker, Adhesive, Spreader, Wetter and Emulsifier).

Aedeagus The male intromittent organ, or penis.

Aerosol A dispersion of spray droplets of diameter 0.1–5.0 μm; usually dispersed from a canister.

Aestivation Dormancy during a hot or dry season.

Agamic Parthenogenetic reproduction; without mating.

Agitator A mechanical device in the spray tank to ensure uniform distribution of toxicant and to prevent sedimentation.

Agroecology The study of ecology in relation to agricultural systems.

Allochthonous Not aboriginal; exotic; introduced; acquired from elsewhere (opp. autochthonous).

Allopatric Having separate and mutually exclusive areas of geographical distribution (opp. sympatric).

Anemophilous Plants which are pollinated by the wind.

Anholocyclic Alternation of generations with suppression of the sexual phase; permanently parthenogenetic.

Anionic surfactant Salt of an organic acid, the structure of which determines its surface activity.

Antibiosis The resistance of a plant to insect attack by having, for example, a thick cuticle, hairy leaves, toxic sap, etc.

Anti-feedant A chemical possessing the property of inhibiting the feeding of certain insect pests.

Anti-frothing agent Material added to prevent frothing of the liquid in the spray tank.

Approved product Proprietary brand of pesticide officially approved by the Ministry of Agriculture, Fisheries and Food, UK.

Arista A large bristle, located on the dorsal edge of the apical antennal segment in the Diptera.

Asymptote The point in the growth of a population at which numerical stability is reached.

Atomiser Device for breaking up a liquid stream into very fine droplets by a stream of air.

Atrophied Reduced in size; rudimentary; vestigial.

Attractant Material with an odour that attracts certain insects; lure. Several proprietory lures are manufactured.

Autoecious Passing the different life-history stages all on (or in) the same host organism.

Autecology The ecological study of a single species.

Autochthonous Aboriginal; native; indigenous; formed where found (opp. allochthonous).

Autocide The control of a pest by the sterile-male technique.

Bait Foodstuff used for attracting pests; usually mixed with a poison to form a poison bait.

Band application Treatment of a band of soil in row-crops, usually covering plant rows, with either sprays or granules.

Biocide A general poison or toxicant.

Boom (spray) Horizontal (or vertical) light frame carrying several spray nozzles.

Brachypterous Having short wings that do not cover the abdomen.

Breaking The separation of the phases from emulsion.

Budworm Common name in the USA for various tortricid larvae.

Calling A virgin female moth releasing sex pheromones to attract males for the purpose of mating.

Carrier Material serving as diluent and vehicle for the active ingredients in a pesticide spray or powder.

Cationic surfactant Material in which surface activity is determined by the basic part of a compound.

Caterpillar Eruciform larva; larva of a moth, butterfly, or sawfly.

Chaetotaxy The arrangement and nomenclature of the bristles on the insect exoskeleton, both adults and larvae.

Chemosterilant Chemical used to render an insect sterile without killing it.

Chrysalis The pupa of a butterfly.

Climatograph A polygonal diagram resulting from plotting temperature means against relative humidity.

Clone A group of identical individuals propagated vegetatively from a single plant.

Coarctate pupa A pupa enclosed inside a hardened shell formed by the previous larval skin.

Cocoon A silken case inside which a pupa is formed.

Colloidal formulation Solution in which the particle size is less than 6 μm in diameter, and the particles stay indefinitely dispersed.

Commensalism Two organisms living together and sharing food, both species usually benefiting from the association; a type of symbiosis.

Community The collection of different species and types of plants and animals, in their respective niches, within the habitat. (1) Closed: the habitat is completely colonized by plants; no areas of bare soil; strong competition for space. (2) Open: habitat is not completely colonized by plants; bare areas of soil; competition for space is thus reduced.

Compatibility The ability to mix different pesticides without physical or chemical interactions which would lead to reduction in biological efficiency or increase in phytotoxicity.

Compressed Flattened from side to side.

Concentrated solution (c.s.) Commercial pesticide preparation before dilution for use.

Concentrate spraying Direct application of the pesticide concentrate without dilution.

Concentration Proportion of active ingredient in a pesticide preparation, before or after dilution.

Contact poison Material killing pests by contact action, presumably by absorption through the cuticle.

Control (noun) Untreated subjects used for comparison with those given a particular crop protection treatment.

Control (verb) To reduce damage or pest density to a level below the economic threshold. (1) Legislative: the use of legislation to control the importation and to prevent any spread of a pest within a country. (2) Physical: the use of mechanical (hand picking, etc.) and physical methods (heat, cold, radiation, etc.) of controlling pests. (3) Cultural: regular farm operations designed to destroy pests. (4) Chemical: the use of chemical pesticides as smokes, gas, dusts, and sprays to poison pests. (5) Biological: the use of natural predators, parasites and disease organisms to reduce pest populations. (6) Integrated: the very carefully reasoned use of several different methods of pest control in conjunction with each other to control pests with a minimum disturbance to the natural situation.

Cosmopolitan A species occurring very widely throughout the major regions of the world.

Costa A longitudinal wing vein, usually forming the anterior margin (leading edge).

Cover Proportion of the surface area of the target plant on which the pesticide has been deposited.

Crawlers The active first instar of a scale insect (Homoptera; Coccoidea).

Cremaster A hooked, or spine-like process at the posterior end of the pupa, often used for attachment (Lepidoptera).

Crepuscular Animals that are active in the twilight; pre-dawn and at dusk in the evenings.

Crochets Hooked spines at the tips of the prolegs of lepidopterous larvae with which they cling to plant foliage.

Crop hygiene (= Phytosanitation) The removal and destruction of heavily infested or diseased plants from a crop so that they do not form sources of reinfestation.

Cutworm Larva of certain Noctuidae that lives in the soil, emerging at night to eat foliage and stems; serious pests of many crops as seedlings, and root crops.

Deflocculating agent Material added to a spray suspension to delay sedimentation.

Defoliant Spray which induces premature leaf-fall.

Deposit (spray) Amount and pattern of spray or dust deposited per unit area of plant surface.

Deposit (dried) Amount and pattern of active ingredient deposited per unit area of plant surface.

Deposition velocity Velocity at which the spray impinges on the target.

Depressed Flattened dorso-ventrally.

Desiccant Chemical which kills plant foliage by inducing excessive water loss.

Diluent Component of spray or dust that reduces the concentration of the active ingredient, and may aid in mechanical application but does not directly affect toxicity.

Disinfect (1) To free from infection by destruction of the pest or pathogen established in or on plants or plant parts. (2) To kill or inactivate pests or pathogens present upon the surface of plants or plant parts, or in the immediate vicinity (e.g. in soil).

Dispersal Movement of individuals out of a population (emigration) or into a population (immigration).

Diurnal Active during the daytime.

Dormant Alive but not growing; buds with an unbroken cover of scales; quiescent; inactive; a resting stage.

Dose; dosage Quantity of pesticide applied per individual, or per unit area, or per unit volume, or per unit weight.

Drift Spray or dust carried by natural air currents beyond the target area.

Drop spectrum Distribution, by number or volume of drops, of spray into different droplet sizes.

Duster Equipment for applying pesticide dusts to a crop.

Ecdysis The moulting (shedding of the skin) of larval arthropods from one stage of development to another, the final moult leading to the formation of the puparium or chrysalis.

Ecesis (= Oikesis) The establishment of an organism in a new habitat; accidental dispersal and establishment in a new area.

Ecoclimate Climate within the plant (crop) community.

Ecology The study of all the living organisms in an area, their interreactions, and their physical environment.

Economic damage The injury (damage) done to a crop which will justify the cost of artificial control measures.

Economic-injury level The lowest population density that will cause economic damage.

Economic pest A pest causing a definite financial crop loss; usually of about 5–10%, according to definition.

Economic threshold The pest population level at which control measures should be started to prevent the pest population from reaching the economic-injury level.

Ecosystem The interacting system of the living organisms in an area and their physical environment.

Efficiency of a pest control measure The more or less fixed reduction of a pest population regardless of the number of pests involved.

Effectiveness of a pest control measure This is shown by the number of pests remaining after control treatment.

Elateriform larva A larva resembling a wireworm with a slender body, heavily sclerotinized, with short thoracic legs and only a few body bristles.

Elytron The thickened forewing of adult beetles (Coleoptera).

Emergence (1) The adult insect leaving the last nymphal skin, or pupal case. (2) Germination of a seed and the appearance of the shoot either though the test, or the soil surface.

Emigration The movement of individuals out of a population.

Emulsifiable concentrate (e.c.) Liquid formulation that when added to water will spontaneously disperse as fine droplets to form an emulsion (= Miscible oil).

Emulsifier Spray additive which permits formation of a stable suspension of oil droplets in aqueous solution, or of aqueous solution in oil.

Emulsion A stable dispersion of oil droplets in aqueous solution, or vice versa.

Emulsion, Invert Suspension of aqueous solution in oil.

Encapsulation Or microencapsulation: the encapsulation of a pesticide in a non-volatile envelope of gelatin, usually of minute size, for delayed release.

Entomophagous An animal (or plant) which feeds upon insects.

Erinium A growth of hairs in dense patches on plant leaves resulting from the attack of certain Eriophyidae (Acarina).

Eruciform larva Caterpillar; a larva with a cylindrical body, well-developed head, and with both thoracic legs and some abdominal prolegs.

Exarate pupa A pupa in which the appendages are free and not glued to the insect body.

Exuvium The cast skin of arthropods after moulting.

Fecundity Capacity to produce offspring (reproduce); power of a species to multiply rapidly.

Filler Inert component of pesticide dust or granule formulation.

Flowability Property of flowing possessed by dusts, colloids, liquids, and some pastes.

Fluorescent tracer Fluorescent material added to a spray to aid the assessment of spray deposits on plants.

Formulation (1) Statement of nature and amount of all constituents of a pesticide concentrate. (2) Method of preparation of a pesticide concentrate.

Fossorial Modified for digging; in the habit of digging or burrowing.

Frass Wood fragments made by a wood-boring insect, usually mixed with the faeces.

Fumigant Pesticide exhibiting toxicity in the vapour phase.

Furrow application Placement of pesticides with seed in the furrow at the time of sowing.

Gall An abnormal growth of plant tissues, caused by the stimulus of an animal or another plant.

Generation The period from any given stage in the life-cycle (usually adult) to the same stage in the offspring.

Granule Coarse particle of inert material (pumice, Fuller's earth, rice husks) impregnated or mixed with a pesticide. Used mainly for soil application, but sometimes for foliar application (pumice formulation).

Granule applicator Machine designed to apply measured quantities of granules.

Grease band Adhesive material (e.g. resin in castor oil, or 'Sticktite') applied as a band around a tree to trap or repel ascending wingless female moths, caterpillars, or ants.

Grub (White) A scarabaeiform larva; thick-bodied, with a well-developed head and thoracic legs, without abdominal prolegs, usually sluggish in behaviour; general term for larvae of Coleoptera.

Habitat The place where plants and animals live, usually with a distinctive boundary (e.g. field, pond, sand-dune, rocky crevice).

Hemelytron The partly thickened forewing of Heteroptera.

Hemimetabolous Insects having a simple metamorphosis, like that in the Orthoptera, Heteroptera and Homoptera.

Herbivorous Feeding on plants (phytophagous).

Hibernation Dormancy during the winter, or cold season.

Hollow-cone Spray jet with a core of air breaking to give drops in an annular pattern.

Holometabolous Insects having a complete metamorphosis, as in the Diptera, Hymenoptera, Coleoptera, Lepidoptera.

Honey-dew Liquid with high sugar content discharged from the anus of some Homoptera.

Hornworm A caterpillar with a dorsal spine or horn on the last abdominal segment: larvae of Sphingidae.

Host The organism in or on which a parasite lives; and the plant on which an insect feeds.

Humectant Material added to a spray to delay evaporation of the water carrier.

Hyaline Transparent; clear.

Hypermetamorphosis A type of complete metamorphosis in which the different larval instars represent two or more different types of larvae.

Hyperparasite A parasite whose host is another parasite.

Hythergraph A polygonal diagram resulting from plotting temperature means against rainfall.

Imago The adult, or reproductive stage of an insect.

Immigration The movement of individuals into a population.

Immune Exempt from infection; resistant to infection.

Incompatible Not compatible; incapable of forming a stable mixture with another chemical.

Indicator Marker; or organism characteristic of a particular habitat.

Inert A material having no biological action.

Infect To enter and establish a pathogenic relationship with a plant (host); to enter and persist in a carrier.

Infest To occupy and cause injury to either a plant, soil or stored products.

Injector A device for ejecting a pesticide below the soil surface, or into the transport system of a tree.

Insecticide A toxin effective against insects.

Instar The form of an insect between successive moults; the first instar being the stage between hatching and the first moult.

Intercropping The growing of two crops simultaneously in the same field.

Jet Liquid stream emitted from a nozzle orifice (in USA = nozzle).

Key pest An important major pest species in the complex of pests attacking a crop, with a dominating effect on control practices.

Lacquer Pesticide incorporated into a lacquer or varnish to achieve slow release over a lengthy period of time.

Larva The immature stages of an insect, between the egg and pupa, having a complete metamorphosis; the six-legged first instar of the Acarina.

Larvicide Toxicant (poison) effective against insect larvae.

LC_{50} Lethal concentration of toxicant required to kill 50% of a large group of individuals of one species.

LD_{50} Lethal dose of toxicant required to kill 50% of a large group of individuals of one species.

Leaf area index (LAI) Ratio of leaf surface to soil surface area, in relation to utilization of solar energy for photosynthesis.

Leaf miner An insect which lives in and feeds upon the cells between the upper and lower epidermis of leaf, these being larvae of Agromyzidae (Diptera), Lyonetiidae and Gracillariidae (Lepidoptera), Hispidae (Coleoptera), etc.

Life table The separation of a pest population into its different age components (e.g. eggs, larvae, pupae, adults).

Looper A caterpillar of the family Geometridae, with only one pair of abdominal prolegs (in addition to the terminal claspers), which moves by looping its body.

Macropterous Large, or long-winged; normally winged.

Maggot A vermiform larva, legless, without a distinct head capsule (Diptera).

Miscible liquid (m.l.) A formulation in which the technical product is dissolved in an organic solvent which is then, on dilution, dissolved in the water carrier.

Mist blower Sprayer producing a fine air-carried spray.

Miticide Preferably called Acaricide.

Molluscicide Toxicant effective against slugs and snails.

Monoculture The extensive cultivation of a single species of plant.

Monophagous An insect restricted to a single host plant species, or genus.

Mortality Population decrease factor; death rate.

Mutualism The symbiotic relationship between two organisms in which both parties derive benefit.

Natality Population 'increase' factor; birth rate.

Necrosis Death of a localized part of a plant.

Nematicide Toxicant effective against nematodes (= eelworms).

Nocturnal Active at night.

Non-ionic surfactant A surfactant that does not ionize in solution and is therefore compatible with both anionic and cationic surfactants.

Nozzle (1) Air blast: nozzle using high velocity air to break up the spray liquid supplied at low pressure. (2) Anvil: nozzle in which the spray liquid jet strikes a smooth, solid surface at a high angle of incidence. (3) Cone (or swirl): nozzle in which the liquid emerges from the orifice with tangential velocity imparted by passage through one or more tangential or helical channels in the swirl chamber. (4) Hollow cone: nozzle in which spray jet has a core of air breaking to give drops in an annular pattern. (5) Fan nozzle: the aperture is an elongate horizontal slit, producing a fan-shaped (linear) spray pattern. (6) Deflector: nozzle in which a fan-shaped sheet of spray is formed by directing the liquid over a sharply inwardly curving surface.

Nymph The immature stage of an insect that does not have a distinct pupal stage; also the immature stages of Acarina that have eight legs; it basically resembles a small adult.

Obtect pupa A pupa in which the appendages are more or less glued to the body surface (Lepidoptera).

Oligophagous (= Stenophagous) An animal feeding upon only a few, closely related, host plants (usually a single plant family); or it may be an animal parasite.

Onisciform larva A flattened platyform larva, like a woodlouse in appearance.

Orifice (nozzle) velocity Velocity at which the spray leaves the nozzle orifice.

Ovicide Toxicant effective against insect or mite eggs.

Oviparous Reproduction by laying eggs.

Paint gun Type of small, hand-carried, air-blast machine.

Pantropical A species occurring widely throughout the tropical and subtropical parts of the world.

Parasite An organism living in intimate association with a living organism (plant or animal) from which it derives material essential for its existence while conferring no obvious benefit in return.

Parasitoid An organism alternately parasitic and free-living; most parasitic Hymenoptera and Diptera fall into this category as usually only the larvae are parasitic.

Parthenogenesis Reproduction without fertilization; usually through eggs but sometimes through viviparity.

Parts per million (ppm) Proportion of toxicant present in relation to that of plant material on which it has been deposited. Usually applied in connection with the edible portion of a crop and its suitability for consumption; otherwise refers to pollution studies.

Pellet Seed coated with inert material, often incorporating pesticides, to ensure uniform size and shape for precision drilling.

Penetrant Oil added to a spray to enable it to penetrate the waxy insect cuticle more effectively.

Persistance The length of time during which a chemical is active or effective, after application.

Persistent The term applied to chemicals that remain active for a long period of time after application.

Pest An animal or plant causing damage to man's crops, animals or possessions.

Pest density The population level of a pest species; especially when economic damage is done.

Pest management The careful manipulation of a pest situation, after extensive consideration of all aspects of the life system as well as ecological and economic factors.

Pest spectrum The complete range of pests attacking a particular crop.

Pesticide A chemical which by virtue of its toxicity (poisonous properties) is used to kill pest organisms. A term of wide application which includes all the more specific applications: insecticide, acaricide, bactericide, fungicide, herbicide, molluscicide, nematicide, rodenticide, etc.

Pheromone (= Ectohormone). A substance secreted by an insect to the exterior causing a specific reaction in the receiving insects.

Phytophagous Herbivorous; plant eating

Phytosanitation Measures requiring the removal or destruction of infected or infested plant material likely to form a source of reinfection or reinfestation. (See Crop Hygiene.) International phytosanitation refers to inspection of plants and seeds to prevent noxious pests and diseases from being brought into a country; often involves plant quarantine.

Phytotoxic A chemical liable to damage or kill plants (especially the higher plants), or plant parts.

Planidium larva A type of first-instar larva in certain Diptera and Hymenoptera which undergoes hypermetamorphosis.

Poison bait An attractant foodstuff for insects, molluscs or rodents, mixed with an appropriate toxicant.

Polyphagous An animal feeding upon a wide range of hosts.

Potentiation The enhancing of the toxic effect of a pesticide by various different means.

Pre-access interval The interval of time between the last application of pesticide to an area and safe access to the area for domestic livestock and man.

Predisposition Making a plant more susceptible to a pest or disease, usually as a result of genetic, cultural or environmental defects.

Preference The factor by which certain plants are more or less attractive to insects by virtue of their texture, colour, aroma or taste.

Pre-harvest interval The interval of time between the last application of pesticide and the safe harvesting of edible crops for immediate consumption.

Pre-oviposition period The period of time between the emergence of an adult female insect and the start of its egg-laying.

Pre-pupa A quiescent stage between the larval period and the pupa; found in some Diptera and Thysanoptera.

Preventative A measure applied in anticipation of pest attack.

Proleg A fleshy abdominal leg found in caterpillars (Lepidoptera, sawflies), bearing a characteristic arrangement of crochets.

Proprietory name Distinguishing name given by the manufacturer to a particular formulated product.

Protective clothing Clothing to protect the spray operator from the toxic effects of crop protection chemicals. This may include rubber gloves, boots, apron, respirator. face mask, etc.

Protonymph The second instar of mites.

Pterostigma A thickened opaque or dark spot along the costal margin of the wing, near the tip (e.g. Odonata, Hymenoptera).

Pupa The stage between larva and adult in insects with complete metamorphosis; a non-feeding and usually inactive stage.

Puparium The case formed by the hardened last larval skin in which the pupa of higher Diptera is formed.

Quarantine All operations associated with the prevention of importation of unwanted organisms into a territory, or their exportation from it.

Recruitment The addition of new individuals to a population, usually by either birth or immigration.

Redistribution Movement of pesticide subsequent to the initial application to other parts of the plant, usually by rain.

Repellant A chemical which has the property of inducing avoidance by a particular pest.

Residue Amount of pesticide remaining in or on plant tissues (or in soil) after a given time, especially at harvest time.

Resistance The natural or induced capacity to avoid or repel attack by pests (or parasites). Also the ability to withstand the toxic effects of a pesticide or a group of pesticides, often by metabolic detoxification.

Rodenticide A toxicant effective against rodents.

Roguing The removal of unhealthy or unwanted plants from a crop.

Rostrum The beak or proboscis of Hemiptera.

Run-off The process of spray shedding from a plant surface during and immediately after application, when droplets coalesce to form a continuous film and surplus liquid drops from the surface.

Scarabaeiform larva A grub-like larva, with a thickened cylindrical body, well-developed head and thoracic legs, without abdominal prolegs, and sluggish in behaviour.

Scavenger An animal that feeds on dead plants and animals, on decaying matter, or on animal faeces.

Secondary pest Species whose numbers are usually controlled by biotic and abiotic factors which sometimes break down, allowing the pest to increase in numbers.

Seed dressing A coating (either dry or wet) of protectant pesticide applied to seeds before planting. Dry seed dressings are often physically stuck to the testa of the seed by a sticker such as methyl cellulose.

Semilooper Caterpillar from the subfamily Plusiinae (Noctuidae) with two or three pairs of prolegs, which locomotes in a somewhat looping manner.

Siphunculi The paired protruding organs near the terminal end of the abdomen of Aphidoidea, also called cornicles, through which a waxy secretion is extruded.

Slurry Paste-like liquid used as a seed coating.

Smoke Aerial dispersal of minute solid particles of pesticides through the use of combustible mixtures.

Soil sterilant Toxicant added to, or injected into, soil for the purpose of killing pests and pathogens in the soil.

Solid cone Jet with air-core reduced to give a cone of spray droplets.

Solvent Carrier solution in which the pesticide (technical product) is dissolved to form the concentrate.

Spray (1) Air-carried: spray propelled to target in a stream of air. (2) Coarse: dispersion of droplets of mass median diameter over 200 μm. (3) Concentrate: undiluted commercial pesticide preparation. (4) Fine: dispersion of droplets of mass median diameter from 50–150 μm. (5) Floor: spray applied to the litter on the ground surface. (6) High-volume: over 1000 l/ha on bushes and trees; over 700 l/ha on ground crops. (7) Low-volume: spray of 200–500 l/ha on bushes and trees; 50–200 l/ha on ground crops. (8) Median-volume: 500–1000 l/ha on bushes and trees; 200–700 l/ha on ground crops. (9) Mist: dispersion of droplets of 50–100 μm in diameter. (10) Ultra-low volume: less than 50 l/ha on trees and bushes; or less than 5 l/ha on ground crops.

Spray angle Angle between the sides of a jet leaving the orifice.

Sprayer Apparatus for applying pesticide sprays; not to be confused with 'Spray operator'.

Spray operator Person operating a sprayer, and applying a spray.

Spread Uniformity and completeness with which a spray deposit covers a continuous surface, such as a leaf or a seed.

Spreader Material added to a spray to lower the surface tension and to improve spread over a given area (= wetter).

Spur An articulated spine, often on a leg segment, usually the tibia. A serrulate tibial spur is characteristic of the Delphacidae (Homoptera).

Stability The ability of a pesticide formulation to resist chemical degradation over a period of time.

Sticker A material of high viscosity used to stick powdered seed dressings on to seeds; two commonly used stickers are paraffin (kerosene) and methyl cellulose. A solution of methyl cellulose can be added to a spray to increase retention on plant foliage.

Stomach poison A toxicant (poison) which operates by absorption through the intestine after having been ingested by the insect, usually on plant material or in a bait.

Supplement (spray) (= Adjuvant).

Surfactant (= Spreader; Wetter).

Susceptible Capable of being easily infested or infected; not resistant.

Swath Width of target area sprayed at one pass.

Symbiosis The general term for two organisms that live together in a partnership, sometimes beneficial; includes commensalism, inquilinism, mutualism, and parasitism.

Sympatric Having the same, or overlapping, areas of geographical distribution.

Synanthropic Living close to human habitation.

Synecology The ecological study of a particular community.

Synergism Increased pesticidal activity of a mixture of pesticides above that of the sum of the values of the individual components.

Systematics The classification of animal and plant species into their higher taxa; sometimes regarded as synonymous with taxonomy.

Systemic A pesticide absorbed through the plant surfaces (usually roots) and translocated through the plant vascular system.

Taint Unwanted flavour in fresh or processed food from a pesticide used on the growing crop.

Target surface The surface intended to receive a spray or dust application.

Taxonomy The laws of classification as applied to natural history; identification of plant and animal species.

Technical product The usual form in which a pesticide is prepared and handled prior to formulation; usually at a high level of purity (95–8%) but not completely pure.

Tegmen The thickened and leathery forewing in the Orthoptera and Dictyoptera.

Tenacity The property of a pesticide deposit or residue to resist removal by weathering.

Tenacity index Ratio of the quantity of residue per unit area at the end of a given period of weathering to that present at the beginning.

Tolerance Ability to endure infestation (or infection) by a particular pest (or pathogen) without showing severe symptoms of distress.

Tolerance, permitted Maximum amount of toxicant allowed in foodstuffs for human consumption.

Toxicant Poison, or chemical exhibiting toxicity.

Toxicity Ability to poison, or to interfere adversely with vital processes of the organism by physico-chemical means.

Tracer Additive to facilitate location of a deposit, by radioactive or fluorescent means.

Translaminar A pesticide which passes through from one surface of a leaf to the other (from lamina to lamina) through the leaf tissue.

Translocation The uptake of a pesticide into part of a plant body and its subsequent dispersal to other parts of the plant body.

Transovarially Transmitted through the egg stage from mother to offspring.

Trap crop Crop of plants (sometimes wild plants) grown especially to attract insect pests, and when infested either sprayed or collected and destroyed. Trap plants usually grown between the rows of the crop plants or else peripherally.

Trapping-out The removal of individuals from a pest population, in significantly large numbers, by means of trapping (often using u.v. light traps or pheromone traps).

Triungulin larva The active first-instar larva of Meloidae (Coleoptera) and Strepsiptera.

Vector Organisms able to transmit viruses or other pathogens either directly or indirectly. Direct virus vectors include insects, mites and nematodes.

Vermiform larva A legless (apodous), headless (acephalic), worm-like larva typical of some Diptera.

Vestigial Poorly developed; degenerate; non-functional.

Viviparous Giving birth to living young (Aphidoidea etc.).

Volunteer Crop plant growing accidentally from shed seed; not deliberately cultivated.

Weathering Generally refers to the deleterious effects of weather on the effectiveness of pesticides, i.e. removal from plant foliage by rainfall and chemical degradation due to insolation, etc.

Wetter Material added to a spray to lower the surface tension and to improve spread over the target area (=Spreader; Surfactant).

Wireworm (Elateriform larva) The larva of Elateridae (Coleoptera); long, slender, well-sclerotized, thoracic legs but no prolegs, and few setae.

This glossary was originally based in part upon one produced for the Horticultural Education Association, UK by Mr R. W. Marsh, OBE, late of Long Ashton Research Station, Bristol, whose assistance is gratefully acknowledged.

An excellent recently published dictionary of relevance is Lincoln, Boxshall & Clark (1982).

C Preservation, shipment, and identification of insect specimens – simple hints

Killing

For specimens to be preserved dry: potassium cyanide, ethyl acetate, chloroform, ether, petrol, etc. Small insects are quickly killed by tobacco smoke. For preservation in fluids: the preservative or very hot water.

Labelling

Specimens without adequate labels are of little value. Data should include place, date and name of collector, clearly stated – e.g. 'Feltwell, Cambs., ENGLAND. 12 Jan. 1972; J. Smith'. In remote districts state latitude and longitude rather than the name of a very small village. Notes such as 'boring apples', 'eating cabbage leaves', 'at light', should be added when relevant; host data are generally very important. When it is desirable to add more on a label, make it quite clear to which specimens the descriptions refer.

Unnecessary words such as 'Insects found in . . .' should be avoided. Labels for tubes should be placed *inside* them, written in Indian ink or soft pencil on paper of good quality.

Dried specimens

Most insects (e.g. grasshoppers, bugs, beetles, butterflies and moths, bees and wasps, flies) are usually pinned and dried. Only stainless steel insect pins, for preference, should be used, even in the UK unsuitable pins soon rust and corrode. If pinning is inconvenient, specimens can be preserved dry in pill-boxes, envelopes, etc., with packing of soft tissue paper, etc. to prevent movement. (Cotton-wool should not touch specimens since claws become entangled in it. Cellulose wool, however, is suitable for packing small Diptera, Hymenoptera, etc.)

After pinning, or packing, specimens should be protected from mould with liberal amounts of naphthalene (e.g. moth balls).

Preservation in liquids

Alcohol 70% to 80%. Substitutes: methylated spirit (slightly diluted), strong rum, gin or whisky. Bottles should not be more than half filled with specimens, to avoid excessive dilution of the preservative. Avoid formalin if possible. 80% alcohol to which 5% glycerine has been added is as good as anything. Soft-bodied insects (caterpillars, maggots, aphids, termites, etc.) should be preserved in liquids. Insects too small for pinning are often most conveniently preserved thus. (Almost all insects, except butterflies and moths, preserve well in liquids, though pinning is the usual practice with most of them.)

Packing

Pins must be pushed deep into the cork of boxes, otherwise many of them come adrift. Large insects must be prevented from rotating on the pins by extra pins touching each side. Naphthalene should be firmly anchored with pins, or melted to adhere to the box. Boxes of pinned insects should be surrounded on all sides by at least 3 cm of thin wood shavings ('excelsior'), packed in a strong box sufficiently firmly to prevent all movement and to absorb shock. If all these precautions are adopted, delicate specimens survive mailing – otherwise considerable damage is almost inevitable.

For specimens in fluids, the only precautions necessary are to push corks firmly into bottles (sealing corks with paraffin wax is desirable, but not essential) and to pack securely put tissue paper in tubes of very delicate specimens. Cigarette and biscuit tins, wrapped in corrugated cardboard, make good containers for mailing bottles. Do not send corked tubes of alcohol by air.

Living specimens

Although it is usually preferable either to dry specimens or preserve them in liquids for shipment, it is sometimes necessary to send living material. Live insects should be sent in suitable strong cardboard, wooden or tin boxes. When tins are used, the lid should be perforated with numerous small holes to prevent undue condensation of moisture. In general, larvae should be supplied with an adequate supply of food material, but if their food is of a succulent nature, it will probably decompose and it may be better to pack them loosely in damp moss. Many adult insects survive several days without food, but a little tissue paper or muslin, or in a dry climate some damp moss, should be placed in the container with them. Larval Noctuidae are often cannibalistic when packed into small containers.

For further information on collecting and preserving insects, with special reference to the tropics (but equally applicable anywhere in the world), the small book by McNutt is recommended (McNutt, D. N. (1976). *Insect Collecting in the Tropics*, 68 pp.; copies available from COPR (now TDRI), London, free to workers in Commonwealth countries, otherwise £1.30).

Identification of insect pests

In many cases the experienced entomologist is able to identify the insects regularly encountered in the field, using both a named collection of local insect pests (which most research/advisory stations and colleges maintain) and local publications. In temperate regions, as distinct from the tropics, most field workers have ready access to both insect collections and to adequate libraries, and so most determinations can be made without difficulty.

However, with some groups of insects and mites, such as aphids, gall midges (Cecidomyiidae), muscoid flies, some Microlepidoptera, Tetranychidae, Tarsonemidae, and the like, the taxonomic criteria are quite esoteric and there may be many closely related species recorded locally. Thus, when accurate determination is required, specimens will have to be sent to the taxonomic experts where the major international museum collections are maintained. In most of the developed countries there will be a national collection of insects, usually in the capital city; but the two largest international organizations for the identification of insect pests are the Commonwealth Institute of Entomology, London, and the Smithsonian Institution, Washington DC, together with Beltsville Agricultural Research Centre. The addresses are given opposite:

The Director
Commonwealth Institute of Entomology,
c/o British Museum (Natural History),
Cromwell Road,
South Kensington,
London, SW7 5BD,
ENGLAND.

The Chairman,
Insect Identification & Beneficial Insect Introduction Institute,
US Department of Agriculture,
Building 003, Room 1,
Beltsville Agricultural Research Centre – West,
Beltsville,
Maryland, 20705,
USA

Both organizations now issue printed booklets and forms for guidance to any would-be users, and they should be contacted prior to any despatch of specimens.

Parcels sent to either of these institutions should be sent by 'Air Parcel Post', and *not* by 'Air Freight' since the latter service involves heavy charges for customs clearance and delivery from the airport, and there is considerable delay before delivery.

D Some standard abbreviations and acronyms

Units and general abbreviations

a.e.	acid equivalent
a.i.	active ingredient
b.p.	boiling point
°C	degrees Celsius
cm	centimetre
cv.	cultivar
e.c.	emulsifiable concentrate
g	gram
h	hour
ha	hectare
h.v.	high volume
i.r.	infra-red
kg	kilogram
km	kilometre
£	pound sterling
l	litre
LC_{50}	median lethal concentration
LD_{50}	median lethal dose
l.v.	low volume
m	metre
mg	milligram
min	minute
ml	millilitre
mm	millimetre
pH	hydrogen ion concentration
post-em	post-emergence
ppm	parts per million
pre-em	pre-emergence
RH	relative humidity
s	second
sp.	species
spp.	species (plural)
ssp.	subspecies
sspp.	subspecies (plural)
$	dollar
u.l.v.	ultra-low volume
u.v.	ultra-violet
var.	variety
vol.	volume
w.p.	wettable powder
w/w	weight for weight
w/v	weight for volume
w.s.p.	water-soluble powder

Miscellaneous abbreviations

BPH	Brown Planthopper of Rice
BC	Biological control
BSI	British Standards Institute
Cda	Controlled droplet application
EAG	Electroantennagram
GV	Granulosis virus
IPM	Integrated pest management
HMSO	Her Majesty's Stationery Office, UK
OC	Organochlorine compounds
OP	Organophosphorous compounds
PHV	Polyhedrosis virus
PM	Pest management
SIRM	Sterile insect release method

Organizations (acronyms)

ADAS	Agricultural Development and Advisory Service (formerly NAAS), MAFF, UK
ARC	Agricultural Research Council, UK
AVRS	Asian Vegetable Research Station, Taiwan
BM(NH)	British Museum (Natural History) London, UK
CAB	Commonwealth Agricultural Bureaux, Slough, UK
CIAT	Centre for International Tropical Agriculture, Cali, Colombia
CIBC	Commonwealth Institute of Biological Control (headquarters), London, UK
CIE	Commonwealth Institute of Entomology, London, UK
CIH	Commonwealth Institute of Helminthology, St Albans, UK
CIMMYT	International Maize and Wheat Improvement Centre, Londres, Mexico
CIP	International Potato Centre, Lima, Perµ
CMI	Commonwealth Mycological Institute, Kew, UK

COPR	Centre for Overseas Pest Research (now TDRI – Tropical Development and Research Institute, London, UK)
CSIRO	Commonwealth Scientific and Industrial Research Organization, Canberra, Australia
EAAFRO	East African Agricultural and Forestry Research Organization, Nairobi, Kenya
EPA	Environmental Protection Agency, Washington, USA
EPPO	European Plant Protection Organization, Paris, France
FAO	Food and Agricultural Organization of the United Nations, Rome, Italy
GCRI	Glasshouse Crops Research Institute, UK
IAC	International Agricultural Centre, Wageningen, Netherlands
ICARDA	International Centre for Agricultural Research in Dry Areas, Aleppo, Syria
ICIPE	International Centre for Insect Physiology and Ecology, Nairobi, Kenya
ICRISAT	International Crops Research Institute for the Semi-arid Tropics, Hyderabad, India
IFIS	International Food Information Service, Reading, UK
IITA	International Institute of Tropical Agriculture, Ibadan, Nigeria
IRRI	International Rice Research Institute, Manila, Philippines
MAFF	Ministry of Agriculture, Fisheries and Food, UK
MARDI	Malaysian Agricultural Research and Development Institute, Selangor, Malaysia
NAAS	National Agricultural Advisory Service (now ADAS), MAFF, UK
NAPPO	North American Plant Protection Organisation, USA
NVRS	National Vegetable Research Station, Wellesbourne, UK
ODA	Overseas Development Administration, London
ODM	Ministry of Overseas Development, UK
PBI	Plant Breeding Institute, Cambridge, UK
PESTDOC	Derwent Pooled Pesticidal Literature Documentation
RESL	Royal Entomological Society of London, UK
RTI	Royal Tropical Institute, Amsterdam, Netherlands
TDRI	Tropical Development and Research Institute, London, UK
TRI	Tropical Products Institute, London, UK (now TDRI)
US AID	United States Aid for International Development, USA
USDA	United States Department of Agriculture, USA
WHO	World Health Organization, Geneva, Switzerland
WICSCBS	West Indies Central Sugarcane Breeding Station, Barbados, West Indies
WRO	Weed Research Organization, Oxford, UK

Index